ENVIRONMENTAL
CHEMISTRY
of
SELENIUM

BOOKS IN SOILS, PLANTS, AND THE ENVIRONMENT

ENVIRONMENTAL CHEMISTRY
of
SELENIUM

edited by

William T. Frankenberger, Jr.
University of California at Riverside
Riverside, California

Richard A. Engberg
U.S. Department of the Interior
Washington, D.C.

CRC Press
Taylor & Francis Group
Boca Raton London New York

CRC Press is an imprint of the
Taylor & Francis Group, an **informa** business

CRC Press
Taylor & Francis Group
6000 Broken Sound Parkway NW, Suite 300
Boca Raton, FL 33487-2742

First issued in paperback 2019

© 1998 by Taylor & Francis Group, LLC
CRC Press is an imprint of Taylor & Francis Group, an Informa business

No claim to original U.S. Government works

ISBN-13: 978-0-8247-0136-9 (hbk)
ISBN-13: 978-0-367-40065-1 (pbk)

Visit the Taylor & Francis Web site at
http://www.taylorandfrancis.com

and the CRC Press Web site at
http://www.crcpress.com

Library of Congress Cataloging-in-Publication Data

Environmental chemistry of selenium/edited by William T. Frankenberger, Jr.,
    Richard A. Engberg.
        p. cm. — (Books in soils, plants, and the environment; v.64)
    Includes bibliographical references and index.
    ISBN 0-8247-0136-4
        1. Selenium—Environmental aspects. I. Frankenberger, W. T. (William T.).
    II. Engberg, R. A. III. Series.
    QH545.S45E58 1998
    577.27'5724—dc21   97–44122
                                                                              CIP

Preface

Agricultural-related selenium investigations shifted focus in the 1980s from domestic animal-crop accumulation issues to resource protection issues following the discovery of selenium poisoning at Kesterson Reservoir, California. The 1994 book *Selenium in the Environment*, edited by W. T. Frankenberger, Jr., and Sally Benson, (Marcel Dekker, Inc.) focused primarily on defining and investigating the processes involved in selenium contamination problems in California and methodologies and techniques for remediating these problems. More recently, research has taken two directions. Pure research in analytical methodologies, feeding supplementation, selenium speciation, volatilization, deselenification, and bioremediation has proceeded in a parallel fashion with applied research in determining other areas of selenium contamination, pathology of fish and wildlife in a field setting, bioaccumulation, synergistic and antagonistic impacts of other constituents, and institutional selenium concerns.

Intended as a complement to *Selenium in the Environment*, this book compiles recent basic research on selenium remediation and extensive information on new developments in applied research, particularly related to field selenium investigations. Its purpose is to provide a blending of knowledge from both these important sources and enhance the fact that they are necessary components in the study of a constituent with the JANUS-like complex behavior of selenium. As we approach the millennium, selenium is arguably the naturally occurring trace constituent of greatest concern worldwide. In overabundant amounts it can lead to toxicosis and teratogenesis in domestic and wild animals and birds, while the impact of selenium deficiency is equally dramatic.

The discovery of selenium problems at Kesterson Reservoir has led to investigations in other areas of the western United States and to the increased involvement of Federal, state, and local governments on the need for environmental remediation and/or regulation of selenium, particularly where anthropogenic enhancement of selenium occurs. Chapter 1 provides a perspective of irrigation drainage-related problems of excess selenium in the western United States. This area may serve as the bellweather for selenium-related problems worldwide when coupled with the need for enhanced sustainable agriculture to meet world food requirements in the 21st century. The chapter presents several examples of real-world attempts to manage selenium issues of remediation and regulation that point to the need for the continuing research described later in this volume.

Chapters 2–4 discuss the challenges in detecting minute levels of selenium with state-of-the-art instrumentation in water, soil, and plants. This field is constantly evolving in determining various inorganic and organic selenium species in salt-enriched matrices. In order to meet regulatory standards extremely low detection limits are needed to prevent a threat to the environment. Fractionation of selenium species is needed to monitor mobility and toxicity.

Chapters 5–9 focus on selenium supplementation in animal diets. Recent discoveries have revealed that selenium can have many beneficial effects in metabolism and development of ruminants and livestock. Selenium deficiencies are now considered a paramount worldwide problem in the nutrition of animals. With supplementation, there is some concern with recycled selenium being released in the food chain, possibly causing deformation and death to wildlife as discovered at Kesterson Reservoir. The critical levels of selenium needed to sustain the health of animals and the environmental implications of selenium metabolites being released are thoroughly discussed. These chapters bring a perspective on the need to maintain adequate nutrition with a trace element that can also cause severe damage to our preserved resources.

Chapters 10–14 deal with field studies of selenium in a variety of environments in the western United States from large areas as the San Joaquin Valley in California to smaller areas as Benton Lake in Montana. These chapters focus on the partitioning of selenium into water, bottom sediment, and the biota of the area and examine the cycling and coexistence of inorganic and organic selenium between these media. They also focus on geological sources of selenium and the processes that allow selenium to become available to the hydrologic and biological cycles of the areas.

Chapter 15 examines the mobility of selenium from coal-mine backfill to groundwater. The chemical and physical properties of backfill material and the chemistry of reclaimed coal-mine groundwater are used together to determine the effects of organic solutes on selenium adsorption and desorption reactions in backfill material.

Chapters 16–19 provide diagnostic information and field research results concerning pathology and toxicology of selenium impacts on fish, waterfowl, and mammalian species. Laboratory studies may not reveal the complete impact of selenium on a variety of species because bioassay testings may not include dietary pathways. Other trace constituents and environmental factors may interact to make species more or less tolerant to elevated selenium concentrations in water or food sources. Numerous case histories support the contention that more meaningful environmental selenium standards and criteria may be established through field-based studies on the impact of selenium.

Chapters 20–21 discuss simulation or predictive techniques to assess the impacts of selenium in areas where little or no data exist. A decision tree and a classification model based on geology, hydrology, and climatic data are designed to aid in the determination of areas susceptible to irrigation-induced selenium contamination. Chapter 22 examines the impact of nitrate-nitrogen from commercial fertilizers on the mobility and reaction kinetics of selenium in groundwater systems. When oxidized forms of selenium occur in a moderately reducing environment, microbial processes quickly transform them to elemental selenium and/or organic selenium compounds. The presence of nitrate inhibits these transformations.

Chapters 23–26 highlight the remedial treatment technology of removing selenium directly from water. This technology involves bioreduction of oxidized soluble inorganic selenium species (SeO_4^{2-} and SeO_3^{2-}) into an insoluble selenium precipitate (Se°) with zero valence. Once precipitated, the colloidal selenium can be removed by filtration or some other extraction technique. The success of this treatment system is highly dependent on microbial uptake mechanisms, metabolic biochemistry, and physiological factors affecting the performance of bioreduction. A number of the microbial strains being considered have advantages over others in this treatment technology. The authors have identified critical factors influencing this microbial transformation, including: selective measures to give a competitive advantage of the inoculum, economic carbon sources, removal of interferences, and determination of optimal conditions to promote microbial growth and sustain high selenium removal efficiency.

Chapter 27 emphasizes the use of blue-green algae to remove selenium from agricultural drainage water. The primary mechanism of removal is volatilization of the soluble selenium into a gaseous form. Chapter 28 describes a new approach in which selenium-enriched drainage water is sprayed onto a land treatment bed to promote volatilization of selenium. Selenium volatility is extensively reviewed in Chapter 29, which describes various species of gaseous selenium generated by microorganisms and plants, while Chapter 30 focuses on particulate selenium in the atmosphere.

Phytoremediation of selenium by specific plants is being investigated as a new remedial technology. Chapter 31 emphasizes factors that affect this technology, including plant species, solute partitioning, microbial interactions, and mechanisms of removal. Chapter 32 describes selenium accumulation and uptake by plants, with emphasis on the biochemical characterization of assimilated seleno-amino acids.

A thorough and complete understanding of the chemistry of selenium is essential to the remediation of selenium problems. Gross concentration data are less important to the understanding of the environmental impact of selenium than is information on partitioning, speciation, volatilization, microbial reduction, and the interaction of other constituents with selenium. Since 1983, there has been an explosion of knowledge regarding selenium chemistry, but each new piece of information seems to point to several other avenues of needed investigation. The information presented herein is a compilation of the latest research into this complex and elusive trace element.

We are indebted to the national and international authors of many disciplines representing academia, industry, and government for sharing their extraordinary knowledge for this compilation. The understanding and hence the solution of environmental selenium problems remain in the hands of these and the many other capable and dedicated scientists engaged in selenium research worldwide.

William T. Frankenberger, Jr.
Richard A. Engberg

William T. Frankenberger, Jr.   **Richard A. Engberg**

Contents

Contributors

D. Jack Adams, Ph.D.   Director, Center for Bioremediation, Weber State University, Ogden, Utah

Georg Alfthan, Ph.D.   Head of Laboratory, Department of Nutrition, National Public Health Institute, Helsinki, Finland

Antti Aro, M.D.   Research Professor, Department of Nutrition, National Public Health Institute, Helsinki, Finland

Sally M. Benson, Ph.D.   Director, Earth Sciences Division, Ernest Orlando Lawrence Berkeley National Laboratory, Berkeley, California

Thomas A. Cahill, Ph.D.   Professor, Department of Physics and Atmospheric Sciences, University of California at Davis, Davis, California

Thomas G. Chasteen, Ph.D.   Associate Professor, Department of Chemistry, Sam Houston State University, Huntsville, Texas

Michael Delamore, B.S.   Supervisory Natural Resources Specialist, Mid-Pacific Region, South-Central California Area Office, United States Bureau of Reclamation, Fresno, California

Päivi Ekholm, D.Sc.   Department of Applied Chemistry and Microbiology, University of Helsinki, Helsinki, Finland

Robert A. Eldred   Crocker Nuclear Laboratory, University of California at Davis, Davis, California

Richard A. Engberg, B.S.   Manager, National Irrigation Water Program, Office of the Secretary, Department of the Interior, Washington, D.C.

Teresa W.-M. Fan, Ph.D. Assistant Research Biochemist, Department of Land, Air and Water Resources, University of California at Davis, Davis, California

William T. Frankenberger, Jr., Ph.D. Professor, Department of Soil and Environmental Sciences, University of California at Riverside, Riverside, California

Gunnar Gissel-Nielsen, Ph.D., Dr. Agro. Professor, Department of Plant Biology and Biogeochemistry, Risø National Laboratory, Roskilde, Denmark

Steven J. Hamilton, Ph.D. Research Fishery Biologist, Biological Resources Division, Department of the Interior, United States Geological Survey, Yankton, South Dakota

George P. Hanna, Jr., Ph.D.† Engineering Research Institute, California State University at Riverside, Riverside, California

Richard M. Higashi, Ph.D. Research Chemist, Crocker Nuclear Laboratory, University of California at Davis, Davis, California

Delmar D. Holz, M.A. Resource Planning Advisor, Program Analysis Office, United States Bureau of Reclamation, Denver, Colorado

Matthew M. Jarman Physical Scientist, Water Resources Division, United States Geological Survey, Salt Lake City, Utah

Ivan T. Kishchak, Ph.D. Head, Animal Husbandry Department, Nikolayev Regional Agricultural Department, Nikolayev Agricultural Institute, Nikolayev, Ukraine

A. Dennis Lemly, Ph.D. Research Fisheries Biologist, Southern Research Station, United States Forest Service, Blacksburg, Virginia

Mark E. Losi Department of Soil and Environmental Sciences, University of California at Riverside, Riverside, California

John Maas, D.V.M., M.S., Diplomate, A.C.V.N. and A.C.V.I.M. Extension Veterinarian, Veterinary Medicine Extension, School of Veterinary Medicine, University of California at Davis, Davis, California

Dean A. Martens, Ph.D. Soil Scientist, United States Department of Agriculture, National Soil Tilth Laboratory, Ames, Iowa

---

†Deceased.

Angus E. McGrath, Ph.D.   Environmental Soil Chemist, Earth Sciences Division, Lawrence Berkeley National Laboratory, Berkeley, California

John B. Milne, Ph.D.   Professor, Department of Chemistry, University of Ottawa, Ottawa, Ontario, Canada

Johnnie N. Moore, Ph.D.   Professor, Department of Geology, University of Montana, Missoula, Montana

David L. Naftz, Ph.D.   Research Hydrologist, Water Resources Division, United States Geological Survey, Salt Lake City, Utah

James E. Oldfield, Ph.D.   Professor Emeritus, Department of Animal Sciences, Oregon State University, Corvallis, Oregon

Dónal O'Toole, M.V.B., Ph.D., M.R.C.V.S., F.R.C.Path.   Veterinary Pathologist, Department of Veterinary Sciences, University of Wyoming, Laramie, Wyoming

Lawrence P. Owens, Ph.D.   Research Manager and Lecturer, Engineering Research Institute, California State University at Fresno, Fresno, California

Ivan S. Palmer, Ph.D.   Professor, Department of Chemistry and Biochemistry, South Dakota State University, Brookings, South Dakota

Tim M. Pickett, B.S.   Program Manager, Center for Bioremediation, Weber State University, Ogden, Utah

David Z. Piper, Ph.D.   Geologist, Minerals Resources Survey Program, United States Geological Survey, Menlo Park, California

Theresa S. Presser   Research Chemist, National Research Program, United States Geological Survey, Menlo Park, California

Merl F. Raisbeck, D.V.M., M.S., Ph.D., D.A.B.V.T.   Professor, Department of Veterinary Sciences, University of Wyoming, Laramie, Wyoming

Katta J. Reddy, Ph.D.   Senior Research Scientist and Adjunct Professor, Wyoming Water Resources Center, University of Wyoming, Laramie, Wyoming

Roy A. Schroeder, Ph.D.   Hydrologist, United States Geological Survey, San Diego, California

Ralph B. See   Hydrologist, Water Resources Division, United States Geological Survey, Louisville, Kentucky

Ralph L. Seiler, M.S. Hydrologist, United States Geological Survey, Carson City, Nevada

James G. Setmire, B.S. Hydrologist, United States Geological Survey and Bureau of Reclamation, Temecula, California

Joseph P. Skorupa, Ph.D. Senior Biologist, Division of Environmental Contaminants, United States Fish and Wildlife Service, Sacramento, California

Doyle W. Stephens, Ph.D. Research Hydrologist, Water Resources Division, United States Geological Survey, Salt Lake City, Utah

Donald L. Suarez, B.A., Ph.D. Research Leader, Soil and Water Chemistry Research, United States Department of Agriculture, United States Salinity Laboratory, Riverside, California

Yuzo Tamari, Ph.D. Associate Professor, Department of Chemistry, Konan University, Kobe, Japan

Norman Terry, B.Sc., M.Sc., Ph.D. Professor, Department of Plant and Microbial Biology, University of California at Berkeley, Berkeley, California

George F. Vance, Ph.D. Associate Professor, Department of Natural Resources, University of Wyoming, Laramie, Wyoming

Pertti Varo, Ph.D. Associate Professor, Department of Applied Chemistry and Microbiology, University of Helsinki, Helsinki, Finland

Bruce Waddell, M.S. Environmental Contaminant Specialist, Ecological Services, United States Fish and Wildlife Service, Salt Lake City, Utah

Dennis W. Westcot Environmental Program Manager, Central Valley Regional Quality Control Board, Sacramento, California

Lin L. Wu, Ph.D. Professor, Department of Environmental Horticulture, University of California at Davis, Davis, California

Peter T. Zawislanski Staff Research Associate, Earth Sciences Division, Lawrence Berkeley National Laboratory, Berkeley, California

Adel Zayed Department of Plant and Microbial Biology, University of California at Berkeley, Berkeley, California

YiQiang Zhang, Ph.D. Research Assistant, Department of Geology, University of Montana, Missoula, Montana

Pioneers of Selenium Research

This book is dedicated to the many investigators worldwide who, since the early 1930s, have built and enhanced the knowledge base of selenium chemistry. While this has been an effort of many hundreds of researchers, the editors in their perusal of the body of literature have identified several researchers whose work has greatly influenced that in which we have been engaged. As a result we wish to recognize these individuals for their contributions.

Frederick Challenger	**Alvin L. Moxon**	**Kurt W. Franke**
The University of Leeds Leeds, England	Ohio State University Wooster, Ohio	South Dakota State College Brookings, South Dakota

Oscar E. Olson
South Dakota
Agricultural
Experiment Station
Brookings,
South Dakota

Orville A. Beath
University of
Wyoming
Laramie, Wyoming

O. H. Muth
Oregon State
University
Corvallis, Oregon

Douglas V. Frost
Nutritional
Biochemistry
Consultant
Brattleboro, Vermont

Peter J. Peterson
Kings College
University of London
London, England

Howard E. Ganther
University of
Wisconsin
Madison, Wisconsin

Alex Shrift
Binghampton
University
Binghampton,
New York

Irene Rosenfeld
University of
Wyoming
Laramie, Wyoming

James E. Oldfield
Oregon State
University
Corvallis, Oregon

**Gunnar Gissel-
Nielson**
Risø National
Laboratory
Roskilde, Denmark

Harry M. Ohlendorf
CHM2 Hill
Sacramento, California

Orville Levander
USDA-ARS
Beltsville Human
Nutrition Research
Center
Beltsville, Maryland

1

Federal and State Perspectives on Regulation and Remediation of Irrigation-Induced Selenium Problems

RICHARD A. ENGBERG
Department of the Interior, Washington, D.C.

DENNIS W. WESTCOT
Central Valley Regional Water Quality Control Board, Sacramento, California

MICHAEL DELAMORE
United States Bureau of Reclamation, Fresno, California

DELMAR D. HOLZ
United States Bureau of Reclamation, Denver, Colorado

I. INTRODUCTION

Domestic animal impacts, crop accumulation, and human health concerns were the principal focus of agriculture-related selenium research in the United States before 1983. The malady "alkali disease," both lethal and nonlethal in large animals and first described by a mid–nineteenth century army surgeon in Nebraska Territory, was identified as caused by selenium in 1931–1934 (Moxon, 1937). Engberg (1973) described a pony in Boyd County, Nebraska, that developed symptoms of alkali disease from eating locally grown hay. Schroeder et al. (1970) described selenium as an essential micronutrient in animal and human nutrition. Miller and Byers (1937) classified plants according to their ability to absorb selenium from soils and to convert selenium to soluble forms. Documentation of human selenium toxicity is rare, but cases include contamination from consumption of home-grown produce (Rosenfeld and Beath, 1964) and from well water (Beath, 1962; Brogden et al., 1979).

The discovery in 1983 that waterfowl deformities at Kesterson Reservoir in California were caused by selenium delivered in irrigation drainage (Ohlendorf et al., 1986) changed agricultural-related selenium investigations from a domestic animal/crop accumulation focus to a resource protection issue. Irrigation project management has been complicated greatly in the western United States in areas where irrigation projects are developed on seleniferous soils. Drainwater or seepage from several irrigation projects has created lakes and wetlands that accumulate selenium and other potentially toxic trace constituents. Migratory waterfowl and endangered species that use these lakes and wetlands are protected by federal legislation. Federal and state officials face the dilemma of determining how to provide resource protection while also protecting the water resources and agricultural infrastructures responsible for the economic development of the rural western United States. Following a history of irrigation development in California and the West and its relationship to water quality and selenium problems, this chapter addresses selenium management issues of regulation and remediation from both state and federal government perspectives. Several case histories on approaches to selenium management are presented.

II. HISTORICAL BACKGROUND OF IRRIGATION DRAINAGE AND SELENIUM PROBLEMS IN WESTERN UNITED STATES

Inadequate drainage and accumulating salts have been persistent problems in the western United States and in parts of the western San Joaquin Valley of California since the first land was irrigated in the nineteenth century. Large-scale use of surface water for irrigation exacerbated the problem by importing more salt into the system, increasing the area of land irrigated, and increasing the amount of water applied. State and federal water resource engineers recognized that water supply development plans need to include provisions for drainage.

Prior to the identification of selenium problems at Kesterson Reservoir the "drainage problem" centered on the impact of poor drainage conditions on crop productivity. The traditional solution to poor drainage conditions was the installation of artificial drains. These drains were designed to collect and carry away water and salts from the root zone, and to intercept shallow groundwater rising toward the root zone. Drainage from these systems was usually discharged to natural or man-made channels, often leading to terminal sinks throughout the West.

The San Luis Act (PL 86-488) passed by Congress in 1960 authorized a plan for the San Luis Unit of the federal Central Valley Project in California. The plan was for joint use facilities providing supplemental water to both the federal and state service areas in the San Joaquin Valley. Because plans for a valley master drain were not fully developed, the act included provision for a drainage outlet for the area.

In 1967 construction was initiated by the Bureau of Reclamation (BOR) on the federal San Luis Drain (SLD). While no final point of discharge in the delta was selected, construction proceeded consistent with alignments and plans developed to that point. The overall plan for SLD called for a 188-mile, concrete-lined channel to carry drainage from approximately 300,000 acres. The plan also included regulating reservoirs to temporarily hold drainage water in order to equalize and control flows in the drain. Initial construction extended from the southern portion of the San Luis Unit service area northward to a regulating reservoir to be located at the Kesterson site near Los Banos, California (see Figs. 1 and 2, Chapter 10).

In 1968 a 5900-acre parcel was acquired at the Kesterson site and construction was initiated on the SLD. By 1970, the first stage of Kesterson Reservoir was completed. It consisted of 12 cells of approximately 100 acres each, separated by earthen berms and averaging about 3 feet deep. In 1970 an agreement was executed between BOR and the U.S. Fish and Wildlife Service (FWS) (then the Bureau of Sport Fisheries and Wildlife) providing for FWS to manage the reservoir for wildlife benefit. Between 1970 and 1978 an average of approximately 2500 acre-feet per year of freshwater was provided to the reservoir and approximately 295 acres of permanent wetland habitat had been developed at the site (U.S. Fish and Wildlife Service, 1988).

By 1975, the initial stage of construction had been completed and the SLD extended approximately 85 miles to Kesterson Reservoir. Construction was halted because continuing questions and concerns about the effects of the discharge in the bay–delta area had prevented the selection of a point of discharge in the delta. Furthermore, questions regarding the authorized funding limits in the San Luis Act had arisen.

In 1975 BOR and the Westland Water District initiated the design and installation of a drainage collector system. By 1978, the collector system encompassing an area of approximately 42,000 acres had been installed in the northwest part of the district. Operation of the collector system began in 1978, and by 1981, 100% of the flow to Kesterson Reservoir was drainage water.

Biologists began to notice the decline and even disappearance of certain species from the San Luis Drain and Kesterson Reservoir in 1981 and initiated studies to determine the cause. In 1982 elevated levels of selenium were detected in fish (Saiki, 1986), and in 1983 there were reports of deformed embryos, adult mortality, and poor reproductive success of water birds. Tests indicated that selenium poisoning was the most probable cause (Ohlendorf et al., 1986). The selenium concentrations in flowing drainage water were found to average about 300 micrograms per liter (μg/L) and the source of the selenium was traced to the shallow groundwater in the drainage collector system area.

An adjacent landowner in 1984 petitioned the California State Water Resource Control Board (CSWRCB) to take action on the problems at Kesterson Reservoir. In February 1985, the CSWRCB issued an order declaring that the Kesterson Reservoir was hazardous and action must be taken to clean up and abate the nuisance conditions created by its operation. On March 15, 1985, the Secretary of the Interior ordered closure of Kesterson Reservoir, citing concern that the selenium poisoning could be causing violations of the Migratory Bird Treaty Act. On April 3, 1985, the Department of the Interior and the Westland Water District entered into an agreement that discharges to the San Luis Drain would cease. By June 1986, the collector system had been completely plugged; the last of the drainage water remaining in the San Luis Drain was discharged to the Reservoir in August 1986.

During planning and initial construction of the SLD, environmental concerns of operation of the regulating reservoirs focused on seepage problems and waterlogging of nearby lands, and wildlife management concerns focused on controlling and managing salinity levels. Concerns over discharges to the bay–delta system focused variously on salinity, boron, and nutrients. Before discovery of problems at Kesterson Reservoir, state and federal wildlife officials recommended that prior to final disposal, the drainage water be used to enhance wetland habitat. Discovery of the selenium problems at Kesterson Reservoir halted the ongoing studies that were aimed at completion of the drain. This interruption persists to the present.

A series of newspaper articles on selenium in the West, published September 8–10, 1985, in the *Sacramento Bee*, focused public attention on selenium problems in locations other than the San Joaquin Valley of California (Engberg, 1992). Assertions of detrimental effects of elevated selenium concentrations at 13 locations in the western states led several members of Congress to ask the Department of the Interior (DOI) to address these concerns. By December 1985, an organizational plan for the DOI Irrigation Drainage Program had been prepared and a management strategy developed. Selenium was the focus of the program, but several other naturally occurring trace constituents were to be studied. The program was to investigate irrigation projects or drainage facilities constructed or managed by DOI that potentially impact national wildlife refuges or other migratory bird/endangered species areas receiving drainwater from the projects. The scope of the program includes not only site investigations but also, when appropriate, remediation.

The program, which subsequently was renamed the National Irrigation Water Quality Program (NIWQP), conducted reconnaissance investigations at 26 irrigation project areas (Fig. 1) during 1986–1994 in 14 of the 17 western states exclusive of the San Joaquin Valley in California where federal irrigation projects are located. Reconnaissance investigations were to determine whether potentially toxic concentrations of trace constituents, particularly selenium, occur in the

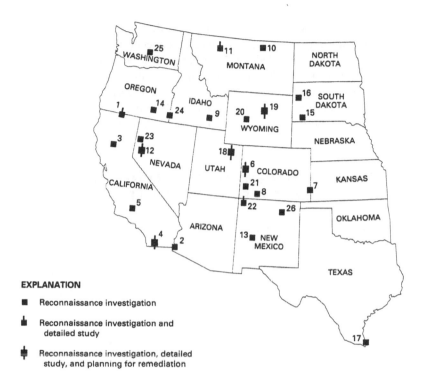

EXPLANATION

■ Reconnaissance investigation

▲ Reconnaissance investigation and detailed study

⬛ Reconnaissance investigation, detailed study, and planning for remediation

STUDY AREA

1. Klamath Basin Refuge Complex, CA-OR
2. Lower Colorado River Valley, CA-AZ
3. Sacramento Refuge Complex, CA
4. Salton Sea Area, CA
5. Tulare Lake Bed Area, CA
6. Gunnison River Basin/Grand Valley Project, CO
7. Middle Arkansas River Basin, CO-KS
8. Pine River Area, CO
9. American Falls Reservoir, ID
10. Milk River Basin, MT
11. Sun River Area, MT
12. Stillwater Wildlife Management Area, NV
13. Middle Rio Grande and Bosque del Apache National Wildlife Refuge, NM

14. Malheur National Wildlife Refuge, OR
15. Angostura Reclamation Unit, SD
16. Belle Fourche Reclamation Project, SD
17. Lower Rio Grande and Laguna Atascosa National Wildlife Refuge, TX
18. Middle Green River Basin, UT
19. Kendrick Reclamation Project Area, WY
20. Riverton Reclamation Project, WY
21. Dolores-Ute Mountain Area, CO
22. San Juan River Area, NM
23. Humboldt River Area, NV
24. Owyhee-Vale Reclamation Project Areas, OR-ID
25. Columbia River Basin, WA
26. Vermejo Project, NM

FIGURE I   Location of study areas, DOI National Irrigation Water Quality Program, 1986–1995. (Modified from Feltz and Engberg, 1994.)

water, sediment, and biota. Sampling at all areas and interpretation of results were guided by a common protocol to allow comparability between study areas.

Analysis of data from all 26 reconnaissance areas indicated potentially serious irrigation and drainage-related water quality problems in 8 areas (Engberg and Sylvester, 1993; Engberg, 1997). Selenium was the principal constituent of concern in 6 of 8 areas. In these areas the median concentration of selenium in water equaled or exceeded the federal Environmental Protection Agency (U.S. EPA) chronic criterion of 5 μg/L for protection of aquatic life (USEPA, 1987). The maximum observed selenium concentration was 27,000 μg/L in a groundwater sample from the Kendrick Reclamation Project, Wyoming. The highest median selenium value was 44 μg/L at the Salton Sea area, California.

III. DEVELOPING MANAGEMENT APPROACHES FOR IRRIGATION-RELATED SELENIUM PROBLEMS

Selenium problems at Kesterson Reservoir and elsewhere in the West led to the identification of selenium as a significant environmental contaminant and highlighted the need for development of approaches to management of irrigation-related selenium problems at the federal, state, and local levels. Research provides the scientific background for the management of selenium problems. Managers and planners must translate the complexities of selenium chemistry and biochemistry and the relationship of selenium to other environmental contaminants into cogent plans. Understanding both the uniqueness of selenium and its sources from irrigated agriculture are necessary to develop management and regulatory approaches. Formulation of clear environmental goals and an understanding of management constraints are required to obtain reasonable and workable management plans.

A.   Selenium as an Environmental Contaminant

I.   Uniqueness of Selenium

Unlike most other naturally occurring trace constituents, selenium has characteristics whose unique combination underscores the difficulty of determining management approaches to solving environmental selenium problems. Whereas other trace constituents may exhibit one or more of these characteristics, selenium is the only trace constituent to exhibit them all.

a.   Selenium Occurs at Low Concentrations in Water

Less than 1% of all surface water samples collected from 26 NIWQP reconnaissance investigation study areas had selenium concentrations exceeding 1.0 milligram per liter (mg/L). Forty percent of all surface water samples had selenium concentrations that were at or below the laboratory reporting level of 1.0 μg/L. The median values of selenium in water samples from three NIWQP study areas where selenium-related waterfowl embryonic deformities were found were 35, 20, and 7 μg/L (Engberg, 1997).

b. Selenium Bioaccumulates

Bioaccumulation of selenium occurred in all NIWQP study areas but was greater in areas where there were significant sources of selenium. In some areas, selenium concentrations as low as 3 μg/L may lead to potentially harmful bioaccumulation in higher tropic levels such as migratory waterfowl or shorebirds (Joseph Skorupa, personal communication, 1994).

c. Information on Long-Term Impacts of Selenium Is Inadequate

The discovery of deformed waterbird embryos at Kesterson Reservoir occurred within a few years of the diversion of irrigation drainwater containing large amounts of selenium into the previously pristine environment of the reservoir. For most waterbird and fish species, however, available information on the chronic effects of long-term exposure to elevated but sublethal concentrations of selenium or on the collective impact of selenium and combinations of other trace environmental contaminants is inadequate.

d. Selenium is Non–Point Source Derived

The principal source of selenium in surface water in the western United States is soils and subsoils beneath irrigated land. Infiltrating irrigation water mobilizes selenium from these deposits and delivers it to receiving streams and wetlands. Nonpoint sources are more difficult to remediate than are point sources.

2. Sources of Selenium from Irrigated Agriculture

In the western United States, irrigated agriculture is the major source of selenium when it is a contaminant in the environment (Engberg, 1997). The sources of selenium from irrigation must be considered in developing management plans and include selenium imported in or introduced into drainwater, mobilized by the practice of irrigation, and concentrated as a result of irrigation practices.

a. Selenium Imported in or Introduced into Irrigation Water

Selenium in the Imperial Valley and Salton Sea area of California is imported in irrigation water diverted from the Colorado River. Selenium concentrations in the imported supply are less than the EPA chronic criterion of 5 μg/L for protection of aquatic life. Because of bioaccumulation, this criterion may not be protective of all beneficial uses including wetlands and the waterfowl that use the wetlands.

Selenium discharged by individual farmers in the Imperial Valley is in itself not a water quality threat. When thousands of neighboring farmers import the same water supply for irrigation and discharge return flows, however, a significant load of selenium is discharged to the Salton Sea and surrounding wetlands.

b. Selenium Mobilized by the Practice of Irrigation

Water supplies used for irrigation in much of the western United States, for the most part, contain selenium at concentrations less than 1.0 μg/L. At this concentration, selenium is not a threat to wildlife or aquatic life. In major parts of the San Joaquin Valley and other western areas, the soils being irrigated contain selenium that is mobilized by infiltrating irrigation water. The selenium is discharged either to surface water by subsurface drainage or to groundwater by deep percolation. Selenium concentrations in the subsurface drainage flows in the San Joaquin River Basin range from less than 1.0 μg/L to 1800 μg/L, depending on the initial concentration in the soil (Chilcott et al., 1988). This represents a continual source of selenium that is discharged to the environment.

c. Selenium Concentrated as a Result of Irrigation Practices

The selenium concentrations in water imported to the Imperial Valley average about 2 μg/L (Setmire et al., 1990). Because of consumptive use during irrigation, selenium concentrations in both subsurface and surface return flows often exceed the EPA criterion and constitute a water quality threat (CRWQCB, Colorado River Region, 1993). Recent efforts to increase the efficiency of water use in the Imperial Valley have reduced the quantity of return flows lost to the drains, which has generally increased selenium concentrations in the collector drains. These efforts have been temporarily halted because of the threat that selenium concentrations will continue to increase in the drains.

A different scenario exists in the San Joaquin Valley, where selenium from subsurface drainage water is being concentrated in evaporation ponds (Westcot et al., 1989, 1993). These ponds pose a threat to wildlife and waterfowl that use them. Techniques to minimize the size of the ponds include increasing the efficiency of water use and reuse of the water. Both techniques increase the selenium concentrations in the ponds and the threat to wildlife and waterfowl that use the ponds.

B. Approaches to Regulating Selenium

1. Environmental Goals

Goals for environmental protection often are defined and implemented through regulatory laws. Water quality regulatory laws at both the state and federal levels initially were written to deal with the massive municipal and industrial discharges to the nation's waterways. In these cases, there was a defined point of discharge, a defined pollutant that entered the water, and, most important, a responsible entity to deal with control of that discharge. These laws are now being used to deal with more difficult discharges, including those from irrigated agriculture. The selenium-laden discharges from irrigated agriculture in the western United States are an example. These discharges are the result of an activity that is a legal practice, and the selenium originates from a natural source and was not added to the water supply. The contaminant, selenium, either is brought in with the water supply or is mobilized or concentrated by the practice of irrigation. The environmental impact is the result not of one discharge but of the discharge from hundreds or thousands of similar nonpoint sources.

Additionally, because the environmental impact is the result of an legal activity, it is unlikely that a responsible entity for control of the contaminant exists. Because of the nature of the selenium sources and the mechanisms that allow it to move into the environment from irrigated areas, defining a control mechanism and a responsible party for implementation is difficult. It requires a new approach to the implementation of existing regulatory laws.

The key element to the success of a regulatory plan is to ensure that environmental goals are clearly defined. The beneficial use to be protected is the foundation of a water quality plan. For selenium, the most critical beneficial uses to be protected are human health, wildlife, and aquatic life. Human health impacts relate primarily to drinking water and to consumption of aquatic life. Direct human health impacts are unlikely to occur when selenium concentrations in drinking water are less than the present federal drinking water standard of 50 μg/L (U.S. EPA, 1995). Animal health impacts are unlikely to occur when animal drinking water concentrations are below 20 μg/L (Ayers and Westcot, 1985).

The level of aquatic life beneficial use is less easy to define. As demonstrated in numerous studies in California, the most sensitive beneficial use to elevated selenium concentrations are wildlife, principally waterfowl, and aquatic life (CVRWQCB, 1995). In 1987 the EPA adopted a 96-hour, 5 μg/L chronic water quality criterion for selenium that is protective of all beneficial uses including wildlife and aquatic life. This criterion was developed using standard protocol for acute and chronic toxicity to aquatic organisms. Selenium, however, is a bioaccumulative substance, as demonstrated at many locations in the western United States.

The EPA water quality criterion does not take bioaccumulation into account and may not be protective of certain aquatic life, including waterfowl. Data developed during a Basin Plan amendment hearing by the CVRWQCB in California led to adoption of a 2 μg/L monthly mean water quality criterion for all uses that involve wetlands and waterfowl in California (CVRWQCB, 1995). Recent bioaccumulation data from three California sites (evaporation ponds in the Tulare Lake Basin, wetlands in the Grassland area of Merced County, and wetlands near the Salton Sea in Imperial County) support this lower water quality criterion. However, recent data from the San Francisco Bay and Estuary and other locations in western states show that even where selenium concentrations are too low to be detected by sensitive analytical techniques, aquatic life bioaccumulation of selenium is occurring. Evidence indicates that rapid bioaccumulation may be related to the chemical form of selenium being discharged. Whether the California water quality criteria are protective of all aquatic life is still open to question. Thus the tasks of developing a control program to provide protection and measuring the success of environmental protection efforts present a huge dilemma.

A new approach to measuring environmental protection is now being used in the Grasslands Watershed as part of the efforts to control selenium in California (U.S. Bureau of Reclamation, 1995). Beginning in the fall of 1996, for selenium leaving the Grassland watershed in Merced County, measurements are focused on more than just selenium concentrations in the water column (Henderson et al., 1995). Three levels of concern have been identified for five environmental indicators (Table 1). Decisions regarding needed changes in selenium management will be based on measurements of all five indicators. Indicators will be updated as data are developed by the program. Certain levels of selenium will trigger testing of additional indicators.

TABLE I   Selenium Levels of Concern for Five Environmental Indicators[a]

Indicator	Normal background	Threshold ranges	Level of concern ranges	Toxicological and reproductive effects a certainty
Water[a]	<0.5–1.5	Avian protection, 2–3, fish protection, 2–5	2–5	>5
Sediment	<2 μg/g	2–4	2–4	>4
Food chain	Usually <2, rarely >5	2–4 in the diet	3–7	>7
Fish	Usually <2, rarely >5		4–12	>12
Avian eggs	Usually <3, max <3	4.2–9.7	3–8 (pop. hatchability)	>8

[a]Units for water are micrograms per liter (μg/L); all other measurements are micrograms per gram (μg/g), dry weight.
Source: Modified from Henderson et al. (1995).

In the Tulare Lake Basin, evaporation pond selenium concentrations range from less than 1 μg/L to 800 μg/L (Westcot et al., 1993). Aquatic life impacts are occurring to waterfowl that use these sites. A mix of alternative and mitigation wetland habitats that are free of selenium is being created with the environmental goal of reducing the overall selenium levels in waterfowl to the no-effect level. Initial results from this site show promise for this approach.

In the San Francisco Bay and Estuary, programs are under way to reduce the total loading of selenium to the Bay/Estuary system with the environmental goal of measuring success of this effort by measuring selenium levels in indicator species of aquatic life. These environmental measurements are being done in addition to measurements of water concentrations for selenium.

2. Management Approaches

The regulatory mechanisms available under the federal Clean Water Act and a variety of state laws stress regulation of individual dischargers. Discharge permits to control point source municipal and industrial discharges have been the principal regulatory tools used. The focus of the permits has been on pollutant reduction by source control and treatment. These actions have succeeded because responsible parties generally are easy to identify and because dischargers generally are successful in identifying sources and controlling the input of pollutants into receiving streams.

The ability to manipulate pollutant input from a nonpoint source, such as selenium from irrigated land, is not as easy. The source often is not well understood, the mechanisms for control are modified agricultural practices that take time to implement, the responses from control actions are not well understood, and measuring success may take several years. In addition, there must be an entity responsible for implementation often does not exist. In the absence of a responsible entity, efforts must be undertaken to form a local entity as present federal law does not give state or federal governments the authority to form such entities.

Control of selenium-containing nonpoint sources requires planners to develop long-term solutions based on local and regional variability (Young and Congdon, 1994). The San Joaquin Valley Drainage Program outlined a selenium control program that focused on three steps (SJVDP, 1990). The first step was to minimize the environmental selenium input from irrigated agriculture. The second step was to capture and reuse the selenium-laden water. The final step was to treat or dispose of the remaining selenium-laden water. This process is one of concentrating the selenium into the smallest volume of wastewater possible. The advantage is that it minimizes the handling, treatment, and disposal costs of the final effluent. A disadvantage is the potential environmental exposure to high concentrations of selenium and the difficulty of finding a way to manage these in the environment without adverse impacts.

The State of California adopted a non–point source management plan (NPS plan) in 1988 that identifies a progressive tiered approach to regulation rather than the traditional permit approach (CSWRCB, 1988). The progressive tiers are (1) voluntary, (2) regulatory encouragement of Best Management Practices (BMP) implementation, and (3) effluent limits established through permits. The least stringent option that obtains compliance is the preferred approach.

Options to manage selenium may be classified as pre-farm, on-farm or post-farm. For each option, the following management steps may be applicable:

1. Control selenium at the source.
2. Reduce the mobilization of selenium.
3. Capture selenium that is mobilized.
4. Utilize, dilute, detoxify, or dispose of the selenium.
5. Mitigate the adverse effects of the selenium.

The initial program of the California NPS plan took place in the Grassland watershed. It emphasized the first two options by stressing improved on-farm water use efficiency to reduce selenium or its mobilization. This was needed as a first step to reduce the existing problem of high concentrations of selenium in the San Joaquin River. Improved water management and land use practices adopted in the Grassland Basin resulted in a 50% reduction in selenium loads to the San Joaquin River and water quality improvement in the river (Karkoski, 1994; CVRWQCB, 1996).

In contrast, selenium load reduction did not improve water quality in the effluent-dominated tributaries within the Grassland Basin. Increased efficiency of water use resulted in higher selenium concentrations in surface and subsurface irrigation return flows and in an almost continuous violation of the adopted water quality objectives for selenium in the effluent-dominated tributary sloughs (Karkoski, 1994; CVRWQCB, 1996). It must be recognized that improved efficiency helps reduce selenium loads but will not make the selenium go away. The connection to on-farm practices and the amount of selenium discharged however is not well understood.

Controlling selenium through improved water management is directed at only one pollutant and using one management technique. Other pollutants including salt and boron also may be concentrated. The increased concentrations may limit options 3 and 4 in the future. For example, improved efficiency in the Imperial Valley of California now limits the options for selenium management and disposal. Regulatory rules are focused on the selenium concentrations in the drains rather than on how to better manage selenium in the environment. A focus on the total load discharged may be a more effective management technique.

The evaporation ponds in the Tulare Lake Basin of California offer a similar example. Improved efficiency would reduce the size of the ponds but would increase the selenium concentrations flowing to the ponds and in the ponds, thus increasing the threat to waterfowl using them.

In the Grassland Basin, consideration of control actions is now being expanded to include distribution and delivery efficiency, on-farm and off-farm reuse of return flows, and regulation of discharge runoff. Because each of these varies on a local and regional basis, selenium reduction mechanisms must be developed on a regional or subwatershed basis. At present, regulatory law is focused on a waterbody-by-waterbody approach. The need to fully meet all water quality criteria has resulted in a dilemma for irrigated agriculture especially in constructed drains and effluent-dominated waterbodies.

Of the five management options presented, mitigation (option 5) may be the only long-term solution to managing selenium. Mitigating the environmental impacts of selenium and managing small concentrations of selenium in the environment may be the most logical steps for dealing with selenium-enriched areas in California and other western states. Initial steps have begun to mitigate the impacts from evaporation ponds in the Tulare Lake Basin of California, and steps are being taken to mitigate the impacts in the Grassland Basin in Merced County, California. These steps do not eliminate selenium, but instead manage it at a reasonable level.

3. Management Constraints

The objective of managing selenium as an environmental contaminant is to improve the general health of the ecosystem in the irrigation project area under investigation. Management can become a tug-of-war between competing interests. One side often comprises irrigators, local residents, and business interests, while the other side includes those charged with enforcing environmental laws and those favoring strict enforcement. These lines are not clearly drawn and may shift radically as scientific information is developed and presented. Very frequently the reaction of owners and irrigators is to question the veracity of the data responsible for delineation of the problem and to question the potential costs of remediation. Members of environmental groups often react by asking that the irrigation project be closed or severely curtailed.

4. Economic and Social Considerations

Economies of areas where irrigation projects are located are closely tied to the projects. When selenium problems are identified in an irrigation project area and a management approach is developed, the approach must include economic and social components. Impacts of proposed changes in project operation or management on individual, local, and institutional economies must be considered in determining management options. Rights of landowners that serve to minimize hardships of proposed selenium management on individuals must be considered.

Management options may impact local economies and successful management may depend on minimal economic or social disruption.

Successful management of selenium problems involving socioeconomic concerns may include the development at no cost to impacted communities of replacement wetlands on federal or state lands and enhanced fishing and hunting in wetland or set-aside areas. Incentives may be provided to landowners in areas whose land is targeted for retirement from agriculture in the form of premium prices for acquisition of the land, water rights exchanges to equivalent land, incentives for reestablishment of irrigation on the alternatives location, or land exchanges. Acquisition of land or water rights should always be on a willing buyer/willing seller basis.

5. Financial Considerations and Funding Sources

Management of selenium problems should be approached in ways that minimize costs and maximize benefits. Costs should never far exceed benefits regardless of cleanup goals. If target concentrations or load limitations are unattainable at reasonable costs, all involved parties must agree that other targets or limitations must be considered within the context of existing laws.

Costs and benefits may best be balanced by the use of multiple-option solutions and by partnerships in determining funding sources to achieve management solutions. Solutions can involve a large variety of possible options, and most frequently the most cost-effective solution involving the greatest benefits is a multiple-option solution. For example, the best fix of a selenium problem may be construction of a treatment facility for drainwater. This solution may be enhanced by on-farm reduction of drainwater production. Drainwater requiring treatment might further be diminished by successive reuse of drainwater on more salt-tolerant crops. In this fashion multiple options may reduce costs and increase benefits while reducing the selenium ultimately available for environmental degradation.

Solutions based on partnerships require cooperation and trust between groups whose overall missions may be different, although all are seeking to solve selenium problems. Federal, state, and local entities including irrigation districts all have stakes in selenium cleanup, and partnerships between these entities can reduce costs and maximize benefits by emphasizing the strengths of the participating entities. For example, irrigation districts could market water saved by canal lining or improved delivery systems to municipal users on a prepayment basis to defray future expenditures.

IV. SELENIUM MANAGEMENT EXAMPLES

Regulating and remediating selenium problems at irrigation project areas in the western United States requires development of management approaches unique to the individual areas. The selenium problems at Kesterson Reservoir were addressed by the following steps: (1) reservoir closure to eliminate further selenium input, (2) reservoir grading and filling to eliminate aquatic and wildlife habitats, hence, minimize wildlife exposure, (3) development of mitigation lands, and (4) long-term monitoring. Individualized management approaches have been developed and are being applied at other areas in the western United States. The western areas discussed in Sections IV.A to IV.D offer examples that highlight the approach to remediation, the difficulties involved in remediation, and the results of these activities as appropriate.

A. San Joaquin Valley Drainage Program

Selenium has been the critical constraint in developing an acceptable solution to drainage problems of the San Joaquin Valley in California ever since the problems associated with selenium became apparent at Kesterson Reservoir. The San Joaquin Valley Drainage Program (SJVDP), established in 1984, was a major federal and state effort to develop comprehensive solutions to agricultural drainage problems of the San Joaquin Valley (see Fig. 1, chapter 10). The SJVDP's oversight committee and management team included a wide variety of public and private entities representing a broad spectrum of interests. SJVDP initiated extensive technical investigations into the various aspects of the drainage problem and its potential solutions. SJVDP also included years of political consensus building, and it was this consensus, more than the development of a final plan for solving the valley's drainage problems, that shaped the SJVDP final report.

Early in the process of consensus development, as a result of the failure to determine a point of discharge in the delta for the SLD, it was decided that the SJVDP would focus only on in-valley solutions to the drainage problems. This decision forced a significant change in the traditional approach to dealing with agricultural drainage by placing the emphasis on minimizing the amount of drainage water. The SJVDP developed the concept of "problem water" and formulated management recommendations to minimize and manage problem water. The SJVDP recognized that these recommendations alone would not ultimately solve the problem but considered them steps prerequisite to long term solution.

The major components of the SJVDP recommended plan (SJVDP, 1990) were as follows:

1. Source control. On-farm and district irrigation system improvements will reduce the amount of applied water and in turn, reduce the amount of potential problem drainage water.
2. Drainage reuse. Drainage water should be used progressively on more salt-tolerant plants to reduce the volume of drainage water and concentrate salts and trace elements for easier containment and disposal.

3. Evaporation systems. Drainage water remaining after reuse on salt-tolerant plants will be stored and evaporated. Four types of evaporation pond were identified: (a) nontoxic ponds in which selenium in the in flowing drainage water is less than 2 $\mu g/L$; (b) selenium-contaminated ponds (inflow containing 2–50 $\mu g/L$ Se) designed with wildlife safeguards and including alternative freshwater habitat; (c) accelerated evaporation ponds requiring minimal pond surface area; and (d) temperature-gradient solar ponds that generate electricity.

4. Land retirement. Irrigation will be discontinued on difficult-to-drain lands or lands overlying shallow groundwater with high levels of selenium.

5. Groundwater management. Planned pumping of the semiconfined aquifer will occur in places where the near-surface water table can be lowered and the pumped water is of suitable quality for irrigation or wildlife habitat.

6. Discharge to the San Joaquin River. Controlled and limited discharge of drainage water will be allowed from the Northern Subarea (see Fig. 1, Chapter 10) to the San Joaquin River, while meeting water quality objectives.

7. Protection, restoration, and provision of substitute water supplies for fish and wildlife habitat. Freshwater supplies will be substituted for drainage water previously used on wetlands and for the protection and restoration of contaminated fisheries and wetland habitat.

8. Institutional change. Institutional measures will be implemented such as tiered water pricing (to encourage water conservation, hence source control), improved irrigation scheduling, water transfers and marketing, and formation of regional drainage management organizations to aid in implementing the other components.

The final report of the SJVDP in 1990 was a joint effort of federal and state governments to put forth a set of effective measures to address drainage issues on which all could agree. The recommendations were limited by both their focus on short-term solutions and by the incomplete technical development of some of the innovative plan components. Nevertheless, some of the SJVDP components have been largely implemented, and others continue to be pursued. Water supply shortages caused by drought, coupled with supply restrictions imposed by environmental requirements and passage by Congress in 1992 of the Central Valley Project Improvement Act (CVPIA), have resulted in implementation of source control measures recommended by the SJVDP. The CVPIA also provided the substitute water supplies recommended for fish and wildlife protection and restoration, and provided some of the institutional changes recommended by the SJVDP. Innovative efforts are under way to develop practical drainage reuse systems. A regional drainage management organization has been formed in the San Joaquin Basin to begin implementation of the river discharge component of the SJVDP plan.

B. Case Study for the Grassland Watershed in Merced County, California

In 1988 the California Central Valley Regional Water Quality Control Board (CVRWQCB) began a program to regulate selenium in subsurface agricultural drainage discharges from a 90,000-acre, selenium-laden irrigated area within the 370,000-acre Grassland watershed, an area on the west side of the San Joaquin River (see Fig. 2, Chapter 10). Discharges were to wetlands of the Grassland watershed and to the San Joaquin River and its tributaries. The approach was to promote adoption of Best Management Practices (BMPs) that would improve on-farm water management to reduce drainage flows and meet established water quality objectives (CVRWQCB, 1994). The farmers and water districts focused their efforts on improved water management. This resulted in improved irrigation efficiency in the selenium problem area; however, water quality goals within the watershed and downstream were not met (Karkoski, 1994).

This lack of compliance prompted the CVRWQCB to reconsider the regulatory direction established in 1988. In May 1996, the CVRWQCB adopted a new approach to regulating non–point source discharges of selenium from the Grassland watershed. The focus continued on source control efforts but emphasized controlling the total load discharged from the watershed and allowing the dischargers to act through a responsible regional entity to implement the load limitations (CVRWQCB, 1996). Under the new approach, obtaining compliance in all areas may be difficult, costly, and time-consuming. Because not all actions can be accomplished at one time, beneficial use attainment was prioritized on a watershed basis. In the Grassland watershed, wetlands and wetland water supply channels have the highest priority, followed by protection of in-stream aquatic life in the San Joaquin River and its tributaries.

Wetlands protection in the Grassland watershed may be achieved by consolidation of drainage flows into channels isolated from the wetland areas (SJVDP, 1990). Consolidation removes selenium-laden drainage flows from the wetlands and from most channels but increases drainage water flows and selenium concentrations in the remaining channels. The benefit is a higher level of beneficial use attainment on a watershed basis as a result of increased wetlands protection. However, meeting final water quality objectives in some channels is difficult or impossible, especially in constructed canals, drains, or natural waterbodies that are strongly effluent-dominated (Westcot, et al., 1996).

Consolidation will be followed by efforts to achieve aquatic life protection in the San Joaquin River. Selenium load reduction to the river should be effective in achieving this goal (Karkoski et al., 1993; Karkoski, 1994; CVRWQCB, 1996).

The regulatory approach for implementing load reductions is through establishment of effluent limits. This allows flexibility for the dischargers to design their own methods of compliance while clearly defining responsibilities for meeting downstream objectives. This approach allows greater participation by dischargers, who will have more responsibility in deciding how to apportion loads among themselves to achieve the most cost-effective protection (Young and Congdon, 1994).

The most difficult step is to achieve aquatic life protection in the effluent-dominated channels. Experience shows that selenium load reductions are not likely to achieve final water quality objectives (Chilcott et al., 1995). The achievement of aquatic life protection in effluent-dominated channels will depend on the ability to develop technology to manage selenium in the drainage flows while also maintaining the economic viability of the watershed. At present, some of that technology is not available. Therefore, a reasonable time for compliance must be given. Where compliance is difficult, consideration must be given to the development of site-specific water quality objectives (CVRWQCB, 1996).

C. Green River Basin Remedial Planning

The Jensen and Vernal Units of the Central Utah Project are located near the Green River in eastern Utah (see inset, Fig. 1, Chapter 11). The Jensen Unit provides irrigation water for about 4000 acres of land and the Vernal Unit for about 14,000 acres (Stephens et al., 1992). Seleniferous soils derived from Mancos Shale underlie nearly all of the Jensen Unit and about 10% of the Vernal Unit. Subsurface drains from about 750 acres of the Jensen Unit convey irrigation drainage to Stewart Lake, which has an outlet to the Green River. Drainage from the Vernal Unit reaches Ashley Creek, whose confluence with the Green River is about 0.5 mile downstream from Stewart Lake.

Median selenium concentrations for samples collected in 1988–1989 from five drains to Stewart Lake range from 5.5 to 79 μg/L. Selenium concentrations reaching Ashley Creek from the Vernal Unit generally are less than 5 μg/L (Stephens et al., 1992).

Sewage lagoons for the city of Vernal, Utah, are located adjacent to Ashley Creek and about 9 miles upstream from its confluence with the Green River. Seepage from the lagoons dissolves selenium from underlying Mancos Shale, and the seepage with its elevated selenium concentrations reaches Ashley Creek. In 1989 a selenium concentration of 16,000 μg/L was observed in a sample of seepage discharged directly to Ashley Creek, down gradient of the lagoons. (Stephens et al., 1992).

The selenium load at the confluence of Ashley Creek and the Green River is derived principally from Ashley Creek (about 80%) and Stewart Lake (about 15%). Backwaters near the confluence are used extensively for spawning by an endangered fish, the razorback sucker *Xyrauchen texanus* (Stephens et al., 1992). To protect the endangered fish, remediation of selenium problems is necessary both for Ashley Creek and Stewart Lake.

Remedial planning for Stewart Lake began in 1992. An NIWQP study team conducted public meetings and developed a large number of remedial options. Evaluation of the options and development of several remedial alternatives occurred in 1993–1995. The preferred remedial alternative selected in 1996 will reroute drains from Stewart Lake directly to the Green River and will provide freshwater flushing of Stewart Lake. Drainwater will be dispersed into the streamflow of the Green River to allow rapid assimilation of selenium. Implementation of the alternative was scheduled to begin in 1997, funded by existing federal monies.

Remediation of selenium problems for the Ashley Creek Basin is progressing slowly. Restructuring or abandonment of the sewage lagoons is necessary to reduce the selenium delivered to Ashley Creek. The local sewage authority has been reluctant to acknowledge the problem or to accept responsibility. Negotiations have taken place for over 2 years between combinations of federal and state agencies to solve the problem and to provide the finances necessary to determine a solution. Some federal and state agencies have pledged financial support, but the pledges alone are not sufficient to meet the estimated costs of a solution; additional financial commitments from local interests are needed. Until a viable solution for the selenium problems related to the sewage lagoons can be effected, the spawning area of the endangered fish cannot be protected.

D. Kendrick Reclamation Project Remedial Planning

A detailed study report of the Kendrick Reclamation Project in central Wyoming (Fig. 2) documented elevated selenium concentrations in samples of water, bottom sediments, plants, and aquatic birds at four wetlands within the Kendrick area (See et al., 1992). Two of the wetlands, Goose Lake and Rasmus Lee Lake (about 100 surface acres each), are closed basins, and two are flow-through wetlands, Thirty-Three Mile Reservoir (about 30 surface acres) and Illco Pond (about 5 surface acres). The report documented poor egg hatchability, embryo mortality, and embryo deformities associated with elevated selenium levels in Canada goose, American avocet, and eared grebe in the closed-basin systems.

The remedial planning effort was initiated by a three-member core team following the release of the detailed study report in February 1992 (See et al., 1992). The Bureau of Reclamation provided the core team leader, and the U.S. Geological Survey and the Fish and Wildlife Service each provided a team member. Initial efforts were aimed at public education, problem recognition, and problem definition. It is imperative that the participants recognize that a problem exists and come to an agreement or consensus on a definition of the problem prior to moving forward in the planning process.

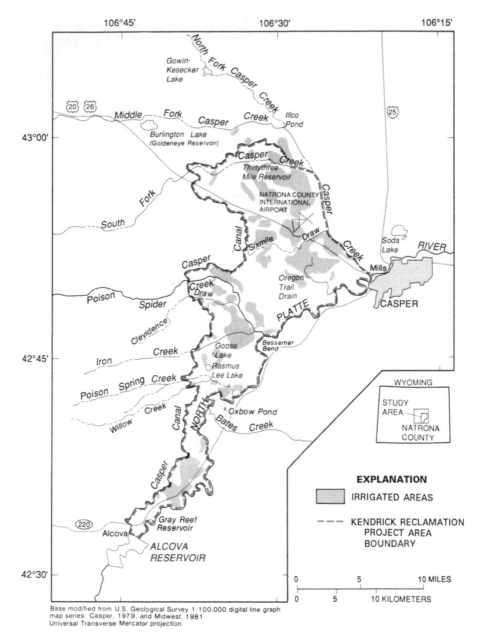

FIGURE 2   Irrigated lands and water bodies, Kendrick Reclamation Project Area, Wyoming.
(From See et al., 1992.)

The problem was defined as follows: surface and groundwater transport of selenium, influenced by irrigation practices, is resulting in adverse impacts to migratory birds. The primary remediation objective and goal was to protect and restore or replace migratory bird habitat at the four wetlands identified as adversely affected by irrigation drainage-induced selenium impacts.

The core team, utilizing an interdisciplinary technical team and the public involvement process, moved into a planning process that included identifying issues, developing options, formulating alternative solutions, and evaluation of alternatives. As the evaluation of alternatives was nearing completion in December 1993, the Casper—Alcova Irrigation District (district) submitted an additional plan to remediate the problem at the four wetlands. The proposal committed the district to the design and construction costs associated with eliminating the irrigation-induced drainage impacts at the wetlands.

The district proposal was based primarily on improved water management and water conservation by eliminating the two closed-basin wetlands and reducing irrigation-induced drainage to the two flow-through wetlands. However, the proposal did not contain all measures for a viable alternative that had been identified by the alternative development process. In an attempt to develop the proposal into a viable alternative, the core team began negotiations with the district and the state of Wyoming. Through negotiation, a viable alternative evolved that will be implemented in 1997 as a partnership effort based on cooperation and shared remediation activities and commitments.

The major remediation commitments of the district and the state include the funding, design, and construction of irrigation delivery system rehabilitation in the four wetland areas and providing water up to the amount conserved for remediation purposes, pursuing retirement of irrigated lands within the closed-basin areas, and committing to fund and implement additional remediation measures if determined necessary following a 5-year monitoring program. The federal government will do the following: coordinate and fund remediation monitoring activities, provide wetland habitat replacement as necessitated by remediation activities at the two closed-basin wetlands, and coordinate and fund activities required for compliance with the National Environmental Policy Act. The federal government will not seek reimbursement or cost-sharing associated with past investigation and planning activities.

It will be possible to move forward with implementation of the remedial measures identified in the partnership without seeking congressional construction authority and funding. The district and the state are providing the design, funding, and construction of the activities that would normally require congressional authority and funding. The project components provided by the federal government will be provided within existing Bureau of Reclamation authorities and funds.

This partnership approach to the remediation of the adverse selenium impacts associated with the Kendrick Reclamation Project will provide timely benefit to the impacted migratory birds. The time-consuming and uncertain process of seeking congressional authority and funds has been avoided with this partnership approach. Not only do the migratory birds received timely benefits, but the stakeholders are contributing financially to the solution.

V. SUMMARY

Selenium problems in biota resulting from the collection of drainwater into a holding reservoir were first observed at Kesterson Reservoir in the San Joaquin Valley, California, in 1982 and soon after in other areas of the western United States. This awareness changed the primary focus of agriculture-related selenium investigations from domestic animal/crop accumulation to resources protection. Managers must translate the complexities of selenium chemistry and biochemistry into cogent management and regulatory approaches, all the while understanding and blending financial, economic, and social constraints into the solutions. Because of selenium's unique properties and its environmental sources, solutions to selenium problems must be tailored specifically for each area.

Examples of management strategy development were provided for several areas in California. Four selenium management plans were discussed from the earliest attempt to develop a broad-based plan for the San Joaquin Valley, California, in the late 1980s to current planning efforts for remediation of selenium problems in the Grasslands Watershed, California, the Green River Basin, Utah, and the Kendrick Reclamation Project, Wyoming. The San Joaquin Valley Drainage Program final report provided several comprehensive recommendations, which if all implemented should provide adequate drainage management in the valley. However, to date, only part of the plan's components have been implemented. For the other three areas, which are substantially smaller, the implementation of the selenium management plans is just beginning. It will take several years to evaluate the success of these plans.

It is evident from these management efforts that the success of selenium management plans depend ultimately on partnership between federal, state, and local entities and on strong commitments by all participants to the management efforts. Without a dedicated and focused approach to these problems, successful management and remediation cannot be assured.

REFERENCES

Ayers, R. S., and D. W. Westcot. 1985. Water Quality for Agriculture. Food and Agricultural Organization of the United Nations, Irrigation and Drainage Paper No. 29, Rev. 1.

Beath, O. A. 1962. Selenium poisons Indians. *Sci. News Lett.* 81:254.

Brogden, R. E., E. C. Huchinson, and D. E. Hillier. 1979. Availability and quality of groundwater, Southern Ute Indian Reservation, southwestern Colorado. U.S. Geological Survey Water-Supply Paper No. 1576-J.

California Regional Water Quality Control Board (CRWQCB), Colorado River Region. 1993. Salton Sea briefing paper. California Regional Water Quality Control Board, Colorado River Region.

California State Water-Resources Control Board (CSWRCB). 1988. Nonpoint Source Management Plan.

Central Valley Regional Water Quality Control Board (CVRWQCB). 1994. Grassland Basin Irrigation and Drainage Study, Vols. 1 and 2. Prepared under contract by Cal Poly Irrigation Training and Research Center, San Luis Obispo. California Regional Water Quality Control Board, Central Valley Region.

Central Valley Regional Water Quality Control Board (CVRWQCB). 1995. Beneficial Use Designations and Water Quality Criteria to be used for the Regulation of Agricultural Subsurface Drainage Discharges in the San Joaquin Basin (5c): Staff Report. California Regional Water Quality Control Board, Central Valley Region.

Central Valley Regional Water Quality Control Board (CVRWQCB). 1996. Amendments to the Water Quality Control Plan for the Sacramento River and San Joaquin River Basins for the Control of Agricultural Subsurface Drainage Discharges: Staff report. California Regional Water Quality Control Board, Central Valley Region.

Chilcott, J. E., D. W. Westcot, K. Werner, and K. K. Belden. 1988. Water Quality Survey of Tile Drainage Discharges in the San Joaquin River Basin. California Regional Water Quality Control Board, Central Valley Region.

Chilcott, J. E., J. Karkoski, M. R. Ryan, and C. Laguna. 1995. Water Quality of the Lower San Joaquin River: Lander Avenue to Vernalis; October 1992 to September 1994 (Water Years 1993 and 1994). California Regional Water Quality Control Board, Central Valley Region.

Engberg, R. A. 1973. Selenium in Nebraska's groundwater and streams. University of Nebraska Conservation and Survey Division, Nebraska Water Survey Paper No. 35.

Engberg, R. A., 1992. Department of the Interior National Irrigation Water Quality Program Summary, FY 1986-91. Department of the Interior National Irrigation Water Quality Program Information Report No. 2.

Engberg, R. A. 1997. Remediation of irrigation-related contamination at Department of the Interior project areas in the western United States. In Dudley, L. M. and J. Guitjens (eds.) "Agroecosystems and the Environment: Sources, Control and Remediation of Potentially Toxic, Trace Element Oxyanions, pp. 57–76. Am Assoc. Adv. Sci.-Pacific Division, San Francisco, CA.

Engberg, R. A., and M. A. Sylvester. 1993. Concentrations, distribution and sources of selenium from irrigated lands in western United States. *ASCE J. Irrig. Drain. Eng.* 119:522–536.

Feltz, H. R., and R. A. Engberg. 1994. Historical perspective of the U.S. Department of the Interior National Irrigation Water Quality Program. In R. A. Marston and V. R. Hasfurther (eds.), *Effects of Human-Induced Changes on Hydrologic Systems*, pp. 1011–1020. American Water Resources Association, Proceedings of the 1994 Summer Symposium, Jackson, WY, June 26–29, 1994.

Henderson, J. D., T. C. Maurer, and S. E. Schwarzbach. 1995. Assessing selenium contamination in two San Joaquin Valley, CA, sloughs. U.S. Fish and Wildlife Service, Region 1, Division of Environmental Contaminants, Sacramento, CA.

Karkoski, J., 1994. A Total Maximum Monthly Load Model for the San Joaquin River. California Regional Water Quality Control Board, Central Valley Region.

Karkoski, J., T. F. Young, C. H. Congdon, and D. A. Haith. 1993. Development of a Selenium TMDL for the San Joaquin River. In "Management of Irrigation and Drainage Systems." ASCE Irrigation and Drainage Systems Division, Park City, UT.

Miller, J. T., and H. G. Byers. 1937. Selenium in plants in relation to its occurrence in soils. *J. Agric. Res.* 55: 59–68.

Moxon, A. L. 1937. Alkali disease or selenium poisoning. South Dakota State College of Agriculture and Mechanical Arts, Agric. Exp. Stn. Bull. No. 311.

Ohlendorf, H. M., D. J. Hoffman, M. K. Saiki, and T. W. Aldrich. 1986. Embryonic mortality and abnormalities of aquatic birds—Apparent impacts of selenium from irrigation drainwater. *Sci. Total Environ.* 52:49–63.

Rosenfeld, I., and O. A. Beath. 1964. *Selenium—Geobotany, Biochemistry, Toxicity, and Nutrition.* New York, Academic Press.

Saiki, M. K. 1986. Concentrations of selenium in aquatic food chain organisms and fish exposed to agricultural tile drainage water. Proc. Second Selenium Symp., pp. 25–33. Bay Institute of San Francisco, Tiburon, CA.

San Joaquin Valley Drainage Program (SJVDP). 1990. A Management Plan for Agricultural Subsurface Drainage and Related Problems on the Westside San Joaquin Valley, California. U.S. Department of the Interior and California Resources Agency, Sacramento, CA.

Schroeder, H. A., D. V. Frost, and J. J. Balassa. 1970. Essential trace metals in man—Selenium. *J. Chronic Dis.* 23:227–243.

See, R. B., D. L. Naftz, D. A. Peterson, J. G. Crock, J. A. Erdman, R. C. Severson, P. Ramirez, Jr., and J. A. Armstrong. 1992. Detailed study of selenium in soil, representative plants, water, bottom sediment, and biota in the Kendrick Reclamation Project Area, Wyoming, 1988–90. U.S. Geological Survey Water Resources Invest. Report No. 91-4131.

Setmire, J. G., J. C. Wolfe, and R. K. Stroud. 1990. Reconnaissance investigation of water quality, bottom sediment and biota associated with irrigation drainage in the Salton Sea area, California, 1986–87. U.S. Geological Survey Water Resources Invest. Report No. 89-4102.

Stephens, D. W., B. Waddell, L. A. Peltz, and J. B. Miller. 1992. Detailed study of selenium and selected elements in water, bottom sediment, and biota associated with irrigation drainage in the Middle Green River Basin, Utah, 1988–90. U.S. Geological Survey Water Resources Invest. Report No. 92-4084.

U.S. Bureau of Reclamation. 1995. Finding of no significant impact and supplemental environmental assessment, Grassland bypass channel project: Interim use of a portion of the San Luis Drain for conveyance of drainage water through the Grassland Water District and adjacent Grassland areas. FONSI No. 96-01-MP. U.S. Bureau of Reclamation, Sacramento, CA.

U.S. Environmental Protection Agency. 1987. Ambient Water Quality Criteria for Selenium—1987. Publication No. 440/5-87-006. EPA, Washington, DC.

U.S. Environmental Protection Agency. 1995. Drinking Water Regulations and Health Advisories. EPA, Office of Water.

U.S. Fish and Wildlife Service. 1988. Final mitigation report for the closure and post closure maintenance of Kesterson Reservoir. Division of Ecological Services, Sacramento, CA, March 1988, Vol. 1.

Westcot, D. W., S. Rosenbaum, and G. Bradford. 1989. Trace element buildup in drainage water evaporation basins, San Joaquin, Valley. In *Toxic Substances in Agricultural Water Supply and Drainage, an International Perspective.* pp. 123–135. Proceedings of the Second Pan American Regional Conference on Irrigation and Drainage. USCID.

Westcot, D. W., J. E. Chilcott, and G. Smith. 1993. Pond water, sediment and crystal chemistry. In *Management of Irrigation and Drainage Systems; Integrated Perspectives,* pp. 587–594. American Society of Civil Engineers, Irrigation and Drainage Division Park City, UT

Westcot, D. W., J. Karkoski, and R. J. Schnagl. 1996. Non-point Source Policies for Agricultural Drainage. ASCE North American Water and Environment Congress '96. Anaheim, CA, June 24–28,1996. Session C-40, CD ROM.

Young, T. F., and C. H. Congdon. 1994. Plowing New Ground—Using Economic Incentives to Control Water Pollution from Agriculture. Environmental Defense Fund Nonpoint Source Pollution Report—1994.

2

Methods of Analysis for the Determination of Selenium in Biological, Geological, and Water Samples

YUZO TAMARI

Konan University, Kobe, Japan

I. INTRODUCTION

Selenium is widely distributed in the hydrosphere, lithosphere, atmosphere, and biosphere of the earth in chemically different oxidation states: selenide (Se^{2-}), elemental selenium (Se^0), selenite (Se^{4+}), selenate (Se^{6+}), and organic selenium (organically complexed species such as selenomethionine or selenocysteine).

Concentrations of selenium in seawater have been reported as follows: as Se^{4+} in coastal seawater of Japan, 20 ng/L (Nakaguchi et al., 1985), 22–24 ng/L (Tamari et al., 1989) and 27–30 ng/L (Tamari et al., 1991); as total selenium, 40–80 ng/kg (Nakaguchi et al., 1995) in surface water and 100 ng/kg (Nakaguchi et al., 1995) in the 1000 m depth-layer water of the North Pacific and 80 ng/L (Cutter and Cutter, 1995) in the Atlantic Ocean. Concentrations of selenium in freshwater have been reported as follows: 30–50 ng/L as total selenium in lake water (Nakaguchi et al., 1985); 7 ng/L as Se^{4+} and 9 ng/L as Se^{6+} in river water in Japan (Nakaguchi et al., 1985); 0.3–54.6 ng/L (mean 11.5 ng/L, $n = 29$) as Se^{4+} and 0.1–20.5 ng/L (mean 3.8 ± 4.7 ng/L, $n = 29$) as Se^{6+} in groundwater used for drinking water in Japan (Tamari et al., 1987); 22.0–60.5 ng/L as Se^{4+} in groundwater for Japanese sake brewing (Tamari et al., 1993); and 19.1–30.0 ng/L (mean 24.4 ng/L, $n = 5$) as Se^{4+} in commercially available mineral water (Tamari et al., 1993). Accordingly, the concentration of total selenium is estimated to range from 10 to 100 ng/L both in seawater and freshwater except in rain and in polluted water.

In geological materials selenium content was estimated at 0.03 to 0.8 mg/kg with an average of 0.09 mg/kg for igneous rocks by Goldschmidt and Strock (1935). The selenium content of rock and rock-forming minerals and the dissolution of selenium by chemical weathering have been investigated; total selenium concentrations of 0.1 to 136 μg/kg with a geometric mean of 8.6 μg/kg for 115 igneous rocks and 2.2 to 1280 μg/kg with a geometric mean of 88.1 μg/kg for 46 sedimentary rocks have been reported (Tamari et al., 1990), indicating the enrichment of selenium both in alkali-feldspar (7.5–149 μg/kg) and in plagioclase (3.6–147 μg/kg) (Tamari et al., 1990).

The selenium content of biological materials can vary widely. For example, content of selenium in food seems to depend on the location and the plants grown there. The following data were reported by Foster and Sumar (1995a) on the selenium content of food in various countries: cereals and cereal products (0.01–0.56 mg/kg), meat, offal, fish, and eggs (0.03–1.33 mg/kg), milk and dairy products (0.002–0.3 mg/kg), and vegetables and fruits (0.001–0.01 mg/kg).

Thus the selenium content of environmental materials has been indicated generally to be in the range of milligrams per kilogram or less except for sediments, seleniferous plants, and water and biological materials in selenium-contaminated areas. For the determination of selenium in environmental samples, methods involving greater sensitivity of selenium are required.

Generally fluorometric analysis (FA), hydride generation (HG) atomic absorption spectrometry (AAS), graphite furnace (GF) AAS, inductively coupled plasma mass spectrometry (ICP-MS), single-column ion chromatography (SC-IC), and neutron activation analysis (NAA) are used for the determination of trace selenium with the following absolute detection limits of selenium: 0.05 ng for FA (Tamari, 1984; Tamari et al., 1986, 1993), 0.2 ng/mL for HG-AAS (Tamari et al., 1992, 1993, 1996a), 1–25 ng for GF-AAS (Kölbl, 1995), 0.1 ng for ICP-MS (Kölbl, 1995), 3–110 ng/mL for SC-IC (Karlson and Frankenberger, Jr., 1986a, 1986b; Mehra and Frankenberger, 1988), and 20 ng for NAA at the neutron flux of 1×10^{12} n/cm^2/s (Suzuki and Hirai, 1980; Suzuki et al., 1982; Okada et al., 1987), although this detection limit depends on the neutron flux of an atomic reactor. For the determination of selenium in real samples, selenium should be separated from the sample matrix, since the matrix interferes with the measurement of selenium of the analysis described above. However, these interferences can be removed, because in FA selenium is measured after the element has been separated by solvent extraction as a selenium complex reacted with fluorescent agents, and in HG-AAS selenium is also determined by separating the selenium as hydrogen selenide.

This chapter discusses FA as a classical and highly sensitive method, HG-AAS as a rapid, convenient, and popular method, and NAA as a nondestructive and reliable method for the determination of trace selenium in biological, geological, and water samples.

II. FLUOROMETRIC ANALYSIS

Analytical reagents of o-phenylenediamine, 3-3'-diaminobenzidine (DAB), and 2,3-diaminonaphthalene (DAN) are among the aromatic diamino compounds that have been used for the spectrometric and fluorometric determination of trace selenium. Selenium reacts with their ortho-diamino groups, forming the selenodiazol five-membered ring system (Parker and Harvey, 1962). Fluorometry generally has greater sensitivity than spectrometry for the detection of selenium. DAN is the most sensitive reagent among the aromatic diamino compounds available and has been widely applied for determining the selenium content in many environmental samples with different matrix components. The reagent DAN selectively reacts with selenite (Se^{4+}) but not with selenate (Se^{6+}), forming the strongly fluorescent complex of 4,5-benzopiazselenol, which is usually extracted with cyclohexane or benzene. Selenite is determined by measurement of the fluorescence of piazselenol (selenite–DAN complex) in an organic solvent. The fluorometric determination of selenium with DAN was first developed by Parker and Harvey (1962), later by Lott et al. (1963), investigating the interferences of foreign ions, and by Watkinson (1966), who applied fluorometry to biological samples and introduced an effective sample digestion treatment.

    The specific reaction of DAN to selenium has been demonstrated by means of radioisotope experiments. The separation factor β of selenium against foreign ions is calculated by the following equation (Tamari, 1979)

$$\beta: \frac{D_1}{D_2} = \frac{E_1(100-E_2)}{E_2(100-E_1)}$$

where D_1 and E_1 are the distribution ratio and the percent extraction, respectively, of ^{75}Se, and D_2 and E_2 are those of another nuclide at the same volume ratio. The value β of selenium has been calculated to be 2×10^6 for Fe (^{59}Fe), 4×10^6 for Cr (^{51}Cr), 2×10^4 for Sc (^{46}Sc), 2×10^5 for Co (^{60}Co), and 2×10^6 for Ce (^{141}Ce). Based on this radiotracer experiment, the resulting selenium separation factor is within the magnitude of 10^4–10^6 for any ion. This indicates that DAN is a highly selective reagent for selenium. Therefore DAN can be used not only as a fluorometric reagent but also as a separation reagent for selenium.

A. Purification of 2,3-Diaminonaphthalene

The purification of DAN is essential for determining nanogram amounts of selenium because the low detection limit of selenium depends on a low reagent-blank value. The DAN solution (0.1 % w/v) used for the fluorometric analysis of selenium is prepared by the following three purification steps (Tamari, 1984).

1.   A 1 g portion of DAN produced by Tokyo-Kasei-Kogyo (Tokyo), or Aldrich Chemical (Milwaukee, WI) is dissolved with 20 mL of ethanol on a hot plate. Impurities are filtered with a Toyo No. 5A filter paper, and the filtrate is cooled to about 0°C. The recrystallized DAN is then filtered with a 3G glass filter, washed with 20 mL of 50% cold ethanol, dried in a desiccator for a few days in the dark, and stored in a freezer. The recrystallization step can be skipped when freshly manufactured DAN reagent is available, since DAN may be oxidized during the recrystallization and drying treatments.

2.   A 0.1 g portion of the dry DAN powder is dissolved on a hot plate with 100 mL of 0.1 mol/L hydrochloric acid containing 0.5 g of hydroxylamine hydrochloride. After cooling, 2 g of sodium acetate anhydride as a salting-out agent is added to the solution to extract the fluorescent impurities effectively to the organic phase (Table 1 and Fig. 1). The DAN solution is adjusted to pH 1 with a few milliliters of hydrochloric acid, transferred to a separatory funnel, and extracted three times using 50 mL of cyclohexane each time. The DAN solution is stored in a freezer.

3.   Before the experiments, the frozen DAN is dissolved by warming at 50°C for more than 20 minutes. After cooling to room temperature, 100 mL of the DAN solution is shaken with 50 mL of cyclohexane, and the aqueous phase is shaken with 25 mL of chloroform. The aqueous DAN solution is then centrifuged at 3000 rpm for 5 minutes to separate the chloroform.

TABLE I   Effect of the Purification of DAN Solutions on the Reagent Blank Values in the Fluorometry of Selenium

Purification procedure of DAN solutions	Reagent blank/RFI[a]
DAN solution (solution No. 1)	26.0
Solution No. 1 purified by cyclohexane extraction (solution No. 2)	18.0
Solution No. 1 containing sodium acetate, purified by cyclohexane extraction (solution No. 3)	12.3
Solution No. 3 purified by chloroform extraction (solution No. 4)	10.0
	10.1
	10.4
	10.5
	10.3 ± 0.6
Determination of 10 ng of selenium with the solution No. 4	45.3

[a]Relative fluorescence intensity measured by DAN fluorometry.
Source: Tamari (1984).

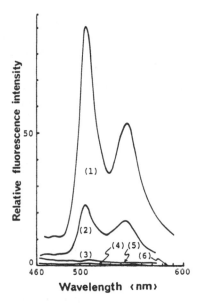

FIGURE 1   Fluorescent spectra of impurities extracted from DAN solutions supplied by Aldrich (A) and Tokyo-Kasei-Kogyo (T): peaks 1 (A) and 2 (T), first extraction, peaks 3 (A) and 4 (T), second extraction; peaks 5 (A) and 6 (T), third extraction. (From Tamari, 1984.)

B. Wet Digestion of Samples

For the determination of trace selenium content in environmental materials, the samples should be digested with an appropriate acid mixture prior to application of the selenium methods such as FA and AAS; predigestion is not necessary for the nondestructive NAA procedure. To decompose the silicate matrix of samples, a hydrofluoric–nitric–perchloric acid mixture is often used for determining the total selenium content of rocks (Tamari, 1984; Tamari et al., 1990) and sediments (Tamari, 1979; Tamari et al., 1978, 1979). To decompose biological samples, the following acids have been used for selenium analysis: nitric acid–urea for blood (Kurahashi et al., 1980); nitric–perchloric acids for urine (Geahchan and Chambon, 1980; Lalonde et al., 1982; Lane et al., 1982); for blood (Lalonde et al., 1982; Lane et al., 1982; Koh and Benson, 1983; Ducros et al., 1994), for hair (Tamari et al., 1986), and for milk (Tamari et al., 1992, 1993, 1995, 1996a, 1996b); sulfuric–nitric acids for blood (Lloyd et al., 1982), for skin (Fairris et al., 1988), and for oily wastes (Campbell and Kanert, 1992); sulfuric acid–nitric acid–hydrogen peroxide for wastewater (Krivan et al., 1985); and sulfuric–nitric–perchloric acids for blood (McLaughlin et al., 1990; Mikac-Devic et al., 1990); and for feeds (Hocquellet and Candillier, 1991).

In experimental attempts to recover selenium from two different digestion treatments with nitric–perchloric acids and sulfuric–nitric acids for a standard reference material (NBS SRM 1549, milk powder), the decomposition temperature was changed from 150°C to 350°C. Almost 100% of the total selenium in the milk was recovered with the nitric–perchloric acid treatment at any reaction temperature, whereas 6 to 11% of the total selenium was recovered with the sulfuric–nitric acid treatment (Tamari et al., 1996a). Decomposition methods for milk selenium analysis have been summarized by Foster and Sumar (1995b), who indicated that sulfuric–nitric acids are effective for the decomposition of the sample, but the presence of residual sulfuric acid interferes with the fluorescence signal of DAN because DAN crystallizes.

Generally wet digestion with nitric–perchloric acids is recommended for selenium analysis to decompose biological samples. In this digestion, 0.5 to 2 g of a sample is placed to a 50 mL beaker, and 2 mL of nitric acid and 2 mL of perchloric acid are added. The mixture is heated on a hot plate at 200°C. Sometimes a few milliliters of nitric acid is added to avoid carbonization. The mixture is heated until white fumes of the perchloric acid appear and continue to evolve for a few minutes. During the digestion, small amounts of perchloric acid are added dropwise to maintain the solution volume at approximately 1 mL and to avoid dryness or near dryness, since the acid is slowly consumed by heating. It is, however, noted that dryness or near-dryness can result in volatilization of selenium due to higher temperature. This volatile selenium is considered to be selenium chloride such as $SeCl_4$ (Tamari et al., 1993) or oxochloride such as $SeO \cdot Cl_2$ and $SeO_2 \cdot 2HCl$ (Dedina and Tsalev, 1995). To prevent this loss of selenium, be sure to keep the solution volume at about 1 mL during wet digestion.

With the nitric–perchloric acid digestion, Tamari et al. (1992) have indicated that not all the selenium is oxidized to selenate (Se^{6+}) during digestion. Rather, 24 to 88% of the total selenium in milk samples is reduced to selenite (Se^{4+}) by the hydrochloric acid produced with the decomposition of perchloric acid, or by chloride ions contained in the milk sample. For the fluorometry of selenium, all the selenium in the digest sample should be reduced to selenite, since the DAN reacts selectively with selenite but not with selenate.

In wet digestion the added nitric acid should be completely removed from the sample solution. Otherwise else the residual nitrate ions can interfere in the fluorometric determination with DAN; that is, the fluorescence intensity of piazselenol (selenium–DAN complex) tends to decrease, producing a negative error in the fluorometry of selenium (Tamari et al., 1986). To remove the nitric acid in the digestion step, continue to heat the almost completely decomposed solution until white fumes of the perchloric acid appear and evolve for a few minutes more. The solution volume needs to be maintained at about 1 mL by adding perchloric acid to avoid dryness during digestion.

C. Reduction of Selenate to Selenite

For the determination of total selenium in a sample, all the selenium must be reduced to selenite, because in FA, the DAN reagent selectively reacts with selenite (Se^{4+}). Also, in HG-AAS, the selenite species are detectable under conditions described later in Section III.

Hydrochloric acid has been mainly used for selenium reduction, since this reduction treatment is rapid and convenient, and there is less contamination during the reduction of geological samples (Tamari et al., 1979, 1990; Tamari, 1984; Dong et al., 1987; Haygarth et al., 1993; Saraswati et al., 1995) and biological samples (Kurahashi et al., 1980; Geahchan and Chambon, 1980; Lane et al., 1982; Lloyd et al., 1982; Koh and Benson, 1983; Krivan et al., 1985; Tamari et al., 1986, 1992, 1995, 1996a, 1996b; Fairris et al., 1988; McLaughlin et al., 1990; Mikac-Davic et al., 1990; Hocquellet and Candillier, 1991; Arikawa and Iwasaki, 1991; Campbell and Kanert, 1992; Ducros et al., 1994).

This reduction depends on the following reaction:

$$SeO_4^{2-} + 2HCl \rightarrow SeO_3^{2-} + H_2O + Cl_2$$

Selenium reduction has been performed by boiling for 3 minutes the sample solution, which is prepared to contain greater than 6 mol/L hydrochloric acid. Figure 2 shows the effect of the reaction temperature on the selenium reduction in the 6 mol/L hydrochloric acid solution for a 10-minute reaction. Figure 3 shows the effect of the concentration of hydrochloric acid when boiling proceeds for 3 minutes. Selenate (Se^{6+}) is quantitatively reduced to selenite (Se^{4+}) by boiling with 6 mol/L hydrochloric acid for 3 minutes, as is seen in Figure 4.

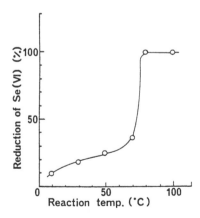

FIGURE 2   Effect of the reaction temperature under 6 mol/L hydrochloric acid for a 10-minute reaction, using standard selenate, 50 ng as Se. (From Tamari et al., 1993.)

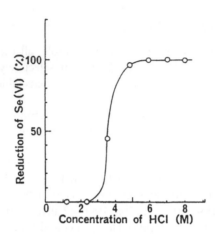

FIGURE 3 Effect of the concentration of hydrochloric acid under boiling for 3 minutes, using standard selenate, 50 ng as Se. (From Tamari et al., 1993.)

When concentrated hydrochloric acid is added to acidify the standard selenate or sample solutions, part or considerable amounts of the selenate in the solutions are reduced to selenite because of the exothermic reaction by adding the acid (Tamari et al., 1987, 1993; Tamari and Ogura, 1997). Therefore use

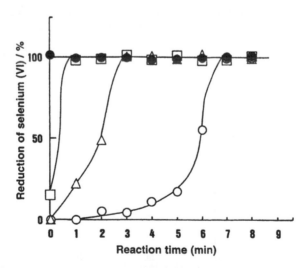

FIGURE 4 Effect of the concentration of hydrochloric acid and reaction time on the reduction of selenium: open circles, 2; triangles, 4; squares, 6; solid circles, 8 mol/L HCl. (From Tamari et al., 1996a.)

caution when adding concentrated hydrochloric acid for the preparation of standard selenate solutions, and for the preparation of natural water samples for the determination of selenate species. To avoid reduction at this stage, hydrochloric acid at a concentration of less than 6 mol/L should be used for acidification.

Potassium bromide has also been used as a reductant of selenate to selenite (Nakaguchi et al., 1985; Tanaka et al., 1986; Tamari et al., 1993; Tamari and Ogura, 1997). It is useful for the selenium reduction with dilute hydrochloric acid (1–1.3 mol/L) and has been applied for the speciation of selenium in natural waters.

D. Fluorometric Interference

The DAN reagent reacts with trace amounts of organic compounds such as aldehyde and α-ketonic acid to yield imidazol and quinoxaline derivatives, respectively. This reagent also has been used for the fluorometry of trace pyruvic acid, cinnamic aldehyde, and benzoaldehyde (Tamari et al., 1993). Note that the DAN reagent can react not only with selenium but also with trace organic compounds derived from biological samples, if the sample is not completely destroyed in the wet digestion step. Figure 5 shows the fluorescent interference of organic compounds produced by the direct reaction of DAN with the digest solutions of samples of cow's milk (2 g used) and human plasma (1 g used). The fluorescent spectrum for piazselenol is greatly affected by the spectra for the reaction products of DAN, with the residual organic compounds at wavelengths of about 470 to 530 nm. Decomposition of the biological sample by wet digestion gives variable results because the degree of decomposition differs from sample to sample, depending on the skill of the analyst and on the type of sample. Accordingly, highly positive errors may occur in the fluorometric determination of selenium if the analyst fails to identify the standard piazselenol spectrum with the correlated sample spectrum. However, this interference can be removed by back-extracting the first extracted organic layer with concentrated nitric acid, redecomposing the nitric layer, and then reextracting with the purified DAN (Tamari et al., 1979, 1986, 1993; Tamari, 1984). This modified DAN fluorometry for biological samples (Tamari et al., 1986) has been applied for the determination of nanogram amounts of selenium in infant formulas and in the colostrum of human milk samples (Tamari et al., 1995).

III. HYDRIDE GENERATION ATOMIC ABSORPTION SPECTROMETRY

Flame atomic absorption spectrometry (F-AAS) is a well-established technique, but its detection limits for selenium were reported to be approximately 0.1 μg

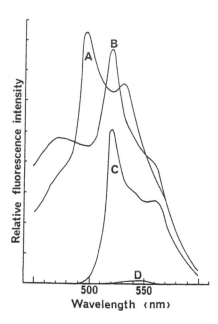

FIGURE 5   Fluorescent spectra of interfering organic compounds reacted with 2,3-diaminonaphthalene. Selenium was determined with DAN after A, the decomposition of 2 g of cow's milk with HNO$_3$; B, the decomposition of 1 g of human plasma with HNO$_3$-HClO$_4$; C, standard piazselenol (Se-DAN complex, 50 ng as Se); D, reagent blank. (From Tamari et al., 1993.)

Se/mL and 10 to 100 μg Se/mL in linear calibration graphs with an air–acetylene flame (Kölbl, 1995). Graphite furnace atomic absorption spectrometry (GF-AAS) offers lower detection limits for the determination of selenium, 0.5 to 50 ng Se/mL with 20 μL sample volume (Kölbl, 1995), and is effective for the assay of small amounts of a sample solution. Generally, however, there is a requirement for matrix modifiers such as copper–magnesium, nickel, nickel–magnesium, palladium, palladium–magnesium, palladium–ascorbic acid, silver–copper–magnesium, rhodium, and others (Hirano et al., 1994; Foster and Sumar, 1995b).

Hydride generation atomic absorption spectrometry (HG-AAS) has been widely used for the determination of trace selenium in many kinds of environmental samples since this technique is rapid and convenient, and has few interferences owing to the separation of selenium as hydrogen selenide from a sample solution containing matrix components. The detection limit for selenium in HG-AAS has been considered to be 0.1 to 0.3 ng Se/mL with a quartz tube atomizer (QTA) (Tamari et al., 1992; Dedina and Tsalev, 1995). The HG technique is effective in avoiding matrix interferences for trace selenium analysis. Accordingly, this technique has been applied to HG inductively coupled plasma atomic emission spectrometry (ICP-AES), of which the detection limit for selenium is about three-fold worse than that in HG-AAS (with QTA), and also has been applied to HG inductively coupled plasma mass spectrometry (ICP-MS), of which the detection limit is as low as 0.1 ng Se (Kölbl , 1995) and 1 to 30 ng Se/L (Dedina and Tsalev, 1995) in the continuous flow mode.

In the HG system, sodium tetrahydroborate ($NaBH_4$) is exclusively used as a reductant to the selenide form, owing to the low selenium impurity contained in the reagent and to the high yield of selenide generated by the reductant. As shown in Figure 6, the hydride generator consists of a peristaltic pump by which all the solutions are introduced and mixed in a manifold. Hydrogen selenide that is generated is introduced to a quartz tube with argon under the conditions shown in Table 2. The hydride formation depends on the following reaction:

$$4H_2SeO_3 + 3BH_4^- + 3H^+ \rightarrow 3H_3BO_3 + 3H_2O + 4H_2Se$$

The hydride generation is usually performed in a hydrochloric acid solution of a few moles per liter. In the presence of hydrochloric acid, most of the tetrahydroborate is hydrolyzed in acid media, producing considerable hydrogen:

$$BH_4^- + H^+ + 3H_2O \rightarrow H_3BO_3 + 4H_2$$

Under acid conditions, selenite species can be reduced to hydrogen selenide, whereas selenate species are not reduced, as indicated by Tamari et al. (1992, 1996a) and Dedina and Tsalev (1995). Therefore, before hydride generation is

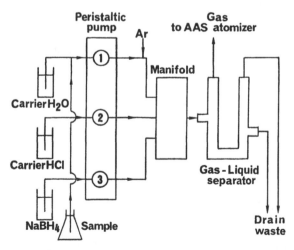

FIGURE 6   Hydride generation system diagram for selenium analysis. (From Tamari et al., 1996a.)

TABLE 2 Instrumental Conditions for Selenium Analysis

AAS (Hitachi, model Z6100)	
Wavelength of hollow cathode lamp (HCL)	196.0
Electric current of HCL	12.5 mA
Acetylene pressure	0.2 kg/cm^2
Air pressure	1.6 kg/cm^2
Hydride generation apparatus (Hitachi, model HFS-2)	
Argon	100 mL/min
Sampling time	0.7 min
Reaction time	0.5 min
Acid carrier	3.6 mol/L HCl
Reducing reagent	0.75 w/v % NaBH4
Sample solution	3.6 mol/L HCl
Detection limit	0.2 ng/mL Se

Source: Tamari et al. (1996a).

undertaken, all selenium must be reduced to selenite by boiling the selenium solutions with 6 mol/L hydrochloric acid for 3 minutes, as described in Section II. In addition to the HG technique, the use of a higher concentration of hydrochloric acid, to increase the acidity of both the acid carrier and the sample solutions, can give rise to higher sensitivity or greater absorbance of selenium, as shown in Table 3. Accordingly, hydrochloric acid (a few mol/L) is often used for HG to generate the steady detectable hydrogen selenide, because the use of more than 6 mol/L hydrochloric acid could cause unstable absorbance due to the

TABLE 3 Effect of the Acidity of Acid Carrier and Sample Solutions

	Relative absorbance (mm in scale)	
Concentration[a]	10 ng/mL selenite	10 ng/mL selenate
6.0 mol/L HCl	104	0
4.8 mol/L HCl	105	0
3.6 mol/L HCl	100	0
2.4 mol/L HCl	98	0
0.9 mol/L HCl	56	0
0.6 mol/L HCl	29	0
0.09 mol/L HCl	30	0
H$_2$O	0	0

[a]The concentration of HCl in a measuring solution is equal to that in an acid carrier solution.
Source: Tamari et al. (1996a).

violent reaction of this acid with tetrahydroborate. To determine selenium in the higher acid concentration in HG in the interest of obtaining high sensitivity, attention is necessary because the different for commercial HG generators have different flow rates.

IV. NEUTRON ACTIVATION ANALYSIS

Neutron activation analysis (NAA) offers a method for determining trace selenium by measuring the γ-radiation intensity of ^{75}Se (half-life 120 days) or ^{77m}Se (half-life 17.5 s). In nondestructive NAA, direct measurements of these selenium nuclides are difficult because of the large interference of other radionuclides, such as ^{24}Na, ^{28}Al, and ^{38}Cl, which are shorter-lived radionuclides than ^{75}Se and are derived from matrix compounds of biological and geological samples. Therefore, the measurements of ^{75}Se should generally be performed on long counting periods after the decay of the large radioactivities of the short-lived nuclides (i.e., more than a week later, at least). Nondestructive NAA has been applied for the determination of hair selenium content (Tamari et al., 1986), since hair selenium is within the milligrams-per-kilogram range and few interfering radionuclides are generated. In contrast to hair samples, the Standard Reference Material 1571 (orchard leaves) was analyzed, and no ^{75}Se (energy of 265 keV) and also no ^{77m}Se (energy of 162 keV) were detected after neutron irradiation had proceeded for 5 hours or 10 seconds at the neutron flux of 10^{12} n/cm^2/sec, because the selenium content is certified as low as 80 μg/kg (Suzuki and Hirai, 1980). Also no ^{75}Se in sediment samples was detected by instrumental NAA with 80 hours of irradiation at the neutron flux of 5 × 10^{13} n/cm^2/sec, because the γ-radiation intensity of the ^{75}Se was at a lower order of magnitude than that of ^{59}Fe, ^{46}Sc, and ^{60}Co even after the decay of ^{24}Na, ^{28}Al, ^{38}Cl, and other short-lived isotopes (Tamari, 1979). Chemical separation techniques are, therefore, essential for the determination of selenium, to eliminate interfering radioactivities, which would result in a lower detection limit of selenium. In radiochemical NAA, techniques such as digestion of samples, reduction of selenium, and extraction of selenium with DAN also are required, as well as in FA and HG-AAS.

Table 4 compares FA, HG-AAS, and NAA for the determination of selenium in U.S. and Japanese standard reference materials. Fairly good agreement of selenium determination has been obtained between the certified values and the data by FA, by HG-AAS, and by NAA. Figure 7 shows the selenium content of formula milk, rice and tea leaf samples determined by FA and by HG-AAS, and also shows hair samples determined by FA and by nondestructive NAA . Between FA and HG-AAS, a coefficient of correlation, $\gamma = 0.99$ for 16 samples, was obtained. Between FA and NAA, the correlation coefficient, $\gamma = 0.91$ for 29 hair samples, may be not statistically outstanding, but it seems to be reasonable in light

TABLE 4   Comparison of Selenium Content (ng/g) of Biological Materials:
Three Methods

Standard reference materials	Fluorometric analysis	Atomic absorption spectrometry	Neutron activation analysis	Certified value
NBS-SRM 1549, Milk powder	101 ± 8	103 ± 6		110 ± 10
NBS-SRM 1571, Orchard leaves	70.5 ± 1.1	88.9 ± 1.7	50 ~90	80
NIES, Rice flour, unpolished				
Low Cd	56.8 ± 1.9	63.7 ± 1.7	61 ± 5	60
Medium Cd	21.5 ± 1.9	18.6 ± 1.9	20 ± 2	20
High Cd	69.5 ± 2.2	64.2 ± 1.0	68 ± 5	70
NIES Tea leaves	20.5 ± 1.1	17.3 ± 1.8	22 ± 6	

[a]NBS-SRM, U.S. National Bureau of Standards (now National Institute of Standards and Technology) Standard Reference Material; NIES, National Institute of Environmental Studies (of Japan).
Source: Tamari et al. (1993).

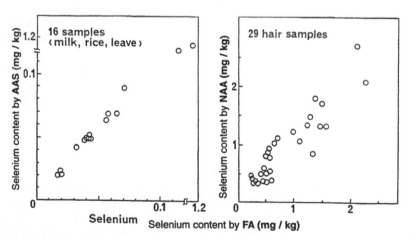

FIGURE 7   Relationships between fluorometric analysis (FA) and hydride generation atomic absorption spectrometry (HG-AAS) and FA and nondestructive neutron activation analysis (NAA). (From Tamari et al., 1993.)

of a mean selenium content with 30% of relative standard deviation determined by the NAA.

V. PRECONCENTRATION AND SPECIFICATION OF SELENIUM

Determination of selenium in natural waters has been investigated at nanogram-per-liter levels, as described at the beginning of this chapter. Accordingly, preconcentration techniques for selenium in water samples are required before analysis by FA and HG-AAS.

Coprecipitation techniques have been most commonly applied for the preconcentration of selenium in natural water samples, including iron(III) hydroxide, which coprecipitates selectively the selenite but not the selenate species in river and sea waters (Hiraki et al., 1973; Yoshii et al., 1977), and tellurium, which collects both selenite and selenate in 1 mol/L hydrochloric acid solution, with reduction of gray metallic tellurium from tellurium(IV) by means of hydrazine sulfate (Yoshii et al., 1977). However, selenium impurities contained in the iron(III) and tellurium(IV) used as coprecipitation carriers are serious problems for the determination of nanogram amounts of selenium. Other coprecipitants containing a negligible amount of selenium impurity following coprecipitation techniques have been reported: zirconium(IV) hydroxide, which collects both the selenite and selenate in groundwater (Tamari et al., 1987) but selectively collects selenite in seawater, probably as a result of the interference of large amounts of seawater sulfate ions (Tamari et al., 1991); lanthanum(III) hydroxide, which collects selenite (Maher, 1982; Tao and Hansen, 1994); and thorium(IV) hydroxide, which coprecipitates selenite in seawater (Tamari et al., 1989).

Compared with coprecipitation techniques, solvent extraction is a rapid and convenient method for the preconcentration of selenium and for the separation of selenium from sample matrices. A method of solvent extraction of selenite in a water sample with DAN was developed and applied for the speciation of selenite, selenate, and organic complexed selenium species in lake water (Nakaguchi et al., 1985). Another method for specification of selenite was developed to collect as a selenite complex with Bismuthiol II on activated carbon from ground- and rainwater samples (Okutani et al., 1991). Both in the former and latter chelating methods, selenate species in water are determined after the selenate has been reduced to selenite by boiling with 1.3 and 1.5 mol/L hydrochloric acid solutions containing potassium bromide, respectively. Speciation of inorganic and organic selenium in marine environments has been reviewed by Nakayama et al. (1989), Dauchy et al. (1994), Olivas and Donard (1994), Kölbal (1995), and Nakaguchi et al. (1995).

With respect to speciation of selenium in geological samples, chemical leaching techniques have been usually applied. An ultrasonic bath has been used to leach the selenium in a river sediment sample with 0.2 or 2 mol/L hydrochloric acid to estimate the selenium derived from the sedimentation with iron(III) hydroxide particulates (Tamari et al., 1978, 1982b). Also CO_2-bubbling water has been used to estimate the leachable selenium dissolved by chemical weathering of rocks and rock-forming minerals (Tamari et al., 1990). To elucidate the vertical distribution of selenium content in a reservoir core sediment, 0.1 mol/L hydrochloric acid has been used to leach the inorganic selenium associated with the sedimentation of iron(III) hydroxide, followed by a 1 mol/L sodium hydroxide solution to leach out organically complexed selenium, and finally by hydrofluoric acid to dissolve the selenium-related silicate residue (Tamari et al., 1990). A separation technique to determine selenide constituents from the powdered mixtures of elemental selenium, selenite, and selenate has also been developed (Tamari et al., 1982a).

Further studies are needed to elucidate the geochemical cycle of selenium on the earth. Trace selenium methods that are accurate, precise, and detectable in amounts as low as nanograms or picograms are required for this purpose. Advancements in the speciation methods for trace selenium should also be developed to clarify the role of selenium in the environment and the biosphere.

REFERENCES

Arikawa, Y., and M. Iwasaki. 1991. Determination of selenium in biological samples by hydride generation–AAS after combustion in high-pressure oxygen (in Japanese). *Nihon Kagaku Kaishi*, pp. 120–124.

Campbell, M. B., and G. A. Kanert. 1992. High-pressure microwave digestion for the determination of arsenic, antimony, selenium and mercury in oily wastes. *Analyst* 117:121–124.

Cutter, G. A., and L. S. Cutter. 1995. Behavior of dissolved antimony, arsenic and selenium in the Atlantic Ocean. *Mar. Chem.* 49:295–306.

Dauchy, H., M. Potin-Gautier, A. Astruc, and M. Astruc. 1994. Analytical methods for the speciation of selenium compounds: A review. *Fresenius J. Anal. Chem. 348*:792–805.

Dedina, J., and D. L. Tsalev. 1995. *Hydride Generation Atomic Absorption Spectrometry*, pp. 308–354. Wiley, Chichester.

Dong, A., V. V. Rendig, R. G. Buran, and G. S. Besga. 1987. Phosphoric acid, nitric acid, and hydrogen peroxide digestion of soil and plant materials for selenium determination. *Anal. Chem.* 59:2728–2730.

Ducros, V., D. Ruffieux, N. Belin, and A. Favier. 1994. Comparison of two digestion methods for the determination of selenium in biological samples. *Analyst* 119:1715–1717.

Fairris, G. M., B. Lloyd, and H. T. Delves. 1988. Skin selenium content measured by hydride generation and atomic absorption spectrometry. *J. Trace Elem. Electrolytes Health Disease* 2:181–184.

Foster, L. H., and S. Sumar. 1995a. Selenium in the environment, food and health. *Nutr. Food Sci.* 5:17-23.

Foster, L. H., and S. Sumar. 1995b. Methods of analysis used for the determination of selenium in milk and infant formulas: A review. *Food Chem.* 53:453–466.

Geahchan, A., and P. Chambon. 1980. Fluorometry of selenium in urine. *Clin. Chem.* 26:1272–1274.

Goldschmidt, V. M., and L. W. Strock. 1935. *Ges. Wiss. Gottingen Math. Physik. Kl.* 1:123. Cited in S. D. Faust, and O. M. Aly. 1981. *Chemistry of Natural Waters*, p. 360. Ann Arbor Science Publishers, Ann Arbor, MI.

Haygarth, P. M., A. P. Rowland, S. Stürup, and K. C. Jones. 1993. Comparison of instrumental methods for the determination of total selenium in environmental samples. *Analyst* 118:1303–1308.

Hiraki, K., O. Yoshii, H. Hirayama, Y. Nishikawa, and T. Shigematsu. 1973. Fluorometric determination of selenium in sea water (in Japanese). *Bunseki Kagaku* 22:712–718.

Hirano, Y., K. Yasuda, and K. Hirokawa. 1994. Relationship between effective atomic vapor temperature of selenium and matrix modifiers in graphite furnace atomic absorption spectrometry (in Japanese). *Bunseki Kagaku* 43:105–110.

Hocquellet, P., and M. P. Candillier. 1991. Evaluation of microwave digestion and solvent extraction for the determination of trace amounts of selenium in feeds and plant and animal tissues by electrothermal atomic absorption spectrometry. *Analyst* 116:505–509.

Karlson, U., and W. T. Frankenberger, Jr. 1986a. Determination of selenate by single-column ion chromatography. *J. Chromatogr.* 368:153–161.

Karlson, U., and W. T. Frankenberger, Jr. 1986b. Single-column ion chromatography of selenite in soil extracts. *Anal. Chem.* 58:2704–2708.

Koh, T. S., and T. H. Benson. 1983. Critical re-appraisal of fluorometric method for determination of selenium in biological materials. *J. Assoc. Off. Anal. Chem.* 66:918–926.

Kölbl, G. 1995. Concepts for the identification and determination of selenium compounds in the aquatic environment. *Mar. Chem.* 48:185–197.

Krivan, V., K. Petrick, V. Welz, and M. Melcher. 1985. Radiotracer error—Diagnostic investigation of selenium determination by hydride-generation atomic absorption spectrometry involving treatment with hydrogen peroxide and hydrochloric acid. *Anal. Chem.* 57:1703–1706.

Kurahashi, K., S. Inoue, S. Yonekawa, Y. Shimoishi, and K. Toei. 1980. Determination of selenium in human blood by gas chromatography with electron-capture detection. *Analyst* 105:690–695.

Lalonde, L., J. Jean, K. D. Roberls, A. Chapdelaine, and G. Bleau. 1982. Fluorometry of selenium in serum and urine. *Clin. Chem.* 28:172–174.

Lane, H. W., R. D. Alberto, O. Barroso, D. Englert, R. N. Stanley, J. Dudrick, and B. S. MacFadyen. 1982. Selenium status of seven chronic intravenous hyperalimentation patients. *J. Parenter. Enteral Nutr.* 6:426–431.

Lloyd, B., P. Holt, and H. T. Delves. 1982. Determination of selenium in biological samples by hydride generation and atomic absorption spectrometry. *Analyst* 107:927–933.

Lott, P. F., P. Cukor, G. Moriber, and J. Solga. 1963. 2,3-Diaminonaphthalene as a reagent for the determination of milligram–submicrogram amounts of selenium. *Anal. Chem.* 35:1159–1163.

Maher, W. A. 1982. Fluorometric determination of selenium in some marine materials after digestion with nitric and perchloric acids and co-precipitation of selenium with lanthanum hydroxide. *Talanta* 29:1117–1118.

McLaughlin, K., D. Dadgar, and M. R. Smyth. 1990. Determination of selenium in blood plasma and serum by flow injection hydride generation atomic absorption spectrometry. *Analyst* 115:275–278.

Mehra, H. C., and W. T. Frankenberger, Jr. 1988. Simultaneous analysis of selenate and selenite by single-column ion chromatography. *Chromatographia* 25:585–588.

Mikac-Devic, M., D. Ferenec, and A. Tiefenbach. 1990. Serum selenium levels in untreated children with acute lymphoblastic leukemia I. *J. Trace Elem. Electrolytes Health Disease.* 4:7–10.

Nakaguchi, Y., K. Hiraki, Y. Tamari, Y. Fukunaga, Y. Nishikawa, and T. Shigematsu. 1985. Fluorometric determination of inorganic selenium(IV), selenium(VI) and organic selenium in natural waters. *Anal. Sci.* 1:247–252.

Nakaguchi, Y., Y. Koike, and K. Hiraki. 1995. Chemical speciation of selenium in natural waters. In H. Sakai and Y. Nozaki (eds.), *Biogeochemical Processes and Ocean Flux in the Western Pacific,* pp. 139–158. Terra Scientific Publishing, Tokyo.

Nakayama, E., Y. Suzuki, K. Fujiwara, and Y. Kitano. 1989. Chemical analyses of seawater for trace elements. Recent progress in Japan on clean sampling and chemical speciation of trace elements: A review. *Anal. Sci.* 5:129–139.

Okada, Y., S. Suzuki, and S. Hiraki. 1987. Determination of selenium in soybean from different regions by instrumental neutron activation analyses (in Japanese). *Bunseki Kagaku* 36:856–861.

Okutani, T., T. Kubota, N. Sugiyama, and Y. Tsuruta. 1991. Determination of selenium(IV) and selenium(VI) in water samples by graphite furnace atomic absorption spectrometry after collection as 3-phenyl-5-mercapto-1,3,4-adiazole-2(3H)-thione-selenium(IV) complex on activated carbon. *Nihon Kagaku Kaishi,* pp. 375–379.

Olivas, R. M., and O. F. X. Donard. 1994. Analytical techniques applied to the speciation of selenium in environmental matrices: Review. *Anal. Chim. Acta* 286:357–370.

Parker, C. A., and L. G. Harvey. 1962. Luminescence of some piazselenols: A new fluorimetric reagent for selenium. *Analyst* 87:558–565.

Saraswati, R., T. W. Vetter, and R. L. Watters. 1995. Comparison of reflux and microwave oven digestion for the determination of arsenic and selenium in sludge reference material using flow injection hydride generation and atomic absorption spectrometry. *Analyst* 120:95–99.

Suzuki, S., and S. Hirai. 1980. Neutron activation analysis of selenium using a coincidence counting method (in Japanese). *Radioisotopes* 29:161–165.

Suzuki, S., S. Hirai, and K. Noda. 1982. Determination of selenium in herb plants by neutron activation analysis using a coincidence counting method (in Japanese). *Bunseki Kagaku* 31:67–71.

Tamari, Y. 1979. Neutron activation analysis of selenium and 17 elements in sediment. *Radioisotopes* 28:1–6.

Tamari, Y. 1984. Fluorometric determination of nanogram amounts of selenium in rocks. *Bunseki Kagaku* 33:E115–E122.

Tamari, Y., and H. Ogura. 1997. Determination of selenium in water samples by hydride generation AAS with the potassium bromide reduction of selenium (in Japanese). *Bunseki Kagaku* 46: 313–317.

Tamari, Y., K. Hiraki, and Y. Nishikawa. 1978. State analysis of selenium in sediment: Investigation of selenium in sediment by leaching method (in Japanese). *Chikyu Kagaku* 12:37–43.

Tamari, Y., K. Hiraki, and Y. Nishikawa. 1979. Fluorometric determination of selenium in sediments with 2,3-diaminonaphthalene (in Japanese). *Bunseki Kagaku* 28:164–169.

Tamari, Y., K. Hiraki, and Y. Nishikawa. 1982a. Separation of selenide constituents from selenide samples containing elemental selenium. *Bull. Chem. Soc. Jpn.* 55:101–103.

Tamari, Y., Y. Inoue, H. Tsuji, and Y. Kusaka. 1982b. Ultrasonic extraction of several cations with dilute nitric acid solutions from soils. *Bunseki Kagaku* 31:E409–E412.

Tamari, Y., S. Ohmori, and K. Hiraki. 1986. Fluorometry of nanogram amounts of selenium in biological samples. *Clin. Chem.* 32:1464–1467.

Tamari, Y., R. Hirai, H. Tsuji, and Y. Kusaka. 1987. Zirconium coprecipitation method for fluorometric determination of ppt level of selenium(IV) and selenium(VI) in groundwaters. *Anal. Sci.* 3:313–317.

Tamari, Y., M. Kitagawa, H. Tsuji, and Y. Kusaka. 1989. Fluorometric determination of ppt level selenium(IV) in sea water by thorium hydroxide coprecipitation method. *Bull. Soc. Sea Water Sci. Jpn.* 42:242–245.

Tamari, Y., H. Ogawa, Y. Fukumoto, H. Tsuji, and Y. Kusaka. 1990. Selenium content and its oxidation state in igneous rocks, rock-forming minerals and a reservoir sediment. *Bull. Chem. Soc. Jpn.,* pp. 2631–2638.

Tamari, Y., H. Tsuji, and Y. Kusaka. 1991. Fluorometric determination of ppt level selenium(IV) in sea water by a coprecipitation method with zirconium hydroxide. *Bull. Soc. Sea Water Sci., Jpn.* 45:83–86.

Tamari, Y., M. Yoshida, S. Takagi, K. Chayama, H. Tsuji, and Y. Kusaka. 1992. Determination of selenium in biological samples by hydride generation AAS (in Japanese). *Bunseki Kagaku* 44:T77–T81.

Tamari, Y., K. Chayama, and H. Tsuji. 1993. Studies on the analysis of ultra trace amounts of selenium (in Japanese). *Biomed. Res. Trace. Elements* 4:263–270.

Tamari, Y., K. Chayama, and H. Tsuji. 1995. Longitudinal study on selenium content in human milk particularly during early lactation compared to that in infant formulas and cow's milk in Japan. *J. Trace Elements Med. Biol.* 9:34–39.

Tamari, Y., H. Ogura, K. Fujimori, and H. Tsuji. 1996a. Determination of selenium in infant formulas by hydride generation atomic absorption spectrometry. *Mem. Konan Univ., Sci. Ser.* 43:37–46.

Tamari, Y., E. S. Kim, and H. Tsuji. 1996b. Selenium distribution in infant formulas and breast milk. *Met. Ions Biol. Med.* 4:508–510.

Tanaka, S., M. Nakamura, H. Yokoi, M. Yumura, and Y. Hashimoto. 1986. Determination of antimony(III), antimony(V), selenium(IV) and selenium(VI) in natural waters by hydride generation atomic absorbance spectrometry combined with a cold trap (in Japanese). *Bunseki Kagaku* 35:116–121.

Tao, G., and E. H. Hansen. 1994. Determination of ultra-trace amounts of selenium(IV) by flow injection hydride generation atomic absorption spectrometry with on-line preconcentration by coprecipitation with lanthanum hydroxide. *Analyst* 119:333–337.

Watkinson, J. H. 1966. Fluorometric determination of selenium in biological material with 2,3-diaminonaphthalene. *Anal. Chem.* 38:92–97.

Yoshii, O., K. Hiraki, Y. Nishikawa, and T. Shigematsu. 1977. Fluorometric determination of selenium(IV) and selenium(VI) in sea water and river water (in Japanese). *Bunseki Kagaku* 26:91–97.

3
Analytical Detection of Selenium in Water, Soil, and Plants

Ivan S. Palmer
South Dakota State University, Brookings, South Dakota

I. INTRODUCTION

The detection of selenium (Se) in water and soil has gained considerable interest in recent years. The Se content in these two types of environmental samples represents the combination of naturally occurring forms as well as the forms of Se put back into the environment by human activity. The Se content of plants represents the amount of the element extracted from the soil and water sources, and the organic forms in the plants are quite available to the animal world.

The detection of Se in water, soil, and plant material can be accomplished by a number of analytical techniques when the materials being studied are from selenium-rich geographic areas that produce samples with relatively high Se content. It is considerably more difficult to detect the element in samples from Se-deficient areas, and the sensitivities of most methods are challenged by the current demands that the element be measured at the level of 1 μg/L or less in waters. The analytical method chosen will depend on whether a nondestructive measure of Se content is desired or partial or total sample destruction is acceptable. If speciation is desired, then additional problems are introduced that are specific to the sample matrices. This chapter deals mainly with the determination of total Se in water, soil, and plants.

Early methods of Se analysis such as gravimetric, volumetric, and colorimetric methods are not very useful for most environmental samples because they lack sufficient sensitivity and/or selectivity. Gravimetric procedures are still used for analyzing high purity elemental standards such as National Institute of Science

47

and Technology's (NIST) Standard Reference Material (SRM) 726 (selenium). The discovery by Schwarz and Foltz (1957) that Se is a nutritionally essential element made it imperative to develop sensitive methods of analysis, capable of measuring the relatively low levels that are common for biological samples. A number of reviews of the developing methodology for Se analysis were published between 1967 and 1976, including those by Watkinson (1967), Alcino and Kowald (1973), Olson et al. (1973), Cooper (1974), Shendrikar (1974), and Olson (1976). More recently, excellent reviews have been published by Lewis and Veillon (1989) and Thomassen et al. (1994), and this chapter draws heavily on the material from the latter two reviews.

Lewis and Veillon (1989) point out distinctions in the quality of methods of analysis. Definitive methods represented by neutron activation analysis (NAA) and mass spectrometry are of the highest quality, but they require instrumentation too complex and are too expensive to permit their common usage by many laboratories. Reference methods or refereed methods are those that have been compared with the definitive methods. Methods such as the Association of Official Analytical Chemists' (AOAC) fluorometric method for plants (Williams, 1984a) and foods (Williams, 1984b), and the hydride generation atomic absorption spectrometry (HGAAS) methods for food (Williams, 1984c) can be considered reference methods for selenium. The official status of these methods applies only to the specified sample matrices. Verification must be done for each new sample matrix that is studied.

In addition to the official methods recognized by the AOAC, the Environmental Protection Agency (EPA) has designated HGAAS and graphite furnace atomic absorption spectrometry (GFAAS) as acceptable for water analysis (U.S. EPA, 1983). The Standard Methods for Examination of Water and Wastewater (1992), published by the American Public Health Association, has listed hydride generation atomic absorption spectrometry, fluorometry, electrothermoatomic absorption spectrometry (EAAS), and inductively coupled plasma (ICP) methods as acceptable methods for the analysis of Se in environmental samples.

Most methods for selenium determination require either considerable sophistication in instrumentation or relatively extensive experience on the part of the analyst in order to provide reliable results, if a wide variety of sample matrices are to be studied. Of the numerous methods of selenium analysis, many are techniques of separation of Se or its derivatives. Examples include gas liquid chromatography (GLC), liquid chromatography (LC), thin-layer chromatography (TLC), ion-exchange chromatography, high performance liquid chromatography (HPLC), and ion chromatography. Since each of these techniques must be used in conjunction with some form of detection such as fluorometry or AAS, they are not discussed in detail in this chapter. For instance, the product of the reaction of selenite [Se (IV)] and 2,3-diaminonaphthalene, which is the product measured in fluorometric methods, may be determined by GLC, HPLC, and conventional fluorometry. The method of measuring the fluorescence often is determined by the instrumentation available in individual laboratories.

This chapter discusses five methods that have been applied to environmental samples: neutron activation analysis (NAA), mass spectrometry (MS), inductively coupled plasma methods (ICP), atomic absorption spectrometry (AAS), and fluorometry.

II. SAMPLE PREPARATION

Most procedures for the destructive methods of Se analysis, involve the treatment of samples with mineral acids followed by conversion of Se to the +4 valence state (Thomassen, 1994). Care must be exercised to prevent loss of selenium and yet ensure the complete freeing of the element from the sample matrix. Gorsuch (1959) has reviewed these precautions. Preferred digestion procedures involve nitric and perchloric acids (Olson et al., 1973; Nève et al., 1982). By the use of ^{75}Se-labeled samples, Olson et al. (1973) showed that no Se was lost with rigorous digestion in a nitric and perchloric acid mixture if care was used to avoid charring. Others have suggested the use of a combination of nitric, sulfuric, and perchloric acids (Watkinson, 1979; Welz et al., 1984a). A comprehensive review of sample digestion techniques was published by Ihnat (1992). Various digestion techniques including dry ashing are discussed by Reamer and Veillon (1981a), and May (1982) has discussed the losses of selenium by dry ashing. Certainly both wet digestion and dry ashing methods have been successfully used on a variety of sample matrices. However, care is needed to avoid losses on dry ashing, and samples must not be allowed to char during wet digestions.

III. NEUTRON ACTIVATION ANALYSIS (NAA)

Neutron activation analysis has been used for the determination of selenium in a wide concentration range and with a wide variety of sample matrices (National Academy of Sciences, 1976; Thomassen et al., 1994). It can be used with or without destruction of the sample. The nondestructive analysis of samples has the advantage of multielement analysis and speed; however, high concentrations of sodium and chloride in biological samples cause interferences, and usually some sort of sample pretreatment and/or separation is needed to avoid the interferences (Cornelis et al., 1982). Absolute detection limits have been reported as 4 and 10 ng for geological and biological materials, respectively (Thomassen et al., 1994). NAA with isotope dilution techniques has been helpful in nutrition studies, particularly in those involving humans where the use of stable isotopes has a distinct advantage (Janghorbani et al., 1981, 1982; Morris et al., 1981; Christensen et al., 1983; Kasper et al., 1984; Sirichakwal et al., 1985). The main disadvantages of NAA are the limited availability and high cost of the required nuclear reactor and counting facilities (Thomassen et al., 1994).

IV. INDUCTIVELY COUPLED PLASMA
SPECTROMETRY (ICP)

Inductively coupled plasma–atomic emission spectrometry (ICP-AES) has the distinct advantage of providing fast multielement analysis with relatively low detection limits for most elements (Hwang et al., 1989). However, elements such as Se may be present in environmental samples in concentrations too low to detect by means of conventional sample introduction or ultrasonic nebulization. Since Se along with arsenic, antimony, bismuth, and tin form volatile hydrides, introduction of the samples by means of the hydrides improves detection limits and reduces matrix interferences. Tracey and Möller (1990) have developed an ICP-AES method that seems to compare well with other methods in determining Se in water and biological samples.

The use of ICP with mass spectrometry detection (ICP-MS) shows promise, but still requires some analyte concentration for the determination of Se in clinical and environmental samples because of low sensitivity for the element (Thomassen et al., 1994). ICP-MS can apparently provide Se isotope ratio measurements in biological samples with the required accuracy and precision (Ting and Janghorbani, 1987; Ting et al. 1989; Thomassen et al., 1994). Hydride generation can serve as an analyte concentration technique for ICP-MS (Ting et al., 1989) and has recently been applied to the analysis of water (McCurdy et al., 1993; Tao et al., 1993). Krushevska et al. (1996), who reviewed various means of eliminating interferences in the determination of Se by ICP-MS, suggest the use of tertiary amines in the analysis of environmental samples to enhance the Se signal and reduce the interferences. Ding and Sturgeon (1997) have recently reported on the use of hydride generation techniques to minimize transition metal interferences. The necessity of resorting to the use of hydride generation for increasing sensitivity and reducing interferences results in the loss of the ability of simultaneous multielement analysis, which is one of the main benefits of ICP methods. In many laboratories, it may be undesirable to limit the use of ICP equipment to the analysis of one element such as Se.

V. MASS SPECTROMETRY

Lewis and Veillon (1989) have given a thorough review of the use of mass spectrometry techniques for Se analysis. Spark source mass spectrometry (SSMS) is one of the methods used by NIST to determine Se in Standard Reference Materials (Lewis and Veillon, 1989) and other reference materials (Olson et al., 1973). Because of the volatility of the element in a high temperature spark, SSMS is not a well suited for the direct analysis of Se.

MS techniques combined with gas chromatography (GCMS) for Se analysis have proven useful. The introduction of the sample by GC has allowed shorter analysis times and good sample throughput. The limitation of such methods is the availability of suitable chelating agents for the analyte (Lewis and Veillon, 1989). Reamer and Veillon (1981b) developed a method using 4-nitro-*o*-phenylenediamine as the derivatizing agent for the subsequent determination of Se by GCMS. The method was further modified to a double isotope dilution GCMS method for the analysis of various biological matrices (Reamer and Veillon, 1983). The method is particularly useful for metabolic studies in which radioisotopes are unacceptable. The method has been used to analyze serum reference materials for NIST (Veillon et al., 1985).

VI. ATOMIC ABSORPTION SPECTROMETRY (AAS)

Since its introduction decades ago by Walsh (1955) for the analysis of elements, AAS has become a popular technique for trace element analysis. Availability of a wide range of instrumentation, potential for automation, and good sensitivity have increased the use of the method. Flame atomic absorption atomization techniques have been used for the analysis of plant materials from areas high in Se (Olson et al., 1983). However, such methods are generally not sensitive enough for most biological and environmental samples (Thomassen et al., 1994). Sensitivity can be improved by generation of volatile selenium hydride and atomization in heated quartz cells (HGAA) or by electrothermal atomization of solutions in graphite tubes (L'vov, 1961) commonly known as graphite furnace atomization atomic absorption spectrometry (GFAAS). Because of its high sensitivity, capability of handling pretreated or nonpretreated samples, and ease of automation, GFAAS has become a very popular method of analysis for trace elements (Thomassen et al., 1994). Since selenium is subject to losses during drying and charring, matrix modification by the addition of metal ions is usually required (Saeed et al., 1979). A number of metal ions will prevent the volatilization of inorganic forms of Se (Alexander et al., 1980), but only nickel and silver appear to adequately stabilize metabolized forms of Se.

In addition to the difficulties with volatilization, the GFAAS method with use of deuterium (D_2) background correction is subject to spectral interferences from iron and phosphate decomposition products (Saeed and Thomassen, 1981). These interferences are reduced by nickel, nickel–platinum, or nickel–palladium matrix modifiers. It has been suggested that a Zeeman-based GFAAS instrument provides the best application of the AA methods for a wide variety of sample matrices (Bauslaugh et al. 1984; Radziuk and Thomassen, 1992). Thomassen et al. (1994) have stated that Zeeman-based GFAAS "provides an acceptable situation for selenium determinations and permits most analyses to be performed against matrix matched calibration graphs". Of course there is some difficulty in matching matrices when working with real-world samples of unknown composition. The same authors conclude that only serum and whole blood can be analyzed without sample pretreatment. All other samples require pretreatment and/or digestion.

The AAS method will undoubtedly grow in popularity because of its sensitivity and good sample throughput. Each new sample matrix type will require some validation to ensure that Se is being accurately determined.

VII. HYDRIDE GENERATION ATOMIC ABSORPTION SPECTROMETRY (HGAAS)

Techniques involving hydride generation with subsequent measurement by atomic absorption spectrometry have the advantages of increased sensitivity and freedom from many interferences caused by sample matrices. As reviewed by Siemer and Koteel (1977), there are essentially two fundamental ways of implementing HGAAS analysis. One is to convey the metallic hydride into the atomizer as soon as it is produced (continuous flow). The other way of implementing HGAAS analysis is to collect the hydride as soon as it is produced, with subsequent introduction into the atomizer (batch-type generation). Batch collection has often been accomplished by freezing the hydride out in a collection tube (McDaniel et al., 1976) or by collecting it in a balloon (Chu et al., 1972).

An HGAAS method for foods developed by Ihnat and Miller (1977a) was subjected to rigorous interlaboratory investigation under the auspices of the AOAC (Ihnat and Miller, 1977b; Ihnat and Thompson, 1980). Although accuracy was acceptable when the results were averaged over all laboratories, variation among laboratories and method precision were considered unacceptable. On the basis of these studies, the method was not adopted as an official method. Holak (1980) developed a closed-system nitric acid digestion method in combination with HGAAS, and this method has been adopted as an official method for foods by AOAC. Apparently HGAAS methods are very dependent on completeness of decomposition of the sample.

Welz and coworkers (1984a, 1984b) have shown HGAAS can be accurate if care is used to ensure complete sample digestion and conversion of all the selenium to selenite (+4 valence). Interferences with the HGAAS method have been noted by many workers. Iron (Goulden and Brooksbank, 1974), copper (Vijan and Wood, 1976), and nickel (Brown et al., 1981) have been shown to decrease hydride formation. Manipulation of the hydrochloric acid concentration in the hydride generation system has been found to eliminate these interferences (Vijan and Leung, 1980).

Hershey and Keliher (1986) studied the possible interferences by 50 elements on the HGAAS method. They report that much less interference was obtained with the steady state hydride generation systems (continuous hydride generation) than with the "batch" systems. They also reported that the interferences observed with the steady state systems could be further reduced by manipulation of the hydrochloric acid concentration. Welz and Schubert-Jacobs (1986) have reported that increasing the hydrochloric acid concentration from 0.5 mol/L to 5 mol/L and decreasing the sodium tetraborate concentration from 3% to 0.5% significantly improved the range of interference-free determination of selenium by HGAAS. Flow injection provides an efficient means of sample introduction for HGAAS and has the advantages of high sample throughput, better tolerance to chemical interferences, and improved detection limits (McLaughlin et al., 1990; Thomassen et al., 1994).

An extensive interlaboratory study was completed in 1988 (Walker et al., 1988) involving 29 laboratories. Study materials included water, soil, sediment, tissues, and vegetation. Analytical methods used included NAA, AAS, continuous flow HGAAS, X-ray fluorescence, GFAAS, ICP, HG-ICP, and fluorometry. The study concluded that laboratories attempting to determine selenium in environmental samples should use either continuous flow HGAAS (with atomic absorption or ICP spectrometry) or fluorometry. Recently a collaborative continuous flow HGAA method for the determination of selenium in feeds and premixes was approved for first action by the AOAC (Palmer and Thiex, 1996). The results obtained with the HGAAS method compared favorably with the results obtained with a fluorometric method developed in the same study.

VIII. FLUOROMETRY

Several reviews are available on the use of fluorometry to determine selenium in biological and environmental samples (Olson et al., 1973; Olson, 1976). Cousins (1960) originally took advantage of the fluorescence of the Se–diaminobenzidine complex (DAB) in hydrocarbon solvents to determine Se. Parker and Harvey (1962) found the product formed from Se and 2,3-diaminonaphthalene (DAN) to be more strongly fluorescent than the Se-DAB complex. Watkinson (1966) conducted a thorough study on the use of DAN to determine Se without previous separation of the element. The sample was digested in a nitric–perchloric acid mixture and the digest heated with HCl to convert all Se to selenite. After pH adjustment, the selenite was reacted with 2,3-diaminonaphthalene and the resulting complex extracted into cyclohexane. Only a few elements caused interference, and their effect could be masked by the use of ethylenediaminetetraacetic acid (Watkinson, 1966; Chen, 1993). Olson (1969) used a modified version of the Watkinson (1966) method and conducted a collaborative study that was approved as an official AOAC method for determining Se in plants. The method has been further modified (Ihnat, 1974a, 1974b; Olson et al., 1975; Horwitz, 1980). In 1979 Watkinson developed a semiautomated fluorometric method that allows more sample throughput.

Finally, Koh and Benson (1983) did a thorough reappraisal of the fluorometric method. They used an aluminum block for the simultaneous digestion of 121 samples. Upon completion of the digestion, they converted all Se to selenite by adding HCl to the tubes in the heating block; then they continued the heating at 100 to 150°C for 30 minutes. Their studies also demonstrated that tedious pH adjustment of the samples prior to reaction with DAN is not necessary. These modifications allow the digestion, reduction, complexation, and extraction of the selenium without any transfers of the sample. Because of the number of samples that could be facilitated by the digestion system, the modifications of Koh and Benson (1983) allowed the treatment of all standards, blanks, and quality control samples under conditions identical to those experienced by the samples. Used in conjunction with an automated fluorometer, the method of sample treatment described by Koh and Benson (1983) allows the processing of a large number of samples per day.

An intralaboratory study comparing fluorometry with HGAAS was conducted on 22 different sample matrices varying in selenium concentration from 0.1 to 380 Tg/g (Chen, 1993; Palmer, 1995). The two methods gave very similar results in the hands of the same analyst. A few soil samples contained enough iron to cause interference with the fluorometric method, but this could be detected and corrected by dilution of the sample or by increasing the amount of ethylenediaminetetraacetic acid used in the procedure. Under uniform conditions, where the same analyst performed the analysis on identical samples, the two methods appeared to be equivalent in performance.

The fluorometric method often is referred to as the simplest, least expensive, and most versatile method for selenium analysis, if close attention is paid to detail (Lewis and Veillon, 1989; Thomassen et al., 1994). Walker et al. (1988) stated that fluorometry was the method of choice for environmental samples, along with continuous generation HGAAS. A recent collaborative study has resulted in the first action approval by the AOAC of a fluorometric method for selenium in feeds and premixes as an official method along with an HGAAS method (Palmer and Thiex, 1996).

IX. SUMMARY

The methods discussed in this chapter are applicable to the determination of total Se. If speciation is desired, additional sample treatment and separation techniques must be used. It is probable that any of the methods of Se analysis can be appropriately customized to determine Se in a specific sample matrix, provided the analyst is attentive to sample preparation and matrix effects. Individual laboratory preferences will undoubtedly be influenced by the availability of equipment and the required sample capacity. It is the opinion of this author that NAA methods are very appropriate if the required nuclear reactor is available. For laboratories that need to analyze a large number of samples with varying and unknown sample matrices, the fluorometric and continuous flow HGAAS methods seem to be most applicable. These methods are discussed in greater detail in Chapter 2.

REFERENCES

Alcino, J. F. and J. A. Kowald. 1973. Analytical methods. In D. L. Klayman and W. H. H. Gunther (eds), *Organic Selenium Compounds: Their Chemistry and Biology*, pp. 1050–1081. Wiley, New York.

Alexander, J., K. Saeed, and Y. Thomassen. 1980. Thermal stabilization of inorganic and organoseleno-compounds for direct electrothermal atomic absorption spectrometry. *Anal. Chim. Acta* 120:377–382.

Bauslaugh, J., B. Radziuk, K. Saeed, and Y. Thomassen. 1984. Reduction of effects of structured non-specific absorption in the determination of arsenic and selenium by electrothermal atomic absorption spectrometry. *Anal. Chim. Acta* 165:149–157.

Brown, R. M., Jr., R. C. Fry, J. L. Moyers, S. J. Northway, M. B. Denton, and G. S. Wilson. 1981. Interference by volatile nitrogen oxides and transition-metal catalysis in the preconcentration of arsenic and selenium as hydrides. *Anal. Chem.* 53:1560–1566.

Chen, T. 1993. Evaluation of the fluorometric and continuous hydride generation atomic absorption method for selenium determination. M.S. thesis, South Dakota State University, Brookings.

Christensen, M. J., M. Janghorbani, F. H. Steinke, N. Istfan, and V. R. Young. 1983. Simultaneous determination of absorption of selenium from poultry meat and selenite in young men: Application of a triple stable isotope method. *Br. J. Nutr.* 50:43–50.

Chu, R. C., G. P. Brown, and P. A. W. Baumgarner. 1972. Arsenic determination at submicrogram levels by arsine generation and flameless atomic absorption spectrophotometric technique. *Anal. Chem.* 44:1476–1479.

Cooper, W. C. 1974. Analytical chemistry of selenium. In R. A. Zingaro, and W. C. Cooper (eds.), *Selenium*, pp. 615–653. Van Nostrand Reinhold, New York,

Cornelis, R., J. Hoste, and J. Versieck. 1982. Potential interferences inherent in neutron activation analysis of trace elements in biological materials. *Talanta* 29:1029–1034.

Cousins, F. B. 1960. A fluorimetric microdetermination of selenium in biological material. *Aust. J. Expl. Biol. Med. Sci.* 38:11–16.

Ding, W.-W., and R. E. Sturgeon. 1997. Minimization of transition metal interferences with hydride generation techniques. *Anal. Chem.* 69:527–531.

Gorsuch, T. T. 1959. Radiochemical investigations on the recovery for analysis of trace elements in organic and biological materials. *Analyst* 84:135–173

Goulden, P. D., and P. Brooksbank. 1974. Automated atomic absorption determination of arsenic, antimony and selenium in natural water. *Anal. Chem.* 46:1431–1436.

Hershey, J. W., and P. N. Keliher. 1986. Some hydride generation inter-element interference studies utilizing atomic absorption and inductively coupled plasma emission spectrometry. *Spectrochim Acta* 41B:713–723.

Holak, W. 1980. Analysis of foods for lead, cadmium, copper, zinc, arsenic and selenium using closed system sample digestion: Collaborative study. *J. Assoc. Off. Anal. Chem.* 63:485–495.

Horwitz, W. 1980. *Official Methods of Analysis of the Association of Official Analytical Chemists.* 13th ed., Sect 3.097-3.101. AOAC, Arlington, VA.

Hwang, J. D., G. D. Guenther, and J. P. Diomiguardi. 1989. A hydride generation system for a 1-kW inductively coupled plasma. *Anal. Chem.* 61:285–288.

Ihnat, M. 1974a. Fluorometric determination of selenium in foods. *J. Assoc. Off. Anal. Chem.* 57:368–372.

Ihnat, M., 1974b. Collaborative study of a fluorometric method for determining selenium in foods. *J. Assoc. Off. Anal. Chem.* 57:373–378.

Ihnat, M. 1992. Selenium. In M. Stoeppler (ed.), *Hazardous Metals in the Environment Techniques and Instrumentation in Analytical Chemistry*, Vol. 12, pp. 475–515. Elsevier, Amsterdam.

Ihnat, M., and H. S. Miller. 1977a. Analysis of foods for arsenic and selenium by acid digestion hydride evolution atomic absorption spectrophotometry. *J. Assoc. Off. Anal. Chem.* 60:813–825.

Ihnat, M., and H. S. Miller. 1977b. Acid digestion, hydride evolution atomic absorption spectrophotometric method for determining arsenic and selenium in foods. Collaborative Study I. *J. Assoc. Off. Anal. Chem.* 60:1414–1433.

Ihnat, M., and B. K. Thompson. 1980. Acid digestion, hydride evolution atomic absorption spectrophotometric method for determining arsenic and selenium in foods. Collaborative Study II. Assessment of collaborative study. *J. Assoc. Off. Anal. Chem.* 63:814–834.

Janghorbani, M., B. T. G. Ting, and V. R. Young. 1981. Use of stable isotopes of selenium in human metabolic studies: Development of analytical methodology, *Am. J. Clin. Nutr.* 34:2816–2830.

Janghorbani, M., M. J. Christensen, A. Nahapetian, and V. R. Young. 1982. Selenium metabolism in healthy adults: Quantitative aspects using the stable isotope $^{74}SeO_3^{2-}$. *Am. J. Clin. Nutr.* 35:647–654.

Kasper, L. J., V. R. Young, and M. Janghorbani. 1984. Short-term dietary selenium restriction in young adults: Quantitative studies with the stable isotope $^{74}SeO_3^{2-}$. *Br. J. Nutr.* 52:443–455.

Koh, T.-S., and T. H. Benson. 1983. Critical re-appraisal of fluorometric method for determination of selenium in biological materials. *J. Assoc. Off. Anal. Chem.* 66:918–926.

Krushevska, A., M. Kotrebai, A. Lásztity, R. Barnes, and D. Amarasiriwardena. 1996. Application of tertiary amines for arsenic and selenium signal enhancement and polyatomic interference reduction in ICP-MS analysis of biological samples. *Fresenius J. Anal. Chem.* 355:793–800.

Lewis, S. A., and C. Veillon. 1989. Inorganic analytical chemistry of selenium. In M. Ihnat (ed.), *Occurrence and Distribution of Selenium*, Chap. 2, pp.13–14. CRC Press, Boca Raton, FL.

L'vov, B. V. 1961. The analytical use of atomic absorption spectra. *Spectrochim. Acta* 17:761–770.

May, T. W. 1982. Recovery of endogenous selenium from fish tissues by open system dry ashing. *J. Assoc. Off. Anal. Chem.* 65:1140–1145.

McCurdy, E. J., J. D. Lange, and P. M. Haygarth. 1993. The determination of selenium in sediment using hydride generation ICP-MS. *Sci. Total Environ.* 135:131–136.

McDaniel, M., A. Shandrikar, K. Reiszner, and P. West. 1976. Concentration and determination of selenium from environmental samples. *Anal. Chem.* 48:2240–2243.

McLaughlin, K., D. Dadgar, M. R. Smyth, and D. McMaster. 1990. Determination of selenium in blood plasma and serum by flow injection hydride generation atomic absorption spectrometry. *Analyst* 115:275–278.

Morris, J. S., D. M. Mckown, H. D. Anderson, M. May, P. Primm, M. Cordts, D. Gebbardt, S. Crowson, and V. Spate. 1981. The determination of selenium in samples having medical and nutritional interest using a fast instrumental neutron activation analysis procedure. In J. E. Spallholz, J. L. Martin and H. E. Ganther (eds.), *Selenium in Biology and Medicine*, pp. 438–448. AVI Publishing, Westport, CT.

National Academy of Sciences. 1976. *Selenium. Medical and Biological Effects of Environmental Pollutants.* NAS, Washington, DC.

Neve, J., M. Hanocq, L. Molle, and G. Lefebure. 1982. Study of some systematic errors during the determination of the total selenium and some of its ionic species in biological materials, *Analyst* 107:934–941.

Olson, O. E. 1969. Fluorometric analysis of selenium in plants *J. Assoc. Off. Anal. Chem.* 52:627–634.

Olson, O. E. 1976. Methods of analysis for selenium, a review. In *Proceedings of the Symposium on Selenium–Tellurium in the Environment*, pp. 67–84. University of Notre Dame, Notre Dame, IN.

Olson, O. E., I. S. Palmer, and E. I. Whitehead. 1973. Determination of selenium in biological materials. In D. Glick (ed.), *Methods of Biochemical Analysis*, Vol. 21, pp. 39–78. Wiley, New York.

Olson, O. E., I. S. Palmer, and E. Carey. 1975. Modification of the official fluorometric method for selenium in plants. *J. Assoc. Off. Anal. Chem.* 58:117–121.

Olson, O. E., R. J. Emerick, and I. S. Palmer. 1983. Measurement of selenium in plants of high selenium content by flame absorption analysis. *At. Spectrosc.* 4:55–58.

Palmer, I. S. 1995. Selenium in water, plants and soil: Analytical methodology. *Proceedings of Symposium on Selenium in the Environment: Essential Nutrient, Potential Toxicant*, pp. 20–37. University of California, Davis, May 31–June 2, 1995.

Palmer, I. S., and N. Thiex. 1996. Selenium in feeds and premixes. Alternative I—fluorometric method , and alternative I—continuous hydride generation atomic absorption method. Adopted as First Action Method 996.16, 3rd Supplement to Official Methods of Analysis, AOAC, 16th ed. Association of Official Analytical Chemists, Arlington, VA.

Parker, C. A., and L. G. Harvey. 1962. Luminescence of some piazselenols, a new fluorimetric agent for selenium. *Analyst* 87:558–565.

Radziuk, B., and Y. Thomassen. 1992. Chemical modification and spectral interferences in selenium determination using Zeeman-effect electrothermal atomic absorption spectrometry. *J. Anal. At. Spectrom.* 7:397–404.

Reamer, D. C., and C. Veillon. 1981a. Preparation of biological materials for determination of selenium by hydride generation–atomic absorption spectrometry. *Anal. Chem.* 53:1192–1195.

Reamer, D. C., and C. Veillon. 1981b. Determination of selenium in biological materials by stable isotope dilution gas chromatography–mass spectrometry. *Anal. Chem.* 53:2166–2169.

Reamer, D. C., and C. Veillon. 1983. A double isotope dilution method for using stable selenium isotopes in metabolic trace studies: analysis by gas chromatography/mass spectrometry (GC/MS). *J. Nutr.* 113:786–792.

Saeed, K., and Y. Thomassen. 1981. Spectral interferences from phosphate matrices in the determination of arsenic, antimony, selenium and tellurium by electrothermal atomic absorption spectrometry. *Anal. Chim. Acta* 130:281–287.

Saeed, K., Y. Thomassen, and F. J. Langmyhe. 1979. Direct electrothermal atomic absorption spectrometric determination of selenium in serum. *Anal. Chim. Acta* 110: 285–289.

Schwarz, K., and C. M. Foltz. 1957. Selenium as an integral part of factor 3 against dietary necrotic liver degeneration. *J. Am. Chem. Soc.* 79:3292–3293.

Shendrikar, A. D. 1974. Critical evaluation of analytical methods for the determination of selenium in air, water and biological materials. *Sci. Total Environ.* 3:155–168.

Siemer, D. D., and P. Koteel. 1977. Comparison of methods of hydride generation atomic absorption spectrometric arsenic and selenium determination. *Anal. Chem.* 49:1096–1099.

Sirichakwal, P. P., V. R. Young, and M. Janghorbani. 1985. Absorption and retention of selenium from intrinsically labeled egg and selenite as determined by stable isotope studies in humans. *Am. J. Clin. Nutr.* 41:264–269.

Standard Methods for Examination of Water and Wastewater. 1992. In Section 3500, Selenium, pp. 3–83 to 3–87. [A. E. Greenberg, L. S. Clesceri, A. D. Easton, and M. A. H. Franson, (eds.), 18th ed.] American Public Health Association, Washington, DC.

Tao, H., J. W. H. Lam, and J. W. McLaren. 1993. Determination of selenium in marine certified reference materials by hydride generation inductively coupled plasma mass spectrometry. *J. Anal. At. Spectrom.* 8:1067–1073.

Thomassen, Y., S. A. Lewis, and D. Veillon. 1994. Trace element analysis in biological specimens. In R. F. Herber (ed.), *Series in Techniques and Instrumentation in Analytical Chemistry*, pp. 489–500. Elsevier, Amsterdam.

Ting, B. T. G., and M. Janghorbani. 1987. Application of ICP-MS to accurate isotopic analysis in human metabolism studies. *Spectrochim. Acta* 42B: 21

Ting, B. T. G., C. S. Mooers, and M. Janghorbani. 1989. Isotopic determination of selenium in biological materials with inductively coupled plasma mass spectrometry. *Analyst* 114:667–674.

Tracey, M. L., and G. Möller. 1990. Continuous flow vapor generation for inductively coupled argon plasma spectrometric analysis. Part I. Selenium. *J. Assoc. Off. Anal. Chem.* 73:404–410.

U.S. Environmental Protection Agency. 1983. Methods for Chemical Analysis of Water and Waste. EPA-600/4-79-020, Methods 270.2 and 270.3.

Veillon, C., S. A. Lewis, K. Y. Patterson, W. R. Wolf, J. M. Harnly, J. Versieck, L. Vanballenberghe, T. C. O'Haver, and R. Cornelis. 1985. Characterization of a bovine serum reference material for major, minor and trace elements. *Anal. Chem.* 57:2106–2109.

Vijan, P. N., and D. Leung. 1980. Reduction of chemical interference and speciation studies in the hydride generation–atomic absorption method for selenium. *Anal. Chim. Acta* 120:141–146.

Vijan, P. N., and G. R. Wood. 1976. An automated submicrogram determination of selenium in vegetation by quartz-tube furnace atomic absorption spectrophotometry. *Talanta* 23:89–94.

Walker, W. J., R. G. Burau, D. Silberman, and A. Jacobson. 1988. The San Joaquin Valley Drainage Program Selenium Round Robin—Final report. Department of Land, Air and Water Resources, University of California, Davis.

Walsh, A. 1955. The application of atomic absorption spectra to chemical analysis. *Spectrochim. Acta* 7:108.

Watkinson J. H. 1966. Fluorometric determination of selenium in biological material with 2,3-diaminonaphthalene. *Anal. Chem.* 38:92–97.

Watkinson, J. H. 1967. Analytical methods for selenium in biological material. In O. H. Muth, J. E. Oldfield, and P. H. Weswig (eds.), *Selenium in Biomedicine*, pp. 97–117. AVI Publishing, Westport, CT.

Watkinson, J. H. 1979. Semi-automated fluorimetric determination of nanogram quantities of selenium in biological material. *Anal. Chim. Acta* 105:319–325.

Welz, B., and M. Schubert-Jacobs. 1986. Mechanisms of transition metal interferences in hydride generation atomic absorption spectrometry. *J. Anal. At. Spectrom.* 1:23–27.

Welz, B., M. Melcher, and J. Nève. 1984a. Determination of selenium in human body fluids by hydride generation atomic absorption spectrometry—Optimization of sample decomposition. *Anal. Chim. Acta* 165:131–140

Welz, B., M. Melcher, and G. Schlemmer. 1984b. Accuracy of the selenium determination in human body fluids using atomic absorption spectrometry. In P. Brätter and P. Schramel (eds.), *Trace Element Analytical Chemistry in Medicine and Biology*, Vol. 3, pp. 207–215. Walter de Gruyter, Berlin.

Williams, S. 1984a. Selenium in plants, fluorometric method. In *Official Methods of Analysis of the Association of Official Analytical Chemists*, 14th ed., pp. 3.102 to 3.107. AOAC, Arlington, VA.

Williams, S. 1984b. Selenium in food, fluorometric method. In *Official Methods of Analysis of the Association of Official Analytical Chemists*, 14th ed., pp. 25.154 to 25.157. AOAC, Arlington, VA.

Williams, S. 1984c. Arsenic, cadmium, lead, selenium and zinc in food, multielement method, first action. In *Official Methods of Analysis of the Association of Official Analytical Chemists*, 14th ed., p. 25.001 to 25.007. AOAC, Arlington, VA.

4

Sequential Extraction of Selenium Oxidation States

DEAN A. MARTENS
United States Department of Agriculture, National Soil Tilth Laboratory, Ames, Iowa

DONALD L. SUAREZ
United States Department of Agriculture, United States Salinity Laboratory, Riverside, California

I. INTRODUCTION

Selenium (Se) is recognized as a micronutrient that is essential to mammals, birds, fish, algae, and many bacteria for growth and survival (Stadtman, 1979; Burau, 1985). In environmental and biological samples, Se can exist in inorganic forms as elemental Se (Se[O]), metal selenides, selenite (Se[IV]), and selenate (Se[VI]), and as volatile or nonvolatile organic species with direct Se–carbon bonds such as methylated compounds, selenoamino acids, and selenoproteins. The forms and concentrations of Se in soil solution are governed by various physical–chemical factors expressed in terms of pH, dissociation constants, solubility products, and oxidation–reduction potentials (Geering et al., 1968). In addition, recent research has indicated that soil microbial populations can affect the concentration and forms of Se (Maiers et al., 1988; Frankenberger and Karlson, 1989). Because of its toxicological effects on waterfowl, Se has become one of the major contaminant problems in irrigation drainage waters in the western portion of the United States. An overview of the Se problem in California evaporation ponds including the Kesterson Reservoir (evaporation ponds) has been presented by Ohlendorf and Santolo (1994).

Recommended dietary Se concentrations have been shown to lie in the range of 0.1 to 0.3 mg/kg^{-1}, whereas levels from 2 to 10 mg Se/kg may give rise to chronic toxicity symptoms (Scott, 1973). Toxic effects of inorganic and organic forms of Se result from the alteration of protein three-dimensional structures and the impairment of enzymatic function in organisms by incorporation of Se amino acids in place of sulfur (S) forms (Frost and Lish, 1975). The structure of proteins is controlled by the sequence of amino acids and the location of disulfide bonding of the cysteine residues (Stryer, 1981). The substitution of Se for S, as in the amino acid, selenocysteine (SeCys), results in the interruption of the protein structure due to pK differences between the sulfhydryl and selenol bridges. The toxic dose of Se is very much dependent on its chemical form, with a different toxicity for organic and inorganic Se compounds (Heinz et al., 1987, 1988, 1989). Thus it is more important to identify and determine of the concentration of Se in each oxidation state in soil, plant, and water samples than to determine of a total Se concentration (Martens and Suarez, 1997a).

Most of our knowledge of the forms of Se in biological samples has been obtained by chemical extraction techniques. Application of extraction methods to selectively remove Se is complicated by the fact that Se may exist in more than one oxidative state, each of which has a unique adsorption behavior (Frost and Griffen, 1977). Since Se chemistry bears little resemblance to transition metal chemistry and geochemical behavior, it is not surprising that the sequential extraction schemes devised for transition metals are not applicable to Se extraction (Weres et al., 1989). In addition, extractants that are oxidizing or reducing in nature may alter the oxidation state of the Se extracted and confound the interpretation of results (Gruebel et al., 1988). However, the principles of sequential extraction techniques, specifically the use of progressively stronger extractants to solubilize the element from different sources, are applicable to Se extraction and can provide meaningful results when carefully interpreted.

Analysis of Se present in evaporation pond sediments from the west side of the San Joaquin Valley of California (contaminated with Se[VI]- and Se[IV]-laden drainage water) revealed that the vast majority of the Se became concentrated in the surface 15 cm of the containment ponds, in forms whose mineralization and mobilization potential are not well understood, and did not percolate into the groundwater (Tokunaga and Benson, 1992; White and Dubrovsky, 1994). The inability of presently employed extraction techniques to identify the distribution of Se oxidation states in Se-contaminated materials has emphasized the need to develop an efficient Se extraction methodology for natural systems (Cary et al., 1967; Chao and Sanzolone, 1989; Tokunaga et al., 1991; Kang et al., 1993; Tokunaga et al., 1994; Zhen-Li et al., 1994; Sharmasarkar and Vance, 1995; Zhang and Moore, 1996). Detailed Se speciation is required to predict Se mobility and leaching from seleniferous soils and evaporation ponds.Treatment and management of these lands to minimize Se leaching and toxicity also require knowledge of Se redox status and speciation. This chapter describes a sequential extraction technique developed by Martens and Suarez (1997a) for determination of the four oxidation states of Se in biological samples.

II. SELENIUM EXTRACTION AND SPECIATION

A. Selenium Extraction

For an in-depth presentation of the sequential extraction scheme used for the analysis of Se[VI], Se[IV], selenide (Se[−II]), and Se[O] present in seleniferous samples, consult Martens and Suarez (1997a).

1. Water and Phosphate Extraction

Material containing unknown levels of Se (5.0 g) was placed into a 40 mL Teflon centrifuge tube and fractionated as follows. A 25 mL aliquot of distilled ionized (DI) water was added to the material and shaken (130 oscillations/min at ambient temperature) on a horizontal shaker for 1 hour. The sample was then centrifuged [10,000 × g, relative centrifugal force (rcf); 20 min], and the supernatant was decanted to a 30 mL polyethylene bottle. A 25 mL aliquot of 0.1 M (pH 7.0) K_2HPO_4–KH_2PO_4 buffer (P-buffer) then was added to the pelleted material and shaken for one hour (130 oscillations/min at ambient temperature). The sample was centrifuged (10,000 × g, rcf; 20 min) and the supernatant was decanted to a 30 mL polyethylene bottle. The P-buffer sample was then shaken for 2 minutes with 5 mL of DI water, centrifuged, and the supernatants combined.

2. NaOH Oxidation

The material remaining after the P-buffer extraction was treated with 25 mL of 0.1 M NaOH (90°C) for 2 hours. The NaOH–soil suspension was centrifuged (10,000 × g, rcf; 20 min), the supernatant decanted, the sample mixed with 5 mL of DI water, shaken, centrifuged, and the supernatants combined in a 30 mL polyethylene bottle.

3. Nitric Acid Oxidation

The material remaining after NaOH extractions was treated with 2.5 mL of 17 M HNO_3, heated to 90°C for 30 minutes and thoroughly cooled. Then 20 mL of DI water was added and the sample heated at 90°C for 1.5 hours. The resulting 2 M HNO_3 solution was cooled to ambient temperature, centrifuged (10,000 × g, rcf; 20 min), and decanted to a 30 mL polypropylene bottle.

B. Selenium Analysis

Selenium was analyzed by hydride generation atomic absorption spectrophotometry (HGAAS). Hydride generation AAS conditions, reagents and standard preparation, and the quality assurance program utilized are described by Martens and Suarez (1997a).

C. Distribution of Selenium Oxidation States

Distribution of Se oxidation states was determined by HGAAS on three treatments of the water or P-buffer extract. This procedure enables speciation of Se[IV], Se[VI], and Se[−II] solubilized by the extractions by selective determination of the Se[IV] oxidation state. To 25 mL graduated glass test tubes, we added from 0.1 to 2.0 mL of the specified extract and (1) 6 M HCl for Se[IV] concentration (25 mL total), (2) 6 M HCl heated at 90°C for 30 minutes for Se[IV] and Se[VI] concentrations (25 mL total), and (3) 1 mL 0.1 M $K_2S_2O_8$ (90°C) for 30 minutes, then 6 M HCl (25 mL total; 90°C) for 30 minutes to determine Se[IV], Se[VI], and Se[−II] concentrations. Persulfate ($K_2S_2O_8$) oxidation results in conversion of all soluble Se species to Se[VI]. The Se[−II] concentration was determined by subtraction of the Se[IV] and Se[VI] concentration (analysis 2) from the Se concentration of the $K_2S_2O_8$ oxidation (analysis 3). The Se[VI] concentration was calculated by subtracting the Se[IV] (analysis 1) from the Se[IV] and Se[VI] (analysis 2). Recovery tests with Se[IV], Se[VI] or selenomethionine (Se[−II]) spiked to a nonseleniferous soil and extracted with the P-buffer resulted in spike recovery rates by the listed HGAAS speciation procedure that exceeded 95%.

From 0.1 to 2.0 mL of the NaOH extract was added to 25 mL graduated glass test tubes and (1) 6 M HCl for Se[IV] concentration (25 mL total) and (2) 1 mL 0.1 M $K_2S_2O_8$ (90°C) for 30 minutes were added, followed by 6 M HCl (25 mL total; 90°C) for 30 minutes, to determine Se[IV] and Se[−II] concentrations. The NaOH extraction has been shown to recover tightly held Se[IV] and inorganic and organic Se[−II] (Martens and Suarez, 1997a).

Selenium analysis was performed on sample aliquots (0.1–1.0 mL) of the HNO_3 extract treated with 6 M HCl (total 25 mL) heated at 90°C for 30 minutes. The HNO_3 extraction accounts for the remaining Se inventory in the sequential extraction method as Se[O] (Martens and Suarez, 1997a). All allotropes of Se[O], insoluble S–Se associations, and possible S amino acid–Se complexes are grouped together in the Se[O] fraction (Weres et al., 1989).

III. SEQUENTIAL EXTRACTION OF SELENIUM OXIDATION STATES FROM BIOLOGICAL MATERIAL

A. Distribution of Selenium Oxidation States in Geologic Materials

A reconnaissance study conducted by the U.S. Geological Survey (Presser et al., 1990) determined that the primary source of Se to the west-central San Joaquin Valley that resulted in Se contamination at Kesterson Reservoir was the nearby

Diablo Range of the California Coastal Range. Presser et al. (1990) established that elevated concentrations of Se were present in the extensive shale exposures of the Upper Cretaceous–Paleocene, Moreno, and the Eocene–Oligocene, Kreyenhagen, formations. These materials were transported over time to the adjacent alluvial soils, and agricultural practices further concentrated the Se in sediments at the Kesterson Reservoir via the San Luis Drain. The shales, alluvial soils, San Luis Drain sediment, and Kesterson soils were analyzed by the sequential extraction methodology to determine whether trends in the distribution of Se oxidation states exist from the Se source (shales) to the Se sink (evaporation ponds) in this drainage series (Martens and Suarez, 1997b).

The speciation data provide much information on the potential mobility and availability of Se present in the tested materials (Tables 1 and 2). The Se[VI] fraction, which is very mobile when in solution, is found only in the water extraction along with low levels of water-soluble Se[IV]. Selenite was found in all samples as water-soluble, ligand-exchangeable, and very tightly held Se[IV]. The sum of water-soluble and exchangeable Se in the materials tested was low (< 15%) in comparison to the total Se contents. The extraction procedure employed here also indicates that a large portion of the Se inventory (> 80%) is present in these materials as organic-associated Se and as Se[O]. Evidence for

TABLE I Selenium Speciation (mg/kg) of the Moreno and Kreyenhagen Shales and Panoche Fan Alluvial Soils[a]

Material	Water		P-buffer		NaOH		HNO$_3$	ΣSe
	Se[IV][b]	Se[VI]	Se[IV]	Se[−II]	Se[IV]	Se[−II]	Se[O]	
Kreyenhagen shale	0.03 (0.00)	0.47 (0.04)	1.35 (0.06)	0.36 (0.00)	3.98 (0.30)	2.98 (0.41)	1.77 (0.05)	10.94
Moreno shale	0.03 (0.01)	0.10 (0.03)	0.49 (0.02)	0.20 (0.01)	0.98 (0.28)	1.57 (0.23)	1.84 (0.28)	5.21
Tumey Gulch soil	ND	0.09 (0.01)	0.02 (0.01)	0.02 (0.01)	0.04 (0.01)	0.33 (0.02)	0.56 (0.05)	1.06
Moreno Gulch soil	ND	0.05 (0.01)	0.01 (0.01)	0.02 (0.01)	0.01 (0.00)	0.18 (0.01)	0.31 (0.02)	0.58
Salt Creek soil	ND	0.01 (0.01)	0.01 (0.01)	0.01 (0.01)	0.01 (0.00)	0.10 (0.01)	0.27 (0.03)	0.41

[a]The shales and soils (5 g) were placed in 40 mL Teflon centrifuge tubes and sequentially extracted with water (25 mL), P-buffer (25 mL), NaOH (25 mL; 90°C), and nitric acid (2.5 mL + 20 mL water; 90°C). Se speciation in the extractants was determined by HGAAS. The value in parentheses indicates relative standard deviation of the mean (n = 4).
[b]ND, not detectable.
Source: Martens and Suarez (1997b).

TABLE 2   Selenium Speciation (mg/kg) of the San Luis Drain Sediment and the Seleniferous Kesterson Evaporation Pond Soils[a]

| Material | Water | | P-buffer | | NaOH | | HNO$_3$ | |
	Se[IV]	Se[VI]	Se[IV]	Se[−II]	Se[IV]	Se[−II]	Se[O]	ΣSe
San Luis	0.11	2.75	3.23	1.71	1.05	42.23	32.16	83.24
Drain	(0.01)	(0.01)	(0.31)	(0.16)	(0.28)	(0.16)	(3.71)	
sediment								
Kesterson	0.49	2.48	1.42	0.77	2.45	12.46	27.73	47.80
pond 4	(0.00)	(0.08)	(0.11)	(0.12)	(0.20)	(0.13)	(1.23)	
Kesterson	0.10	0.51	0.51	0.24	0.50	2.56	2.29	6.70
pond 7	(0.01)	(0.01)	(0.01)	(0.00)	(0.12)	(0.04)	(0.45)	
Kesterson	0.11	0.92	0.43	0.29	1.34	0.46	1.80	5.35
pond 11	(0.01)	(0.06)	(0.01)	(0.06)	(0.08)	(0.03)	(0.23)	

[a]The soils (5 g) were placed in 40 mL Teflon centrifuge tubes and sequentially extracted with water (25 mL), P-buffer (25 mL), NaOH (25 mL; 90°C), and nitric acid (2.5 mL + 20 mL water). Se speciation in the extractants was determined by HGAAS. The value in parentheses indicates relative standard deviation of the mean ($n = 4$).
Source: Martens and Suarez (1997b).

the presence of Se[O] in the Kesterson soils was determined by X-ray absorption spectroscopy based on synchrotron radiation (Pickering et al., 1995).

Selenium is chemically similar to S, and many of the reactions involving Se species are assumed to follow the reactivity of S analogs. Research has shown that soil organic matter is the predominant factor for retention of S in nongypsiferous soils (Tabatabai and Bremner, 1972). Weres et al. (1989) noted a linear relationship for seleniferous Kesterson soils (total Se content) and organic C contents and theorized that Se was initially deposited with the organic fraction of the sediment. In the San Luis Drain sediment and Kesterson soils, the Se[O] content, the Se[−II] content, and the sum of (Se[−II] and Se[O])–Se were linear with organic C content ($r^2 = 0.90*$, $0.86*$, and $0.96*$, respectively). Other research (Elsokkary, 1980; Naftz and Rice, 1989; Abrams et al., 1990) also reported a correlation of total soil Se content with organic matter content. An exponential function ($r^2 = 0.96**$) rather than a linear relationship was noted when the sum of (Se[−II] and Se[O])–Se was plotted against the organic C content of nine soil materials investigated (Martens and Suarez, 1997b).

To further test the observed relationship between Se and organic C, an additional 12 Se and organic C content values (evaporation pond sediments and alluvial soil data from the western San Joaquin Valley obtained by Martens and Suarez, 1997b, and values taken from Karlson and Frankenberger, 1989, and

Weres et al., 1989) were included in the data set, shown in Figure 1. The close approximation of the additional data points to the determined exponential relationship suggests that Se concentrations rapidly increased as organic C contents increased above 20 g organic C/kg (2% organic C) in evaporation pond sediments. This increase in the ratio of Se to organic matter may be related to the change from aerobic conditions (nearly all values with less than 20 g organic C/kg) to anaerobic conditions (all samples with more than 20 g organic C/kg) reported to occur in evaporation pond sediments. Such a relationship is expected to be found only when the materials have similar inputs of Se (i.e., C content is not a general predictor of Se content).

B. Carbon/Selenium Mineralization

Carbon-bonded S, a major S component of photosynthetic leaves may represent an important source of sulfate for terrestrial systems (Likens and Bormann, 1974).

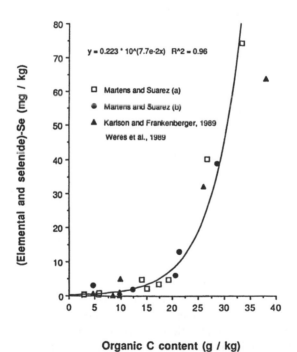

Organic C content (g / kg)

FIGURE I   The relationship between (a) the sum of the Se[−II] and Se[O] concentrations and the organic C content in the nine soil materials tested, (b) additional evaporation pond Se and organic C values (data not reported here), and Se and organic C values from previous research with western San Joaquin Valley alluvial soils (Frankenberger and Karlson, 1989) and the Kesterson evaporation ponds (Weres et al., 1989). (Reprinted with permission, Martens and Suarez, 1997b.)

The sulfur-containing amino acids methionine (Met) and cysteine constitute a large portion of the plant C-bonded S and may account for 5 to 35% of the S present in soil (Freney et al., 1970). In contrast, the importance of C-bonded Se in the environment is not well known. Plants have been reported to assimilate soluble inorganic Se into analogs of the sulfur amino acids from seleniferous soils (Gupta et al., 1993). Abrams and Burau (1989) and Rael and Frankenberger (1995) have presented unequivocal mass spectrum data proving that the Se-containing amino acid selenomethionine (SeMet) is present in seleniferous soils and sediments. The origin of this selenoamino acid in the soil materials was not identified as either plant or microbial, nor were concentrations provided. In contrast to our knowledge of the transformations of S-amino acids, the mobility and mineralization of SeMet and selenocystine (SeCys) in soil have been neglected research topics (Haygarth, 1994).

To determine the mineralization of selenoamino acids in soil, three rates of SeMet and three rates of SeCys were added to samples of a Panoche and a Panhill soil, and the disappearance of the Se-amino acids was followed by amino acid analysis and the distribution of resulting Se oxidation states by sequential extraction methodology (Martens and Suarez, 1997c).

I.   Selenomethionine Mineralization

SeMet additions to the Panoche soil were mineralized faster than SeMet additions to the Panhill soil (Martens and Suarez, 1997c). The time required for 50% mineralization of the SeMet addition for the Panoche and Panhill soils, respectively, was calculated as follows: 5 mg, 3.2 and 24.5 hours; 25 mg, 15.5 and 41.6 hours; 50 mg, 36.0 and 47.1 hours. The SeMet additions followed pseudo-first-order decomposition kinetics, with the order dependent on SeMet concentration. Both the Panoche and the Panhill soils showed apparent first-order kinetics at the lowest addition rate of 5 mg and apparent zero-order kinetics at the higher addition rates of 25 and 50 mg.

A time series of high performance ion chromatographic traces detailing the disappearance of the 50 mg SeMet-Se/kg soil application rate in the Panoche soil at the 6, 24, and 72 hour sampling periods is shown in Figure 2 (Martens and Suarez, 1997c). The Panhill traces were nearly identical to the Panoche traces except for the remaining SeMet concentrations (data not shown). The faster SeMet mineralization rates of the Panoche soil compared to the Panhill soil may be partially explained by the cultivation history of the two soils. The Panoche soil when sampled was under intensive vegetable cultivation in the central San Joaquin Valley of California, but the nearby Panhill soil had not been subjected to recent cultivation. Monreal and McGill (1989) found that Cys mineralization ($^{14}CO_2$ production) was greater in heavily cultivated soils compared to the respective noncultivated soils, and the differences were attributed to different C allocation patterns among microbially mediated processes.

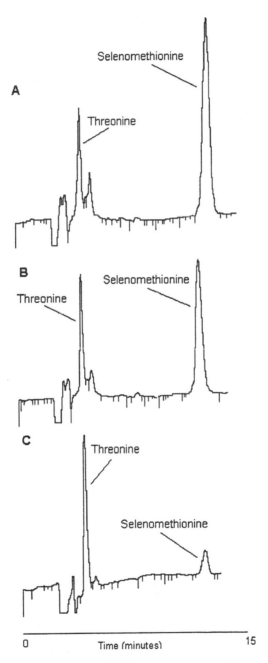

FIGURE 2 Chromatographic traces of SeMet concentrations (50 mg SeMet-Se/kg soil) remaining in the Panoche soil after incubation for (A) 6 hour, (B) 24 hour, and (C) 72 hours. (Reprinted with permission, Martens and Suarez, 1997c.)

Karlson and Frankenberger (1989) reported SeMet additions to soil resulted in evolution of volatile Se compounds. The amounts of Se volatilized were different for the two soils tested. The Panhill soil evolved 70, 60, and 40% of the added SeMet at the 5, 25, and 50 mg SeMet-Se/kg soil addition rates, respectively. In contrast, the Panoche soil evolved 50, 76, and 80% of the added Se at the same rates of addition (50 mg addition rate presented in Fig. 3). The decrease in Se volatilized from the Panhill soil with increasing SeMet levels may be due to the accumulation of a P-buffer-soluble organic Se compound (Se[−II]) as determined by HGAAS analysis (Martens and Suarez, 1997c). This unidentified compound reached a maximum concentration of 21.5 mg/kg (43%) of the 50 SeMet mg/kg rate applied to the Panhill soil (Figurge 3). Extensive chromatographic analysis determined that this accumulated organic Se compound was not SeMet or an Se derivative of homoserine.

Volatile compounds also are reported from incubation of Met additions to soil (Frankenberger and Karlson, 1989; Martens and Suarez, 1997c). The intermediates in the metabolism of Met, dimethylsulfoniopropionate (ambient incubation) and S-methylmethionine (100°C incubation) can be detected by formation of dimethyl sulfide with application of a two-step alkaline hydrolysis

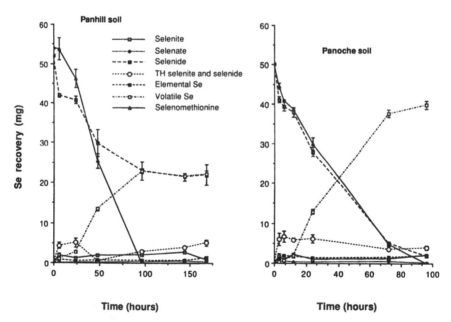

FIGURE 3   Recovery of SeMet-Se (mean ± standard deviation) from the Panhill soil and the Panoche soil amended with 50 mg SeMet-Se/kg soil and the HGAAS speciation analysis; TH, tightly held. (Reprinted with permission, Martens and Suarez, 1997c.)

method (White, 1982). Application of the alkaline hydrolysis procedure also released dimethyl selenide (DMSe) from the Panoche soil treated with SeMet, but DMSe was not released from the SeMet-treated Panhill soil. Evolution of DMSe was measured with ambient NaOH incubation, indicating the presence of dimethylselenopropionate (DMSeP) and evolution of DMSe with the 100°C NaOH incubation indicates the presence of Se-methylselenomethionine (Se-methyl-SeMet). The detection of the possible intermediates of Se volatilization from SeMet additions to soils has not been reported, but Cooke and Bruland (1987) reported evidence of Se-methyl-SeMet in seleniferous surface waters. This evidence suggests that several pathways for Se volatilization are active in the soils tested, a methylation and a demethylation pathway, similar to the two pathways found for S volatilization (Taylor and Gilchrist, 1991). The failure to detect DMSeP or Se-methyl-SeMet evolution from the alkaline-treated Panhill soil extracts suggests that the organic Se compound that accumulated in the SeMet-treated Panhill soil was 3-selenol-propionate ($HSeCH_2CH_2CO_2H$). This Se compound is analogous to a 3-mercaptopropionate ($HSCH_2CH_2CO_2H$) compound found to accumulate with aerobic marine bacteria that metabolize the demethylation pathway intermediate 3-methiolpropionate ($CH_3SCH_2CH_2CO_2H$) (Taylor and Gilchrist, 1991).

2. Selenocystine Mineralization

In contrast to the extensive Se volatilization noted with application of SeMet to the Panoche and Panhill soils, no measurable volatile Se was observed for SeCys applications to the Panhill soil, and much lower levels (maximum of 18% of SeCys-Se applied) of volatile Se were measured with SeCys applications to the Panoche soil (Martens and Suarez, 1997c). Challenger (1951) suggested that dimethyl diselenide formation in natural systems was the result of SeCys metabolism with concomitant release of the amino acid serine. Headspace analysis of the Panoche soil treated with SeCys determined that only DMSe was present and extensive amino acid analysis failed to detect the release of serine. After incubation for 6 hours, SeCys was not detected in the P-buffer extracts of either soil, yet the majority of the Se remaining in both soils was determined to be Se[−II] by HGAAS (Fig. 4). The organic Se compound(s) formed in the soils appears to be 3-selenol-propionate ($HSeCH_2CH_2CO_2H$), the same compound that was proposed to accumulate in the Panhill soil treated with SeMet. The rapid oxidation of the SeCys (Se[−II]) to Se[O] and Se[IV] is in contrast to the very slow conversion of soil amended Se[O] to Se[IV] reported by Geering et al. (1968). We conclude that SeCys is very unstable in natural systems, that SeCys present in biological materials oxidizes rapidly, and that the majority of the Se[−II] released by oxidation of SeCys is slowly transformed to inorganic Se. The quantity and distribution of the two selenoamino acids, SeMet and SeCys, in seleniferous plant

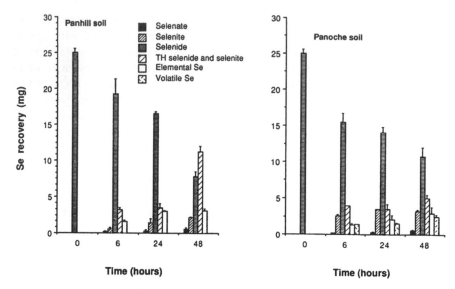

FIGURE 4 Distribution of Se (mean ± standard deviation) in the Panhill and Panoche soils amended with 25 mg SeCys-Se/kg soil: TH, tightly held. (Reprinted with permission, Martens and Suarez, 1997c.)

material have not been completely described. Wheat, soybeans, and other grains have been reported to contain SeMet (Beilstein and Whanger, 1986; Yasumoto et al., 1988), while *Allium* and *Brassica* spp. reportedly are rich sources of the amino acids SeCys and methylselenocysteine (M-SeCys) (Cai et al., 1995).

3.   Seleniferous Alfalfa Mineralization

To determine the distribution of Se in seleniferous plant tissue before and after mineralization, seleniferous alfalfa was produced by growing alfalfa in 0.5 M Hoaglands nutrient solution with and without 2 mg/L Se[VI] (Martens and Suarez, 1997d). The Se accumulated in plant tissue has been reported to be in the Se[−II] form as selenoamino acids. The sequential extraction procedure was used to analyze the alfalfa samples. Results indicate that plants such as alfalfa show a wide range in solubility of Se compounds (Fig. 5). The alfalfa (per-kilogram dry weight basis ±standard deviation) assimilated 2.6 ± 0.31 mg Se[VI], 13.0 ± 0.60 mg Se[−II], and a trace of Se[IV] as P-buffer-extractable Se. A rapid incorporation of Se[VI] into organic Se compounds was evidenced by the low levels of inorganic Se[IV] and Se[VI] (8.8%) determined in the alfalfa tissue. Phosphate-extractable components have been shown to be associated with the cytoplasm and soluble cellular components of the plant tissue (Michaud and Asselin, 1995). The remaining alfalfa Se appears to be incorporated into insoluble proteins and/

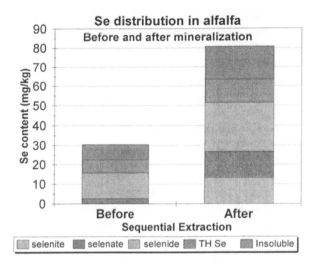

FIGURE 5 Distribution of assimilated alfalfa Se as determined by sequential extractions before and after mineralization by soil microorganisms for 60 days and HGAAS speciation analysis; TH Se, tightly held Se.

or cell walls due to the need for stronger oxidation steps for tightly held (TH) Se (8.0 ± 0.58 mg Se) and acid-soluble Se (6.7 ± 0.37 mg Se). Amino acid analysis with Se amino acid peak fraction collection, digestion, and HGAAS analysis confirmed the presence of Se-M-SeMet, SeMet, SeCys, and M-SeCys in the soluble phase (data not presented). Of the Se content determined by the captured amino acid fractions, 75% of the amino acid-Se was present as SeCys or M-SeCys.

After mineralization of the alfalfa by soil microorganisms for 60 days, the Se distribution in the alfalfa showed an increase in Se concentration in the remaining plant residue (Fig. 5). Sequential extractions found the following concentrations: 13.1 ± 1.2 mg Se[IV], 13.6 ± 1.35 mg Se[VI], 24.8 ± 2.15 mg Se[−II], 11.9 ± 1.02 mg TH selenium, and 17.4 ± 0.96 mg refractory Se for the mineralized alfalfa residue (Se/kg dry weight basis ±standard deviation). Only about 10% of the alfalfa residue-Se was captured as volatile Se. The lack of extensive Se volatilization is perhaps due to the low levels of SeMet compared to the concentrations of SeCys compounds determined in the alfalfa tissue. The pattern of Se oxidation states in the decomposed alfalfa tissue (Fig. 5) resembled the pattern of Se oxidation states present in the same soils treated with SeCys (Fig. 4).

The ratio of C to Se determined for the fresh alfalfa was 13,400. After microbial decomposition, C/Se decreased to 3800, indicating that the relative Se content in the residue increases as C mineralization proceeds. Analysis of the alfalfa tissue remaining after the 60-day mineralization experiment found that the tissue was devoid of the extractable amino acids that had been determined before the experiment. Only low levels of SeMet and phenylalanine were identified in the alfalfa extractions, indicating that most of the P-buffer-extractable selenide-Se remaining in the mineralized alfalfa residue was present as non–amino acid selenide-Se, as was also noted in the SeCys experiments. A total protein digestion (Martens and Frankenberger, 1992) of the alfalfa tissue remaining after mineralization indicated that all the amino acids measured before decomposition were still accounted for in the mineralized residue, but the concentrations had been diminished by approximately 50% (data not presented). The concentration of the inorganic Se species, Se[VI] and Se[IV], accounted for just 32.5% of the Se inventory after incubation, indicating that the majority of selenide-Se present in the plant residues (ca. 10% inorganic forms initially) will not be mineralized to inorganic forms of Se. The decrease in the C/Se ratio after mineralization suggests that the soil organisms are mineralizing the nonseleniferous organics at a faster rate than the seleniferous organic compounds.

The large increase in Se concentrations with mineralization may help to explain how elevated levels of water-insoluble Se can occur in aquatic systems exposed to soluble inorganic Se. To test whether this mechanism for Se enrichment is occurring in contaminated soil, plant residues (washed free of soil and lyophilized) present in the Kesterson pond 4 ecosystem (sampled May 1996) were analyzed for C and Se; the results indicated that the C/Se ratio in the organic litter had decreased to near 1000. The Se enrichment mechanism noted in the alfalfa C mineralization experiment appeared to be occurring in natural systems (data not presented).

C. Seleniferous Irrigation Drainage Water Management

I.   Organic Matter Management

The slower mineralization rate of Se-containing organic compounds compared to nonseleniferous compounds may help explain why concentrations of Se compounds can rapidly increase in the environment. Martens and Suarez (1997b), after evaluating a number of seleniferous soils and evaporation pond sediments from the San Joaquin Valley of California, suggested that the distribution of the Se oxidation states present in the evaporation ponds can be managed by limiting organic C accumulation. Martens and Suarez (1997b) found that organic-associated Se present in evaporation pond soils increased rapidly when the organic C content increased above 20.0 g organic C per kilogram of soil.

The importance of the relationship between Se content and organic C content in evaporation pond management is demonstrated by analyzing the results of two strategies for dealing with irrigation drainage waters. An aerated drainage water evaporation pond located on the Sumner Peck ranch (Peck ponds, west San Joaquin Valley, CA), which limited plant growth (4.6 g organic C/kg sediment), assimilated less organic-associated Se (18%) from irrigation drainage waters compared to organic-associated Se (> 80%) in Kesterson evaporation ponds (Kesterson pond 4) with uncontrolled plant growth (26.7 g organic C/kg sediment). At present, no management plans are in place for dealing with seleniferous irrigation drainage wastewater, but this research suggests that management of vegetation in evaporation ponds can affect the distribution of Se and may possibly limit the formation of toxic organic Se compounds.

2. Bioremediation Success Interpreted by the Distribution of Selenium Oxidation States

Frankenberger and his coworkers (Frankenberger and Karlson, 1989) demonstrated that stimulation of microorganisms that volatilize Se may effectively remove Se from Se-contaminated soils. Successful alkylselenide production in soil is controlled by several factors (Doran and Alexander, 1977; Frankenberger and Karlson, 1989, 1994). The oxidation state of the Se present in the contaminated material is the predominant factor implicated in Se volatilization (Doran and Alexander, 1977). The oxidation state order determined by Doran and Alexander (1977) for Se volatilization was Se[−II] (as SeMet) ⩾ Se[IV] > Se[VI] ⩾ Se[O]. Knowledge of the distribution of Se oxidation states present in a contaminated soil enables prediction of the efficiency and rate of Se removal from the system.

Frankenberger and his coworkers implemented bioremediation practices at the Peck ranch and Kesterson evaporation pond systems during the late 1980s and early 1990s (Frankenberger and Karlson, 1995; Flury et al., 1997). They obtained different results at the two sites. The Peck ponds (range of 7.1–18.1 mg total Se/kg sediment; mean, 11.4 mg Se) could rapidly be detoxified by microbial bioremediation to meet the cleanup goal of a 4 mg total Se/kg soil in as short of a time frame as 2.6 years. However, Se concentrations at the Kesterson pond 4 test site (1987 Se content, 47.8 mg total Se/kg sediment) (with the same practices as at the Peck ponds) did not decrease to the desired 4 mg total Se/kg sediment in more than 8 years of remediation. The remediation success at the Peck evaporation site could have been predicted by evaluation of the distribution of the Se oxidation states in the soil by use of the outlined sequential extraction methodology. A Peck evaporation pond sediment had an oxidation state distribution (12.2 mg total Se/kg sediment) of 0.81 mg Se[VI], 8.24 mg Se[IV], 2.24 mg Se[O] (18%), and 0.89 mg Se[−II]/kg sediment (7.2%). In contrast, the Kesterson pond 4 had an Se oxidation state distribution (mg Se/kg sediment) of 2.48 mg Se[VI], 4.36 mg Se[IV], 27.73 mg Se[O] (58%), and 13.23 mg Se[−II]/kg sediment (27.7%). The Kesterson soil had a much greater percentage of the Se inventory in Se forms that are known to be poor substrates for microbial volatilization, such as Se[O] and Se[−II] (86% of inventory) when compared to the Peck pond soil levels of Se[O] and Se[-II] (25% of inventory).

D. Conclusions

Recent research has been aimed at understanding the reactions of inorganic forms of selenium. However, we now know that the most toxic Se and also the least understood Se forms are the organic Se compounds. The research presented here illustrates a small portion of perhaps the very important role of carbon-bonded Se in the cycling of Se in biological systems.

REFERENCES

Abrams, M. M., and R. G. Burau. 1989. Fractionation of selenium and detection of selenomethionine in a soil extract. *Commun. Soil Sci. Plant Anal.* 20:221–237.

Abrams, M. M., R. G. Burau, and R. J. Zasoski. 1990. Organic selenium distribution in selected California soils. *Soil Sci. Soc. Am. J.* 54:979–982.

Burau, R. G. 1985. Environmental chemistry of selenium. *California Agric.* 39:16–18.

Beilstein, M. A., and P. D. Whanger. 1986. Deposition of dietary organic and inorganic selenium in rat erythrocyte proteins. *J. Nutr.* 116:1701–1710.

Cai, X. J., E. Block, P. C. Uden, X. Zhang, B. D. Quimby, and J. J. Sullivan. 1995. *Allium* chemistry: Identification of selenoamino acids in ordinary and selenium-enriched garlic, onion, and broccoli using gas chromatography with atomic emission detection. *J. Agric. Food Chem.* 43:1754–1757.

Cary, E. E., G. A. Wieczorek, and W. H. Allaway. 1967. Reactions of selenite-selenium added to soils that produce low-selenium forages. *Soil Sci. Soc. Am. Proc.* 31:21–26.

Challenger, F. 1951. Biological methylation. *Adv. Enzymol.* 12:429–491.

Chao, T. T., and R. F. Sanzolone. 1989. Fractionation of soil selenium by sequential partial dissolution. *Soil Sci. Soc. Am. J.* 53:385–392.

Cooke, T. D., and K. W. Bruland. 1987. Aquatic chemistry of selenium: evidence of biomethylation. *Environ. Sci. Technol.* 21:2114–2119.

Doran, W. T., and M. Alexander. 1977. Microbial transformations of selenium. *Appl. Environ. Microbiol.* 33:31–37.

Elsokkary, I. H. 1980. Selenium distribution, chemical fractionation and adsorption in some Egyptian alluvial and lacustrine soils. *Z. Pflanzenernaehr. Bodenkd.* 143:74–83.

Flury, M., W. T. Frankenberger, Jr., and W. A. Jury. 1997. Long-term depletion of selenium from Kesterson dewatered sediments. *Sci. Total Environ.* 198:259–270.

Frankenberger, Jr., W. T., and U. Karlson. 1989. Environmental factors affecting microbial production of dimethylselenide in a selenium-contaminated sediment. *Soil Sci. Soc. Am. J.* 53:1435–1442.

Frankenberger, Jr., W. T,. and U. Karlson. 1994. Soil management factors affecting volatilization of selenium from dewatered sediments. *Geomicrobiol. J.* 12:265–278.

Frankenberger, Jr., W. T., and U. Karlson. 1995. Volatilization of selenium from a dewatered seleniferous sediment: A field study. *J. Ind. Microbiol.* 14:226–232.

Freney, J. R., G. E. Melville, and C. H. Williams. 1970. The determination of carbon bonded sulfur in soils. *Soil Sci.* 114:310–318.

Frost, D. V., and P. M. Lish. 1975. Selenium in biology. *Annu. Rev. Pharmacol.* 15:259–284.

Frost, R. R., and R. A. Griffen. 1977. Effect of pH on adsorption of arsenic and selenium from landfill leachate by clay minerals. *Soil Sci. Soc. Am. J.* 41:53–57.

Geering, H. R., E. E. Cary, L. H. P. Jones, and W. H. Allaway. 1968. Solubility and redox criteria for the possible forms of selenium in soils. *Soil Sci. Soc. Am. Proc.* 32:35–40.

Greubel, K. A., J. A Davis, and Leckie J. O. 1988. The feasibility of using sequential extraction techniques for arsenic and selenium in soils and sediments. *Soil Sci. Soc. Am. J.* 52:390–397.

Gupta, U. C., K. A. Winter, and J. B. Sanderson. 1993. Selenium content of barley as influenced by selenite- and selenate-enriched fertilizers. *Commun. Soil Sci. Plant Anal.* 24:1165–1170.

Haygarth, P. M. 1994. Global importance and global cycling of selenium. In W. T. Frankenberger, Jr., and S. Benson (eds.), *Selenium in the Environment*, pp.1–28 New York.

Heinz, G. H., D. J. Hoffman, A. J. Krynitsky, and D. M. G. Weller. 1987. Reproduction in mallards fed selenium. *Environ. Toxicol. Chem.* 6:423–433.

Heinz, G. H., D. J. Hoffman, and L. G. Gold. 1988. Toxicity of organic and inorganic selenium to mallard ducklings. *Arch. Environ. Contam. Toxicol.* 17:561–568.

Heinz, G. H., D. J. Hoffman, and L. G. Gold. 1989. Impaired reproduction of mallards fed an organic form of selenium. *J. Wildl. Manage.* 53:418–428.

Kang, Y., H. Yamada, K. Kyuma, and T. Hattori. 1993. Speciation of selenium in soil. *Soil Sci. Plant Nutr.* 39:331–337.

Karlson, U., and W. T. Frankenberger, Jr. 1989. Accelerated rates of selenium volatilization from California soils. *Soil Sci. Soc. Am. J.* 53:749–753.

Likens, G. E., and F. H. Bormann. 1974. Acid rain: A serious regional environmental problem. *Science* 184:1176–1179.

Maiers, D. T., P. L. Wichlacz, D. L. Thompson, and D. F. Bruhn. 1988. Selenate reduction by bacteria from a selenium-rich environment. *Appl. Environ. Microbiol.* 54:2591–2593.

Martens, D. A., and W. T. Frankenberger, Jr.. 1992. Pulsed amperometric detection of amino acids separated by anion exchange chromatography. *J. Liquid Chromatog.* 15:423–439.

Martens, D. A., and D. L. Suarez. 1997a. Selenium speciation of soil/sediment determined with sequential extractions and hydride generation atomic absorption spectrophotometry. *Environ. Sci. Technol.* 31:171–177.

Martens, D. A., and D. L. Suarez. 1997b. Selenium speciation of marine shales, alluvial soils, and evaporation basin soils of California. *J. Environ. Qual.* 26:424–432.

Martens, D. A., and D. L. Suarez. 1997c. Mineralization of selenium-containing amino acids in two California soils. *Soil Sci. Soc. Am. J.* (in press).

Michaud, D., and A. Asselin. 1995. Application to plant proteins of gel electrophoretic methods. A review. *J. Chromatogr.* 698:263–279.

Monreal, C. M., and W. B. McGill. 1989. The dynamics of free cystine cycling at steady-state through the solutions of selected cultivated and uncultivated Chernozemic and Luvisolic soils. *Soil Biol. Biochem.* 21:689–694.

Naftz, D. L., and J. A. Rice. 1989. Geochemical processes controlling selenium in ground water after mining, Powder River Basin, Wyoming, USA. *Appl. Geochem.* 4:565–575.

Ohlendorf, H. M., and G. M. Santolo. 1994. Kesterson reservoir—Past, present and future: An ecological risk assessment. In W. T. Frankenberger, Jr., and S. M. Benson (eds.), *Selenium in the Environment,* pp. 69–117. Dekker, New York.

Pickering, I. J., G. E. Brown, Jr., and T. K. Tokunaga. 1995. Quantitative speciation of selenium in soils using X-ray absorption spectroscopy. *Environ. Sci. Technol.* 29:2456–2459.

Presser, T. S., W. C. Swain, R. R. Tidball, and R. C. Severson. 1990. Geologic sources, mobilization, and transport of selenium from the California Coastal ranges to the western San Joaquin Valley: A reconnaissance study. U.S. Geological Survey, Water Resources Invest. Report No. 90-4070.

Rael, R. M., and W. T. Frankenberger, Jr. 1995. Detection of selenomethionine in the fulvic fraction of a seleniferous sediment. *Soil Biol. Biochem.* 27:241–242.

Scott, M. L. 1973. Selenium dilemma. *J. Nutr.* 103:803–808.

Sharmasarkar, S., and G. F. Vance. 1995. Fractional partitioning for assessing solid-phase speciation and geochemical transformations of soil selenium. *Soil Sci.* 160:43–55.

Stadtman, T. C. 1979. Some selenium-dependent biochemical processes. *Adv. Enzymol. Relat. Areas Mol. Biol.* 48:1–28.

Stryer, L. 1981. Introduction to protein structure and function. In (ed.), *Biochemistry,* p. 32. Freeman, San Francisco.

Tabatabai, M. I., and J. M. Bremner. 1972. Distribution of total and available sulfur in selected soils and soil profiles. *Agron. J.* 64:40–44.

Taylor, B. F., and D. C. Gilchrist. 1991. New routes for aerobic biodegradation of dimethyl-sulfoniopropionate. *Appl. Environ. Microbiol.* 57:3581–3584.

Tokunaga, T. K., and S. M. Benson. 1992. Selenium in Kesterson reservoir ephemeral pools formed by groundwater rise: I. A field study. *J. Environ. Qual.* 21:246–251.

Tokunaga, T. K., D. S. Lipton, S. M. Benson, A. W. Yee, J. O. Oldfather, E. C. Duckart, P. W. Johanis, and K. E. Halvorsen. 1991. Soil selenium fractionation, depth profiles, and time trends in a vegetated upland at Kesterson Reservoir. *Water Air Soil Pollut.* 57:31–41.

Tokunaga, T. K., S. R. Sutton, and S. Bajt. 1994. Mapping of selenium concentrations in soil aggregates with synchrotron X-ray fluorescence microprobe. *Soil Sci.* 158:421–434.

Weres, O., A.-R. Jaouni, and L. Tsao. 1989. The distribution, speciation and geochemical cycling of selenium in a sedimentary environment, Kesterson Reservoir, California, U.S.A. *Appl. Geochem.* 4:543–563.

White, R. H. J. 1982. Analysis of dimethyl sulfonium compounds in marine algae. *Mar. Res.* 40:529–535.

White, A. F., and N. M. Dubrovsky. 1994. Chemical oxidation–reduction controls on selenium mobility in groundwater systems. In W. T. Frankenberger, Jr., and S. M. Benson (eds.), *Selenium in the Environment,* pp. 185–222. Dekker, New York.

Yasumoto, K., T. Suzuki, and M. Yoshida. 1988. Identification of selenomethionine in soybean protein. *J. Agric. Food Chem.* 46:463–467.

Zhang, Y., and J. N. Moore. 1996. Selenium fractionation and speciation in a wetland system. *Environ. Sci. Technol.* 30:2613–2619.

Zhen-Li, H., Y. Xiao-E, Z. Zu-Xiang, X. Wei-Ping, P. Jian-Ming, and L. Xiao-Ya. 1994. Fractionation of soil selenium with relation to Se availability to plants. *Pedosphere* 4:209–216.

5

Effects of Selenium Supplementation of Fertilizers on Human Nutrition and Selenium Status

ANTTI ARO and GEORG ALFTHAN
National Public Health Institute, Helsinki, Finland

PÄIVI EKHOLM and PERTTI VARO
University of Helsinki, Helsinki, Finland

I. INTRODUCTION

In Finland the selenium content of agricultural soils is relatively low (0.2–0.3 mg/kg dry matter). The climatic conditions and the low pH and high iron content of the soil favor the deposition of reduced selenium compounds such as selenites and selenides, which form insoluble complexes and are poorly available for plants. Thus, only about 5% of total selenium in the soil is in a soluble form and available for plants (Koljonen, 1975; Sippola, 1979).

Muscular dystrophy and other disorders associated with insufficient selenium intake were found to be a problem in animal husbandry during the 1950s and 1960s. Since 1969, commercial animal feeds in Finland have been supplemented with 0.1 mg of sodium selenite per kilogram of product, dry weight. This reduced the incidence of selenium deficiency problems but did not entirely remove the need for selenium medication of domestic animals, and did not appreciably affect human nutrition. In a large study of the mineral composition of Finnish foods conducted in the late 1970s it was shown that all agricultural products grown and produced in Finland contained very low amounts of selenium (Varo and Koivistoinen, 1980).

During the 1970s, the dietary selenium intake was relatively stable and was estimated to be as low as 25 μg/10 megajoules (10 MJ = 2400 kcal). During periods of poor domestic crop yields, when high-selenium wheat from North America was imported, as was the case in the early 1980s, the selenium content of cereal products increased. Dietary selenium intakes and serum selenium concentrations increased in 1979, 1980, and 1982 (Varo and Koivistoinen, 1981; Alfthan, 1988).

In certain areas of the People's Republic of China, children and young women with extremely low selenium intakes (≤ 10 μg/day) have been affected by an endemic, often fatal cardiomyopathy called Keshan disease (Keshan Disease Research Group, 1979). In Finland no apparent disorders caused by selenium deficiency have been observed in human subjects. However, the results of some Finnish prospective epidemiological studies suggested that particularly low selenium intakes, reflected as low serum selenium levels, were associated with increased risk of cardiovascular death (Salonen et al., 1982) and certain types of cancer (Salonen et al., 1984). These findings, together with the known problems caused by selenium deficiency in domestic animals and the increasing popularity of self-medication with commercial selenium preparations, persuaded the Finnish authorities in 1983 to consider means for increasing the selenium content of domestic agricultural products (Varo et al., 1994).

II. SELENIUM SUPPLEMENTATION OF FERTILIZERS IN FINLAND

The Ministry of Agriculture and Forestry of Finland decided in 1983 to supplement multimineral fertilizers with selenium in the form of sodium selenate starting in the fall of 1984. Direct enrichment of foods was considered dangerous because of the toxicity of selenium compounds. Supplementation of fertilizers was preferred, because the plants form a biological barrier that would protect the target population from the effects of accidental overdosage of selenium. The decision was based on studies conducted in Denmark (Gissel-Nielsen and Bisbjerg, 1970) and Finland (Yläranta, 1983) showing that supplementary selenate, in contrast to selenite, stays in the soil in a form that is available for plants for several months. It was also shown that the amount necessary to increase the selenium content of grass is lower than the amount needed for increasing the selenium content of grains (Yläranta, 1984a, 1984b). Initially the fertilizers for grain production were supplemented with 16 mg/kg of selenium as sodium selenate and those for the production of hay and fodder with 6 mg/kg. The primary goal was to increase the selenium content of cereal grains by tenfold. Although selenium is probably not an essential element for plants, they absorb the mineral and transform it into organic compounds. Therefore it could be foreseen that animal nutrition and foods of animal origin would be affected as well.

An expert group was appointed by the government to monitor the effects of the supplementation on soils, waters, animal feeds, grain crops, foods, and the human selenium status. Since 1990, all fertilizers have been supplemented with the same amount of selenium (Varo et al., 1994). The expert group recommended the amount of 8 mg/kg for all fertilizers, but the ministry decided to reduce the level to 6 mg/kg.

III. EFFECTS OF FERTILIZATION ON HUMAN NUTRITION

A. Animal Feeds

The selenium content of animal feeds is determined mainly via agricultural practices. Animal farms usually produce their own grassy and cereal feeds whose selenium level is determined by the supplemented fertilizers. However, commercial animal feed mixtures and mineral concentrates for pigs contain added selenite of, at most, 0.1 mg Se/kg, so that the total selenium content of feed does not exceed 0.4 mg/kg (Ministry of Agriculture and Forestry, 1994). Selenium supplementation of commercial feeds is not allowed for dairy cattle or in beef production. During their life span, animals may also undergo selenium medication. The selenium content of domestic animal products is the sum of all these factors.

Although the range of selenium concentrations in foods remained narrow and safe throughout the supplementation period (1985–1990), some peak concentrations exceeding 1 mg/kg dry matter were detected in fodder and hay samples from a few individual farms. The high values were most probably caused by liberal use of high-selenium fertilizers (16 mg/kg) in the production of grassy feeds and hay. Elimination of these unnecessarily high concentrations was the main reason for lowering the level of supplemented selenium in fertilizers after 1990.

Oats and barley for animal feed are grown in large quantities in Finland. They are sown in the spring like spring wheat. The farm-to-farm variation in the selenium concentration of cereal feeds has remained within acceptable limits after adoptation of the practice of supplementing all fertilizers with the same amount of selenium (Fig. 1), and high values (>0.5 mg/kg) have been eliminated from grassy feeds as well.

B. Foods

The selenium concentration of spring cereals responded dramatically to the changes in the supplementation level of fertilizers both in 1985 and 1991 (Table 1). The highest selenium concentrations, of about 0.25 mg/kg wheat in 1987–1990, decreased to the level of 0.1 mg/kg and below in 1991–1996 (Fig. 2). The decline in selenium concentration has been smaller in other foods, because

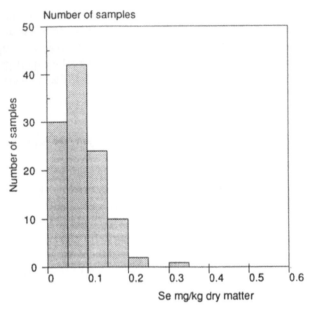

FIGURE I   Farm-to-farm variation in the selenium concentration of barley grown in Finland in 1993.

the lower selenium level fertilizers (6 mg Se/kg) were already used in grassy feed production before 1990.

The overall effect of selenium supplementation on winter cereals (winter wheat and rye) has been small, and their selenium concentrations never exceeded

TABLE I   Selenium Concentration (mg/kg dry matter, mean ± SD) of Finnish Foods Between 1984 and 1996

	n	1984	n	1990	n	1996[a]
Wheat bread, French loaf	24	0.05 ± 0.04	16	0.23 ± 0.02	12	0.10 ± 0.02
Rye bread, whole grain	24	0.07 ± 0.05	16	0.06 ± 0.02	12	0.05 ± 0.03
Potato	2	<0.01	16	0.11 ± 0.03	12	0.03 ± 0.01
Beef, steak	24	0.17 ± 0.06	16	0.64 ± 0.08	12	0.37 ± 0.05
Pork, chop	24	0.35 ± 0.07	16	1.09 ± 0.09	12	0.56 ± 0.07
Milk, whole	24	0.06 ± 0.01	16	0.23 ± 0.02	12	0.13 ± 0.01
Cheese, Edam type	24	0.09 ± 0.02	16	0.42 ± 0.04	12	0.23 ± 0.01
Eggs	24	0.69 ± 0.15	16	1.27 ± 0.13	12	0.89 ± 0.06

[a]March–September.

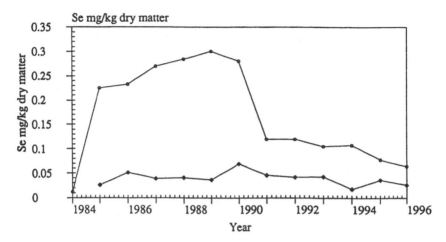

FIGURE 2   Selenium concentrations of spring (circles) and winter (diamonds) wheat, 1984–1996.

0.1 mg/kg. The difference from spring cereals is mainly due to the time difference between application of fertilizer for sowing and the growing season during the next summer.

Finland occasionally imports grain because of an inadequate or low quality harvest, and domestic and imported grain is mixed for milling in ratios stipulated by the Ministry of Agriculture. A substantial part of the imported grain comes from North America. The imported grain slightly raised the selenium content of flour and bread even during the years of selenium fertilization in 1989 and 1990, as the selenium content of North American grain was higher than the grain produced in Finland with selenium-enriched fertilizers (Ekholm et al., 1995).

Selenium concentrations have declined in all Finnish agricultural products since 1991. Milk has been the most sensitive indicator food throughout the monitoring period. The selenium concentration of milk is known to be closely dependent on that of feeds (Jacobsen et al., 1965; Conrad and Moxon, 1979; Aspila, 1991). Thus the present change in the selenium concentrations of foods was first observed in milk, in the summers of 1985 and 1991.

The selenium concentrations of milk, cheese, and eggs increased two- to fivefold between 1984 and 1990 and decreased by 30 to 45% during 1991 to 1996 (Table 1). The use of selenite-supplemented commercial feeds was already common in egg production in the 1970s and 1980s. Consequently, in 1996 the selenium concentration of eggs was only 30% higher than that in 1984, before selenium was added to fertilizers.

Selenite raises the selenium content of muscle tissues less effectively than does organic selenium (Ekholm et al., 1991). This explains the effectiveness of the selenium-supplemented fertilization on the selenium content of meat. The selenium concentrations of beef and pork increased by two- to threefold between 1984 and 1990, and the increase from the 1970s was about tenfold.

The change to a single level of selenium in supplemented fertilizers decreased the selenium concentration of beef by more than 40% and that of pork by 50% (Table 1). The changes have been less marked in bovine and porcine livers than in the corresponding muscle tissues. The selenium content of skeletal muscles and other soft tissues is known to be linearly dependent on the selenium concentration of the diet, reaching a plateau level with the increasing selenium (Mahan and Moxon, 1978; Sankari, 1985; Echevarria et al., 1988). The selenium content of liver reaches its plateau at a lower dietary selenium level (0.25 mg/kg fodder) than muscle tissue (>0.40 mg/kg fodder) (Ekholm et al., 1991). The present selenium concentration of feeds is still high enough to keep the selenium concentration of bovine and porcine liver near its saturation level (Ekholm et al., 1990).

The selenium content of fish has remained unchanged, that is, between 0.50 and 1.50 mg/kg dry matter, depending on the species and the fat content of fish (Eurola et al., 1991). The fat content varies seasonally and therefore there is a small but consistent seasonal variation in fish selenium.

C. Dietary Intake of Selenium

The estimated daily selenium intake per person in Finland in the mid-1970s, when grain was not imported, was 25 μg/10 MJ. Grain imports occasionally raised the selenium intake by 10 to 20 μg/10 MJ, but most of the time the selenium intake was below the lower limit of the safe and adequate intake for selenium as defined in the United States by the National Academy of Sciences (1980), namely, 50 μg/day.

In the estimated selenium intake, a plateau of 110 to 120 μg/10 MJ was reached in 1987, (Fig. 3) and remained constant until 1990 (Varo et al., 1994). The level of intake was confirmed by a 3-day double-portion analysis in 41 subjects in 1991, indicating a mean intake of 95 μg/day: 120 μg/day in men and 85 μg/day in women.

The selenium concentration of human milk correlates well with the estimated selenium intake (Fig. 4). During the period of selenium supplementation, the selenium concentration of breast milk increased from 0.05 mg/kg dry matter in 1977 (Koivistoinen, 1980) to about 0.11 mg/kg dry matter in 1990. The selenium concentration in human milk is low in comparison with cow's milk, because of the lower protein content of human milk. The effect of the decreased dietary selenium intake on the selenium concentration of breast milk was evident in 1992. No further decrease was noted in 1995–1996, indicating that the selenium intake has reached a new plateau.

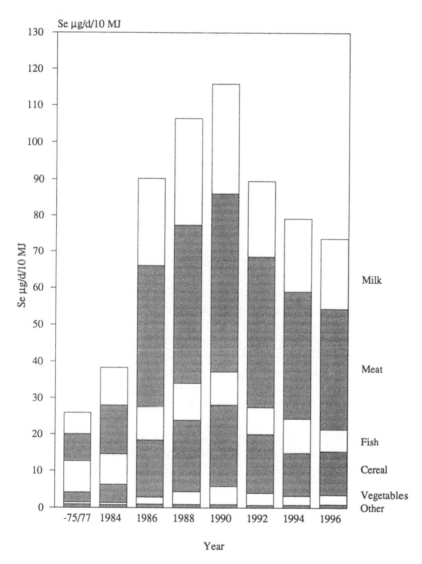

FIGURE 3 The estimated dietary selenium intake (energy level of 10 MJ) in Finland before and during selenium fertilization.

The decrease in the supplemented selenium in fertilizers lowered the daily human intake of selenium to 70 µg/day in 1995, as calculated from Finnish food balance sheets at an energy intake of 10 MJ (Information Centre of Ministry of Agriculture and Forestry, 1996) (Fig. 3). This is well within the current U.S.

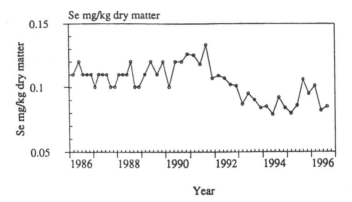

FIGURE 4   Selenium concentrations of human milk, 1986–1996.

recommendations (55 μg/day for women, and 70 μg/day for men) (National Academy of Sciences, 1989) and the Nordic countries (30–60 μg/day) (Nordic Council of Ministers, 1989). The average selenium intake is still higher in Finland than in most other European countries, and is at almost the same level reported in some parts of the United States and Canada (Levander and Morris, 1984; Dokkum et al., 1989; Oster and Prellwitz, 1989; Bratakos and Ioannou, 1991; Pennington and Young, 1991). At present, about 40% of the selenium intake comes from meat, 24% from dairy products and eggs, and 11% from fish. Animal protein is thus the main source of dietary selenium. Cereal products account for 19% of total intake. The overall selenium intake may still be decreasing slightly (Fig. 3).

The present level of selenium in foods guarantees a safe and adequate intake with diets of all kinds. Excessive food-based intakes are most unlikely, even in exceptional dietary compositions.

IV. EFFECTS OF FERTILIZATION ON HUMAN SELENIUM STATUS

A.   Serum and Whole Blood Selenium Concentrations

Serum selenium concentrations in healthy Finnish adults have been monitored since the early 1970s. The low intake of selenium, 25 μg/day in the 1970s (Mutanen and Koivistoinen, 1983), was reflected in a correspondingly low serum selenium level of 0.63 to 0.76 μmol/L (Alfthan, 1988). This level was among the lowest values reported in the world (Alfthan and Nève, 1996). In the early 1980s, the mean serum selenium concentration varied between 0.75 and 1.23 μmol/L depending on the amount of imported high-selenium wheat (Alfthan,

1988). Since 1985 the serum selenium concentration of the same healthy urban male and female, Helsinki ($n = 24-35$) and rural Leppävirta ($n = 35-45$) groups have been followed systematically. The combined annual serum selenium means for 1984 to 1996 are shown in Figure 5. Before selenium supplementation of fertilizers started, the mean serum selenium concentration was 0.89 μmol/L, and it reached a level of 1.5 μmol/L, among the highest values in Europe, 4 years later. After the decrease in fertilizer selenium, the serum selenium decreased to a new level of 1.25 μmol/L in 1996. This serum selenium level is still among the highest in Europe, but lower than is generally found in North America (Alfthan and Nève, 1996).

Serum selenium concentrations in certain population groups thought to be at risk for suboptimal selenium intake also have been studied (Varo et al., 1994). Table 2 shows that the mean serum selenium concentrations of mothers at delivery and their neonates during 1983 to 1996 have paralleled adult serum selenium levels on a lower level. Neither exceptionally low or high individual values have been observed.

Whole-blood selenium has been monitored systematically only in the urban group since 1985. The relative effect of selenium supplementation was larger in whole blood compared with serum (125% vs. 70%, respectively: Fig. 5). Whole-blood selenium reached a maximal level approximately a year later than serum. Before supplementation of fertilizers with selenium it was 1.15 μmol/L, and it

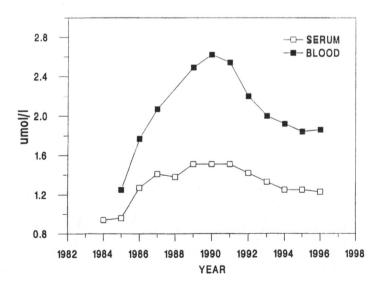

FIGURE 5   Annual mean serum (open squares) and whole-blood (solid squares) selenium concentration in healthy Finns before and during selenium fertilization (starting in 1985).

TABLE 2 Mean Serum Selenium Concentration (μmol/L) in Finnish Mothers at Delivery and Their Neonates Before and During Supplementation of Selenium to Fertilizers

	Before		During			
	n	1983–1985	n	1988–1991	n	1996
Mothers at delivery	15	0.76	21	1.40	20	0.89
Neonate	15	0.59	21	0.86	20	0.71

was at a maximum of 2.60 μmol/L at peak supplementation. After 1992 the decrease paralleled that of serum selenium, and in 1996 the mean level was 1.86 μmol/L.

B. Tissue Selenium Concentrations

Toenails reflect the integrated intake of selenium over a period from 6 months to a year (Longnecker et al., 1993). Unlike serum and whole-blood data, the selenium concentration of toenails has not been studied systematically annually. Data over the period 1984–1995 have been compiled from Finnish studies on different groups of healthy subjects in addition to data from the follow-up groups (Alfthan et al., 1991a; Ovaskainen et al., 1993) (Fig. 6). Before the selenium

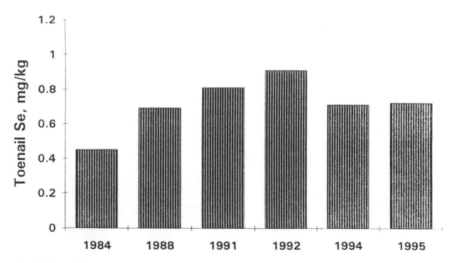

FIGURE 6 Mean toenail selenium concentration in healthy Finnish subjects before and during selenium fertilization (starting in 1985).

supplementation of fertilizers started, the mean toenail selenium concentration was 0.45 mg/kg. The maximum level, 0.91 ± 0.10 mg/kg, was observed in 1992, about 2 years later than for whole blood. In accordance with serum and whole blood, a clear decrease was seen, and the latest value from 1995 is 0.72 ± 0.08 mg/kg. In a recent European multicenter study comprising eight countries, the toenail selenium concentration of 59 middle-aged Finnish men sampled during 1990–1992 was the highest, 0.84 mg/kg (Virtanen et al., 1996), in support of our toenail data.

It has been shown in rat experiments that the major fraction of the body stores of selenium are located in the liver. Liver selenium is mobile and reflects dietary selenium intake over a relatively short period (i.e., weeks) (Levander et al., 1983a). Selenium has been determined in human liver samples obtained at autopsy from men who had died in traffic accidents both before (1983–1985) and during (1988–1989) the program of selenium supplementation of fertilizers. Initially, the mean value was 0.95 mg/kg dry weight ($n = 53$). The increased intake of selenium increased the mean selenium concentration of liver tissue by 60% in samples obtained 3 to 4 years later to 1.58 mg/kg (Varo et al., 1994). The maximal value of adult liver selenium was slightly lower than the values available from other European countries: the Netherlands, 1.75 mg/kg (Aalbers et al., 1987), and Scotland, 1.80 mg/kg (Lyon et al., 1989). The selenium concentration of fetal liver and heart tissue increased by 20 and 40%, respectively, during the same time period (Varo et al., 1994).

C. Serum, Red Blood Cell, and Platelet Glutathione Peroxidase Activity

In serum, whole blood, and red blood cells, activity of the selenium-dependent enzyme glutathione peroxidase (GSHPx) is associated with selenium within moderate intake levels, namely, for serum, below 50 μg/day, and for whole blood, below approximately 60 to 80 μg/day (Yang et al., 1987; Levander, 1989; Alfthan et al., 1991b). At higher intake levels, the activity of the enzyme in serum and whole blood reaches a plateau and cannot be stimulated further. Saturation of serum GSHPx activity has been regarded as a measure of optimal selenium intake and has been the basis of the current U.S. Recommended Dietary Allowance (Levander, 1989).

To find out whether platelet GSHPx activity could be increased by selenium supplementation, and to determine the qualitative effect of organic/inorganic selenium supplementation on GSHPx activity, two placebo-controlled supplementation studies were carried out in central Finland in the same group of healthy middle-aged blood donors (Levander et al., 1983b; Alfthan et al., 1991b).

The first study (supplementation for 11 weeks) was done in 1981, before selenium fertilization, and the other (supplementation for 16 weeks), was done during the program. In both studies, 10 men were supplemented with 200 μg of organic selenium as selenium-enriched yeast or with 200 μg inorganic selenate. The third group received a placebo. At baseline in 1981, the mean plasma selenium concentration was 0.89 μmol/L and in 1987 it was 1.40 μmol/L, reflecting selenium intakes of 39 and 100 μg/day, respectively.

Figure 7 shows the percentage increase of platelet GSHPx activity related to the activity of the placebo groups. Before the addition of selenium to fertilizers, selenate and yeast-selenium increased the activity by 104 and 75%, respectively. During fertilization the effects of selenate and yeast-selenium were much lower: 41 and 6%, respectively. The results suggested that upon the intake of selenium in 1987, 100 μg/day was still not sufficient to completely saturate GSHPx activity in platelets. At the higher intake level, however, both plasma and red blood cell GSHPx was maximally stimulated. An extrapolation of platelet data including the two Finnish studies and five other studies having a similar design suggested that maximal stimulation of platelet GSHPx activity would occur at a plasma selenium level of 1.25 to 1.45 μmol/L (Alfthan et al., 1991b).

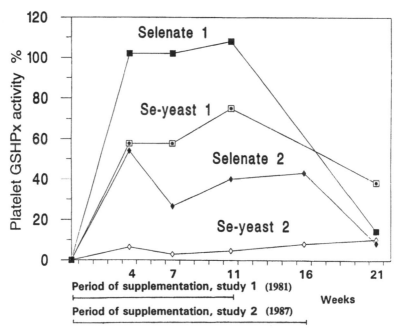

FIGURE 7   Activity of platelet glutathione peroxidase in response to organic and inorganic selenium supplementation (200 μg/day) in healthy Finnish subjects before (in 1981) and during (in 1987) selenium fertilization.

V. HUMAN HEALTH AND ENVIRONMENT

The supplementation of fertilizers in Finland is a nationwide experiment affecting all residents; it is not a placebo-controlled clinical trial. Therefore it is impossible to measure in exact terms the health outcomes of the intervention. Apart from Keshan disease, which can be prevented by prophylactic administration of sodium selenite (Keshan Disease Research Group, 1979), the role of selenium in human disease in unclear. Judging from the results of epidemiological studies, mortality and morbidity in cardiovascular diseases (Salonen et al., 1982) and cancer (Salonen et al., 1984; Knekt et al., 1990) are the end points that might have been affected by increasing the selenium intake of Finnish people. If it is anticipated that selenium exerts its effects via the antioxidant activity of the selenium-dependent enzyme GSHPx, then the intervention should have been successful, since it resulted in near-maximal stimulation of the GSHPx activity in all blood components (Alfthan et al., 1991b).

The age-adjusted mortality from coronary heart disease has declined continuously in Finland since the end of the 1960s. Most of the decline can be attributed to favorable changes in the levels of classical risk factors such as serum cholesterol and blood pressure, and after 1985 it is impossible to find any change in the declining trend that could be attributed to selenium supplementation (Pietinen et al., 1996). Neither do the data on cancer mortality in Finland suggest any specific effects due to increased dietary selenium intake in the late 1980s (Varo et al., 1994). These multifactorial diseases are affected by numerous different factors, and it will be difficult and probably impossible to determine whether the increased selenium intake resulting from supplementation of fertilizers has influenced the health of the nation. Conclusions on the effects of increased selenium intakes should be based on controlled clinical trials.

No health hazards from possible environmental effects of selenium supplementation have been observed. Studies on the selenium levels of tap water, groundwaters, lake and river waters, and lake sediments have disclosed no environmental effects that could be ascribed to the supplemented fertilization (Wang et al., 1991, 1994, 1995; Alfthan et al., 1995). The amounts of selenium that have been used annually in fertilizers (20 tons during 1985–1990, 7.6 tons since 1991) are comparable with the total fallout of selenium from precipitation, estimated to be 18 tons in 1989 (Wang et al., 1993).

VI. SUMMARY AND CONCLUSIONS

For geochemical reasons, Finland has been a low-selenium area. Based on studies of the selenium content of foods, epidemiological associations, and field trials of selenium supplementation, it was decided in 1984 to increase the selenium content of foods and the selenium intake of the population by supplementing multimineral fertilizers with selenium in the form of sodium selenate. Within 2 years, a threefold increase of selenium intake was observed to levels of 100 to 120 μg/day in 1989–1991. The supplementation affected the selenium content of all major food groups with the exception of fish. The serum and whole-blood selenium concentrations of people increased concomitantly by 70 and 125%, respectively. In 1990 the amount of selenium added to fertilizers for grain production was reduced by about 60%. This reduced the selenium intake by 30% and mean serum and whole-blood selenium levels by 20 and 40%, respectively, from the highest levels observed in 1989 and 1991. Plants take up part of the supplemented selenate and convert it into organic selenium compounds, mainly selenomethionine. This modification in plant content affects human nutrition by increasing the selenium content of foods of both animal and vegetable origin. It also reduces the need to supplement animal feeds with inorganic selenium.

In Finland, where the geochemical conditions are relatively uniform, the supplementation of fertilizers with selenium has proved to be an effective, safe, and controlled way of bringing the selenium intake of the whole population to the recommended level (Ekholm, 1997). Since most of the supplemented selenium is derived from foods of animal origin, supplementation of animal feeds with organic selenium would be an alternative way of increasing the selenium concentration in foods and the selenium intake of people (Aro, 1996).

REFERENCES

Aalbers, T. G., J. P. W. Houtman, and B. Makkink. 1987. Trace-element concentrations in human autopsy tissue. *Clin. Chem.* 33:2057–2064.

Alfthan, G. 1988. Longitudinal study on the selenium status of healthy adults in Finland during 1975–1984. *Nutr. Res.* 8:467–476.

Alfthan, G., and J. Nève. 1996. Reference values for serum selenium in various areas evaluated according to the TRACY protocol. *J. Trace Elem. Biol. Med.* 10:77–87.

Alfthan, G., A. Aro, H. Arvilommi, and J. K. Huttunen. 1991a. Selenium metabolism and platelet glutathione peroxidase activity in healthy Finnish men: Effects of selenium yeast, selenite, and selenate. *Am. J. Clin. Nutr.* 53:120–125.

Alfthan, G., A. Aro, H. Arvilommi, and J. K. Huttunen. 1991b. Deposition of selenium in toenails is dependent on the form of dietary selenium. In F. J. Kok and P. van't Veer (eds), *Biomarkers of Dietary Exposure,* p. 110. Proceedings of the Third Meeting on Nutritional Epidemiology, Rotterdam, Jan. 23–25, 1991.

Alfthan, G., D. Wang, A. Aro, and J. Soveri. 1995. The geochemistry of selenium in groundwaters in Finland. *Sci. Total Environ.* 162:93–103.

Aro, A. 1996. Various forms and methods of selenium supplementation. In J. T. Kumpulainen and J. T. Salonen (eds.), *Natural Antioxidants and Food Quality in Atherosclerosis and Cancer Prevention,* pp. 168–171. Royal Society of Chemistry, Cambridge.

Aspila, P. 1991. Metabolism of selenite, selenomethionine and feed-incorporated selenium in lactating goats and dairy cows. *J. Agric. Sci. Finland* 63:1–74.

Bratakos, M. S., and P. V. Ioannou. 1991. Selenium in human milk and dietary selenium intake by Greeks. *Sci. Total Environ.* 105:101–107.

Conrad, H. R., and A. L. Moxon. 1979. Transfer of dietary selenium to milk. *J. Dairy Sci.* 62:404–411.

Dokkum, W. van, R. H. Vos, T. H. de Muys, and J. A. Wesstra. 1989. Minerals and trace elements in total diets in the Netherlands. *Br. J. Nutr.* 61:7–15.

Echevarria, M. G., P. R. Henry, C. B. Ammerman, and P. V. Rao. 1988. Effects of time and dietary selenium concentration as sodium selenite on tissue selenium uptake by sheep. *J. Anim. Sci.* 66:2299–2305.

Ekholm, P. 1997. Effects of selenium supplemented commercial fertilizers on food selenium contents and selenium intake in Finland. Dissertation, EKT-Series No. 1047, University of Helsinki.

Ekholm, P., M. Ylinen, P. Koivistoinen, and P. Varo. 1990. Effects of general soil fertilization with sodium selenate in Finland on the selenium content of meat and fish. *J. Agric. Food Chem.* 38:695–698.

Ekholm, P., P. Varo, P. Aspila, P. Koivistoinen, and L. Syrjälä-Qvist. 1991. Transport of feed selenium to different tissues of bulls. *Br. J. Nutr.* 66:49–55.

Ekholm, P., M. Ylinen, P. Koivistoinen, and P. Varo. 1995. Selenium concentration of Finnish foods: Effects of reducing the amount of selenate in fertilizers. *Agric. Sci. Finland* 4:377–384.

Furola, M., P. Ekholm, M. Ylinen, P. Koivistoinen, and P. Varo. 1991. Selenium in Finnish foods after beginning the use of selenate-supplemented fertilizers. *J. Sci. Food Agric.* 56:57–70.

Gissel-Nielsen, G., and B. Bisbjerg. 1970. The uptake of applied selenium by agricultural plants. 2. The utilization of various selenium compounds. *Plant Soil* 32.382–396.

Information Centre of Ministry of Agriculture and Forestry. 1996. Balance Sheet of Food Commodities, 1995, preliminary. Ministry of Agriculture and Forestry, Helsinki.

Jacobsen, S. O., H. E. Oksanen, and E. Hansson. 1965. Excretion of selenium in the milk of sheep. *Acta Vet. Scand.* 6:299–312.

Keshan Disease Research Group. 1979. Epidemiologic studies on the etiologic relationship of selenium and Keshan disease. *Chin. Med. J.* 92:477–482.

Knekt, P., A. Aromaa, J. Maatela, G. Alfthan, R-K. Aaran, M. Hakama, T. Hakulinen, R. Peto, and L. Teppo. 1990. Serum selenium and subsequent risk of cancer among Finnish men and women. *J. Natl. Cancer Inst.* 82:864–868.

Koivistoinen, P. 1980. Mineral element compositions of Finnish foods. I–XI. *Acta Agric. Scand. Suppl.* 22:1–164.

Koljonen, T. 1975. The behavior of selenium in Finnish soils. *Ann. Agric. Fenn.* 14:240–247.

Levander, O. A. 1989. Progress in establishing human nutritional requirements and dietary recommendations for selenium. In A. Wendel (ed.), *Selenium in Biology and Health*, pp. 205–220. Springer-Verlag, New York.

Levander, O. A., and V. C. Morris. 1984. Dietary selenium levels needed to maintain balance in North American adults consuming self-selected diets. *Am. J. Clin. Nutr.* 39:809–815.

Levander, O. A., D. P. DeLoach, V. C. Morris, and P. B. Moser. 1983a. Platelet glutathione peroxidase activity as an index of selenium status in rats. *J. Nutr.* 113:55–63.

Levander O. A, G. Alfthan, H. Arvilommi, C. G. Gref, J. K. Huttunen, M. Kataja, P. Koivistoinen, and J. Pikkarainen. 1983b. Bioavailability of selenium to Finnish men as assessed by platelet glutathione peroxidase activity and other blood parameters. *Am. J. Clin. Nutr.* 37:887–897.

Longnecker, M. P., M. J. Stampfer, J. S. Morris, V. Spate, C. Baskett, M. Mason, and W. C. Willett. 1993. A 1-y trial of the effect of high-selenium bread on selenium levels in blood and toenails. *Am. J. Clin. Nutr.* 57:408–413.

Lyon, T. D. B., G. S. Fell, D. J. Halls, J. Clark, and F. McKenna. 1989. Determination of nine inorganic elements in human autopsy tissue. *J. Trace Elem. Electrolytes Health Disease* 3:109–118.

Mahan, D. C., and A. L. Moxon. 1978. Effects of adding inorganic and organic selenium sources to the diets of young swine. *J. Anim. Sci.* 47:456–466.

Ministry of Agriculture and Forestry. 1994. Annual Report. Working Group Report No. 2 (in Finnish). Helsinki.

Mutanen, M., and P. Koivistoinen. 1983. The role of imported grain in the selenium intake of Finnish population 1941–1981. *Int. J. Vitam. Nutr. Res.* 53:34–38.

National Academy of Sciences. 1980. *Recommended Dietary Allowances,* 9th ed. National Research Council, Washington, DC.

National Academy of Sciences. 1989. *Recommended Dietary Allowances,* 10th ed. National Research Council, Washington, DC.

Nordic Council of Ministers. 1989. Nordic Dietary Recommendations. Report 2. Copenhagen.

Oster, O., and W. Prellwitz. 1989. The daily selenium intake of West German adults. *Biol. Trace Elem. Res.* 20:1–14.

Ovaskainen, M. L., J. Virtamo, G. Alfthan, J. Haukka, P. Pietinen, P. Taylor, and J. K. Huttunen. 1993. Toenail selenium as an indicator of selenium intake among middle-aged men in a low-soil selenium area. *Am. J. Clin. Nutr.* 57:662–665.

Pennington, J. A. T., and B. E. Young. 1991. Total diet study of nutritional elements, 1982–1989. *J. Am. Diet. Assoc.* 91:179–183.

Pietinen, P., E. Vartiainen, R. Seppänen, A. Aro, and P. Puska. 1996. Changes in diet in Finland from 1972 to 1992. Impact on coronary heart disease risk. *Prev. Med.* 25:243–250.

Salonen, J. T., G. Alfthan, J. K. Huttunen, J. Pikkarainen, and P. Puska. 1982. Association between cardiovascular death and myocardial infarction and serum selenium in a matched-pair longitudinal study. *Lancet* ii:175–179.

Salonen, J. T., G. Alfthan, J. K. Huttunen, and P. Puska. 1984. Association between serum selenium and the risk of cancer. *Am. J. Epidemiol.* 120:342–349.

Sankari, S. 1985. Plasma glutathione peroxidase and tissue selenium response to selenium supplementation in swine. *Acta Vet. Scand.* 81:1–127.

Sippola, J. 1979. Selenium content of soils and timothy (*Phleum pratense* L.) in Finland. *Ann. Agric. Fenn.* 18:182–187.

Varo, P., and P. Koivistoinen. 1980. Mineral element composition of Finnish foods. XII. General discussion and nutritional evaluation. *Acta Agric. Scand. Suppl.* 22:165–171.

Varo, P., and P. Koivistoinen. 1981. Annual variations in the average selenium intake in Finland: Cereal products and milk as sources of selenium in 1979/1980. *Int. J. Vitam. Nutr. Res.* 51:62–65.

Varo, P., G. Alfthan, J. K. Huttunen, and A. Aro. 1994. Nationwide selenium supplementation in Finland—Effects on diet, blood and tissue levels, and health. In R. F. Burk (ed.), *Selenium in Biology and Human Health,* pp. 198–218. Springer-Verlag, New York.

Virtanen, S. M., P. van't Veer, F. Kok, A. F. M. Kardinaal, and A. Aro, for the Euramic Study Group. 1996. Predictors of adipose tissue tocopherol and toenail selenium levels in nine countries: The EURAMIC Study. *Eur. J. Clin. Nutr.* 50:599–606.

Wang, D., G. Alfthan, A. Aro, L. Kauppi, and J. Soveri. 1991. Selenium in tap water and natural water ecosystems in Finland. In A. Aitio, A. Aro, J. Järvisalo, and H. Vainio (eds.), *Trace Elements in Health and Disease,* pp. 49–56. Royal Society of Chemistry, Cambridge.

Wang D., G. Alfthan, A. Aro, and J. Soveri. 1993. Anthropogenic emissions of Se in Finland. *Appl. Geochem. Suppl.* 2:87–93.

Wang, D., G. Alfthan, A. Aro, P. Lahermo, and P. Väänänen. 1994. The impact of selenium fertili sation on the distribution of selenium in rivers in Finland. *Agric. Ecosyst. Environ.* 50:133–149.

Wang, D., G. Alfthan, A. Aro, A. Mäkelä, S. Knuuttila, and T. Hammar. 1995. The impact of selenium supplemented fertilization on selenium in lake ecosystems in Finland. *Agric. Ecosyst. Environ.* 54:137–148.

Yang, G. Q., L. Z. Zhu, S. J. Liu, L. Z Gu, P. C. Huang, and M. O. Lu. 1987. Human selenium requirements in China. In J. F. Combs, J. E. Spallholz, O. A. Levander, and J. E. Oldfield (eds.), *Proceedings of the Third International Symposium on Selenium in Biology and Medicine,* pp. 589–607. AVI Press, Westport, CT.

Ylaranta, T. 1983. Sorption of selenite and selenate in the soil. *Ann. Agric. Fenn.* 21:103–113.

Yläranta, T. 1984a. Raising the selenium content of spring wheat and barley using selenite and selenate. *Ann. Agric. Fenn.* 23:75–84 .

Yläranta, T. 1984b. Effect of selenite and selenate fertilization and foliar spraying on selenium content of timothy grass. *Ann. Agric. Fenn.* 23:96–108.

6

Effects of Selenium Supplementation of Field Crops

GUNNAR GISSEL-NIELSEN
Risø National Laboratory, Roskilde, Denmark

I. INTRODUCTION

In recent discussions about the quality of the food products of modern agriculture, lack of selenium (Se) in food has often been viewed as an example of a mineral imbalance related to intensive plant production. Selenium has been recognized as essential for human and livestock nutrition for more than 30 years, and several attempts have been made to demonstrate its essentiality for plants by depleting Se in their growth medium, but so far without success. Furthermore, none of the many Se-containing enzymes active in animals are found in plants. Consequently, interest of Se in plants is related to the quality of the plants as animal fodder and as human food. The main source of Se for animal fodder and human food is the soil–plant system. Therefore field treatment with Se is a possible way of improving the Se nutrition of livestock and people.

II. GEOGRAPHICAL DISTRIBUTION OF SELENIUM

The Se concentration of the total fodder needed to meet the minimum requirements of livestock is considered to be in the range of 0.05 to 0.1 mg/kg dry matter, and toxic effects can be expected at chronic intakes of fodder that exceed about 1 mg Se/kg. According to these limits, different areas of the world are characterized as Se-deficient, Se-adequate, and Se-toxic. Outside Europe, crops containing toxic Se concentrations are found in the midwestern regions of the

United States and Canada, and in Venezuela, India, and China. Areas of Se deficiency are more widespread. This condition is reported from both the western and the eastern coastal areas of North America, and from Venezuela, Australia, New Zealand, Japan, and China (Gissel-Nielsen et al., 1984). However no information is available about the Se status of most countries of the world .

The situation in Europe is illustrated in Figure 1. There is a geographical pattern in this map showing Scandinavia as a natural low-Se area, while central Europe balances between deficiency and sufficiency. Information from southern Europe is sparse. A few samples from Italy point to a range spanning from somewhat deficient to sufficient (Bordini et al., 1985). Results presented at the International Symposium on Selenium in Belgrade in 1991, indicated that the Se status of the former Yugoslavia ranges between adequate and inadequate. Selenium toxicity has been observed only spotwise in Wales, Eire, and Russia.

The Se concentrations in plants are reflected in the daily Se intake by humans and in their blood Se content. Table 1 gives some data on this matter

☐ Inadequate   ▨ Adequate
▦ Spotwise toxic   ☐ No information

FIGURE 1   Distribution of Se in fodder crops in Europe; upper and lower symbols for Finland refer to before and after 1985, respectively.

TABLE I   Human Dietary Se Intakes and Blood Se Levels

Country	Se intake (μg/day)	Blood Se (μg/L)
Belgium	55	123
Canada	98–224	182
China	11–4990	8–3180
Denmark	40	86
Finland before 1985	30	56–87
New Zealand	28–56	59–83
United States	62–216	157–265

Source: Gissel-Nielsen et al. (1984).

from a few countries. These Se values indicate a relationship between the Se content of food produced in different countries and the blood Se content of the inhabitants. A multitude of publications indicate a similar correlation between the Se content of animal food and animal blood Se.

The sources for the intake of Se are many, and many publications provide the results of surveys on the Se content of foodstuffs. One of the first comprehensive surveys was that of Koivostoinen (1980), giving the results from the low-Se area of Finland. Robberecht et al. (1982) from Belgium provided information on more Se-sufficient countries. The differences in the Se content of the same foodstuff from different surveys are obvious when the reported results are compared, but when the data are arranged according to decreasing Se content, similarities are also obvious. Consequently, no exact values can be given for the Se content of the different foodstuffs, but the foodstuff groups can be listed in a relative order of Se content, as is done in Table 2.

Because of the correlation between Se intake and blood Se shown in Table 1, the Se intake of humans and livestock living in a certain area can be estimated roughly by evaluating the Se uptake by the fodder crops grown in the area in

TABLE 2   Groups of Foodstuffs in Order of Relative Se Content

Relatively high	Seafood
	Meat
	High-sulfur vegetables
	Dairy products
	Cereals
	Other vegetables
Relatively low	Fruits

question. This is because cereals and vegetables are dominant in the human diet and, along with pasture plants, also in livestock fodder. Therefore crop plants are responsible for the greater part of the Se intake of humans and livestock even if crop plants are relatively low in Se, as seen in Table 2. As a consequence, an increase in the Se concentration of crops is an obvious way to remedy Se deficiency in humans or livestock.

III. SELENIUM UPTAKE BY PLANTS

Only when data from areas with extreme differences in Se content are compared can a straight, positive correlation between soil Se and plant content be seen. Areas with low to moderate Se content show no general correlation because of the large number of factors influencing the availability of soil Se to plants (Gissel-Nielsen et al., 1984).

In soils with a high pH, inorganic Se will occur mainly as selenate (Se^{6+}), which is hardly fixed at all in the soil. Under conditions of low precipitation and low leaching, most of the Se will be available to the plants, and the crops will be rich in Se. This is known in some places in India and in South Dakota. Contrary to this, a low pH favors the selenite form (Se^{4+}), which is fixed strongly to the soil clay particles and iron hydroxides (Gissel-Nielsen, 1977). The same total Se content of such soils as in the example with high-pH soils might result in the Se concentration of the crops being 10 times lower.

The factors contributing to this situation are presented in Figure 2, which depicts the adsorption of selenite by clay minerals and organic matter, along with a very strong fixation by iron hydroxides. Volatile Se is lost to the atmosphere through microbial activity. However, Se also returns to the soil from the atmosphere. All this leaves only a minor part of the Se in the cycle to pass through the plant–animal system. The Se concentration of plant samples depends furthermore on the time of sampling, as illustrated in Figure 3 (Gissel-Nielsen, 1975b). The release of Se from clay and from organic matter is a very slow process. The Se released during the winter is available for the crops in the spring, when the yield of grass is low, giving a relatively high Se concentration. However, when the growth of the grass increases during the summer, less Se is taken up and distributed over a far greater yield, and therefore the concentration drops as a result of dilution.

IV. FIELD TREATMENT WITH SELENIUM

To remedy the low natural Se content of crops in some areas, different ways of raising the concentrations have been investigated over the years, and the subject

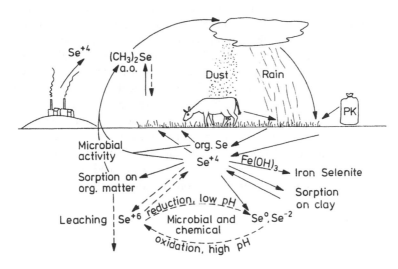

FIGURE 2 Possible cycles of Se under field conditions. (From Gissel-Nielsen, 1977.)

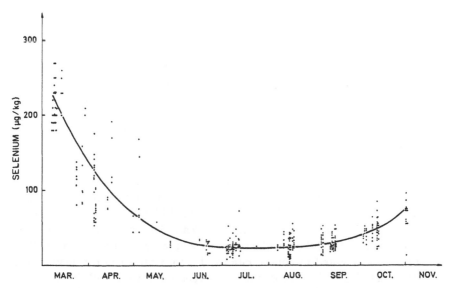

FIGURE 3 Selenium concentrations in pasture samples for eight months in year 1973. (From Gissel-Nielsen, 1975b.)

has been discussed in a number of reviews (e.g., Gissel-Nielsen et al., 1984; Mikkelsen et al., 1989). Of all the different ways of increasing the Se concentration in crops, soil and foliar application of selenate or selenite are those of practical importance. Presowing treatment of barley seeds with selenite has also been tested (Gissel-Nielsen, 1975a). However, the amount of selenite needed to obtain a desirable Se concentration in the harvested grain was the same as that indicated when selenite was added through the fertilizer. The presowing treatment with selenite resulted in a concentration of several hundred milligrams of Se per kilogram of seed, making it highly toxic to animals and humans. Yläranta (1984b) treated seed of barley and wheat with selenate before sowing. As in the case of selenite, the same amount of selenate was needed for seed pretreatment as for soil application to obtain a desirable Se concentration in the crop.

A. Soil Application

The first studies on the soil application of Se involved the spraying of selenite or selenate solutions onto the soil surface (Cary et al., 1967; Watkinson and Davies, 1967). These relatively simple experiments showed the possibilities of the method, and they have been followed by comprehensive studies of the method worldwide. Moreover, soil application of elemental Se (Se^0) was tested by Grant (1965), Peterson and Butler (1966), Cary and Allaway (1969), and Gissel-Nielsen and Bisbjerg (1970). This is a slow-release form of Se and was intended as a treatment having a long-term effect. However, Se^0 has to be oxidized to Se^{4+} or Se^{6+} before it becomes available to the plants, and many environmental factors (e.g., pH, humidity, microorganisms) have an impact on the rate of oxidation. Consequently, it is very difficult to predict the effect of soil supplementation using elemental Se on the Se concentration of the plants.

Different Se salts were tested in pot and field experiments of Gissel-Nielsen and Bisbjerg (1970). Selenates had in general a 20 to 50 times greater availability than selenites in the first year, but in the second year the concentration was lower in the selenate-treated plants. The slightly soluble $CuSeO_4$ and $BaSeO_4$ had longer-lasting effects than did K_2SeO_4.

Typical of many of the soil application experiments in the early years was the addition of quite large amounts of Se salts in an attempt to find the optimum annual amount and to study the residual effects of higher amounts. Thus, Cary and Allaway (1973) added 2.2 and 4.5 kg Se/ha as Na_2SeO_3 to a number of silt loam soils and cropped them with corn, oats, and forage crops for 4 years. Both amounts of Se resulted in concentrations of 1 to 3 mg/kg in dry matter. The authors concluded at the time that these concentrations were probably not toxic to animals, but today they are considered to be in the low range of toxicity for animals. In the following 3 years the 2.2 kg treatment gave Se concentrations in pasture crops in the desirable range, whereas after 3 years the effect disappeared in the annual crops.

Whelan (1989) tested the addition of sodium selenite to subterranean clover in Western Australia using up to 800 g Se/ha. He found the residual value of a single application to be in the order of 25% in year 2 and 15% in year 3, when compared to plots with fresh applications in years 2 and 3. A large-scale field experiment given an annual addition for 5 years of 60 and 120 g Se/ha as Na_2SeO_3 incorporated in an NPK compound fertilizer was carried out at 21 farms covering the common Danish soil types (Gissel-Nielsen, 1977). The soils differed in their content of clay, organic matter, previous cropping, and so on but were all glacial deposit mineral soils with a pH of 5 to 7. The 120 g treatment raised the native Se concentration of 0.02 to 0.04 mg/kg of barley, wheat, ryegrass, clover, and fodder beets (0.09 mg/kg in the beet top) to 0.08 to 0.13 mg Se/kg, which is considered a sufficient and safe level for animal nutrition.

Gupta and Winter (1981) and Gupta et al. (1982) also found that a small annual addition of selenite was preferable to a single, large application for annual crops. For perennial pasture, a single application giving a residual effect for 4 to 5 years might be a good measure. Butler and Peterson (1963) reported a similar uptake of Se^{4+} and Se^{6+} into three pasture plant species when added as ^{75}Se-labeled compounds to a nutrient solution. When Se-fortified fertilizers were top-dressed to a pasture crop in the field, Grant (1965) found a 20 times greater effect of selenate than of selenite, and this result was confirmed in all later experiments with soil application of Se^{4+} and Se^{6+}. The reason for this discrepancy in potency is the stronger adsorption of selenite in the soil as discussed earlier in this chapter.

Hupkens van der Elst and Watkinson (1977) investigated the importance of an even distribution of the Se on a paddock, using fertilizer prills as a carrier for sodium selenate. They treated a strip of the paddock, covering 5% of the area, with 340 g Se/ha, which was equivalent to 17 g Se/ha for the total area. The strip treatment was expected to give greater variation in the blood Se content of the grazing animals, but this was not the case. All animals had the same Se blood content as animals grazing a paddock with even distribution of the selenized prills. This demonstrates that the free movement of the grazing animals compensates for an uneven distribution of the Se.

Van Dorst and Peterson (1984) studied the effect on the availability of selenate of raising a natural soil pH of 4.5 to a pH value of 7 by liming. At the low pH, selenate disappeared much faster from the soil solution than at pH 7, so the high pH resulted in a higher Se concentration in the plant tissue as well as a longer lasting effect.

Coutts et al. (1990) used a British-produced fertilizer prill enriched with 1% Se as selenate on a grass/clover sward. With 10 g Se/ha, Se levels in all cuts were within the adequate range for livestock nutrition.

Yläranta (1984a, 1984b, 1984c) made a comprehensive study of all the above-mentioned Se treatments of crops in the field. On this basis, he recommended the use of compound fertilizers enriched with selenate in amounts that resulted in the addition of 10 g Se/ha for normal fertilization procedures. This recommendation was adopted by the Finnish authorities for the compulsory selenization of all Finnish-grown food and fodder crops.

In a pot experiment, Singh (1994) compared the ability of $Ca(NO_3)_2$ and an NPK fertilizer as carriers for selenate. He found twice as much Se in wheat at maturity when $Ca(NO_3)_2$ served as a carrier, depending on the time of application. The difference in the Se concentration is probably an effect of the time of application: the NPK was added to the soil before sowing, whereas the $Ca(NO_3)_2$ was added at tillering and at heading growth stages.

Gupta (1995) tested Selcote, a commercial product from New Zealand, on cereals and forage crops in field and pot experiments. Per gram of Se, it had the same effect as sodium selenate.

Flue gas desulfurization waste, fly ash, and sewage sludge are examples of alternative sources of Se for field crops (e.g., Logan et al., 1987; Gissel-Nielsen and Bertelsen, 1988). They all caused an increase in the Se concentration of the crops to an extent depending on the Se concentration of the waste, which again depended on the sources behind the waste products. In most cases, the amount of waste needed for remedying an Se deficiency was so high that other problems occurred, such as excessive Cd concentrations or improper levels of salinity.

These waste products, therefore, should be acknowledged to have a positive effect on the Se concentration of crops when deposited in the field anyway, but they should not be concidered Se fertilizers as such.

B. Foliar Application

To avoid the complication of soil influence on the availability of added Se, foliar application was considered at an early stage. Davies and Watkinson (1966) discovered as long ago as in the mid-1960s that selenite sprayed onto pasture plants has a much greater effect than when it is sprayed onto the soil surface. In 1972–1973, Gissel-Nielsen (1975a) tested foliar applications of up to 50 g Se/ha as selenite in field experiments with barley. A linear correlation was found between added Se and the concentration in the plants; and 5 to 10 g Se/ha resulted in the desired level of 0.05 to 0.1 mg/kg Se in the grain. Further, the increase in Se concentration was greater in the mature grain than in the straw, indicating an effective translocation of the absorbed Se to the edible part of the plant. To improve the efficiency of the foliar application, Gissel-Nielsen (1981) added a detergent (Lissapol-N) to the Se solution in field experiments with barley at 26 Danish farms. The result was a twofold increase in the Se concentration, giving 0.05 to 0.1 mg/kg in the grain using 2 to 5 g Se/ha. In the same experiments, the foliar application was performed at two growth stages: Feekes 4 and 6 (end of May and end of June). The late application also resulted in a twofold increase of the effect. The likely explanation is a better plant cover of the soil, allowing a greater part of the solution to wet the leaves.

In a field experiment, Archer (1983) sprayed pasture crops with 70 and 140 g Se/ha as sodium selenite and as sodium selenate. Selenate had a more pronounced but shorter lasting effect on the Se concentration of the plants than did the selenite. The large amounts used by Archer resulted in concentrations of about 10 mg Se/kg dry matter in the first cut, an amount highly toxic to animals. Gissel-Nielsen (1984) compared different ways of adding small amounts of selenite to pasture crops. Foliar application in the spring of 10 g Se/ha as sodium selenite to an established pasture resulted in concentrations within the desirable range for all cuts during the growing season. The recovery in the crops of the added Se was up to 35%.

Yläranta's comprehensive studies (1984a, 1984b, 1984c) included the foliar application of selenite and selenate. His conclusions are in accordance with the earlier findings that selenate is somewhat more effective than selenite, that 5 to 10 g Se/ha is sufficient, and that the effect depends on the stage of development of the crop at the time of spraying.

C. Experience on a Larger Scale

As mentioned above, Se treatment of field crops is carried out in some countries. Only in Finland, however, is it performed to an extent that enables study of overall long-term effects. In 1984 Finnish authorities decided that all commercial fertilizers for food and fodder crops should be enriched with sodium selenate: 16 mg Se/kg fertilizer for cereals and horticulture crops, and 6 mg Se/kg fertilizers for fodder beet and grass production. It was further decided to monitor the Se concentrations in the crops carefully and decrease the applications when or if concentrations became higher than necessary for preventing Se deficiency in livestock and man. A number of studies (e.g., Eurola et al., 1989) reported an elevated Se content in all crops, typically an increase from about 0.01 mg/kg to 0.1–0.3 mg/kg in dry matter.

As a result of an increasing residual effect, it was decided to remove from the market the fertilizers containing 16 mg Se/kg, leaving only those having 6 mg/kg for all crops. This change stabilized the average Se concentrations at a somewhat lower level: Paivi et al. (1995) reported about 0.1 mg/kg in cereal grain, while Jukola et al. (1996) reported about 0.2 mg/kg in barley and oats, but only 0.13 and 0.17 mg/kg in hay and grass silage, respectively. In any case, Finland has become a country producing crops with adequate Se levels for livestock and human nutrition instead of a Se-deficient area.

D. Main Guidelines

Considering all the experiments reported here, the overall conclusion is that the foliar application of about 5 g Se/ha as selenite or selenate and fertilization using about 10 g Se/ha as selenate or about 120 g Se/ha as selenite are effective annual treatments for raising the Se content of annual crops to a level desirable for human and animal nutrition. The effect of Se is increased when it is used together with a detergent for foliar application, and for all treatments the effect is greatest when carried out on a well-established crop. The residual effects of these treatments are very small. A somewhat higher amount is needed for pasture crops, but this gives a residual effect lasting 2 to 3 years.

V. BIOAVAILABILITY OF SELENIUM

Another factor to be considered in Se field treatment as a means of preventing deficiency in livestock is the chemical form in which Se is administered to animals. Whether the Se is offered in premixed fodder, in concentrates, mineral supplements, water, lickstones, or ruminant pellets, it occurs as inorganic selenite, whereas the Se in crops occurs predominantly as free selenoamino acids or in proteins. The significance of organic versus inorganic Se in animal nutrition is still subject to much discussion, so one cannot recommend a single Se compound as the best in all situations. However, there is no doubt that the bioavailability of different Se compounds varies strongly, and this has been shown in several experiments. Table 3 gives the results from just one of them.

Laws et al. (1986) fed Se-deficient chicks with a number of Se sources, and Table 3 shows the percentage of surviving birds fed the same amount of Se but from the different sources. This table clearly demonstrates a marked difference in bioavailability between the different organic sources of Se as well as between

TABLE 3   Relative Biological Values of Se[a]

Se source	Survival (%)
Chick diet, controls	34
+ Dried fish	38
+ Bread dough	53
+ Raw beef	69
+ Baked bread	72
+ Selenite	81

[a]Measured as survival of Se-deficient chicks given different diets containing 0.02 mg Se/kg.
Source: Laws et al. (1986).

organic and inorganic forms. Such differences stress the importance of the form of Se in animal and human foodstuffs, and research has been carried out to evaluate the possibilities of influencing the chemical form of the Se in plants. Earlier short-term experiments carried out with tomato roots (Asher et al. 1977) and with maize (Gissel-Nielsen, 1979) showed that when Se is added to the nutrient solution as selenate, it is translocated in the xylem sap as selenate, whereas selenite is metabolized immediately to Se-amino acids and translocated as such (Fig. 4).

These results indicated a possible way of affecting the metabolic pathway of Se in plants. Consequently, ryegrass and barley were given Se as selenite or selenate through the roots or by foliar application in a pot experiment (Gissel-Nielsen, 1987). A series of fractionations of the Se compounds of the grass and the barley grain separated the compounds into total and water-soluble Se proteins, Se-amino acids, selenite, and selenate. Neither the oxidation state of the added Se nor the method of application had any significant effect on the distribution of the Se over the possible compounds. In all four cases only about 10% of the Se was present in an inorganic form. This implies that any method of supplementing crops with inorganic Se leads to the same Se compounds in the mature plants. Therefore the effect of the different methods can be evaluated purely on the basis of the total uptake in percent of the total added.

VI. CONCLUSION

Relatively few cases of Se toxicity in man and livestock have been observed over the years, and only China has reported severe cases of deficiency leading to the

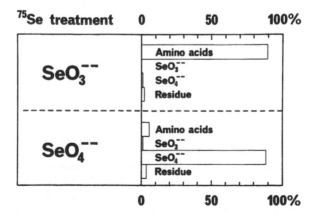

FIGURE 4 The distribution of Se in fractions of xylem sap from maize when added as selenate or selenite. (From Gissel-Nielsen, 1979.)

death of many humans. However moderate deficiencies leading to problems in animals are seen in many countries, and these deficiencies might be more dangerous to human health than often considered. Consequently greater action should be taken to prevent Se deficiency in man and livestock. As described earlier, concern over high rates of cardiovascular disease and certain forms of cancer in Finland prompted the Finnish authorities to legislate the introduction of selenate in all compound fertilizers from 1985. Selenized compound fertilizers are available in New Zealand, and in some countries (e.g., Sweden, the United Kingdom) micronutrient solutions including Se are on the market for foliar application. In many countries premixed fodder enriched with inorganic Se is available for livestock production.

Even if Se is available in tablet form for human consumption in most countries, and even if in general a much more varied diet including relatively Se-rich foodstuffs such as fish and some vegetables is consumed, there is far more attention paid to the Se nutrition of animals than of the human population. In future it is likely that increased attention will be paid to the importance of sufficient Se in the human diet, as well as in livestock fodder. This chapter shows that a reasonable Se supplementation of our crops is an inexpensive, safe, and easy way of ensuring a desirable human and animal intake of Se.

REFERENCES

Archer, F. C. 1983. Agriculture group symposium. Trace elements in soils, crops and forages. *J. Sci. Food Agric.* 34:49–61.

Asher, C. J., G. W. Butler, and P. J. Peterson. 1977. Selenium transport in root systems of tomato. *J. Exp. Bot.* 28:279–291.

Bordini, A., M. Di Vincenzo, C. Sinisalchi, and A. Valeriani. 1985. Evaluation of the selenium content of foods commonly consumed in Italy. *Riv. Soc. Ital. Sci. Aliment.* 14:357–360.

Butler, G. W., and P. J. Peterson. 1963. Availability of selenium in forage to ruminants. *Proc. N. Z. Soc. Anim. Prod.* 23:13–19.

Cary, E. E., and W. H. Allaway. 1969. The stability of different forms of selenium applied to low-selenium soils. *Soil Sci. Soc. Am. Proc.* 33:571–574.

Cary, E. E., and W. H. Allaway. 1973. Selenium content of field crops grown on selenite-treated soils. *Agron. J.* 65:922–925.

Cary, E. E., G. A. Wieczorek, and W. H. Allaway. 1967. Reactions of selenite-selenium added to soils that produce low-selenium forage. *Soil Sci. Soc. Am. Proc.* 31:21–26.

Coutts, G., D. Atkinson, and S. Cooke. 1990. Application of selenium prills to improve the selenium supply to a grass/clover sward. *Commun. Soil Sci. Plant Anal.* 21:951–963.

Davies, E. B., and J. H. Watkinson. 1966. Uptake of native and applied selenium by pasture species. I. Uptake of Se by browntop, ryegrass, cocksfoot, and white clover from Atiamuri sand. *N. Z. J. Agric. Res.* 9:317–327.

Eurola, M., P. Ekholm, M. Ylinen, P. Koivistoinen, and P. Varo. 1989. Effects of selenium fertilization on the selenium content of selected Finnish fruits and vegetables. *Acta Agric. Scand.* 39:345–350.

Gissel-Nielsen, G. 1975a. Foliar application and pre-sowing treatment of cereals with selenite. *Z. Pflanzenern. Bodenkd.* 138:97–105.

Gissel-Nielsen, G. 1975b. Selenium concentration in Danish forage crops. *Acta Agric. Scand.* 25:216–220.

Gissel-Nielsen, G. 1977. Control of selenium in plants. Risø Report No. 370, pp. 1–42. Roskilde, Denmark.

Gissel-Nielsen, G. 1979. Uptake and translocation of ^{75}Se in *Zea mays*. In *Isotopes and Radiation in Research on Soil–Plant Relationships*, pp. 427–436. International Atomic Energy Agency, Vienna.

Gissel-Nielsen, G. 1981. Foliar application of selenite to barley plants low in selenium. *Commun. Soil Sci. Plant Anal.* 12:631–642.

Gissel-Nielsen, G. 1984. Improvement of selenium status of pasture crops. *Biol. Trace Elem. Res.* 6:281–288.

Gissel-Nielsen, G. 1987. Fractionation of selenium in barley and rye-grass. *J. Plant Nutr.* 10:2147–2152.

Gissel-Nielsen, G., and F. Bertelsen. 1988. Inorganic element uptake by barley from soil supplemented with flue gas desulphurization waste and fly ash. *Environ. Geochem. Health* 10:21–25.

Gissel-Nielsen, G., and B. Bisbjerg. 1970. The uptake of applied selenium by agricultural plants. 2. The utilization of various selenium compounds. *Plant Soil* 32:382–396.

Gissel-Nielsen, G., U. C. Gupta, M. Lamand, and T. Westermarck. 1984. Selenium in soils and plants and its importance in livestock and human nutrition. *Adv. Agron.* 37:397–460.

Grant, A. B. 1965. Pasture top-dressing with selenium. *N. Z. J. Agric. Res.* 8:681–690.

Gupta, U. C. 1995. Effects of Selcote® ultra and sodium selenate (laboratory versus commercial grade) on selenium concentration in feed crops. *J. Plant Nutr.* 18:1629–1636.

Gupta, U. C., and K. A. Winter. 1981. Long-term residual effects of applied selenium on the selenium uptake by plants. *J. Plant Nutri.* 3:493–502.

Gupta, U. C., K. B. McRae, and K. A. Winter. 1982. Effect of applied selenium on the selenium content of barley and forages and soil selenium depletion rates. *Can. J. Soil Sci.* 62:145–154.

Hupkens van der Elst, F. C. C., and J. H. Watkinson. 1977. Effect of topdressing pasture with selenium prills on selenium concentration in blood of stock. *N. Z. J. Exp. Agric.* 5:79–83.

Jukola, E., J. Hakkarainen, H. Saloniemi, and S. Sankari. 1996. Effect of selenium fertilization on selenium in feedstuffs and selenium, vitamin E, and beta-carotene concentrations in blood of cattle. *J. Dairy Sci.* 79:831–837.

Koivistoinen, P. 1980. Mineral element composition of Finnish foods: N, K, Ca, Mg, P, S, Fe, Cu, Mn, Zn, Mo, Co, Ni, Cr, F, Se, Si, Rb, Al, B, Br, Hg, As, Cd, Pb and ash. *Acta Agric. Scand. Suppl.* 22:1–171.

Laws, J. E., J. D. Latshaw, and M. Biggert. 1986. Selenium bioavailability in foods and feeds. *Nutr. Rep. Int.* 33:13–24.

Logan, T. J., A. C. Chang, A. L. Page, and T. J. Ganje. 1987. Accumulation of selenium in crops grown on sludge-treated soil. *J. Environ. Qual.* 16:349–352.

Mikkelsen, R. L., A. L. Page, and F. T. Bingham. 1989. Factors affecting selenium accumulation by agricultural crops. Soil Science Society of America Special Publication No. 23:65–94.

Paivi, E., Y. Maija, K. Pekka, and V. Pertti. 1995. Selenium concentration of Finnish foods: Effects of reducing the amount of selenate in fertilizers. *Agric. Sci. Finland* 4:377–384.

Peterson, P. J., and G. W. Butler. 1966. Colloidal selenium availability to three pasture species in pot culture. *Nature* 212:961–962.

Robberecht, H., H. Deelstra, D. Vanden Berghe, and R. van Grieken. 1982. Selenium gehalten van levensmiddelen gekonsumeerd in Belgie. *Rev. Ferment. Ind. Aliment.* 37:188–201.

Singh, B. R. 1994. Effect of selenium-enriched calcium nitrate, top-dressed at different growth stages, on the selenium concentration in wheat. *Fert. Res.* 38(3):199–203.

Van Dorst, S. H., and P. J. Peterson. 1984. Selenium speciation in the soil solution and its relevance to plant uptake. *J. Sci. Food Agric.* 35:601–605.

Watkinson, J. H., and E. B. Davies. 1967. Uptake of native and applied selenium by pasture species. III. Uptake of selenium from various carriers. *N. Z. J. Agric. Res.* 10:116–121.

Whelan, B. R. 1989. Uptake of selenite fertiliser by subterranean clover pasture in Western Australia. *Aust. J. Exp. Agric.* 29:517–522.

Yläranta, T. 1984a. Raising the selenium content of spring wheat and barley using selenite and selenate. *Ann. Agric. Fenn.* 23:75–84.

Yläranta, T. 1984b. Effect of selenium fertilization and foliar spraying at different growth stages on the selenium content of spring wheat and barley. *Ann. Agric. Fenn.* 23:85–95.

Yläranta, T. 1984c. Effect of selenite and selenate fertilization and foliar spraying on selenium content of timothy grass. *Ann. Agric. Fenn.* 23:96–108.

7

Selenium Metabolism in Grazing Ruminants: Deficiency, Supplementation, and Environmental Implications

JOHN MAAS
University of California at Davis, Davis, California

I. BIOCHEMISTRY

The classic nutritional research of Schwarz and Foltz (1957), showing that selenium (Se) is the critical element in factor 3 that prevents liver necrosis in rats, began the work that subsequently proved Se to be an essential nutrient for man, cattle, other grazing ruminants, and all vertebrates examined to date. Subsequently, Muth (1963) and Hogue et al. (1962) reported that Se and vitamin E administration can prevent white muscle disease in young ruminants. In recent years, several Se-responsive conditions have been described in cattle, and these have been reviewed (Maas, 1983). The Se-responsive conditions in cattle include nutritional myodegeneration (white muscle disease), retained placenta, abortions, neonatal weakness, diarrhea, ill thrift, infertility, and immune system deficits. Relationships between vitamin E and Se were observed; however, the biochemical basis for the action of Se was only speculated upon. Minimum dietary requirements of cattle for Se were developed from feeding and response trials. In 1973 Rotruck et al. published work outlining a basic biochemical mechanism that accounts for the role of Se as an essential nutrient. That work (Rotruck et al.,1973) showed that Se is a component of glutathione peroxidase (GSH-Px; EC 1.11.1.9) in erythrocytes. Selenium-deficient rats were found to have low GSH-Px activity and signs of Se deficiency. Most of a dose of radioactive ^{75}Se was present in the GSH-Px activity of erythrocytes of test animals. The same workers purified Se-containing GSH-Px from sheep erythrocytes.

It was later shown that GSH-Px contains 4 gram-atoms of Se per mole of GSH-Px. For many years this was the only known biochemical role of Se, and while GSH-Px was found to have important antioxidant activity, it was recognized that several tissues are high in Se concentrations and low in GSH-Px activity. A number of other Se-containing proteins (selenoproteins) with biologic activity have been identified (Burk and Hill, 1993). In addition to the original GSH-Px, now referred to as cellular glutathione peroxidase (cGSH-Px), three other glutathione peroxidases have been characterized (Ursini et al., 1985; Takahashi et al., 1987; Chu et al., 1993). Both type I and type II iodothyronine deiodinases that contain Se have been characterized (Arthur et al., 1990; Behne et al., 1990; Berry et al., 1991; Croteau et al., 1995). Additionally, three other selenoproteins have been sequenced and characterized, but their biological role is unknown (Hill et al., 1991; Karimpour et al., 1992; Vendeland et al., 1993). Recently, a new 18 kDa, membrane-bound selenoprotein has been reported (Kyriakopoulos et al., 1996). All the Se-containing proteins described contain selenocysteine and all appear to be genetically controlled. While the various biochemical and physiologic functions of all these proteins are not presently clear, it is evident that Se is an important antioxidant, and this helps property to explain its role in prevention of a number of disease conditions. At the present time, the only clinical use of Se-containing enzymes is the analysis of blood GSH-Px activity for diagnosis of Se status by some laboratories.

II. SELENIUM DEFICIENCY SYNDROMES IN GRAZING RUMINANTS

Several disease syndromes in cattle have been shown to be Se-responsive; that is, Se administration will reverse the condition or prior Se supplementation will prevent the condition. It must be emphasized that the precise pathophysiologic mechanism(s) of many of these conditions can only be theorized at present. In the original work by Muth (1963), not all the Se-deficient forages produced white muscle disease, and while Se administration would successfully prevent or treat white muscle disease, it was apparent that factor(s) other than Se deficiency were involved. The other Se-responsive syndromes are similar in that Se deficiency is the underlying problem to be addressed, while other factors involved with pregnancy, infectious diseases, exercise, or stress are important for the condition to become manifest. The Se-responsive syndromes can be put into four major disease categories: musculoskeletal, reproductive, gastrointestinal, and immunologic.

The musculoskeletal conditions include nutritional myodegeneration (NMD, e.g., white muscle disease; nutritional muscular dystrophy), neonatal weakness, and myodegeneration of adult cattle. These diseases, particularly NMD, are widespread throughout the world and affect both domestic and wild ruminants on every continent except Antarctica, where grazing ruminants do not occur. NMD is a particular problem in the United States, Canada, Australia, New Zealand, and Europe. Its cardiac form (Fig. 1) can occur within 2 to 3 days of birth and often is associated with severe myocardial lesions and peracute to acute death. Calves or lambs affected at 1 to 4 weeks of age often appear lame or stiff and are reluctant to move. A substantial number of these calves will die if not treated promptly with injectable Se.

Elevated serum enzymes of muscle origin such as creatinine kinase or aspartate aminotransferase are helpful in diagnosing the myopathy. Creatinine kinase is most specific for muscle necrosis. On post mortem examination, pale streaks are seen in the muscles, and hyaline degeneration and calcification are common observations on histopathologic examination. The cardiac form of NMD is more severe, with necrosis and calcification of the heart muscle and the intercostal muscles. Neonatal weakness due to Se deficiency is a less severe clinical manifestation. Myodegeneration of adult cattle is often associated with exercise or parturition, and common clinical signs include paresis and myoglobinuria.

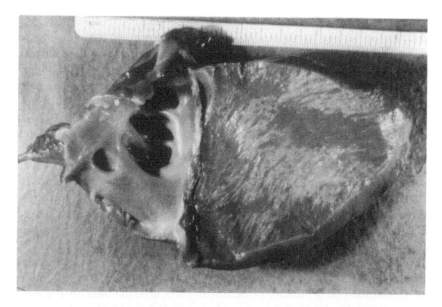

FIGURE 1   Heart of a calf with the cardiac form of nutritional myodegeneration (NMD; white muscle disease). The interventricular septum shows large areas (white) of cardiac muscle necrosis and calcification.

The role of Se as an antioxidant is a key factor in these musculoskeletal conditions. Grazing ruminants are particularly susceptible to Se deficiency diseases because their diets often consist of feeds from a small geographic area; if the soil and plants are low in Se in that area, animals are particularly predisposed.

Selenium deficiency also can cause reproductive pathologies such as abortion, retained placenta, and infertility in cattle. Cattle, sheep, and other grazing animals are susceptible to a number of infectious agents that can cause abortion, such as *Brucella* spp., *Campylobacter* spp., *Leptospira* spp., and numerous viral agents. Selenium deficiency has also been documented as a direct cause of abortion in cattle and sheep (Hedstrom et al., 1986). These aborted fetuses often have lesions similar to NMD and may represent an in utero form of myodegeneration. Retained placenta is a multifactorial condition that is not specific to Se deficiency even though in some parts of the United States Se-responsive retained placenta is common. Infertility in cattle can be due to Se deficiency; however, infertility can also be caused by infectious agents (various protozoa, bacteria, and viruses), trauma, hormonal imbalances, and other nutritional or metabolic diseases.

The most common presenting signs of Se deficiency in cattle involve the gastrointestinal tract as diarrhea and/or "ill thrift," both of which are nonspecific clinical signs. The diarrhea must be differentiated from the many causes of diarrhea in cattle, which include bacteria (*Salmonella* spp., *Mycobacterium paratuberculosis*, etc.), viruses (bovine virus diarrhea, rotavirus, etc.), parasites (*Trichostrongylus* spp., *Ostertagia* spp.), or other nutritional and metabolic conditions. In addition to preventing Se-responsive diarrhea, Se added to the diet of Se-deficient ruminants will allow for normal weight gains and can increase feed efficiency (Nunn et al., 1995). The fact that normal Se nutrition optimizes feed efficiency in ruminants is a very important phenomenon. There are resource management implications both for domestic grazing ruminants and wild ruminants with respect to this aspect of Se nutrition. Marked Se deficiency could decrease feed efficiency by as much as 30%, and this could have a major impact on carrying capacity and range utilization.

The effect of Se deficiency on the immune system of laboratory animals and man has been examined in detail, and it has generally been concluded that Se deficiency decreases both the cellular and the humoral immune response to specific antigens. Additionally, genetic mechanisms that utilize Se to maintain the activity of immune system functions, such as lymphocyte proliferation and T-lymphocyte function, have been characterized (Kiremidjian-Schumacher et al., 1994; Roy et al., 1994). The research data for domestic ruminants is not as clear. Research on the immune system responses to Se deficiency in ruminants has been confounded by a number of factors, in particular, the vitamin E status of the experimental animals. Selenium and vitamin E have similar effects on the immune response and in many cases the responses overlap or are synergistic. Therefore, in situations with high levels of dietary vitamin E, such as would occur in animals grazing lush pastures, a decrease in immune function due to Se deficiency may not be observed because the very high vitamin E status preserves normal immune function. Conversely, if the experimental animals are deficient in both Se and vitamin E, adding nutritive levels of Se may not be enough to support the immune system for normal function.

Other important experimental variables have been the dose of Se, the relative ability to deplete experimental ruminants of Se (or vitamin E) during studies, and inadvertent addition of Se or vitamin E in serum products (fetal bovine serum) used for in vitro immunologic tests, in essence correcting the Se deficiency of the cells in the test system. In published data (Finch and Turner, 1996), selenium deficiency in ruminant species has been shown to have little effect on specific antibody response. Selenium deficiency has not been shown to have much effect on the ability of neutrophils to engulf a variety of microorganisms (Finch and Turner, 1996). However, Se has been shown to be an important factor in the ability of neutrophils to exhibit microbicidal activity on ingested microorganisms. Selenium deficiency decreases the oxidative capacity of neutrophils (Finch and Turner, 1996). Also, normal Se status increases the mitogen response of lymphocytes in vitro versus lymphocytes from Se-deficient ruminants (Finch and Turner, 1996). Selenium deficiency has been shown to be a major risk factor for mastitis in dairy cattle (Smith et al., 1984; Erskine et al., 1987, 1989), and while the precise immune mechanisms may not be currently known, Se deficiency seems to be a very important cause of the heightened susceptibility of grazing ruminants to a number of disease conditions.

III. SELENIUM DEFICIENCY IN HUMANS

Selenium deficiency is a major factor in the human disease referred to as Keshan disease, which is endemic in certain regions of China. Keshan disease is a juvenile cardiomyopathy that presents as congestive heart failure in infants and young children. In fact, Keshan disease has been virtually eliminated by supplying sodium selenite pills to those at risk. One of the confounding observations regarding Keshan disease is its seasonal occurrence. Chinese researchers isolated a Coxsackievirus, B4, from Keshan disease patients, and a series of interesting experiments were subsequently performed. Using coxsackievirus B3, Beck et al. (1994a, 1994b, 1995) found that Se-deficient mice developed cardiomyopathy when exposed to a nonvirulent strain of coxsackie virus B3. Additionally, they reported the following: (1) a normally resistant strain of mice that were Se deficient exhibited cardiac lesions when infected, (2) a nonvirulent form of coxsackievirus B3 mutated to a virulent form when passed through Se-deficient mice, (3) the age-related resistance that mice normally possessed against viral infection was lost when they were Se deficient, and (4) six of the seven base pairs of the genetic code of Coxsackievirus B3, thought to determine virulence, consistently mutated from the avirulent form to the virulent form when passed through Se-deficient mice (Beck et al., 1994a, 1994b, 1995). This biologic phenomenon is of unknown clinical significance; however, the fact that Se status might alter the host–parasite relationship in a way that permits virulent mutation of viruses on such a fundamental level is extremely important. The possibility that this type of mutation could occur in Se-deficient ruminants would help explain observations regarding the frequency of viral mutations in both domestic and wildlife species.

IV. DETERMINATION OF NUTRITIONAL SELENIUM STATUS

A central consideration for Se deficiency, or Se toxicity, is accuracy in diagnosing Se status. While recognition of clinical disease is important, many losses can result from subclinical disease, decreased weight gains, or decreased feed efficiency. Most areas of the United States and Canada (Fig. 2) are at risk of having Se-deficient grazing livestock and wildlife. The Pacific Coast, including most of California, the Intermountain West, the states and provinces bordering the Great Lakes, the northeastern United States, and the eastern coastal states are Se deficient. The central plains tend to have normal Se levels in the soils with localized areas of excess Se. Selenium concentration of plants, feeds, soil, and water can be determined; however, because of numerous dietary interactions, the Se status of animals is difficult to predict from these data. Some of the nutrient interactions will be discussed later.

Nutritional Se status in cattle can be determined by measuring Se concentration (Olson, 1969; Tracy and Moller, 1990) or GSH-Px activity (Agergaard and Jensen, 1982; Maas et al., 1993) in a number of tissues. The tissues most commonly used are blood, liver, and kidney. In a clinical setting, tissues of convenience include whole blood (EDTA or heparin tubes) and serum or plasma. Blood Se concentrations from 0.1 ppm to 1.0 μg/mL are considered normal, with most supplemented or Se normal ruminants having values of 0.1 to 0.3 μg/mL (Maas, 1983). Blood Se concentrations below 0.05 μg/mL are frankly deficient and usually associated with clinical disease. Blood Se concentration of 0.05 to 0.1 μg/mL are considered marginal, and subclinical disease can be common in these instances. While serum and/or plasma can be analyzed for Se concentration, it has recently been shown that diagnostic interpretation of these serum Se values is severely limited (Maas et al., 1992). Serum Se concentrations that are 0.01 μg/mL or less are diagnostic of Se deficiency and serum Se concentrations of 0.10 μg/mL or greater are diagnostic of Se adequacy (Maas et al., 1992). However, serum Se values between 0.01 and 0.10 μg/mL (the vast majority of samples) cannot be diagnostically interpreted with enough accuracy to make clinical deci-

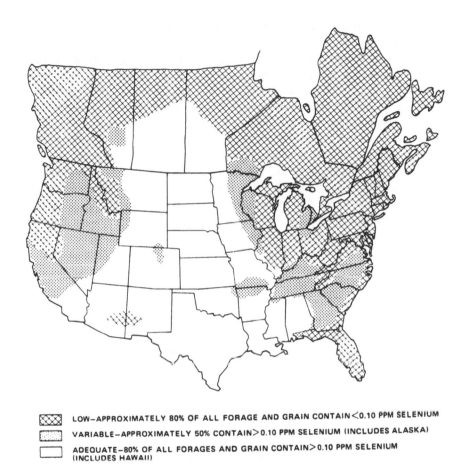

LOW—APPROXIMATELY 80% OF ALL FORAGE AND GRAIN CONTAIN<0.10 PPM SELENIUM

VARIABLE—APPROXIMATELY 50% CONTAIN>0.10 PPM SELENIUM (INCLUDES ALASKA)

ADEQUATE—80% OF ALL FORAGES AND GRAIN CONTAIN>0.10 PPM SELENIUM
(INCLUDES HAWAII)

FIGURE 2 Regional distribution of forages and grains containing low, variable, or adequate levels of selenium in the United States and Canada. (From National Research Council, *Selenium in Nutrition*, rev. ed. National Academy Press, Washington, DC, 1983, p. 24.)

sions (Maas et al., 1992). Blood Se concentrations of 5.0 μg/mL or greater are diagnostic of Se toxicosis. (Se toxicosis is discussed in detail by O'Toole and Raisbeck in Chapter 19). Blood GSH-Px activity levels (IU/mg hemoglobin/min) of 0 to 15, 15 to 25, and 25 to 100 correspond to blood Se levels of 0.01 to 0.05, 0.05 to 0.1, and 0.1 μg/mL and above, respectively, in our laboratory (Maas et al., 1993). Methods for GSH-Px are rarely standardized, however, and each GSH-Px method must be validated for each laboratory and for each species to be of diagnostic value. The luxury of having control animals that are fed standardized diets under controlled conditions does not exist in a diagnostic setting, and therefore the rigor applied to diagnostic testing and interpretation must be higher than in experimental situations.

There have been numerous interactions documented between Se and other dietary components. The most significant of these interactions is between Se and vitamin E. Both these essential nutrients are potent antioxidants and tend to work in concert to protect cells from biological damage. Vitamin E and Se have an apparent sparing effect on each other. This sparing effect is not without limits in grazing ruminants, and there are reports in which Se status was normal, but because of vitamin E deficiency, additional Se was not effective in treatment or prevention of NMD (Maas et al., 1984). Also, much of the experimental work with Se deficiency or vitamin E deficiency in ruminants has involved variable interactions between the two nutrients and has made interpretation difficult. In a duckling myopathy model, Van Vleet reported that the following elements interfered with Se and vitamin E metabolism: silver, zinc, cadmium, tellurium, cobalt, copper, mercury, tin, lead, arsenic, iron, and sulfur, either alone or in combination (Van Vleet, 1982a, 1992b). Some of these elements are potent oxidants (Cd, Hg, Fe), while others interfere directly with Se metabolism (As, S, Te). Selenium has been found to be protective against toxic doses of compounds such as monensin, a polyether ionophore, which can induce myopathy.

One of the more important interactions on a clinical basis is that between Se and sulfur. Sulfur interferes with Se uptake by plants (Severson and Gough, 1992) and also with Se metabolism in ruminants (Jones et al., 1987). Therefore, sulfur fertilization of pastures and rangelands could significantly increase the interference with Se metabolism in animals and predispose them to Se deficiency. Another concern has been potential Se interference with absorption and/or retention of copper. A recent report examined this possibility in preweaned and weaned beef heifers (Maas et al., 1994). These young heifers were supplemented with (Se 3 mg/day), and no measurable interference with Cu metabolism, as characterized by hepatic and serum copper concentrations, was reported (Maas et al., 1994). A similar study in lactating dairy cattle (Buckley et al., 1986) had similar results, and thus no apparent interference with copper metabolism seems to occur even with high levels of Se supplementation. Yet numerous interactions between Se, other nutrients, drugs, other elements, and diseases can occur, and this increases the importance of accurate diagnostic criteria and careful interpretation of data related to Se deficiency or Se toxicity.

V. SELENIUM SUPPLEMENTATION

A. Regulatory Aspects of Supplementation

Selenium deficiency is a widespread problem in North American grazing ruminants, in both domestic and wildlife species. A survey of state veterinary diagnostic laboratories found that Se deficiency was diagnosed in 46 states, constituting a major problem in livestock or wildlife in 37 states, and that natural Se toxicosis was a rare problem (Edmondson et al., 1993). When Se toxicosis does occur naturally, it can be associated with aquatic environments in some regions, while in others the soils and plants have high levels of Se. In California, a series of evaporation ponds in the San Joaquin Valley (Kesterson Reservoir) downstream from Se-rich soils and irrigated lands accumulated high levels of Se, resulting in Se toxicosis in certain wildlife species. Data from this episode were used by the Department of the Interior and others to request the Food and Drug Administration (FDA) to reduce the level of allowable Se that could be fed to livestock and poultry (U.S. FDA, 1993). This action sparked controversy throughout the scientific and regulatory communities. No evidence was presented that Se supplemented to animals resulted in accumulations of toxic amounts of Se at Kesterson or other sites. Indeed, there was no historical practice of supplementing Se to animals upstream from Kesterson, as this area is not Se deficient.

The FDA eventually began regulatory action to decrease the level of Se that could be added to animal diets. Selenium is unique among essential nutrients in that its use in animals is regulated by the FDA. This is not the case for other nutrients, as Se, in all its forms and products, is treated as a drug. In 1994 two laws were passed by the 103rd Congress and signed by the president, affecting Se supplementation to animals. Title VI of Public Law 103-330 provides for suspension of the stay published in the 1993 final rule of the 1987 food additive regulation relating to Se until December 31, 1995 (U.S. FDA, 1995). The second act, Public Law 103-354, under Subtitle G—Food Safety Section 262, "Conditions for Implementation of Alteration in the Level of Additives Allowed in Animal Diets," prohibits the implementation or enforcement of the 1993 final rule that stayed the 1987 amendments unless certain determinations are made by the FDA (U.S. FDA, 1995). These determinations can be stated as follows: (1) selenium additives are not essential at levels authorized in the absence of such final rule, to maintain animal nutrition and protect animal health (i.e., Se is not an essential nutrient for animals up to 0.3 mg/kg of the diet, dry matter basis); (2) selenium at such levels is not safe to the animals consuming the additive; (3) selenium at such levels is not safe to individuals consuming edible portions of animals that receive the additive; (4) selenium at such levels does not achieve its intended effect of promoting normal growth and reproduction of livestock and poultry; and (5) the manufacture and use of selenium at such levels cannot reasonably be controlled by adherence to current good manufacturing practice requirements (U.S. FDA, 1995). If none of these items is determined, the 1987 amendment will continue in force. In general terms, the 1987 amendment to the food additive regulation allows for supplemental dietary Se levels (concentrations) up to 0.3 mg/kg dry matter basis (DMB), not to exceed 3.0 mg Se per animal per day, for beef and dairy cattle. Salt–mineral mixes for cattle may contain a maximum of 120 mg/kg Se, and mixtures of these types could contain a maximum of 90 mg/kg Se for sheep. The pre-1987 supplemental Se levels are 0.1 mg/kg DMB in the diet, and 20 mg/kg Se in salt–mineral mixes for cattle and 30 mg/kg Se in salt–mineral mixes for sheep. It is important to realize that the National Research Council of the National Academy of Sciences has determined the nutrient requirements for Se to be between 0.1 and 0.5 mg/kg for cattle and most other species (CAST, 1994). Therefore, 0.1 mg/kg is not adequate in all circumstances, and interference by other dietary components can increase the requirements greatly. The presence of sulfates in the feed or water, for example, can markedly interfere with Se absorption and utilization. The use of blood Se concentration as a diagnostic and management tool is very helpful in assessing the nutritional program for grazing livestock.

B.   Methods of Supplementation

The need to supplement Se in grazing situations is common in the United States and Canada. Current laws and regulations recognize the scientific facts regarding Se as an essential nutrient for grazing animals and other species. In 1989, the FDA regulation was amended to allow Se supplementation by an intraruminal bolus at levels provided by the food additive regulation. This petition, filed by Schering-Plough Animal Health Corporation, also relied on the 1987 amendment. While this bolus, which provided 3 mg Se/day for 120 days, is legal, it is off the market indefinitely as a result of the FDA actions. Therefore, one of the tools for Se supplementation was lost.

One of the first methods of supplementation of Se used was via injection of sodium selenite solutions. This form of Se was particularly useful for treatment of NMD in calves, lambs, and other species. Recently, new data regarding pharmacodynamic aspects of Se injections in cattle has been reported (Maas et al., 1993). It was found that peak blood Se concentrations, after using the label dose (2.5 mg/100 lb body weight), occurred 5 hours postinjection. The peak blood Se concentration was above 0.1 μg/mL for 10 hours postinjection; however, the blood Se concentration continued to decrease rapidly and was less than 0.05 μg/mL by 28 days postinjection. The blood GSH-Px activity showed a significant increase by day 28 postinjection; however, the blood GSH-Px activity achieved a level only about 40% of that considered to be adequate before declining sharply (Maas et al., 1993). Some of the other significant conclusions of this study included the following (Maas et al., 1993):

1.   The label dose of injectable Se, as sodium selenite, does not achieve blood Se or blood GSH-Px activity considered to be adequate.
2.   Serum Se or serum GSH-Px is not an accurate predictor of Se status or current Se injection history.

3. Injectable Se is an excellent therapeutic agent and can be used for strategic supplementation, but should not be the sole method of long-term supplementation unless repeated often.
4. The use of blood Se and blood GSH-Px can serve retrospectively in the diagnosis of Se status within 14 days of the time of an injection.
5. Approximately 30% of the injected Se dose is eliminated via the kidney within the first 48 hours postinjection.

Selenium injections are best considered as therapeutic agents or short-term supplements. For grazing ruminants, the use of sustained-release Se boluses is an excellent alternative. These products have been shown to be effective and safe for grazing animals (Campbell et al., 1990; Coe et al., 1993; Maas et al., 1994). Their use is particularly advantageous in extensive grazing or range situations. Another form of supplementation is the use of salt–mineral mixes with added Se. This can be a very effective method, but it relies on voluntary consumption, which can vary greatly with season and individual animal. For feeding under more intensive situations, Se can be added to a premix and fed as part of a total mixed feed or a portion of a grain feeding program. Fertilization with Se is done in countries such as New Zealand, Australia, and Finland. This method has been used experimentally in California on range utilized by mule deer (Oliver et al., 1990a). The need for Se supplementation is not limited to grazing domestic ruminants. Does (Columbia black-tailed deer) given intraruminal Se boluses exhibited a 2.6-fold increase in fawn survival over a 3-year period (Flueck, 1989) versus controls. Blood samples (1695) were collected from mule deer in 15 geographical regions of California and analyzed for Se content (Oliver et al., 1990b). Two-thirds of the herd groups had first-quartile Se concentrations less than 0.05 ppm, indicating widespread Se deficiency in this species of grazing wildlife (Oliver et al., 1990b).

VI. ENVIRONMENTAL IMPLICATIONS OF SELENIUM SUPPLEMENTATION

Selenium supplemented to grazing ruminants is excreted in various forms and by three major routes, the feces, urine, and rumen gases. Oral Se is excreted mostly in the feces (Cousins and Cairney, 1961), whereas parenteral administration of Se (sodium selenate or sodium selenite) results in the majority being excreted in the urine (Wright and Bell, 1966; Maas et al., 1993). Trimethyl selenide is the major urinary metabolite (Palmer et al., 1969), and trimethyl selenide has been shown to be essentially unavailable nutritionally for the prevention of liver necrosis in rats (Tsay et al., 1970). The predominant forms of Se in the feces of sheep, and probably in that of other ruminants, is the highly insoluble elemental Se or the metal selenides (Cousins and Cairney, 1961; Peterson and Spedding, 1963; Obermeyer et al., 1971). According to the report by the Council of Agricultural Science and Technology (CAST), excretion of dietary Se by livestock supplemented at or near 0.3 mg/kg dietary Se is about 60% of intake; furthermore, excretion by all U.S. animals that legally could be supplemented with the maximum allowable amount of Se could contribute 28.5 tons of Se to the environment annually (CAST, 1994). The latter amount would constitute 0.2% of the Se entering the environment from anthropogenic sources and natural sources, including movement into soil and water from water eluviation of Se-containing rocks. Thus, Se from supplemented grazing ruminants contributes an insignificant amount to the environment and furthermore, the excreted Se is present in forms with limited or no bioavailability. This has been the practical experience of many livestock operations in Se-deficient areas. After years of supplementing Se to cattle and sheep, the pasture and forage concentrations of Se have not increased. Also, a survey of aquatic ecosystems in northern California downstream from ranches that had been supplementing Se for 3 to 8 years showed no accumulation of Se in water, sediment, algae, aquatic plants, invertebrates, or fish versus upstream control sites (Norman et al., 1992). Work presently under way by this author indicates that grazing ruminants can be useful in safely removing Se from Se-contaminated grasslands and may represent an efficient, low cost method of remediation of such areas.

VII. SUMMARY

Selenium is an essential trace element for all vertebrates, including grazing ruminants. Selenium deficiency is widespread in the United States and Canada. Significant advances have been made in understanding Se deficiency problems and in our ability to diagnose Se deficiency and to monitor Se supplementation. Grazing ruminants do not represent an environmental concern with regard to Se; however, it is likely that large domestic ruminants could be helpful in remediation of Se-contaminated areas.

REFERENCES

Agergaard, N., and P. T. Jensen. 1982. Procedure for blood glutathione peroxidase determination in cattle and swine. Acta Vet. Scand. 23:515–527.

Arthur, J. R., F. Nicol, and G. J. Beckett. 1990. Hepatic iodothyronine deiodinase: The role of selenium. Biochem. J. 272:537–540.

Beck, M. A., P. C. Kolbeck, Qing Shi, L. H. Rohr, V. C. Morris, and O. A. Levander. 1994a. Increased virulence of a human enterovirus (Coxsackie B3) in selenium-deficient mice. J. Infect. Dis. 170:351–357.

Beck, M. A., P. C. Kolbeck, L. H. Rohr, Qing Shi, V. C. Morris, and O. A. Levander. 1994b. Benign human enterovirus becomes virulent in selenium-deficient mice. *J. Med. Virol.* 43:166–170.

Beck, M. A., Qing Shi, V. C. Morris, and O. A. Levander. 1995. Rapid genomic evolution of a non-virulent Coxsackie B3 in selenium-deficient mice results in selection of identical virulent isolates. *Nature Med.* 1:433–436.

Behne, D., A. Kyriakopoulos, H. Meinhold, and J. Kohrle. 1990. Identification of type I iodothyronine 5'-deiodinase as a selenoenzyme. *Biochem. Biophys. Res. Commun.* 173:1143–1149.

Berry, M. J., L. Banu, and P. R. Larsen. 1991. Type I iodothyronine deiodinase is a selenocysteine-containing enzyme. *Nature* 349:438–440.

Buckley, W. T., S. N. Hucklin, L. J. Fisher, and G. K. Eigendorf. 1986. Effect of selenium supplementation on copper metabolism in dairy cows. *Can. J. Anim. Sci.* 66:1009–1018.

Burk, R. F., and K. E. Hill. 1993. Regulation of selenoproteins. *Annu. Rev. Nutr.* 13:65–81.

Campbell, D. T., J. Maas, D. W. Weber, O. R. Hedstrom, and B. B. Norman. 1990. Safety and efficacy of two sustained-release intrareticular selenium supplements and the associated placental and colostral transfer of selenium in beef cattle. *Am. J. Vet. Res.* 51:813–817.

CAST (Council for Agricultural Science and Technology). 1994. *Risks and Benefits of Selenium in Agriculture.* Ames, IA. 35 pp.

Chu, F. F., J. H. Doroshow, and R. S. Esworthy. 1993. Expression, characterization, and tissue distribution of a new cellular selenium-dependent glutathione peroxidase, GSHPx-GI. *J. Biol. Chem.* 268:2571–2576.

Coe, P. H., J. Maas, J. Reynolds, and I. Gardner. 1993. Randomized field trial to determine the effects of oral selenium supplementation on milk production and reproductive performance of Holstein heifers. *J. Am. Vet. Med. Assoc.* 202:875–881.

Cousins, F. B., and I. M. Cairney. 1961. Some aspects of selenium metabolism in sheep. *Aust. J. Agric. Res.* 12:927–943.

Croteau, W., S. L. Whittemore, M. J. Schneider, and D. L. St. Germain. 1995. Cloning and expression of cDNA for a mammalian type II iodothyronine deiodinase. *J. Biol. Chem.* 270:16569–16575.

Edmondson, A. J., B. B. Norman, and D. Suther. 1993. Survey of state veterinarians and state veterinary diagnostic laboratories for selenium deficiency and toxicosis in animals. *J. Am. Vet. Med. Assoc.* 202:865–872.

Erskine, R. J., R. J. Eberhart, L. J. Hutchinson, and R. W. Scholz. 1987. Blood selenium concentrations and glutathione peroxidase activities in dairy herds with high and low somatic cell counts. *J. Am. Vet. Med. Assoc.* 190:1417–1421.

Erskine, R. J., R. J. Eberhart, P. J. Grasso, and R. W. Scholz. 1989. Induction of *Escherichia coli* mastitis in cows fed selenium-deficient or selenium-supplemented diets. *Am. J. Vet. Res.* 50:2093–2100.

Finch, J. M., and R. J. Turner. 1996. Effects of selenium and vitamin E on the immune responses of domestic animals. *Res. Vet. Sci.* 60:97-106.

Flueck, W. 1989. The effect of selenium on reproduction of black-tailed deer (*Odocoileus hemionus columbianus*) in Shasta County, California. Ph.D. thesis, University of California, Davis. 284 pp.

Hedstrom, O. R., J. P. Maas, B. D. Hultgren, E. D. Lassen, E. A. Wallner-Pendleton, and S. R. Snyder. 1986. Selenium deficiency in bovine, equine, and ovine with emphasis on its association with chronic diseases. *29th Annual Proceedings American Association of Veterinary Laboratory Diagnosticians,* pp. 101–126.

Hill, K. E., R. S. Lloyd, J.-G. Yang, R. Read, and R. F. Burk. 1991. The cDNA for rat selenoprotein P contains 10 TGA codons in the open reading frame. *J. Biol. Chem.* 266:10050–10053.

Hogue, D. E., J. F. Proctor, R. G. Warner, and J. K. Loosli. 1962. Relation of selenium, vitamin E and an unidentified factor to muscular dystrophy (stiff-lamb or white muscle disease) in the lamb. *J. Anim. Sci.* 21:25–29.

Jones, M. B., D. M. Center, V. V. Rendig, M. R. Daly, B. B. Norman, and W. A. Williams. 1987. Selenium enhances lamb gains on sulfur-fertilized pastures. *California Agric.* May–June, 14–16.

Karimpour, I., M. Cutler, D. Shih, J. Smith, and K. C. Kleene. 1992. Sequence of the gene encoding the mitochondrial capsule selenoprotein of mouse sperm: Identification of in-phase TGA selenocysteine codons. *DNA Cell Biol.* 11:693–699.

Kiremidjian-Schumacher, L., M. Roy, H. I. Wishe, M. W. Cohen, and G. Stotzky. 1994. Supplementation with selenium and human immune cell functions. II. Effect on cytotoxic lymphocytes and natural killer cells. *Biol. Trace Elem. Res.* 41:115–127.

Kyriakopoulous, A, C. Hammel, H. Gessner, and D. Behne. 1996. Characterization of an 18-kD selenium-containing protein in several tissues of the rat. *Am. Biotech. Lab.* July, p. 22.

Maas, J. P. 1983. Diagnosis and management of selenium-responsive diseases in cattle. *Compend. Contin. Educ. Pract. Vet.* 5:S393–S399.

Maas, J., M. S. Bulgin, B. C. Anderson, and T. M. Frye. 1984. Nutritional myodegeneration associated with vitamin E deficiency and normal selenium status in lambs. *J. Am. Vet. Med. Assoc.* 184:201–204.

Maas, J. , F. D. Galey, J. R. Peauroi, J. T. Case, E. S. Littlefield, C. C. Gay, L. D. Koller, R. O. Crisman, D. W. Weber, D. W. Warner, and M. L. Tracy. 1992. The correlation between serum selenium and blood selenium in cattle. *J. Vet. Diagn. Invest.* 4:48–52.

Maas, J., J. R. Peauroi, T. Tonjes, J. Karlonas, F. D. Galey, and B. Han. 1993. Intramuscular selenium administration in selenium-deficient cattle. *J. Vet. Intern. Med.* 7:342–348.

Maas, J, J. R. Peauroi, D. W. Weber, and F. W. Adams. 1994. Safety, efficacy, and effects on copper metabolism of intrareticularly placed selenium boluses in beef heifer calves. *Am. J. Vet. Res.* 55:247–250.

Muth, O. H. 1963. White muscle disease, a selenium responsive myopathy. *J. Am. Vet. Med. Assoc.* 142:272–277.

Norman, B. B., G. Nader, M. Oliver, R. Delmas, D. Drake, and H. George. 1992. Effects of selenium supplementation in cattle on aquatic ecosystems in California. *J. Am. Vet. Med. Assoc.* 201:869–872.

Nunn, C. L., H. A. Turner, D. J. Drake. 1995. Effect of selenium boluses on weight gain and feed efficiency of wintering beef steers. *Proc. Am. Anim. Sci. Soc.* (Abstr.), p. 749.

Obermeyer, B. D., I. S. Palmer, O. E. Olson, and A. W. Halverson. 1971. Toxicity of trimethylselenonium chloride in the rat with and without arsenite. *Toxicol. Appl. Pharmacol.* 20:135–146.

Oliver, M. N., D. A. Jessup, and B. B. Norman. 1990a. Selenium supplementation of mule deer in California. *Trans. West. Sec. Wildl. Soc.* 26:87–90.

Oliver, M. N., G. Ros-Mcgauar, D. A. Jessup, B. B. Norman, and C. E. Franti. 1990b. Selenium concentrations in blood of free-ranging mule deer in California. *Trans. West. Sec. Wildl. Soc.* 26:80–86.

Olson, O. E. 1969. Fluorometric analysis of selenium in plants. *J. Assoc. Off. Anal. Chem.* 42:627–634.

Palmer, I. S., D. D. Fischer, A. W. Halverson, and O. E. Olson. 1969. Identification of a major selenium excretory product in rat urine. *Biochim. Biophys. Acta* 177:336–341.

Peterson, P. J., and D. J. Spedding. 1963. The excretion by sheep of Se-75 incorporated in red clover. The chemical nature of the excreted selenium and its uptake by three plant species. *N. Z. J. Agric. Res.* 6:13–22.

Rotruck, J. T., A. L. Pope, H. E. Ganther, A. B. Swanson, D. G. Hafeman, and W. G. Hoekstra. 1973. Selenium: Biochemical role as a component of glutathione peroxidase. *Science* 179:588–590.

Roy, M., L. Kiremidjian-Schumacher, H. I. Wishe, M. W. Cohen, and G. Stotzky. 1994. Supplementation with selenium and human immune cell functions. I. Effect on lymphocyte proliferation and interleukin 2 receptor expression. *Biol. Trace Elem. Res.* 41:103–114.

Schwarz, K., and C. M. Foltz. 1957. Selenium as an integral part of factor 3 against dietary necrotic liver degeneration. *J. Am. Chem. Soc.* 79:3292–3293.

Severson, R. C., and L. P. Gough. 1992. Selenium and sulfur relationships in alfalfa and soil under field conditions, San Joaquin Valley, California. *J. Environ. Qual.* 21:353–358.

Smith, K. L., J. H. Harrison, D. D. Hancock, D. A. Todhunter, and H. R. Conrad. 1984. Effect of vitamin E and selenium supplementation on incidence of clinical mastitis and duration of clinical symptoms. *J. Dairy Sci.* 67:1293–1300.

Takahashi, K., N. Avissar, J. Whitin, and H. Cohen. 1987. Purification and characterization of human plasma glutathione peroxidase: A selenoglycoprotein distinct from the known cellular enzyme. *Arch. Biochem. Biophys.* 256:677–686.

Tracy, M. L., and G. Moller. 1990. Continuous flow generation for inductively coupled argon plasma spectrometric analysis. Part I: Selenium. *J. Assoc. Off. Anal. Chem.* 73:404–410.

Tsay, D. T., A. W. Halverson, and I. S. Palmer. 1970. Inactivity of dietary trimethylselenonium chloride against the necrogenic syndrome of the rat. *Nutr. Rep. Int.* 2:203–207.

Ursini, F., M. Maiorino, and C. Gregolin. 1985. The selenoenzyme phospholipid hydroperoxide glutathione peroxidase. *Biochim. Biophys. Acta* 839:62–70.

U.S. Food and Drug Administration (FDA). 1993. Food additives permitted in feed and drinking water of animals; selenium; stay of the 1987 amendments. *Fed. Regist.* 58:47962–47973.

U.S. Food and Drug Administration (FDA). 1995. Food additives permitted in feed and drinking water of animals; selenium. *Fed. Regist.* 60:53702–53704.

Van Vleet, J. F. 1982a. Amounts of twelve elements to induce selenium-vitamin E deficiency in ducklings. *Am. J. Vet. Res.* 43:851-857.

Van Vleet, J. F. 1982b. Amounts of eight combined elements required to induce selenium–vitamin E deficiency and protection by supplements of selenium and vitamin E. *Am. J. Vet. Res.* 43:1049–1055.

Vendeland, S. C., M. A. Beilstein, C. L. Chen, O. N. Jensen, E. Barofsky, and P. D. Whanger. 1993. Purification and properties of selenoprotein W from rat muscle. *J. Biol. Chem.* 268:17103–17107.

Wright, P. L., and M. C. Bell. 1966. Comparative metabolism of selenium and tellurium by sheep and swine. *Am. J. Physiol.* 211:6–34.

8
Environmental Implications of Uses of Selenium with Animals

JAMES E. OLDFIELD
Oregon State University, Corvallis, Oregon

Agricultural producers worldwide are becoming increasingly concerned about the effects their practices may have on the maintenance of environmental quality; in fact, the impact of animal and crop production technologies on soil and groundwater is a primary concern of modern-day agriculture (Zebarth, 1996). Attention has focused first on nitrogen and phosphorus, which are major elements in animal wastes, and concerns have been sufficiently strong that in some areas, such as the Netherlands, nationwide reductions in animal numbers have been mandated (Cheeke, 1995). From these problem areas there has been a natural extension of interest to various other substances that are added to livestock diets, and particularly to those, such as selenium, that may contribute to a toxicity burden.

Selenium's use in agriculture dates to the early 1960s, after it had been demonstrated to be an essential micronutrient (Schwarz and Foltz, 1957). Right from the beginning, its useful applications were tempered by concerns relating to its long-understood toxicity in excess. There was another troublesome issue, too, that weighed heavily in the development of regulations for its use, and this was the matter of potential carcinogenicity, which if confirmed would bring selenium under the purview of the so-called Delaney clause of the 1962 amendments (Public Law 87-781) of the federal Food, Drug and Cosmetics Act (Ullrey, 1992). Taken together, these concerns about toxicity and possible carcinogenicity meant that intense scrutiny was directed toward selenium to ensure that its uses would not leave unacceptable levels of it in the environment.

I. SELENIUM TOXICITY

Knowledge of selenium's toxicity dates right to its discovery by Berzelius, as he investigated worker illnesses in a Swedish sulfuric acid plant in 1817 (Oldfield, 1995). Originally suspecting arsenic poisoning, Berzelius was surprised to find, instead, a new element in slimes from the acid vats to which workers were exposed. He named his discovery selenium, after Selene, the Greek goddess of the moon. Confirmation of its toxicity came with the observation that livestock poisonings in the north-central range regions of the United States, could be related to high levels of selenium in native forages (Franke, 1934). This connection was first noted by Madison (1860), an army surgeon at Fort Randall, in what is now South Dakota. The problem was first called "alkali disease" on the assumption that it was caused by animals drinking from alkaline ponds (Moxon, 1937).

Selenium toxicity has been studied, intermittently, down through the years, but the early investigations, many of which were done at the University of Wyoming, have been cited as references so many times that they are practically a part of the range lore of the north-central states. Livestock poisonings have understandably received widespread publicity, and at least one anecdotal book has been written about them (Harris, 1986). Selenium toxicity gained new prominence a decade ago, through a highly publicized report of death and deformities among waterfowl nesting at the Kesterson Reservoir, in California's San Joaquin Valley (Ohlendorf et al., 1986). It seems clear that selenium was, indeed, involved in the earlier range poisonings in South Dakota and in the more recent avian incidents in California, but there is mounting evidence, in retrospect, that other substances may have been implicated, too. O'Toole points out differences in symptomatology between experimentally induced selenium toxicosis and the historic instances of poisoning on the range (O'Toole and Raisbeck, 1995), while Ullrey (1992) describes the Kesterson Reservoir as "essentially a wastewater sump receiving irrigation drainage waters from the San Luis drain in Western Fresno county." Such waters contain many other elements, beside selenium, some of them in toxic quantities.

The issue of selenium and cancer merits mention here because of its implications for the regulatory control of selenium's use in agriculture. Selenium was first proposed as a carcinogen when liver tumors were reported in rats after induction of cirrhosis by feeding selenium (Nelson et al., 1943), and this caused the federal Food and Drug Administration (FDA) to withhold authority to add supplementary selenium to livestock rations for a number of years. A research effort was mounted to explore the issue of selenium's proposed carcinogenicity, and six published studies from it have been reviewed (Hegsted, 1970). Three studies identified selenium as a carcinogen, while the other three came to the opposite view. However, reviewers concluded there were flaws in the former set, including inadequate experimental design, unnaturally high levels of dietary selenium, and presence of infectious disease. The latter three studies were judged to be properly planned and controlled, and on this basis the FDA concluded that "judicious administration of selenium derivatives to domestic animals would not constitute a carcinogenic risk" (U.S. FDA, 1973).

II. NUTRIENT ESSENTIALITY OF SELENIUM

The major area of use for selenium in agriculture is as an essential micronutrient, provided to animals whose diets would otherwise be deficient in it. Following Schwarz and Foltz's (1957) disclosure of selenium's essentiality as a result of experiments with laboratory rats, animal scientists tested its usefulness against a number of problems affecting domestic animals for which no cure was then available. In short order, a number of conditions were identified, which became known as "selenium-responsive diseases" (Oldfield, 1987). Signs of selenium deficiency have been reported in a remarkable number of species, including among domestic food-producing animals, cattle, chickens, ducks, fish, goats, pheasants, quail, sheep, swine, and turkeys. These are widespread, and one, "white muscle disease," a myopathy affecting both cardiac and skeletal muscles in young animals, has been diagnosed in 30 of the 50 states (Wolf et al., 1963).

Mortalities from these selenium deficiency diseases, though significant, were not as important economically as the lowered production efficiency of affected animals, occasioned by lowered reproductivity and growth. Selenium supplementation was demonstrated to enhance both reproduction (Andrews et al., 1968) and growth efficiency (Oldfield et al., 1960)—two bases of successful livestock production—so animal producers worldwide were quick to adopt the practice, which has since become, where needed, an accepted animal production technology. Continuing research with selenium, too, has suggested that the additive may increase animals' resistance to certain infections (Smith et al., 1984; Finch and Turner, 1989, 1996; Nemec et al., 1990), and this has strengthened further its commercial application. It has been suggested that selenium deficiency can compromise the immune system, with adverse effects on production and performance appearing before clinical symptoms are observed.

Selenium can be made available to animals in two ways: direct supplementation, in which the selenium is given to the animals orally or parenterally, and indirectly (i.e., added to soil as a fertilizer amendment, thereby enhancing the selenium content of livestock feeds and forages). Oral treatments include an ingenious heavy pellet, originating in Australia (Kuchel and Buckley, 1969), which lodges in the forestomach of ruminant animals, and dispenses selenium over extended periods of time. A more recent application of this principle employs a heavy bolus of phosphate-based soluble glass as the selenium carrier (Telfer et al., 1983). Experience indicates that different forms of selenium may be most appropriate for these different methods of administration. In general, feed supplementation has favored the use of sodium selenite; elemental selenium has been used, with iron for added weight in the heavy rumen pellets; and the fertilizer applications have involved sodium selenate or barium selenate.

Methods of selenium supplementation vary, by area, worldwide. In the United States, most supplementary selenium is given in feed; in Australia, the heavy pellets are widely used, and in New Zealand and Finland—two small countries with extensive land areas of selenium deficiency—most of the supplementary selenium is added to fertilizers. None of these methods is 100% efficient, which means there is always some residual selenium returned to the environment, and it is appropriate to examine and to quantify such residues, where possible.

III. ENVIRONMENTAL IMPACTS

A. Range Animals

Conditions that have come to be recognized as symptoms of selenium deficiency were noted very early, specifically in the United States and New Zealand, in animals grazing on range- or pastureland. The reason was obvious: these animals frequently derived their entire diet from forage herbage, and when this was grown on selenium-deficient soil, it did not provide sufficient selenium to meet the animals' needs. Animals most frequently affected were beef cattle and sheep. In contrast, dairy cattle, pigs, and poultry were more likely to be fed diets supplemented with feedstuffs from other areas, which could make up the deficiency in local feeds. However, selenium deficiency is not uncommon among these species when their diets are drawn largely from a single source, such as grain grown in an area of selenium deficiency.

When the poisonings of wild waterfowl at the Kesterson Reservoir first occurred, the source of the demonstrated high levels of selenium was not immediately identifiable. Any uses of selenium, including nutritional supplementation, became suspect, and public opinion demanded research that would monitor just what happened to the selenium given to animals as a feed supplement. Norman et al. (1992) at the University of California at Davis quickly supplied data from an investigation of beef cattle operations in areas of known selenium deficiency in northern California. The researchers measured selenium levels in stream water above and below ranches on which selenium supplementation of beef cattle had been practiced for 3 to 8 years, reporting no significant differences in the upstream and downstream contents of selenium in water, stream sediment, algae, invertebrates, and fish (Table 1).

Meanwhile, studies were continuing in areas adjacent to the Kesterson Reservoir. The source of the high selenium was determined to be seleniferous outcrops of rocks on the west side of the San Joaquin Valley, from which over

TABLE I   Selenium Values (μg/kg, wet basis) for Specimens Obtained on Four Cattle Ranches: Upstream (control) and Below (treated) Where Se-Supplemented Animals Were Held

	Water	Sediment	Algae	Invertebrates	Fish[a]
Ranch 1					
Control					
Mean	3.3	29.0	13.7	110.7	145.0
SD	0.5	0.0	0.9	12.7	6.4
Treated					
Mean	2.3	122.3	10.3	112.3	199.7
SD	0.5	43.7	2.4	9.0	34.8
Ranch 2					
Control					
Mean	0.9	163.7	58.5	35.3	319.0
SD	0.0	49.4	5.5	21.9	26.4
Treated					
Mean	1.7	103.3	49.3	493.0	289.0
SD	0.5	58.5	0.9	0.0	19.4
Ranch 3					
Control					
Mean	1.7	73.3	25.3	112.7	N/S
SD	0.5	9.7	1.2	4.0	
Treated					
Mean	2.3	53.3	26.0	84.7	N/S
SD	1.0	4.5	14.4	12.6	
Ranch 4					
Control					
Mean	1.7	197.0	11.3	236.7	N/S
SD	0.5	13.9	1.2	3.6	
Treated					
Mean	1.7	140.7	19.0	179.7	N/S
SD	0.5	28.8	0.8	5.8	

[a]N/S, not studied.
Source: Norman et al. (1992).

time infiltrating irrigation water had leached selenium downward to the valley floor, where it was picked up in deep runoff drains and carried to shallow evaporation ponds at Kesterson. No evidence was found to implicate animal supplementation practices in any residual accumulation, and interestingly, positive responses of cattle, in terms of increased weight gains, were reported in an adjacent area of the San Joaquin Valley (Nelson and Miller, 1987), indicating absence of any significant soil buildup. Further, indirect evidence of the safety of agricultural uses of selenium was provided by an extensive survey of vegetable crops exposed to varying selenium environments in California; these results confirmed that agricultural supplements posed no selenium hazard when consumed at normal levels of intake (Grattan et al., 1990).

B. Feedlot Animals

The dispersion of animals grazing on rangelands represented an area of low potential selenium hazard. However, the question has been logically raised about the effects of selenium supplementation when large numbers of livestock are concentrated together, as in feedlots. Here, two items are significant: the selenium levels in the excreta and the levels at which the manures are applied to soils.

One empirical piece of evidence about the possible accumulation of selenium in soils following application of manures from selenium-supplemented animals was provided by the ongoing records from the Morrow Field plots, at the University of Illinois. These plots, established in 1876, represent the longest, continually cropped land in the world (Odell et al., 1982), and they offered a unique opportunity to study selenium levels in soils fertilized with manures before and after selenium supplementation of animals was practiced (Table 2). The mean selenium content of the corn crop from these plots rose from 26 to 30 μg/kg when manures from selenium-supplemented animals were applied, an insignificant increase. Moreover, the higher level in the fertilized plot only equaled that in the crop that had received no fertilizer at all (Oldfield, 1994).

Other studies of the effects of applying to croplands, manures from feedlot animals of different species, which had been supplemented with selenium, have

TABLE 2   Selenium in Corn (μg/kg, dry basis) Grown on the Morrow Plots[a] (University of Illinois, Urbana-Champaign) Fertilized with Manure Before and After FDA Approval of Se Supplementation of Livestock Diets

	Year	No fertilizer	Commercial fertilizer[b]	Manure[c]
Before FDA approval	1970	36	20	28
	1971	25	10	25
	Mean	30	15	26
After FDA approval	1986	25	11	24
	1987	40	12	36
	Mean	32	12	30

[a]Corn grown continuously since 1876.
[b]Applied as recommended based on soil test. Limestone applied when soil pH fell below 5.8–5.9.
[c]Most of the manure from beef cattle, but some from dairy cattle and swine. Six tons applied per acre in even-numbered years. Limestone applied when soil pH fell below 5.8–5.9.
Source: Ullrey (1988).

been reported from Michigan State University. Stowe (1992) documented selenium contents of the manure-pit liquid under slotted-floor feedlot pens filled to capacity with beef cattle, dairy cows or heifers, or pigs, and calculated the effects these might have on crops at normal levels of manure application (Tables 3 and 4). From the beef cattle pens, the liquid manure contained an average of 62 μg/kg selenium, and if applied to the soil at a rate of 4000 gallons per acre (37,400 L/ha), which is sufficient to support production of 150 bushels of shelled corn per acre, it would return only 0.9 g of selenium per acre, or 2.3 g/ha. From

TABLE 3   Selenium Concentrations in Manure-Pit Liquid Below Slatted-Floor Pens Filled with Beef Cattle

Sample	Position	Dry matter (%)	Se (μg/L)
1	Top	10.2	59
	Middle	10.4	60
	Bottom	10.6	55
2	Top	10.1	60
	Middle	10.5	63
	Bottom	11.1	67
3	Top	10.4	61
	Middle	10.3	63
	Bottom	10.5	71
	Mean	10.5	62

Source: Stowe (1992).

TABLE 4   Selenium Concentrations in Manure-Pit Liquid Below Slatted-Floor Pens Filled with Swine, Replacement Dairy Heifers, or Lactating Dairy Cows, or in Liquid in a Swine Manure Lagoon (diets supplemented with 0.3 mg/kg Se from sodium selenite)

Species	Sample	Dry matter (%)	Se (μg/L)
Swine	Pit 1	7.3	133
	Pit 2	4.0	108
	Pit 3	2.3	88
	Pit 4	2.3	85
	Lagoon W	0.2	5
	Lagoon E	0.2	6
Dairy	Heifer pit	10.0	88
	Cow pit	6.7	70

Source: Stowe (1992).

the pens of pigs and dairy cows or heifers, the selenium content of the effluent ranged from 70 to 86 $\mu g/kg$. If this were applied to soil at appropriate rates, it would add 1.0 to 1.3 g of selenium per acre (2.5–3.2 g/ha).

It is pertinent that New Zealanders have been applying more than five times as much selenium (10 g/ha) in a governmentally sanctioned program for more than a decade, with no sign of significant accumulations in the soil (Watkinson, 1992). The possibility always exists of overapplication of animal wastes on restricted acreages, but the likelihood of this happening is greatly reduced by current regulations in many areas, which limit allowable levels of manure to control soil nitrogen and phosphorus.

C. Selenium in Animal Feed

Despite its widespread and useful application, the total amount of selenium given directly to animals is small compared with other environmental sources. The annual livestock use of selenium in the United States is about 47.5 tons (Oldfield, 1990), while industrial uses, in electronic and photocopier components, glass manufacturing, and chemical and pigment applications, approximate 700 tons annually. Beyond these direct use figures, some 2650 tons of selenium is released annually from combustion of coal and liquid fuels (Ullrey, 1992), while other anthropogenic sources, including refuse burning, metal smelting, and refining and industrial production (Herring, 1991), bring the total up to 4670 tons. Natural emissions of selenium, both gaseous and particulate, have been estimated at 10,300 tons worldwide—about 1.7 times as much as contributed by human activities. Since the United States is an industrialized country, its activities would be expected to release more selenium into the environment than less-industrial areas, so assuming that natural and industrial emissions would contribute equally in this country, the total annual releases of selenium would be about 9340 tons (Ullrey, 1992). In comparison, if all the selenium incorporated into animal feeds were eventually excreted into the environment, it would only be 47.5 tons, or about 0.5% of that available from other sources—certainly a very small proportion of the total.

There is additional evidence against the likelihood of any dangerous residues from this supplementary use of selenium. New Zealand investigators showed some years ago that the biological availability of selenium administered to ruminant animals was lessened by its transformation to elemental selenium and selenides in the feces (Peterson and Spedding, 1963). A similar diminished availability occurs in monogastrics with the formation of selenonium ion or selenides, which are excreted in the urine (Joblin and Pritchard, 1983; Ganther et al., 1990). It is relevant, too, that manures from animals whose diets were supplemented with copper at many times the level of selenium supplementation have been applied to soils without creating any environmental hazard (CAST, 1996). One may thus feel reasonably assured that in areas of soil deficiency, supplementation of livestock with selenium, at governmentally approved levels will cause no environmental hazards. The long history of successful use of selenium with livestock, in this country and in New Zealand, provides anecdotal confirmation.

D. Selenium in Fertilizer

As experience was gained in the nutritional use of selenium with animals, research continued on the most effective means of administration. Where young animals are routinely handled, and treated with other materials, the addition of selenium is simple and effective. This practice does not work as well in extensive livestock operations with range-grazing animals, which are seldom handled and often receive no supplementary feeds in which the selenium might be mixed. An alternative method, first developed in New Zealand (Watkinson, 1983), is to apply selenium to pasture or range areas in amounts sufficient to raise the selenium content of the herbage to levels compatible with good animal health. In two small countries, Finland and New Zealand, both of which have significant proportions of the landmass deficient in selenium, this method has been practiced extensively— with governmental sanction. Because this technology employs substantially greater amounts of selenium than direct animal supplementation, it is appropriate to examine its impact on the environment.

In New Zealand, application of selenium as a fertilizer amendment has been permitted since 1982. Originally, the form used was sodium selenate, which was applied in granules containing 1% Se, at a rate of 1 kg of granules per hectare. In comparison trials, application of sodium selenite produced significantly higher forage Se levels in ryegrass and clover for 350 days after the tenth annual application than were observed in untreated areas; sodium selenate did likewise for 550 days. In a separate experiment, concentrations of Se in whole wheat and wheat flour adequate for normal health in livestock and humans were obtained by applying sodium selenate, to supply 10 g Se/ha (Stephen et al., 1989). More recently, slow-release fertilizer products containing barium selenate* have been successfully used. The slow-release technology, using barium selenate has been assessed in experiments with sheep in West Australia (Whelan and Barrow, 1994). A single application of the selenized fertilizer (1 kg/ha = 10 g Se and 17.4 g Ba) maintained the animals' plasma Se concentration at adequate levels for 4 years. There was no significant increase in barium in either the pasture forage or the sheep (Whelan, 1993). This slow-release principle, since it reduces the total amount of selenium needed, can be regarded as environmentally desirable. Watkinson (1992) concluded that "there was only a small residual effect, even after

---

*Produced by ICI Crop Care, and marketed as **Selcote Ultra**.

the equivalent of 20 years of topdressing, and that seasonal differences never resulted in excessive initial levels."

While the governmental decision to allow amendment of fertilizers with selenium in New Zealand was made solely for the benefit of livestock production, in Finland the decision represented an added concern for human health. In Finland, all agricultural fertilizers have been supplemented with sodium selenate since 1984 because of concerns about the very low levels of selenium in indigenous foods (Koivistoinen, 1980) and about the low selenium status of the Finnish population and possible implications of this condition for human health (Salonen et al., 1982, 1984).

Initially, two levels of selenium supplementation of fertilizers were implemented in Finland: 16 mg Se/kg in fertilizers used on grain fields and 6 mg Se/kg on those used to produce hay and fodder. The intent was to raise the selenium levels in cereals and in animal products so that the Finnish food intake would approximate the adequate range of 0.05 to 0.20 mg/day. The results were carefully monitored, and in the spring of 1990 the Ministry of Agriculture and Forestry mandated a switch to the single, lower level of 6 mg Se/kg of fertilizer.

The average daily Se intake by Finns in the mid-1970s was 25 μg/day. In 1987, after fertilization at the higher level of selenium supplementation had been practiced for 3 years, intake plateaued at 110 to 120 μg/day. The reduction in Se addition from 16 mg/kg to 6 mg/kg fertilizer caused a drop in dietary selenium intake to about 90 μg/day in 1992, which is considered adequate (Varo, 1993). Extensive analyses have been conducted on a wide range of Finnish foods, after the beginning of use of selenate-supplemented fertilizers, and these have been published (Eurola et al., 1991). It is seldom that such a study, encompassing an entire country and its population, is made available, and we are indebted to the Finnish investigators for their comprehensive data. Although there has been interest in selenium fertilization in other areas, including Denmark (Gissel-Nielsen, 1993), Sweden (Johnsson, 1991), and Australia (Whelan et al., 1994), it has been largely research-oriented and has not resulted, to date, in widespread commercial applications.

IV. SUMMARY

The first, and still the major biological use for selenium is as a dietary supplement for food-producing domestic animals and poultry that would otherwise have insufficient supplies of this essential micronutrient. Such supplementation has not only effectively prevented symptoms of metabolic diseases, including specific myopathies, and degeneration of the liver and pancreas, but has also enhanced growth and reproductive performance in a cost-effective manner. The obvious advantages of such results have made selenium supplementation an attractive animal production technology in areas of natural deficiency worldwide.

Selenium is also well recognized as a toxicant, and regulatory agencies in various countries have carefully monitored its use and have established levels at which supplementary selenium use is authorized. These levels were originally set with the safety of the target animals in mind, but more recently attention has also been paid to the effects of treatment residuals on quality of the environment (CAST, 1996). The concern here has been that with continued usage over extended periods of time, treatment residues might accumulate to levels that are toxic. There is evidence that this has not happened, even in feedlot animals, where animal concentrations produce larger amounts of excreted selenium than is the case of free-ranging, pastured livestock. An alternate method of selenium use, where it is added to fertilizer as an amendment to increase the selenium content of pasture forage or other feed crops, has been found to be environmentally innocuous, as well as nutritionally effective.

Perhaps the most compelling indicator that continued supplementary use of selenium poses no environmental threats is the observation from user countries like Finland and New Zealand that its application over a decade or more has not lowered the need. Animals in these countries still respond positively to added selenium, which would not be expected if any significant environmental buildup had occurred over time. Food animal producers, and the consuming public, can be assured that selenium, at governmentally approved levels, can be profitably administered to food animals and that such use will be environmentally acceptable.

REFERENCES

Andrews, E. D., W. J. Hartley, and A. B. Grant. 1968. Selenium-responsive diseases of animals in New Zealand. *N. Z. Vet. J.* 16:3–17.

CAST (Council for Agricultural Science & Technology). 1996. *Integrated Animal Waste Management.* Ames, IA. 87.

Cheeke, P. R. 1995. Environmental issues in animal production. *Prof. Anim. Sci.* 11:1–8.

Eurola, M. H., P. I. Ekholm, M. E. Ylinen, P. E. Koivistoinen, and P. T. Varo. 1991. Selenium in Finnish foods after beginning the use of selenate-supplemented fertilizers. *J. Sci. Food Agric.* 56:57–70.

Finch, J. M., and R. J. Turner. 1989. Enhancement of ovine lymphocyte responses: A comparison of selenium and vitamin E supplementation. *Vet. Immunol. Immunopathol.* 23:245–256.

Finch, J. M., and R. J. Turner. 1996. Effects of selenium and vitamin E on the immune responses of domestic animals. *Res. Vet. Sci.* 60:97–106.

Franke, K. W. 1934. A new toxicant occurring naturally in certain samples of plant foodstuffs. I. Results obtained in preliminary feeding trials. *J. Nutr.* 8:597–608.

Ganther, H. E., S. Vadhanavikit, and C. Ip. 1990. Selenium methylation, demethylation and biological activity in the rat. *Fed. Am. Soc. Exp. Biol. J.* 4 Part I, A 372 (Abst.).

Gissel-Nielsen, G. 1993. General aspects of selenium fertilization. In Arne Froslie (ed.), *Problems on Selenium in Animal Nutrition, Norwegian Journal of Agricultural Science,* Supplement No. 11, pp. 135–140.

Grattan, S. R., C. Shennan, C. M. May, and R. G. Burau. 1990. Effect of saline drainage water reuse on yield and accumulation of selenium in melon and tomato. In K. K. Tanji (ed.), *Selenium Contents in Animal and Human Food Crops Grown in California,* pp. 81–83. University of California (Davis) Extension Publication No. 3380.

Harris, T. 1986. *Death in the Marsh.* Island Press, Washington DC. 245 pp.

Hegsted, D. M. 1970. Selenium and cancer. *Nutr. Rev.* 28:75.

Herring, J. R. 1991. Selenium geochemistry—A conspectus. *Proceedings of the 1990 Billings Land Reclamation Symposium on Selenium in Arid and Semi-arid Environments,* pp. 7–24. U.S. Geological Survey, Denver.

Joblin, K., and M. W. Pritchard. 1983. Urinary effects of variations in the selenium and sulphur contents of ryegrass from pasture. *Plant Soil* 70:69–76.

Johnsson, L. 1991. Selenium uptake by plants as a function of soil type, organic matter content, and pH. *Plant Soil* 133:57–64.

Koivistoinen, P. (ed.). 1980. Mineral element composition of Finnish foods, I–XII. *Acta Agric. Scand. Suppl.* 22:1–171.

Kuchel, R. E., and R. A. Buckley. 1969. The provision of selenium to sheep by means of heavy pellets. *Aust. J. Agric. Res.* 20:1099–1107.

Madison, T. C. 1860. Sanitary Report—Fort Randall. In R. H. Coolidge (ed.), Statistical Report on the Sickness and Mortality in the Army in the United States. Senate Exchange Doc. 52:37.

Moxon, A. L. 1937. Alkali Disease, or Selenium Poisoning. South Dakota Agric. Exp. Stn. Bull. No. 311. Brookings. 91 pp.

Nelson, A. A., O. G. Fitzhugh, and H. O. Calvery. 1943. Liver tumors following cirrhosis caused by selenium in rats. *Cancer Res.* 3:230.

Nelson, A. O., and R. F. Miller. 1987. Responses to selenium in a range beef herd. *California Agric.* 41:4–5.

Nemec, M., M. Hidiroglou, K. Neilsen, and J. Proulx. 1990. Effect of vitamin E and selenium supplementation on some immune parameters following vaccination against brucellosis in cattle. *J. Anim. Sci.* 68:4303–4309.

Norman, B. B., G. Nader, M. Oliver, R. Delmas, D. Drake, and H. George. 1992. Effects of selenium supplementation in cattle on aquatic ecosystems in northern California. *J. Am. Vet. Med. Assoc.* 201:869–872.

O'Toole, D., and M. F. Raisbeck. 1995. Pathology of experimentally-induced chronic selenosis (alkali disease) in yearling cattle. *J. Vet. Diagn. Invest.* 7:364–373.

Odell, R. T., W. M. Walker, L. V. Boone, and M. G. Oldham. 1982. The Morrow Plots: A century of learning. Illinois Agric. Exp. Stn. Bull. No. 775. Urbana-Champaign. 22 pp.

Ohlendorf, H. M., D. J. Hoffman, M. K. Saiki, and T. W. Aldrich. 1986. Embryonic mortality and abnormalities of aquatic birds: Apparent impacts of selenium from irrigation drainage water. *Sci. Total Environ.* 52:49–63.

Oldfield, J. E. 1987. Contributions of animals to nutrition research with selenium. In J. F. Combs, J. E. Spallholz, O. A. Levander, and J. E. Oldfield (eds.), *Selenium in Biology and Medicine*, pp. 33–46. Van Nostrand Reinhold, New York.

Oldfield, J. E. 1990. *Selenium: Its Uses in Agriculture, Nutrition and Health, and the Environment*. Selenium Tellurium Development Association, Grimbergen, Belgium. 8 pp.

Oldfield, J. E. 1994. Impacts of agricultural uses of selenium on the environment. *Proceedings of the Fifth International Symposium on Uses of Selenium and Tellurium, Brussels*, pp. 81–84. Selenium Tellurium Development Association, Grimbergen, Belgium.

Oldfield, J. E. 1995. The history of selenium as an essential trace element for animals. *Proceedings of a Symposium on Selenium in the Environment, Sacramento, California*, pp. 1–6. School of Veterinary Medicine, University of California, Davis.

Oldfield, J. E., J. R. Schubert, and O. H. Muth. 1960. Selenium and vitamin E as related to growth and white muscle disease in lambs. *Proc. Soc. Exp. Biol. Med.* 103:799–800.

Peterson, P. J., and D. J. Spedding. 1963. The excretion by sheep of Se-75 incorporated in red clover. The chemical nature of the excreted selenium and its uptake by three plant species. *N. Z. J. Agric. Res.* 6:13–22.

Salonen, J. T., G. Alfthan, J. K. Huttunen, J. Pikkarainon, and P. Puska. 1982. Association between cardiovascular death and myocardial infarction and serum selenium, in a matched-pair longitudinal study. *Lancet* 2:175–179.

Salonen, J. T., G. Alfthan, J. K. Huttunen, and P. Puska. 1984. Association between serum selenium and risk of cancer. *Am. J. Epidemiol.* 120:342–349.

Schwarz, K., and C. M. Foltz. 1957. Selenium as an integral part of factor 3 against dietary, necrotic liver degeneration. *J. Am. Chem. Soc.* 79:3292–3293.

Smith, K. L., J. H. Harrison, D. D. Hancock, D. A. Todhunter, and H. R. Conrad. 1984. Effect of vitamin E and selenium supplementation on incidence of clinical mastitis and duration of clinical symptoms. *J. Dairy Sci.* 67:1293–1301.

Stephen, P. C., L. J. Seville, and J. H. Watkinson. 1989. The effects of sodium selenate applications on growth and selenium concentrations in wheat. *N. Z. J. Crop Hortic. Sci.* 17:229–237.

Stowe, H. D. 1992. In Risks and Benefits of Selenium in Agriculture. Issue Paper No. 3, Supplement, pp. 14–16. Council for Agricultural Science and Technology, Ames, IA.

Telfer, S. B., G. Zervas, and P. Knott. 1983. U.K. Patent Application 2116424 A.

Ullrey, D. E. 1992. Basis for regulation of selenium supplements in animal diets. *J. Anim Sci.* 70:3922–3927.

Ullrey, D. E. 1988. Wildlife as indicators of enzootic selenium deficiency. Pp. 155–159. In L. S. Hurley, C. L. Keen, B. Lonnerdal, and R. B. Rucker (Eds.). *Trace Elements in Man and Animals 6*. Plenum Press, New York.

U.S. Food and Drug Administration (FDA). 1973. Draft environmental impact statement: Rule making on selenium in animal feeds. FDA, Rockville, MD.

Varo, P. 1993. Selenium fertilization in Finland: Selenium content in feed and foods. *Norw. J. Agric. Sci. Suppl.* 11:151–158.

Watkinson, J. H. 1983. Prevention of selenium deficiency in grazing animals by annual topdressing of pasture with sodium selenate. *N. Z. Vet. J.* 31:78–85.

Watkinson, J. H. 1992. Personal communication to the author, 24 January. Ministry of Agriculture and Fisheries, Ruakura Agricultural Research Center, Hamilton, NZ.

Whelan, B. R. 1993. Effect of barium selenate fertilizer on the concentration of barium in pasture and sheep tissues. *J. Agric. Food Chem.* 41:768–770.

Whelan, B. R., and N. J. Barrow. 1994. Slow-release selenium fertilizers to correct selenium deficiency in sheep in West Australia. *Fert. Res.* 39:183–188.

Whelan, B. R., D. W. Peter, and N. J. Barrow. 1994. Selenium fertilizers for pastures graded by sheep. I. Selenium concentrations in whole blood and plasma. *Aust. J. Agric. Res.* 45:863–875.

Wolf, E., V. Kollonitsch, and C. H. Kline. 1963. A survey of selenium treatment in livestock production. *J. Agric. Food Chem.* 11:355–360.

Zebarth, B. J. 1996. Waste management impacts on groundwater issues and options in the lower Fraser Valley. *Proceedings of the 31st Annual Pacific Northwest Animal Nutrition Conference, Seattle*, pp. 65–76.

9

Supplementation of Selenium in the Diets of Domestic Animals

Ivan T. Kishchak
Nikolayev Agricultural Institute, Nikolayev, Ukraine

I. DISTRIBUTION OF SELENIUM IN ANIMALS AND PLANTS

A nutritionally balanced diet is important in obtaining highly productive domestic animals. Feed additives may be used to provide optimum amounts of mineral elements. Selenium (Se) is in the group of mineral elements of vital necessity, since it, together with vitamin E, is indispensable for preserving animals' health and maximizing their productivity.

Selenium is widespread throughout the biosphere and often associated with natural sulfur compounds. The average Se content of the earth's crust is 1×10^{-5} %, in soils it is 10^{-6} %, and in freshwater, 10^{-9}%. Trace amounts also occur in air and rainwater.

Selenium is distributed unequally in the bodies of domestic animals: in general, muscle tissue has 50 to 52% of the total; skin, wool, horns, and horn formations, 14 to 15%; skeleton, 10%; liver, 8%; and the rest of the tissues, 15 to 18%. The concentration of Se in tissue increases with increased levels of Se in an animal's diet. In the whole blood of animals of different species, Se content ranges from 5 to 100 $\mu g/100$ mL. Its concentration in erythrocytes is twofold higher than in plasma in which Se is bound with albumin γ- and β-globin functions. The transport function of Se is performed by the albumin fraction. Rumen bacteria are Se concentrators in ruminants, and their quantity is sometimes two- to thirtyfold higher than that contained in the diet.

Selenium is widely distributed in the plant kingdom. Its quantity depends on the ability of plants to accumulate Se into their biomass and on the degree of its fixation in soils. In reality, Se absorption by plants does not depend on its total concentration in soils. Plants absorb Se easily from alkaline soils, where it often exists in water-soluble forms. In such regions, acute (blind staggers) and chronic (alkali disease) poisoning of animals has been observed when Se occurs in the diet at the rate of 10 to 20 mg/per kilogram of dry matter (Bereshtein, 1966; Samohin, 1981; Chumachenko et al., 1989). Although acid soils may contain high Se concentrations, plants assimilate only small amounts because the Se is bound by insoluble iron compounds. White muscle disease may occur in calves and lambs pastured in acid soil areas as a result of Se deficiency (<0.1 mg/kg dry matter) in the forage.

Low concentrations of Se are found in dernopodzolic soils, and higher concentrations are found in peat and mountain soils. Selenium concentrations in soils tend to diminish from February to April in the latter regions. Soils rich in sulfate retard the assimilation of Se by plants.

Plants are divided into three groups according to their ability to accumulate Se: grasses that are poor in Se accumulation (<5 mg/kg), grain crops capable of accumulating Se (5–30 mg/kg), and leguminous or cruciferous plants containing Se at more than 1000 mg/kg. The Se content in plants depends on many factors, including the season of year, and on the method of harvesting and preserving. Feeds not harvested at the optimum time and preserved in open air or in rainy weather often contain less Se. Typically the Se content of most plants ranges from 0.06 to 1 mg/kg of air-dried matter.

Selenium deficiency in animals often leads to E-avitaminosis and the emergence of other diseases. A correlation has been established between white muscle disease in animals and low Se content in plants. This Se deficiency is accompanied by overloading (more intensive use) of some organs and muscle tissue (heart), skeletal degeneration, toxic dystrophy of the liver in pigs and poultry (hens), liver necrosis in cattle, exudative diathesis in poultry, and crazy chick disease and other pathologues in chickens. Lambs, calves, pigs, and poultry are the most susceptible to these conditions. The mortality in younger animals can reach as high as 60%. The activity of glutathione peroxidase is considered to be the basic criterion for estimating Se availability to animals, since this enzyme rises abruptly when Se increases in the blood. The optimum activity of glutathione peroxidase in tissues is observed at a dietary Se level of 0.12 mg/kg dry matter (Kalnitsky, 1985).

Modern intensive use of genetic engineering and feed supplementation of domestic animals leads to optimum loading of their body tissue. To understand the impacts of Se on domestic animals, it is necessary to study and understand the complex interrelationships of regional conditions; the ecological status of water, plants, and feeds; the biological features of the animals; and the nature and effects of selenium itself.

Animals assimilate Se mainly from feeds. To provide the animals with Se and for prophylaxis, Se may be included in feed preparations, in mineral mixes and intraruminal boluses. Treatment of pastures with Se fertilization, use of Se additives in water, and injections of Se and vitamin E are other ways of delivering Se to animals.

The Se content in various feeds may range from 0.023 mg/kg in maize ears to 2.663 mg/kg in dried egg albumin. The highest concentrations of Se are found in hay, mixed feeds, and wheat straw (0.164–0.283 mg/kg of air-dried matter) and in sour chips or maize silage (0.028–0.077 mg/kg).

Analysis of feeds given to animals in southern Ukraine indicate that the Se content of dry matter varies between 0.045 and 0.248 mg/kg Se (Table 1). The minimum Se requirement for pigs is 0.1 to 0.5 mg/kg of Se dry matter. Table 1 compares the Se content in separate feeds used for agricultural animal feeding in southern Ukraine and northern England.

Selenium in animal organs and tissues is closely correlated with its level and availability in feeds. When Se occurs in feeds at adequate levels and is highly available, the Se concentration may range from 5 to 8 μg/100 mL in blood and

TABLE 1   Selenium Content in Feed[a]

Feeds	Northern England	Southern Ukraine
Barley	0.032(49)	0.095(5)
Oats	0.030(15)	0.084(3)
Wheat	0.035(5)	0.045(5)
Maize	0.042(6)	0.068(4)
Wheat bran	1.093(1)	0.088(2)
Sunflower oilseed meal	—	0.135(2)
Soybean oilseed meal	0.153(13)	—
Sunflower seed meal	—	0.150(3)
Cottonseed meal	0.360(3)	—
Bacterial protein	0.389(8)	0.77(3)
Feed yeast	—	0.112(3)
Sugar beet pulp, dried	0.304(11)	0.073(3)
Meat and bone meal	0.356(6)	0.248(2)
Dried fat-free milk	0.109(3)	0.113(3)
Mangel (wurzel)	0.035(3)	0.056(5)
Alfalfa pellets	0.075(1)	0.072(2)
Cereal straw	0.068(8)	0.059(7)
Mixed feeds for cattle	—	0.078(14)
Running water (mg/L)	—	0.0035(5)
Water (snow) (mg/L)	—	0.023(5)

[a]Unless otherwise noted, units are milligrams per kilogram dry matter; quantity of analyzed samples given in parentheses.

from 12 to 16 μg/100 g in the liver. Selenium concentrations of 0.1 to 0.8 and 3 to 4 μg/100 g in the blood and liver, respectively, may indicate deficiencies, which may in turn lead to white muscle disease.

II. THE EFFECT OF SELENIUM ON DOMESTIC ANIMALS

Selenium's biological effect is similar to that of vitamin E. Selenium is 250 times more active than the amino acid L-cystine. It is established that though Se and vitamin E perform separate functions, they act synergistically, supplementing each other in the prevention of metabolic damage. It is well known that Se affects protein metabolism in animals, especially the sulfur-containing amino acids, impacting the tissue aspiration and the immunological reactivity of the organism. During activation of sulfur-containing amino acids observed in a number of pathological processes, one Se atom is sufficient to make 350,000 molecules of sulfur-containing amino acids biologically active.

Selenium also functions as a catalyst. At low doses it activates nonspecific phosphatases, ATPase, and pyruvate decarboxylation by means of catalytic oxidation of lipoic acid and thio groups, and raises total Se activity and the activity of α-ketoglutarate oxidase.

High doses of Se inhibit ATPase and succinate dehydrogenase activity, thus leading to inhibition of carbohydrate metabolism. This property explains to some extent the toxic effect of Se, which is accompanied by decreasing serum aspartate amino transferase, tissue respiration rate, and the immunological activity of the organism.

Selenium regulates assimilation and losses of vitamins A, C, and E and influences the rate of oxidation–reduction reactions. Selenium is known to decrease adrenaline activity, thus reinforcing insulin activity and promoting glycogen accumulation in muscle tissue, as well as retarding aerobic oxidation. Selenium is a factor 3 of the reduced glutathione phosphates that prevent the development of dystrophic processes. Selenium deficiency leads to the accumulation of peroxide compounds and to the development of degenerative changes in tissues.

The biological activity of Se and its impact on various functions of the organism depend to a great extent on its interaction with sulfur in tissues. Chemically, Se is similar to sulfur, but it is more active and toxic than sulfur. Sulfur neutralizes Se toxic effects to some extent, but Se is able to substitute for sulfur in specific proteins. Arsenic is antagonistic to Se, and its compounds effectively inhibit the toxic impact of Se.

The duodenum absorbs Se slowly, and absorbed Se also accumulates in the liver and kidneys. The presence of calcium, cadmium, arsenic, cobalt, lead, tellurium, silver, or sulfur can diminish Se assimilation in the organism.

Selenium in whole grain is an antioxidant and permits the organism to inhibit the chemical activity of free radicals that participate in the destruction of living cells. Free radicals attack the cell membranes, and when they are able to penetrate the cells, they can disrupt the calcium balance, leading to a number of problems such as destruction of the genetic information coding in the cells. Such injuries can lead to the synthesis of proteins capable of resulting in serious misfunctions of the organism. Antioxidants including Se and vitamins A, C, and E can protect living cells by aiding certain enzymes in the regulation of free electrons that are converted into free radicals, thus controlling the amounts of these substances produced.

The antioxidant function of Se is based on the partial interchangeability of vitamin E in the prophylaxis of some diseases and is explained by the element being a constituent of the enzyme glutathione peroxidase, which helps vitamin E prevent cell injuries.

Selenium acts as an inhibitor to proteolytic enzymes, phosphatization, and tissue breathing, and it activates succinic dehydrogenase in all tissues, especially in the liver. The complex effect of Se and vitamin E activates the immune system of the organism and stimulates both specific and nonspecific responses.

Selenium provides the organism with protective means against toxic elements such as mercury and cadmium. Under the action of mercury, the immunity of the organism decreases because mercury diminishes the quantity of white blood cells, which kill foreign bodies in the organism. Cadmium is a cumulative toxic substance that promotes hypertension. It accumulates in the kidneys and inhibits the immune system. Possible effects of excess cadmium include anemia, compromised zinc metabolism, lung diseases, and diminished T-cell counts in the organism. Total immunity is decreased because organs (liver, kidneys) that are vital for the immune system are impaired as a result of the dimination of the T-cell counts.

Figure 1 illustrates the wide variation of Se impacts on growth and development and highlights the necessity for conducting further studies on the role of Se on domestic agricultural animals. Both positive and negative impacts of Se are shown.

III. BIOLOGICAL AVAILABILITY OF SELENIUM FOR FARM ANIMALS

Selenium availability from milk, some vegetable feeds, organic compounds, selenate, and $HSeO_3$ is rather high, but it varies from low to moderate for a majority of plant feeds. The more reduced inorganic forms such as elemental selenium are of low solubility and relatively unavailable to animals.

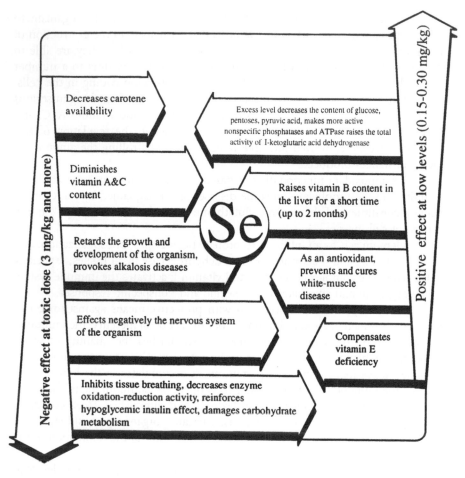

FIGURE 1   Positive and negative impacts of Se on domestic animals.

The total concentration of a mineral does not reveal the impact of feeds and feed additives as sources of macronutrients and trace minerals, since only the bioavailable parts can be absorbed and changed into metabolically active forms. Biological availability of Se relates to the effectiveness of its assimilation from different sources by farm animals, and their use of the substance in various physiological states.

The alimentary canal, the respiratory tract, and the skin are potential sites of Se absorption from the environment. Selenium consumed with animal feeds and feed additives is quickly absorbed both by ruminant and monogastric animals. The low part of small intestine of swine and sheep is the place of peak Se absorption, and endogenous secretion takes place in the duodenum. There is practically no absorption from the stomach or abomasum. Absorption of Se is higher in monogastric animals than in ruminants, in which some Se is converted to insoluble forms in the rumen.

Selenite is absorbed through brush fringe membranes of the small intestine, and selenate is absorbed in its middle and distal part of the intestine at the expense of active transport mechanisms. Selenate absorption is depressed by inorganic ions (sulfate, thiosulfate, molybdate, chromate) and organic anions (oxalate, oxaloacetate), relative to its physical and chemical properties. Cysteine and glutathione stimulate Se absorption from selenite but not from selenate, and inhibit absorption of methionine and its analogs.

Amino acids containing Se and their corresponding sulfur analogs (cysteine, methionine) have common mechanisms of absorption In ruminants, selenate bioavailability is higher than that of selenite. In poultry selenate bioavailability is lower than that of selenite.

Efficiency of Se use for exudative diathesic prophylaxis in chickens, expressed as percentage of selenite (Na_2SeO_3), was as follows: from selenate, 58 to 90; seleno-D-cystine, 68 to 78; seleno-DL-methionine, 18 to 61; seleno-DL-ethionine, 44; sodium selenide, 42; elemental selenium, 8; alfalfa meal, 210; brewer's yeast, 89; corn, 86; cottonseed meal, 86; brewer's grain, 80; wheat, 71; dried grain distillate, 68; soybean meal, 60; milk, >100; herring meal, 25; tuna meal, 22; poultry residue meal, 18; American herring meal, 18; and fish extract, 9 (Kuznetsov, 1992; Kishchak, 1995). The bioavailability of the above-mentioned chemical compounds for turkey poults is similar.

In experiments on nonpregnant cows and ewes, Se bioavailability from organic compounds such as selenomethionine, yeast products, and high Se wheat was higher than from sodium selenite (National Academy of Science, 1989; Spain, 1994).

Upon acid hydrolysis of fish meal or at its lyophilization of fish meal, bioavailability of Se is increased. Researchers connect low bioavailability of Se with feeds (except milk) from animal origin as a result of the formation of complex compounds with purine bases, mercury, and other elements (Spain, 1993).

Selenium bioavailability depends to a great extent on the chosen criterion that is related to characteristics of the element and its sources. Selenium bioavailability estimated by chicken survival rate and blood glutathione peroxidase activity, respectively (in percent compared to sodium selenite), was as follows: from baked bread, 80 and 62; from sodden bread, 40 and 51; raw beef, 58 and 73; roast beef, 35 and 60.

Selenium bioavailability from sodium selenite and calcium selenite estimated by the microelement content in tissues is the same both for chickens and lambs. Judging by blood Se content, the efficiency of the element's use in sheep from different compounds is arranged in the following decreasing order: $HSeO_3$ > sodium selenate > sodium selenite > cadmium selenide. Selenious acids also had a positive influence on live weight gains and wool production.

Feeding of soluble-glass boluses containing 14% of copper, 0.63% of cobalt, and 0.265% Se to pregnant ewes or Se soluble-glass boluses to pregnant cows increased Se to ewes, cows, and their offspring (Dyachenko, 1988; Kishchak, 1994).

IV. SELENIUM IN FEEDING PIGS

Until recently in the former USSR, Se was considered to be toxic only for farm animals. Plants containing even low concentrations of Se were looked upon as potentially toxic. In reference books on agricultural animal feeding in Ukraine, Se was not discussed as an essential nutrient. However, in the most recent edition of reference book *Detailed Rates of Agricultural Animals Feeding* (Nozdrin et al., 1991), Se is discussed as an essential nutrient for young cattle and sheep in Ukraine.

Selenium demand in ruminants has been well studied. Rickaby (1981) indicates that it is necessary to study the problem of Se demand in pigs, specifically the mechanism of interaction between D-α-tocopherol and Se and their effects on pigs.

Selenium deficiency develops in pigs fed musty grains containing Se in quantities less than 0.1 mg/kg dry matter. The optimum Se level in pigs diet is 0.1 to 0.5 mg/kg of dry matter, but higher levels of Se are toxic.

The supplementation of Se into the diet of pigs that previously had a diet low in Se promoted their recovery from white muscle disease. In Sweden, where white muscle disease occurs rather frequently in pigs, 50 mg of D-α-tocopherol or 2 mg of sodium selenite are added to the diet of pigs to prevent this disease. In many cases, Se is more effective than vitamin E in the prevention of white muscle disease. When Se is added to the feed mixture in an amount of 0.5 mg/kg of feed, one often observes improved reproductive function of sows, increased birthrate of viable piglets, and increased weight before weaning.

It has been established that supplementation works best when Se is added to feeds in combination with vitamin E. Introduction of 1 kg of vitamin E and 10 g of sodium selenite dissolved in distilled water in one ton of premixed KS-1 (a complex mixture of vitamins and microelements used in feed for sows and adult pigs) promotes improvement of breeding boars, sperm production, and impregating property of the sperm. The best effect on morphological and biochemical blood indexes of piglets was reached with the addition of vitamin E in the amount 0.5 mg in combination with 0.15 mg/kg Se dry matter.

The growth-promoting effect of sodium selenite in the amount of 0.03 to 1 mg/kg of the animal's live weight allowed an increase of 15% average daily gain of a test group of piglets during a 2-month fattening period as compared to a control group.

When Se is added to milking sows' diet, its content in their milk is raised, but piglets do not receive a sufficient amount of Se from this source to satisfy their requirements. Breeding of pigs on a diet containing Se at the rate 0.14 mg/kg of feed is sufficient to prevent selenium deficiency, provided 4.5 mg of D-α-tocopherol is added to 1 kg of feed.

It has been established that sodium selenite has a less positive effect on live weight gain when added to the diet of feeder pigs in the southern Ukraine.

Intramuscular injection to pregnant swine of barium selenate suspension in the dose of 0.5 to 1.0 mg of Se per kilogram of live weight increased Se concentrations significantly in newborn piglets. Intramuscular injection of selenite to suckling pigs increased their resistance to alimentary tract diseases. Application of sodium selenite by spray in a dose of 0.4 mg/kg of live weight to piglets increased nonspecific resistance of the organisms.

We examined the performance index of feeder pig stock whose diet included a preparation of sodium selenite and Nutril-Se [vitamins and selenoamino acids added in feed of farm animals produced in Slovenia (former Yugoslavia)]. There was a positive Se impact on maximum metabolism stabilization. The feeds promoted an average daily gain of stocker pigs of 14% for the group that received Nutril-Se in feeds. The growth-promoting Se impact was due to both nutrients and biologically active substances that were constituents of the preparations.

It is evident that further research is needed in this direction because of the limited amount of the published data on the mechanisms of Se effects separately and the complex interaction of Se, vitamin E, and other compounds on agricultural animals.

REFERENCES

Bereshtein, F. 1966. *Microelements in Physiology and Pathology of Animals*. Urozhay, Minsk.

Chumachenko, J., S. Stoyanovsky, and P. Lagodvk. 1989. *Reference Book in Using Biologically Active Matter in Live-stock Breeding*. Urozhay, Kiev.

Dyachenko, L. S. 1988. Effect of different selenium sources in ration on productivity and metabolism in yearling ewes. *Sci. Tech. Biol. Ukr. Sci. Res. Inst. Animal Husb. Steppe Regions*. Kherson 1:45–50.

Kalnitsky, B. 1985. *Minerals in Animal Feeding*. Agropromizent, Leningrad.

Kishchak, I. T. 1994. Biological availability of selenium for farm animals. In *Proceedings of the Fifth International Symposium on Uses of Selenium and Tellurium*, pp. 357–358. Brussels.

Kishchak, I. T. 1995. *Production and Use of Premixes*. Kiev, Urozhai.

Kuznetsov, S. G. 1992 *Bioavailability of Minerals for Animals*. VNNIITTRagroprom (Review information), Moscow, pp. 41–43.

National Academy of Science. 1989. *Nutrient Requirements and Signs of Deficiency. Nutrient Requirements of Dairy Cattle*, 6th rev. ed. National Academy Press, Washington, D C.

Nozdrin, M. T., M. M. Karpus, V. F. Kirovaschenko, and others. 1991. *Detailed Rates of Agricultural Animals Feeding*. Urazhai, Kiev.

Rickaby, D. C. 1981. Se demand in ruminants. The latest achievements in the investigation of animals feeding. Ministry of Agriculture, Fisheries and Food. Ken. for Bar. Newcastle upon Tyne. 3:207.

Samohin, V. 1981. *Prophylaxis of Violation of Metabolism of Microelements in Animals*. Kolos, Moscow.

Spain, J. 1993. Trace mineral supplementation strategies for lactating dairy cow diets. In *Proceedings Tune Tuning Rations: Is the Benefit Worth the Cost?*, pp. 26–32. University of Missouri Cow College.

Spain, J. 1994. Nutrition and immune function: Defining the relationship. In *Proceedings on Successfully Meeting the Challenge: Focus on Forages, Feeding, and Health*. University of Missouri Cow College.

10

Mass Balance Approach to Selenium Cycling Through the San Joaquin Valley: From Source to River to Bay

THERESA S. PRESSER and DAVID Z. PIPER
United States Geological Survey, Menlo Park, California

I. INTRODUCTION

Surface and ground waters of the Central Valley of California (e.g., rivers, dams, off-stream storage reservoirs, pumping facilities, irrigation and drinking water supply canals, agricultural drainage canals) are part of a hydrologic system that makes up a complex ecosystem extending from the riparian wetlands of the Sacramento and San Joaquin Rivers through the San Francisco Bay/Delta Estuary to the Pacific Ocean (Fig. 1). Water quality concerns center on elevated selenium (Se) and salt concentrations in irrigation drainage water discharged into the waterways of the relatively arid San Joaquin Valley (SJV), including the San Joaquin River (SJR). These waters are made unique by dissolved Se, weathered from marine sedimentary rocks of the Coast Ranges to the west, being ultimately concentrated to toxic levels in aquatic wildlife in the wetlands of the SJV/SJR trough (Figs. 1 and 2) (Presser and Ohlendorf, 1987; Presser, 1994). Scientific and environmental concerns focus on the bioreactive properties of Se and its partitioning among biota, water, and sediment, and on whether simple dilution models can be applied to an element that bioaccumulates. Because of state and federal commitments to provide water for irrigation, as well as drainage of irrigation wastewater by the year 2000 drainage from over 180,000 ha of seleniferous, salinized farmland within the western SJV will create approximately 387 million cubic meters of potentially toxic drainage water annually (i.e., "problem water," as defined by the San Joaquin Valley Drainage Program, 1990), thus lending urgency to an understanding of the biogeochemistry of Se in this environment.

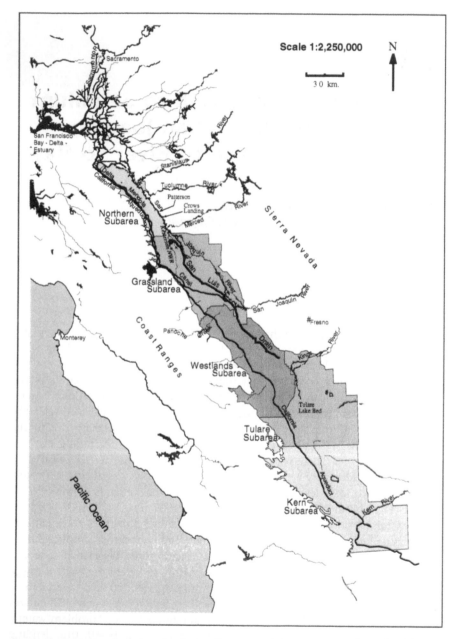

FIGURE 1   The San Francisco Bay/Delta Estuary and San Joaquin Valley of California, showing the major waterways and the five agricultural drainage management subareas (Northern, Grassland, Westlands, Tulare, and Kern).

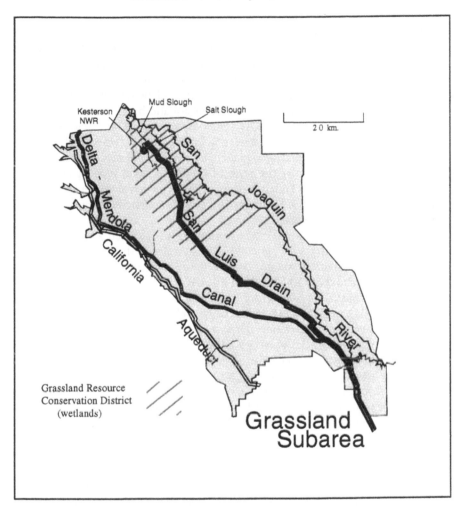

Figure 2 The Grasslands drainage management subarea showing the San Joaquin River, riparian wetlands, and drainage system, including the San Luis Drain.

Commencing in 1968, a 137 km section of the San Luis Drain (SLD) was built to Kesterson National Wildlife Refuge, in what was then seen as the first step of a plan to transport agricultural drainage to the San Francisco Bay/Delta Estuary (Fig. 1). To minimize the cost of agricultural drainage disposal in this interim project (i.e., prior to completion of the drain), wastewater that was drained from surface and subsurface agricultural lands and transported by drainage canals, including the SLD, was used as water supplies for aquatic wildlife habitat on the Pacific Flyway in the SJV (San Joaquin Valley Interagency Drainage Program,

1979). However, questions were raised beginning in 1983 about the practice of disposal of irrigation drainage in wetlands, owing to the appearance of deformities among embryos and hatchlings of waterfowl at Kesterson National Wildlife Refuge (Ohlendorf et al., 1986), the main recipient of SLD wastewater (Fig. 2). Selenium was found to bioaccumulate in the food chain and hence, in the proteins of higher level species, causing teratogenicity in aquatic birds. In 1985, the U.S. Department of the Interior (USDOI) shut down the SLD. In 1988 Kesterson Reservoir, representing approximately one-fourth of the refuge and its most contaminated part, was filled and graded by the U.S. Bureau of Reclamation (USBR, 1986), as ordered by the state of California, as a type of what some consider "on-site disposal." Today, Kesterson Reservoir remains a terrestrial habitat monitored for return of Se to the surface, Se accumulation in plants and biota, and Se seepage to groundwater.

The SJR, the only natural outlet from the SJV, has been and continues to be used as a de facto drain for agricultural wastewater to the San Francisco Bay/ Delta Estuary distribution system, despite the closure of the SLD. The SJR is California's largest river south of the San Francisco Bay/Delta Estuary. Its headwaters are in the Sierra Nevada; in its lower reaches, the river flows northward through the SJV trough for 110 miles to the Delta (Fig. 1). Currently, 98% of the water is diverted for irrigation purposes. This leaves the lower reach of the river at certain times of the year as a "dry reach" that is totally dependent on disposal of agricultural drainage for its flow. This usage of the SJR water is the subject of a lawsuit filed in 1988 by The Natural Resources Defense Council in the U.S. District Court in Sacramento, of recent complaints (Smith, 1995; Bard, 1996) to the California State Water Resources Control Board (CSWRCB), and of congressional legislation (The Central Valley Project Improvement Act of 1992) because of purported lack of protection (1) of fish, wildlife, and habitat of the river, (2) of the river as a public trust resource, (3) of endangered species using the river, and (4) of water quality (due to disposal of agricultural drainage). Its riparian wetlands contain the largest tract of natural wetlands remaining in the SJV, the Grassland Resource Conservation District (Fig. 2). Nearly 95% of wetland habitat in the SJV has been lost. Historically, varying amounts of drainage have been used to supplement contracted water supplies in these wetlands tributary to the SJR, and varying lengths of the complex channel system within the wetlands have been employed to convey agricultural drainage to the river.

The SJR has been monitored for Se concentrations since 1985, mainly by the California Central Valley Regional Water Quality Control Board (CCVRWQCB, 1996a, 1996b). In 1991 and 1992 the state acknowledged elevated levels of Se in the SJR and parts of the San Francisco Bay/Delta Estuary. It declared the lower reach of river "impaired" (CCVRWQCB, 1991) and the Se levels in the bay "of concern" (CSFBRWQCB, 1992). The major source of Se to the river is agricultural drainage; sources to the San Francisco Bay/Delta Estuary include both agricultural drainage under certain flow conditions and oil refinery waste that originates from oil refineries around the bay. Flow within the SJR is complex owing to climatic variability (i.e., alternating drought and flood conditions) and agricultural management variability (i.e., preirrigation, irrigation, drainage). The juncture of the SJR and the estuary is further complicated by the need to (1) maintain a balance between water supply and demand that permits achievement of salinity standards in the estuary and (2) recycle a certain percentage of the SJR back to the south via supply canals (e.g., the Delta Mendota Canal, Fig. 2) for irrigation purposes.

Since the closure of Kesterson Reservoir, management and treatment options have been researched and debated for disposal of the potentially toxic soil leachate mobilized by irrigation. With finite water supplies, however, environmental effects and benefit/cost ratios have prevented the achievement of an effective balance between maintaining agricultural production and supporting environmental resources. Legal challenges to the use and/or disposal of agricultural drainage and regulatory efforts at the federal and state levels generally come under the Clean Water Act (1972), including the National Pollutant Discharge Elimination System permit program, as well as under the Endangered Species Act (1973), the National Environmental Policy Act (1969), the Resource Conservation and Recovery Act (1988), the Porter–Cologne Water Quality Act (California State Water Resources Control Board, 1969), the California Toxic Pits Cleanup Act (California Health and Safety Code, 1984), the California Environmental Quality Act (California Public Resources Code, 1970), and the Public Trust Doctrine (Dunning, 1980). The Migratory Bird Treaty Act of 1918 (Margolin, 1979), which might address concerns about bird deaths at Kesterson Reservoir, has not been the subject of a specific legal action. An exhaustive review of enacted statutes and legal actions is beyond the scope of this chapter. Particular concerns, however, have centered on (1) protection of aquifers, human health, and aquatic life, (2) definition and disposal of hazardous waste, (3) definition of toxicity on a functional basis, (4) evidence of risk and potential harm, (5) application of a dilution credit for bioaccumulable substances, (6) adoption of waste discharge requirements, and (7) exemption of agricultural drains and waters from regulation. Regulation of agricultural drainage is hampered by the lack of authority on the part of the U.S. Environmental Protection Agency (USEPA) to enforce water quality standards for either point source or non–point source pollution from irrigated agriculture. This is because agricultural runoff (irrigation return flow) has been exempted from point source regulation through the Clean Water Act, and adherence to federal non–point source regulation programs is voluntary (Environmental Defense Fund, 1994).

A cooperative project among agricultural, government, and environmental parties enacted by the USBR (1995) reopened the SLD on an interim 5-year basis to transport drainage to the SJR. Renamed the Grassland Bypass Channel Project, the project focuses on the Grassland drainage management subarea of the western SJV (Figs. 1 and 2). This project is a regional management effort to improve water quality by regulating Se loads, to protect wildlife habitat by assuring the wetlands of an adequate supply of clean water through removal of drainage from the wetlands and, at the same time, to examine possible adverse effects that may result from this more direct routing of drainage through the SLD and Mud Slough to the SJR (Fig. 2). As we discuss below, drainwater entering the river will now be deprived of the effects of dilution by wetland flows and loss of Se to sediment and biota due to bioaccumulation.

The Grassland Bypass Channel Project proposes the following goals: regulation of Se based on total loads (rather than water quality standards), commitment to meet and further define environmental concerns for wetlands and the SJR, creation of a regional drainage entity to assign responsibility for pollution, agreement on monetary penalties for exceedence of loads, and development of a long-term management strategy to achieve water quality objectives. Although load targets have been developed mainly through a simple averaging of data from 1986 to 1994, as a requirement for reuse of the SLD, conditions of the SJR "will not worsen" over yet-to-be-defined historical input loads (USBR, 1995). Recent adoption by the state of a water quality objective of less than 2 micrograms per liter (μg/L) Se for the Grassland wetland channels (CCVRWQCB, 1996c), has essentially removed these channels as alternative flow paths for drainage water (approximate average 62 μg/L Se). This regulation will make it difficult to reuse the wetland channels, for example, during flood runoff events as a means of circumventing project objectives, even if loads to the SLD are exceeded.

A decision by the U.S. District Court in Fresno, California, announced on December 16, 1994, also addresses agricultural drainage relief through use of the SLD. The court order calls for the USBR to explore the possibility of obtaining a state permit to discharge agricultural drainage directly into the San Francisco Bay/Delta Estuary (Fig. 1) by extending the SLD to the bay as originally designed in 1968. The decision is under appeal. During the 1950s, agriculturalists and water purveyors proposed a master drain to the San Francisco Bay, and this proposal was enacted as part of the federal San Luis Act in 1960. More recently, the state (CCVRWQCB, 1994a, 1995; CSWRCB, 1995) has emphasized that an extension of an export drain is the only "technically feasible solution" for management of agricultural drainage in the western SJV, although no recipient of the drainage has been officially selected. Some proponents envision expansion of the drainage facilities to include other types of discharge (e.g., municipal and industrial). If these out-of-valley disposal plans are enacted, Se-enriched agricultural drainage from Grassland, Westlands, and Tulare drainage management subareas (Fig. 1) could be collected in the future in a valley-wide agricultural drain, transferring the waste load from the SJV directly to the San Francisco Bay/Delta Estuary or the Pacific Ocean.

This chapter presents a retrospective view of Kesterson Reservoir as it applies to a mass balance approach for current and future monitoring of Se loadings in the SJV due to agricultural drainage. Table 1 provides Se concentrations and loads in the western SJV where man-induced perturbations have enhanced or diminished natural levels. These concentrations are compared to primary source, background, and concern levels in soil, sediment, water, and wildlife. We discuss the status of drainage disposal after 10 years of management, regulation, and legal challenges in two drainage management areas of the SJV, Grassland, and Tulare subareas (Fig. 1). Our main focus, however, is on the reopened SLD in the Grassland subarea and the relation of the Se load from this subarea to the SJR (Fig. 2). The areas from which data are reported represent essentially two types of environment, differentiated by flow. The described Kesterson and Tulare systems, consisting of basin evaporation ponds, are static regimes. The Grassland system, consisting of many miles of wetland channels in addition to Mud and Salt Sloughs, is a flow-through regime. The San Francisco Bay/Delta Estuary may be an example of a system that is intermediate in terms of flow. This estuary system now has a clear and significant Se load due to oil refinery waste and some agricultural waste and may have an additional Se load due to extension of the SLD.

Whether the detrimental effects of Se observed in closed systems will develop in open or intermediate systems remains to be proven for different Se pathways (i.e., SJR or SLD) and repositories (bay or ocean). Our mass balance calculations from the small amount of available data, however, suggest long-term Se mass loading in the SJR and in the highly productive San Francisco Bay/Delta Estuary. Monitoring protocols must be developed to evaluate the exact magnitude and rate of loss/gain of Se from the water column and the precise mechanisms that drive Se's pathway from source rock through biota to its accumulation in bed sediment. As opposed to historical data, data collected during current (e.g., Grassland Bypass Channel Project) and future projects may allow improvement in these types of mass balance calculations.

II. DRAINAGE SUBAREAS

Five drainage management subareas (Fig. 1) have been designated by the joint federal–state San Joaquin Valley Drainage Program (1990) that developed recommendations for a management plan for agricultural subsurface drainage from 1985 to 1990. The subareas are Northern, Grassland, Westlands, Tulare, and Kern. These subareas are important for determining management options, which include source control, drainage reuse, evaporation systems, land retirement, groundwater management, discharge to the San Joaquin River, and institutional change (e.g., tiered water pricing). The main disposal option in each subarea is:

1.   Grassland subarea drains irrigation subsurface wastewater into the SJR.

Moreno Formations)	
g)	
Nonseleniferous rocks (e.g., Panoche and Lodo Formations)	1.1
San Joaquin Valley (western slope)	0.14
San Joaquin Valley (Panoche Fan)	0.68
Western United States	0.34
Conterminous U.S.	0.26
t (μg/g)	
Kesterson Reservoir	
Top 15 cm	5.0
Top 5 cm, "muck"	55
Organic detritus	165
Cleanup criterion	4.0
San Luis Drain	84
Tulare evaporation ponds	0.05–15
Grassland wetlands	<2.0
San Francisco Bay	0.1–0.6
Deep sea	0.17
μg/L)	
Westlands drainage sumps	140–1400
Grassland drainage sumps	8–4200
San Luis Drain (inflow to pond 2)	330
Tulare evaporation ponds	<1–6300
Mud and Salt Sloughs	16/2–100
Panoche Creek	2–60
Coast Ranges (seeps/ephemeral streams)	<2–3500
U.S. rivers	0.2

	San Joaquin River water quality objective	5.0
	San Francisco Bay	0.07–0.4
	Pacific Ocean	0.13
	Aquatic maximum (drinking water standard)	50
level—solids (μg/g)[c]	Sediment	2.0–4.0
	Aquatic vegetation and invertebrates	3.0–7.0
	Fish (whole body)	4.0–12.0
	Avian eggs	3.0–8.0
level—liquids (μg/L)[c]	Aquatic level (for protection of wildlife)	<2.3–5.0
us waste	Solid (μg/g)	100
	Liquid (μg/L)	1000
el (kst)[a]	San Luis Drain (bed sediment)	0.26–0.85
	San Luis Drain (reuse annual target)	0.38[d]
	San Joaquin River (model calculation)	0.067–0.15
	San Joaquin River (measured, WY 1995)	0.90
	San Joaquin River (total, 1986–1995)	5.85
	Kesterson Reservoir (total, 1981–1985)	1.0
	Grassland wetlands (cumulative loss, 1986–1995)	0.95
	Annual human dietary intake	$10^{-8}–10^{-8.6}$

ations for solids are reported on a dry weight basis. Load levels are in Kestersons (kst): 1 kst = 7900 kg, or 17,400 lb, Se. See discu
designate the following references, as listed at the end of the chapter: 1, Presser et al. (1990); 2, Tidball et al. (1989); 3, Shacklette an
(1986); 5, CCVRWQCB (1990); 6, Henderson et al. (1995); 7, Luoma (personal communication 1997); 8, Turekian and Wedepoh

2. Westlands subarea, prior to 1986, drained to Kesterson Reservoir, but since has a "no discharge" policy, that is, storage of drainage in a ground water aquifer and use of the aquifer for dilution.
3. Tulare and Kern subareas, as hydrologically closed basins, continue to drain to evaporation ponds.
4. No action was recommended for drainage originating in the Northern subarea.

The natural landscape dictated, to some extent, which option was implemented in each subarea. This subdivision has allowed for mitigation of some drainage problems (e.g., migration of drainage from upslope to downslope) by assigning ownership of subsurface drainage to regional drainage entities such as that instituted within the Grassland Bypass Channel Project.

Identification of contaminated lands that produce a high percentage of the overall Se load has been listed as a priority step in determining a long-term, in-valley solution for drainage management (National Research Council, 1989). A memorandum of understanding was signed by state and federal agencies and agricultural water managers in 1991, to implement the in-valley recommendations of San Joaquin Valley Drainage Program (CCVRWQCB,1994a). This agreement was reinforced in the 1995 Water Quality Control Plan for the San Francisco Bay/Sacramento–San Joaquin Delta Estuary, which called for implementation of the program to help achieve water quality objectives (CSWRCB, 1995). A coordinated effort to oversee implementation of the management plan was instituted in 1991, but the program was never funded (SJVDIP, 1994). No comprehensive monitoring has been done to document the overall effectiveness of steps taken to manage drainage after the baseline studies during the late 1980s. Pilot land retirement programs on both Federal and State levels have been funded (the federal Central Valley Project Improvement Act of 1992 and the California State San Joaquin Valley Drainage Relief Act of 1992), but delays in the development of guidelines for the project have slowed acquisition.

III. HISTORICAL PERSPECTIVE FOR WETLAND AND SAN JOAQUIN RIVER PROTECTION

A significant regulatory action for subsurface drainage was taken in 1985 by the California State Water Resources Control Board when it adopted Order No. WQ 85-1 (CSWRCB, 1985). This order was passed to protect the habitat of the Grassland Resource Conservation District and the SJR from a fate similar to that of Kesterson Reservoir. The order required development of plans for (1) cleanup and abatement of Kesterson Reservoir, (2) regulation of agricultural drainage to the San Joaquin River, including waste discharge requirements, and (3) control of agricultural subsurface drainage in the San Joaquin Basin (CSWRCB, 1987). Included in these plans were extensive monitoring, adoption and implementation of improved irrigation practices and drainage reduction strategies, and the development, recommendation, and implementation of water quality objectives for Se. In 1988 the state adopted Se water quality objectives for the San Joaquin Basin, calling for compliance by 1991 for the lower reaches of the SJR and compliance by 1993 for the upper reaches. To date, no waste discharge requirements or permits for disposal of agricultural drainage in the SJR have been issued by the state, and full compliance to state objectives is now not required until the year 2010 (CCVRWQCB, 1996c).

In an effort to increase protection of the SJR under the federal and state antidegradation policy, the USEPA in 1992 rejected part of the state-adopted Se water quality objective for protection of aquatic life (USEPA, 1992a). The USEPA promulgated lower allowable water concentrations (5 μg/L Se instead of 8–10 μg/L; 4-day average instead of a monthly mean) in some tributaries and reaches of the SJR. The lower standards were based on national data from field observations of impacts to aquatic life. Research based on ecological risk assessment to wildlife, mainly avian egg bioaccumulation, suggests the need for a level even lower, namely, less than 2.3 μg/L Se for water (1.0 μg/L dissolved Se, or 2.3 μg/L total recoverable Se) to protect wildlife (Skorupa and Ohlendorf, 1991; Peterson and Nebeker, 1992; Lemly, 1993). However, federal and state water quality objectives for the lower SJR and some of its tributaries were exceeded more than 50% of the time from water years 1986 to 1995 (a water year begins in October), and for some months continuously (CCVRWQCB, 1996a, 1996b).

Because the lower reach of the SJR is a quality-impaired water body (CCVRWQCB, 1991), the federal Clean Water Act requires that water quality standards be translated into total maximum daily loads (TMDLs) through modeling that focuses on sources (Environmental Defense Fund, 1994). The model developed for the SJR, named the Total Maximum Monthly Load (TMML) model, is a conservative element dilution model for assimilative capacity (i.e., dilution capacity) of the river that does not take into account bioaccumulation (CCVRWQCB, 1994b). Additional assumptions used in the model to derive loads result in calculated SJR annual loads significantly higher (an additional 1817–1449 kg) than that derived using more stringent, hence more protective criteria (Table 1). However, load targets for the reuse of the SLD to deliver drainage to the SJR were eventually negotiated among federal and state agencies, farmers, and environmentalists with only an environmental commitment that the input loads to the SJR "will not worsen" over historical loads (USBR, 1995). Annual load targets for the 5-year SLD reuse project are 3000 kg for each of the first two years (Table 1), with at least a 5% load reduction for each succeeding year of a 3-year period (USBR, 1995). This level essentially maintains a status quo that includes extensive violation of USEPA Se standards, since the targets were based on the greater of either a historical average of limited data collected between 1986 and 1994 or the TMML calculation.

The compliance point for the reuse of the SLD is the drain terminus at Mud Slough, located at Kesterson Reservoir terminal ponds (Fig. 2). Although the former Kesterson Reservoir ponds no longer support aquatic habitat, the area remains as a grim reminder of the loss of California's natural resources, at a time when Se has once again advanced to elevated levels in the drainage water of the SLD. An additional compliance point for state regulators is the SJR at Crows Landing (Fig. 1). A possible extension of the SLD to bypass Mud Slough and the lower reach of the SJR above Crows Landing has been discussed (CCVRWQCB, 1996c). This alternative would avoid having drainage in these parts of the slough and river in the future. Such planning again provides for more direct routing of drainage to the SJR and more control by management to comply with objectives that would be dependent mainly on dilution by the Merced River.

Recent management and regulatory approach strategies in the Grassland subarea have lacked a scientific conceptual basis. For example, in the past 6 years, no causal connection between drainage management (i.e., pollution load reduction) and improved water quality has been documented by monitoring efforts (Presser et al., 1996; Westcot et al., 1996). Source control efforts such as improvements in irrigation efficiency and recycling of surface and subsurface drainage may have resulted in reduction of deep percolation, but not in an overall reduction in water quality loads. The reuse of the SLD is being adopted by the USBR as an effort for long-term drainage management despite the lack of data and conceptualization just indicated.

IV. ENVIRONMENTAL FRAMEWORK
SELENIUM VALUES

The difference between essential and toxic levels for Se is quite narrow. Concern levels, which are between levels considered safe (no effect) and levels considered harmful (toxicity threshold), are intended to provide protection for the environment. These values are listed in Table 1 for comparison with ranges for primary geologic sources and background levels and with a compilation of Se concentration means and ranges in soil, sediment, and water for Kesterson Reservoir and the Tulare and Grassland subareas. Ranges of concentrations for sources of agricultural drainage (i.e., collector sumps) and drainage canals for 1985 are given; recent detailed data are not available. Until long-term studies have better defined contamination and cumulative risk due to Se, these ecological risk guidelines must be used in the assessment of environmental conditions. Historical loading to the SJR is also given, along with estimated load attenuation in the Grassland wetlands in units of Kestersons (kst), for which 1 kst equals 7900 kg of Se. The 7900 kg amount is based on the mass loading at Kesterson Reservoir (USBR, 1986, see below). Thus, we propose the Kesterson as a unit to represent a measure of hazard to wildlife, where 1 kst was the cause of the ecotoxicity problem at Kesterson Reservoir.

V.   SELENIUM LOADS AND FIELD EVIDENCE FOR LOAD ATTENUATION

A.   Kesterson Reservoir

Beginning in 1978, agricultural subsurface drainage flowed through the SLD from the Westlands drainage management subarea into Kesterson National Wildlife Refuge (Figs. 1 and 2). Introduction of Se into the food chain from these waters was first demonstrated in 1983 at Kesterson Reservoir, a closed series of evaporation ponds that covered approximately one-fourth of the refuge (Presser and Barnes, 1985; Ohlendorf et al., 1986). Selenium in the inlet pond, as soluble selenate (SeO_4^{2-}) was at a concentration of 330 $\mu g/L$ (Table 1). The water was classified as a sodium–sulfate water with a sulfate concentration of 5550 mg/L and sodium concentration of 2750 mg/L. The correlation between dissolved Se and both sodium and sulfate was +0.93 in water samples from agricultural drainage collector sumps that discharged into the SLD, and thus into the reservoir ponds.

Concentrations of these soluble constituents were expected to increase progressively as the water flowed through a series of 12 interconnected evaporation ponds. In the terminal evaporation pond at Kesterson Reservoir, however, the concentration of Se in water was very low (14 $\mu g/l$), whereas the sodium and sulfate concentrations, as expected, had increased (6250 and 11,500 mg/L, respectively) (Presser and Barnes, 1984, 1985). These data showed a significant loss of dissolved Se between input and terminal pond. Selenium was later found to have been taken into an algal mat that covered the almost dry terminal pond. The concentration of Se in the algal mat approached a level of 13 $\mu g/g$, dry weight in May 1983, and 24 $\mu g/g$ in August 1983. These levels exceed the concern level for selenium in aquatic vegetation (Table 1) and thus were in the toxic range for the diet of birds.

In later studies, organisms showed extensive biomagnification of Se through the food chain (Sakai and Lowe, 1987). By contrast, salt crusts accumulating on the pond, identified as the mineral thenardite (sodium sulfate), contained an order of magnitude less Se, 1.8 $\mu g/g$. Soil Se averaged 55 $\mu g/g$, dry weight, in the depth range of 0 to 5 cm, with 76 to 96% of the Se present in this uppermost layer of sediment (USBR, 1986). Organic detritus from six of the ponds averaged 165 mg/g dry weight. Although this early sampling at Kesterson Reservoir was limited, the data demonstrated that (1) Se lost from solution in the ponds entered the food chain through uptake by biota and (2) organic processes were more effective in removing Se from surface water than were inorganic processes (Presser and Ohlendorf, 1987).

Discrepancies between the chemical balances of Se versus those of sodium and sulfate elucidated the different processes undergone by a nonconservative element (Se) and a conservative element (sodium). A general Se mass balance for Kesterson Reservoir, for the years 1981 to 1985 (USBR, 1986), gives the total input of Se into the SLD and Kesterson Reservoir at 10,300 kg. Of that amount, 2400 kg was estimated to have been retained by the sediment of the SLD. The remaining 7900 kg, or 1 kst (Table 1), was distributed in the biota, water, and sediment at Kesterson Reservoir. Ecotoxic levels were achieved in this environment by high Se mobility, recycling, reactivity, and efficiency of transfer of Se among biota, water, and sediment.

A complete inventory of Se in water, sediment, and biota was not made, but might have enabled prediction of Se behavior through detailed mass balance models using various factors for load attenuation and ecotoxicity. The low level of Se in the terminal pond further demonstrates that waterborne Se alone may not be a good indicator of its level of toxicity; low levels of Se can occur in the water column even though food chain exposure of fish and wildlife may be substantial (Skorupa et al., 1996). According to Skorupa et al., low waterborne Se concentrations can indicate low mass loading (low risk) or high biotic uptake (high risk).

Since its burial, Kesterson Reservoir is no longer subjected to drainage or wetland flows. However, winter rains in this otherwise dry habitat create ephemeral pools whose waters contain up to 1600 μg/L Se (USBR, 1996), a concentration more than sufficient to classify it as as a hazardous waste (USEPA, 1980; USBR, 1986; Table 1). The concentrations of Se in samples of aquatic invertebrates from these ponds (geometric means of 8.5–12.5 μg/g Se, dry weight, from 1988 to 1995) exceeded toxicity thresholds for adverse hatchability and teratogenic effects to birds feeding on these biota (Skorupa and Ohlendorf, 1991; Heinz, 1996; Table 1). The Se concentrations in invertebrates sampled in 1996 indicated some of the highest values of Se bioaccumulated in aquatic organisms (up to 60 μg/g Se, dry weight) since the original sampling at Kesterson Reservoir (CH2MHill, 1996). Geometric mean Se concentrations in samples of vegetation have equaled or exceeded concern levels (2-6 μg/g, dry weight) since monitoring began in 1988. Since 1990, when a maximum mean of 5.2 μg/g, dry weight, was reached, mean Se levels in plants have not changed significantly. Maximum Se concentrations increased though, to 90 μg/g, dry weight, in 1990. Organic detritus from three habitats (filled, open, or grassland) sampled in 1996 all showed extremely high maxima of Se (230–340 μg/g, dry weight). Although these food chain pathways are thought to represent relative degrees of risk, there has been no evidence of recent selenotoxic effects to higher species in the terrestrial ecosystem.

The appearance of these elevated concentrations of Se raise questions about the persistence of Se in the environment and the future remediation of wetlands contamination by Se. For example, how is the cumulative depuration or loss of Se from contaminated ecosystems to be monitored and what is the operational definition of "inert" (i.e., unavailable) form(s) of Se? Recent experiments (Zawislanski and Zavarin, 1996; Zawislanski et al., 1996) have shown that (1) up to 50% of "refractory" Se was actually oxidized to soluble Se(VI) and (2) rates of volatilization from present-day sediments at Kesterson Reservoir are low (1–5%).

B. Tulare Subarea

Tulare drainage management subarea (Fig. 1) is a hydrologically closed basin where intensively farmed lands are located next to managed wetlands on the Pacific Flyway that remain as part of the relict Tulare Lake (Presser et al., 1995). Subsurface drains are widely used to manage groundwater salinity and water levels, with disposal in privately owned evaporation basins. In 1992, 2700 ha of evaporation ponds, divided into 21 separate basins, existed as part of the management alternative for disposal of subsurface drainage in the larger Tulare Basin. Most pond systems consisted of a series of connected evaporation ponds, each representing a closed system (CCVRWQCB, 1990).

Dry climatic conditions and runoff into this subarea, from both the Coast Ranges on the west and the Sierra Nevada on the east, complicate and intensify effects in the Tulare Basin associated with soil salinization over those seen at Westlands subarea and consequently seen at Kesterson Reservoir. Elemental concentration patterns in the waters from different geographic zones of the basin (alluvial fan, trough, and lake bed) are discernible (CCVRWQCB, 1990; Fujii and Swain, 1995). In the alluvial fan zone, the geometric mean Se concentration in ponds is 320 $\mu g/L$ and in inlet flow it is 250 $\mu g/L$ (CCVRWQCB, 1990), comparable to the average inflow seen at Kesterson Reservoir of 330 $\mu g/L$. The geometric means for pond inflows for the lake bed and trough zones are low (11 and 2 $\mu g/L$ Se, respectively) as are concentrations in ponds of these zones. Inlet concentrations for ponds in these two zones reached a maximum of 62 $\mu g/L$ Se. Other trace elements, mainly oxyanions, also show elevated concentrations in certain zones. Concentration maxima ($\mu g/L$) in ponds of the alluvial fan, trough, and lake bed zones are Se 6300, U 8000, As 420, Mo 12,000, V 112, and B 700,000 (CCVRWQCB, 1990). Samples of shallow groundwater reach the following maxima ($\mu g/L$): Se 1000, U 5400, As 2600, Mo 15,000, and B 73,000, showing the close linkages of the surface water and groundwater systems (Fujii and Swain, 1995; Presser et al., 1995).

Intensive management of the constructed ponds in the Tulare Basin, popularly labeled as a series of "mini-Kestersans," was initiated to minimize bird use and, therefore, lessen potential adverse effects. Unfortunately, the Kesterson analogy turned out to be all too prophetic and management efforts inadequate. From 1987 to 1990, greater than 50% of bird eggs in 11 of 17 basins studied were documented to have Se means in species-specific reproductive effect ranges (Skorupa et al., 1996; Skorupa, this volume, Chapter 18). Selenium levels in at least 13 of the basins were elevated enough to lead to the prediction that waterbird populations breeding at these ponds would experience either reduced hatchling success or increased teratogenic levels (CH2MHill et al., 1993). In Chapter 18 Skorupa documents a 10–50% rate of embryo teratogenesis in one or more species of waterbirds at four sites in the Tulare Basin ponds, comparable in both type and rate of deformity to Kesterson Reservoir birds (Presser and Ohlendorf, 1987). Because avian toxicity, in fact, occurred at ponds containing relatively low Se concentration (7–18 μg/L), (Bay Institute, 1993; Skorupa et al., 1996; Skorupa, this volume, Chapter 18), the adequacy and applicability of traditionally defined Se toxicity to the ponds (i.e., 1000 μg/L as a hazardous Se waste, based on an allowable concentration that is two orders of magnitude above that defined for the drinking water standard—USEPA, 1980), as opposed to functional toxicity, was questioned by the U.S. Fish and Wildlife Service (USFWS) and environmentalists before the SWRCB during 1995 (CSWRCB, 1996). In view of those studies and of data presented at the evidentiary hearing process, pond acreage was decreased and alternative and compensating aquatic habitats were created to mitigate and remediate "unavoidable losses" of aquatic birds.

At two recently constructed experimental pond sites, accelerated evaporation of concentrated drainage water is employed to reduce aquatic bird use and thereby bird loss. At these sites, the rate of evaporation is increased by spraying brine into the air at a rate that prevents ponding. For future management, construction of numerous 0.80 ha sites containing this type of "pond" has been suggested as part of an overall drainage reduction strategy. The management objective is to maintain bird-free ponds. However, shorebirds (i.e., stilts, avocets, killdeer) nested after a winter storm in 1996 when ponding of water occurred at these sites and adjacent experimental areas, where extremely salt-tolerant plants (halophytes, including trees) are grown . At the Red Rock Ranch site, Se concentrations in the inlet water (average 1400 μg/L Se) were greater than 1000 μg/L (a hazardous Se waste); moreover, pond Se concentrations measured up to 18,000 μg/L, and inviability of eggs was 67% (J. P. Skorupa, personal communications April 11, July 1, 1996; Chapter 18, this volume). The level of teratogenicity (56.7%) surpasses the level found at Kesterson Reservoir. Although these pools containing high Se concentrations were deemed a management problem that could be corrected, the question remains as to how biomagnification that led to this extent of deformity could have occurred in the food chain over a relatively short period of time.

One of the Tulare Basin series of ponds is experiencing an apparent loss, or load attenuation, of Se between inlet water and the terminal evaporation pond.

This series of ponds contained an approximate total of 3.9 kst in solution in the summer of 1995 (Tulare Lake Drainage District, 1995). Dissolved Se concentrations showed a continuous decrease from cell 1 to cell 7, whereas salt concentrations (represented by electrical conductivity) increased. Fan et al. (1996) attributed depletion from water to possible Se volatilization as well as precipitation in evaporation ponds and recommended such pond systems for "in situ Se removal." Because the trend was based on Se concentrations for the water column alone, rather than a loss from the entire system, the conjectured loss of Se via volatilization cannot be quantified until a complete inventory of major Se reservoirs is made. A mass balance monitoring approach that includes, in addition to the amount of Se volatilized, Se in water, suspended matter, precipitated material, bottom sediment, and biota would document the partitioning of Se between its different reservoirs, thus permitting an evaluation of the relative importance of removal mechanisms. Use of pond systems and of plants (phytoremediation) for depleting Se from the water, although promising, may still present risks, since Se in certain phases of the ecosystem concurrently increases and subsequent exposures to wildlife occur. As at Kesterson Reservoir, elevated concentrations of Se in the precipitated salts, sediment, and vegetation accumulating from these ponds may persist for years, or require exorbitant sums for cleanup.

Location of evaporation ponds that represent oxygen-depleted depositional environments juxtaposed to trace-element-enriched geologic formations and farmlands that represent alkaline-oxidized sources, leads to the repetitive cycling of trace elements in arid hydrologically closed basins like the Tulare Basin. With this type of land use and disposal system, the concentration of trace elements in pond sediment, surface waters, and groundwaters seems inevitable. The economic analysis of whether "evaporation pond farming" is sustainable in view of the need to provide compensating and alternative habitat for wildlife was not definitive (CSWRCB, 1996). Connecting the Tulare subarea to the existing 137 km SLD is viewed by state planners (CCVRWQCB, 1994a, 1995) as the answer to water quality problems in the Tulare Basin. This export drain could potentially remove salt- and Se-laden agricultural waste as well as industrial and municipal waste from the basin. No recipient of the drainage has yet been selected.

C. Grassland Subarea

Agricultural drainage from the Grassland subarea has historically been disposed of in the SJR (Fig. 2). Subsurface drains, originally called "deep drains," were installed in the 1950s. The route of the drainage water to the SJR changed following the demise of Kesterson Reservoir. Prior to 1985, drainage water was mixed with freshwater in wildlife habitats, including duck clubs and wetland channels of the Grassland Resource Conservation District, which is tributary to the SJR. From 1985 to 1996, drainage water was alternated with freshwater in fewer wetland channels to reduce exposure of the entire habitat, with the drainage water eventually traveling through Mud and Salt Sloughs to the SJR. In September 1996, the drainage water was routed into the northernmost 45 km section of the SLD and 10.6 km of Mud Slough and then into the SJR, largely bypassing the wetland habitat (USBR, 1995). Unless otherwise noted the data presented in the discussion below of the loads discharged along the pathway the drainage takes, from agricultural drainage collector sump through earthen canals or drains and wetlands to the SJR, are referenced in state documents CCVRWQCB (1996a, 1996b, 1996c).

Average annual Se concentrations in the water from source agricultural drainage canals in the Grassland subarea (Figs. 1 and 2) were as high as 82 μg/L for water year (WY) 1986 to WY 1995, lower than that in water delivered to Kesterson Reservoir by the SLD (330 μg/L). Water from individual farm field collector sumps contained Se concentrations as high as 4200 μg/L Se, three times greater than the maximum collector sump for the SLD of 1400 μg/L (Presser and Barnes, 1985) (Table 1). A recent limited assessment of the mean Se concentration for drainage sumps showed 211 μg/L, with a median concentration of 134 μg/L (CCVRWQCB, 1996c). Annual loads originating in the subarea at the beginning of management in 1986 and 1987 were 0.56 and 0.62 kst, respectively. Annual loads did vary with drought and flood conditions in subsequent years, but in WY 1995, which was 10 years after institution of the state drainage control program and 5 years after completion of the federal–state drainage management plan, the annual Se source load was the highest ever recorded, 0.65 kst.

Except for WY 1990, Se loads discharged from source sumps and drains have been higher than loads measured downstream of Grassland Resource Conservation District wetland area (i.e., at Mud and Salt Sloughs; Fig. 2). The 1986 Se load in Mud and Salt Sloughs decreased by 0.18 kst between the source load and the load discharged after traveling through the wetland channels. This pattern of load attenuation has been repeated in successive years. For WYs 1986–1994, the cumulative difference between the input drains and Mud and-Salt Slough outputs was 0.95 kst (CCVRWQCB,1996a, 1996b, 1996c). The USBR (1995) data also showed Se input loads higher than exported loads with a comparable difference of 0.89 kst and an annual maximum attenuation of 50%. This loss of Se approximately equals the mass loading of Se that created the ecotoxicity at Kesterson Reservoir (1 kst). Failure to account for this Se loss in the Grasslands (0.89–0.95 kst) could be due to several factor, but in the absence of comprehensive monitoring data, we can only speculate on the reasons. Possible factors include errors in flow and concentration measurements owing to temporal and spatial extrapolations used to define calculated load, and failure to document accurately the partitioning and bioaccumlation of Se among water, sediment, and biota.

Elevated levels of Se in fish and invertebrates taken from Mud and Salt Sloughs during 1992 and 1993 indicate that bioaccumulation accounts for some of the loss of Se. Approximately 77 % of whole-body fish samples in Mud Slough and 85% in Salt Slough were in the concern range (Henderson et al., 1995, Table 1). Selenium concentrations in invertebrates were also in the concern range (Table 1), with 15% of Mud Slough samples exceeding the toxicity threshold of 7 μg/ g Se. The most recent Se levels in water birds in the Grassland wetlands (Fig. 2) were measured from 1984 to 1988 (Ohlendorf and Hothem, 1987; Paveglio et al., 1992). In the southern wetlands, species averages of Se decreased overall for livers of both coots (from 23 μg/g to 10 μg/g) and ducks (from 24 μg/g to 14 μg/g). These Se levels are elevated when compared to mean background levels in coots (5.4 μg/g) and ducks (8.4 μg/g) but below levels associated with adverse effects on reproduction in coots (82 μg/g) and ducks (20 μg/g) at Kesterson Reservoir (Paveglio et al., 1992). Eggs collected in WY 1995 by the USFWS from Grassland birds, as background data for the reuse of the SLD project, have not been analyzed. USFWS's overall objectives for reuse of the SLD are to maintain Se concentrations below toxicity threshold levels in biota in the area of Mud Slough (where drainage has been increased) and to lower Se concentrations in biota in the area of Salt Slough (where drainage has been removed) to background levels (Henderson et al., 1995).

VI. POTENTIAL FOR LOAD ATTENUATION IN GRASSLAND BYPASS CHANNEL PROJECT

Because the 5 μg/L Se objectives in the lower SJR have been violated a majority of the time since Se concentrations have been monitored in the river, the cumulative Se load discharged from the Grassland subarea to the SJR from WY 1986 through WY 1994 (4.3 kst) is greater than the allowable loads calculated using a conservative element, standard steady state TMDL model for the SJR (1.35–0.60 kst). Annual load targets for the SLD reuse project are 0.38 kst (Table 1) for each of the first two years, with a 5% load reduction for each succeeding year of a 3-year period (USBR, 1995). The cumulative load for the 5-year project, if load targets are met, is 1.80 kst. However, the impact to the SJR will likely increase because drainage collected in the SLD and Mud Slough will discharge directly into the SJR without traveling through wetland channels containing freshwater dilution flows. If the same logic of in-transit loss of Se through uptake in biota and sediment of evaporation ponds and distribution channels observed at the Kesterson, Tulare, and Grassland wetland areas is applied to the pathway now taken by the drainage, a significant potential for load attenuation exists for the SLD, Mud Slough, and the SJR. The cumulative load of Se discharged to the SJR system from WY 1986 to WY 1995, including the in-transit loss of Se, was

5.85 kst. Despite this large load compared to that at Kesterson Reservoir, few samples of sediment and biota have been taken to document the Se inventory in the SJR system, although plankton and clam samples show levels of contamination (up to 5 μg/g, dry weight) (CSWRCB, 1991) that could adversely impact fish and birds (Table 1).

Except for WY 1989, Se loads measured for the SJR at Crows Landing (Fig. 1), a state compliance site for Se further downstream in the SJR below the Merced River (CCVRWQCB, 1994a), were higher than those discharged from Mud and Salt Sloughs. The cumulative difference between annual loads for Mud and Salt Sloughs and annual loads measured at Crows Landing from WY 1986 to WY 1994 is 0.58 kst, with an annual maximum difference of 0.25 kst in 1986. The 1995 load measured at Patterson (Fig. 1), yet further downstream than the Crows Landing site and the only SJR load data available that year, was 0.90 kst, a historical maximum. This value compares to 0.65 kst for the drains and 0.61 kst from Mud and Salt Sloughs.

The relation among the agricultural drainage area source loads, the Mud and Salt Slough loads, and the SJR loads "remains unexplained" (CCVRWQCB, 1996b), although Mud and Salt Sloughs have been identified as contributing most of the total Se load drained into the SJR (e.g., 81% in WY 1988 and 86% in WYs 1993 and 1994). In addition to load attenuation of Se in the wetlands, these data suggest a significant second source of Se (0.58 kst) added to the SJR that may be perceived as worsening yet further the SJR condition. From a mass balance perspective, these data show that 5.85 kst has been loaded into the SJR system between 1986 and 1995, and the current annual rate may be as great as 1 kst (Table 1). This comes 12 years after the Kesterson disaster and 10 years after identification and implementation of management and treatment strategies intended to control subsurface agricultural drainage. Because no causal connection is seen between pollution load reduction and improvements in water quality in the SJR, there would seem to be a rational and even urgent need to identify monitoring requirements, including accurate flow measurements, that adequately determine a mass balance for Se and the processes controlling it. Despite variable system inputs (rainfall and applied water) and outputs (drainage), these large loads would seem to be driven, in part, by increased agricultural production levels. In the Grassland subarea, production is now at a maximum for the region (38,000 ha) as a result of availability of full allotments of irrigation water.

VII. SELENIUM ENRICHMENT IN SAN LUIS DRAIN SEDIMENT

During its initial operation, from 1978 to 1986, the SLD conveyed subsurface agricultural drainage to Kesterson Reservoir. During its closure from 1987 to

1996, the SLD acted as an evaporator and seepage collector, although it briefly acted as a conduit for flood and drainage water during SJV flooding in WY 1995. Therefore, analyses of bed sediment of the SLD may provide a continuing history of Se partitioning and flux occurring in the sediment during static, controlled flow, and flooding regimes between 1978 and 1996.

Data from a survey in 1994 (45 km segment) indicated that Se concentrations in bed sediment in the SLD averaged 44 μg/g dry weight, with a maximum of 146 μg/g (Presser et al., 1996). Earlier surveys (137 km segment), made when the SLD was in operation, found a maximum of 210 μg/g Se in sediment samples, with an average concentration of 84 μg/g (USBR, 1986). These data sets are difficult to compare because variation at single sampling sites is large and trends with depth are opposite for the 1985 and the 1994 surveys (Presser et al., 1996). However, the elevated Se levels cannot be explained by geologic source material. Seleniferous rocks in the California Coast Ranges average 8.9 μg/g Se (Presser et al., 1990; Table 1). Western SJV soils average 0.14 μg/g Se. Soils from the most contaminated alluvial fan, Panoche Fan, average 0.68 μg/g, with a maximum of 4.5 μg/g (Tidball et al., 1989; Table 1). It seems reasonable to assume that bioaccumulation is the mechanism whereby Se concentrations increased in the sediment under flowing conditions when the drain contained irrigation drainage water.

Because of the high levels of Se documented in the SLD in 1985, the sediment residing in the drain was classified for regulatory purposes (USBR, 1986). SWRCB Order No. WQ 85-1 found the soils and wastewater associated with Kesterson Reservoir to be a "designated waste" that posed a hazard to the environment, and, as such, should be handled, stored, or disposed of in a manner consistent with hazardous waste management provisions. This concern, prompted a recommendation for complete sediment removal from the portion of the SLD to be reopened as part of the Grassland Bypass Channel Project as the project was originally conceived by the San Joaquin Valley Drainage Program (1990). However, this recommendation was never carried out. Rather, management of the sediment in the drain is to take place (USBR, 1995; Presser et al., 1996) even though levels of concern and hazard have been exceeded (Table 1). California's criterion for a solid hazardous Se waste is defined as 100 μg/g wet weight (California Code of Regulation, 1979; USBR, 1986; Table 1).

The Kesterson Reservoir cleanup criterion called for filling and grading of pond sediments (and associated vegetation) if Se concentrations were greater than 4 μg/g. The sediment Se toxicity threshold for ecological risk developed by the USFWS for the reuse of the SLD is 4 μg/g, dry weight (Henderson et al., 1995). A concern level of 2 to 4 μg/g Se, dry weight, is recommended for bottom sediment as a performance guideline for remediation of contaminated irrigation drainage sites across the western United States studied under USDOI's National Irrigation Water Quality Drainage Program (Gober, 1994). Luoma et al. (1992)

further identified the surficial layer of bottom sediment and suspended material as important indicators of toxicity based on bivalve uptake, defining 1.0 to 1.5 μg/g Se, dry weight, for particulates as a limit for protection of aquatic life.

A mass balance estimation made during initial use of the SLD showed 0.30 kst of Se residing in the accumulated sediment in the drain (137 km segment) (USBR, 1986). Within the 45 km section of the SLD reopened in September 1996, an estimated 42,650 m^3 of sediment was present. In terms of mass loading, an average concentration of 44 μg/g Se represents 0.26 kst contained in the sediment (Table 1). A statistically significant inventory of Se in SLD sediment to establish a baseline would reveal the degree to which Se is being gained or lost from the water column and concentrated or removed from the sediment during future projects. Even without the data necessary to quantify this transfer, sediment has proven to support a substantial food chain (Moore, 1990) that further contributes to Se bioavailability.

In contrast to a Se accumulation mechanism in SLD sediment during its initial operation, mobilization or depuration of Se from sediment was hypothesized during an emergency discharge of water to the SLD during flooding in WY 1995. The SLD acted as a conduit for floodwater into the SJR that included runoff from the Coast Ranges, which was expected to contain elevated Se concentrations. Panoche Creek (Fig. 1), the source of the major loading of Se to the western SJV, was sampled where it issues from the Coast Ranges onto the alluvial fan of the SJV to represent drainage from the approximately 81,000 ha upper watershed. The creek waters contained low levels of dissolved Se (approximately 5 μg/L) during nonstorm periods and approximately 60 μg/L dissolved Se in first-flush runoff events during storms in 1988 as evaporated salts were mobilized and debris flows occurred (Presser et al., 1990). At the same location during the large-magnitude flood of March 1995 (estimated as a 50 to 100-year event), dissolved Se concentrations in composited runoff samples of water and sediment from Panoche Creek showed a maximum of 45 μg/L at a flow rate of 15.6 m^3/s and 19 μg/L at a flow rate of 283 m^3/s (USEPA, personal communication, 1996). The Se concentration in the sediment portion of the runoff was estimated as approximately 10 times that of the dissolved fraction.

A short-duration maximum concentration of 120 μg/L Se was seen in the SLD outfall during that event, with a downstream concentration of 33 μg/kg in Mud Slough. Altogether, approximately 0.1 kst was discharged to the SJR during the 15 days the SLD was open during this flood event, mainly from the SLD, Mud Slough, and Salt Slough. Agricultural drainage collector sumps were not shut off during this flood period, and flooded fields were pumped into the SLD. The amount of Se attributable to each potential source of Se (subsurface drainage, runoff, or that mobilized from sediment) discharged to the SJR during the WY 1995 flooding is still unknown, in part, because no conclusive data were collected to differentiate sediment load from dissolved load. The section of the SLD reopened during the flood event (approximately 105 km) was estimated to hold 133,800 m³ of sediment (USBR, 1986). If the average Se concentration in the sediment was 44 μg/g, then Se in the sediment on the floor of the SLD contained roughly 0.85 kst. Mobilization or resuspension and subsequent transport of about 12% of this sediment could account for the discharge of the 0.1 kst measured at the check point on the SJR.

VIII. DISCUSSION

Wetlands and agriculture in the SJV are inexorably linked via irrigation drainage. As a consequence of this linkage, elevated Se concentrations occur, leading to avian reproductive toxicity as at Kesterson Reservoir, Tulare Basin, and Red Rock Ranch (see also Skorupa, this volume, Chapter 18). In a recent evaluation, a load reduction of 47–80% of the amount of agricultural drainage disposed of in the SJR is necessary to meet the state compliance schedule (CCVWQCB, 1996c). A model using dynamic drainage effluent limits based on the real-time assimilative capacity of the SJR was suggested recently as a future management tool (Karkoski, 1996). This "real-time" model would allow approximately 140% more Se to be disposed of in the SJR than the amounts calculated by previous quasi-static models (TMML model) and those adopted as load targets by the Grassland Bypass Channel Project (Table 1). Hence, the "real-time" model amount is substantially greater (293–603%) than that allowed if a standard steady state TMDL method with a 5 μg/L Se objective is used for determining the load to the SJR. With "real time" Se management, ponds for flow regulation would be necessary to maximize release of Se loads during different flow conditions in the river. Models such as this, and the site-specific TMDL and TMML models developed for the SJR, address, to some degree, the complexities of flow, but fail to consider the complexities of Se chemistry and biology. In these models, the behavior of an element is defined solely on the basis of dilution, thus treating Se as a conservative element. If real-time management is enacted, Se disposal would be maximized and the SJR kept continuously at a 5 μg/L concentration based on flow conditions; it is unknown what overall effect this regime would have on the SJR. For example, this scheme would eliminate beneficial natural flushing of the system during high flow events and would create a concentration that could be thought of, to some degree, as "static."

The available data suggest that the definition of Se contamination must be formulated on an ecosystem level and a mass balance basis. For future management, the lessons learned at Kesterson Reservoir may be applicable to the Tulare evaporation ponds and those learned at Grassland wetlands, applicable to the SLD/Mud Slough/SJR system. A final repository for Se, if the SLD is extended, is the San Francisco Bay/Delta Estuary and/or the Pacific Ocean. Based on flow rates and transport, residence time and exposure pathway for Se in the ecosystems associated with these repository water bodies may be important variables to measure to define Se uptake and retention and thus, ecotoxicity. The bay may represent a system intermediate between static (evaporation ponds) and flowing (SLD, Mud Slough, and SJR) systems.

Assimilative capacity for Se in a receiving water cannot be based on a dilution model. Allowable Se loads need to be determined using a mass balance approach that recognizes the cumulative loading of Se in water, sediment, and biota, including past loading (e.g., in bed sediment). Although not all the ramifications of Se cycling are known, a mass balance approach to understanding Se transport and fate would contribute to establishing limits of bioaccumulation of Se in relation to such important variables as flow and speciation. These data are essential for the design of management strategies that optimize Se concentrations and loads and also comply with regulatory and environmental commitments that adequately protect the environment.

The efficiency of Se transfer through an environment may center on the dominance of biological versus chemical processes that may be a function of flow. Oremland et al. (1989) have shown that for reactions involving Se, biological reactions are much the faster. High rates of exchange are expected among Se species (selenate, selenite, elemental Se, and organic and inorganic selenide) and between intercellular pools. Such exchange makes for an efficient transfer of Se among water, sediment, and biota. Selenium is known to bioaccumulate very efficiently in terminal (static) sink evaporation ponds (e.g., Kesterson Reservoir, Tulare evaporation ponds). Flushing flows, either natural or managed, in some wetland areas (e.g., Grassland) may mitigate the effects of Se mass loadings. To this end, a management strategy could be developed whereby flow is slow enough to prevent sediment movement but fast enough to minimize Se bioaccumulation. At present, the data necessary to evaluate this scheme, or to determine an optimum flow condition have not been collected. Bioconcentration of waterborne dissolved Se and bioaccumulation (possibly biomagnification) of food-borne particulate Se are required to quantify potential toxicity levels. These exposure pathways require monitoring strategies that include measurement of Se concentrations in water (dissolved) and in organic and inorganic suspended particulate matter, in addition to those in bed sediment and biota. Speciation considerations also lead to two types of Se assessment for which amounts and ranges of risk need to be quantified: (1) ecological hazard created by the high mobility of dissolved selenate species, and (2) bioassay toxicity created by the high toxicity of the dissolved organic selenide species (e.g., selenomethionine).

Waters and wildlife of the San Francisco Bay/Delta Estuary are already at risk (Nichols et al., 1986). The bay, which is the largest estuary on the west coast of North America, has lost 80% of its historic marshes to urban encroachment (i.e., filling and diking). Preservation and restoration of the bay wetlands are long-term goals of a new cooperative effort (CALFED Bay–Delta Program, 1996) among federal and state governments and the general public to ensure a healthy ecosystem and reliable, high quality water supplies. As stated previously, the levels of Se in the San Francisco Bay/Delta Estuary had been listed as "of concern" (CSFBRWQCB, 1992) mainly as a result of six petroleum refinery point sources of Se pollution, but the agricultural drainage component of the Se load has not been quantified. Samples collected in 1987–1989 from sites in the bay, principally Suisun Bay and San Pablo Bay, showed site-mean-Se levels in scoter and scaup (waterfowl) liver tissue of 40 to 209 and 14 to 83 $\mu g/g$, dry weight, respectively (CSFBRWQCB, 1992), surpassing levels in the waterfowl (46 to 82 $\mu g/g$, dry weight Se) at Kesterson Reservoir that showed deformities.

More significantly, deformed embryos have been recovered from a small marsh in the northern bay specifically designed for remediation of Se-enriched oil refinery effluent after 4 years of receiving effluent containing elevated Se concentrations (20 $\mu g/L$). Selenium levels in bird eggs similar to those found at Kesterson Reservoir have been detected, and 30% of mallard nests and 10% of coot nests contained deformed embryos (Skorupa et al., 1996; Skorupa, this volume, Chapter 18). An additional source of Se due to agricultural drainage into the bay via an extension of the SLD can only exacerbate these problems for this now multipurpose aquatic environment. To measure impacts in the future, comprehensive environmental monitoring on a mass balance basis is needed to relate effluent limits to Se bioaccumlative potential, rates of transfer, and effects in the ecosystem.

ACKNOWLEDGMENTS

We wish to acknowledge the technical and policy reviews by Marc A. Sylvester and E. C. Callender of the Water Resources Division of the U.S. Geological Survey and the support of the U.S. Fish and Wildlife Service, Division of Environmental Contaminants, Sacramento Field Office, especially senior biologists Joseph P. Skorupa and Steven E. Schwarzbach.

REFERENCES

Bard, C. 1996. Public Trust Complaint (San Joaquin River) against the U.S. Bureau of Reclamation and California State Water Resources Control Board. CSWRCB, Sacramento.

Bay Institute of San Francisco. 1993. *Death in the Ponds: Selenium Induced Waterbird Deaths and Deformities at Agricultural Evaporation Ponds.* Bay Institute of San Francisco, San Rafael, CA.

Bruland, K. W. 1983. Trace elements in seawater. In J. P. Riley and R. Chester (eds.), *Chemical Oceanography*, Vol. 8, pp. 157–220. Academic Press, San Diego, CA.

CALFED Bay–Delta Program. 1996. Phase I Progress Report. CALFED, Sacramento.

California Central Valley Regional Water Quality Control Board. 1990. Sediment and water quality in evaporation basins used for the disposal of agricultural subsurface drainage water in the San Joaquin Valley, California, 1988 and 1989. CCVRWQCB, Sacramento.

California Central Valley Regional Water Quality Control Board. 1991. Water quality assessment. CCVRWQCB, Sacramento.

California Central Valley Regional Water Quality Control Board. 1994a. The water quality control plan (basin plan) for the Sacramento River Basin and San Joaquin River Basin, 3rd ed. CCVRWQCB, Sacramento.

California Central Valley Regional Water Quality Control Board. 1994b. A total maximum monthly load model for the San Joaquin River. CCVRWQCB, Sacramento.

California Central Valley Regional Water Quality Control Board. 1995. Tulare Basin Plan. CCVRWQCB, Fresno.

California Central Valley Regional Water Quality Control Board. 1996a. Water quality of the lower San Joaquin River: Lander Avenue to Vernalis. CCVRWQCB, Sacramento.

California Central Valley Regional Water Quality Control Board. 1996b. Agricultural drainage contribution to water quality in the Grassland Area of Western Merced County, California. CCVRWQCB, Sacramento.

California Central Valley Regional Water Quality Control Board. 1996c. Amendments to the water quality control plan for the Sacramento and San Joaquin River Basins for control of agricultural subsurface drainage discharges. CCVRWQCB, Sacramento.

California Code of Regulations. 1979 and as amended. Title 22, Social Security, Division 4, Environmental Health, Chapter 30. Minimum standards for management of hazardous and extremely hazardous wastes. Article 66001–66699, s-161, 1121–0649.

California Health and Safety Code. 1984. Section 25208 et seq.

California Public Resources Code. 1970. Section 21000 et seq.

California San Francisco Bay Regional Water Quality Control Board. 1992. Mass emissions reduction strategy for selenium for San Francisco Bay. CSFBRWCB, San Francisco.

California State Water Resources Control Board. 1969. California Porter–Cologne Water Quality Act. California Water Code, Division 7, Water Quality, Chapters 1–10 and as amended. CSWRCB, Sacramento.

California State Water Resources Control Board. 1985. Order No.WQ 85–1, In the matter of the petition of R. J. Claus for the review of inaction of the California Central Valley Regional Water Resources Control Board. File No. A-354, adopted 2/5/85. CSWRCB, Sacramento.

California State Water Resources Control Board. 1987. Regulation of agricultural drainage to the San Joaquin River. SWRCB Order No. W.Q. 85–1 Technical Committee Report. CSWRCB, Sacramento.

California State Water Resources Control Board. 1991. Selenium verification study 1988–1990. No. 91–2-WQ. CSWRCB, Sacramento.

California State Water Resources Control Board. 1995. Water quality control plan for the San Francisco Bay/Sacramento–San Joaquin Delta Estuary. No. 95–1WR (Resolution 95–24, Bay-Delta Plan). CSWRCB, Sacramento.

California State Water Resources Control Board. 1996. Staff technical report on petitions regarding Tulare Lake evaporation ponds. CSWRCB, Sacramento.

CH2MHill. 1996. Kesterson Reservoir Biological Monitoring Report. U.S. Bureau of Reclamation, Mid-Pacific Region, Sacramento.

CH2M Hill, Harvey and Associates, and G. L. Horner. 1993. Cumulative impacts of agriculture evaporation basins on wildlife. California State Department of Water Resources, Fresno.

Cutter, G. A. 1989. The estuarine behavior of selenium in San Francisco Bay. *Estuarine Coastal Shelf Sci.* 28:13–34.

Cutter, G. A., and M. L. C. San Diego-McGlone. 1990. Temporal variability of selenium fluxes in San Francisco Bay. *Sci. Total Environ.* 97/98:235–250.

Dunning, H. C. 1980. The public trust doctrine in natural resources law and management: A symposium. *Univ. California Davis Law Rev.* 14(2):181–496.

Environmental Defense Fund. 1994. *Plowing New Ground, Using Economic Incentives to Control Water Pollution from Agriculture.* Oakland, CA.

Fan, T. W-M., R. M. Higashi, and A. N. Lane. 1996. In situ selenium bioremediation by aquatic plants. In Abstracts of the Society for Environmental Toxicology and Chemistry, Fall 1996, Washington, DC.

Fujii, R., and W. C. Swain. 1995. Areal distribution of selected trace elements, salinity, and major ions in shallow ground water, Tulare Basin, southern San Joaquin Valley, California U.S. Geological Survey Water Resources Invest. Report No. 95–4048.

Gober, J. 1994. The relative importance of factors influencing selenium toxicity. p. 1021–1031. In R. A. Marston and V. R. Hasfurther (eds.), *Proceedings of the Annual Summer Symposium of the American Water Resources Association: Effects of Human-Induced Changes on Hydrologic Systems.* Jackson Hole, WY, June 1994.

Heinz, G. H. 1996. Selenium in birds. In W. N. Beyer, G. H. Heinz, and A. W. Redmon-Norwood (eds.), *Environmental Contaminants in Wildlife: Interpreting Tissue Concentrations.* Lewis, New York.

Henderson, J. D., T. C. Maurer, and S. E. Schwarzbach. 1995. Assessing selenium contamination in two San Joaquin Valley, California, sloughs, an update of monitoring for interim re-use of the San Luis Drain. U.S. Fish and Wildlife Service, Sacramento.

Karkoski, J. 1996. Dynamic versus quasi-static effluent limits. In C. Bathala (ed.), *Proceedings of the North American Water and Environment Congress, 1996.* American Society of Civil Engineers, Environmental Engineering Division, Orange CA.

Lemly, A. D. 1993. Guidelines for evaluating selenium data from aquatic monitoring and assessment studies. *Environ. Monit. Assess.* 28:83–100.

Luoma, S. N., C. Johns, N. Fisher, N. A. Steinberg, R. S. Oremland and J. R. Reinfelder. 1992. Determination of selenium bioavailablity to a benthic bivalve from particulate and solute pathways. *Environ. Sci. Technol.* 26:485–491.

Margolin, S. 1979. Liability under the Migratory Bird Treaty Act. *Ecol. Law Q.* 7:989–1010.

Moore, S., and others. 1990. *Fish and Wildlife Resources and Agricultural Drainage in the San Joaquin Valley, California,* Vol. 1. San Joaquin Valley Drainage Program, Sacramento.

National Research Council. 1980. *Recommended Dietary Allowances,* 9th ed. National Academy of Sciences, Washington, DC.

National Research Council. 1989. *Irrigation-Induced Water Quality Problems. What Can Be Learned from the San Joaquin Valley Experience?* National Academy of Sciences, Washington, DC. 157 pp.

Nichols, F. H., J. E. Cloern, S. N. Luoma, and D. H. Peterson. 1986. The modification of an estuary. *Science* 231:567–573.

Ohlendorf, H. M., and R. L. Hothem. 1987. Selenium contamination of the Grasslands, a major California waterfowl area. *Sci. Total Environ.* 66:169–183.

Ohlendorf, H. M., D. J. Hoffman, M. K. Saiki, and T. W. Aldrich. 1986. Embryonic mortality and abnormalities of aquatic birds: Apparent impacts by selenium from irrigation drainwater. *Sci. Total Environ.* 52:49–63.

Oremland, R. S., J. T. Hollibaugh, A. S. Maest, T. S. Presser, L. G. Miller, and C. W. Culbertson. 1989. Selenate reduction to elemental selenium by anaerobic bacteria in sediments and culture: Biogeochemical significance of a novel, sulfate-independent respiration. *Appl. Environ. Microbiol.* 55:2333–2343.

Paveglio, F. L., C. M. Bunck, and G. H. Heinz. 1992. Selenium and boron in aquatic birds from Central California. *J. Wildl. Manage.* 56:31–42.

Peterson, J. A., and A. V. Nebeker. 1992. Estimation of waterborne selenium concentrations that are toxicity thresholds for wildlife. *Arch. Environ. Contam. Toxicol.* 23:154–162.

Presser, T. S. 1994. The Kesterson effect. *Environ. Manage.* 18:437–454.

Presser, T. S., and I. Barnes. 1984. Selenium concentrations in water tributary to and in the vicinity of Kesterson National Wildlife Refuge, Fresno and Merced, California. U. S. Geological Survey Water Resources Invest. Report No. 85–4220.

Presser, T. S. and I. Barnes. 1985. Dissolved constituents including selenium in the vicinity of the Kesterson National Wildlife Refuge and the West Grassland, Fresno and Merced Counties, California. U. S. Geological Survey Water Resources Invest. Report No. 85–4220.

Presser, T. S. and H. M. Ohlendorf. 1987. Biogeochemical cycling of selenium in the San Joaquin Valley, California, USA. *Environ. Manage.* 11:805–821.

Presser, T. S., W. C. Swain, R. R. Tidball, and R. C. Severson. 1990. Geologic sources, mobilization, and transport of Se from the California Coast Ranges to the western San Joaquin Valley: A reconnaissance study. U.S. Geological Survey Water Resources Invest. Report No. 90–4070.

Presser, T. S., W. C. Swain, and R. Fujii. 1995. Surface water/ground-water linkages in relation to trace element and salinity distributions, Tulare Basin, southern San Joaquin Valley, California. *EOS*, Abstracts of the American Geophysical Union Transactions, Fall 1995, pF241.

Presser, T. S., M. A. Sylvester, N. M. Dubrovsky, and R. J. Hoffman. 1996. Review of the Grassland Bypass Channel Project monitoring program. U.S. Geological Survey Admin. Report, Menlo Park and Sacramento.

Sakai, M. K., and T. P. Lowe. 1987. Selenium in aquatic organisms from subsurface agricultural drainage water, San Joaquin Valley, *California Arch. Environ. Contam. Toxicol.* 16:657–670.

San Joaquin Valley Drainage Program. 1990. A management plan for agricultural subsurface drainage and related problems on the westside San Joaquin Valley. San Joaquin Valley Drainage Program, Sacramento.

San Joaquin Valley Drainage Implementation Program. 1994. Annual progress report. California Department of Water Resources, Sacramento.

San Joaquin Valley Interagency Drainage Program. 1979. Agricultural drainage and salt management in the San Joaquin Valley. California State Department of Water Resources, U.S. Bur. of Reclamation, California State Water Resources Control Board, Sacramento.

Shacklette, H. T., and J. G. Boerngen. 1984. Element concentrations in soils and other surficial material of the conterminous U.S. U.S. Geological Survey Prof. Paper No. 1270.

Skorupa, J. P., and H. M. Ohlendorf. 1991. Contaminants in drainage water and avain risk threshold. In A. Dinar and D. Zilberman (eds.), *The Economics and Management of Water and Drainage in Agriculture*, pp. 345–368. Kluwer Academic Publishers, Boston.

Skorupa, J. P., S. P. Morman, and J. S. Sefchick-Edwards. 1996. Guidelines for interpreting selenium exposures of biota associated with non-marine aquatic habitats. U.S. Fish and Wildlife Service, Sacramento.

Smith, Felix. 1995. Unreasonable use of water, public trust, and nuisance complaint (San Joaquin River) against the U.S. Bureau of Reclamation and California State Water Resources Control Board. CSWRCB, Sacramento.

Tidball, R. R., R. C. Severson, J. M. McNeal, and S. A. Wilson. 1989. Distribution of selenium, mercury, and other elements in soils of the San Joaquin Valley and parts of the San Luis Drain Service Area, California. In A. Howard (ed.), *Selenium and Agricultural Drainage: Implications for San Francisco Bay and the California Environ-ment. Proceedings of the Third Annual Symposium*, 1986, pp. 71–82. Bay Institute of San Francisco, San Rafael, CA.

Tulare Lake Drainage District. 1995. Tulare Lake Drainage District summary of activities. Tulare Lake Drainage District, Corcoran, CA.

Turekian, K. K., and K. H. Wedepohl. 1961. Distribution of the elements in some major units of the earth's crust. *Geol. Soc. Am. Bull.* 72:175–192.

U.S. Bureau of Reclamation. 1986. Final environmental impact statement, Kesterson pro-gram, Vol. II. USBR, Mid-Pacific Region, Sacramento.

U.S. Bureau of Reclamation. 1995. Finding of no significant impact and supplemental environmental assessment, Grassland Bypass Channel Project, interim use of a portion of the San Luis Drain for conveyance of drainage water through the Grassland Water District and adjacent Grassland Areas. USBR, Mid-Pacific Region, Sacramento.

U.S. Bureau of Reclamation. 1996. Annual report for Kesterson Reservoir, Merced County, California. Revised Monitoring and Reporting Program, pp. 87–149, USBR, Mid-Pacific Region, Sacramento.

U.S Environmental Protection Agency. 1980. Hazardous waste management system. *Fed. Regist.* 45:33,063–33,285.

U.S Environmental Protection Agency. 1992a. Rulemaking: Water quality standards: Estab-lishment of numeric criteria for priority toxic pollutants: States' compliance. Final rule. *Fed. Regist.* 57:60848, Dec. 22, 1992.

U.S Environmental Protection Agency. 1992b. Water quality criteria. USEPA, Washing-ton, DC.

Wedepohl, K. H. (ed.). 1969–1978. *Handbook of Geochemistry*, Vol. II, Parts 1–5. Springer-Verlag, New York.

Westcot, D. W., J. Karkoski, and R. Schnagl. 1996. Non–point source policies for agricul-tural drainage. In p.83–88 C. Bathala (ed.), *Proceedings of the North American Water and Environment Congress, 1996*. American Society of Civil Engineers, Environmental Engineering Division, Orange CA.

Zawislanski, P. T., and M. Zavarin. 1996. Nature and rates of selenium transformations: A laboratory study of Kesterson Reservoir soils. *Soil Sci. Soc. Am. J.* 60:791–800.

Zawislanki, P. T., G. R. Jayaweera, J. W. Biggar, W. T. Frankenberger, and L. Wu. 1996. The pond 2 selenium volatilization study: A synthesis of five years of experimental results, 1990–1995. A joint report to the U.S. Bureau of Reclamation, Lawrence Berkeley National Laboratory and University of California, Davis and Riverside.

11

Selenium Sources and Effects on Biota in the Green River Basin of Wyoming, Colorado, and Utah

Doyle W. Stephens
United States Geological Survey, Salt Lake City, Utah

Bruce Waddell
United States Fish and Wildlife Service, Salt Lake City, Utah

I. INTRODUCTION

The Green River Basin of Wyoming, Colorado, and Utah is a part of the Upper Colorado River Basin and has many scenic and mineral resources. High concentrations of salts, selenium, and other trace elements are associated with the sedimentary formations of the basin, and mobilization of these salts and elements has been and is a problem for irrigation in the region. The Upper Colorado River Basin is regarded as the greatest contributor of salinity to the Colorado River (U.S. Environmental Protection Agency, 1971). As early as 1935, drainage from irrigated soils high in selenium was identified as a source of elevated selenium concentrations in the Colorado River and its tributaries by Williams and Byers (1935).

This chapter summarizes data collected from 1976 to 1996, when selenium was studied extensively in the Green River Basin. The principal data sources were the National Water Data Storage and Retrieval System (WATSTORE) database of the U.S. Geological Survey (USGS), publications of the Department of Interior (DOI) National Irrigation Water Quality Program (NIWQP), reconnaissance studies of the DOI Floodplain Habitat Restoration Program, and contaminant assessment reports of the U.S. Fish and Wildlife Service (USFWS). Geologic maps in digital format for Wyoming (Love and Christiansen, 1985), Colorado (Tweto,

1979), and Utah (Hintze, 1980) were used to identify the geologic formations. Data for dissolved selenium in water were limited to results of samples analyzed only by the USGS using an oxidation and reduction digestion procedure prior to hydride generation. Analyses of biological tissues were done under supervision of the Patuxent Analytical Control Facility of the USFWS and values are reported as dry weight.

The Green River, known by the Shoshone Indians as the Seeds-kee-dee-Agie, or Prairie Hen River, originates at altitudes near 4300 m in the Wind River Mountains of western Wyoming and drains a basin of 117,000 km^2 (Fig. 1). It flows 1175 km to join the Colorado River 190 km south of Green River, Utah, and provides 5.5 billion cubic meters of water annually to the Colorado. The basin is noted for its rugged lands and mineral and wildlife resources; within its drainage are five state-managed wildlife or waterfowl areas, three national wildlife refuges, and one Bureau of Land Management (BLM) wetland complex. The basin is sparsely populated, and human population centers are closely associated with irrigated agriculture.

II. SOURCES OF SELENIUM TO THE GREEN RIVER AND ITS TRIBUTARIES

A. Geologic Sources

The river alluvium in headwater areas of the Green River is derived from granitic rocks that contain few soluble salts and are resistant to weathering. Alluvium in downstream areas is composed of local rock as well as rock transported from higher areas in the drainage. Where the local rock is composed primarily of Cretaceous or some Tertiary deposits, the resultant alluvium contains an abundance of soluble minerals and trace elements such as arsenic, boron, selenium, uranium, vanadium, and zinc. However, because of its ubiquity within the area, the element of greatest concern for biota is selenium. Although selenium occurs in rocks formed during the Carboniferous to Quaternary Periods, most studies have focused on the Cretaceous Period, and more is known about the distribution of seleniferous plants on soils derived from rocks of Cretaceous age than on soils from any other period (Berrow and Ure, 1989). Generally, areas with alkaline soils (pH > 7.5) that receive less than 64 cm of precipitation annually have the greatest potential of mobilizing selenium as a water contaminant (Lakin, 1961). Although precipitation in the Green River Basin varies from 15 cm to 152 cm, much of the basin receives less than 64 cm annually, and evaporation rates as high as 103 cm in the lower valleys of the river (Iorns et al., 1965) can greatly concentrate dissolved minerals in impounded water.

Selenium occurs in numerous formations deposited during the Cretaceous and Tertiary Periods. For example, Figure 1 was prepared with information from

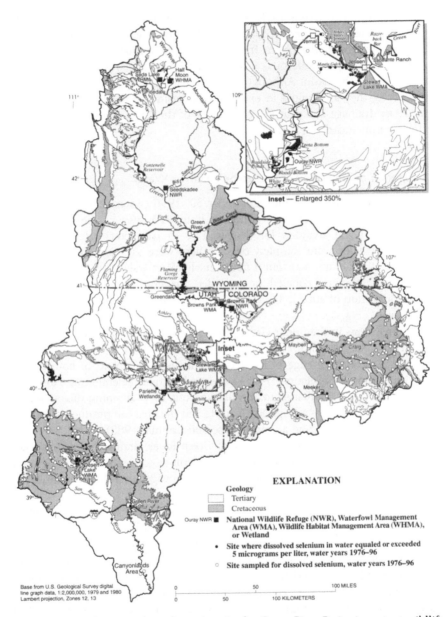

EXPLANATION

Geology

☐ Tertiary

▨ Cretaceous

Ouray NWR ■ National Wildlife Refuge (NWR), Waterfowl Management Area (WMA), Wildlife Habitat Management Area (WHMA), or Wetland

● Site where dissolved selenium in water equaled or exceeded 5 micrograms per liter, water years 1976–96

○ Site sampled for dissolved selenium, water years 1976–96

Base from U.S. Geological Survey digital line graph data, 1:2,000,000, 1979 and 1980 Lambert projection, Zones 12, 13

FIGURE I   Seleniferous geologic formations in the Green River Basin, important wildlife areas, sites where water samples for selenium were collected, and sites where selenium concentration equaled or exceeded the EPA chronic criterion of 5 μg/L for protection of aquatic wildlife.

21 formations of Cretaceous age identified as seleniferous by Rosenfeld and Beath (1964); however, 14 of them account for 84% of the area of all surficial deposits of Cretaceous age in the basin. In addition, eight formations of Tertiary age, identified as partly seleniferous, account for 97% of the area of all deposits of Tertiary age in the basin. Although most formations of Upper Jurassic age are seleniferous, they are limited in extent in the basin and occur primarily in the Price River drainage and near the confluence of the Green and Colorado Rivers and are not included in Figure 1.

B.   Relation of Geologic Sources to Selenium in Water

Between October 1, 1976, and September 30, 1996 (water years 1976–1996), samples of surface water were collected from 473 sites in the Green River Basin (Fig. 1) and analyzed for dissolved (0.45 μm filter) selenium by the USGS. Many sites were sampled once, others were sampled multiple times, particularly if selenium was found to adversely affect water use. Concentrations of dissolved selenium equaled or exceeded the chronic criterion of 5 μg/L for the protection of aquatic life (U.S. Environmental Protection Agency, 1987) at 111 sites (Fig. 1).

Three areas in the upper Green River Basin of Wyoming are underlain by deposits of Cretaceous age:

1.   The western edge of the basin contains a thin band of both Hillard Shale and Frontier Formation (Fm) known to be highly seleniferous (Rosenfeld and Beath, 1964). Tributaries to the Blacks Fork River draining this area have selenium concentrations as high as 12 μg/L, and the area can produce vegetation moderately toxic to grazing animals (Case and Cannia, 1988).

2.   Bitter Creek, east of the city of Green River, drains an area underlain by Baxter Shale and Rock Springs Fm, and selenium concentrations in the creek have been as high as 10 μg/L. The area produces vegetation that may be highly toxic to grazing animals (Case and Cannia, 1988).

3.   The Little Snake River, near the eastern edge of the basin, drains areas underlain by Lewis Shale, Lance Fm, Steele Shale, and Mesa Verde Group, all formations of Cretaceous age, but only the Steele Shale is regarded as highly seleniferous (Rosenfeld and Beath, 1964). Few water samples from streams immediately underlain by the Cretaceous formations have been analyzed for selenium in this area. However, Naftz and Barclay (1991) analyzed 59 surface water samples from the Savery Creek drainage immediately east of the Cretaceous outcrop and found the maximum concentration of selenium was 3 μg/L. The Little Snake River downgradient of these sources typically contains less than 5 μg/L.

A large area of Tertiary age deposits, primarily the Laney Member of the Green River Fm, occurs in the Big Sandy River drainage where selenium concentrations have been as high as 5 μg/L upstream of Seedskadee National Wildlife Refuge (NWR). Members of the Green River Fm appear to be only minor sources of selenium to the Big Sandy River, as concentrations have not exceeded 5 μg/L in numerous samples collected during 1975–1991.

The headwaters of the Yampa River in Colorado are underlain by Precambrian age gneiss and schist and Tertiary age basaltic formations that are not seleniferous. However, the area within a radius of about 50 km of the town of Craig contains extensive Cretaceous deposits (Williams Fork Fm, Iles Fm, and Laramie Fm), all of which are seleniferous. The largest sources are several small streams entering the Yampa River about 34 km east of Craig. Concentrations in these streams ranged from 24 to 300 μg/L and likely account for much of the 13 to 17 μg/L measured in the Yampa River downstream of Craig. Goodspring Creek drains formations of Cretaceous age between Craig and the town of Maybell, and selenium concentrations in the drainage have commonly ranged from 5 to 11 μg/L. The Cretaceous age deposits through which the Yampa River flows upstream of Maybell result in selenium concentrations in the river that range from 0 to 11 μg/L with a mean of about 2 μg/L near Maybell. The Little Snake River drains a large area of Tertiary age Wasatch Fm and Bridger Fm, both of which may be seleniferous in areas but contribute only a small amount of selenium to the Little Snake River.

The bedrock in the headwaters of the White River in Colorado primarily consists of non-selenium-contributing Quaternary age deposits and Tertiary age Weber Sandstone. Areas north and south of the town of Meeker are underlain by Iles and Williams Fork Fm that are seleniferous; concentrations of selenium in the White River near Meeker occasionally exceeded 5 μg/L. Piceance and Yellow Creeks drain an area of Tertiary age Uinta Fm that is partly seleniferous, and concentrations of selenium occasionally exceeded 5 μg/L in both creeks. Evacuation and Bitter Creeks drain an area underlain by the Parachute Creek Member of the Green River Fm, a Tertiary age deposit that is seleniferous in areas. Selenium concentrations in these creeks ranged from 5 to 14 μg/L. Selenium concentrations as high as 8 μg/L have been measured in the White River near its confluence with the Green River.

The Utah part of the Green River Basin contains several extensive areas of Cretaceous age deposits, notably the Mancos Shale, which is about 1500 m thick and extends east of Vernal, and a broad area of Mesa Verde Group and Mancos Shale that extends from the southwest border of the basin through the town of Green River, Utah, and extending 150 km into Colorado. A small area of Mesa Verde Group occurs near the headwaters of the Duchesne River. These source areas are discussed in Section III.

III. SELENIUM IN WETLANDS WITHIN THE GREEN RIVER BASIN

A. Wyoming

Seedskadee NWR, located 32 km north of Green River, Wyoming, near the confluence of the Big Sandy and Green Rivers, is underlain by the Bridger Fm of Tertiary age, a formation that is seleniferous in areas and can support highly seleniferous vegetation (Rosenfeld and Beath, 1964). Ponds and wetlands receive water from both rivers. In 1975 and 1977, two water samples from the Big Sandy River contained selenium concentrations of 5 μg/L, but concentrations in the Green River upstream of Seedskadee NWR were generally less than the reporting limit of 1 μg/L. A study of ponds at Seedskadee NWR completed by Ramirez and Armstrong (1992a) showed that samples of water, bottom sediment, and biota taken in 1988–1989 did not contain selenium in concentrations hazardous to fish or waterfowl (Table 1). Ramirez and Armstrong (1992b) sampled bottom sediment and plants from six sites along the Big Sandy River, 45 km upstream of Seedskadee NWR, and reported that the highest concentration of selenium found was 5.3 μg/g in bottom sediment near a source of seepage to the river.

B. Colorado and Utah

1. Browns Park National Wildlife Refuge

Browns Park NWR (federal) and the adjacent Browns Park Waterfowl Management Area (WMA) (state of Utah) are located on the Green River near the Colorado–Utah state line and provide about 3400 ha of waterfowl area. The wetlands are located on Quaternary age gravels transported by the river, but several of the tributary drainages cut through the potentially seleniferous Browns Park Fm of Tertiary age. However, selenium concentrations in water collected in 1987 from two streams and five ponds at Browns Park NWR did not exceed 1 μg/L, and selenium concentrations in fish from the Green River near the refuge were less than the value of 4 μg/g recognized as a level of concern for cold-water fish (Lemly, 1993). Selenium concentrations in the livers of 10 American coots (*Fulica americana*) from three marshes were also low, ranging from 1.4 to 5.8 μg/g (Table 1).

2. Backwater Areas Along the Green River in Utah

Fourteen sites have been identified along the Green River as potential areas for restoration or enhancement of floodplain functions needed to support recovery of endangered fish in the Upper Colorado River Basin. Investigations of many of the sites (Table 2) have been completed under the Floodplain Habitat Restoration

	Sample period (years)	Surface water (μg/L)	Bottom sediment (μg/g)	Aquatic plants/algae (μg/g)	Invertebrates (μg/g)	Whole-body fish (μg/g)	Bird eggs (μg/g)	Bird livers (μg/g)	
	1988–1989	<1–<3	<0.2–1.9	0.65–2.13	<0.53–2.7	NA	NA	2.1–5.1	Ramirez an (1992a)
'R	1987	<1–1	NA	NA	NA	1.9–3.7	NA	1.4–5.8	Peltz and W
1A	1986–1989	<1–140	1–720	1.2–73	10.4–37.4	4–83.9	2.0–33	1.9–74.7	Peltz and W
-27 km ver)	1986–1989	<1–16,000	7.1	1.6–13.3	36.9–51.4	1.6–122	4.1–71[b]	30.8–50.3	Peltz and W Stephens et
	1986–1990	<1–93	<0.1–26	0.3–91	0.7–71.7	0.58–104	1.2–120	3–125	Peltz and W Stephens et
3asin	1995	<1–2	<1–15.3	<0.5–0.9	0.7–10.7	1.2–8.3	1.2–6	NA	Stephens an unpublish
	1987–1988	<1–9	0.4–1.5	0.3–15.3	3.2–10.6	1.8–16.9	1.4–16.9	5.1–47	Peltz and W
San Rafael	1994	<1–46	<1	<0.5–8	1.1–29	5.5–16.2	1.9–22.6	3.5–72.5	Stephens et Waddell an (1992)
a	1995	<1–<2	NA	NA	9.6	1.3–10.7	NA	NA	Stephens an unpublish

TABLE 2 Concentrations of Selenium in Water, Bottom Sediment, and Biota from Flooded Bottomlands Adjacent to the Green River in Utah, 1993–1996

Area	Distance from mouth (km)	Selenium in Water (μg/L)	Selenium in Sediment (μg/g, dry wt)	Selenium in Biota (μg/g, dry wt)
Vernal area				
Escalante Ranch	491	1–2400	—	7.0 invertebrates, 5.0 avian eggs
Bonanza Bridge	464	<2	<1	2.7 invertebrates
Baeser Bend	438	<2	<1	2.3 invertebrates
Brennan Bottom	430	<2	<2	3.3 carp, 4.8 green sunfish
Ouray NWR Area				
Johnson Bottom	425	<1	—	2.2 chironomids
Wyasket Bottom	410	<1	—	1.6 rooted plants
Canyonlands Area				
Spring Canyon	109	<2	—	1.3 small fish
Millard Canyon	54	<1	—	—
Anderson Bottom	50	<2	—	10.6 carp, 10.7 channel catfish, 4.8 small fish
Holman Canyon	45	<1	—	4.5 small fish, 9.6 crayfish

Program of DOI (K. Holley, U.S. Bureau of Reclamation, written communications, 1995, 1996).

Escalante Ranch, a wetland of about 250 ha adjacent to the Green River 10 km upstream from Stewart Lake WMA, is partly underlain by Mancos Shale. Its proximity to documented spawning areas for the endangered razorback sucker (*Xyrauchen texanus*) makes it desirable habitat for rearing the young fish. During 1993, selenium concentrations in water from wetland ponds were as high as 5 μg/L; concentrations in invertebrates were near 7 μg/g, but concentrations in avian eggs did not exceed 5 μg/g (B. Waddell and D. Stephens, unpublished). Additional measurement of water sources by Cooper and Severn (1994) found that seepage on the bluff east of the wetland contained selenium concentrations of 3 to 1260 μg/L, but selenium contamination was apparently not widespread and was less than detection level in the wetland ponds. Studies of seepage sources to the wetlands in 1994 (D. Stephens and J. Miller, unpublished) provided quantification of 15 seeps containing 1 to 2400 μg/L with a combined loading potential of greater than 200 g of selenium per day into the northern area of the wetlands. Because of the potential for large amounts of selenium to contaminate the wetlands, the area was not considered for development as a rearing site.

Areas considered for habitat restoration receive most of their water from the Green River, which usually contains little selenium. Backwater areas, however, may receive groundwater or surface drainage that could contain selenium in quantities harmful to young fish. Table 2 summarizes contaminant assessments in many of the areas considered for development as backwater habitat. Wetland sites at Ouray NWR and Pariette Wetlands are discussed in Section III.B.5 and III.B.7, respectively. In addition to Escalante Ranch, concentrations of selenium that exceed a level of concern for biota were found at Sheppard Bottom in the Ouray NWR, Pariette Wetlands, and Anderson Bottom in the Canyonlands Area.

3.   Stewart Lake Waterfowl Management Area

Extensive studies have been completed at Stewart Lake WMA, located adjacent to the Green River 19 km southeast of Vernal, Utah. Stewart Lake, managed by the Utah Division of Wildlife Resources, has a surface area of 100 ha and occupies most of the WMA. The lake is flooded every few years by the Green River and occasionally (1995, 1997) by Ashley Creek. Since 1979, the primary source of surface water to Stewart Lake has been discharge from five subsurface drains installed by the Bureau of Reclamation (BR) to replace streamflow diverted to a local reservoir. The drainage system services about 300 ha of irrigated farmland underlain by Mancos Shale.

Studies from 1986 to 1990 (Stephens and Waddell, 1992; Stephens et al., 1992) showed that median concentrations of selenium entering Stewart Lake from the drains exceeded 5 μg/L and were as high as 140 μg/L (Table 1). Concentrations of total selenium in bottom sediment samples collected in 1991 near the point of discharge from the drains were as high as 720 μg/g but declined to 4 to 7 μg/g near the outflow to the Green River. During the summers of 1986–1989, about 252 g of selenium entered Stewart Lake daily from the drains, but only 60 g/day was discharged to the Green River. The lake functioned as a biological treatment facility, retaining about 75% of the selenium discharged from the drains.

During 1986–1989, nesting success by waterbirds was poor, there were few benthic insects, and selenium concentrations in biota were high. Mean selenium concentrations in most whole-body fish substantially exceeded the value of 4 μg/g associated with adverse effects on fish populations (Lemly, 1993), and a razorback sucker captured in Stewart Lake in 1994 had a concentration of selenium in a muscle plug of 13 μg/g. Selenium concentrations in 26 of 49 waterbird eggs from Canada goose (*Branta canadensis*), black-crowned night heron (*Nycticorax nycticorax*), gadwall (*Anas strepera*), redhead (*Aythya americana*), teal (*Anas dis-*

cors), western grebe (*Aechmophorus occidentalis*), and common snipe (*Gallinago gallinago*) exceeded the 8 μg/g value associated with reproductive impairment in populations of some species (Skorupa and Ohlendorf, 1991). Deformities consistent with selenium toxicosis in redhead and teal embryos were found in the area. Planning for remediation of selenium contamination at Stewart Lake WMA was begun in 1991, and a pipeline to deliver a small quantity of water for dilution of selenium in the lake was completed in 1995. Pending completion of environmental documents, construction activities to route the subsurface drains to the Green River and provide dikes for water management are anticipated to begin in the summer of 1997.

4. Ashley Creek

Selenium contamination of Ashley Creek was first documented by Stephens et al. (1988), who reported that concentrations in the lower reach of the creek during 1986–1987 ranged from 25 to 73 μg/L. Stephens et al. (1992), using data from June 1986, calculated that the creek discharged 8 kg of selenium per day to the Green River. Sampling of tributaries and seeps in 1988–1989 identified water sources containing selenium concentrations ranging less than 1 μg/L to 16,000 μg/L (Table 1) within 27 km of the confluence with the Green River. Seventy percent of the water samples from 23 tributaries and seeps contained selenium concentrations exceeding 5 μg/L, and the average concentration of selenium in Ashley Creek near the confluence with the Green River was 52 μg/L.

The area contributing the highest concentrations of selenium was immediately downgradient of the Ashley Valley sewer lagoons, a complex of 5 ponds originally constructed as a total containment facility on a bluff of Mancos Shale underlain by the Split Mountain Anticline. Inflow to the lagoons totaled 4.37 $\times$ 10^6 m^3 in 1987, and potential evaporation for the same period could have removed about 1.36 $\times$ 10^6 m^3, indicating that about 3 $\times$ 10^6 m^3 of water (0.09 m^3/s) could have leaked from the lagoons in 1987 (Stephens et al., 1992). Synoptic measurements of all inflows and outflows to Ashley Creek in a 6.8 km reach downgradient of the lagoons in 1991 and 1992 indicated that about 0.04 to 0.09 m^3/ s of seepage from the lagoons reached Ashley Creek. Water in the lagoons, which initially does not contain selenium, leaks from the facility and seeps through fractured Mancos Shale, mobilizing high concentrations of selenium. The seleniferous water discharges to Ashley Creek. Smaller amounts of selenium are also contributed by 11 BR agricultural drains in the Vernal area, canal seepage and runoff, and drainage from Mantle Gulch. Based on only a few measurements of sources of flow to Ashley Creek, mass balance calculations show that about 85% of the selenium loading to Ashley Creek is due to leakage from the lagoons, 6% to Mantle Gulch, 5% to BR drains, and 4% to miscellaneous sources.

Samples of biota from Ashley Creek had selenium concentrations that rank among the highest found in the Green River Basin. Selenium concentrations in fish from four sites on Ashley Creek ranged from 1.6 μg/g to 122 μg/g with a geometric mean of 32.7 μg/g and were significantly higher at sites downgradient from the lagoons. Selenium concentrations in waterfowl livers from the Winter Storage Pond adjacent to the lagoons ranged from 30.8 to 50.3 μg/g, waterbird eggs were as high as 71 μg/g, and deformed avian embryos were found (Peltz and Waddell, 1991; Stephens et al., 1992). In 1996 the USFWS approached the Ashley Valley Water Management District, the Utah Department of Environmental Quality, and the U.S. Environmental Protection Agency (EPA) urging that a solution be implemented to prevent seepage from the lagoons from contaminating Ashley Creek. Planning is under way to replace the facility with a traditional mechanical treatment plant.

5. Ouray National Wildlife Refuge

Ouray NWR is adjacent to the Green River 48 km south of Vernal and was developed in 1960 as part of the mitigation for construction of Flaming Gorge Reservoir. Quaternary age gravels and recent alluvial deposits characterize the surface geology near the refuge. Fluvial and lake deposits of the Uinta Fm and fluvial sandstone and mudstone of the Duchesne River Fm of Tertiary age, both partly seleniferous, make up most of the upland areas surrounding the refuge. Studies done at Ouray NWR during 1986–1990 (Stephens et al., 1988, 1992) showed that selenium concentrations in refuge ponds were variable and depended on the source of water supplied to the ponds. The selenium concentration in irrigation water supplied to the refuge and water from the Green River rarely exceeded 1 μg/L, but water from two ponds on the western part of the refuge had median concentrations near 40 μg/L. The source of the selenium is shallow groundwater that leaches selenium from the underlying formations and moves southerly across the western edge of the refuge. Less than 10% of the refuge has been affected by selenium contamination, and only rarely is any surface water discharged from the refuge to the Green River.

Geometric mean selenium concentrations in biota from the Roadside Ponds of 54.7 μg/g in bird eggs and 82.6 μg/g in fathead minnows (*Pimephales promelas*) exceeded concentrations at which reproduction and survival are seriously affected (Skorupa and Ohlendorf, 1991; Lemly, 1993). Geometric mean concentrations in most plants (30.3 μg/g) and invertebrates (26.9 μg/g) were higher than values known to cause reproductive failure in mallards (Heinz et al., 1987). Deformed embryos of mallards, American coots, and redheads were found in and near the ponds. Currently (1997), efforts are underway at Ouray NWR to reduce selenium contamination entering the refuge, manage contaminated seepage, and eliminate use of the Roadside Ponds by waterfowl. Ranges of selenium concentrations in all media are reported in Table 1.

6.   Duchesne River Basin

Water quality and biological studies under the NIWQP program in cooperation with the Ute Indian Tribe were done in 1995 on a large part of the Duchesne River Basin to determine whether selenium or other contaminants associated with irrigation had adverse effects on tribal wetlands (D. Stephens and B. Waddell, unpublished). The geology of the basin is mixed but many of the rocks are of early Tertiary age: Uinta Fm, Green River Fm, and Duchesne River Fm, all of which are considered partly seleniferous. Twenty sites were sampled, and although concentrations of dissolved boron in water from several sites were near 10,000 $\mu g/L$, concentrations of selenium in water did not exceed 2 $\mu g/L$. Concentrations of selenium in biota generally were small, but several samples of biota exceeded values of 3 $\mu g/g$ for dietary items and 4 $\mu g/g$ for fish, both regarded as concentrations with potentially toxic effects (Lemly, 1993). Selenium was not a major contaminant in the basin, and the mean concentration of selenium discharged by the Duchesne River to the Green River was about 1 $\mu g/L$. Ranges of selenium concentrations in all media are reported in Table 1.

7.   Pariette Wetlands

Pariette Wetlands, a BLM complex of 3600 ha located adjacent to the Green River 56 km south of Vernal, is underlain by the Uinta Fm of Tertiary age, which is seleniferous in areas. The principal source of water is agricultural diversions from the Duchesne River that enter the complex as tailwater and drainwater from the agricultural area upgradient. Studies done during 1987–1988 (Stephens et al., 1992) showed that water entering the wetlands typically had a selenium concentration of 5 $\mu g/L$, which declined to near 2 $\mu g/L$ at the point of discharge to the Green River. Geometric mean concentrations of selenium in whole-body fish samples from three pond complexes ranged from 5.8 to 9.8 $\mu g/g$ and exceeded a reference value of 4 $\mu g/g$ associated with adverse effects on some fish species (Lemly, 1993). Concentrations of selenium in eggs and livers of American coots in one pond were greater than the toxicity threshold of 3 $\mu g/g$ (eggs) and 10 $\mu g/g$ (liver) regarded as potentially hazardous to waterfowl populations (Lemly, 1993). No deformed waterfowl embryos were found during the study. Ranges of selenium concentrations in all media are reported in Table 1.

8.   Price and Upper San Rafael River Drainages

In 1988 and 1990, concentrations of selenium in waterbird eggs at Desert Lake WMA in the Price River drainage ranged from 6.8 to 22.6 $\mu g/g$ and were associated with concentrations in water of 4 to 12 μg/L (Waddell and Coyner, 1990; Waddell and Stephensen, 1992). The extent of selenium contamination of other wetlands in the area was investigated under the NIWQP program in 1994 (Stephens et al., 1997). Headwaters areas of the Price River and small streams such as Cottonwood and Huntington Creeks that flow to the San Rafael River are underlain by the Cretaceous age Mesa Verde Fm. The streams flow through Mancos Shale in the upper benches of the valley, and several irrigation reservoirs and Desert Lake WMA are located near Mancos Shale deposits. Selenium concentrations exceeded 5 μg/L at several sites in the Price River drainage, and several drains and an open channel near Desert Lake WMA contained 12 to 46 μg/L. Concentrations of selenium entering Desert Lake were typically less than 4 μg/L but contained about 5000 mg/L of dissolved solids. Selenium concentrations in water leaving the wetlands were near 1 to 2 μg/L.

Selenium concentrations in aquatic invertebrates exceeded the dietary toxicity threshold for birds in several ponds at Desert Lake WMA, an open drain near Desert Lake, and several small reservoirs in the Price River drainage (Stephens et al., 1997). Selenium concentrations were greater than the toxicity threshold value of 4 μg/g in all fish samples collected from the Price River drainage and were greater than 13 μg/g in all fish from Desert Lake WMA. Ten of the 18 waterbird eggs collected throughout the study area had selenium concentrations within the level of concern of 3 to 8 μg/g (Lemly, 1993; Skorupa and Ohlendorf, 1991), 6 exceeded the toxicity threshold value of 8 μg/g, and 5 of the 6 highest concentrations were from Desert Lake WMA. No deformities of fish or waterbirds were observed. Selenium concentrations for all media are reported in Table 1.

IV. EFFECTS OF SELENIUM ON THE GREEN RIVER AND ITS FISHERY

A. Selenium Loading to the Green River

Concentrations of selenium in the Green River from the headwaters to downstream of Flaming Gorge Reservoir typically remain less than 1 μg/L with only infrequent spikes to 4 μg/L (Fig. 2). The Yampa River is the first substantial contributor of selenium, and downstream of its confluence, the Green River near Jensen, Utah, maintains a median concentration near 1 μg/L but occasionally spikes to 7 μg/L. Between Jensen and Green River, Utah, relatively high concentrations of selenium enter from Stewart Lake, Ashley Creek, Pariette Wetlands, and the Price River. At Green River, Utah, the median concentration of selenium in the river is 2 μg/L with infrequent spikes as high as 8 μg/L.

Although concentrations are of interest in identifying sources of selenium and for regulatory purposes, the effect of each source is best viewed by the total load it contributes. Daily loading to the Green River from each of the major

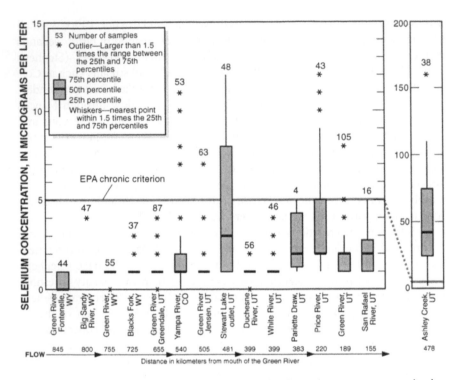

FIGURE 2 Selenium concentrations in water samples from the Green River and select tributaries, water years 1976–1996. Site names correspond to site locations shown in Figure 3.

sources calculated using the mean flow for the period of record and the mean selenium concentration for water years 1976–1996 is shown in Figure 3. The major sources of loading under mean flow conditions are the Yampa River (8 kg/day) and Ashley Creek (7.8 kg/day), which when combined account for nearly 60% of the calculated loading in the river at Green River, Utah. Cumulative contributions of minor sources total about 5.5 kg/day. Because of unmeasured sources and error inherent in calculations using mean flows and mean concentrations, however, summation of source loading does not equal calculated loading at Green River, Utah (27 kg/day), nor does it include unmeasured biological uptake or volatilization of selenium.

B. Effects on Endangered Fish

The Green River historically has supported populations of the Colorado squawfish (*Ptychocheilus lucius*), razorback sucker (*Xyrauchen texanus*), bonytail chub (*Gila*

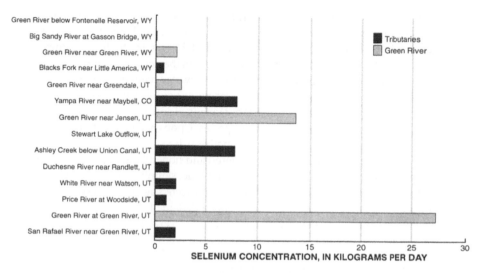

FIGURE 3 Daily selenium loading in the Green River and select tributaries, water years 1976–1996.

elegans), and humpback chub (*G. cypha*). These species are listed as endangered and their critical habitat has been designated as the 100-year floodplain from Craig, Colorado, on the Yampa River to the confluence with the Colorado River, the lower 200 km of the White River, and the lower 4 km of the Duchesne River (Maddux et al., 1993). The spawning areas for Colorado squawfish and razorback suckers are limited (Tyus and Karp, 1990), and critical life stages are from fertilized eggs through the first year of development (Miller et al., 1982). The Green River in the Jensen area is a known spawning location for razorback suckers, with staging in Ashley Creek and the outlet for Stewart Lake (Tyus and Karp, 1990). After hatching, larval fish drift downstream until they reach backwater habitats, where they feed and mature.

The effect of selenium (or other contaminants) on fish populations in the Green River is difficult to distinguish from habitat changes caused by human manipulation of the hydrology of the river. Prior to the construction of dams in the basin, flow in the Green River would begin to increase in March as a result of snowmelt and would peak in June. Prior to construction of Flaming Gorge Reservoir, the average peak at Green River, Utah was 900 m³/s for 1895–1962, and in 1917 the peak reached 1900 m³/s (Collier et al., 1996). The spring floods were cold and carried large quantities of suspended sediment. Summer flows were clear and warm. With the filling of Flaming Gorge Reservoir in 1966, the timing of flows was altered: winter flows increased and the spring flood rarely

occurred. Hypolimnetic releases from the reservoir caused mean water temperatures to drop from a predam average of 19°C to 6°C (Vanicek and Kramer, 1969),
which inhibited spawning of many native fish. Changes in the flow regime and
water temperature were accompanied by changes in the deposition and suspension
of bottom sediment that greatly altered the backwater habitats favored by many
of the native species for spawning and rearing of young fish. These changes in
the environment have favored the establishment of introduced fish species that
compete with, and prey upon, the native species. Upon this background, selenium
is only another agent of stress upon the native fish.

Effects of trace elements on endangered and nonendangered fish have been
investigated by collection and analysis of fish tissue, sampling of food chain items,
and bioassays. Because razorback suckers are protected against harvest, nonlethal
techniques for sampling tissues were developed (Waddell and May, 1995). Muscle
plugs taken from adult razorbacks collected at several locations in the Green
River during 1991–1995 show large concentrations of selenium (median near
34 μg/g) associated with fish collected near the outflows of Ashley Creek and
Stewart Lake (Fig. 4). All fish sampled had muscle plug concentrations greater
than 4 μg/g, a whole-body threshold associated with adverse effects on reproduc-

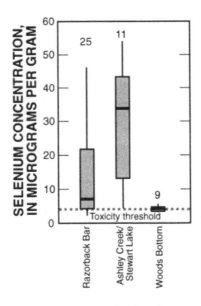

FIGURE 4   Concentrations of selenium in muscle plugs from razorback suckers sampled
at three areas on the Green River, 1991–1995. The toxicity threshold refers to adverse
effects on reproduction at 4 μg/g dry weight in whole-body fish samples (Lemly, 1993).
(Unpublished data from B. Waddell.)

tion (Lemly, 1993). Concentrations of selenium in muscle plugs from fish collected near Razorback Bar, a known spawning site 15 km upstream from Ashley Creek, were elevated relative to samples from Woods Bottom at Ouray NWR (a generally uncontaminated site). Contamination of fish collected near the Razorback Bar is likely because of its close proximity to feeding areas near Ashley Creek. Sensitivity of fish to contaminants is dependent on the ionic form of the contaminant and the life stage most susceptable to the contaminant. Recent studies of three endangered fish species in the Green River by Hamilton (1995) demonstrated that Colorado squawfish were two to five times more sensitive to selenate and selenite at the swimup life state (17–31 days after hatching) than at older stages. Bonytail chub was 5 times more sensitive to selenate at the swimup stage than at older stages, and razorback suckers were equally sensitive at both swimup and juvenile stages. Overall, the study concluded that concentrations of boron, selenium, and zinc in Ashley Creek may be adversely affecting the early life stages of three endangered fish.

In situ bioassays showed that water samples from Ashley Creek and Stewart Lake were directly lethal to larval razorback suckers during 10-day exposures (Finger et al., 1995). Although lethality was not wholly attributable to selenium, the tests showed that selenium in combination with other trace elements caused mortality of larval razorbacks. Additionally, bioassays done using *Daphnia magna*, (fathead minnows) and razorback sucker larvae showed that selenium was the only element that exhibited a statistically significant relation to reproductive failure or mortality. The magnitude of contamination of food sources eaten by larval razorbacks was shown for plankton collected from the backwaters of the Stewart Lake/Ashley Creek area. The plankton contained selenium concentrations of 13.6 to 24.9 $\mu g/g$, or about four to eight times a value of 3 $\mu g/g$ associated with toxic effects through the aquatic food chain (B. Waddell, unpublished). The data indicate that larvae drifting downstream after hatching may encounter lethal conditions in the backwater area created by the outflows of Stewart Lake and Ashley Creek.

C. Effects on Other Fish

Additional studies were done to further define the extent of contamination of nonendangered fish in the Green River that may be associated with the Ashley Creek/Stewart Lake area (B. Waddell and C. Wiens, written commununication, 1994). Analysis of whole-body samples of carp (*Cyprinus carpio*) from the Green River shows that the largest concentrations of selenium in tissues were observed near the confluence of Ashley Creek/Stewart Lake and concentrations were lower both upstream and downstream (Fig. 5). Carp and razorback suckers in the Green River have some common food sources, but it is unknown whether carp and razorback suckers concentrate selenium to the same degree. However, the

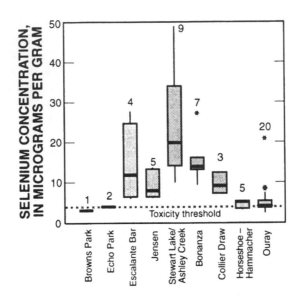

FIGURE 5 Concentrations of selenium in whole-body samples of carp sampled at nine areas on the Green River, 1978–1995. The toxicity threshold refers to adverse effects on reproduction at 4 μg/g dry weight in whole-body fish samples (Lemly, 1993). (Data from Schmitt and Brumbaugh, 1990; Peltz and Waddell, 1991; Waddell, unpublished.)

range of selenium values but not the median of tissue concentration data is similar in both species collected from the same sites.

V. SUMMARY

Sedimentary deposits of Cretaceous and sometimes Tertiary age are the principal sources of selenium in the Green River Basin. Water passing through soils derived from these deposits leaches variable amounts of soluble selenium and transports it to streams and wetlands. Selenium contamination of water and wetlands is not substantial upstream of Flaming Gorge Reservoir, but downstream, relatively large loads of selenium are discharged to the Green River by the Yampa River and Ashley Creek. Median concentrations of selenium in the Green River range from less than 1 μg/L above Flaming Gorge Reservoir to about 2 μg/L at Green River, Utah, with only occasional spikes that exceed the EPA chronic criterion of 5 μg/L. Adverse effects on biota in wetlands adjacent to the Green River generally are caused by localized sources and are almost always associated with shallow groundwater and not with the river. Areas where biota have been adversely affected include Stewart Lake WMA, lower Ashley Creek, Pariette Wetlands, part of Ouray NWR, and Desert Lake WMA. Although direct exposure to dissolved selenium in the Green River does not appear to result in acute lethality to endangered fish, waterborne dietary exposure has resulted in selenium concentrations in razorback suckers that are sufficient to inhibit reproduction.

REFERENCES

Berrow, M. L., and A. M. Ure. 1989. Geological materials and soils. In M. Ihnat (ed.), *Occurrence and Distribution of Selenium*, pp. 213–242. CRC Press, Boca Raton, FL.

Case, J. C., and J. C. Cannia. 1988. Guide to potentially seleniferous areas in Wyoming. Geological Survey of Wyoming, Open-File Report No. 88–1.

Collier, M., R. H. Webb, and J. C. Schmidt. 1996. Dams and rivers—Primer on the downstream effects of dams. U.S. Geological Survey Circular No. 1126.

Cooper, D. J., and C. Severn. 1994. Wetlands of the Escalante Ranch Area, Utah: Hydrology, water chemistry, vegetation, invertebrate communities, and restoration potential. Unnumbered report for the Recovery Program for the Endangered Fishes of the Upper Colorado. Colorado State University, Fort Collins.

Finger, S. E., A. C. Allert, S. J. Olson, and E. C. Callahan. 1995. Toxicity of irrigation drainage and associated waters in the middle Green River Basin, Utah. Report to U.S. Fish and Wildlife Service. National Biological Survey, Columbia, MO.

Hamilton, S. J. 1995. Hazard assessment of inorganics to three endangered fish in the Green River, Utah. *Ecotoxicol. Environ. Saf.* 30:134–142.

Heinz, G. H., D. J. Hoffman, A. J. Krynitsky, and D. M. G. Weller. 1987. Reproduction in mallards fed selenium. *Environ. Toxicol. Chem.* 6:423–433.

Hintze, L. F. Geologic Map of Utah. 1980. Utah Geological and Mineral Survey map. Scale 1 : 500,000.

Iorns, W. V., C. Hembre, and G. Oakland. 1965. Water resources of the Upper Colorado River Basin-. Technical report. U.S. Geological Survey Prof. Paper No. 441.

Lakin, H. W. 1961. Geochemistry of selenium in relation to agriculture. In M. Anderson, H. Lakin, K. Beeson, F. Smith, and E. Thacker (eds.), *Selenium in Agriculture*, pp. 1–12. U.S. Department of Agriculture Handbook No. 200. USDA, Washington, DC.

Lemly, A. D. 1993. Guidelines for evaluating selenium data from aquatic monitoring and assessment studies. *Environ. Monit. Assess.* 28:83–100.

Love, J. D., and A. C. Christiansen. 1985. Geologic Map of Wyoming. U.S. Geological Survey special geologic map. Scale 1 : 500,000.

Maddux, H. A., L. A. Fitzpatrick, and W. R. Noonan. 1993. Colorado River endangered fishes critical habitat. Draft biological support document. U.S. Fish and Wildlife Service, Salt Lake City, UT.

Miller, W. H., J. J. Valentine, D. L. Archer, H. M. Tyus, R. A. Valdez, and L. Kaeding. 1982. Colorado River Fishery Project. Part 1. Final Report, Summary. U.S. Fish and Wildlife Service, Salt Lake City, UT.

Naftz, D. L., and C. Barclay. 1991. Selenium and associated trace elements in soil, rock, water and streambed sediment of the proposed Sandstone Reservoir, south-central Wyoming. U.S. Geological Survey, Water Resources Invest. Report No. 91–4000.

Peltz, L. A., and B. Waddell. 1991. Physical, chemical, and biological data for detailed study of irrigation drainage in the middle Green River Basin, Utah, 1988–89. U.S. Geological Survey Open-File Report No. 91–530.

Ramirez, P., and J. Armstrong. 1992a. Environmental contaminant surveys in three national wildlife refuges in Wyoming. U.S. Fish and Wildlife Service Contaminant Report No. R6/702C/92. USFWS, Cheyenne.

Ramirez, P., and J. Armstrong. 1992b. Environmental contaminants monitoring in selected wetlands of Wyoming. U.S. Fish and Wildlife Service Contaminant Report No. R6/701C/92. USFWS, Cheyenne.

Rosenfeld, I., and O. Beath. 1964. *Selenium—Geobotany, Biochemistry, Toxicity, and Nutrition.* Academic Press, New York.

Schmitt, C. J., and W. G. Brumbaugh. 1990. National contaminant biomonitoring program: Concentrations of arsenic, cadmium, copper, lead, mercury, selenium, and zinc in U.S. freshwater fish, 1976–1984. *Arch. Environ. Contam. Toxicol.* 19:731–747.

Skorupa, J. P., and H. M. Ohlendorf. 1991. Contaminants in drainage water and avian risk thresholds. In A. Dinar and D. Ziberman (eds.), *The Economics and Management of Water and Drainage in Agriculture*, pp. 345–368. Kluwer Academic Publishers, Boston.

Stephens, D. W., and B. Waddell. 1992. Selenium contamination of waterfowl areas in Utah and options for management. In R. Robarts and M. Bothwell (eds.), *Aquatic Ecosystems in Semi-Arid Regions: Implications for Resource Management*, pp. 301–311. National Hydrology Research Institute Saskatoon, Saskatchewan Symposium Series 7. Environment Canada.

Stephens, D. W., B. Waddell, and J. B. Miller. 1988. Reconnaissance investigation of water quality, bottom sediment, and biota associated with irrigation drainage in the middle Green River Basin, Utah, 1986–87. U.S. Geological Survey Water Resources Invest. Report No. 88–4011.

Stephens, D. W., B. Waddell, and J. B. Miller. 1992. Detailed study of selenium and selected elements in water, bottom sediment, and biota associated with irrigation drainage in the middle Green River Basin, Utah, 1988–90. U.S. Geological Survey Water Resources Invest. Report No. 92–4084.

Stephens, D. W., B. Waddell, K. Dubois, and E. Peterson. 1997. Field screening of water quality, bottom sediment, and biota associated with the Emery and Scofield Project areas, central Utah, 1994. U.S. Geological Survey Water Resources Invest. Report No. 96–4298.

Tweto, O. 1979. Geologic Map of Colorado. U.S. Geological Survey special geologic map. Scale 1 : 500,000.

Tyus, H. M., and C. A. Karp. 1990. Spawning and movements of razorback sucker, *Xyrauchen texanus*, in the Green River Basin of Colorado and Utah. *Southwest. Nat.* 35:427–433.

U.S. Environmental Protection Agency. 1971. The mineral quality problem in the Colorado River Basin. Summary Report Regions VIII and IX. EPA, Washington, DC.

U.S. Environmental Protection Agency. 1987. Ambient water quality criteria for selenium—1987. Publication No. EPA 440/5-87-006. EPA, Washington, DC.

Vanicek, C. D., and R. H. Kramer. 1969. Life history of the Colorado squawfish and Colorado chub in the Green River in Dinosaur National Monument, 1964–1966. *Trans. Am. Fish. Soc.* 98:193–208.

Waddell, B., and J. Coyner. 1990. Screening evaluation for inorganic contaminants at Desert Lake Wetlands Waterfowl Management Area. U.S. Fish and Wildlife Service Contaminants Report, Salt Lake City, UT.

Waddell, B., and T. May. 1995. Selenium concentrations in the razorback sucker (*Xyrauchen texanus*): Substitution of non-lethal muscle plugs for muscle tissue in contaminant assessment. *Arch. Environ. Contam. Toxicol.* 28:321–326.

Waddell, B., and S. Stephensen. 1992. Supplemental sampling for inorganic elements in eggs at Tamarisk Lake, Desert Lake Waterfowl Management Area, 1990. U.S. Fish and Wildlife Service Contaminants Info. Bull., Salt Lake City, UT.

Williams, K. T., and H. G. Byers. 1935. Occurrence of selenium in the Colorado River and some of its tributaries. *Ind. Eng. Chem.* 7:431–432.

12

Selenium and Salinity Concerns in the Salton Sea Area of California

JAMES G. SETMIRE
United States Geological Survey and Bureau of Reclamation,
Temecula, California

ROY A. SCHROEDER
United States Geological Survey, San Diego, California

I. BACKGROUND

Selenium concentrations are elevated in subsurface drainwater in the Imperial Valley of California and in surface drains and rivers conveying irrigation drainage (Setmire et al., 1990, 1993). Selenium also is at levels of concern in biota utilizing these resources and in the Salton Sea. The Salton Sea area occupies a topographic and structural trough (Fig. 1). The trough is a landward extension of the depression filled by the Gulf of California, from which it is separated by the broad fan of the Colorado River Delta (Loeltz et al., 1975). The Salton Sea is a closed inland lake occupying about 930 km² whose shoreline is currently at an elevation of about 78.3 m below sea level. At its southern end, the Salton Sea is the terminus of the New and Alamo Rivers, which receive irrigation drainwater from a 2250 km drainage network in the Imperial Valley. The New River also receives municipal and industrial waste from the city of Mexicali, Mexico, along with agricultural return from the Mexicali Valley. The Salton Sea National Wildlife Refuge, located in the southern end of the Salton Sea, is a major waterfowl stopover on the Pacific Flyway. At its northern end, the Salton Sea receives agricultural return flow and runoff from the Coachella Valley via the Whitewater River.

The Imperial Valley is typical of a desert area, with a maximum temperature more than 37.8°C more than 110 days per year. Evapotranspiration from a

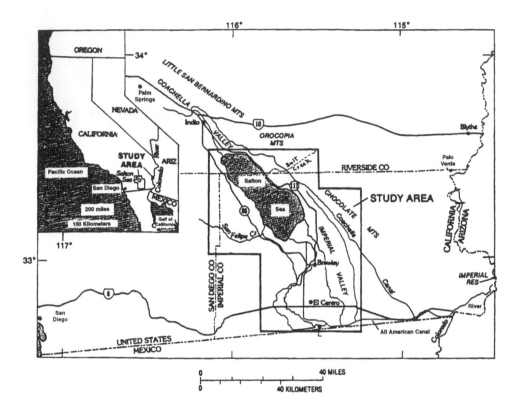

FIGURE I   Location of study area.

growing crop can exceed 1.80 m/year and during hot summer months exceeds 0.8 cm of water per day. The frost-free period is greater than 300 days for 9 of 10 years and 350 days for 3 of 10 years (Zimmerman, 1981). Imported water has transformed this desert environment into a highly productive agricultural area, permitting the growth of crops year-round.

Colorado River water is diverted at Imperial Dam and delivered via the All-American Canal and a 2700 km network of canals and laterals to 194,717 ha (1995) of farmland in the Imperial Valley. Agriculture and livestock grossed $1 billion in revenue for 1995. Of the agricultural use, field crops totaled 173,406 ha (alfalfa, 75,075 ha; Sudan grass, 31,316 ha; wheat, 25,138 ha; and sugar beets, 12,793 ha), and garden crops totaled 36,471 ha (lettuce, 6800 ha; carrots, 5906 ha). The total land in crops was 218,332 ha, with 32,396 ha having duplicate crops (Imperial Irrigation District, personal communication, 1996).

Water ordered by the farmers is delivered by Imperial Irrigation District to the fields with delivery available to each 65 ha. Irrigation water is accounted for in the following manner. "Operational loss" is the excess water needed to convey the requested water to the field. The excess water is discharged to a surface drain. This loss represents about 15% of the flow in the Alamo River at the outlet to the Salton Sea. Water delivered via the lateral canals is applied to the head of the field using flood irrigation (furrow and border strip) and flows downgradient to the lower or tail end. More water than a crop requires is applied to maintain sufficient wetting time at both ends of the field for effective irrigation. This excess water, commonly called tailwater, is collected at the lower end of the field and discharged to the surface drains. Tailwater is similar in concentrations of total dissolved solids and selenium to Colorado River water used for irrigation, although it has a higher silt load and wash-off of pesticides and nutrients.

Irrigation water infiltrates through the soil. Some of this water is intercepted by shallow tile lines installed at a depth of 2 to 3 m to prevent salt accumulation in the root zone. Spacing of these tile lines depends on the soil type. The distance varies from about 15 m in the northern part of Imperial Valley, where Imperial Formation clays are present to 120 m on the east and west sides of the valley, and to the south where Rositas sands are present (Zimmerman, 1981; Imperial Irrigation District, personal communication, 1989). In clayey soils, water applied to the surface flows more freely through the permeable backfill material used to fill trenches dug for installation of the tile lines than through the adjacent soil. This "trench flow" produces increased flows to the tile drains for a shorter duration than is observed in areas having loamy or sandy soil (Tod and Grismer, 1988). Of the water percolating through the soil, only the water within about 3 m (horizontal distance) of the tile drain is affected. Water at greater distances moves more slowly to the drains, and some of it recharges the aquifer beneath the field. Water in the tile lines is collected and discharged via sumps or gravity tile outlets to the surface drains. This discharge is termed subsurface drainwater or "tile" water. Drainwater is collected by a 2370 km drainage network that discharges either to the New River or the Alamo River, or directly to the Salton Sea.

In 1986 irrigation drainwater studies began in the Salton Sea area as part of the National Irrigation Water Quality Program (NIWQP) of the Department of Interior. The NIWQP originated as an outgrowth of the concern over the selenium contamination problems at Kesterson Reservoir, California. Other irrigation drainage projects in the western United States that had the potential for selenium contamination were investigated. Investigations into selenium contamination problems in the Salton Sea and planning studies for selenium remediation are ongoing to date (1997).

II. EFFECT OF IRRIGATED AGRICULTURE ON WATER QUALITY

A. Irrigation Water

Colorado River water is used to irrigate crops in the Imperial Valley. The quality of this irrigation water varies yearly as natural events (floods and droughts) and

anthropogenic factors (land development and management, including irrigation activities) control the quality of water in the Colorado River. Water in the
East Highline Canal (Colorado River water) was sampled monthly from August
1988 to August 1989. The median dissolved solids concentration was 686 mg/L,
with a selenium concentration of 2 μg/L, and a Se/Cl ratio of 2.2 $\times$ 10^{-5} (Setmire et al., 1993). Hydrogen and oxygen isotopes and tritium concentrations
were analyzed in one water sample. Concentrations were the following: δD (deuterium) = -103 permil, δ^{18}O (Oxygen) = -13 permil, and tritium concentration = 30 tritium units (TU) (Schroeder et al., 1993). These concentrations
and ratios will be used in Section II.C later as a reference to show changes that
occur as water moves through the agricultural system in the Imperial Valley. The
Se/Cl ratio will show relative sources and sinks of selenium, the hydrogen and
oxygen isotopes will show the evaporative history of the water, and the tritium
will show the relative composite age of the water.

B. Surface Drainwater

Surface drainwater in the Imperial Valley is found in the 2250 km network of
drains, in the New and Alamo Rivers, and in the Salton Sea. The Alamo River
consists almost entirely of surface drainwater. The median selenium concentration
for monthly water samples collected from the Alamo River at its outlet to the
Salton Sea during August 1988 to August 1989 was 8 μg/L, and the median
dissolved solids concentration was 2170 mg/L (Setmire et al., 1993). Total water
discharge in the Alamo River for the sampling period was 74,100 hectare-meters
(ha $\cdot$ m), yielding a total selenium load of 5.9 metric tons discharged to the
Salton Sea. In comparison, the New River contributed only 2.3 metric tons of
selenium for the same sampling period. Both water discharge and selenium
concentrations were lower for the New River.

    Surface drainwater is composed of subsurface drainwater, operational loss,
canal seepage, tailwater runoff, and occasionally storm water. These sources can
be divided into two categories: the first category is subsurface drainwater; the
second category incorporates the remaining components of surface water that
have major ion and selenium concentrations similar to the Colorado River (Setmire
et al., 1993, Michel and Schroeder, 1994) and are referred to as dilution water.
Median concentrations from the 1988–1989 monthly sampling performed by
the U.S. Geological Survey (USGS) can be used in a simple two-component
mixing equation to calculate the percentage composition of water in the Alamo
River at its outlet to the Salton Sea. The two components are described above:
subsurface drainwater and dilution water. The water in the Alamo River and/or
in surface drains is the result of that mixing.

    The equation is:

$$X A + (1 - X) B = C$$

where A = concentration of Colorado River water (dilution water), B = concentration of subsurface drainwater, and C = concentration in Alamo River at the outlet (also of surface drains).

For chloride the equation is:

$$X(92 \text{ mg/L}) + (1 - X)(1200 \text{ mg/L}) = 420 \text{ mg/L}$$

Solving the equation, the fraction $X = 0.70$. In the Alamo River, 70% of the chloride in the water comes from "dilution water" and 30% comes from subsurface drainwater, while 15% of the chloride comes from dilution water and 85% comes from subsurface drainwater. Solving the equation using median concentrations for dissolved solids yields a value for dilution water (x) of 77%. These values show that about three-fourths of the water in the Alamo River comes from "dilution water" and about one-fourth from subsurface drainwater.

Concentrations in water change rapidly as subsurface drainwater is discharged to the drains and as tailwater and/or operational loss is discharged. Bottom sediment also is affected by these changes. Selenium concentrations in the water column were not good predictors of selenium concentrations in bottom sediments. Rapid changes in the flow rate can resuspend previously deposited material, or deposition can occur as flow velocities in the drain decrease. Either case will produce variable selenium concentrations in bottom sediment. According to Salomons and Forstner (1984), dissolved metal concentrations are partly controlled by the interactions between metals and particulates. Adsorption is a key factor in the removal of trace metals from hydrologic systems. The particle size of the bottom sediment is a major factor controlling metal adsorption. As grain size decreases, metal concentration increases (Horowitz, 1991). Selenium concentrations in bottom sediment were compared with the percentage of bed material finer than 0.062 mm (silt and clay) at about 50 sites from 18 drains to determine whether selenium predominantly occurs on fine sediment (Setmire, unpublished data, 1994). Figure 2 shows that bottom sediment samples consisting mostly of fine material generally have higher selenium concentrations ($r^2 = 0.55$). The bottom sediment sample having the highest selenium concentration of 1.7 μg/g also has the highest portion finer than 0.062 mm at 98%.

Selenium concentrations in 260 soil samples from 15 fields (Schroeder et al., 1993) were compared with selenium concentrations in 48 bottom sediment samples to determine whether selenium concentrations in bottom sediments mirror selenium concentrations in the soil. The median selenium concentration from the soil samples is 0.2 μg/g compared with 0.5 μg/g for the bottom sediment samples. Results of t-tests suggest a very low probability that the two data sets represent the same population. The median concentrations show more than a doubling in selenium concentration from the field soils to bottom sediment in the drains. This increase in selenium concentration in the drain bottom sediments is accompanied by a decrease in the ratio of Se to Cl in the water. The Se/Cl ratio is 2.2×10^{-5} in irrigation water, 2.0×10^{-5} in subsurface drainwater, and

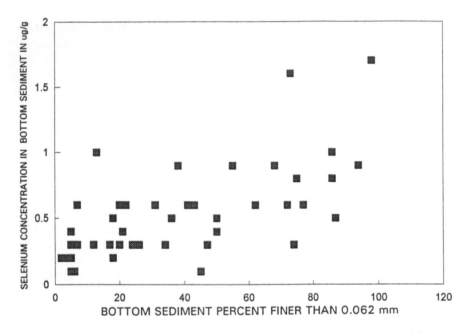

FIGURE 2 Regression plot of selenium concentration in bottom sediment and bottom sediment percentage finer than 0.062 mm for bottom sediment samples collected from selected surface drains in the Imperial Valley, California, August 1994.

1.5×10^{-5} in surface drainwater. Compared with chloride, it appears that some selenium is lost to the drain sediments. Furthermore, it is likely that some selenium reduction is occurring in the drains. Selenium speciation analysis for one drainwater sample showed that selenite comprised 1.4 μg/L of a total selenium concentration of 3.31 μg/L (A. Maest, personal communication, 1989). The effects of reducing conditions have been observed in numerous drains throughout the Imperial Valley. These effects include the presence of black sediment and the generation of hydrogen sulfide gas. Reduction of sulfate to hydrogen sulfide shows that conditions are present whereby selenate also could be reduced. Reduced species of selenium, including selenite, can adsorb on finer sediments that settle, increasing the selenium concentration in bottom sediments.

C. Subsurface Drainwater

Selenium concentrations in 119 sumps or gravity tile outlets sampled during May 1988 ranged from 3 to 300 μg/L, with a median concentration of 24 μg/L and a standard deviation of 58 μg/L (Setmire et al., 1993), reflecting results heavily skewed toward low concentrations. The same sites were sampled in 1986 by the California Regional Water Quality Control Board. Regressions suggest that selenium concentrations were about the same in 1988 and in 1986 ($r^2 = 0.787$, $a < 0.01$, slope $= 0.966$). The lack of change in concentration confirms that processes controlling selenium concentrations in subsurface drainwater are uniform over long periods. The similarity in concentrations was expected, given the comparative constancy in the source water quality (Colorado River) and the long transit times through the delivery and drainage system.

During August 1994 through January 1995, 820 sumps and gravity tile outlets were sampled for specific conductance and discharge, and 304 water samples were collected to determine selenium concentrations. Selenium concentrations ranged from 1 to 311 $\mu g/L$, with a median of 28 $\mu g/L$ and a standard deviation of 52 $\mu g/L$. There was no regression analysis between these data and the 1986 and 1988 data because not all the same sites were sampled. All three sets of data show an area of high selenium and dissolved solids concentrations southeast of the Salton Sea National Wildlife Refuge where land surface altitudes are among the lowest in the Imperial Valley.

Dissolved solids concentrations and specific conductance measurements show a distribution similar to that of selenium, except in the area immediately bordering the southern end of the Salton Sea, where reducing conditions are believed to result in selenium removal. The correlation between specific conductance and selenium concentration for the 1988 data ($r^2 = 0.77, a = 0.01$) (Setmire et al., 1993) is about the same as the correlation between dissolved solids and selenium concentrations for the 1986 data ($r^2 = 0.704$) (California Regional Water Quality Control Board, personal communication, 1986). However, the correlation between selenium and dissolved solids concentrations for the 1994–1995 data is only $r^2 = 0.28$ ($a = 0.01$). Sites for the 1994–1995 samples were selected to represent specific areas and as such caused tighter clustering of the data and weakening the correlation between selenium and dissolved solids.

Hydrogen and oxygen isotopes and tritium concentrations were analyzed from subsurface drainwater samples to show the source of the water and the processes affecting its concentration. Hydrogen and oxygen isotopes, which are chemically conservative, provide the ability to distinguish between increases in dissolved solids concentrations owing to leaching without evaporation and increases in dissolved solids owing to evaporation (Fontes, 1980). Evaporation in an arid environment such as the Imperial Valley results in enrichment of the heavier isotopes in the water that remains. Deuterium and oxygen-18 (^{18}O) are enriched in relation to the common forms or hydrogen (^{1}H) and oxygen (^{16}O). Values are reported as δD and $\delta^{18}O$ relative to the Vienna Standard for mean ocean water (V-SMOW).

For subsurface drainwater in the Imperial Valley, $\delta D = 5.4 \, \delta^{18}O - 34$ (Fig. 3). The standard error of the slope is 0.16 (Schroeder et al., 1991). Fontes (1980) found that $\delta D = 5.8 \, \delta^{18}O - 21$ for irrigation drainage wells in the Juarez

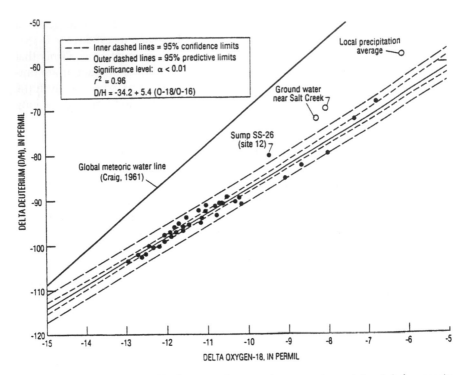

FIGURE 3 Regression plot of hydrogen and oxygen isotope ratios and the global meteoric water line for subsurface drainwater samples collected in the Imperial Valley, California, May 1988.

Valley of Mexico, where the range in dissolved solids concentration was controlled by evaporative concentration as indicated by the global meteoric water line in Figure 3. The comparability of the two slopes, 5.8 and 5.4, shows that similar evaporative processes control the range in dissolved solids concentration in subsurface drainwater from the Imperial Valley. The r^2 of 0.96, $a < 0.01$ (Setmire et al., 1993), suggests a single source of the subsurface drainwater. As indicated earlier, Colorado River water in the East Highline Canal used to irrigate the Imperial Valley has a δD of about −103 permil and a δ¹⁸O of about −13 permil (Schroeder et al., 1993). The regression line showing the relation between hydrogen and oxygen isotopes for Imperial Valley drainwater passes through that point at the lower end and through the isotopic composition of Salton Sea water when extrapolated on the upper end. Therefore, the Colorado River is the source of all subsurface drainwater and surface water in the Imperial Valley, including the water in the Salton Sea. Points not plotting on the regression line lie outside the Imperial Valley or are influenced by water from external sources such as San Felipe Creek, Salt Creek, or local precipitation.

The median tritium concentration in pore water collected at a depth of 1 m near drains in six fields was 32 TU, slightly higher than the tritium concentration in recently analyzed Colorado River water. Tritium concentrations of samples from a depth of 2 m, near the depth of the tile lines, have a median concentration of 52 TU, indicating water with an average transit time of about 5 years (Michel and Schroeder, 1994).

Evaporative concentration of Colorado River water controls the range of dissolved solids concentration in subsurface drainwater from the Imperial Valley (Setmire et al., 1993). U.S. Geological Survey scientists hypothesized that evaporative concentration also controls the range of selenium concentrations. In other areas, such as the San Joaquin Valley of California, oxidation of reduced seleniferous deposits and evaporative concentration produce high levels of selenium in drainwater (Deverel and Fujii, 1987). Data from 270 soil cores collected from 15 fields show that the soil is not a major source of selenium in the Imperial Valley (Schroeder et al., 1993). Regression of δD against selenium initially had a low correlation. Six sites draining fields in a narrow band along the southern end of the Salton Sea strongly affected the regression. These fields had high chloride and low selenium concentrations similar to the Salton Sea and had comparatively reducing conditions. They are atypical of the remainder of the Imperial Valley and were deleted from the regression, giving $r^2 = 0.62$. Similarly high correlations result from regressions between δD and chloride concentrations, δD and dissolved-solids concentrations, and selenium and chloride concentrations. The high correlation between these variables shows that selenium concentrations in subsurface drainwater are controlled by evaporation of Colorado River water.

Selenium is compared with other elements such as chloride to show relative changes that occur as water moves through the agricultural system. These elemental mass (weight) ratios show sources and sinks for selenium. Chloride is used for comparison because it is chemically conservative and highly soluble. Colorado River water in the East Highline Canal had a selenium to chloride ratio (Se/Cl) of 2.2×10^{-5}. This ratio changes when one or the other ion is affected by physical processes, solubility, oxidation–reduction reactions, and/or biological reactions. In the Imperial Valley, the physical process of evaporation affects both elements similarly. The median Se/Cl ratio for subsurface drainwater sampled during May 1988 is 2×10^{-5}. The minimum ratio is 0.027×10^{-5}, the maximum ratio is 7.3×10^{-5}, and 50% of the ratios fall between 1.0 and 3.0×10^{-5} (Setmire et al., 1993). Also, one of the highest detected selenium concentrations, 340 $\mu g/L$, had an associated chloride concentration of 15,000 mg/L and a Se/Cl ratio of 2.3×10^{-5}. The similarity of this Se/Cl ratio with the ratio for Colorado River water (East Highline Canal) shows that evaporative concentration of irrigation water is the most important process producing the elevated levels of selenium present in the Imperial Valley.

Crop type and soil characteristics defining where different crops are grown control the amount of evaporative concentration that occurs in a field. For example, Bermuda grass tends to be grown in clayey, more saline soils of lower permeability, while higher value crops, such as lettuce, tend to be grown in more permeable soils (S. Knell, personal communication, 1997). Hence, the soils and crop type affect conductance and selenium in the subsurface drainage from the respective fields. The regression equation describing the relationship between specific conductance and selenium concentration ($r^2 = 0.28$) for the 304 water samples was used to estimate selenium concentrations from specific conductance values measured during August 1994 through January 1995 at 820 gravity tile outlets and sumps for which crop type was also known. The median modeled selenium concentration for the 820 sites of 36 μg/L is slightly higher than the median of 28 μg/L for the 304 water samples on which selenium was measured. Modeled selenium concentrations ranged from a low of 28 μg/L for fields where lettuce is grown to a high of 52 μg/L for fields where Bermuda grass is grown (Table 1).

D. Quality Changes During Water Transport

Concentrations of several key constituents in Colorado River water are compared with concentrations in surface and subsurface drainwater to show the overall

TABLE I   Modeled Selenium Concentrations for Selected Crops[a] from Selenium to Specific Conductance Regression for Subsurface/Drainwater Samples Collected in the Imperial Valley, California, August 1994 to January 1995

| Crop | Measured selenium (μg/L) | | | Measured conductance (μS/cm) | | | Modeled selenium median (μg/L) |
	Min	Max	n	Min	Max	n	
Asparagus	13	27	(5)	2,750	37.600	(13)	29
Beds and rows	11	284	(8)	1,970	28,500	(62)	36
Between crops	1	10.5	(4)	4,450	34,000	(16)	40
Bermuda grass	1	167	(29)	1,310	35,500	(80)	52
Disked	2	311	(17)	2,010	24,400	(32)	34
Fallow	5	311	(46)	2,090	42,300	(73)	37
Lettuce	5	31	(5)	2,210	15,950	(26)	28
Sudan	3	50.5	(32)	1,080	22,010	(62)	42
Sugar beets	5	173	(10)	4,540	22,500	(41)	46

[a]n = number of sites or samples.

water quality changes due to irrigated agriculture. Median concentrations for monthly samples collected in the East Highline Canal from August 1988 to August 1989 along with concentrations of selected constituents in subsurface water and surface drainwater are shown in Table 2. Concentrations of these selected constituents increase from Colorado River water to irrigation drainwater. Total dissolved solids increases about 3-fold from the source (Colorado River) to output (surface drainwater), nitrate about 20-fold (reflecting use of fertilizers), chloride 4.5-fold, and selenium 3-fold. Some dilution of subsurface drainwater then occurs when fresher water is mixed in the surface drains.

E. Salton Sea

The Salton Sea is the terminus for irrigation drainage originating in the Coachella Valley, the Imperial Valley, and the Mexicali Valley. The Se/Cl ratio in the Salton Sea is 0.007×10^{-5}, showing that the Salton Sea is a major sink for selenium. Evaporation is the major physical process controlling the major ion chemistry of the Salton Sea. Median chloride concentrations increase from 520 mg/L in the Alamo River to about 15,000 mg/L in the Salton Sea. In contrast, median selenium concentrations decrease from 8 μg/L in the Alamo River to 1 μg/L in the Salton Sea (Setmire et al., 1993). Selenium is biologically reduced, eventually ending up in the bottom sediment and the biomass of the Salton Sea, or in the wildlife feeding in the Salton Sea.

A narrow zone of mixing is present between high-salinity Salton Sea water and low-salinity Alamo River water. The Alamo River extends about 500 m into the southern end of the Salton Sea. The river, with levees on both sides, is dredged as needed to maintain adequate flow to the Salton Sea. The mixing zone occurs about 60 m past the end of the levee. The specific conductance in the mixing zone at a depth of 0.4 m was 5000 μS and 51,000 μS at a depth of 1 m (bottom). The selenium concentration was 8 μg/L at the upper depth and only 1.0 μg/L at the lower depth. Samples collected to determine speciation indicate that selenium in the Alamo River, on the river side of the interface, is a

TABLE 2   Median Concentrations of Selected Constituents in Water Samples Representing the East Highline Canal, Subsurface Drainwater, and Surface Drainwater in the Imperial Valley, California

Site	Dissolved solids (mg/L)	Nitrogen (mg/L)		Chloride (mg/L)	Selenium (μg/L)
		As NO$_3$	As NH$_4$		
East Highline Canal	686	0.22	0.03	98	2
Subsurface drainwater	6448	1.1	0.07	1200	28
Surface drainwater	2025	4.95	0.19	420	6

mixture of selenate and selenite. The total selenium concentration measured at this site on June 1989 was 6.35 μg/L, with 2.56 μg/L as selenite and 3.79 μg/L as selenate (A. Maest, personal communication, 1989). These concentrations, along with those from the surface drains, show that selenium in the Alamo River flowing into the Salton Sea is fairly evenly divided between the intermediate (+4)(selenite) and the oxidized (+6)(selenate) oxidation states. Samples collected at the seaside of the interface show that for a total selenium concentration of less than 2.4 μg/L (method- specific reporting limit for June 1989 sample), 1.79 μg/L was as Se(+4) and less than 0.2 μg/L as Se(+6). These results show that virtually none of the selenium on the Salton Sea side of the interface is in the (+6) oxidation state (A. Maest, personal communication, 1989).

Selenium concentrations in bottom sediments of the Alamo River delta ranged from 0.2 to 2.5 mg/kg, with no readily apparent spatial pattern in their distribution. A composite sample collected during 1986 contained a selenium concentration of 3.3 mg/kg and had a dissolved organic content of 1%. A sediment core collected in 1996 near the south buoy (deepest location in the Salton Sea) had a selenium concentration of 9.3 mg/kg and a corresponding dissolved organic carbon content of 9.2 percent (R. Schroeder, unpublished data, 1996). This core was composed of very low density material. The high selenium concentration and the high dissolved organic carbon content of this sample show that selenium is likely incorporated into biomass, which degrades and concentrates in the deepest parts of the Salton Sea.

F. Groundwater

Groundwater in the central part of the Imperial Valley is much too saline for consumption, although there has been limited use in isolated areas for domestic livestock. Analyses from shallow monitoring wells from the early 1960s (Loeltz et al., 1975) indicated the presence of some freshwater which, based on the well's location near rivers or unlined canals, probably results from seepage (Michel and Schroeder, 1994). Multidepth piezometer wells installed at three locations in the Imperial Valley showed that regional groundwater is free of influence from irrigation drainage (Setmire et al., 1993). This water has a salinity at about 15 g/L, about one-third the salinity of the Salton Sea. The shallow groundwater system consists of two zones—an upper zone, extending no deeper than about 10 m, and a lower zone that predates the development of irrigated agriculture. The upper zone, produced by infiltrating irrigation drainwater, has highly variable salinity and selenium concentrations. The lower zone is of relatively constant salinity and contains no selenium. The selenium is believed to have been removed by microbial reduction. Environmental conditions, which are toxic near the surface, become increasingly reducing at depth, resulting in several other characteristic geochemical changes as well—removal of nitrate by denitrification, generation of soluble iron and manganese (both in the +2 oxidation state), and at greater depth, sulfate reduction. Denitrification was confirmed by isotopic studies; nitrogen-15 is enriched (up to 100 permil) in low-nitrate groundwater at intermediate depths. Ammoniam-N in water also is present at concentrations as high as 26 mg/L (Schroeder et al., 1993). It is postulated that if new analytical methods become available, similar isotopic data for selenium (80:76 mass ratios) will confirm microbial selenate reduction in the groundwater.

Groundwater in the Coachella Valley north of the Salton Sea has long been an important source of municipal and domestic supplies. Recharge is from precipitation in the high mountains north of the valley and also from imported Colorado River and northern California water. Groundwater that is suitable for human consumption also is found in the sparsely populated areas west and east of the irrigated low-lying central part of the Imperial Valley. Recharge is from local precipitation in the surrounding mountains, from the Colorado River, and from Colorado River underflow (on the east side of the valley).

III. EFFECTS ON BIOTA

The Salton Sea area has become an increasingly important habitat for resident fish and wildlife and for migratory waterfowl on the Pacific Flyway as wetland acreage in California has decreased. More than 90% of California's wetlands have been lost to other uses in the last 150 years, mostly to agricultural development in the Central Valley (J. Bennett, personal communication, 1996). Almost 400 bird species have been documented to date, and more than a million individual birds use the Salton Sea area annually. The area also contains several endangered species, including the desert pupfish (Cyprinodon macularius), the Yuma clapper rail (Rallus longirostris yumanensis), and the brown pelican (Pelecanus occidentalis). Based on information from other areas, selenium concentrations are high enough in the wildlife of the Salton Sea area to present some hazard to reproductive success.

Selenium bioaccumulation in lower trophic organisms and biomagnification in the food chain causes highest residues to occur in the upper trophic levels. Bioaccumulation and biomagnification are apparent in rivers and drains (Fig. 4) and in the Salton Sea itself (Fig. 5) (Schroeder et al., 1993; Setmire et al., 1993). The figures show that concentrations at the highest freshwater trophic level are only half those in the highest trophic level in the Salton Sea. Selenium concentrations in these upper food chain sources represent a possible threat to the long-term survival of fish-eating birds feeding in the Salton Sea area, especially those feeding on fish from the Salton Sea. The shaded areas in Figures 4 and 5 represent a level of concern where selenium concentrations are elevated above background levels but rarely cause discernible adverse effects. High selenium concentrations

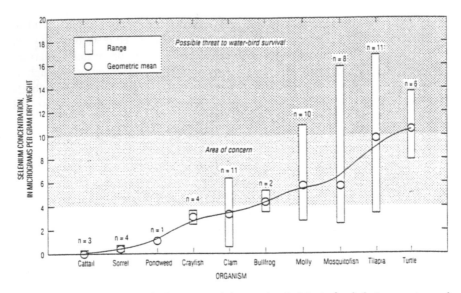

Figure 4 Concentration of selenium, and dietary thresholds, in food chain organisms of rivers and drains in the Imperial Valley, California 1988–1990: light shaded area, thresholds for area of concern; darker shaded area, possible threat to waterbird survival. (Modified from Heinz et al., 1987, 1989.)

also have been responsible for a California State Office of Health and Hazardous Assessment advisory limiting the consumption of fish caught in the Salton Sea to 113 g per 2-week period and prohibiting consumption altogether for pregnant women.

Livers from 145 individuals representing 10 waterfowl species were analyzed in the 1988–1990 study, and 10 (about 7%) of the samples were found to have concentrations exceeding the 30 μg/g dry weight threshold that indicates heavy exposure and a high risk of reproductive impairment (U.S. Fish and Wildlife Service, 1990; J. Skorupa, personal communication, 1992). More than 100 eggs of black-necked stilt (*Himantopus mexicanus*), a resident shorebird species, also were analyzed and only about 7% were found to exceed the 6 μg/g selenium threshold that predicts a 10% probability of embryotoxicity (death or deformity). Because of the low percentages, the likelihood of detecting teratogenesis, such as embryo and developmental abnormalities characteristic of classical selenium responses, is extremely low.

Additional data were collected by the U.S. Fish and Wildlife Service in 1992–1994 to further quantify biological effects of selenium on biota: 38 stilt eggs in 1992 and 40 additional stilt eggs in 1993. The geometric mean selenium

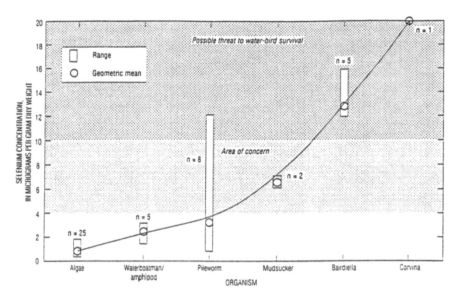

Figure 5 Concentration of selenium and dietary thresholds in food chain organisms of the Salton Sea, 1988–1990: light shaded area, thresholds for area of concern; darker shaded area, possible threat to waterbird survival. (Modified from Heinz et al., 1987, 1989.)

concentrations of 7 and 6 μg/g, respectively, are similar to those obtained during the 1988 to 1990 study (J. Bennett, personal communication, 1996).

Toxicological relationships established from studies in other areas, especially from the San Joaquin Valley, show that 15% of Salton Sea stilt nests were expected to be affected by hatching failure—close to the 13% that was actually observed (J. Skorupa, personal communication, 1994). Normally, about 8% of stilt nests with less than 4.1 μg/g selenium in their eggs have one or more fail-to-hatch eggs; hence there is only about a 5% reproductive depression in the Salton Sea area (J. Bennett, personal communication, 1996).

In addition to studies on waterfowl, sailfin mollies (*Poecilia latipinna*) were collected from 13 drains in 1994 to represent chemical body burdens of the endangered desert pupfish. The selenium levels in 1994 were found to be about the same as those in five mollies and four mosquitofish collected for the 1988–1990 study (Schroeder et al., 1993). For 10 of the 13 drains, selenium concentrations were in the range of 3 to 6 μg/g dry weight for whole-body fish, identified as a level of concern for warm-water fish (J. Bennett, personal communication, 1996). Mollies from two other drains had concentrations exceeding the toxicity threshold of 6 μg/g, above which there is increased risk of teratogenesis and embryo mortality (J. Bennett, personal communication, 1996).

IV. SUMMARY

Selenium is elevated in water, sediment, and biota in the Salton Sea area of California. Selenium in Colorado River water used for irrigation in the Imperial Valley has an average concentration of 2 μg/L. Irrigation drainwater in the Alamo River at its outlet to the Salton Sea has a median selenium concentration of 8 μg/L. Overall, selenium concentrations increase fourfold as a result of irrigated agriculture. Subsurface drainwater in the Imperial Valley had a median selenium concentration of 25 μg/L for 119 samples collected in May 1988 and 28 μg/L for 304 samples collected from August 1994 to January 1995, a 12- to 14-fold increase from the concentrations in the Colorado River. Ratios of selenium to chloride and comparisons between selenium concentrations in fields and in bottom sediments indicate that some selenium likely is lost to the bottom sediments of the surface drains. The Salton Sea, with a very low Se/Cl ratio, is a major sink for selenium, with only 1 μg/L selenium in the water column of the sea compared to the median of 8 μg/L selenium in the inflowing water of the Alamo River. Selenium contents in biota of the rivers and drains and in the Salton Sea also are elevated and are at levels of concern. Selenium bioaccumulates in food chain organisms and biomagnifies from lower to higher trophic levels. At greatest risk of reproductive impairment are the larger piscivorous birds feeding on fish in the Salton Sea.

REFERENCES

Craig, H., 1961. Isotopic variations in meteoric waters. *Science* 113:1702–1703.

Deverel, S. J., and R. Fujii. 1987. Processes affecting the distribution of selenium in shallow ground water of agricultural areas, western San Joaquin Valley, California. U.S. Geological Survey, Open File Report No. 87–220.

Fontes, J. C. 1980. Environmental isotopes in groundwater hydrology. In P. Fritz and J. C. Fontes (eds.), *Handbook of Environmental Isotope Geochemistry. The Terrestrial Environment*, Part A, pp. 75–140. New York, Elsevier Scientific.

Heinz, G. H., D. J. Hoffman, A. J. Krynitsky, and D. M. G. Weller. 1987. Reproduction in mallards fed selenium. *Environ. Toxicol. Chem.* 6:423–433.

Heinz, G. H., D. J. Hoffman, and L. G. Gold. 1989. Impaired reproduction of mallards fed an organic form of selenium. *J. Wildl. Manage.* 53:418–428.

Horowitz, A. J. 1991. *A Primer on Sediment–Trace Element Chemistry*. Lewis Publishers, Chelsea, MI.

Loeltz, O. J., B. Irelan, J. H. Robison, and F. H. Olmsted. 1975. Geohydrologic reconnaissance of the Imperial Valley. U.S. Geological Survey Prof. Paper No. 486-K.

Michel, R. L., and R. A. Schroeder. 1994. Use of long-term tritium records from the Colorado River to determine timescales for hydrologic processes associated with irrigation in the Imperial Valley, California. *Appl. Geochem.* 9:387–401.

Salomons, W., and U. Forstner. 1984. *Metals in the Hydrocycle*. Springer-Verlag, Berlin, Heidelberg, New York, Tokyo.

Schroeder, R. A., J. G. Setmire, and J. N. Densmore. 1991. Use of stable isotopes, tritium, soluble salts, and redox-sensitive elements to distinguish ground water from irrigation water in the Salton Sea basin. In W. J. Ritter (ed.), *Irrigation and Drainage*, pp. 524–530. American Sociey of Civil Engineers, National Irrigation and Drainage Division Conference, Honolulu, HA, July 22–26.

Schroeder, R. A., M. Rivera, and others. 1993. Physical, chemical, and biological data for detailed study of irrigation drainage in the Salton Sea area, California, 1988 to 1990. U.S. Geological Survey Open File Report No. 93-083.

Setmire, J. G., J. C. Wolfe, and R. K. Stroud. 1990. Reconnaissance investigation of water quality, bottom sediment, and biota associated with irrigation drainage in the Salton Sea area, California, 1986 to 1987. U.S. Geological Survey Water Resources Invest. Report No. 89-4102.

Setmire, J. G., R. A. Schroeder, J. N. Densmore, S. L. Goodbred, D. J. Audet, and W. R. Radke. 1993. Detailed study of water quality, bottom sediment, and biota associated with irrigation drainage in the Salton Sea area, California, 1988 to 1989. U.S. Geological Survey Water Resources Invest. Report No. 93-4014.

Tod, I. C., and M. E. Grismer. 1988. Drainage efficiency and cracking clay soils. Meeting paper No. 88-2588, American Society Agricultural Engineers, Winter Meeting, Chicago, Dec. 13–16, 1988.

U.S. Fish and Wildlife Service. 1990. Effects of irrigation drainwater contaminants on wildlife. Laurel, MD, Patuxent Wildlife Research Center.

Zimmerman, R. P. 1981. Soil survey of Imperial County, California, Imperial Valley area. U.S. Department of Agriculture, Soil Conservation Service.

13

Selenium Cycling in Estuarine Wetlands: Overview and New Results from the San Francisco Bay

Peter T. Zawislanski and Angus E. McGrath
Lawrence Berkeley National Laboratory, Berkeley, California

I. INTRODUCTION

Studies of selenium (Se) speciation in ocean waters (Measures and Burton, 1978; Burton et al., 1980; Takayanagi and Wong, 1984, 1985) and in estuarine waters (Measures and Burton, 1978; Takayanagi and Cossa, 1985; Cutter, 1989; Cutter and San Diego-McGlone, 1990) have significantly improved the understanding of Se chemical dynamics in the aqueous state. Far fewer studies deal with Se cycling in estuarine sediment, suspended sediment, and plankton, resulting in a less than complete estuarine Se cycling model. Several key observations have been made that relate to the common coexistence of at least three aqueous Se species: Se(IV), Se(VI), and organo-Se. Data indicate that conversion between Se species is limited and generally difficult to discern on a large spatial scale (e.g., that of an estuary). Most studies found that Se(IV) and Se(VI) remain conservative over a wide range of salinity (Measures and Burton, 1978; Takayanagi and Cossa, 1985) as water moves from the freshwater upper reaches to the saline lower reaches of an estuary. There are exceptions, such as the estuary of the Scheldt River in Belgium (van der Sloot et al., 1985) and the San Francisco Estuary in California (Cutter, 1989), where anthropogenic activities result in significant Se inputs. Although the total loading of Se, or any other contaminant, affects the overall health of an estuary, it is the interaction of Se with the suspended and underlying sediments that may have the most dramatic and immediate effect on the biological system. This is because some of the most important components of the food web are often found in the shallow sediment and tend to be filter feeders. Luoma et al. (1992) investigated Se uptake by bivalves (*Macoma balthica*) and found that all particulate forms of Se were assimilated to some degree, with particulate organo-Se being assimilated with 86% efficiency and particulate Se(0) with 22% efficiency. They also found that 98 to 99% of the Se found in *M. balthica* tissue came from particulate and not dissolved Se. Such findings further underscore the need to understand the dynamics of Se associations with suspended and bottom sediments.

This chapter evaluates the results of past and ongoing research to arrive at a refinement of the Se cycle in estuarine wetlands. Specifically, data from tidal marshes in the northern reach of the San Francisco Estuary are used to define likely physical paths and chemical transformations in marshes and mudflats.

II. SELENIUM CYCLING IN ESTUARINE WETLANDS: PAST EFFORTS

Numerous studies of Se speciation and transformation in terrestrial sediments have been conducted in which special attention was given to adsorption (Hamdy and Gissel-Nielsen, 1977; Balistrieri and Chao, 1987; Neal et al., 1987); reduction (Doran, 1982; Oremland et al., 1990; Tokunaga et al., 1996), oxidation (Sarathchandra and Watkinson, 1981; Tokunaga et al., 1994; Zawislanski and Zavarin, 1996), volatilization (Frankenberger and Karlson, 1989; Karlson and Frankenberger, 1988; 1989), and transport (Long et al., 1990; Benson et al., 1991; Zawislanski et al., 1992). Most of this work was done on sediments with elevated Se concentrations, usually in settings where Se was introduced into the system via Se-contaminated water, with aqueous concentrations as high as 10,000 μg/ L, resulting in sediment levels sometimes exceeding 100 mg/kg (e.g., Weres et al., 1989). In contrast, aqueous Se concentrations in estuarine wetlands rarely exceed 1 μg/L and sediment levels generally are less than 1 mg/kg. Although the underlying geochemical processes may be similar, the Se cycle in estuarine wetlands is complicated further by the potentially dominant role of Se on suspended sediments and by the influence of tidal dynamics on sediment redistribution and redox conditions.

Studies of Se cycling in estuarine wetlands have been relatively limited. The only comprehensive study of this kind was performed by Velinsky and Cutter (1991), who focused their investigation on the annual cycle of Se in a *Spartina alterniflora*–dominated coastal salt marsh on Delaware Bay. They found Se(0) to be the dominant form in sediments and observed oxidation and reduction of 10 to 20% of the total Se inventory due to seasonal redox cycles. Through modeling, they postulated that the sources of Se at this site were the neighboring creek (75%) and the atmosphere (25%). Tidal contributions were neglected, thereby eliminating the influence of estuary waters altogether. This particular site, which was located within approximately 50 m of a creek and was seldom inundated, was probably not ideal for the study of estuarine water–wetlands interaction. To the best of our knowledge, no such studies have been performed in tidal mudflats.

Takayanagi and Belzile (1988) and Belzile and Lebel (1988) investigated Se distributions in sediment cores from the central (300–500 m deep) part of the St. Lawrence Estuary in Canada. A trend of increasing dissolved Se with depth was observed and was hypothesized to be a result of iron (Fe) oxyhydroxide reduction, presumably concomitant with organic matter oxidation. Selenium concentrations in the interstitial water were two- to sevenfold higher than in the overlying water. As a result of such a Se gradient, Takayanagi and Belzile (1988) estimated an upward flux of soluble Se equivalent to approximately 90 $\mu g/m^2$/year, an amount equal to all the dissolved Se in the top 30 cm of the sediment. They suggest that the loss of soluble Se to the overlying water is balanced by the solubilization of Se adsorbed to Fe and/or manganese (Mn) oxyhydroxides.

Selenium association with suspended matter (referred to as total suspended matter, TSM) in estuaries has been considered (Cutter, 1989) but probably has been underestimated in its effect on Se fate. Belzile and Lebel (1988) identified a correlation between Se content of suspended sediments in the Laurentian Trough with marine organisms, based on carbon/nitrogen (C/N) ratios. They hypothesize that Se enriched biogenic material settles and is subsequently degraded, resulting in Se solubilization and eventual precipitation. One of the important considerations is whether the Se that is sampled as "dissolved" is actually associated with colloids. Takayanagi and Wong (1984), in their studies of the southern Chesapeake Bay, found that "dissolved" inorganic Se, as defined by passing a 0.45 μm filter, comprised 77, 40, and 0% colloidal Se in river water (James River), estuarine water, and coastal water, respectively. The organic "dissolved" Se fraction comprised 70, 64, and 35% colloidal Se in those environments, respectively. The decrease in colloidal Se relative to total Se with salinity implies flocculation of colloids and deposition. The association of Se with particulates and colloids takes on further significance in environments where point releases of Se may result in local increases in sediment Se levels.

III. SELENIUM IN ESTUARINE WETLANDS OF THE SAN FRANCISCO ESTUARY

Releases of Se by oil refineries located in the northern reach of the San Francisco Estuary have prompted concern over potential Se accumulation in surrounding wetlands. These concerns were fueled further by findings of elevated tissue Se in clams (Johns et al., 1988) and diving ducks (Fan and Lipsett, 1988). Although no adverse effects were identified, research on the biogeochemical cycling of Se is considered critical to the understanding of the future health of the estuary. Selenium concentrations in the San Francisco Estuary vary depending on location along the net flow direction and correlate inversely with salinity. The two main sources of water to the estuary, which join to form the delta, are the Sacramento River and the San Joaquin River (Fig. 1), which have average Se concentrations of 0.07 and 0.8 $\mu g/L$, respectively (Cutter, 1989). In both rivers, Se(VI) is the dominant form. Since the Sacramento River contributes by far the greater volume of water, by the time the water reaches the Suisun Bay, total Se concentrations are around 0.25 $\mu g/L$. Selenium concentrations at the mouth of the estuary, by the Golden Gate Bridge, are generally below 0.1 $\mu g/L$ (Cutter, 1989). Cutter (1989) identified and Cutter and San Diego-McGlone (1990) confirmed a marked increase in dissolved Se(IV) in the Carquinez Strait (Fig. 1), corresponding to the location of oil refinery discharge points. This increase in Se(IV) was not accompanied by a decrease of other Se forms, suggesting an anthropogenic source, which was further confirmed by the observation that refineries discharge Se predominantly in the Se(IV) form. Furthermore, Cutter (1989) found that suspended particulates were a significant pool of Se, contributing roughly 0.02 to 0.04 $\mu g/L$ in the Carquinez Strait but as little as 0.01 $\mu g/L$ at the Golden Gate Bridge. The decrease is due largely to a decrease in TSM. According to sediment

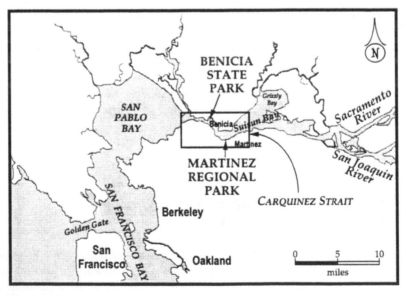

FIGURE I   The San Francisco Estuary (central and northern) and location of field sites.

transport studies, about 50% of suspended particulates are deposited in the estuary (Ogden Beeman & Associates, 1992).

Luoma et al. (1992) attribute particulate Se concentrations that influence clam uptake to phytoplankton bioconcentration of Se(IV) through metabolic transformation into selenocysteine or selenomethionine. Several studies have shown that Se(IV) can be bioconcentrated by phytoplankton (Zhang et al., 1990). Zhang et al. (1990) also found that in both live cell and biocide-treated control samples, selenite concentrations decreased rapidly immediately upon addition. They attributed this phenomenon to adsorption onto cell surfaces. Data presented by Cutter (1989) for solution concentrations of TSM versus chlorophyll-*a* showed a correlation with $r^2 = 0.50$, suggesting that phytoplankton is an important contributor of Se in TSM.

IV. ONGOING STUDIES OF SELENIUM CYCLING IN THE CARQUINEZ STRAIT

Research at Lawrence Berkeley National Laboratory (LBNL) has focused on determining the Se cycle as it pertains to the interaction between estuarine waters and intertidal sediments. At the initiation of this work, the degree to which wetland sediments accumulate Se from the overlying water and the geochemical path Se takes once incorporated were not well understood. The goals of this research may be broadly summarized as follows:

> to define and measure the modes of Se incorporation into wetlands sediment
> to define and measure Se transformations in sediment
> to estimate the potential for Se remobilization

To accomplish these goals, investigators performed detailed sampling and fractionation of sediments, analysis of plants and measurements of plant turnover rates, and measurements of suspended particulate Se and sedimentation dynamics, as well as studies of the influence of tides on redox cycles in soil and sediment. This chapter presents recent data (collected from December 1995 through December 1996) on sediment- Se, TSM- Se, sedimentation rates, and redox rates and uses these in constructing an Se cycling hypothesis.

A. Selenium Distribution and Fractionation in Sediments

Mudflat and marsh sediments were sampled from two sites in the Carquinez Strait: the Martinez Regional Park (MRP) shoreline in Martinez and the Benicia State Park (BSP) shoreline in Benicia (Fig. 1). These sites were chosen because of their proximity to two major refineries that are known to release approximately 1500 kg of Se per year (SFBRWQCB, 1992). Each site contains an extensive mudflat, which is exposed during low tide to a distance of around 150 to 200 m from shore; a *Spartina* zone, generally no wider than 10 m; a *Scirpus/Typha* zone, on the order of 30 m wide; and a *Distichlis/Salicornia*-dominated marsh plain. The *Spartina* and *Scirpus/Typha* zones are subsequently referred to as the "lower marsh." The main difference between the two sites is the relative energy of the depositional environment. At MRP, the entire width of the mudflats is exposed to strong wave action, both natural and due to passing ships, whereas the BSP site is located within Southampton Bay, an inlet that effectively dampens out wave action and results in a very low energy depositional environment. Thus sediments from the MRP mudflats are sandy, whereas those from BSP are composed of fine silt.

1. Sampling, Extraction, Analysis

Sediment cores 20 cm deep were sampled at both sites in December 1995. Prior to being frozen, bulk sediment values for Eh (redox potential) and pH were measured using platinum (Pt) and antimony (Sb) electrodes, respectively. After being frozen, the cores were sectioned into intervals of varying thickness, from 1 cm at the surface, to 5 cm at the bottom. After removal of interstitial water (via centrifugation), the samples were homogenized and subsamples were sequentially extracted and analyzed for Se species. In addition, a subsample was dried and ground for total Se extraction and analysis, as well as for total organic carbon (TOC) determination.

A sequential extraction procedure was developed from techniques used earlier for Se fractionation and speciation (Weres et al., 1989; Velinsky and Cutter, 1990; Lipton, 1991; Tokunaga et al., 1991). Table 1 contains the sequence of extractions and the operationally defined target Se fractions each extraction is designed to remove. Testing of these methods has demonstrated that Se(IV)/Se(VI) spikes are extracted with 93% efficiency by the first three extractants. Selenomethionine is extracted with 65 to 70% efficiency and Se(0) with 78% efficiency. Samples were extracted without drying after removal of interstitial

TABLE I   Sequential Extraction Procedure for Estuarine Wetland Soils and Sediments

Extract name	Solution	Target Se species
Distilled water (Dx)	Distilled water	Soluble
Phosphate (Px)	0.001 M Na_2PHO_4	Adsorbed
Sodium hydroxide (OHx)	0.02 M NaOH	Organically associated
Sulfite (Sx)[a]	1.0 M Na2SO3 (pH 7.0)	Elemental
HCl-Cr^{2+} (Crx)[a]	1.0 M HCl, 1.0 m Cr^{2+}	Pyrite
Total acid digest (TAD)	Conc. HNO_3, H_2O_2, HCl, Urea	Total Se

[a]Method developed by Velinsky and Cutter (1990).

water and determination of water content. Sulfite and HCl-Cr^{2+} extracts were done after NaOH extraction, drying, and grinding. Residual Se is defined as the difference between total acid digest Se (TAD-Se, see Table 1) and the sum of sequentially extracted Se.

Extracts and waters were analyzed by hydride generation flow injected atomic adsorption spectroscopy (HG-AAS) (Weres et al. 1989). Samples with solution concentrations below 0.7 μg/kg or with volumes below 5 ml were analyzed using either a cold vapor method adapted from Cutter (1978) or a flow injection method developed by Tao and Hansen (1994). Total organic carbon (TOC) was measured on a Carlo Erba NA 1500 C and N analyzer.

2. Selenium in the Water Column and on Suspended Sediment

Water and suspended particulate matter within the Carquinez Strait were sampled biweekly at the MRP site in the mudflats and the dock from December 1995 to December 1996. From the field, 21 L volumes were taken to the laboratory, where they were centrifuged (12,000 rpm) and filtered using a 0.45 μm Millipore nitrocellulose filter. Water was double filtered and acidified (4 ml of concentrated HCl per liter of solution) for analysis, while sediments were freeze-dried and stored for analysis.

The high degree of resuspension in the mudflats resulted in high TSM concentrations in those samples, while dock samples contained sediment concentrations that correlated well with deep channel surface samples. TSM-Se concentrations, calculated on a dry weight basis, agree with measurements by Cutter (1989) for surface water sampled in the Carquinez Strait (0.5–1.5 mg/kg). Total Se in TSM appears to increase with decreasing TSM concentration (Fig. 2). In

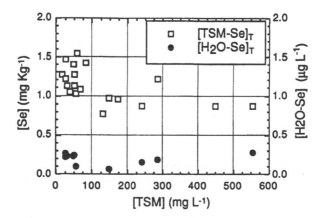

FIGURE 2   TSM-Se (based on oven-dry weight) and dissolved total Se concentrations versus TSM concentration in surface waters at the MRP site.

general, TSM concentrations increase with increasing wind/wave action and storm drainage, and these forces govern the maximum particle size of suspended sediments. Water concentrations for total Se and Se(IV) are in the same range as those previously reported for the Carquinez Strait (Cutter, 1989; Cutter and San Diego-McGlone, 1990).

TSM-Se correlates with TOC ($r = 0.661$, Fig. 3). To understand the mechanisms that control Se accumulation in TSM, it is also necessary both to speciate TSM-Se and to correlate concentrations with other properties. Surprisingly, adsorbed Se was found to be a minor fraction (<5%) of TSM-Se. Preliminary results show that TSM-Se is inefficiently extracted by the sequence developed for soils (78% residual Se). While the reasons for such low efficiency are under investigation, it may be hypothesized that the inability of the procedure to disrupt live cells is responsible. If this is the case, then biogenic Se may dominate TSM-Se.

3. Sediment Selenium Distribution: Eh, pH, and TOC

The distributions of total Se in sediment at MRP and BSP are shown in Figure 4. In both cases Se concentrations increase landward (i.e. are higher on the marsh plain than the mudflats). The concentration gradient is steeper at MRP than BSP, probably because mudflat sediment at MRP is much coarser in texture than at BSP. Except for a slight decline in the mudflats, there are no consistent intersite trends with depth. The most notable feature of this distribution is that sediment Se concentrations in the mudflats are lower than Se concentrations on suspended sediment in the water column (cf. Fig. 2), while those in the marshes are higher. There are several possible explanations for this pattern. Sediment deposited onto

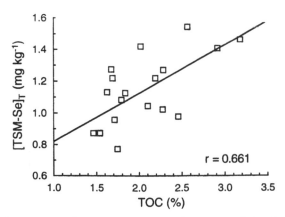

Figure 3   TSM-Se versus total organic carbon (TOC) in suspended sediments from the MRP site.

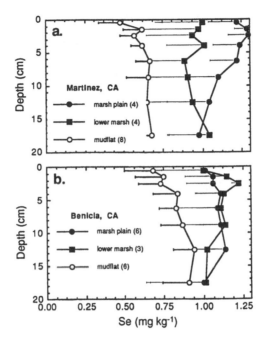

FIGURE 4   Total Se distributions in three sediment–soil environments at (a) the MRP site, and (b) the BSP site. Each point is a mean of the number of samples shown in the legend. Error bars, shown on only one side of the mean for clarity, indicate one standard deviation.

the mudflats may be depleted subsequently with respect to Se as a result of Se dissolution and loss to the water column, while marsh sediment becomes enriched with Se because biochemical reduction takes place in the water column. A more physically based explanation is that the suspended sediment becomes fractionated along particle size; coarser sediment tends to drop out of the water column in the higher energy mudflats, while the finer sediment is deposited in the more quiescent marsh plain, where water is shallow, vegetation is thick, and the effects of wind are diminished. Not only would more Se be expected to adsorb onto finer sediments, but suspended marine organisms also would be more likely to settle out in this low energy environment.

Other properties of the sediment cores were measured, including Eh, pH, and TOC. The results of Eh measurements are shown in Figure 5. The commonality between the two sites are the relatively more oxic conditions in the marsh plain, likely due to a shorter time of inundation. As indicated by the standard deviations, the largest spatial variability was observed in the lower marsh, where the decay of plant material results in local reduced zones. The steepest depth gradient was observed in the mudflat cores, with more oxic sediments near the surface. This

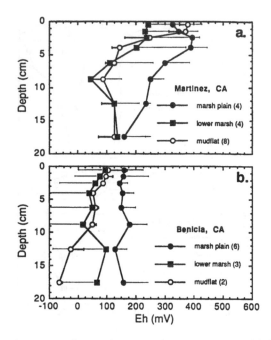

FIGURE 5   Eh distributions in three sediment–soil environments at (a) the MRP site, and (b) the BSP site. Each point is a mean of the number of samples shown in the legend. Error bars, shown on only one side of the mean for clarity, indicate one standard deviation.

is clearly due to the near constant water saturation of these sediments, except at the surface. Nonetheless, almost all samples from all environments showed positive Eh, suggesting that even under these wet and periodically saturated conditions, the redox state of the sediments does not fall within the stability field of Se(0), or below 0 mV. This has important implications for the immobilization of Se, because reduction of Se(IV) and Se(VI) to Se(0) is believed to be one of the main mechanisms by which Se is immobilized in sediments. Interestingly, there is no correlation between Eh and total Se, strongly implying that most of the Se is allogenic.

The distribution of TOC for both sites is shown in Figure 6. A very steep gradient is observed from the marsh plain, where TOC content is as high as 15%, to the mudflats, where TOC is as low as 1%. A positive correlation exists between total Se and TOC ($r = 0.74$), which could indicate that Se is being preferentially reduced or sorbed in sediments with high TOC. On the other hand, this correlation may indicate that Se is associated with suspended particles that are higher in TOC (i.e. finer textured and plankton-dominated), which in turn settle in the lower energy parts of the system, such as the upper marsh. Sediment pH (not

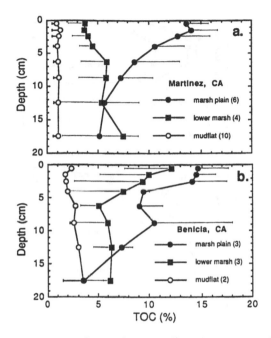

FIGURE 6 TOC distributions in three sediment–soil environments at (a) the MRP site, and (b) the BSP site. Each point is a mean of the number of samples shown in the legend. Error bars, shown on only one side of the mean for clarity, indicate one standard deviation.

shown) increases from the upper marsh (median pH 7.4) to the mudflats (median pH 8.1), but its correlation with total Se is less significant ($r = -0.56$).

4.   Selenium in Interstitial Water

Interstitial water removed via centrifugation is most available to physical exchange at the sediment–water interface and contains most of the soluble Se, referred to hereafter as "interstitial Se." The results of interstitial Se analysis are shown in Figure 7. Selenium concentrations exceed those found in overlying waters by one to two orders of magnitude, with the highest values in the top 1 to 2 cm of the mudflat sediment and in the marsh plain. Together with the speciation results from suspended sediment, this suggests that Se is being solubilized in situ. Such an upward soluble Se gradient may mean that Se will diffuse into the overlying bay water. A rough estimate of Se loss due to diffusive flux, based on Fick's law, falls in the range of 300–500 μg Se/m²/year, which is about a 0.5 to 1 order of magnitude less than the sedimentary Se input (see Section IV.B on trapped sediments).

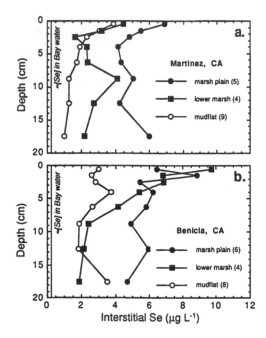

FIGURE 7 Interstitial Se distributions in three sediment–soil environments at (a) the MRP site, and (b) the BSP site. Each point is a mean of the number of samples shown in the legend. The average coefficient of variance for sites MRP and BSP is 0.61 and 0.50, respectively.

5.   Selenium Fractionation

Selenium fractionation was performed on sediments from all three environments. Elemental Se was found to be the dominant fraction in all three environments, comprising around 50% of total Se (Fig. 8). Organically associated Se comprised the second largest fraction (ca. 25%), followed by adsorbed Se (3–10%) and soluble Se (<1%). The most significant difference among the three environments was with respect to the fraction that was *not* extracted, or residual Se. This fraction decreases from 26% in the marsh plain to 7% in the mudflats. This decrease, which is correlated with TOC ($r = 0.51$), may be indicative of the presence of an organo-Se form that is not removed during the sequential extraction process.

No depth trends in Se fractionation were observed in the top 20 cm of sediment, even though the depth intervals near the sediment surface were only 1 cm, suggesting that Se transformations take place very shortly after Se is deposited, probably within millimeters of the surface. There was also no correlation between various Se species and the Eh–pH regime as measured on the solid cores. Evidently, Se transformations, especially oxidation (Zawislanski and

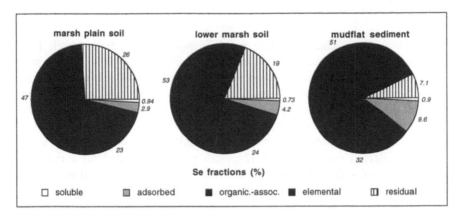

FIGURE 8 Results of soil–sediment Se fractionation at the MRP site.

Zavarin, 1996), occur on much longer time scales than Eh changes associated with tidal cycles. The absence of a distinct soluble Se gradient indicates that Se diffusion below a depth of about 5 cm is limited.

B. Selenium Concentrations on Trapped Sediment

One of the Se cycling hypotheses suggested above invokes the dominance of sedimentation in the observed distribution of Se in the intertidal zone. A study of sedimentation dynamics is ongoing at the MRP and BSP sites to test this hypothesis. The goals are to measure the patterns and rates of sediment and sediment Se deposition.

1. Study Design

Twelve erosion pins and sediment traps were installed at evenly spaced intervals along transects at both sites. Each transect is perpendicular to the shoreline and spans the region from the marsh plain to the mudflats. Erosion pins (2 ft long stainless steel rods) were inserted into the sediment so that only the top 30 cm of each pin remains unburied. These are used to measure either sedimentation or erosion. Sediment traps consist of $10 \times 10 \text{ cm}^2$ squares of plastic with long bolts at each corner to serve as anchors. The plastic surfaces were roughened by gluing fine sediment on them. The traps were installed as flush to the ground surface as possible. Sediment collected in the traps is weighed to determine a net sedimentation rate, analyzed to determine the percentage of organic particulates, and analyzed for total Se. Samples, collected at low tides, are removed using distilled water washes and the wet sediments are freeze-dried prior to analysis.

2. Results

Currently available data, including monthly measurements from March 1996 until August 1996, are summarized in Fig. 9. Sedimentation rates are highest in the mudflats and at the mudflat–lower marsh boundary, specifically in the *Spartina* zone. Selenium concentrations show the opposite trend, with concentrations above 1 mg/kg in the marsh plain and around 0.5 mg/kg in the mudflats. The marked increase in TOC in the particulates trapped in the marsh plain indicates a high organic component, possibly phytoplankton. A comprehensive set of samples collected over at least a 12-month period is needed for a complete interpretation of sediment Se dynamics, but the implication so far is that sediment deposition patterns are very important in Se distribution.

C. Selenium Reduction Rates in Laboratory Microcosms

A laboratory microcosm study was conducted to assess the relative significance of Se reduction from the water column on Se accumulation on marsh and mudflat sediments. The main goal of this study was to measure rates of Se reduction under ponded conditions. Given that different parts of the intertidal zone are flooded for variably long times, appropriate ponding periods were chosen.

1. Experimental Design

The experimental setup consists of a Lucite box with a 10×20 cm base, an open top, and a water reservoir. Platinum electrodes, porous ceramic cup samplers, and

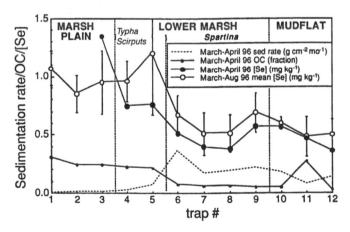

FIGURE 9   Sedimentation rate, TOC, and Se levels of sediments trapped at the MRP site; Se levels are shown for both a single sampling period ($n = 1$; March–April 1996) and as an average for a six month period ($n = 6$; March–August 1996; $\pm$ 1 SD).

water samplers were installed on one side of the box. In a preliminary experiment, 1 × 20 × 10 cm monolith of sediment was sampled at low tide from the mudflats and another from the marsh plain at MRP. These samples were immediately placed in microcosm boxes and transported to the laboratory, where probes and samplers were installed.

After the measurement of initial conditions, 10 cm of bay water, containing 0.218 μg of Se per liter, was ponded on top of the sediment, and reduction trends, measured by means of the Pt electrodes, were observed. Samples of surface and pore waters were collected. Water was drained after 7 and 15 hours from the marsh and mudflat sample, respectively. These ponding times were chosen to represent the maximum amount of time each environment might be flooded during high tide. On average, the marsh plain is inundated for no more than 4 hours, twice a day, while the part of the mudflats from which the sample was taken is usually flooded for about 6 to 8 hours, twice a day. After the marsh sample became reoxidized close to initial values, it was flooded with bay water spiked with Se(IV) (2 μg/L) and remained flooded for nearly a month.

2. Results

Figure 10 shows the Eh trends at the marsh sediment–water interface, where the most notable changes were observed. Experiment 1 in the marsh soil (short-term ponding, drying), conducted in November 1995, is shown on the same time scale as experiment 2 (long-term ponding). During the ponding phase of experiment 1, Eh did not drop below 0 mV in either environment. The redox potential of water within 1 cm of the sediment–water interface did not decrease.

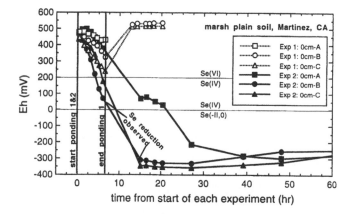

Figure 10 Eh changes observed in a soil–water microcosm of marsh plain soils from MRP. Theoretical Se stability fields (based on work by Masscheleyn et al., 1991) are shown for comparison.

During long-term ponding on the marsh soil (experiment 2), Eh at the sediment–water interface dropped below 0 mV after 10 to 20 hours. A 50% decrease in dissolved Se concentration was observed at the boundary at 16 hours, though no changes in Se were observed deeper in the soil or in the overlying water at that time.

The observed Eh trends show that neither the marsh nor the mudflat sediment become anoxic at the sediment–water interface within one tidal flooding cycle. These results suggest that reduction of Se in the water column is not significant on the time scale of tidal flooding.

V. DISCUSSION

All the available data tend to favor a hypothesis for Se cycling in which the accumulation of Se in estuarine wetlands is influenced by sedimentation dynamics, suspended particulates, diffusive fluxes, and to a lesser degree, reduction–oxidation cycles in very shallow soils and sediments. This hypothetical cycle does not address the fate of Se in geologic time (i.e., changes resulting from diagenesis), although the effect of diagenetic processes on present-day Se concentrations, specifically in interstitial water, should not be disregarded. The predominant processes are shown schematically in Figure 11.

The transformations shown in Figure 11 are, of course, only partial, since most of the Se in the sediment remains in elemental form. However, the relative fraction of Se needed to be solubilized to raise interstitial Se concentrations by 1 $\mu g/L$ is very minor, on the order of 0.1% of the total inventory. Although the processes are portrayed as occurring in the top 1 cm of sediment, this refers

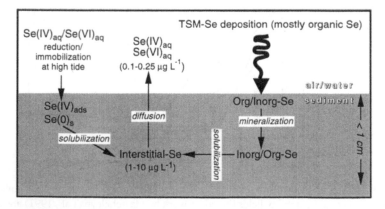

FIGURE 11   Conceptual diagram of the Se cycle near the water–sediment interface in an estuarine wetland.

primarily to the short-term mineralization of biogenic Se. As seen from data from this and previous research (Takayanagi and Belzile, 1988), relatively high interstitial Se concentrations can be found at depths of tens of centimeters and possibly deeper, suggesting that solubilization of Se, either via desorption of Se(IV) or by way of oxidation of organically associated or elemental Se, may occur deeper in the sediment profile. Solubilization of Se has also been observed to occur through dissolution of Se(0) by thiols, sulfides, and polysulfides in saline environments (Weres et al., 1989).

The ecological significance of such a cycle in an anthropogenically affected environment is considerable, given the primary exposure pathways with respect to Se in estuarine wetlands. As discussed earlier, tissue Se concentrations in bivalves, which are a major part of waterfowl diet, are closely related to particulate Se levels. The relatively lower sediment Se concentrations in mudflats are encouraging, since the mudflats are primary feeding grounds for birds. Sediment Se concentrations in both the mudflats and the marsh are fairly uniform with depth, suggesting that recent fluctuations in Se concentrations in the Carquinez Strait have not made a mark in the sediment record. Conversely, it is doubtful that moderate changes in Se levels in the Carquinez Strait would result in a noticeable decrease in sedimentary Se in estuarine wetlands.

The mechanism by which Se becomes associated with suspended particulates in the Carquinez Strait requires further investigation. Given that adsorbed Se(IV) is a minor fraction of the extractable TSM-Se, some transformation must occur, presumably on the particulate surface or, in the case of phytoplankton, within live cells. Since TSM appears to play an important role in the Se cycle, it is essential that these processes be elucidated.

Finally, the implications of the upward gradient in soluble Se from the interstitial water into bay water need to be considered. If Se is continually solubilized in the sediment column, there may be a long-term source for soluble Se to the overlying water. The assumptions made in the hypothesis discussed herein are rather conservative, assuming no advective mixing of tidal waters with interstitial waters. Inasmuch as some mixing is inevitable in this highly dynamic hydrologic system, exchange of soluble Se may occur on a more rapid scale, and diffusion may be a flux of secondary importance. Since much of the biological activity of concern transpires at the sediment–water interface, it is there that Se levels need to be monitored, and, once again, should not be expected to be affected by minor changes in dissolved Se levels in the Carquinez Strait, but rather controlled by the solubilization of sedimentary Se.

ACKNOWLEDGMENTS

The authors thank Emmanuel Gabet for collecting sediment trap data; A. Brownfield, H. Wong, A. Haxo, T. Sears, and D. King for assistance in the field and laboratory; and S.

Mountford and J. Oldfather for analytical support. We also thank S. Benson, T. Tokunaga, S. Luoma, and G. Cutter for valuable comments and suggestions. We acknowledge the support of the Department of Energy (A. Hartstein, E. Zuech) and the San Francisco Bay Regional Water Quality Control Board (M. Carlin, K. Taylor) in cofunding this research.

REFERENCES

Balistrieri, L. S., and T. T. Chao. 1987. Selenium adsorption by goethite. *Soil Sci. Soc. Am. J.* 51:1145–1151.

Benson, S. M., A. F. White, S. Halfman, S. Flexser, and M. Alavi. 1991. Groundwater contamination at Kesterson Reservoir, California 1. Hydrogeologic setting and conservative solute transport. *Water Resour. Res.* 27:1071–1084.

Belzile, N., and J. Lebel. 1988. Selenium profiles in the sediments of the Laurentian Trough (northwest North Atlantic). *Chem. Geol.* 68:99–103.

Burton, J. D., W. A. Maher, C. I. Measures, and P. J. Statham. 1980. Aspects of the distribution and chemical form of selenium and arsenic in ocean waters and marine organisms. *Thalassia Jugosl.* 16:155–164.

Cutter, G. A. 1978. Species determination of selenium natural waters. *Analy. Chim. Acta* 98:59–66.

Cutter, G. A. 1989. The estuarine behaviour of selenium in San Francisco Bay. *Estuarine Coastal Shelf Sci.* 28:13–34.

Cutter, G. A., and M. L. C. San Diego-McGlone. 1990. Temporal variability of selenium fluxes in San Francisco Bay. *Sci. Total Environ.* 97/98:235–250.

Doran, J. W. 1982. Microorganisms and the biological cycling of selenium. *Adv. Microb. Ecol.* 6:1–32.

Fan, A. M., and M. J. Lipsett. 1988. Human health evaluation of findings on selenium in ducks obtained from the verification study, 1986–1987. California Department of Health Services, Health Advisory, June 17, 1988.

Frankenberger, W. T., Jr., and U. Karlson. 1989. Environmental factors affecting microbial production of dimethylselenide in a selenium-contaminated sediment. *Soil Sci. Soc. Am. J.* 53:1435–1442.

Hamdy, A. A., and G. Gissel-Nielsen. 1977. Fixation of selenium by clay minerals and iron oxides. *Z. Pflanzenernaehr. Bodenkd.* 140:63–70.

Johns, C., S. N. Luoma, and V. Elrod. 1988. Selenium accumulation in benthic bivalves and fine sediments of San Francisco Bay, the Sacramento–San Joaquin Delta, and selected tributaries. *Estuarine Coastal Shelf Sci.* 27:381–396.

Karlson, U., and W. T. Frankenberger, Jr. 1988. Determination of gaseous selenium-75 evolved from soil. *Soil Sci. Soc. Am. J.* 52(3):678–681.

Karlson, U., and W. T. Frankenberger, Jr. 1989. Accelerated rates of selenium volatilization from California soils. *Soil Sci. Soc. Am. J.* 53:749–753.

Lipton, D. S. 1991. Associations of selenium with inorganic and organic constituents in soils of a semi-arid region. Ph.D. thesis. University of California, Berkeley.

Long, R. H. B., S. M. Benson, T. K. Tokunaga, and A. Yee. 1990. Selenium immobilization in a pond sediment at Kesterson Reservoir. *J. Environ. Qual.* 19:302–311.

Luoma, S., C. Johns, N. Fisher, N. Steinberg, R. Oremland, and J. Reinfelder. 1992. Determination of selenium bioavailability to a benthic bivalve from particulate and solute pathways. *Environ. Sci. Technol.* 26:485–491.

Masscheleyn, P. H., R. D. Delaune, and W. H. Patrick, Jr. 1991. Biogeochemical behavior of selenium in anoxic soils and sediments: An equilibrium thermodynamics approach. *J. Environ. Sci. Health A* 26(4):555–573.

Measures, C. I., and J. D. Burton. 1978. Behaviour and speciation of dissolved selenium in estuarine waters. *Nature* 273:293–295.

Neal, R. H., G. Sposito, K. M. Holtzclaw, and S. J. Traina. 1987. Selenite adsorption on alluvial soils: I. Soil composition and pH effects. *Soil Sci. Soc. Am. J.* 51:1161–1165.

Ogden Beeman and Associates, Inc. 1992. Sediment budget study for San Francisco Bay. Prepared for Corps of Engineers, San Francisco District.

Oremland, R. S., N. A. Steinberg, A. S. Maest, L. G. Miller, and J. T. Hollibaugh. 1990. Measurement of in-situ rates of selenate removal by dissimilatory bacterial reduction in sediments. *Environ. Sci. Technol.* 24:1157–1164.

Sarathchandra, S. U., and J. H. Watkinson. 1981. Oxidation of elemental selenium to selenite by *Bacillus megaterium. Science* 211:600–601.

SFBRWQCB (San Francisco Bay Region Water Quality Control Board). 1992. Mass emissions reduction strategy for selenium. Staff Report.

Takayanagi, K., and N. Belzile. 1988. Profiles of dissolved and acid-leachable selenium in the upper St. Lawrence Estuary. *Mar. Chem.* 24:307–314.

Takayanagi, K., and D. Cossa. 1985. Speciation of dissolved selenium in the upper St. Lawrence Estuary. In A. C. Sigleo and A. Hattori (eds.), *Mari Estuarine Geochemistry*, pp. 275–284. Lewis Publishers, Chelsea, MI.

Takayanagi, K., and G. T. F. Wong. 1984. Organic and colloidal selenium in southern Chesapeake Bay and adjacent waters. *Mar. Chem.* 14:141–148.

Takayanagi, K., and G. T. F. Wong. 1985. Dissolved inorganic and organic selenium in the Orca Basin. *Geochim. Cosmochim. Acta* 49:539–546.

Tao, G., and E. H. Hansen. 1994. Determination of ultra-trace amounts of selenium (IV) by flow injection hydride generation atomic absorption spectrometry with on-line preconcentration by coprecipitation with lanthanum hydroxide. *Analyst* 119: 333–337.

Tokunaga, T. K., D. S. Lipton, S. M. Benson, A. Y. Yee, J. M. Oldfather, E. C. Duckart, P. W. Johannis, and K. H. Halvorsen. 1991. Soil selenium fractionation, depth profiles and time trends in a vegetated site at Kesterson Reservoir. *Water Air Soil Pollut.* 57/58:31–41.

Tokunaga, T. K., P. T. Zawislanski, P. W. Johannis, S. M. Benson, and D. S. Lipton. 1994. Field investigations of selenium speciation, transformation, and transport in soils from Kesterson Reservoir and Lahontan Valley. In W. T. Frankenberger, Jr., and S. M. Benson, (eds.), *Selenium in the Environment*, pp. 119–138. Dekker. New York.

Tokunaga, T. K, I. J. Pickering, and G. E. Brown. 1996. Selenium transformations in ponded sediments. *Soil Sci. Soc. Am. J.* 60:781–790.

Van der Sloot, H. A., D. Hoede, J. Wijkstra, J. C. Duinker, and R. F. Nolting. 1985. Anionic species of V, As, Se, Mo, Sb, Te, and W in the Scheldt and Rhine Estuaries and the Southern Bight (North Sea). *Estuarine Coastal Shelf Sci.* 21:633–651.

Velinsky, D. J., and G. A. Cutter. 1990. Determination of elemental Se and pyrite-Se in sediments. *Anal. Chim. Acta* 235:419–425.

Velinsky, D. J., and G. A. Cutter. 1991. Geochemistry of selenium in a coastal salt marsh. *Geochim. Cosmochim. Acta* 55:179–191.

Weres, O., A.-R. Jaouni, and L. Tsao. 1989. The distribution, speciation and geochemical cycling of selenium in a sedimentary environment, Kesterson Reservoir, California, U.S.A. *Appl. Geochem.* 4:543–563.

Zawislanski, P. T., and M. Zavarin. 1996. Nature and rates of selenium transformations in Kesterson Reservoir soils: A laboratory study. *Soil Sci. Soc. Am. J.* 60:791–800.

Zawislanski, P. T., T. K. Tokunaga, S. M. Benson, J. M. Oldfather, and T. N. Narasimhan. 1992. Bare soil evaporation and solute movement in selenium-contaminated soils of Kesterson Reservoir. *J. Environ. Qual.* 21:447–457.

Zhang, G. H., M. H. Hu, Y. P. Huang, and P. J. Harrison. 1990. Se uptake and accumulation in marine phytoplankton and transfer of Se to the clam *Puditapes philippnarum.* *Mar. Environ. Res.* 30:179–190.

14

Selenium Accumulation in a Wetland Channel, Benton Lake, Montana

YIQIANG ZHANG and JOHNNIE N. MOORE
University of Montana, Missoula, Montana

I. INTRODUCTION

Selenium (Se) accumulation in wetlands has been studied since the discovery of irrigation-induced selenium poisoning of waterfowl (Ohlendorf, 1989; Weres et al., 1989; Presser et al., 1994; Zhang and Moore, 1996a, 1997). These studies show that selenium accumulation in wetlands is very complex and is controlled by various environmental conditions. Most studies of selenium accumulation in wetland systems have focused on Kesterson Reservoir in California, where selenium bioaccumulation was much higher than most other wetlands of the western United States. Bioaccumulation of selenium in wetland organisms in Kesterson Reservoir created serious hazards to fish and waterfowl (Ohlendorf, 1989), finally resulting in closure of the reservoir in 1986. The problems at Kesterson Reservoir led to an increasing concern about the hazards of selenium bioaccumulation at other wetlands (Presser et al., 1994).

U.S. Department of the Interior reconnaissance investigations of selenium contamination in many wetlands of the western United States showed that most of these wetlands were impacted by contaminated drainage water (Presser et al., 1994). Although there are no documented deleterious effects on waterfowl in most of these wetlands, concentration of selenium may continue to rise, possibly posing a threat to wildlife in the future. Therefore, information about the complex natural processes controlling selenium accumulation in wetlands is important for those responsible for managing and remediating these selenium-impacted wetlands.

The selenium problem at Benton Lake, Montana, has been studied for several years (Nimick et al., 1996; Zhang and Moore, 1996a, 1997). The studies

243

show that selenium-containing drainage water is the major source of selenium entering the lake, and selenium fixation in sediment is the major process removing selenium from the water column. In this study, one channel (about 10 m wide by 1200 m long) within Benton Lake was chosen for detailed examination of selenium accumulation because of its steep selenium concentration gradient. The channel is a ditch excavated in the early 1960s to form the dike on the southeast side of a pond (P4C, Fig. 1). A saline seep at the southwest end of the channel contributes acidic water with high concentrations of selenium and dissolved solids. Discharge from the seep is intermittent and very small. However, large and sudden pulses of selenium probably come from the seep during summer thunderstorms when accumulated selenium-rich efflorescent salts are dissolved and washed into the channel. The northeast end of the channel receives water

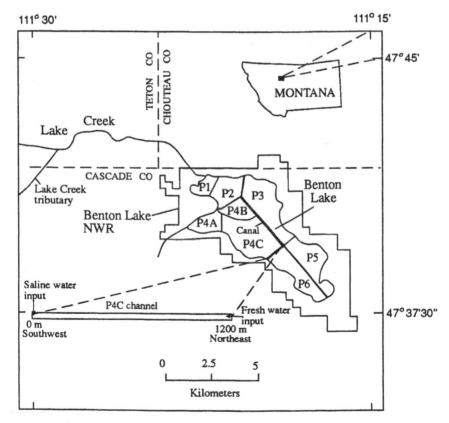

FIGURE 1   Map of Benton Lake showing location of the P4C channel and water sources of the channel; P1, P2, P4C, etc, are designations for pool 1, pool 2, pool 4C, etc.

whenever water is added to P4C. Water delivered to P4C comes from P1 and P2 via the canal between P3 and P4 and a headgate in the northeast corner of the P4C. The selenium concentration in this water is low as a result of selenium-removal processes occurring in P1 and P2 (Zhang and Moore, 1996a; 1997). The P4C channel provides an excellent natural laboratory for studying selenium accumulation in a channel for the purpose of biogeochemical processes to remove selenium from water in wetlands areas. The objectives of this study were to examine the selenium accumulation in the channel receiving selenium-rich water and to quantify the biogeochemical processes compartmentalizing selenium.

II. MATERIALS AND METHODS

Samples were collected from the P4C channel in 1994–1995. Water samples were filtered through a 0.45 μm membrane filter into clean polyethylene bottles. Samples for major cations and trace metals were acidified to a pH of less than 2 with concentrated hydrochloric acid (HCl). Dissolved oxygen (DO) and pH in water were measured in the field using a portable DO meter (Orion 820) and a pH meter (Orion 290A). Surface sediments (upper 8 cm) were collected using a plastic spoon. Plant samples—watermilfoil (*Myrophyllum exculenta*), algae, and alkali bulrush (*Scirpus paludosus*)—were collected by hand. All samples were transported to the laboratory on ice. Bulrush samples were washed to remove all sediment from samples and then sectioned into different parts (leaf and stem, rhizome, roots). Aquatic plant roots were collected from samples of sediment in the laboratory by sieving. Three subsamples of sediment were kept wet and stored in sealed plastic bags at 4°C for selenium fractionation analysis. Other sediment samples, plant roots, watermilfoil, algae, and alkali bulrush were dried at room temperature (about 23°C), ground, and stored in sealed polyethylene bottles for total selenium analysis.

　　To permit quantitative estimations of the amount of selenium associated with organic materials (Se_{OMS}) versus that not associated with organic materials in sediment (Se_{NOMS}), the selenium concentration in all organic material in the sediment was assumed to be the same as the selenium concentration measured separately in small plant roots collected from sediment. Based on this assumption, the selenium concentration associated with organic material in sediment was calculated by using the equation $Se_{OMS} = C_{OMS}/C_{OM} \times Se_{OM}$, in micrograms, where Se_{OMS} is the selenium concentration in organic material incorporated in the sediment of selenium in organic material per gram of sediment; C_{OMS} is the concentration of organic carbon in the sediment, in milligrams of organic carbon per gram of sediment; C_{OM} is the concentration of organic carbon in small plant roots removed from the sediment by sieving, in milligrams of organic carbon per gram of small plant roots; and Se_{OM} is the concentration of selenium in small plant roots removed from the sediment by sieving, in micrograms of selenium in plant roots per gram of plant roots. Se$_{NOMS}$ was estimated to be the difference between the total selenium in the sediment and Se$_{OMS}$. Even though the selenium concentration in the small plant roots removed from sediment may not perfectly represent the selenium concentration in different organic material pools in the sediment, it probably provides a good estimate of Se$_{OMS}$ and Se$_{NOMS}$.

Total selenium concentrations in sediments, small plant roots, watermilfoil, algae, and alkali bulrush were determined using a hydrogen peroxide–hydrochloric acid digestion procedure (Zhang and Moore, 1997). An XAD–resin separation method developed by Fio and Fujii (1990) was used to determine selenium species (selenate, selenite, and organic selenium) in water and adsorbed fractions in sediment. A sequential extraction technique was used to determine *selenium fractionation* in sediment (Zhang 1996; Zhang and Moore, 1996a). The dry weight concentration of the selenium fractions was determined by correcting for water content in sediment samples dried at room temperature. Selenium concentrations in all samples were determined using hydride generation atomic absorption spectrometry (Zhang and Moore, 1996b). Total carbon was determined by combustion of the sediment samples at 1050°C and coulometric titration of CO_2 from the combustion tube. Inorganic carbon was determined by coulometric titration of CO_2 released from an acid–sediment reaction tube. Organic carbon was estimated to be the difference between total carbon and inorganic carbon. Concentrations of dissolved major cations and trace metals were determined by means of inductively coupled plasma optical emission spectrometry (AtomComp Series 800). Sulfate and chloride were determined by ion chromatography (Dionex 2000i). Because concentrations of nitrate in the channel surface water were very low (<0.01 mg/L), its effect on selenium accumulation was not examined in this study. All selenium concentrations in sediment, algae, plant, and plant roots were reported as dry weight (air-dried at about 23°C).

III. RESULTS

A. Water Chemistry and Selenium Removal

Saline seep water collected near the southwest end of the P4C channel was a water of the acidic Na- or $MgSO_4$ type (Table 1). Saline seep water contained high concentrations of Al, Ca, Na, Mg, Cl, and SO_4 and relatively low concentrations of B, Fe, Mn, Si, and Sr. Concentrations of most constituents differed significantly among the three sampling dates. Concentrations were the highest in the September 1994 sample collected during a dry period, when seep water probably had been evapoconcentrated. Concentrations were lower in May 1995 during a rainy period. The concentrations of total dissolved selenium were very high (299–1880 μg/

Table I   Concentrations of Selected Constituents (mg/L), and Selenium Species (μg/L), and pH of the Saline Seep Near the Southwest End of the P4C Channel

Parameter	July 7, 1994	Sept. 22, 1994	May 11, 1995
pH	3.9	4.3	3.7
Al	409	560	108
B	6.15	8.6	1.61
Ca	525	520	442
Fe	5.72	24.6	0.64
Mg	1.09×10^4	1.54×10^4	2.64×10^3
Mn	91.3	137	24.5
Na	9.8×10^3	1.58×10^4	2.66×10^3
Si	43.7	47.6	10.7
Sr	4.47	6.84	3.17
Cl	992	1.31×10^3	268
SO$_4$	6.74×10^4	1.1×10^5	1.91×10^4
Total dissolved Se	1.22×10^3	1.88×10^3	299
Se (VI) (selenate)	967	1.55×10^3	197
Se (IV) (selenite)	19.8	27.4	9.5
Organic Se	232	302	93

L). Selenate accounted for 66 to 82% of the total dissolved selenium, followed by organic selenium, at 16 to 31% and selenite at 1.5 to 3.2%.

The distribution of total dissolved selenium in the channel appears to be controlled by the mixing of seep water and channel water (Table 2). Concentrations of all elements decreased with increasing distance from the southwest end of the channel. The major constituents in the channel water were Ca, Mg, Na, Cl, and SO$_4$. Concentrations of Al, B, Mn, and Sr were low. The concentration ratios of Cl to Se and Na to Se in channel water increased significantly in the first 20 m and then decreased with distance to about 550 m, where they stabilized (Fig. 2). Concentrations of dissolved oxygen differed among the sites. In sites rich in organic material, DO concentrations were relatively low, ranging from 1 to 2 mg/L on the surface water to less than 1 mg/L at the water–sediment interface. In sites containing less organic material, DO concentrations were relatively high, ranging between 5 and 12 mg/L in the surface water and from 3 to 8 mg/L at the water–sediment interface.

Concentrations of selenium differed among the three sampling dates (Fig. 3). In the dry period of August 1994, the water depth in the channel was very shallow and the first 60 m reach at the southwest end of the channel was dry. The concentration of total dissolved selenium was 9.2 μg/L (pH 7.3) at 60 m, decreasing to about 2 μg/L (pH 9.3) at 200 m and remaining near 2 μg/L

Concentrations of Selected Constituents (mg/L) and Total Dissolved Selenium (μg/L) in the P4C Channel Water (Ma...

(m)[a]	Al	B	Ca	Mg	Mn	Na	Sr	Cl	SO$_4$
	76.2	1.67	393	1980	20.7	2620	2.69	233	1.64×10^4
	33.8	1.64	407	1560	15.0	2360	2.57	216	1.33×10^4
	0.641	0.766	203	652	4.58	1130	1.53	126	5.18×10^3
	0.341	0.496	141	377	1.83	723	1.25	81.7	2.85×10^3
	0.228	0.458	132	351	1.48	661	1.22	75.9	2.48×10^3
	0.195	0.464	134	356	1.74	673	1.22	78.1	2.58×10^3
	0.124	0.434	125	329	1.46	610	1.16	65.8	2.31×10^3
	0.067	0.383	115	294	1.15	534	1.07	66.4	2.00×10^3
	0.036	0.337	102	254	0.843	442	0.982	48.1	1.67×10^3
	0.025	0.208	62.7	128	<0.005	162	0.688	27.2	598
	0.023	0.204	63.4	128	<0.005	159	0.696	26.3	607
	0.026	0.204	63.2	128	<0.005	158	0.700	25.8	593

...om SW end of the P4C channel.
...lved selenium.

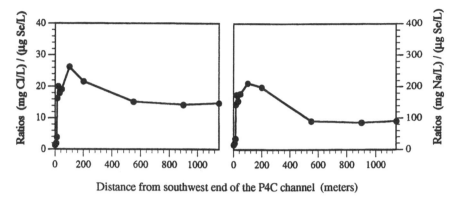

FIGURE 2   Concentration ratios of chloride and sodium to total dissolved selenium in channel water sampled in May 1995.

throughout the remainder of the channel. In September 1994, P4C was filled with a large volume of low-selenium water delivered from P1 and P2 via the canal. The channel reach with elevated selenium (>2 μg/L, pH 4.3–8.8) was only about 25 m long. In the rainy period of May 1995, elevated selenium concentrations (>2 μg/L, pH 4.9–8.6) extended about 100 m from the southwest end of the channel.

Selenate and organic selenium were the main selenium species in channel water and accounted for more than 90% of the total selenium (Fig. 4). Concentrations of selenate and organic selenium decreased with distance away from the southwest end of the channel. The decrease in selenate concentration occurred faster than the decrease in organic selenium concentration.

B.   Selenium Accumulation in Sediment and Plants

Concentrations of total selenium in surface sediment and plant roots collected from the P4C channel in August 1994 (a dry period) decreased away from the southwest end of the channel (Fig. 5). From 0 to 92 m from the southwest end, total selenium concentrations ranged from 9.04 to 20.1 μg/g in sediment and from 15.5 to 52.5 μg/g in small plant roots. Concentrations of total selenium decreased to about 1 μg/g in sediment and 3 μg/g in plant roots at about 200 m. From about 200 m to the northeast end of the channel, total selenium concentration was constant in both sediment and plant roots. The ratios of total selenium concentration in plant roots to sediment for samples collected at the same sampling site ranged from 1.36 to 8.26, with a mean of 3.44.

Watermilfoil accumulated selenium to a high level in the southwest end of the channel (Table 3). Selenium concentrations in watermilfoil were 14.5 μg/

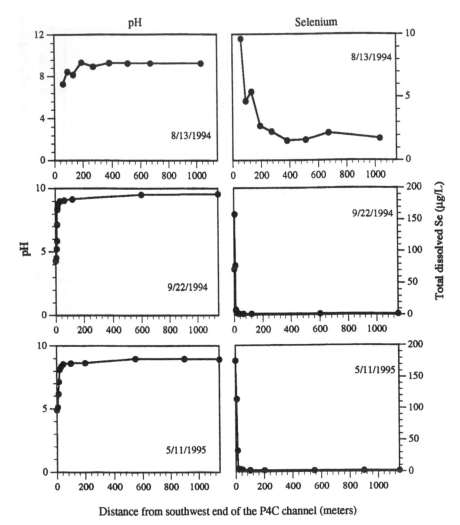

FIGURE 3   Distribution of pH and total dissolved selenium in the P4C channel water.

g at 20 m, 6.45 μg/g at 30 m, and 4.89 μg/g at 60 m from the southwest end. These samples were collected in May 1995 when watermilfoil were only about 5 cm in length. In September 1995, when watermilfoil was fully grown, concentrations of selenium in watermilfoil also was high near the southwest end of the channel and low in the northeast end of the channel. Similarly, higher selenium concentrations in algae also were found near the southwest end of the channel. Selenium concentrations differed in the different parts of bulrush. It increased from leaf and stem to rhizome, and to roots (Table 3). Selenium concentrations

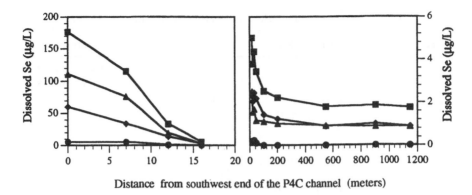

FIGURE 4 Distribution of dissolved selenium species in the P4C channel water sampled in May 1995. *Right:* data from 16 to 1150 m (note different *y*-axis scales). Squares, total selenium; triangles, selenate; circles, selenite; diamonds, organic selenium.

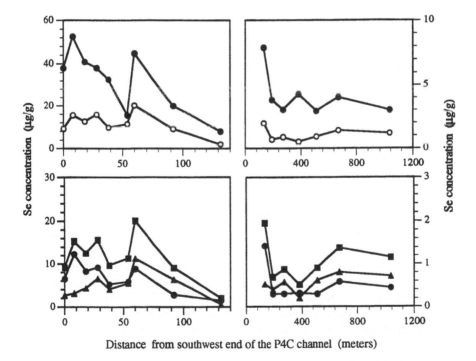

FIGURE 5 *Top:* distribution of selenium in sediment (open circles) and plant roots (solid circles) sampled (August 1994) from the P4C channel. *Bottom:* distribution of selenium fractions in sediment sampled (August 1994) from the P4C channel. Squares, total selenium; circles, selenium in organic material; triangles, selenium in nonorganic material. (Note different *x*- and *y*-scale axes.)

Table 3   Concentrations (μg/g) of Selenium in Plants Within the P4C channel

Distance (m)[a]	Watermilfoil	Algae	Bulrush
14	—	—	1.24, 1.43, 2.29, 22.3[d]
14	—	—	1.32, 1.50, 2.50, 15.7[d]
17	8.38	4.71	—
20	8.39 (14.52)[b]	3.70 (4.92)[c]	—
22	5.26	—	—
30	5.17 (6.45)	— (2.40)	—
60	0.053 (4.89)	0.651	—
500	—	0.019	—
1150	0.177 (0.716)	—	—

[a]Distance from SW end of the P4C channel.
[b]Samples were collected in close sites in May 1995.
[c]Algae samples were collected in close sites when part of channel was dried in September 1995.
[d]Selenium concentrations in upper part, lower part of the leaf and stem, rhizome, and roots, respectively.

were relatively lower in leaf and stem, ranging from 1.24 to 1.50 μg/g, and were much higher in roots (15.7–22.3 μg/g).

Concentrations of Se_{OMS} and Se_{NOMS} in the sediment were highest near the southwest end of the channel and lower farther away from the southwest end of the channel (Fig. 5). At distances less than 60 m from the southwest end of the channel, the concentration of total selenium in sediment was dominated by Se_{OMS}. At 60 m and distances greater than 60 m, Se_{NOMS} was the main form of selenium in the sediment. For all sediment samples in this channel, the mean concentrations of Se_{OMS} and Se_{NOMS} were very similar, accounting for 52 and 48% of the total selenium in sediment, respectively.

C. Selenium Fractionation in Sediment

The soluble and adsorbed selenium fractions were low (Tables 4 and 5), making up only 0.9 to 1.8% and 3.6 to 8.7% of the total selenium in sediment, respectively. Within the adsorbed fraction, selenate and organic selenium were the major selenium species. In contrast, elemental selenium and selenium associated with organic material was higher, accounting for 33.2 to 57.1% and 34.3 to 53.6% of the total selenium in sediment, respectively. Selenium in oxides was also low, comprising from 0.5 to 3.9% of the total selenium in the sediment.

IV. DISCUSSION

Selenium contamination of wetlands has resulted from agricultural drainage containing high concentrations of selenium (Lemly et al., 1993; Presser et al., 1994;

TABLE 4 Selenium (Se) Fractions (μg/g) in Surface Sediments Sampled Near the Bank of the P4C Channel

Distance (m)[a]	Soluble Se	Adsorbed Se	Elemental Se	OM Se[b]	Se-oxides	AAS Se[c]	Σ Se fractions	Total Se
30	0.084	0.304	2.730	1.639	0.023	0.006	4.786	4.739
60	0.027	0.110	1.326	1.547	0.028	0.011	3.049	2.944
1500	0.008	0.044	0.162	0.269	0.019	0.010	0.512	0.493

[a]Distance from SW end of the P4C channel.
[b]Se associated with organic material.
[c]Se in amorphous aluminosilicates.

Zhang and Moore, 1997). When very selenium-rich water entered the channel (Table 2), concentrations of all elements decreased with distance, apparently in response to mixing with low-selenium channel water. Concentration ratios of "conservative" Cl and Na to selenium in channel water showed that the ratios were much higher in the channel mixing area than in the saline input water (Fig. 2). Concentration of selenium decreased much faster than that of "conservative" Cl and Na. Determination of selenium species showed that the decrease in selenate concentration in channel water was faster than that of organic selenium (Fig. 4). This phenomenon also was found at Kesterson Reservoir, California (Cooke and Bruland, 1987), and in the main pond system at Benton Lake, Montana (Zhang and Moore, 1996a). The percentage of selenate decreased dramatically as water moved through the pond system, while the percentages of dissolved organic selenium increased. These studies suggest that selenate is preferentially removed from the water column over other species and that dilution mixing between very

TABLE 5 Selenium (Se) Species (μg/g) in Adsorbed Fractions in Surface Sediments Sampled Near the Bank of the P4C Channel

Distance (m)[a]	Selenium species			
	Se (IV) (selenite)[b]	Se (VI) (selenate)	Organic	Total
30	0.023	0.141	0.140	0.304
60	BDL	0.049	0.061	0.110
1500	BDL	0.025	0.019	0.044

[a]Distance from SW end of the P4C channel.
[b]BDL, below the detection limit (0.007 μg/g).

selenium-rich water and channel water is not the only process reducing selenium concentrations in channel water.

Adsorption of selenium on sediments can remove selenium from water. The degree of adsorption is dependent on the selenium species present (Frize and Hall, 1988; Neal et al., 1987a, 1987b; Neal and Sposito, 1989) and various environmental conditions (Neal et al., 1987a, Goldberg and Glaubig, 1988; Barrow and Whelan, 1989). Selenate adsorption generally is not considered an important process for removing selenium from water because the adsorption of selenate is weak (Neal and Sposito, 1989). Studies on selenium fractionation and species in soil and sediment showed that adsorbed selenate was an important part of adsorbed inorganic selenium in soils and sediment, accounting for 12 to 54% of total adsorbed selenium in soil in San Joaquin Valley, California (Fujii et al., 1988), 53% of total adsorbed selenium in sediment at Benton Lake, Montana (Zhang and Moore, 1996a), and about 50% of total adsorbed selenium in this channel sediment. These studies suggest that selenate can be adsorbed in the sediment and soil. Selenite has a relatively strong affinity for sorption sites (Neal et al., 1987a, 1987b; Frize and Hall, 1988) and readily adsorbs to sediment and soil. This did not happen in the channel sediment because the major selenium species from saline seeps input into the channel was selenate (Table 1). Selenite concentrations were very low, accounting for less than 5% of the total dissolved selenium. Therefore, there was less selenite to compete with selenate for adsorption sites.

The small proportion of total adsorbed selenium in the sediment could be due to the relatively high pH in the channel water and high concentrations of other ions. Barrow and Whelan (1989), Goldberg and Glaubig (1988), and Neal et al. (1987a) all found that increase in solution pH (from 4 to 9) decreased selenite and selenate adsorption on soils. Water pH in the channel was high (8–9) at most sites (Fig. 3), which would substantially reduce selenium adsorption in sediment. Other constituents in channel water also were much higher than selenium (Table 2), so they may have competed for adsorption sites. In this study, total adsorbed selenium accounted for less than 10% of the total selenium accumulated in sediments, showing that selenium adsorption was a minor process removing selenium from the water column.

Reduction of selenium to elemental selenium is the major process in wetlands which transfers selenium from water to sediment. Oremland et al. (1989) found that selenate reduction to elemental selenium proceeded quite rapidly in anoxic sediment. Elemental selenium is a dominant form of selenium in wetland systems (Oremland et al., 1989; Weres et al., 1989; Velinsky and Cutter, 1991; Tokunaga et al., 1996; Zhang and Moore, 1996a). The same pattern was confirmed in this study, in which elemental selenium was an important sink for selenate removal from the water. The formation of large amounts of elemental selenium may be related to the redox potential in the surface sediment. Generally, the redox potential is low in flooding sediment (Gambrell and Patrick, 1978; Koch-Rose et al., 1994), so it supports the reduction of electron acceptors, such as O_2, NO_3 and selenate (Se-[VI]). In this vegetation-rich channel, decomposition of channel plants, such as algae, watermilfoil and sago pondweed, consumes dissolved oxygen and decreases redox potential in surface sediment (low DO values at the water–sediment interface and very low concentrations of nitrate in surface water), resulting in an increased rate of selenate reduction. This cycle was confirmed by recent laboratory studies in which selenate reduction to elemental selenium was shown to be more efficient in sediment with organic amendments than in sediment without organic matter treatment (Oremland et al., 1989; Tokunaga et al., 1996). Selenate reduction to elemental selenium also is supported by the presence of sulfate reduction (very strong smell of H_2S in sediment). Because the redox potential of SO_4/H_2S is lower than Se (VI)/Se^0 (Masscheleyn et al., 1990), selenate reduction is more favorable energetically than SO_4 reduction. This relationship suggests that the wetland sediment environment facilitates selenate reduction to elemental selenium.

Selenium uptake by wetland organisms is an important removal process. Among wetland organisms, aquatic plants may play a more important role in the process of selenium uptake because they dominate the makeup of organic material in the wetland. Arvy (1993) found that selenium uptake by bean plants can transfer selenium from roots to leaves. Concentrations of selenium in roots were higher than those in other parts of the bean plant. The different distribution of selenium concentration in plants also is shown in this study. Selenium concentration in small plant roots was higher than that in watermilfoil sampled at similar sites. Selenium concentration in alkali bulrush roots also was much higher than that in other plant parts (leaf and stem, rhizome). At Kesterson Reservoir, California, the mean selenium concentration in the roots and rhizomes was found to be 64 $\mu g/g$, but it was only 30 $\mu g/g$ in the stem and leaf materials (Ohlendorf, 1989). Cattails are known to accumulate selenium in their roots under wetland habitats (W. T. Frankenberger, Jr., personal communication). These studies suggest that the selenium distribution in wetland aquatic plant roots may reflect maximum selenium uptake by aquatic plants. Selenium accumulation in plant roots, watermilfoil, and algae was also related to the dissolved selenium concentrations in the channel. Highest accumulation occurred at the southwest end of the channel near the region of high dissolved-selenium input. In the rest of the channel, selenium accumulation was low.

Deposition and incorporation of channel organisms (especially wetland plants) to the sediment is an important process accumulating selenium. This study showed that concentrations of root-accumulated selenium are much higher than those of sediment containing these plant roots and that a high percentage of the selenium associated with organic material in sediment is in organic material.

V. CONCLUSION

Results of this study showed that the distribution of selenium in water, sediment, and plants is related to the proximity of the source of the selenium-rich input water, mixing between selenium-rich water and channel water, and biogeochemical processes occurring in the channel. Elevated selenium concentrations in water, sediment, and plants in the channel were found as far as 100 m from the selenium source. Plant roots appeared to accumulate selenium more strongly than sediment in the channel. Elemental selenium and selenium associated with organic material were the major fractions of selenium in sediment. In contrast, adsorbed selenium was low. A quantitative estimation of selenium associated with organic material in channel sediment showed that the mean percentages of organic material selenium and nonorganic material selenium were similar (52 and 48%, respectively). In this wetland channel, selenium *reduction* and selenium binding to organic material are the major processes removing selenium from water.

ACKNOWLEDGMENTS

We thank Erich Gilbert, Mindy Meade, Steve Martin, and Jim McCollum from Benton Lake National Wildlife Refuge and Donald Palawski from the U.S. Fish and Wildlife Service. Special thanks to David Nimick for helpful discussions and assistance with field work. Funding for this study was provided by the U.S. Fish and Wildlife Service and the Department of the Interior's National Irrigation Water Quality Program. Thanks to the M. J. Murdock Charitable Trust for funding a major laboratory upgrade to aid in analyses.

REFERENCES

Arvy, M. P. 1993. Selenate and selenite uptake and translocation in bean plants (*Phaseolus vulgaris*). *J. Exp. Bot.* 44:1083–1087.
Barrow, N. J., and B. R. Whelan. 1989. Testing a mechanistic model: VII. The effects of pH and of electrolyte on the reaction of selenite and selenate with a soil. *J. Soil Sci.* 40:17–28.
Cooke, T. D., and K. W. Bruland. 1987. Aquatic chemistry of selenium: Evidence of biomethylation. *Environ. Sci. Technol.* 21:1214–1219.
Fio, J. L., and R. Fujii. 1990. Selenium speciation methods and application to soil saturation extracts. *Soil Sci. Soc. Am. J.* 54:363–369.
Frize, R. J., and S. D. Hall. 1988. Efficacy of various media in attenuation of selenium. *J. Environ. Qual.* 17:480–484.
Fujii, R., S. J. Deverel, and D. B. Hatfield. 1988. Distribution of selenium in soils of agricultural fields, western San Joaquin Valley, California. *Soil Sci. Soc. Am. J.* 52:1274–1283.

Gambrell, R. P., and J. W. H. Patrick. 1978. Chemical and microbiological properties of anaerobic soils and sediments. In D. D. Hook and R. M. Crawford (eds.), *Plant Life in Anaerobic Environments*, pp. 375–423. Ann Arbor Science, Ann Arbor, MI.

Goldberg, S., and R. A. Glaubig. 1988. Anion sorption on a calcareous, Montmorillonitic soil-selenium. *Soil Sci. Soc. Am. J.* 52:954–958.

Koch-Rose, M. S., K. R. Reddy, and J. P. Chanton. 1994. Factors controlling seasonal nutrient profiles in a subtropical peatland of the Florida Everglades. *J. Environ. Qual.* 23:526–533.

Lemly, A. D., S. E. Finger, and M. K. Nelson. 1993. Sources and impacts of irrigation drainage contaminants in arid wetlands. *Environ. Toxicol. Chem.* 12:2265–2279.

Masscheleyn, P. H., R. D. Delaune, and W. H. J. Patrick. 1990. Transformation of selenium as affected by sediment oxidation–reduction potential and pH. *Environ. Sci. Technol.* 24:91–96.

Neal, R. H., and G. Sposito. 1989. Selenate adsorption on alluvial soils. *Soil Sci. Soc. Am. J.* 53:70–74.

Neal, R. H., G. Sposito, K. M. Holtzclaw, and S. J. Traina. 1987a. Selenite adsorption on alluvial soils: I. Soil composition and pH effects. *Soil Sci. Soc. Am. J.* 51:1161–1165.

Neal, R. H., G. Sposito, K. M. Holtzclaw, and S. J. Traina. 1987b. Selenite adsorption on alluvial soils: II. Solution composition effects. *Soil Sci. Soc. Am. J.* 51:1165–1169.

Nimick, D. A., J. H. Lambing, D. U. Palawski, and J. C. Malloy. 1996. Detailed study of selenium in soil, water, bottom sediment, and biota in the Sun River Irrigation Project, Freezout Lake Wildlife Management Area, and Benton Lake National Wildlife Refuge, West-Central Montana, 1990–1992. U.S. Geological Survey Water Resources Invest. Report No. 95-4170.

Ohlendorf, H. M. 1989. Bioaccumulation and effects of selenium in wildlife. In L. W. Jacobs (ed.), *Selenium in Agriculture and the Environment.* pp. 133–177. American Soil Association and Soil Science Society of America, Madison, WI.

Oremland, R. S., J. T. Hollibaugh, A. S. Maest, T. S. Presser, L. G. Miller, and C. W. Culbertson. 1989. Selenate reduction to elemental selenium by anaerobic bacteria in sediments and culture: Biogeochemical significance of a novel, sulfate-independent respiration. *Appl. Environ. Microbiol.* 55:2333–2343.

Presser, T. S., M. A. Sylvester, and W. H. Low. 1994. Bioaccumulation of selenium from natural geologic sources in western states and its potential consequences. *Environ. Manage.* 18:423–436.

Tokunaga, T. K., I. J. Pickering, and G. E. J. Brown. 1996. Selenium transformation in ponded sediment. *Soil Sci. Soc. Am. J.* 60:791–790.

Velinsky, D. J., and G. A. Cutter. 1991. Geochemistry of selenium in a coastal salt marsh. *Geochim. Cosmochim. Acta* 55:179–191.

Weres, O., A. R. Jaouni, and L. Tsao. 1989. The distribution, speciation and geochemical cycling of selenium in a sedimentary environment, Kesterson Reservoir, California, U.S.A. *Appl. Geochem.* 4:543–563.

Zhang, Y. Q. 1997. Biogeochemical cycling of selenium in Benton Lake, Montana. Ph. D. thesis. University of Montana, Missoula.

Zhang, Y. Q., and J. N. Moore. 1996a. Selenium speciation and fractionation in a wetland system. *Environ. Sci. Technol.* 30:2613–2619.

Zhang, Y. Q., and J. N. Moore. 1997. Controls of selenium distribution in wetland sediment, Benton Lake, Montana. *Water, Air, Soil Pollut.* 97:323–340.

15

Selenite Sorption by Coal Mine Backfill Materials in the Presence of Organic Solutes

GEORGE F. VANCE
University of Wyoming, Laramie, Wyoming

RANDOLPH B. SEE
United States Geological Survey, Louisville, Kentucky

KATTA J. REDDY
University of Wyoming, Laramie, Wyoming

I. INTRODUCTION

The chemistry of selenium (Se) in coal overburden materials can be influenced by mining and reclamation operations (Dreher and Finkelman, 1992). During surface coal mining, rock material overlying the coal is redistributed from its original stratigraphic position. Exposure of previously buried material to surface oxidizing conditions can decrease the stability of Se-containing sulfides and organic matter, increasing the potential solubility of Se species (Lindner-Lunsford and Wilson, 1992; White and Dubrovsky, 1994).

In a study of the cumulative hydrologic impacts of surface coal mining in the eastern Powder River structural basin (PRB) of northeastern Wyoming, Martin et al. (1988) found that the water-soluble Se content of overburden material was not representative of the total quantity of Se that could be released to groundwater after mining. Their study indicated total Se content in overburden material from two mines in the PRB ranged from 0.1 to 3.8 mg/kg. Effluent derived after about one pore volume of water was passed through columns packed with backfill material had Se concentrations generally exceeding 50 μg/L (Martin et al., 1988). These Se concentrations are of concern, particularly if consumed by livestock, wildlife, and humans. Currently, the Wyoming class III groundwater standard for total Se (suitable for livestock) is 50 μg/L (Wyoming Department of Environmental Quality, 1993).

Concentrations of Se ranging from 3.4 μg/L (Martin et al., 1988) to 330 μg/L (Naftz and Rice, 1989) have been detected in shallow postmining groundwaters from coal mines in the PRB. The groundwater contribution to surface water flow may provide a mechanism for the transport of Se from groundwater environments to surface water resources, after which various biogeochemical processes can influence the environmental chemistry of Se (Masscheleyn and Patrick, 1993). For example, surface water Se species concentrations ranging from 2 to 13 μg/L have been reported to cause reproductive problems in aquatic birds (Skorupa and Ohlendorf, 1991). A better knowledge of the geochemical processes affecting Se would therefore be useful to the mining industry, as well as state and federal agencies, for developing effective reclamation techniques for minimizing Se contamination of surface and groundwaters (Fio and Fujii, 1990).

The mobility of Se from coal mine backfill to groundwater is controlled by a number of geochemical processes including the formation of inorganic and organic complexes, sorption/desorption, and precipitation/dissolution reactions. The oxidation of Se compounds can transform reduced and less mobile Se species such as selenide (Se^{2-}) and elemental selenium (Se^0) into oxidized Se species, selenite (SeO_3^{2-}) and selenate (SeO_4^{2-}), which are more soluble in alkaline and oxidizing groundwaters (Naftz and Rice, 1989; Tokunaga et al., 1996; Zawislanski and Zavarin, 1996). Thermodynamic calculations indicate that selenide and elemental Se should be found in reducing environments, selenite in moderately oxidizing environments, and selenate in oxidizing environments (Elrashidi et al., 1989).

Dreher and Finkelman (1992) suggested the source, occurrence, and fate of Se in overburden deposits and backfill water are important in understanding Se chemistry in coal mine environments. The results of their study indicated decreasing soluble Se concentrations were due, in part, to microbially assisted reduction of selenate to selenite that was followed by sorption onto clay minerals in the coal mine backfill materials. Thus, selenite sorption is expected to be one of the major processes controlling the solubility of Se in subsurface environments. Sorption of selenite results from electrostatic attraction or ligand exchange mechanisms and is affected by pH, redox condition, and competition of other ions (Balistrieri and Chao, 1987; Bar-Yosef and Meek, 1987; Neal et al., 1987a, 1987b; Blaylock et al., 1995; Jayaweera and Biggar, 1996). In addition, kinetic studies involving soils have suggested selenite sorption onto oxide surfaces occurs by a ligand-exchange mechanism (Goldberg, 1985; Balistrieri and Chao, 1987; Neal et al., 1987a, 1987b).

Studies by Balistrieri and Chao (1987), Naftz and Rice (1989), and Abrams et al. (1990) have shown that natural organic solutes and organic Se compounds play an important role in the mobility of Se in soil and coal mine overburden. For example, Naftz and Rice (1989) indicated that natural organic solutes in coal mine backfill groundwater systems have the potential to compete for adsorption sites on metal oxides, thus increasing the solubility of Se. They found a statistically significant positive correlation between Se and organic carbon. Using a sequential extraction technique to evaluate solid phase Se, Naftz and Rice (1989) found that overburden samples from two coal mines in the PRB had nearly 100% of the total Se in an operationally defined organic phase.

Dissolved organic carbon (DOC) concentrations in most groundwaters are generally below 2 mg/L (Thurman, 1985). However, DOC concentrations in groundwater associated with organic-rich environments can be extremely large. For example, groundwater receiving recharge from marshes and swamps can contain DOC levels over 10 mg/L, oil field brines over 1000 mg/L, and trona water over 40,000 mg/L DOC (Thurman, 1985). The DOC in groundwater associated with reclaimed coal backfill materials can vary as a function of backfill composition and time after reclamation. Studies involving backfill and overburden materials indicate that organic solutes are important to Se chemistry because of the accompanying redox effects and competitive processes (Naftz and Rice, 1989; Dreher and Finkelman, 1992; Sharmasarkar and Vance, 1997).

The primary objective of this study was to evaluate the role of natural organic solutes in effecting selenite sorption in reclaimed surface coal mines. Specific objectives included the characterization of the chemical and physical properties of backfill core materials, the evaluation of reclaimed coal mine ground water chemistry and DOC composition, and the determination of organic solute effects on Se sorption/desorption reactions involving backfill core materials.

II. MATERIALS AND METHODS

A. Coal Mine Backfill Samples

Two backfill cores were collected from within a 5 m radius of existing monitoring wells at two reclaimed coal mine field sites (Fig. 1). Core holes ranged in depth from 6.8 to 8.4 m and were at least as deep as the associated well. Backfill cores were collected using a rotary-driven, split-spoon auger that allowed the collection of 8.3 cm diameter cores. The core barrel within the auger extended beyond the auger head and rested on a bearing assembly, which prevented the core barrel sampler from rotating, thus permitting the collection of core samples that were only slightly disturbed.

Acrylic liners were used inside the core barrel to collect and store the backfill core samples. Void space in the liners was purged with prepurified argon

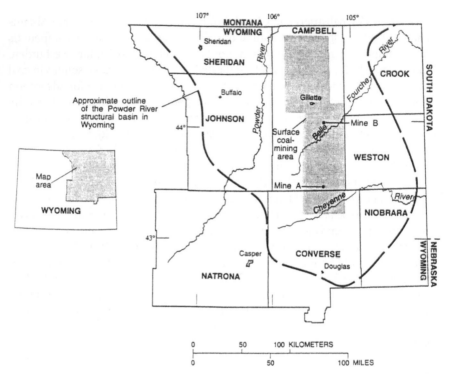

FIGURE I   Location of surface coal mining area, coal mine sampling sites, and the Powder River structural basin in Wyoming.

gas, and the liners were sealed with plastic caps and tape to prevent oxidation of the samples during transportation and storage. Backfill cores were chilled in the field and stored under refrigeration until needed for laboratory experiments. The cores were removed from the acrylic liners and partitioned into subsamples; splits of the core samples were air-dried and sieved through a 10 mesh (2 mm) sieve. A total of 157 backfill sample splits were separated from the six cores.

I.   X-Ray Diffraction and Chemical Analyses

Several core samples were analyzed by X-ray diffraction (XRD) to determine bulk mineralogy following the methods of Moore and Reynolds (1989). Samples (10–20 μm) were wetted using acetone, placed on glass slides, dried at room temperature, and analyzed using a Scintag X-ray diffractometer.* Portions of the core samples also were examined under a microscope and by visual analysis for the presence of carbonates (positive if effervesced) using 0.1 M HCl.

2. Fractionation of Backfill Core Selenium by Sequential Partial Dissolution

A sequential partial dissolution technique developed by Chao and Sanzolone (1989) was modified for partitioning backfill core Se into operationally defined fractions. The procedure partitions Se into six fractions:

Fraction 1: 0.1 M KH_2PO_4 adjusted to a pH 8 with KOH
Fraction 2: 1 M $NaC_2H_3O_2$ adjusted to pH 5 with acetic acid
Fraction 3: NaOCl (5% w/v), adjusted to pH 9.5
Fraction 4: 0.25 M $NH_2(OH) \cdot HCl$
Fraction 5: 4 M HCl
Fraction 6: concentrated HF + HNO_3 + $HClO_4$ total digestion of the residue

After sequential partitioning, the Se in each fraction was determined by atomic absorption spectrophotometry with hydride generation (AAS-HG) (Blaylock and James, 1993; Spackman et al., 1994).

B. Groundwater Samples

Groundwater samples were collected from two wells at field site A and one well at field site B. Samples were collected after pH, specific conductance, and temperature of continuously pumped well water solutions had stabilized. The groundwater samples were filtered on-site through 0.45 μm glass fiber filters into 250 mL, high density polyethylene (HDP) bottles for major and trace ion analyses, and into 4 L HDP bottles for DOC fractionation and isolation studies. Subsamples used for cation analyses were acidified to pH 2 with HNO_3; subsamples for DOC analysis were purged with argon and filled under a stream of argon gas to minimize oxidation. All groundwater samples were chilled in the field with ice and subsequently stored under refrigeration (3°C) until time of analysis.

On-site analyses of groundwater solutions were conducted in a sealed, flow-through chamber that minimized sample contact with the atmosphere. Specific conductance was measured using a Lab Line Instruments Lectro Mho Meter; pH (Orion combination electrode); Eh (Orion platinum electrode with an $Ag/AgCl_2$ reference electrode) was measured using an Orion Research Model 407A Ion

---

* Use of brand, firm, or trade names is for identification purposes only and does not constitute endorsement by the University of Wyoming, the U.S. Geological Survey, or the Wyoming Water Resources Center.

Analyzer; and dissolved oxygen was measured using a YSI Incorporated Model 57 dissolved oxygen meter.

Concentrations of cations (Al, As, Cd, Ca, Cr, Cu, Fe, Pb, Mn, Mg, Mo, Si, Zn) in filtered samples were measured using inductively coupled plasma (ICP) spectrophotometry. Sodium and potassium were measured using atomic emission spectrophotometry (AES). Unacidified filtrates were analyzed for dissolved inorganic carbon (DIC), SO_4, Cl, F, NO_3, and total Se. Concentrations of SO_4, Cl, F, and NO_3 were measured using a Dionex Ion Chromatograph (Sunnyvale, CA), DOC using a Shimadzu TOC-5000 carbon analyzer, and total Se by AAS-HG.

1. Fractionation of Dissolved Organic Carbon

Dissolved organic carbon in groundwater samples was fractionated using the method of Vance and David (1991). The method separates DOC into six fractions (hydrophobic bases, acids, and neutrals, and hydrophilic bases, acids, and neutrals). The percentage of DOC in the different fractions was determined, and the dominant fractions were identified to facilitate their isolation and use in backfill core Se sorption/desorption experiments.

2. Isolation of Dissolved Organic Carbon

Hydrophobic and hydrophilic acids in groundwater samples were isolated and concentrated in accordance with the methods described by Vance and David (1991). The methods involve the substitution of XAD-4 resin for the anion-exchange resin to facilitate the recovery of hydrophilic acids (Malcolm and MacCarthy, 1992). Groundwater samples from wells A-1 and B-1 were used in the isolation procedure; groundwater DOC from well A-2 was not isolated because of low DOC concentrations. The isolated fractions were used later in the sorption/desorption studies to determine their effect on Se sorption.

The following samples were used for the DOC isolation study: 16 L of water from well A-1 (88 mg/L DOC) and 18 L from well B-1 (87 mg/L DOC). After isolation of the hydrophobic and hydrophilic acids, the solutions were passed through a column containing cation-exchange resin. For the well A-1 water sample, 3.3 L of hydrophobic acid (280 mg/L DOC) and 0.8 L of hydrophilic acid (145 mg/L DOC), and for water from well B-1, 3.1 L of hydrophobic acid (300 mg/L DOC) and 0.8 L of hydrophilic acid (130 mg/L DOC) were isolated.

C. Sorption Studies

Selenium sorption experiments were conducted by equilibrating 2 g backfill core samples in polyethylene centrifuge tubes and 20 mL solution containing known amounts of Na_2SeO_3 ranging from 0 to 1000 μg SeO_3/L. Samples were capped and placed on a reciprocating shaker for 24 hours, a reaction time that was chosen based on results of kinetic experiments by Balistrieri and Chao (1987). After equilibrating, samples were centrifuged for 15 minutes at 2000 rpm and the supernatant was filtered through 0.7 μm Whatman glass microfiber filters prerinsed with distilled deionized (DDI) water. The core samples were then extracted with 1 M KH_2PO_4 and the extracted Se was measured to determine Se desorption. Total Se in sorption/desorption solutions was determined using the AAS-HG method reported by Spackman et al. (1994).

Three selenite sorption experiments were conducted using background solutions of DDI water, 0.1 M $CaCl_2$, and solutions containing isolated hydrophobic or hydrophilic acids. Concentration of the isolated DOC from well A-1 used in the competitive sorption experiments was 157 mg/L for hydrophobic acids or 107 mg/L for hydrophilic acids. Competitive sorption experiments using isolated DOC from well B-1 contained 110 mg/L hydrophobic or 88 mg/L hydrophilic acids. The isolated DOC fractions from each groundwater sample were added to the corresponding backfill core samples collected from the sites near each well, except for the backfill core samples corresponding to site A-2, where hydrophobic acids from well A-1 were used. No sorption studies were conducted using hydrophilic acid and backfill core samples from site A-2 because the amount of isolated hydrophilic acids was insufficient.

III.   RESULTS AND DISCUSSION

A.   Backfill Core Sample Analysis

The XRD analyses indicated that the following minerals were present in backfill core samples: quartz, potassium feldspar, kaolinite, illite, and muscovite. Additional minerals detected in some samples included smectite, gypsum, calcite, dolomite, carbonate apatite, and goethite (Table 1). The presence of large-surface-area clay minerals (kaolinite, illite, muscovite, and smectite) indicates that the core samples have potentially large sorption capacities.

Portions of the backfill core samples also were examined under a microscope. Although goethite, an iron oxide, was detected by XRD in only one sample, microscopic analysis revealed that iron oxide coatings and coal particles were present in all the samples. Iron oxides were estimated to occur in amounts less than 1% in all samples. X-ray diffraction peaks of other iron oxides (e.g., hematite) were not detected, which suggests the iron oxides were amorphous rather than crystalline, or the concentrations were below the detection limits of the XRD analysis (about 2%) (See et al., 1995). Samples also were tested for the presence of carbonates, using 0.1 M HCl. All the samples effervesced when hydrochloric acid was added. This is consistent with XRD results indicating the presence of calcite or dolomite or both, in all but one sample. When HCl was added to the

Ray Diffraction Analyses of Minerals[a] in Select Backfill Core Samples from Reclaimed Surface Coal Mines in the Powd[er]
[...]ing

[C]ore and sample no.	Quartz	Potassium feldspar	Kaolinite	Illite	Muscovite	Smectite	Gypsum	Calcit[e]
136C	XX	XX	XXX	XX	X	XXX	−	X
147A	XX	X	XX	X	X	X	−	XXX
246C	XX	X	XX	X	XX	−	X	X
257A	XX	X	XXX	XXX	X	X	−	XX
188A	nd	nd	nd	nd	nd	nd	nd	nd
277A	XX	XXX	X	XX	X	−	−	X
288A	XX	X	X	X	XXX	−	−	XXX
146C	X	X	XX	XXX	X	XXX	−	XX
157A	X	X	X	X	X	XX	−	XXX
256A	X	X	XXX	XXX	X	XX	−	XX

[...]dance of each mineral is indicated as high (XXX) to low (X) based on peak height, or as no discernible peak (−) for the mineral [...]
[...]d.

samples, the iron oxide coatings became more visible as the aggregate particles dispersed.

The saturated paste pH values were variable and ranged from 4.2 to 7.5 (Table 2). The higher pH values were probably due to the buffering effect of carbonates, which were identified in several backfill core samples. The saturated paste pH for backfill core samples collected near the groundwater table were similar to those obtained from on-site pH measurements of groundwater.

Saturated paste Eh measurements were somewhat variable (Table 2). Values of Eh near the groundwater table were also similar to those obtained for the groundwater samples. All saturated paste extract specific conductance values, EC, were greater than 2 decisiemens per meter (dS/m), suggesting a large soluble salt content in the core samples (Table 2). Saturated paste extractable Se concentrations ranged between 2 and 93 μg/kg, with Se generally decreasing in samples below the groundwater table. These results indicate groundwater dilution of Se concentrations or possibly reduction of selenate to selenite, which is subsequently sorbed onto the backfill core material (See et al., 1995).

Evaluation of backfill core samples by sequential dissolution (Chao and Sanzolone, 1989) was conducted to determine the content of Se in different fractions. Selenium concentrations in each of the six fractions and total Se concentrations for the samples studied are shown in Table 3. The fractions represent the following Se characteristics:

Water-soluble and specifically sorbed Se; primarily selenite that would be readily available for leaching and/or plant uptake.

Se in carbonate minerals; potentially available through acidification processes.

Se in organic matter; becomes available through microbial decomposition.

Se in manganese oxides and amorphous iron oxides; oxidation and reduction reactions may release Se.

Se in crystalline iron oxides and acid-volatile sulfides.

Se primarily in crystalline sulfides, occluded and embedded in siliceous materials, silicates, or refractory accessory minerals; relatively resistant to weathering.

Naftz and Rice (1989) found that the organic phase contained 100% of the total Se in 10 of 11 undisturbed, overburden rock samples collected from mine sites in the PRB, Wyoming. Selenium in the organic phase (fraction 3) of this study accounted for about 5 to 50% of the total Se in backfill core samples analyzed. The differences observed may be a result of dissolution potentially due to redox reactions and the redistribution of Se from the organic phase to other fractions in the backfill material during the mixing and disturbance associated with mining and subsequent reclamation processes.

Analyses of Saturated Pastes and Saturated Paste Extracts from Select Backfill Core Samples, Powder River Basin, Wyo

Core and sample no.	Sample depth (m)	Saturated paste		Saturated paste extracts			
		pH	Eh (mV)	EC (dS/m)	Se (μg/L)	DOC (mg/L)	SO_4^{2-}
136C	4.3	7.4	342	5.9	88	15.3	4
147A	4.8	7.5	344	5.4	15	13.5	4
246C	4.5	7.0	348	4.4	89	18.3	9
257A	4.7	6.8	373	6.3	52	25.8	10
188A	5.5	4.2	611	3.2	4	9.1	6
277A	4.7	4.6	405	4.7	15	5.6	10
288A	5.5	5.9	486	3.7	2	15.7	
146C	4.5	7.5	314	3.7	93	35	10
157A	4.7	7.4	310	3.5	36	45.2	
256A	3.9	7.0	450	3.4	62	61.1	9

elenium Concentrations (μg/kg) in Various Fractions of Select Backfill Core Samples from the Powder River Basin,
see text for explanation of the sequential partial dissolution method)

Core and sample no.	Fractions[a,b]						Total sele...
	1	2	3	4	5	6	Sum
136C	160	20	90	20	90	100	480
147A	50	<20	280	20	130	100	580
246C	90	<20	330	<20	150	100	670
257A	120	<20	70	20	130	130	470
_88A	50	<20	50	<20	70	100	270
277A	30	<20	90	20	70	100	330
288A	<20	<20	<20	<20	60	50	110
146C	110	20	<20	<20	110	160	400
157A	nd	nd	nd	nd	nd	nd	nd
256A	120	30	80	<20	140	420	790

0.1 M KH$_2$PO$_4$ adjusted to a pH 8 with KOH; fraction 2, 1 M NaC$_2$H$_3$O$_2$ adjusted to pH 5 with acetic acid; fraction 3, 5% NaOCl b
djusted to pH 9.5; fraction 4, 0.25 M NH$_2$(OH) · HCl; fraction 5, 4 M HCl; fraction 6, concentrated HF + HNO$_3$ + HClO$_4$ total

B. Groundwater Analysis

The pH of the groundwater samples ranged from 5.5 to 7.3 (Table 4). Results of other parameters measured in the groundwater solutions included the following: Se from 3 $\mu g/L$ (well A-2) to 125 $\mu g/L$ (well A-1); DOC from 11 mg/L (well A-2) to 88 mg/L (well A-1); specific conductance (EC) from 3.6 (well B-1) to 10 dS/m (well A-1). Groundwater samples were dominated by sulfate concentrations (2960–11,300 mg/L) (Table 5). Although groundwater Se speciation was not determined in this study, analyses of groundwater from mine B indicated 99% of the Se present was selenate and only a negligible amount (<1%) existed as selenite (Naftz and Rice, 1989).

Hydrogeologic monitoring at mines A and B indicates that dissolved Se concentrations and water levels follow an inverse trend (Fadlelmawla, 1996). Generally there was a decline in dissolved Se concentrations with a concurrent increase in groundwater levels as backfill materials were resaturated. The increasing water levels result from resaturation of the backfill materials via precipitation, water pumped from the active mine pits, and lateral and underflow from adjacent areas.

The results of the DOC fractionation analysis (Table 6) indicate that DOC in the groundwater samples was dominated by hydrophobic and hydrophilic acids (74–84% for well A-1, 38–69% for well A-2, 69–83% for well B-1). Hydrophobic and hydrophilic neutrals in groundwater samples ranged from 16 to 23% for well A-1, 21 to 55% for well A-2, and 10 to 25% for well B-1. Hydrophobic and hydrophilic bases in groundwater samples were generally less than 10%.

TABLE 4 Characteristics of Groundwater Samples Collected from Monitoring Wells in Reclaimed Coal Mine Areas in the Power River Basin, Wyoming

Characteristic	Well no.					
	A-1	A-1	A-2	A-2	B-1	B-1
Sampling date	9-17-91	7-7-92	9-17-91	7-7-92	9-19-91	8-19-92
Water table depth (m)	3.7	3.6	4.7	4.6	3.5	3.5
EC (dS/m)	10	10	5.0	4.2	3.6	3.8
pH	7.1	7.3	5.5	5.6	6.3	6.4
Eh (mV)	350	370	450	450	400	370
Temperature	11	11	10	11	10	14
			Laboratory analyses			
Total	125	121	3	13	88	88
DOC (mg/L)	88	84	11	14	87	79

TABLE 5 Elemental Analysis of Groundwater Samples Collected in 1992 from Monitoring Wells in Reclaimed Coal Mine Areas in the Powder River Basin, Wyoming

Element	Well number		
	A-1	A-2	B-1
Major element (mg/L)			
Ca	427	470	545
Mg	2,080	693	391
Na	1,180	339	347
K	38	18	33
SO$_4$	11,300	4,980	2,960
Cl	340	46	65
Si	7.2	6.0	6.9
NO$_3$	68	41	51
Trace elements (mg/L)			
Al	0.03	—	0.01
Cd	<0.001	0.028	<0.001
Cu	0.017	0.012	0.009
Fe	0.001	0.007	0.006
Mn	0.03	0.01	0.16
Mo	0.005	0.006	0.006

C. Selenium Sorption/Desorption Studies

Sorption/desorption studies were evaluated based on (1) direct comparison of the selenite sorbed and desorbed as a function of the amount of selenite added (i.e., percentage of selenite removed or released) and (2) evaluation of the equilibrium data using several sorption isotherms (e.g., Freundlich, Langmuir, and initial mass isotherms). The direct comparison method of characterizing selenite sorption on backfill materials provides a general description of how selenite removal is dependent on solution selenite concentrations. Sorption isotherms, on the other hand, have been used extensively to study sorption reactions in soils, sediments, and other geological materials, and are capable of providing additional information into the sorption process (Goldberg, 1985). Parameters such as empirical constants, binding affinities, sorption maxima, indigenous soil pools, and distribution coefficients may be determined by the different adsorption isotherms. However, depending on the data, some sorption isotherm approaches fail to provide useful results because of the presence of indigenous concentrations of solutes in the material studied.

sults of the DOC Fractionation Analysis of Groundwater Samples Collected from Monitoring Wells in Reclaimed Coa
Powder River Basin, Wyoming

Date sampled	Hydrophobic solutes (%)				Hydrophobic solutes (%)		
	Acids	Bases	Neutrals	Total	Acids	Bases	Neutrals
9-17-91	42	2	15	59	32	2	8
7-7-92	42	<1	15	57	42	0	1
9-17-91	43	<1	6	49	26	10	15
7-7-92	22	<1	20	42	16	7	35
9-18-91	53	<1	17	70	16	6	8
8-19-92	53	<1	10	63	30	7	0

Results obtained by the direct comparison method using a DDI water background solution indicated that selenite sorption on A-1 backfill core samples ranged from 65 to 92%, on A-2 core samples from 92 to 98%, and on B-1 core samples from 56 to 97%. Sorption of selenite was found to increase with decreasing pH, as was expected because of the increase of positively charged sorption sites (Neal et al., 1987a; Blaylock et al., 1995).

Figure 2 shows selenite sorbed on the backfill core samples as a function of the equilibrium selenite concentration. Selenite sorption increased with increasing concentration of the added selenite. The A-2 samples had the largest and B-1 samples the smallest selenite sorption capacities, which may be a function of the B-1 samples having the greatest DOC concentration (A-2 the lowest) in the associated groundwater and saturated paste extracts. This would imply there is an inverse relation between DOC concentration and sorption capacity for Se, as shown by Naftz and Rice (1989). The phosphate extraction recovered about 50% of the added Se. The difference between the sorbed and recovered Se was probably due to Se precipitation or formation of strong ligand-exchange reactions involving Se.

The 1 M $CaCl_2$ background solution resulted in selenite sorption generally greater than 90%. Increasing sorption with decreasing pH and the increase in sorption capacity with increasing selenite added were also observed in this study. A comparison of experiments using DDI water and $CaCl_2$ background solutions suggests an increase in selenite sorption with the 0.1 M $CaCl_2$ (Fig. 2), due perhaps to precipitation reactions and/or to an increase in positively charged surfaces from calcium sorption (Neal et al., 1987b). Selenite sorption in the presence of hydrophobic or hydrophilic acids was lower than in the DDI water and $CaCl_2$ studies. Hydrophilic acids, however, were not tested in studies involving the A-2 backfill core samples because amounts of isolated hydrophilic acid material were too small.

Freundlich and Langmuir adsorption isotherms were unsuccessful in describing the relations between equilibrium Se concentrations and selenite sorption. Similarly, Blaylock et al. (1995) found that the initial mass isotherm could be used to develop predictive equations for describing selenite sorption on coal mine backfill materials. Initial mass isotherm parameters and distribution coefficients for the experiments involving the backfill materials and different background solutions are listed in Table 7.

The slope (m) and intercept (b) values determined in the initial mass isotherm are important parameters for describing selenite sorption relations. The slope defines the fraction of selenite retained by the backfill material: that is, the greater the slope, the larger the sorption. Intercept values indicate the amount of Se released when the initial solution contains no Se (i.e., negative intercept values signify Se desorption). From slope and intercept values a reserve backfill Se pool (RBSP) can be computed: RBSP = $b/(1 - m)$. The RBSP is the amount of Se present in the backfill that can be readily exchanged with substances in solution under the conditions of the experiment. Distribution coefficients (K_d),

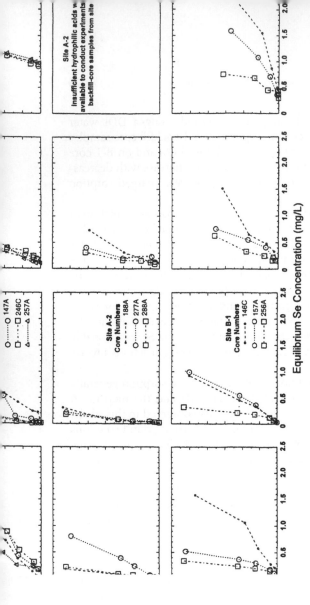

Site A-2
Insufficient hydrophilic acids w...
available to conduct experiments...
backfill-core samples from site

Site A-2
Core Numbers
- ● 147A
- ○ 246C
- □ 257A
- △

Site A-2
Core Numbers
- ● 188A
- ○ 277A
- □ 288A

Site B-1
Core Numbers
- ● 146C
- ○ 157A
- □ 256A

Equilibrium Se Concentration (mg/L)

: 2 Selenite sorption as a function of equilibrium selenium concentrations of backfill core samples with DDI water, and hydrophobic and hydrophilic acids added to background solutions.

TABLE 7 Selenite Sorption Parameters and Distribution Coefficients (K_d) Determined from Initial Mass Isotherms Under Different Experimental Conditions[a]

Site no.	Core and sample no.	Slope	Intercept	RBSP	K_d
		DDI Water			
A-1	136C	0.878	−0.028	0.231	143
A-1	147A	0.871	−0.130	1.01	135
A-1	246C	0.868	−0.154	1.17	132
A-1	257A	0.918	−0.099	1.21	224
A-2	188A	0.985	−0.004	0.301	1340
A-2	277A	0.926	−0.014	0.191	249
A-2	288A	0.982	−0.020	1.08	1080
B-1	146C	0.864	−0.184	1.36	127
B-1	157A	0.962	−0.136	3.56	504
B-1	256A	0.980	−0.127	6.26	966
		0.1 M CaCl$_2$			
A-1	136C	0.907	−0.106	1.15	196
A-1	147A	0.934	−0.057	0.865	283
A-1	246C	0.987	−0.009	0.737	1570
A-1	257A	0.976	0.001	−0.055	804
A-2	188A	0.975	0.004	−0.143	791
A-2	277A	0.982	−0.018	0.988	1090
A-2	288A	0.985	−0.014	0.910	1290
B-1	146C	0.920	−0.057	0.710	229
B-1	157A	0.960	−0.067	1.67	475
B-1	256A	0.979	−0.092	4.29	912
		Hydrophobic acid addition			
A-1	136C	0.956	−0.117	2.67	437
A-1	147A	0.959	−0.163	4.01	471
A-1	246C	0.917	−0.127	1.52	220
A-1	257A	0.915	−0.052	0.607	215
A-2	188A	0.926	−0.057	0.766	250
A-2	277A	0.974	−0.203	7.96	763
A-2	288A	0.978	−0.099	4.40	873
B-1	146C	0.837	−0.154	0.940	102
B-1	157A	0.922	−0.186	2.39	237
B-1	256A	0.938	−0.122	1.95	301
		Hydrophilic acid addition			
A-1	136C	0.942	−0.725	12.6	327
A-1	147A	0.942	−0.955	16.6	327
A-1	246C	0.983	−1.13	65.1	1130
A-1	257A	0.957	−0.928	21.4	442
B-1	146C	0.744	−0.452	1.77	58
B-1	157A	0.822	−0.374	2.10	92
B-1	256A	0.937	−0.356	5.67	298

[a]Correlation coefficients (r^2) for all experiments were >0.99. RSBP represents the reserve backfill selenium pool.

which describe the affinity of Se for the backfill, also can be calculated from initial mass isotherm results (Nodvin et al., 1986; Vance and David, 1991).

Results of the initial mass isotherm slope (0.744 to 0.987) and intercept values (−1.13 to 0.004) indicate that the backfill materials have a large retention capacity for selenite and that generally Se is released from the backfill when background solutions containing little or no selenite are mixed with the backfill core samples. Addition of $CaCl_2$ to background solutions increased sorption of selenite sorption, as demonstrated by the greater slope values in the $CaCl_2$ experiments compared to experiments containing only DDI water. Addition of hydrophobic acid to background solutions had only a minor affect on selenite sorption characteristics, except for shallow core samples collected at site A-1. For these samples, sorption increased similarly to the $CaCl_2$ study, which may be a function of enhanced sorption due to cations associated with the hydrophobic acid solutions.

The greatest impact on selenite sorption occurred with the hydrophilic acid additions. Because of the limited quantities of hydrophilic acids, only backfill core samples from sites A-1 and B-1 were studied. For A-1 samples, large initial mass isotherm slopes were determined along with low intercept values, whereas B-1 samples were found to have low slopes and intercept values. These results suggest that hydrophilic acids play a role in reducing the retention of Se by backfill materials by enhancing the desorption of indigenous Se or by displacing Se through competitive sorption (Weres et al., 1990).

Information derived from RBSP and K_d values also can be used to describe selenite sorption processes. The larger RBSPs associated with backfill core sample depth are indicative of greater amounts of Se present in the backfill that can be readily exchanged by substances such as natural organic solutes. The RBSP values were largest in studies involving hydrophilic acid. Larger RBSP values suggest that greater amounts of Se could be released by additions of hydrophilic acid to backfill materials. In addition, K_d values were generally greater in samples collected at greater depths for each of the backfill cores. A noticeable decrease in the K_d values was found in B-1 samples when results from studies involving DDI water or 0.1 M $CaCl_2$ background solutions were compared to data from studies with additions of hydrophilic acid, further indicating that hydrophilic acid can impact selenite sorption.

Sorption generally decreased in the presence of dissolved organic carbon, which indicates that DOC influences the selenite sorption process. This interference could be due to the formation of neutral complexes with Se that are not sorbed as strongly as selenite, or by competitive sorption processes involving Se and DOC. Addition of hydrophobic acid decreased selenite sorption capacities (vs. sorption with distilled-deionized water) of the A-2 and B-1 samples, while the sorption capacities remained almost constant for the A-1 backfill core samples. The addition of the hydrophilic acid decreased selenite sorption capacity for both

A-1 and B-1 backfill core samples and decreased selenite sorption more than did the addition of hydrophobic acid. The effect of DOC on selenite sorption was generally greater at lower concentrations of added selenite. A comparison of selenite sorption in the presence of hydrophobic and hydrophilic acids would suggest that the latter is capable of enhancing the potential movement of Se in reclaimed coal mine aquifers in the PRB, Wyoming, or within shallow groundwater systems (Deverel and Millard, 1988).

IV. CONCLUSIONS

Selenium in groundwater of reclaimed surface coal mines in the Powder River Basin, Wyoming, can exceed the Wyoming Department of Environmental Quality groundwater standard of 50 μg/L. A study was conducted to examine Se sorption processes and the effects of natural organic solutes on the solubility of Se using backfill core samples from three field sites in reclaimed areas at two large surface coal mines in the PRB. Backfill cores were collected from within a 5 m radius of existing monitoring wells. Samples subjected to X-ray diffraction analyses indicated that the backfill core samples contained quartz, kaolinite, potassium feldspar, illite, and muscovite, as well as other secondary minerals. Selenium concentrations in saturated paste extracts of backfill core samples ranged from 2 to 93 μg/L. A sequential partial dissolution analysis indicated that 5 to 50% of the total Se in the backfill core samples was in an operationally defined organic phase. Groundwater samples had Se concentrations of 3 to 125 μg/L. Dissolved organic carbon in groundwater samples ranged from 11 to 88 mg/L and was generally dominated by hydrophobic and hydrophilic acids (38–84%).

Isolated hydrophobic and hydrophilic acids from the groundwater samples were used in competitive sorption/desorption studies to determine the effect of DOC on selenite sorption. Three sorption/desorption experiments were conducted using background solutions of distilled deionized water, 0.1 M CaCl$_2$, and isolated hydrophobic and hydrophilic acids added to DDI water. An increase in selenite sorption occurred with 0.1 M CaCl$_2$ as compared to DDI water, whereas selenite sorption generally decreased in the presence of hydrophobic and hydrophilic acids, particularly with the latter. Results of this study suggest that when DOC, and particularly hydrophilic acids, are present in groundwater from reclaimed surface coal mines there is potential for an increase in the mobility of selenite in surface coal mine backfill materials.

ACKNOWLEDGMENTS

This work was supported in part by both the National Science Foundation EPSCoR and Abandoned Coal Mine Lands Research Program (ACMLRP) at the University of Wyoming.

Support for the ACMLRP was administered by the Land Quality Division of the Wyoming Department of Environmental Quality from funds returned to Wyoming from the Office of Surface Mining of the U.S. Department of Interior. Additional funding was provided by the U.S. Geological Survey, the Wyoming Water Resource Center, and the Department of Plant, Soil, and Insect Sciences at the University of Wyoming. Amr Fadlelmawla and Richard Allen provided valuable assistance with the laboratory analyses.

REFERENCES

Abrams, M. M., R. G. Berau, and R. J. Zasoki. 1990. Organic selenium distribution in selected California soils. *Soil Sci. Soc. Am. J.* 54:979–982.

Balistrieri, L. S., and T. T. Chao. 1987. Selenium adsorption by goethite. *Soil Sci. Soc. Am. J.* 51:1145–1151.

Bar-Yosef, B., and D. Meek. 1987. Selenium sorption by kaolinite and montmorillonite. *Soil Sci.* 144:11–18.

Blaylock, M. J., and B. R. James. 1993. Selenite and selenate quantification by hydride generation-atomic absorption spectrometry, ion chromatography, and colorimetry. *J. Environ. Qual.* 22:851–857.

Blaylock, M. J., T. A. Tawfic, and G. F. Vance. 1995. Modeling selenite sorption in reclaimed coal mine soil materials. *Soil Sci.* 159:43–48.

Chao, T. T., and R. F. Sanzolone. 1989. Fractionation of soil selenium by sequential partial dissolution. *Soil Sci. Soc. Am. J.* 53:385–392.

Deverel, S. J., and S. P. Millard. 1988. Distribution and mobility of selenium and other trace elements in shallow groundwater of the western San Joaquin Valley, California. *Environ. Sci. Technol.* 22:697–702.

Dreher, G. B., and R. B. Finkelman. 1992. Selenium mobilization in a surface coal mine, Powder River Basin, Wyoming, U.S.A. *Environ. Geol. Water Sci.* 19:155–167.

Elrashidi, M. A., D. C. Adriano, and W. L. Lindsay. 1989. Solubility, speciation, and transformation of selenium in soils. In L. W. Jacobs (ed.) *Selenium in Agricultural and the Environment*, pp. 51–63. Soil Science Society of America. Special Publication No. 23. SSSA, Madison, WI.

Fadlelmawla, A. A. 1996. Characterization and modeling hydrogeochemical processes controlling selenium transport in aquifers associated with reclaimed coal mines in the Powder River Basin of Wyoming. Ph.D. dissertation, University of Wyoming, Laramie.

Goldberg, S. 1985. Chemical modeling of anion competition on goethite using the constant capacitance model. *Soil Sci. Soc. Am. J.* 49:851–856.

Fio, J. L., and R. Fujii. 1990. Selenium speciation methods and application to soil saturation extracts from San Joaquin Valley, California. *Soil Sci. Soc. Am. J.* 54:363–369.

Jayaweera, G. R., and J. W. Biggar. 1996. Role of redox potential in chemical transformation of selenium in soils. *Soil Sci. Soc. Am. J.* 60:1056–1063.

Lindner-Lunsford, J. B., and J. F. Wilson, Jr. 1992. Shallow ground water in the Powder River Basin, northeastern Wyoming—Description of selected publications, 1950–

91, and indicators for further study. U.S. Geological Survey Water Resources Invest. Report No. 91–4067.

Malcolm, R. L., and P. MacCarthy. 1992. Quantitative evaluation of XAD-8 and XAD-4 resins used in tandem for removing organic solutes from water. *Environ. Int.* 18:597–607.

Martin, L. J., D. L. Naftz, H. W. Lowham, and J. G. Rankl. 1988. Cumulative potential hydrologic impacts of surface coal mining in the eastern Powder River structural basin, northeastern, Wyoming. U.S. Geological Survey Water Resources Invest. Report No. 88–4046.

Masscheleyn, P. H., and W. H. Patrick, Jr. 1993. Biogeochemical processes affecting selenium cycling in wetlands. *Environ. Toxicol. Chem.* 12:2235–2243.

Moore, D. M., and R. C. Reynolds, Jr. 1989. *X-Ray Diffraction and the Identification and Analysis of Clay Minerals*. Oxford University Press, Oxford.

Naftz, D. L., and J. Rice. 1989. Geochemical processes controlling selenium in ground water after mining, Powder River Basin, Wyoming, U.S.A. *Appl. Geochem.* 4:565–576.

Neal, R. H., G. Sposito, K. M. Holtzclaw, and S. J. Traina. 1987a. Selenite adsorption on alluvial soils: I. Soil composition and pH effects. *Soil Sci. Soc. Am. J.* 51:1161–1165.

Neal, R. H., G. Sposito, K. M. Holtzclaw, and S. J. Traina. 1987b. Selenite adsorption on alluvial soils: II. Solution composition effects. *Soil Sci. Soc. Am. J.* 51:1165–1169.

Nodvin, S. C., C. T. Driscoll, and G. E. Likens. 1986. Simple partitioning of anions and dissolved organic carbon in a forest soil. *Soil Sci.* 142:27–35.

See, R. B., K. J. Reddy, G. F. Vance, A. A. Fadlelmawla, and M. J. Blaylock. 1995. Geochemical processes and the effects of natural organic solutes on the solubility of selenium in coal-mine backfill samples from the Powder River Basin, Wyoming. U.S. Geological Survey Water Resources Invest. Report No. 95–4200.

Sharmasarkar, S., and G. F. Vance. 1997. Extraction and distribution of soil organic and inorganic selenium in coal mine environments of Wyoming, USA. *Environ. Geol.* 29:17–22.

Skorupa, J. P., and H. M. Ohlendorf. 1991. Contaminants in drainage water and avian risk thresholds. In A. Dinar and D. Zilberman (eds.), *The Economics and Management of Water and Drainage in Agriculture*, pp. 345–368. Kluwer Academic Publishers, Boston.

Spackman, L. K., G. F. Vance, L. E. Vickland, P. K. Carroll, D. G. Steward, and J. G. Luther. 1994. Standard operating procedures for the sampling and analysis of selenium in soil and overburden/spoil material. University of Wyoming Agric. Exp. Stn. Publ. No. MP-82.

Thurman, E. M. 1985. *Organic Geochemistry of Natural Waters*. Martinus Nijhoff/Dr. W. Junk, Dordrecht, Netherlands.

Tokunaga, T. K., I. J. Pickering, and G. E. Brown, Jr. 1996. Selenium transformations in ponded sediments. *Soil Sci. Soc. Am. J.* 60:781–790.

Vance, G. F., and M. B. David. 1991. Chemical characteristics and acidity of soluble organic substances from a northern hardwood forest floor, central Maine, USA. *Geochim. Cosmochim. Acta* 55:3611–3625.

Weres, O., H. R. Bowman, A. Goldstein, E. C. Smith, L. Tsao, and W. Harnden. 1990. The effect of nitrate and organic matter upon mobility of selenium in groundwater and in a water treatment process. *Water, Air, Soil Pollut.* 49:251–272.

White, A. F., and N. M. Dubrovsky. 1994. Chemical oxidation–reduction controls on selenium mobility in groundwater systems. In W. T. Frankenberger, Jr., and S. Benson (eds.), *Selenium in the Environment*, pp. 185–221. Dekker, New York.

Wyoming Department of Environmental Quality. 1993. Water quality rules and regulations, Chapter VIII. Wyoming Department of Environmental Quality, Water Quality Division.

Zawislanski, P. T., and M. Zavarin. 1996. Nature and rates of selenium transformations: A laboratory study of Kesterson Reservoir soils. *Soil Sci. Soc. Am. J.* 60:791–800.

16

Pathology of Selenium Poisoning in Fish

A. DENNIS LEMLY

United States Forest Service, Blacksburg, Virginia

I. INTRODUCTION

Selenium presents an interesting paradox in the field of aquatic toxicology because it is both a nutrient and a poison. As a nutrient, it is required in the diet of fish at concentrations of about 0.1 to 0.5 μg/g dry weight (Hodson and Hilton, 1983; Gatlin and Wilson, 1984) (In this chapter, all dietary and tissue concentrations are reported on a dry weight basis.) Selenium is necessary for proper formation and functioning of glutathione peroxidase, which is a major cellular antioxidant enzyme (Heisinger and Dawson, 1983; Bell et al., 1986). This enzyme protects cell membranes from damage or lysis due to lipid peroxidation. Without adequate selenium, normal cellular and organ metabolism break down because of peroxides produced as a by-product of digestion. Symptoms of selenium deficiency in fish include reduced growth, anemia, exudative diathesis, muscular dystrophy, and increased mortality (Poston et al., 1976; Bell and Cowey, 1985; Bell et al., 1985; Gatlin et al., 1986). Thus, the beneficial effects of proper selenium in the diet of fish are firmly established.

At dietary concentrations of only 7 to 30 times those required (i.e., >3 μg/g), selenium becomes a poison. Some of the major toxic effects are due to a simple principle of cell biology. From a biochemical perspective, selenium is very similar to sulfur, and cells do not discriminate well between the two when carrying out one of their key functions—protein synthesis. When present in excessive amounts, selenium is erroneously substituted for sulfur in proteins that are being formed inside the cells. Sulfur-to-sulfur linkages (ionic disulfide bonds) are necessary for protein molecules to coil into their tertiary (helix) structure which, in turn, is necessary for proper functioning of the protein, either as a cellular building block or as a component of enzymes. Substitution of selenium for sulfur disrupts the normal chemical bonding, resulting in improperly formed and dysfunctional proteins or enzymes (Diplock and Hoekstra, 1976; Reddy and Massaro, 1983; Sunde, 1984).

Thresholds for dietary selenium toxicity in fish are easily reached and exceeded in contaminated aquatic systems. For example, selenium released in wastewater from a coal-fired electric generating station contaminated Belews Lake, North Carolina, to the extent that fish were consuming 20 to 80 $\mu g/g$ selenium (Cumbie and Van Horn, 1978; Lemly, 1985). Naturally occurring selenium leached from soils as a result of agricultural irrigation in California bioaccumulated in wetlands to concentrations of over 100 $\mu g/g$ in fish food organisms (Lemly et al., 1993; Lemly, 1994). Both these sites experienced massive poisoning of fish and wildlife (Lemly, 1997a). These events emphasize how severe the environmental impacts of excessive selenium can be. Moreover, in a field setting fish accumulate some selenium from water through their gills, which increases the risk that concentrations in tissues may reach toxic levels.

Excessive selenium can cause a wide variety of toxic effects at the biochemical, cellular, organ, and system levels (e.g., Sorensen, 1986). This chapter examines the most prominent outward manifestation of selenium toxicosis—teratogenic deformities—and discusses the use of this pathological symptom as a diagnostic tool for evaluating reproductive impairment and assessing impacts to fish populations in contaminated aquatic habitats.

II. OCCURRENCE AND PERSISTENCE OF TERATOGENIC EFFECTS

Teratogenic deformities in fish are a permanent pathological marker of selenium poisoning. They are congenital malformations that are due to excessive selenium in eggs. The process begins with the diet of parent fish. Excess dietary selenium (>3 $\mu g/g$) causes elevated concentrations of selenium to be deposited in developing eggs, particularly the yolk. When eggs hatch, larval fish rapidly utilize the selenium-contaminated yolk, both as an energy supply and as a source of protein for building new body tissues. Hard and soft tissues may be deformed if the molecular structure of the protein building blocks has been distorted as a result of substitution of selenium for sulfur. Some tissues may not be generated at all, resulting in missing body parts.

The prevalence of teratogenic deformities increases rapidly when selenium concentrations in eggs exceeds 10 $\mu g/g$. Hatchability of eggs is not affected by elevated selenium even though there may be a high incidence of deformities in resultant larvae and fry, and many may fail to survive (Gillespie and Baumann, 1986; Coyle et al., 1993). Teratogenesis is induced when larval fish are relying on their attached yolk sac for nourishment and development. Once external feeding has begun, the potential for teratogenic effects declines and is soon lost. Feeding excessive selenium (up to lethal levels) to fry or juvenile fish as they are growing will not cause teratogenic malformations to occur (Hamilton et al., 1990; Cleveland et al., 1993). Moreover, dietary selenium levels sufficient to load eggs beyond teratogenic thresholds (diet of 5–20 μg/g) do not cause teratogenesis in parent fish, or otherwise generally affect their health or survival (Coyle et al., 1993). Thus, the teratogenic process is strictly an egg–larvae phenomenon. Because of these relationships, teratogenesis can be a very subtle, but important cause of reproductive failure in fish. Entire populations may disappear with little evidence of "toxicity," since major impacts to early life stages can be taking place at the same time that adult fish appear healthy (Cumbie and Van Horn, 1978; Lemly, 1985).

Mortality of larval fish can be high if the teratogenic defects are severe enough to impair critical body functions (Woock et al., 1987). Not all abnormalities are life-threatening, however, and in some cases the malformations can persist into juvenile and adult life stages (Lemly, 1993). Such relatively benign cases are likely restricted to locations where there is little threat from predators, since all but the most subtle deformities would probably compromise a fish's ability to feed and avoid predators. Thus, in assessing the prevalence of teratogenic defects, it is important to focus on the earliest life stages (i.e., newly emerging larvae and young fry).

III. SYMPTOMS OF TERATOGENESIS

Teratogenic deformities can occur in most, if not all, hard or soft tissues of the body. However, some of the most conspicuous (consequently, the most diagnostic) are found in the skeleton, fins, head, and mouth. Typical examples include (1) lordosis—concave curvature of the lumbar region of the spine, (2) scoliosis—lateral curvature of the spine, (3) kyphosis—convex curvature of the thoracic region of the spine, resulting in "humpback" condition, (4) missing or deformed fins, (5) missing or deformed gills or gill covers (opercle), (6) abnormally shaped head, (7) missing or deformed eyes, and (8) deformed mouth. Several of these symptoms are shown in Figures 1 to 3.

In general, a careful fish-in-hand inspection is sufficient to diagnose any of the major teratogenic deformities. However, careful examination with the aid of a dissection microscope is needed to make the diagnosis for larvae and fry, small species (e.g., small cyprinids, poeciliids), or when the subtle, less overt

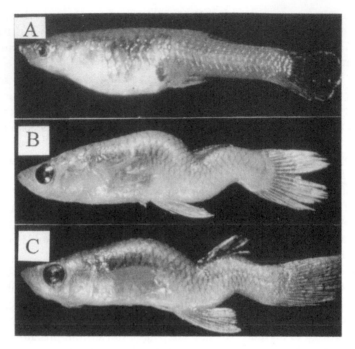

FIGURE 1   Normal (A) and teratogenic (B, C) adult mosquitofish (*Gambusia affinis*) exhibiting dorsoventral deformation (kyphosis and lordosis) of the spine.

symptoms (e.g., slightly deformed fins, opercles, etc.) must be tabulated. This is particularly true for larval fish. Some of their undeveloped features could erroneously be considered a defect when, in fact, they are a consequence of a premature life stage, not selenium teratogenesis. However, this is not a serious concern because larval fish have distinctive patterns of development that quickly become apparent to the investigator looking for teratogenesis. With a bit of hands-on experience, the true teratogenic defects are easily distinguishable, even in young fish.

Certain other symptoms of selenium poisoning may be confused with teratogenic effects. These are generally thought to represent acute toxic responses to high doses or tissue concentrations of selenium—that is, they are not true teratogenic effects. The most common of these symptoms are (1) edema—swollen and distended abdomen due to accumulation of fluid in the visceral cavity, (2) exophthalmus (bulging or protruding eyes) due to accumulation of fluid in the eye sockets, and (3) cataracts, which appear as a white coating on the eyes. All these symptoms may be present concurrently, along with the true teratogenic

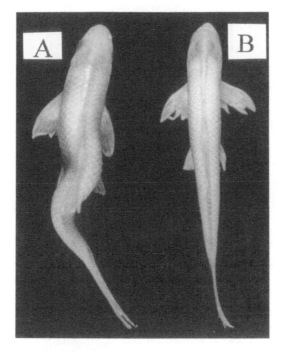

FIGURE 2   Teratogenic (A) and normal (B) juvenile red shiners (*Notropis lutrensis*). The teratogenic effect is scoliosis, or lateral curvature of the spine.

effects. Particular care must be exercised when examining larval fish. Edematous larvae with distended abdomens are a common occurrence (e.g., Bryson et al., 1984; Gillespie and Baumann, 1986; Pyron and Beitinger, 1989). This condition may progress to, or be associated with, the expression of terata, but the edema itself does not constitute a teratogenic defect. However, severe edema is usually accompanied by deformity of the spine (most often lordosis) or soft tissues in the abdomen (Fig. 4B). The prevalence of edema and terata can be virtually the same (e.g., Schultz and Hermanutz, 1990), or quite different (e.g., Hermanutz et al., 1992). Thus, one should not assume a 1 : 1 relationship. Reasonable caution (i.e., close inspection and comparison with normal larvae, as in Fig. 4A) will prevent inaccurate diagnoses.

   To draw a conclusion of selenium-induced teratogenesis, the visual indicators and symptoms (deformities) must be corroborated with the presence of elevated concentrations of selenium in tissues. Concentrations in the range of 10 to 20 μg/g or greater (whole-body homogenate) would be sufficient to confirm the diagnosis. This amount corresponds to concentrations of about 6 to

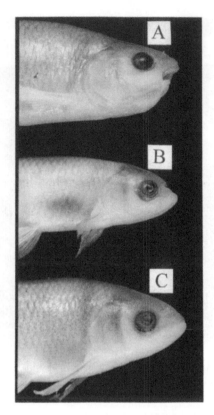

FIGURE 3 Teratogenic (A, B) and normal (C) juvenile red shiners (*Notropis lutrensis*). Teratogenic effects shown are deformity of the mouth and lower jaw (A), upper portion of the head (B), and pectoral fins (B).

12 μg/g in muscle (fillets), or 20 to 40 μg/g in visceral tissues, including the liver. Although measurement of tissue concentrations is essential, it is not necessary to conduct extensive surveys on hundreds of fish. Analysis of six samples per fish species (e.g., six adults or juveniles with teratogenic deformities or six composites for teratogenic larvae/fry) is sufficient.

IV. IMPLICATIONS FOR ASSESSING IMPACTS TO FISH POPULATIONS

Resource managers dealing with aquatic systems known or suspected of being contaminated with selenium usually face a dilemma over how to proceed with

A

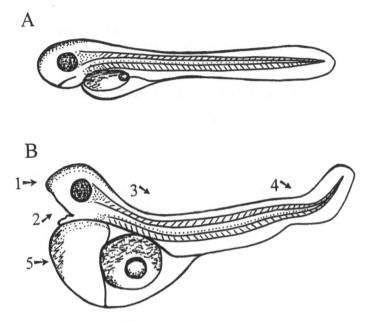

B

FIGURE 4   Typical appearance of larval fish at about 2 to 4 days after hatching. (A) Normal larvae showing yolk absorption nearing completion and straight, developing spine. (B) Abnormal development due to selenium-induced terata: 1, deformed, pointed head; 2, deformed, gaping lower jaw; 3, kyphosis (curvature of the thoracic region of the spine); 4, lordosis (concave curvature of the lumbar and/or caudal regions of the spine). Other symptoms of selenium poisoning that usually accompany terata include edema (swollen, fluid-filled abdomen: see 5) and delayed yolk absorption.

hazard assessment. A typical course of action is to initiate a monitoring program for measuring selenium in water, sediments, and biota. Such efforts will reveal the level and extent of contamination, but the information gained will also raise two very important questions: What do the concentrations of selenium mean with regard to potential biological impacts? And is there any evidence of effects at the site? Unless these questions are answered, the resource manager cannot accurately assess risks or develop a prudent plan for reducing or mitigating hazards.

The value of using teratogenic deformities as a tool for assessing impacts to fish populations becomes apparent when one considers the direct linkage with mortality and reproductive impairment. Concentrations of selenium in tissues can be used to suggest a cause–effect linkage, but there must be other evidence to confirm that actual toxic impacts have occurred. Teratogenesis is a direct expression of selenium toxicity, and it is a clear marker of a cause–effect relation.

It can be used to evaluate and predict impacts to fish populations without the expenditure of large amounts of time and money on contaminant monitoring which, in itself, leaves important questions unanswered.

V. AN INDEX FOR TERATA-BASED ASSESSMENT

A considerable amount of data are available for assessing or predicting the impact of selenium-induced teratogenesis on fish populations. Laboratory studies provide important information on the relationships between egg concentrations of selenium, prevalence of teratogenic deformities in larvae, and associated mortality. These studies are of three types: (1) those in which captive adult fish were fed selenium-laden diets or exposed to high-selenium water and then allowed to spawn in indoor tanks (Bryson et al., 1984, 1985a, 1985b; Woock et al., 1987; Pyron and Beitinger, 1989), (2) those in which outdoor artificial streams were dosed with waterborne selenium, providing for exposure of adult fish to natural food chain selenium prior to spawning (Schultz and Hermanutz, 1990; Hermanutz, 1992; Hermanutz et al., 1992), and (3) those in which adult fish were taken from selenium-contaminated aquatic habitats and spawned artificially (that is, eggs and milt were removed and mixed, and the resultant hatch was monitored; Gillespie and Baumann, 1986).

The field data come from Belews Lake, North Carolina. This lake was impounded in the early 1970s to serve as a cooling reservoir for a large coal-fired electric generating station (2250 MW generating capacity). Fly ash produced by the power plant was disposed in a settling basin, which released selenium-laden effluent containing 100 to 200 μg/L; (about 80% selenite) in return flows to the lake. This selenium bioaccumulated in aquatic food chains, and within 2 years the fishery of Belews Lake began to decline because of reproductive failure. Of the 20 species of fish originally present in the reservoir, 16 were entirely eliminated, including all the primary sport fish. Two species were rendered effectively sterile but persisted as aging adults; one was eliminated, although adults managed to recolonize to a limited extent from a relatively uncontaminated headwater area (but did not reproduce); and one was unaffected. This pattern of selenium contamination from the power plant and resultant poisoning of fish persisted from 1974 to 1985 (Lemly, 1985). In late 1985, under mandates from the state of North Carolina, the power company changed operations for fly ash disposal, and selenium-laden effluent no longer entered the lake. Since that time, selenium levels have fallen and the fishery has begun to recover, but sediments and associated aquatic food chains remain moderately contaminated, and there are residual effects on the fishery, including persistent teratogenic deformities.

Teratogenic assessment was used to evaluate the reproductive success of fish and determine the degree of impact in Belews Lake in 1975, 1978, 1982,

1992, and 1996 (Lemly, 1993, 1997b). The 1975 survey was conducted during the period of initial selenium contamination of the reservoir, before the fishery experienced serious decline. This was the only survey made when the original assemblage of fish species was still present in the lake. This data set is quite informative because it documents levels of teratogenesis in a wide range of species that represent various feeding modes and trophic positions. Selenium concentrations in whole-body fish samples (juveniles and adults) were high (40–65 $\mu g/g$), as was the prevalence of teratogenic deformities (up to 55%). Teratogenesis was present in all 19 species examined. Surveys conducted in 1978 and 1982 yielded similar results, although only four fish species remained in 1978, and six in 1982. Very high selenium concentrations (up to 130 $\mu g/g$) were closely paralleled by the occurrence and prevalence of teratogenic deformities, which ranged up to 70% (juvenile and adult fish). The survey in 1992 indicated that gradual recovery was taking place, but there were still only 9 of the original 20 species present, and numerical abundance was quite low. Concentrations of selenium in fish had fallen to 11 to 20 $\mu g/g$, and the incidence of teratogenesis did not exceed 11% (juveniles and adults). Fish were sucessfully reproducing, and it was soon possible to collect larval fish for examination. Further recovery of the fishery was evident in 1996. All the major sport fish had reestablished and were successfully reproducing. Tissue concentrations of selenium had fallen to 5 to 10 $\mu g/g$, and teratogenic deformities were 6% or lower (larvae, fry, juveniles, and adults).

Figures 5 and 6, which show relationships between the amount of selenium in fish tissues, prevalence of teratogenesis, and associated mortality, represent a compilation of all field and laboratory data on teratogenic effects (e.g., field studies such as Lemly, 1993, laboratory studies such as Woock et al., 1987). The prevalence of teratogenic deformities is dependent on tissue concentrations of selenium—more selenium results in more frequent terata. However, the association follows an exponential function rather than a linear relationship. In natural populations of juvenile and adult centrarchids (i.e., not laboratory studies), the inflection point for the function occurs in the range of 40 to 50 $\mu g/g$. At these concentrations, about one-fourth of the fish exhibit teratogenesis (Fig. 5). Beyond the inflection point, relatively small increases in selenium cause substantial increases in terata. The maximum observed frequency is 70% for individuals with body burdens of selenium in the 70 to 90 $\mu g/g$ range. The exponential function holds for larval centrarchids as well, but the selenium concentrations for the inflection point and maximum are much lower. For example, up to 80% deformities result at tissue concentrations of only about 30 to 40 $\mu g/g$.

The relationship between teratogenesis and mortality is of primary importance in developing an assessment index. Whereas the prevalence of terata is influenced by tissue concentrations of selenium, the degree of mortality from terata is not; that is, about 80% of teratogenic larval fish die regardless of their

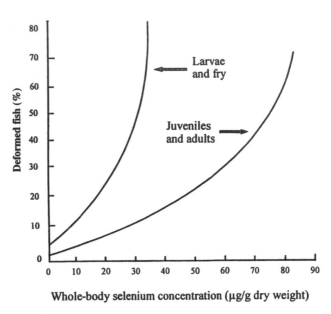

FIGURE 5   Relationship between whole-body concentrations of selenium and prevalence of teratogenic deformities in fish.

body burden of selenium (Fig. 6). This suggests that there is a maximum body burden for generation of lethal terata. Saturation beyond this maximum by additional selenium has little impact. Mortality is nearly constant for juvenile and adult fish as well, but the magnitude is not nearly as great as for larvae—only about 25% of teratogenic juvenile and adult fish die. This is probably a reflection of simple mathematics and, to some extent, the severity of terata (i.e., the 20% or so of teratogenic larvae that survive will make up the teratogenic juvenile/adult population). Although terata persist, they may no longer be as life-threatening as in younger fish.

The difference in mortality between life stages indicates that larval fish should be the priority for assessing or predicting population level impacts of selenium because it is more likely that teratogenic mortality will be expressed in this life stage. Moreover, persistence of deformities into the juvenile and adult life stages may occur only under special circumstances, where natural predation has been sharply reduced or eliminated (Lemly, 1993). Ideally, all life stages should be examined, with the focus placed on larval fish.

Both the laboratory and field data indicate a close parallel between selenium concentrations, incidence of teratogenic deformities, and magnitude of reproductive failure in fish. Using these relationships, an index was developed for teratogenic-based assessment of impacts to fish populations (Table 1). This index is

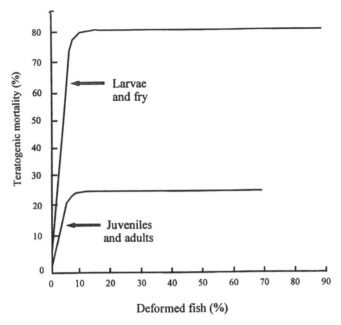

FIGURE 6 Relationship between prevalence of deformities and incidence of teratogenic mortality in fish.

composed of three ratings that signify increasing levels of terata-induced population mortality: 1, negligible impact (<5% population mortality); 2, slight to moderate impact (5–20% population mortality); and 3, major impact (>20%

TABLE I Index for Evaluating the Impact of Selenium-Induced Teratogenic Mortality on Fish Populations

Fish life stage	Proportion with terata (%)	Population mortality (%)[a]	Index rating	Anticipated impact
Larvae or fry	<6	<5	1	Negligible
	6–25	5–20	2	Slight to moderate
	>25	>20	3	Major
Juveniles or adults	<20	<5	1	Negligible
	20–80	5–20	2	Slight to moderate
	>80	>20	3	Major

[a] Mortality expressed as a percentage of the total fish population, not teratogenic mortality. For example, 20% larvae with terata translates to 16% population mortality because up to 20% of those with terata would be expected to survive to adulthood (e.g., only about 80% of teratogenic larvae die).

population mortality). Each rating is based on the anticipated population level impact of the corresponding degree of mortality; thus little effect is expected when mortality is less than 5%, but substantial effects may occur when mortality exceeds 20%. Population mortality is calculated in four simple steps:

1. Determine the percentage of teratogenic fish and the percentage of normal fish in the total sample.
2. Multiply the percentage of teratogenic fish times the expected mortality rate (80% for larvae, 25% for juveniles and adults) to estimate the percentage of fish that will survive.
3. Add the percentage of normal fish and the percentage of surviving teratogenic fish.
4. Subtract this sum from 100%. The result is population mortality, which will be less than teratogenic mortality.

For example, 20% teratogenic larvae with 80% mortality translates to 16% population mortality; 20% teratogenic juveniles/adults with 25% mortality translates to 5% population mortality. Because of the differences in teratogenic mortality between larval and juvenile/adult fish, age-specific indices were developed. As discussed previously, persistence of terata in older life stages (and thus accurate evaluation) can be heavily influenced by predation, thus the index for juveniles and adults may have limited application.

The terata–mortality relationships are based on data for two fish families: Centrarchidae (bass, sunfish) and Cyprinidae (minnows). The resultant index for impacts may or may not be directly applicable to cold- or cool-water families such as Salmonidae (e.g., trout, salmon) or Esocidae (e.g., pike and muskellunge). However, extrapolation to other families of fish is probably not necessary, since centrarchids and cyprinids have characteristics that make them a good indicator or sentinel for other species. Specifically, they are sensitive to selenium, include nationally important sport fish species, and they occupy most of the aquatic habitats in the continental United States (Lee et al., 1980; Lemly, 1993).

Because the index consists of impact-based assessment, it can be applied to virtually any aquatic habitat. Impacts (terata) are a function of selenium concentrations in fish eggs. Conditions responsible for getting selenium into fish eggs—bioaccumulation in aquatic food chains and consumption of contaminated diets by parent fish—can be highly variable from location to location and are influenced by such factors as hydrology and landform (amount and timing of precipitation; stream, lake, or wetland), chemical form of selenium (selenate, selenite, organoselenium), and timing and amount of selenium inputs relative to spawning periods (Lemly and Smith, 1987). Consequently, the potential hazard (likelihood of toxic impacts) of selenium to fish and wildlife is also highly variable (Lemly, 1995, 1996). However, the index is based on a measure of existing impact (terata), not potential hazard. As such, terata are an expression of the sum total of parental exposure, regardless of the temporal, spatial, or chemical variations that may exist from site to site. Thus, the applicability of the index is not influenced by local environmental conditions that affect selenium dynamics and biological uptake. It makes no difference whether the system is a fast-flowing stream in which selenate predominates and bioaccumulation is low, or a terminal wetland experiencing high bioaccumulation from selenite. Population level impacts are indicated only if a sufficient amount of terata exists.

VI. EXAMPLE ASSESSMENTS

1. Collect and examine 500 larval fish using ichthyoplankton sampling techniques; assess the prevalence of teratogenic deformities, and measure selenium concentrations in six composite samples of teratogenic individuals. The investigation reveals that 15% have terata and the associated selenium concentrations are 10 to 15 μg/g. The expected population level mortality is 12% (85% normal + 3% surviving teratogenic = 88% total survival), resulting in an index rating of 2. Conclusion: slight to moderate impact on the population due to teratogenic effects of selenium.

2. Collect and examine 300 juvenile and 200 adult fish. Assess the prevalence of teratagenic deformities and measure selenium concentrations in individuals with terata (whole-body; 6 juveniles and 6 adults). The investigation reveals that 8% have terata and the associated selenium concentrations are 20 to 30 μg/g. The expected population level mortality is 2% (92% normal + 6% surviving teratogenic = 98% total survival), resulting in an index rating of 1. Conclusion: negligible impact on the population due to teratogenic effects of selenium.

3. Sample 1000 larval, 200 juvenile, and 100 adult fish. Determine the prevalence of teratogenic deformities and measure selenium concentrations in teratogenic individuals (6 composite samples for larvae, 6 individuals for juveniles and adults). The investigation reveals that 35% of larvae have terata and 3% of juveniles and adults have terata; selenium concentrations are 10 to 30 μg/g. The expected population level mortality is 28% for larvae (65% normal + 7% surviving teratogenic = 72% total survival) and 0.6% for juveniles and adults (97% normal + 2.4% surviving teratogenic = 99.4% total survival). Resulting index ratings are 3 for larvae and 1 for juveniles/adults. Conclusion: major impact on the population due to teratogenic effects of selenium on larvae.

VII. CONCLUSIONS

Teratogenic deformities are reliable bioindicators of selenium toxicosis in fish. They are produced in response to dietary exposure of parent fish and subsequent deposition of selenium in eggs. Toxicity of waterborne and dietary selenium to adult fish is variable, with organic forms such as selenomethionine being most readily bioaccumulated and toxic, followed by selenite and selenate. However, once ingested, selenium is biochemically processed and incorporated into egg proteins, primarily as selenoamino acids (selenomethionine, selenocystine, etc.). If concentrations are sufficiently high, deformed embryos develop as a result of dysfunctional proteins and enzymes. Thus, terata are produced by a rather uniform process regardless of what chemical form(s) of selenium the parent fish were exposed to.

Teratogenic-based impact assessment provides a conclusive cause–effect linkage between the contaminant and the fish. It is particularly useful for verifying selenium-induced impacts on reproductive success because poor reproduction can be caused by many things (fluctuating water levels, nest predation, food shortages, poor recruitment etc.). The index given here should be a useful tool for evaluating the effect of selenium on fish populations. Moreover, application of this technique may save considerable time and money by identifying the most efficient use of manpower and funds early in the assessment process.

REFERENCES

Bell, J. G., and C. B. Cowey. 1985. Roles of vitamin E and selenium in the prevention of pathologies related to fatty acid oxidation in salmonids. In C. B. Cowey, A. M. Mackie, and J. G. Bell (eds.), *Nutrition and Feeding in Fish*, pp. 333–347. Academic Press, New York.

Bell, J. G., C. B. Cowey, J. W. Adron, and A. M. Shanks. 1985. Some effects of vitamin E and selenium deprivation on tissue enzyme levels and indices of tissue peroxidation in rainbow trout (*Salmo gairdneri*). B. J. Nutr. 53:149–157.

Bell, J. G., B. J. S. Pirie, J. W. Adron, and C. B. Cowey. 1986. Some effects of selenium deficiency on glutathione peroxidase (EC 1.11.1.9) activity and tissue pathology in rainbow trout (*Salmo gairdneri*). B. J. Nutr. 55:305–311.

Bryson, W. T., W. R. Garrett, M. A. Mallin, K. A. MacPherson, W. E. Partin, and S. E. Woock. 1984. Roxboro Steam Electric Plant 1982 Environmental Monitoring Studies, Vol. 2. Hyco Reservoir Bioassay Studies. Technical Report. Carolina Power and Light Company, New Hill, NC.

Bryson, W. T., W. R. Garrett, M. A. Mallin, K. A. MacPherson, W. E. Partin, and S. E. Woock. 1985a. Roxboro Steam Electric Plant—Hyco Reservoir 1983 Bioassay Report. Technical Report. Carolina Power and Light Company, New Hill, NC.

Bryson, W. T., M. A. Mallin, K. A. MacPherson, W. E. Partin, and S. E. Woock. 1985b. Roxboro Steam Electric Plant—Hyco Reservoir 1984 Bioassay Report. Technical Report. Carolina Power and Light Company, New Hill, NC.

Cleveland, L., E. E. Little, D. R. Buckler, and R. H. Wiedmeyer. 1993. Toxicity and bioaccumulation of waterborne and dietary selenium in juvenile bluegill (*Lepomis macrochirus*). Aquat. Toxicol. 27:265–280.

Coyle, J. J., D. R. Buckler, C. G. Ingersoll, J. F. Fairchild, and T. W. May. 1993. Effect of dietary selenium on the reproductive success of bluegills (Lepomis macrochirus). Environ. Toxicol. Chem. 12:551–565.

Cumbie, P. M., and S. L. Van Horn. 1978. Selenium accumulation associated with fish mortality and reproductive failure. Proc. Annu. Conf. Southeast. Assoc. Fish Wildl. Agency 32:612–624.

Diplock, A. T., and W. G. Hoekstra. 1976. Metabolic aspects of selenium action and toxicity. CRC Crit. Rev. Toxicol. 5:271–329.

Gatlin, D. M., III, and R. P. Wilson. 1984. Dietary selenium requirement of fingerling channel catfish. J. Nutr. 114:627–633.

Gatlin, D. M., III, W. E. Poe, and R. P. Wilson. 1986. Effects of singular and combined dietary deficiencies of selenium and vitamin E on fingerling channel catfish (Ictalurus punctatus). J. Nutr. 116:1061–1067.

Gillespie, R. B., and P. C. Baumann. 1986. Effects of high tissue concentrations of selenium on reproduction by bluegills. Trans. Am. Fish. Soc. 115:208–213.

Hamilton, S. J., K. J. Buhl, N. L. Faerber, R. H. Wiedmeyer, and F. A. Bullard. 1990. Toxicity of organic selenium in the diet to chinook salmon. Environ. Toxicol. Chem. 9:347–358.

Heisinger, J. F., and S. M. Dawson. 1983. Effect of selenium deficiency on liver and blood glutathione peroxidase activity in the black bullhead. J. Exp. Zool. 225:325–327.

Hermanutz, R. O. 1992. Malformation of the fathead minnow (Pimephales promelas) in an ecosystem with elevated selenium concentrations. Bull. Environ. Contam. Toxicol. 49:290–294.

Hermanutz, R. O., K. N. Allen, T. H. Roush, and S. F. Hedtke. 1992. Effects of elevated selenium concentrations on bluegills (Lepomis macrochirus) in outdoor experimental streams. Environ. Toxicol. Chem. 11:217–224.

Hodson, P. V., and J. W. Hilton. 1983. The nutritional requirements and toxicity to fish of dietary and waterborne selenium. Ecol. Bull. 35:335–340.

Lee, D. S., C. R. Gilbert, C. H. Hocutt, R. E. Jenkins, D. E. McAllister, and J. R. Stauffer, Jr. 1980. Atlas of North American Freshwater Fishes. North Carolina State Museum of Natural History, Raleigh.

Lemly, A. D. 1985. Toxicology of selenium in a freshwater reservoir: Implications for environmental hazard evaluation and safety. Ecotoxicol. Environ. Saf. 10:314–338.

Lemly, A. D. 1993. Teratogenic effects of selenium in natural populations of freshwater fish. Ecotoxicol. Environ. Saf. 26:181–204.

Lemly, A. D. 1994. Irrigated agriculture and freshwater wetlands: A struggle for coexistence in the western United States. Wetlands Ecol. Manage. 3:3–15.

Lemly, A. D. 1995. A protocol for aquatic hazard assessment of selenium. Ecotoxicol. Environ. Saf. 32:280–288.

Lemly, A. D. 1996. Evaluation of the hazard quotient method for risk assessment of selenium. Ecotoxicol. Environ. Saf. 35:156–162.

Lemly, A. D. 1997a. Environmental implications of excessive selenium. Biomed. Environ. Sci. (in press).

Lemly, A. D. 1997b. Ecosystem recovery following selenium contamination in a freshwater reservoir. Ecotoxicol. Environ. Saf. 36:275–281.

Lemly, A. D., and G. J. Smith. 1987. Aquatic cycling of selenium: Implications for fish and wildlife. Fish and Wildlife Leaflet No. 12. U.S. Fish and Wildlife Service, Washington, DC.

Lemly, A. D., S. E. Finger, and M. K. Nelson. 1993. Sources and impacts of irrigation drainwater contaminants in arid wetlands. *Environ. Toxicol. Chem.* 12:2265–2279.

Poston, H. A., G. F. Combs, Jr., and L. Leibovitz. 1976. Vitamin E and selenium interrelations in the diet of Atlantic salmon (*Salmo salar*): Gross, histological, and biochemical deficiency signs. *J. Nutr.* 106:892–904.

Pyron, M., and T. L. Beitinger. 1989. Effect of selenium on reproductive behavior and fry of fathead minnows. *Bull. Environ. Contam. Toxicol.* 42:609–613.

Reddy, C. C., and E. J. Massaro. 1983. Biochemistry of selenium: An overview. *Fundam. Appl. Toxicol.* 3:431–436.

Schultz, R., and R. Hermanutz. 1990. Transfer of toxic concentrations of selenium from parent to progeny in the fathead minnow (*Pimephales promelas*). *Bull. Environ. Contam. Toxicol.* 45:568–573.

Sorensen, E. M. B. 1986. The effects of selenium on freshwater teleosts. In E. Hodgson (ed.), *Reviews of Environmental Toxicology*, Vol. 2, pp. 59–117. Elsevier Biomedical Press, New York.

Sunde, R. A. 1984. The biochemistry of selenoproteins. *J. Am. Org. Chem. Soc.* 61:1891–1900.

Woock, S. E., W. R. Garrett, W. R. Partin, and W. T. Bryson. 1987. Decreased survival and teratogenesis during laboratory selenium exposures to bluegill, *Lepomis macrochirus*. *Bull. Environ. Contam. Toxicol.* 39:998–1005.

17

Selenium Effects on Endangered Fish in the Colorado River Basin

STEVEN J. HAMILTON
United States Geological Survey, Yankton, South Dakota

I. INTRODUCTION

As a result of its long isolation, the Colorado River basin (Fig. 1) supports some of the most distinctive ichthyofauna in North America. Several fish species have declined in the basin, which originally contained 32 native species of which 75% were endemic (Minckley, 1991). The first two listed as endangered were the humpback chub (*Gila cypha*) and the Colorado squawfish (*Ptychocheilus lucius*). Both were listed in 1967 (USDOI, 1967) as part of the Endangered Species Preservation Act of 1966, six years prior to the passage of the Endangered Species Act of 1973. Passage of this act relisted the humpback chub and Colorado squawfish (USFWS, 1974); later, in 1980, the bonytail (*Gila elegans*) was listed (USFWS, 1980), and in 1991 the razorback sucker (*Xyrauchen texanus*) (USFWS, 1991a) (Fig. 2). Several other fish in the Colorado River basin are now listed as species of concern (formerly category 2 species), including flannelmouth sucker (*Catostomus latipinnis*) and roundtail chub (*Gila robusta*) (USFWS, 1991b) (Fig. 2).

Each of these species has or will have recovery plans, which list activities the U.S. Fish and Wildlife Service will undertake to establish viable, self-sustaining populations that will lead to downlisting or delisting of the species. A combined approach for recovery of the four endangered fish in the upper Colorado River basin has been undertaken by the Recovery Implementation Program for Endangered Fish Species in the Upper Colorado River Basin, which was initiated in 1987 (USFWS, 1987). The goal of the 15-year program is to reestablish self-sustaining populations of the four species while allowing continued development of water.

FIGURE 1   The Colorado River basin. (Modified from Carlson and Muth, 1989.)

Because the remaining populations of humpback chub and Colorado squaw-fish seem to be stable, the recovery program has put most of its effort into recovering bonytail and razorback sucker (F. Pfeifer, personal communication, 1996). The remaining population of razorback sucker in the Green River has been estimated at about 1000 individuals (Lanigan and Tyus, 1989) and at 300 to 600 (Modde et al., 1996). Razorback sucker are very rare in the upper Colorado River, where only 10 fish were found in the river between 1989 and1996 (C.

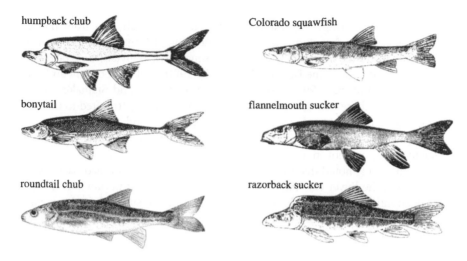

FIGURE 2   Endangered fish and species of concern in the Colorado River basin. (From Lee et al., 1980.)

McAda, personal communication, 1997). Similarly, the population of Colorado squawfish is considered low relative to historical levels and is estimated at 600 to 650 individuals in the upper Colorado River (Osmundson and Burnham, 1996).

The reasons for the decline of these species are related to a combination of factors including stream alteration (dams, irrigation, dewatering, channelization), loss of habitat (spawning sites and backwater nursery areas), changes in flow regime, blockage of migration routes, water temperature changes, competition with and predation from introduced species, parasitism, and changes in food base (USFWS, 1987). Although irrigation and pollution were suggested in 1976 as possible contributing factors to the decline of the endangered fish (Seethaler et al., 1979), the recovery program did not seriously consider contaminant concerns until 1994.

Historically, Colorado squawfish and razorback sucker were very abundant in the upper Colorado River basin, as observed by Jordan (1891) during a survey of streams in 1889. Quartarone (1993) interviewed 111 people (average age about 77) who recounted numerous experiences from the 1920s to early 1950s of easily catching these fish. Many noted how good Colorado "whitefish" (Colorado squawfish) tasted, and recalled that they had been easy to catch using a variety of baits including mice, rabbits, birds, grubs, and other more traditional baits, as well as "hardware" (lures). However, in the late 1940s and early 1950s, many of those with fishing experience dating back to the 1920s stated that these fish were becoming rare in the upper Colorado River basin. They believed this rarity was the result of "pollution" in the rivers from dumping of raw sewage from cities like Rock Springs, Wyoming, and Green River, Utah, and from railroad maintenance yards that dumped oil and scale cleaned from steam locomotives. Historical accounts showed that Colorado squawfish, bonytail, and razorback sucker were present in the Dolores River in the 1920s and were caught and used for food (Quartarone, 1993). However, in 1971 Holden and Stalnaker (1975) found none of these fish in the Dolores River, a result they attributed to the river being "far from its natural state due to irrigation uses and its record of rather severe pollution." Four Colorado squawfish were found in the lower 2 km of the river in 1992, but not in a 1982 survey effort (Valdez et al., 1992).

Dill (1944) noted that Colorado squawfish, bonytail, and razorback sucker were rare in the lower Colorado River basin after 1930, but in previous decades had been numerous. Colorado squawfish and razorback sucker have been extirpated from the Gila River system in Arizona, where they were once abundant (Minckley, 1973).

II. ENDANGERED FISH AND IRRIGATION DRAINS

One historical observation overlooked until recent times goes back to the 1940s and early 1950s, when, according to widespread accounts, Colorado squawfish, bonytail, and razorback sucker were found in irrigation drainage ditches throughout the basin (Quartarone, 1993). Some of these ditches were adjacent to the Green River near Green River, Utah, and Linwood, Wyoming (now under waters of the Flaming Gorge Reservoir); the Hartland Ditch on the Gunnison River near Delta, Colorado, was also included.

Prior to making a spawning run, adult razorback sucker typically stage in backwater (areas of low velocity, high temperature water, that are high in productivity), usually found at stream mouths. Two of these staging sites in the Green River basin are the lower 0.8 km of Ashley Creek and Stewart Lake outfall (Tyus, 1987; Tyus and Karp, 1990), which are adjacent to each other. Both these sites are known to have elevated selenium concentrations in water and food organisms. Tyus and Karp (1990) and personal communications from other researchers reported ripe razorback sucker in Ashley Creek/Stewart Lake outfall area during 1978–1989. Tyus (1987) found that 73% of the ripe razorback sucker captured in 1981–1986 were in the Jensen area, which includes Ashley Creek and Stewart Lake outfall. Stephens et al. (1988) reported that selenium concentrations in water were 25 to 73 μg/L in Ashley Creek and 6 to 10 μg/L in Stewart Lake outflow in 1986–1987.

Also, in recent years, backwater sites on the mainstem of the Colorado River where razorback suckers have been found, have been documented to have elevated selenium concentrations. McAda (1977) reported that 22 razorback sucker were collected in fall 1974 and 27 in spring 1975 from a gravel pit at

Walter Walker State Wildlife Area (WWSWA) on the Colorado River near Grand Junction, Colorado. Approximately 40 razorback sucker were captured in a back-water area about 26 km upstream from WWSWA, referred to as Clifton Pond, which was flooded by irrigation flows in the spring of 1976. There was no evidence of successful spawning of razorback sucker or other native species in the gravel pit, even though suspected spawning behavior was observed in razorback sucker, and flannelmouth sucker was abundant in the area. Kidd (1977) reported that 28 razorback suckers were captured at WWSWA in 1974, 75 in 1975, and 16 in 1976, and 60 were captured at Clifton Bridge backwater (i.e., Clifton Pond) in 1975, but none were captured in 1974 or 1976. Kidd also reported 14 Colorado squawfish captured at WWSWA in 1974, 28 in 1975, and 2 in 1976. Valdez et al. (1982) reported that 9 Colorado squawfish and 37 adult razorback sucker were collected in 1979–1981 from the gravel pit at WWSWA and Clifton Pond. All the razorback sucker collected by McAda and Valdez were considered to be old fish; no juveniles or subadults were collected (C. McAda and R. Valdez, personal communication, 1997).

Although the investigators did not measure concentrations of selenium and other inorganics at the sites just mentioned, the sites probably had elevated selenium concentrations. Selenium concentrations in the channel area of WWSWA [former gravel pit site used by razorback sucker in 1974–1975 (McAda, 1977) and 1979–1981 (Valdez et al., 1982)], at two sites where Colorado squaw-fish were captured in 1994–1996, were measured as 12 and 60 μg/L in May 1995 (unpublished data). Another site adjacent to the channel area, North Pond, contained 133 μg/L in May 1995. The channel area and North Pond both receive groundwater through a cobble aquifer (Phillips, 1986) that is recharged almost exclusively by irrigation activities (Butler et al., 1989). Virtually all this groundwa-ter flows into the Colorado River. Selenium concentrations measured in 1991–1992 in water from a drainage ditch flowing into Clifton Pond contained 86 to 100 μg/L in the winter and 18 μg/L in the summer (Holley and Weston, 1995).

Valdez et al. (1982) reported that Colorado squawfish congregated at the mouths of irrigation drainage ditches that empty into backwater areas of the Colorado River near Grand Junction, including Reed Wash, [Little] Salt Wash, and Big Salt Wash. Butler et al. (1994) reported that in 1991, samples from Reed Wash contained 120 μg/L of selenium, Little Salt Wash 16 μg/L, and Big Salt Wash 24 μg/L. Average whole-body selenium concentrations in several fish species at these sites were 11, 8, and 9 μg/g, respectively. (All sediment and tissue concentrations are given as dry weight.) Consequently, adult Colorado squawfish that congregated at these three sites may have been exposed to substan-tial amounts of selenium in water and in prey fish. In the summer of 1995, two Colorado squawfish were caught in an irrigation drain that emptied into the Colorado River just upstream of Adobe Creek near Grand Junction (B. Osmundson, personal communication, 1996), demonstrating that these fish continue to use drains (no samples were collected).

III. DOCUMENTED SELENIUM PROBLEMS IN THE COLORADO RIVER BASIN

The National Contaminant Biomonitoring Program has documented temporal and geographic trends in concentrations of persistent environmental contaminants, including selenium, that may threaten fish and wildlife. Of 99 to 112 stations nationwide where fish were collected every other year between 1972 and 1984, selenium concentrations at about 13 to 16 stations have exceeded the 85th percentile, which is an arbitrary point distinguishing "high" concentrations. Selenium concentrations in whole-body fish in the Colorado River basin have been among the highest in the nation (Walsh et al., 1977; Lowe et al., 1985; Schmitt and Brumbaugh, 1990). They exceeded the 85th percentile in fish at five out of six Colorado River basin stations: Green River at Vernal, Utah (the only upper basin station) and Colorado River at four sites in Arizona (Imperial Reservoir, Lake Havasu, Lake Mead, and Lake Powell). Samples were collected in 1972–1973. Selenium concentrations in whole-body fish also exceeded the 85th percentile in 1978–1981 and 1984 at six out of seven stations: the five just named, plus Colorado River at Yuma, Arizona. Only the station on the Gila River (San Carlos Reservoir, AZ) has not exceeded the 85th percentile. The long-term selenium contamination of the lower Colorado River basin may have been one of the factors contributing to the disappearance of endangered fish in the early 1930s reported by Dill (1944).

Several National Irrigation Water Quality Program (NIWQP) studies in the Colorado River basin have reported elevated concentrations of selenium and other elements in rivers where endangered or threatened fish occurred historically, but are rare now. Prior to the NIWQP studies under the auspices of the U.S. Department of the Interior (DOI), a few studies in the 1930s by the U.S. Department of Agriculture reported elevated selenium concentrations in the upper Colorado River basin (Anderson et al., 1961).

An NIWQP study of federal irrigation projects in the Uncompahgre area, which includes the Uncompahgre and Gunnison Rivers, found elevated selenium in water and biota. Butler et al. (1991) reported that water samples from the Uncompahgre River, a tributary of the Gunnison River at Delta, Colorado, contained a median selenium concentration of 15 μg/L (maximum 52 μg/L) during 1968–1988, and the Gunnison River at Grand Junction, contained 10 μg/L (maximum 21 μg/L) during 1975–88. These investigators surmised that the Uncompahgre Irrigation Project was a major source of selenium to the Uncompahgre and Gunnison Rivers because the maximum selenium concentration in water from upstream of the project was 2 μg/L. Butler et al. (1991) reported that selenium concentrations in 1988–1989 were less than 1 to 8 μg/L in water, 6 to 7 μg/g in invertebrates, and 7 to 10 μg/g in small omnivorous fish in the Gunnison River at Delta, (upstream of confluence with Uncompahgre River) and 2 to 34 μg/L in water, 4 μg/g in invertebrates, and 6 μg/g in fish in the Uncompahgre River at Delta.

In 1992 Butler et al. (1994) found that selenium concentrations in samples from the Gunnison River at Delta were 6 μg/L in water and 11 to 17 μg/g in speckled dace (Rhinichthys osculus). Selenium concentrations in the Uncompahgre River at Delta were 8 to 25 μg/L in water and 7 to 9 μg/g in speckled dace. Historically, Anderson et al. (1961) reported that in 1936 the Gunnison River at its mouth contained a mean selenium concentration of 22 μg/L in water (maximum 55 μg/L). High selenium concentrations have apparently occurred in the Gunnison River over a long period of time, probably since at least 1936. Seniors interviewed by Quartarone (1993) recalled that the endangered fish were becoming rare in the late 1940s and early 1950s in the upper Colorado River basin.

An NIWQP study of federal irrigation projects in the Grand Valley (i.e., Colorado River in the Grand Junction area) found elevated selenium concentrations in water and biota. Butler et al. (1994) reported that in 1991–1992 the Colorado River at Cameo (above irrigation influences) contained an average of less than 1 μg/L, but at the Colorado–Utah state line contained 4 to 7 μg/L (mean 4 μg/L). The increase in selenium at the state line was a result of inputs from 970 km of irrigation drains in the Grand Valley and the Gunnison River basin (Butler et al., 1989). Elevated selenium concentrations in samples collected in 1991–1992 from drains emptying into the Colorado River, where endangered fish have been observed to congregate, were reported for Little Salt Wash (6–24 μg/L), Big Salt Wash (8–46 μg/L), and Reed Wash (120 μg/L) (Butler et al., 1994). Other drains emptying into the Colorado River also contained selenium: Adobe Creek, 9–92 μg/L; Hunter Wash, 7 to 35 μg/L; Pritchard Wash, 12 to 23 μg/L; and Persigo Wash, 15 to 86 μg/L (Butler et al., 1994).

Endangered fish in the upper Colorado River near Grand Junction have apparently been exposed to elevated selenium concentrations since the first major irrigation project delivered water in 1915 (Butler et al., 1989). Initial selenium concentrations in drainwater from newly irrigated soils were probably extremely high—2000$^+$ μg/L (Anderson et al., 1961) in the Grand Valley. Yet concentrations currently measured in drainwater, decades after irrigation was initiated, are still substantially elevated. Historically, selenium concentrations measured in four irrigation drains in the Grand Valley in the 1930s were 320 to 1050 μg/L, whereas two new drains near Mack, Colorado, (about 32 km downstream from Grand Junction) contained 1980 μg/L in 1934 and 2680 μg/L in 1935 (Anderson et al., 1961). In 1936 selenium concentrations in the Colorado River at Cameo, Colorado, contained 0 to 1 μg/L, whereas downstream of irrigation influences, but above the Gunnison River, the Colorado contained 0 to 10 μg/L (mean 4 μg/L). At the same time, the Gunnison River at its mouth contained 5 to 55 μg/L (mean 22.5 μg/L). Anderson et al. (1961) also reported that at one point in the 1930s the Gunnison River at Grand Junction contained 80 μg/L and the Colorado River at Grand Junction contained 30 μg/L.

An NIWQP investigation of irrigation projects on the San Juan River below Navajo Reservoir in northwestern New Mexico found elevated selenium concentrations in water and biota. Blanchard et al. (1993) reported that selenium was elevated in a drain located on the east part of the Hogback Irrigation Project near Shiprock, New Mexico, that is the water source for a backwater of the San Juan River. Concentrations of selenium in the east Hogback Drain were 11 to 21 μg/L in water, 11 to 17 μg/g in invertebrates, and 27 to 42 μg/g in fish. The San Juan River currently has a small population of Colorado squawfish, but no razorback sucker. Both species were common in the river historically. In spite of intensive sampling efforts, few larval or young-of-year Colorado squawfish have been collected in recent years (18 in 1987; 1 each in 1988, 1990, 1992; 13 in 1993; 7 in 1994; 2 in 1995; 1 in 1996; none in 1989 and 1991) (Holden and Masslich, 1997). An unusually high incidence of abnormal lesions on fish in the San Juan River, especially flannelmouth sucker, which is a species of concern, has been attributed to pathogens requiring inducement by stressors such as high contaminant concentrations, malnutrition, or poor water quality (Abell, 1994). The highest incidence of abnormalities was found in the river section just below the east Hogback Drain.

Butler et al. (1995) reported high selenium concentrations in two tributaries of the San Juan River in southwestern Colorado. Concentrations in the Mancos River near its confluence with the San Juan River in the Four Corners area were 6 to 10 μg/L in water, 3 μg/g in sediment, 3 to 7 μg/g in invertebrates, and 3 to 14 μg/g in fish. Concentrations in McElmo Creek (18 km downstream of Four Corners) above its confluence with the San Juan River were 3 to 9 μg/L in water, 1 to 2 μg/g in invertebrates, and 1 to 6 μg/g in fish. Flow in both tributaries was primarily irrigation drainwater except for spring freshets (R. Krueger, personal communication, 1995), and both have backwater areas associated with their confluence with the San Juan River. Three young-of-year Colorado squawfish (2 in 1987 and 1 in 1994) were found in the San Juan River just below the confluence of the Mancos River (Holden and Masslich, 1997).

An NIWQP investigation of irrigation projects in the middle Green River basin in northeastern Utah found elevated selenium concentrations in water, sediment, and biota. Stephens et al. (1988) reported that concentrations of selenium in water in 1986–1987 were 25 to 73 μg/L in Ashley Creek , 6 to 10 μg/L in Stewart Lake outflow, and 31 μg/L in Marsh 4720. In a follow-up study, Peltz and Waddell (1991) and Stephens et al. (1992) reported that selenium concentrations in 1988–1989 in Ashley Creek were 59 to 78 μg/L in water and 40 to 122 μg/g in fish, whereas Stewart Lake outfall had 2 to 11 μg/L in water and 11 to 25 μg/g in fish. Endangered razorback suckers were found staging at Ashley Creek and at the outfall of Stewart Lake during 1978–1989 prior to making spawning runs to upstream spawning sites (Tyus, 1987; Tyus and Karp, 1990). One of the two known spawning sites for razorback sucker is located only about 20 km upstream from the Ashley Creek/Stewart Lake area near Jensen, Utah, in the Green River at a location commonly known as "Razorback Bar."

Peltz and Waddell (1991) and Stephens et al. (1992) are the only NIWQP reports to include information on selenium concentrations in endangered fish. These sources report that one sample of eggs from a razorback sucker collected in 1988 near Jensen, Utah, contained selenium concentrations of 4.9 μg/g, whereas muscle, liver, and gonads from a Colorado squawfish collected in 1982 from near the confluence of Ashley Creek had 2.7, 9.2, and 6.5 μg/g, respectively. The same tissues from a humpback chub collected in 1986 in the Green River (specific site undetermined) had selenium concentrations of 4.3, 6, and 7 μg/g, respectively. One sample of eggs from a razorback sucker caught near Razorback Bar in 1991 contained 28 μg/g of selenium (B. Waddell, personal communication, 1992).

At the lower end of the upper Colorado River basin, elevated selenium concentrations have been found in fish from Lake Powell, and also in fish from the upper end of the lower basin near Lee's Ferry. Selenium concentrations in a wide variety of fish collected from several locations in Lake Powell ranged from 2.5 to 14 μg/g (Waddell and Wiens, 1993). Selenium concentrations were 4.1 to 18 μg/g in headless whole-body samples and 7.8 to 73 μg/g in eggs of rainbow trout (*Oncorhynchus mykiss*) collected in the 24 km river reach between Glen Canyon Dam and Lee's Ferry (A. Ayres, personal communication, 1997).

An NIWQP investigation in the lower Colorado River of 11 locations between Davis Dam and the United States–Mexico border in 1986–1987 revealed elevated selenium concentrations in water, sediment, and biota. Radtke et al. (1988) reported finding consistently higher selenium concentrations in backwaters and oxbow lakes that received water from the Colorado River than in canals returning irrigation drainwater to the river. They suggested that selenium contamination in the lower Colorado River was not from local agricultural practices, but rather from upstream sources (i.e., the upper Colorado River basin). Three other studies have also documented elevated selenium in water, sediment, invertebrates including clams, and fish in the lower Colorado River (Bell-McCaulou, 1993; Lusk, 1993; Welsh and Maughan, 1994).

IV. SELENIUM INFLUENCES HABITAT RESTORATION EFFORTS

One component of the Recovery Implementation Program for the Endangered Fish Species of the Upper Colorado River Basin is the Floodplain Habitat Restoration

Program. Its goal is to restore a large number of floodplain sites for use by adults for staging prior to spawning runs and for use by larval fish as nursery areas. An early evaluation of five sites by Cooper and Severn (1994a–1994e) focused on site suitability based primarily on hydrology and food abundance (aquatic and benthic invertebrates), and secondarily on vegetation and water chemistry including measurement of selenium and other inorganics. A major shortcoming of this early effort was the paucity of information on selenium that was considered in assessing and evaluating the sites. In one report, the investigators recommended restoration of Sheppard Bottoms at Ouray National Wildlife Refuge (NWR), Ouray, Utah, based on only one selenium value in water. They failed to consider three NIWQP reports that showed elevated selenium concentrations in water, sediment, and biota at Ouray NWR, which could adversely affect fish and wildlife (Peltz and Waddell, 1991; Stephens et al., 1988, 1992).

Because of concerns in the Floodplain Habitat Restoration Program about adverse effects from selenium and other inorganics on endangered fish, screening for contaminants was the primary focus of site assessments initiated in late 1994 by Holley and others (Holley and Weston, 1995; Stephens et al., 1995; Archuleta and Holley, 1996). To date, six site screenings on the Colorado River, three on the Gunnison River, and 21 on the Green River have been completed. Ten sites have been approved for restoration, 10 more were put on temporary hold pending outcome of ongoing selenium research with endangered fish, and 10 sites were identified that will not be restored. Of the sites that will not be restored, four are on the Green River, and three each are on the Colorado and Gunnison Rivers. Seven of these sites were dropped because of elevated selenium concentrations in water or biota or both, and the other three sites were dropped because of concerns about other inorganic constituents.

V. SELENIUM STUDIES WITH ENDANGERED FISH

The Colorado River Fisheries Project, another component of the Recovery Implementation Program for the Endangered Fish Species of the Upper Colorado River Basin, contracted in 1981 to have short-term tests conducted to determine the toxicity of selenium and several other inorganic and organic chemical constituents to Colorado squawfish and humpback chub (Beleau and Bartosz, 1982). Those tests seemed to indicate at the time that selenium and other inorganics and organics were not contributing to the decline of the species. However, the release of NIWQP reconnaissance investigations of federal irrigation projects on the middle Green, upper Colorado, Gunnison, Uncompahgre, and San Juan Rivers reported elevated selenium concentrations in water and biota. Consequently, additional toxicity tests were warranted and conducted with several life stages of Colorado squawfish, razorback sucker, bonytail, and flannelmouth sucker in water qualities simulating environmental conditions (Hamilton, 1995; Buhl and Hamilton, 1996; Hamilton and Buhl, 1996, 1997). These tests evaluated, several individual inorganic constituents and inorganic mixtures simulating environmental conditions where water quality was strongly influenced by irrigation drainwater. The results showed that waterborne selenium itself was not a hazard to the fish. However, six mixtures that included selenium from the San Juan River and two that included selenium from the Green River (Ashley Creek and Stewart Lake outfall) showed high hazards to endangered fish because the ratio of biological effects concentration (acute toxicity data) to environmental concentration was <100.

Long-term tests with larval razorback sucker and bonytail exposed to a mixture of inorganics in water including selenium that simulated Ashley Creek and the Green River showed adverse effects at concentrations less than 10 times of the environmental concentrations, which suggests a high hazard (Hamilton and Buhl, 1994). Razorback sucker larvae had reduced survival after 40 days exposure to the 16X treatment (1X was the environmental concentration) and after 60 days at 8X, reduced growth after 30 days at 8X and after 60 days at 4X, and reduced critical swimming speed after 60 days at 8X. Bonytail had reduced survival after 20 days of exposure to the 16X treatment, reduced growth after 60 days at 8X, and reduced critical swimming speed after 60 days at 16X.

In another long-term study, larval razorback sucker were fed live zooplankton collected from six sites at Ouray NWR that contained different amounts of selenium. Survival was 0 to 20% in four studies initiated with larvae aged 5, 10, 24, and 28 days old (Hamilton et al., 1996). This study suggested that selenium concentrations of 2.3 $\mu g/g$ or greater in food organisms, in combination with other inorganics in the water or food organisms, caused reduced survival in larval razorback sucker. Both the Ashley Creek and Ouray NWR studies showed adverse effects on larval razorback sucker at selenium concentrations present in current environmental conditions in the upper Colorado River basin.

A reproduction study with adult razorback suckers at three sites near Grand Junction (North Pond at WWSWA, diked tertiary channel near Adobe Creek, and reference ponds at Horsethief State Wildlife Area), with different selenium concentrations in water and biota, showed that selenium was taken up readily in tissue and eggs (unpublished data). After 9 months of exposure to selenium in water and food organisms, selenium concentrations in muscle plugs of adults at the reference site were unchanged at 4.5 $\mu g/g$. At Adobe Creek, muscle plugs increased from 3.9 $\mu g/g$ to 12 $\mu g/g$; and muscle plugs at North Pond increased from 4.1 $\mu g/g$ to 17 $\mu g/g$. Selenium concentrations in eggs collected after spawning was induced were 6.5 $\mu g/g$ in reference fish, 44 $\mu g/g$ in Adobe Creek fish, and 38 $\mu g/g$ in North Pond fish. In a 30-day grow-out study with larvae hatched from those eggs, survival was low for all three sites. A severe infestation of anchor worm in the reference fish may have compromised the health of those fish during oogenesis. However, larvae from brood stock held at the reference site had high survival. This study is being repeated, and results should be available in 1998.

Elevated selenium concentrations in wild razorback sucker and Colorado squawfish have been documented. Selenium concentrations in muscle plugs of eight adult razorback sucker from Ashley Creek, one from Stewart Lake outfall, and three from Razorback Bar exceeded 11.5 μg/g (maximum 54 μg/g), whereas two from the Escalante spawning area in the Green River (about 10 km upstream of Ashley Creek), nine from Razorback Bar, and one from Old Charlie Wash (about 76 km downstream of Ashley Creek) contained less than 7.4 μg/g (Waddell and May, 1995). Eggs from three of these fish had selenium concentrations of 3.7 to 10.6 μg/g, with the highest selenium concentration from a fish with 32 μg/g in its muscle plug (Hamilton and Waddell, 1994). The selenium concentrations in muscle plugs from the Ashley Creek fish were similar to those in fish used in the reproduction study conducted near Grand Junction with the same species. However, the selenium concentrations in eggs in the study by Hamilton and Waddell (1994) were about one-quarter those in the Grand Junction study, which suggests that the Grand Junction fish probably accumulated more selenium during oogenesis because their exposure time was longer than that of the Green River fish, which were free to move at will. However, the nine fish at Ashley Creek/ Stewart Lake outfall area and three at Razorback Bar with high selenium concentrations in their muscle tissue (Waddell and May, 1995) apparently were exposed to elevated selenium concentrations long enough to accumulate a substantial amount of selenium in their muscle tissue, and presumedly in their eggs.

A similar pattern of selenium residues in muscle plugs from Colorado squawfish was found in the Colorado River. About 7 km above or below WWSWA near Grand Junction, selenium in muscle plugs averaged less than 5 μg/g, but at WWSWA the average was 16 μg/g (maximum 31 μg/g) (B. Osmundson, personal communication, 1995). These selenium residues in razorback sucker and Colorado squawfish show that some endangered fish in high selenium areas are accumulating selenium to concentrations that could harm their reproductive success.

VI. SUMMARY

Historically, fish in the Colorado River basin that are now endangered were widespread and abundant. Many physical changes to the rivers in the basin have resulted from dams and water diversions for domestic and agricultural use, as well as biotic changes from the introduction of nonnative fish. Chemical changes in the basin's rivers have occurred as a result of widespread use of irrigation and the concomitant disposal of drainwater. Documentation of very high selenium concentrations in drainwater in the mid-1930s, combined with the disappearance of endangered fish from both the upper and lower basins in the late 1940s and early 1950s, suggests a causal relation. NIWQP studies in the late 1980s and early 1990s documented elevated selenium concentrations in water, sediment, and biota throughout the entire basin.

Recent studies, designed to determine the biological effects of selenium on larval and adult endangered fish in the upper Colorado River basin, showed effects at concentrations found in water and biota in the NIWQP studies and by others. Although the decline of endangered fish has been closely tied to physical and biotic changes, the link to chemical changes associated with selenium seems to be another important factor. The recovery programs for endangered fish in the upper Colorado River and the San Juan River are indicative of recent acknowledgment of the possibility of contaminant impacts on endangered fish. Selenium research has been incorporated into the two recovery programs. Amelioration of selenium effects may be essential to the recovery of endangered fish in the Colorado River basin.

ACKNOWLEDGMENTS

I thank R. Engberg, W. Frankenberger, K. Holley, F. Pfeifer, L. Sappington, and D. Woodward for reviewing this chapter.

REFERENCES

Abell, R. 1994. *San Juan River Basin Water Quality and Contaminants Review*, Vols. 1 and 2, Museum of Southwestern Biology, University of New Mexico, Albuquerque.

Anderson, M. S., H. W. Lakin, K. C. Beeson, F. F. Smith, and E. Thacker. 1961. *Selenium in Agriculture*. Agriculture Handbook No. 200, U.S. Department of Agriculture, Washington, DC.

Archuleta, A., and K. Holley. 1996. Contaminant screening results. Floodplain Habitat Restoration Program, Green River, Utah. U.S. Bureau of Reclamation, Grand Junction, CO.

Beleau, M. H., and J. A. Bartosz. 1982. Acute toxicity of selected chemicals: Data base. Colorado River Fishery Project Final Report Contracted Studies, Report No. 6, pp. 242–254. U.S. Fish and Wildlife Service, Salt Lake City, UT.

Bell-McCaulou, T. M. 1993. *Corbicula fluminea* as a bioindicator on the lower Colorado River. M.S. thesis. University of Arizona, Tucson.

Blanchard, P. J., R. R. Roy, and T. F. O'Brien. 1993. Reconnaissance investigation of water quality, bottom sediment, and biota associated with irrigation drainage in the San Juan River area, San Juan County, northwestern New Mexico, 1990–91. U.S. Geological Survey, Albuquerque, NM. Water-Resources Investi. Report No. 93-4065.

Buhl, K. J., and S. J. Hamilton. 1996. Toxicity of inorganic contaminants, individually and in environmental mixtures, to three endangered fishes (Colorado squawfish, bonytail, and razorback sucker). *Arch. Environ. Contam. Toxicol.* 30:84–92.

Butler, D. L., B. C. Osmundson, and S. McCall. 1989. Review of water quality, sediment, and biota associated with the Grand Valley Project, Colorado River basin, Colorado. U.S. Geological Survey, Grand Junction, CO.

Butler, D. L., R. P. Krueger, B. C. Osmundson, A. L. Thompson, and S. K. McCall. 1991. Reconnaissance investigation of water quality, bottom sediment, and biota associated with irrigation drainage in the Gunnison and Uncompahgre river basins and at Sweitzer Lake, west-central Colorado, 1988–89. U.S. Geological Survey, Denver. Water-Resources Investi. Report No. 91-4103.

Butler, D. L., W. G. Wright, D. A. Hahn, R. P. Krueger, and B. C. Osmundson. 1994. Physical, chemical, and biological data for detailed study of irrigation drainage in the Uncompahgre project area and in the Grand Valley, west-central Colorado, 1991–92. U.S. Geological Survey, Denver. Open File Report No. 94-110.

Butler, D. L., R. P. Krueger, B. C. Osmundson, and E. G. Jensen. 1995. Reconnaissance investigation of water quality, bottom sediment, and biota associated with irrigation drainage in the Dolores Project area, southwestern Colorado and southeastern Utah, 1990–91. U.S. Geological Survey, Denver. Water-Resources Investi. Report No. 94-4041.

Carlson, C. A., and R. T. Muth. 1989. The Colorado River: Lifeline of the American Southwest. *Proceedings of the International Large River Symposium*, pp. 220–239. Can. Spec. Pub Fish. Aquat. Sci. 106.

Cooper, D. J., and C. Severn. 1994a. Wetlands of the Escalante Ranch area, Utah: Hydrology, water chemistry, vegetation, invertebrate communities, and restoration potential. Final Report to Recovery Implementation Program for Endangered Fish Species in the Upper Colorado River Basin, Denver.

Cooper, D. J., and C. Severn. 1994b. Evaluation of the 29 5/8 mile pond near Grand Junction, Colorado. Final Report to Recovery Implementation Program for Endangered Fish Species in the Upper Colorado River Basin, Denver.

Cooper, D. J., and C. Severn. 1994c. Wetlands of the Escalante State Wildlife Area on the Gunnison River, near Delta, Colorado: Hydrology, water chemistry, vegetation, invertebrate communities, and restoration potential. Final Report to Recovery Implementation Program for Endangered Fish Species in the Upper Colorado River Basin, Denver.

Cooper, D. J., and C. Severn. 1994d. Wetlands of the Ouray National Wildlife Refuge, Utah: Hydrology, water chemistry, vegetation, invertebrate communities, and restoration potential. Final Report to Recovery Implementation Program for Endangered Fish Species in the Upper Colorado River Basin, Denver.

Cooper, D. J., and C. Severn. 1994e. Ecological characteristics of wetlands at the Moab Slough, Moab, Utah. Final Report to Recovery Implementation Program for Endangered Fish Species in the Upper Colorado River Basin, Denver.

Dill, W. A. 1944. The fishery of the lower Colorado River. *California Fish Game* 30:109–211.

Hamilton, S. J. 1995. Hazard assessment of inorganics to three endangered fish in the Green River, Utah. *Ecotoxicol. Environ. Saf.* 30:134–142.

Hamilton, S. J., and K. J. Buhl. 1994. Irrigation drainwater effects on the endangered larval razorback sucker and bonytail in the middle Green River. Abstract 415, p. 75. Society of Environmental Toxicology and Chemistry, Oct. 30–Nov. 3, Denver.

Hamilton, S. J., and K. J. Buhl. 1996. Hazard assessment of inorganics, singly and in mixtures, to flannelmouth sucker in the San Juan River, New Mexico. Final Report to San Juan River Recovery Implementation Program. National Biological Service, Yankton, SD.

Hamilton, S. J., and K. J. Buhl. 1997. Hazard assessment of inorganics, individually and in mixtures, to two endangered fish in the San Juan River, New Mexico. *Environ. Toxicol. Water Qual.* 12(2):195–209.

Hamilton, S. J., and B. Waddell. 1994. Selenium in eggs and milt of razorback sucker (*Xyrauchen texanus*) in the middle Green River, Utah. *Arch. Environ. Contam. Toxicol.* 27:195–201.

Hamilton, S. J., K. J. Buhl, F. A. Bullard, and S. F. McDonald. 1996. Evaluation of toxicity to larval razorback sucker of selenium-laden food organisms for Ouray NWR on the Green River, Utah. Final Report to Recovery Implementation Program for Endangered Fish Species in the Upper Colorado River Basin. National Biological Service, Yankton, SD.

Holden, P. B., and W. Masslich. 1997. Summary Report 1991–1996. San Juan River Recovery Implementation Program. BIO/WEST, Inc., Logan, UT.

Holden, P. B., and C. B. Stalnaker. 1975. Distribution of fishes in the Dolores and Yampa river systems of the upper Colorado basin. *Southwest. Nat.* 19:403–412.

Holley, K., and K. Weston. 1995. Contaminant screening results. Floodplain Habitat Restoration Program, Colorado and Gunnison rivers, Colorado. U.S. Bureau of Reclamation, Grand Junction, CO.

Jordan, D. S. 1891. Report of explorations in Colorado and Utah during the summer of 1889, with an account of the fishes found in each of the river basins examined. *Bull. U.S. Fish. Comm.* 9:1–40.

Kidd, G. 1977. An investigation of endangered and threatened fish species in the upper Colorado River as related to Bureau of Reclamation projects. Final Report to U.S. Bureau of Reclamation, Grand Junction, CO.

Lanigan, S. H., and H. M. Tyus. 1989. Population size and status of the razorback sucker in the Green River basin, Utah and Colorado. *North Am. J. Fish. Manage.* 9:68–73.

Lee, D. S., C. R. Gilbert, C. H. Hocutt, R. E. Jenkins, D. E. McAllister, and J. R. Stauffer, Jr. 1980. *Atlas of North American Freshwater Fishes.* North Carolina Museum of Natural History, Raleigh.

Lowe, T. P., T. W. May, W. G. Brumbaugh, and D. A. Kane. 1985. National contaminant biomonitoring program: Concentrations of seven elements in freshwater fish, 1978–1981. *Arch. Environ. Contam. Toxicol.* 14:363–388.

Lusk, J. D. 1993. Selenium in aquatic habitats at Imperial National Wildlife Refuge. M.S. thesis, University of Arizona, Tucson.

McAda, C. W. 1977. Aspects of the life history of three catostomids native to the upper Colorado River basin. M.S. thesis. Utah State University, Logan.

Minckley, W. L. 1973. *Fishes of Arizona.* Arizona Fish and Game Department, Phoenix.

Minckley, W. L. 1991. Native fishes of the Grand Canyon region: An obituary. In *Colorado River Ecology and Dam Management*, pp. 124–177. National Academy Press, Washington, DC.

Modde, T., K. P. Burnham, and E. J. Wick. 1996. Population status of the razorback sucker in the middle Green River (U.S.A.). *Conserv. Biol.* 10:110–119.

Osmundson, D. B., and K. P. Burnham. 1996. Status and trends of the Colorado squawfish in the upper Colorado River. Final Report to Recovery Implementation Program for Endangered Fish Species in the Upper Colorado River Basin, Denver.

Peltz, L. A., and B. Waddell. 1991. Physical, chemical, and biological data for detailed study of irrigation drainage in the middle Green River basin, Utah, 1988–89, with selected data for 1982–87. U.S. Geological Survey, Salt Lake City, UT, Open File Report No. 91-530.

Phillips, W. A. 1986. Cobble aquifer investigation. Colorado River Basin Salinity Control Program, U.S. Bureau of Reclamation, Grand Junction, CO.

Quartarone, F. 1993. Historical accounts of upper Colorado River basin endangered fish. Final Report to Recovery Implementation Program for Endangered Fish Species in the Upper Colorado River Basin, Denver.

Radtke, D. B., W. G. Kepner, and R. J. Effertz. 1988. Reconnaissance investigation of water quality, bottom sediment, and biota associated with irrigation drainage in the lower Colorado River Valley, Arizona, California, and Nevada, 1986–87. U.S. Geological Survey, Tuscon, AZ, Water-Resources Investi. Report No. 88-4002.

Schmitt, C. J., and W. G. Brumbaugh. 1990. National contaminant biomonitoring program: Concentrations of arsenic, cadmium, copper, lead, mercury, selenium, and zinc in U.S. freshwater fish, 1976–1984. *Arch. Environ. Contam. Toxicol.* 19:731–747.

Seethaler, K. H., C. W. McAda, and R. S. Wydoski. 1979. Endangered and threatened fish in the Yampa and Green rivers of Dinosaur National Monument, *U.S. Natl. Park Serv. Trans. Proc. Ser.* 5:605–612.

Stephens, D. W., B. Waddell, and J. B. Miller. 1988. Reconnaissance investigation of water quality, bottom sediment, and biota associated with irrigation drainage in the middle Green River basin, Utah, 1986–87. U.S. Geological Survey, Salt Lake City, UT, Water-Resources Investi. Report No. 88-4011.

Stephens, D. W., B. Waddell, L. A. Peltz, and J. B. Miller. 1992. Detailed study of selenium and selected elements in water, bottom sediment, and biota associated with irrigation drainage in the middle Green River basin, Utah, 1988–90. U.S. Geological Survey, Salt Lake City, UT, Water-Resources Investi. Report No. 92-4084.

Stephens, D., B. Waddell, and K. Holley. 1995. Contaminant screening results. Floodplain Habitat Restoration Program, Green River, Utah. U.S. Bureau of Reclamation, Grand Junction, CO.

Tyus, H. M. 1987. Distribution, reproduction, and habitat use of the razorback sucker in the Green River, Utah, 1979–1986. *Trans. Am. Fish. Soc.* 116:111–116.

Tyus, H. M., and C. A. Karp. 1990. Spawning and movements of razorback sucker, *Xyrauchen texanus,* in the Green River basin of Colorado and Utah. *Southwest. Nat.* 35:427–433.

U.S. Department of the Interior. 1967. Native fish and wildlife endangered species. *Fed. Regist.* 48:4001.

U.S. Fish and Wildlife Service. 1974. Endangered native wildlife. *Fed. Regist.* 39:1175.

U.S. Fish and Wildlife Service. 1980. Determination that the bonytail chub (*Gila elegans*) is an endangered species. *Fed. Regist.* 45:27710–27713.

U.S. Fish and Wildlife Service. 1987. Recovery implementation program for endangered fish species in the upper Colorado River basin. USFWS, Denver.

U.S. Fish and Wildlife Service. 1991a. Endangered and threatened wildlife and plants; the razorback sucker (*Xyrauchen texanus*) determined to be an endangered species. *Fed. Regist.* 56:54957–54967.

U.S. Fish and Wildlife Service. 1991b. Endangered and threatened wildlife and plants; animal candidate review for listing as endangered or threatened species, proposed rule. *Fed. Regist.* 56:58804–58836.

Valdez, R., P. Mangan, R. Smith, and B. Nilson. 1982. Upper Colorado River investigation (Rifle, Colorado to Lake Powell, Utah). In Colorado River Fishery Project Final Report Field Investigations, Part 2, pp. 100–279. U.S. Fish and Wildlife Service, Salt Lake City.

Valdez, R. A., W. J. Masslich, and A. Wasowicz. 1992. Dolores River native fish habitat suitability study. Final Report No. TR-272-02 to Utah Division of Wildlife Resources, Salt Lake City.

Waddell, B., and T. May. 1995. Selenium concentrations in the razorback sucker (*Xyrauchen texanus*): Substitution of non-lethal muscle plugs for muscle tissue in contaminant assessment. *Arch. Environ. Contam. Toxicol.* 28:321–326.

Waddell, B., and C. Wiens. 1993. Reconnaissance study of trace elements in water, sediment, and biota of Lake Powell. Interim Report. U.S. Fish and Wildlife Service, Salt Lake City, UT.

Walsh, D. F., B. L. Berger, and J. R. Bean. 1977. Residues in fish, wildlife, and estuaries: Mercury, arsenic, lead, cadmium, and selenium residues in fish, 1971–73— National pesticide monitoring program. *Pestic. Monit. J.* 11:5–34.

Welsh, D., and O. E. Maughan. 1994. Concentrations of selenium in biota, sediments, and water at Cibola National Wildlife Refuge. *Arch. Environ. Contam. Toxicol.* 26:452–458.

18

Selenium Poisoning of Fish and Wildlife in Nature: Lessons from Twelve Real-World Examples

JOSEPH P. SKORUPA

United States Fish and Wildlife Service, Sacramento, California

I. INTRODUCTION

Ecotoxicologists ultimately endeavor to understand the attributes of toxicants in the real world. Although controlled studies can contribute to that understanding, they are often plagued by an inability to translate laboratory results to real ecosystems: the so-called lab-to-field dilemma (Landis and Yu, 1995). The lab-to-field dilemma is of particular concern for bioaccumulative toxicants because standardized aquatic bioassay testing rarely includes a dietary pathway, and dietary exposure of fish and wildlife to bioaccumulative toxicants is usually the primary risk factor. That critical flaw was appropriately recognized by the U.S. Environmental Protection Agency (EPA) when the freshwater chronic criterion for selenium was established at 5 micrograms per liter (μg/L) in 1987. EPA established the 5 μg/L criterion based largely on a single well-documented episode of selenium poisoning in nature rather than on a larger accumulation of data from bioassay toxicity testing (USEPA, 1987). Although that choice was clearly prudent, a national water quality criterion based largely on one real-world case study is an easy target for criticism. A single study does not provide sufficient basis for assessing a criterion's applicability across a variety of aquatic ecosystems and site-specific environmental conditions.

During the decade since EPA's publication of the 5 μg/L freshwater chronic criterion for selenium (USEPA, 1987), there has been much research on the toxicity of selenium to fish and wildlife populations, as well as numerous reviews of available information (Lillebo et al., 1988; UC Committee, 1988; DuBowy, 1989; Ohlendorf, 1989; Beyer, 1990; Moore et al., 1990; USFWS, 1990a, 1990b; Skorupa and Ohlendorf, 1991; Sorensen, 1991; Peterson and Nebeker, 1992; CH2M HILL et al., 1993; Lemly, 1993a, 1995; Council for Agricultural Science and Technology, 1994; Gober, 1994; Maier and Knight, 1994; Heinz, 1996; O'Toole et al., 1996). None of these recent reviews, however, systematically presents an updated inventory of real-world case studies and the associated comparative results. In this chapter I will endeavor to provide such an inventory and, in some cases, provide the first detailed documentation to be found outside government reports and regulatory environmental assessment documents. This is not trivial: there are at least a dozen real-world case studies of clearly confirmed or highly probable selenium poisoning in nature (Fig. 1). Finally, I will examine this new abundance of real-world information for practical insights and the applicability of EPA's 5 μg/L criterion.

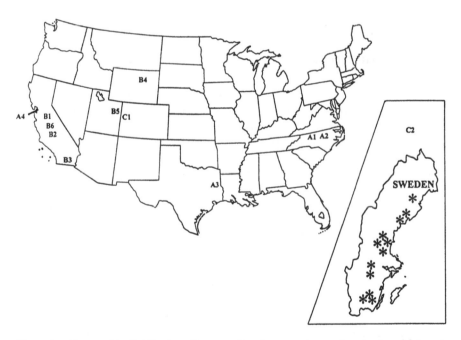

FIGURE I   Geographic distribution of case studies documenting toxic episodes of fish and wildlife exposure to selenium. Site numbering follows text section divisions of Section II Thus site A1 = Belews Lake (subsection A.1), site B1 = Kesterson Reservoir (subsection B.1), and so on. Asterisks indicate the distribution of individual lakes included in the Swedish lakes study.

II. REAL-WORLD TOXIC EPISODES

A. Case Studies Associated with Coal and Petroleum

1. Belews Lake, North Carolina: Power Plant Cooling Basin

Belews Lake (A1 in Fig. 1), a man-made reservoir, was constructed to provide condenser cooling water for a large coal-fired electric power plant. The reservoir began filling in 1970 and biological monitoring also commenced immediately. The first unit of the power station began operating in 1974. Within a year (1975) juvenile recruitment among most species of fish was very low, and by 1977 the fish community had been drastically altered. Water entering Belews Lake from a fly ash settling basin contained 150 to 200 μg/L selenium, and of 16 trace elements studied, only selenium was highly elevated in water and biota. By 1978, it was clear that a severe episode of selenium poisoning had occurred at Belews Lake (Cumbie and Van Horn, 1978; Lemly, 1985a, 1985b; 1993b).

Although the selenium concentration in the main portion of Belews Lake had been elevated only to values on the order of 8 to 22 μg/L, and an average value of 10 μg/L, highly elevated rates of teratogenic fish (10–70% vs. normal baseline of 1–3%) were documented, and populations of 16 species of fish completely collapsed. Only 4 species of fish remained by 1978. As summarized in Table 1, this level of waterborne selenium resulted in biotic tissues that sometimes exceeded 100 mg/kg selenium. (All tissue concentrations in this chapter are expressed on a dry weight basis unless otherwise noted.) Compared to normal reference values, which are uniformly in single digits, the tissue data were extremely elevated and were consistent with the severe adverse effects data. Both the exposure and the response data from the main reservoir at Belews Lake indicated that 10 μg/L waterborne selenium was well above the toxicity threshold (Lemly, 1985a, 1985b, 1993b).

In contrast to the main reservoir, a semi-isolated reach known as the Highway 158 arm contained less than 5 μg/L waterborne selenium, and an overtly normal fish community persisted there (Cumbie and Van Horn, 1978). A subsequent histopathological and hematological study of green sunfish (Lepomis cyanellus) measured 3 to 4 μg/L waterborne selenium in the Highway 158 arm and detected in some of the fish sublethal toxic effects such as generalized edema and abnormal ovarian tissue damage (Sorensen et al., 1984). Thus, the threshold region for toxicity at Belews Lake appeared to be in the neighborhood of 2 to 5 μg/L waterborne selenium. By the mid-1980s the Belews Lake episode had been closely studied and documented for more than a decade. Based largely on that body of documentation, EPA revised the national chronic criterion for selenium downward from 35 μg/L to 5 μg/L (USEPA, 1987).

A dry ash system was implemented in late 1985 at Belews Lake, curtailing the input of selenium to the reservoir. Fish sampling 7 years postremediation,

	(µg/L)	(µg/L)	(dw)	(dw)	body	Eggs	Muscle	Hepatic	Eggs	Muscle	Hepatic	Resources
Range:	N/A	0.1–1	0.2–2	0.4–4.5	<1–4	<1–4	<1–4	2–8	0.5–4.5	1–3	2–15	None
Typical:	N/A	<0.5	<1	<2	<2	<3	<2	<5	<3	<2	<10	
	150–200	10	4–30	20–50	40–125	20–170	25–200	ND	ND	ND	ND	16 Fish spp. extirpated, 10–70% rates of terata
	N/A	3–4	0.7–3	4–8	ND	ND	7–9	25–30	ND	ND	ND	Sublethal effects; ovarian damage, generalized edema
	N/A	ND	ND	ND	10–20	ND	ND	ND	ND	ND	ND	5–10% fingerling rates of terata (1–3% normal)
	N/A	<1	1–4	2–5	ND	3–20	ND	ND	ND	ND	6–15	3–6% fry rates of terata (0% fingerling terata)
	50–200	10	3–40	10–30	ND	30–50	35–50	ND	ND	ND	ND	Fish densities reduced 38–75%, reproductive failure
	2200–2700	5	0.8–68	ND	ND	ND	10–40	25–100	ND	ND	ND	Collapse (>90%) of planktivorous fish biomass

N/A	ND	ND	1–15	5–10	17	ND	20–50	<0.5–12+	ND	10–35	Blackbird hatch decline (54%), fish ovary damage
N/A	ND	ND	ND	5—10	ND	5–10	15–25	ND	ND	ND	Not monitored
10–60	5–20	3	10–45	ND	ND	ND	ND	1.9–57	ND	ND	10–30% teratogenic waterfowl nests (vs. <1% normal)
230–420	15–430	0.3–67	9.9–290	69–430	ND	ND	ND	2.3–180	3.5–40	3.1–360	4–49% inviable waterbird eggs, adult & juvenile toxicosis
463–943	323–6280	0.8–15	4.7–250	ND	ND	ND	ND	15–148	ND	ND	33–50% teratogenic shorebird eggs (vs. <0.5% normal)
83–671	29–603	0.2–4.1	9.4–140	ND	ND	ND	ND	2.1–164	5.5–36	6.7–120	2–16% teratogenic shorebird eggs (vs. <0.5% normal)

ce	source (µg/L)	Water system (µg/L)	(mg/kg dw)	(mg/kg dw)	Whole body	Eggs	Muscle	Hepatic[c]	Eggs	Muscle	Hepatic[c]	and Wildlife Resources
	2–10	1.5	3.3	0.8–12.1	6.1–16	ND	7.9–14	ND	1.6–35	2.7–7.2	2.7–42	5% reduction of black-necked stilt nesting proficiency
	ND	7–1300	4–17	87–166	ND	ND	ND	ND	2.4–135	17–35	2.6–170	38–52% inviable avocet and Canada goose eggs
	ND	44–70	20–43	36–65	ND	ND	ND	ND	39–160	23–26	22–134	8–26% inviable avocet and eared grebe eggs
	<1–830	9–93	8–26	10–71	22–104	ND	ND	ND	3.8–120	6–84	19–213	10% teratogenic coot nests, >85% inviable coot eggs
	1151–2114	1600–11,300	ND	ND	ND	ND	ND	ND	7.2–81	ND	ND	14–57% teratogenic shorebird eggs (vs. <0.5% normal)
	ND	96–160	ND	14–20	ND	ND	ND	<2–40	ND	ND	ND	Progressive mortality of stocked game fishes

ND	<5–80	305	ND	ND	ND	<48–243	<74–255	ND	ND	ND	Human health warnings posted
ND	10–70	8.6–41	27–30	15–50	31–32	22–31	40–78	5.6–18	ND	9.0–84	Possible reproductive failure among catfish
ND	24	31	12–26	29–62	ND	ND	ND	23–32	ND	ND	Not monitored
N/A	3–5	ND	ND	ND	ND	14–52	53–135	ND	ND	ND	No catastrophic effects evident
N/A	1–5	ND	ND	ND	ND	0.9–36	ND	ND	ND	ND	Collapse of perch populations in five lakes

water and sediment are less standardized between studies than are sampling methods for biotic tissues.
: applicable; dw, dry weight

80; 2, Lemly, 1985b; 3, Lillebo et al., 1988; 4, Maier and Knight, 1994; 5, Irwin, 1996; 6, Moore et al., 1990; 7, Martin and Ha
5; 9, USFWS, 1990a; 10, Bezer, 1990; 11, Birkner, 1978; 12, Saiki and Lowe, 1987; 13, Hothem and Ohlendorf, 1989; 14, Schule
1992; 16, Saiki et al, 1993; 17, Welsh and Maughan, 1994; 18, Walsh et al., 1977; 19, Schmitt and Brumbaugh, 1990; 20, Jenkins
. Saiki, 1989; 23, Ogle and Knight, 1989; 24, Hamilton et al., 1990; 25, USFWS, 1990b; 26, Cleveland et al., 1993; 27, Coyle et al.
994; 29, Coughlan and Velte, 1989; 30, Hermanutz et al., 1992; 31, Hamilton and Waddell, 1994; 32, Sorensen, 1988; 33, Whit
, 1988; 35, White et al., 1987; 36, Ohlendorf et al., 1990; 37, Barnum, 1994; 38, Skorupa et al., unpublished data—bird, hepa
rf, 1991; 40, Ohlendorf et al., 1993; 41, Lemly, 1985a; 42, Lemly, 1993c; 43, Sorensen et al., 1984; 44, Great Lakes Science A
ck and Summers, 1984; 46, Gillespie and Baumann, 1986; 47, Cutter, 1986; 48, Cumbie and Van Horn, 1978; 49, Garrett and

in 1992, still revealed slightly elevated rates of teratogenic fish (5–10%) and elevated tissue selenium (Table 1). It was not reported how far the waterborne concentrations of selenium in the main reservoir had declined between initiation of remediation in 1985 and the 1992 sampling of fish. An exposure–response curve relating frequency of deformities in centrarchids (sunfish family) to whole-body selenium concentrations showed an excellent fit to an exponential function ($r^2 = 0.88$; Lemly, 1993b). The EC_{10} and EC_{50} estimates from the fitted exponential function are about 30 and 70 mg/kg whole-body selenium, respectively (visual estimates from the published curve), where EC_{10} and EC_{50} represent the 10 and 50% effect concentrations, respectively.

A more comprehensive study of selenium in the Belews Lake ecosystem was conducted in 1996, a full decade after the dry ash disposal system was implemented (Lemly, 1997). Although the waterborne concentration of selenium had relaxed to less than 1 μg/L, concentrations of selenium in sediment, inverte-brates, and fish ovaries (= eggs) were still slightly to moderately elevated (Table 1). The 3 to 6% incidence of terata among fish fry also exceeded normal back-ground levels (1–3%; Lemly, 1993b, 1997). Long-term residual exposure of biota to elevated levels of selenium from short-term selenium inputs has not been unique to Belews Lake (see Martin Reservoir and Kesterson Reservoir case studies, Sections II.A.3 and II.B.1), and suggests that once selenium has entered biotic pathways (e.g., via algal uptake), it is very efficiently recycled over time. In some systems, the peak waterborne concentration of selenium that was reached may be more relevant to assessing risk than longer term average concentrations. The 1996 data for Belews Lake clearly illustrate at least one set of circumstances under which measures of waterborne selenium alone would be inadequate for assessing toxic risk.

2.   Hyco Reservoir, North Carolina: Power Plant Cooling Basin

Hyco Reservoir (A2 in Fig. 1), like Belews Lake, served as a cooling water impoundment for a large coal-fired electric power plant. Although operation of the Hyco units dated back to at least 1973, biological monitoring was not initiated until 1978. The impetus for a monitoring program included reports from anglers of declining bass catches at Hyco Reservoir, and the heightened concern of state regulatory officials caused by events at Belews Lake. Water entering Hyco Reservoir from fly ash settling basins contained 50 to 200 μg/L selenium, and as at Belews Lake, the selenium concentration in the reservoir was elevated to an average of about 10 μg/L. Again, consistent with data from Belews Lake, this level of waterborne selenium was sufficient to result in highly contaminated biotic tissues (Table 1).

Adverse effects on fish populations were pronounced: censuses in four separate coves of Hyco Reservoir indicated 38 to 75% reductions in densities of adult fish between 1979 and 1980 and severe (>95%) reductions in densities of larval fish along three transects. Potentially toxic trace elements other than selenium were not notably elevated in fish tissues, and clinical demonstration of selenium-associated inviability of Hyco fish larvae led to the conclusion that selenium was the causative agent responsible for fish declines in the reservoir (Woock and Summers, 1984; Gillespie and Baumann, 1986). Data generated from studies at Hyco Reservoir are well removed from toxicity threshold regions and exposure-response insights were limited to circumstances of severe exposure.

3. Martin Reservoir, Texas: Power Plant Cooling Basin

Martin Reservoir (A3 in Fig. 1) was constructed in 1974 to provide a source of cooling water for a large coal-fired electric power plant. The plant began operating in 1977, and in 1978–79 there were unauthorized discharges from two fly ash settling ponds into Martin Reservoir. Water in the fly ash ponds contained 2200 to 2700 $\mu g/L$ selenium. Measures of waterborne selenium in Martin Reservoir varied from 1 to 34 $\mu g/L$, with an overall average of 2.6 $\mu g/L$, and the contaminant was thought to average about 5 $\mu g/L$ in the primary impact areas. This was sufficient to cause highly elevated tissue selenium in fish and birds, and elevated tissue selenium persisted for at least a decade after the discharge episode (Table 1).

Fish die-offs were noted within 2 months of the initial fly ash pond discharges, and in mid-1979 the Texas Parks and Wildlife Department began an investigation. Fish populations had been monitored in 1977, prior to initiation of fly ash pond discharges, and the 1977 monitoring effort was closely replicated in 1979 and 1980 to reveal a decline in biomass of planktivorous fish that exceeded 90%. Chemical analyses of multiple chemical elements revealed that only selenium concentrations were elevated sufficiently to explain the collapse of planktivorous fish populations. About 7 to 8 years after the discharge episode, high selenium concentrations in red-winged blackbird (Agelaius phoeniceus) eggs (mean value of 11.1 mg/kg) were documented and associated with greater than 50% depression in egg hatchability, while barn swallows (Hirundo rustica) had nearly normal concentrations of selenium in their eggs and normal egg hatchability. Similar to Hyco Reservoir, data collected at Martin Reservoir have not been sufficient to permit the construction of exposure–response curves. The data did reveal, however, that at or below 5 $\mu g/L$ waterborne selenium in Martin Reservoir, biotic responses were relatively severe in magnitude—suggesting that the hazard threshold lies below 5 $\mu g/L$ (Garrett and Inman, 1984; Lemly, 1985a; Sorensen, 1986, 1988; King, 1988; Texas Parks and Wildlife Department, 1990; King et al., 1994).

4. Chevron Richmond Oil Refinery, California: Constructed Wetland

In 1988–1989 Chevron USA initiated an experimental program to route process wastewater from their Richmond (California) oil refinery (A4 in Fig. 1) through a small (36 ha) constructed wetland prior to discharge into the San Francisco Bay estuary. Initial (1989–91) water monitoring revealed that outflow from the constructed wetland typically contained substantively less selenium (ca. 10 $\mu g/$ L) than the inflow had contained (ca. 20 $\mu g/L$). By 1994 the marsh was attracting substantive use by waterbirds, prompting officials at the San Francisco Bay Regional Water Quality Control Board to request that Chevron USA conduct a study of selenium exposure and reproductive performance among birds nesting at the marsh. The board requested an additional study of food chain bioaccumulation of selenium for 1995.

During 1995, wastewater inflow averaged about 20 $\mu g/L$ selenium and after flowing through the three segments of the marsh averaged about 5 $\mu g/L$ in the outflow. Aquatic invertebrates and birds using this flow-through system accumulated highly elevated tissue concentrations of selenium (Table 1). Black-necked stilts (Himantopus mexicanus) were chosen as a focal bird species for random sampling of eggs in 1994 and 1995. Nests of other species of waterbirds were also monitored in 1994, and some fail-to-hatch eggs were nonrandomly collected. In both 1994 and 1995, random stilt eggs averaged about 20 to 30 mg/kg selenium and concentrations exceeding 50 mg/kg were observed among fail-to-hatch eggs of stilts and other bird species. Deformed embryos were recovered from about 30% of mallard (Anas platyrhynchos) nests and about 10% of American coot (Fulica americana) nests that yielded one or more assessable embryos. Normally, fewer than 1% of assessable nests should yield a deformed embryo. Selenium exposure at Chevron Marsh was sufficient to expect a 6.7% deformity rate in stilt eggs (based on the exposure–response data for stilts exposed to seleniferous agricultural drainage water; see Tulare Basin data presented in Section II.B.2), but none of 16 assessable stilt embryos were deformed. At an expected 6.7% deformity rate, a sample size of 16 has a power of only 69% for detecting a deformed embryo. Doubling the sample size would have provided a more acceptable power of 90%.

Although selenium poisoning of mallards and coots was confirmed, and poisoning of stilts seems likely, net effects on the local breeding populations of these species is unknown. Chevron USA operates a predator control program at the marsh, and the benefits of that protection could counterbalance losses caused by selenium poisoning. To date, neither the predator control benefits nor the selenium poisoning losses (especially posthatch) have been quantified to a level of scientific certainty that would permit a reliable cost–benefit evaluation.

Stilt eggs averaged about the same selenium exposure at Chevron Marsh (20–30 mg/kg) as that observed at Kesterson Reservoir, California (25–37 mg/kg), but the source water at Chevron Marsh averaged less than 10% as much selenium as the source water at Kesterson (20 vs. 300 $\mu g/L$). This unexpected result prompted the 1995 follow-up bioaccumulation study at Chevron Marsh, which revealed that transfer of selenium from water to aquatic invertebrates was greatly enhanced compared to Kesterson and other agricultural drainage water sites, while transfer of selenium from aquatic invertebrates to black-necked stilt eggs was comparable. Thus, the unexpectedly high level of selenium in stilt eggs at Chevron Marsh appears to be a function of selenium chemistry in the water (primarily selenite at Chevron Marsh vs. selenate at Kesterson), not in the food chain. A 5-year remedial management and monitoring plan is being implemented in 1997 by Chevron USA and the San Francisco Bay Regional Water Quality Control Board, but the plan will not include any further monitoring of avian reproductive performance beyond the collection of small numbers of eggs in 1997, 1999, and 2001 (CH2M HILL, 1994, 1995; Chevron USA, 1996; San Francisco Bay Regional Water Quality Control Board file data).

B. Agricultural Drainage Water Associated Case Studies

1. Kesterson Reservoir, California: Drainage Water Evaporation Impoundment

Kesterson Reservoir (B1 in Fig. 1), a 500 ha shallow impoundment (1–1.5 m deep) subdivided into 12 interconnected cells, was constructed in the northern San Joaquin Valley as part of a federal irrigation project. Located at the terminus of the San Luis Drain, Kesterson Reservoir served dual roles as an evaporation basin for agricultural drainage water and a managed wetland intended to benefit fish and wildlife populations. Following an initial period (1972–78) of receiving high quality agricultural spill water, during which time robust marsh vegetation and animal populations were established, Kesterson Reservoir began receiving highly saline subsurface drainage water. By 1981, virtually all inflow to the reservoir was saline drainage water. By the spring of 1982, federal biologists and resource managers noted an apparent deterioration of the aquatic ecosystem. Detailed ecotoxicological research was conducted during 1983–85 (Zahm, 1986; Ohlendorf, 1989).

Saline drainage water discharged from the San Luis Drain to Kesterson Reservoir averaged about 300 μg/L selenium. The cells at Kesterson Reservoir were operated in series, such that concentrations of total dissolved solids, boron, and some other elements increased as water moved down series. In contrast, selenium concentrations in impounded water decreased as water moved down series from the receiving cells. Even in down-series cells, however, selenium concentrations still usually exceeded 50 μg/L. This resulted in a severely contaminated aquatic habitat, with selenium concentrations in some food chain fauna, fish, and wildlife samples as high as 100 to 400 mg/kg (Table 1).

Although there are no hard data on the species composition of the fish fauna at Kesterson Reservoir prior to inflows of saline drainage water, as late as September, 1983, a multispecies warmwater fish assemblage was sampled in the San Luis Drain near its discharge point to Kesterson Reservoir (M. K. Saiki,

U.S. Geological Survey, personal communication). Therefore, there is at least a circumstantial basis to suspect that Kesterson Reservoir initially also contained a similar multispecies assemblage of warm-water fish. Shortly thereafter, however, only the pollution-tolerant mosquitofish (Gambusia affinis) persisted in the San Luis Drain and Kesterson Reservoir. The role of selenium in the disappearance of all other fish species, if selenium played any role, is unknown. During 1984 and 1985, samples of mosquitofish from the San Luis Drain averaged whole-body concentrations of about 120 mg/kg selenium and were documented to have a 20 to 30% incidence of stillborn fry compared to 1 to 3% in reference samples. The high incidence of stillborn fry was believed to be attributable to the high exposures of these fish to selenium (as opposed to a salinity effect) because many of the stillborn fry exhibited superficial signs of teratogenesis, and the salinity of San Luis Drain water, although elevated, was well below the upper limits known to be tolerated by mosquitofish (Saiki and Ogle, 1995; M. K. Saiki, personal communication).

The geometric mean selenium content of waterbird eggs at Kesterson Reservoir during 1983–85 varied by species and ranged from means of about 4 to 70 mg/kg (Ohlendorf and Hothem, 1994). Kesterson was surrounded by a rich landscape mosaic of cleaner wetlands, and wide-ranging species of birds, such as ducks, exhibited lower average selenium exposures than the more sedentary species did. Depressed egg viability (hatchability) and elevated incidence of embryo deformities were documented for several species of waterbirds. Upon pooling data for all species, it was found that at least 39% of 578 nests contained one or more inviable eggs and 26% of 2281 fully incubated eggs were inviable (vs. 1.2% for pooled reference eggs). When the reproductive data were segregated by species, findings were as follows: 4 to 49% of the fully incubated Kesterson eggs failed to hatch, and 0 to 15% contained deformed embryos. Embryo deformities were often multiple and typically involved the eyes, beak, and limbs (Ohlendorf et al., 1988). This distinctive pattern of multiple embryo deformities associated with highly elevated egg selenium is hereafter referred to as the "Kesterson syndrome." Complete posthatch juvenile mortality was reported for several species. Signs of acute poisoning of adults, such as carcasses that exhibited alopecia, also were evident (Ohlendorf et al., 1986a; Ohlendorf, 1989; Ohlendorf and Skorupa, 1989; Skorupa and Ohlendorf, 1991).

The avian reproductive data collected at Kesterson Reservoir and nearby comparison sites were sufficient to permit the construction of several exposure–response curves. The curves clearly indicated that reproductive toxicity was exposure responsive and, combined with data from experimental feeding studies using captive mallards, confirmed the causative link between selenium exposure and embryotoxicity. However, because most data came from the highly contaminated Kesterson Reservoir, the lower end of the exposure axis (x axis) was poorly represented and threshold points for adverse effects could not be identified very precisely (Ohlendorf et al., 1986b; Heinz, 1996).

All agricultural drainage discharges to Kesterson Reservoir were halted in 1986. By the end of 1988, Kesterson Reservoir had been dried up and the low-lying areas within the ponds filled with soil to at least 15 cm above the expected average seasonal rise of groundwater. Although ephemeral pooling of rainwater still occurs, Kesterson Reservoir has been transformed to a mosaic of primarily terrestrial habitats that are much less contaminated than the aquatic habitats they replaced. Postclosure biological monitoring has been conducted annually since 1987. Selenium concentrations in animal tissues have stabilized at slightly elevated levels, and no toxic effects are apparent (Ohlendorf and Santolo, 1994).

2. Tulare Basin, California: Drainage Water Evaporation Impoundments

The Tulare Basin (B2 in Fig. 1), located in the southern San Joaquin Valley about 160 km south of Kesterson Reservoir, has no natural drainage to the ocean. To meet the demand for agricultural drainage generated by extensive irrigation within the basin, more than 20 shallow impoundments were constructed during 1972–85 to provide for evaporative disposal of saline drainage water. These facilities varied from large multiple-celled systems (similar in design to Kesterson Reservoir) to small single-celled "ponds." None of these facilities were intended to provide fish or wildlife benefits and, unlike Kesterson, they are devoid of emergent marsh vegetation. Nonetheless, these impoundments proved very attractive to waterbirds, including populations of nesting birds. Although all the facilities received saline drainage water, there was a wide span of ionic and trace element composition of impounded water. The selenium content of water discharged to these impoundments varied from less than 1 μg/L to more than 1000 μg/L. Detailed studies of avian exposure to contaminants and reproductive performance, methodologically matched to the Kesterson studies, were conducted during 1987–89. Additional wildlife monitoring and research at these facilities has occurred from 1982 to present (Moore et al., 1990; CH2M HILL et al., 1993; Robinson et al., 1997).

During 1987–89, four sites were found to exhibit highly elevated rates of embryo teratogenesis (10–50%) in one or more species of waterbirds (Table 1). The embryo deformity types at these sites matched the Kesterson syndrome, and maximum selenium concentrations in bird eggs at all four sites were highly elevated (80–164 mg/kg). The least contaminated of the four sites, the Tulare Lake Drainage District South facility, averaged about 15 μg/L selenium in impounded water and about 20 mg/kg in duck eggs (Skorupa and Ohlendorf, 1991). This confirmed that reproductive toxicity could occur at agricultural drainage water sites with far less contamination than at Kesterson Reservoir. The true research value of the Tulare Basin, however, was the wide range of selenium exposures associated with different evaporation facilities and the ubiquitous abundance of two species of birds, black-necked stilts and American avocets (Recurvirostra americana). Stilt and avocet eggs could be sampled across an exposure risk of four orders of magnitude (i.e., <1 to >100 mg/kg selenium in eggs). Thus, the Tulare Basin presented ideal circumstances for examining exposure–response relationships and threshold points for avian reproductive toxicity.

It is possible to determine unambiguously the overt teratogenic status (e.g., presence or absence of eyes) of any fresh stilt or avocet embryo incubated to at least 6 to 8 days of age. Consequently, teratogenic response is a very precise variable for examining species' relative sensitivity to selenium exposure. Sufficient field data have now been collected for stilts and avocets to allow delineation of species-specific teratogenic response curves (Fig. 2). Surprisingly, even though stilts and avocets are each other's closest phylogenetic relatives (Sibley et al., 1988), their response coefficients for selenium (coefficient b_1 in Fig. 2) are significantly different ($Z = 2.70$, $P < 0.01$; Afifi and Clark, 1996). Based on predicted 50% effect concentrations (EC_{50} values in Fig. 2), stilt embryos are about twice as sensitive as avocets to in ovo selenium exposure. Stilts, therefore, are the more sensitive model species for further examination of threshold points.

Although embryo teratogenesis is a precise response variable, it is also relatively insensitive because teratogenesis is a severe response. Subtle disruptions of embryonic physiology and development are likely to cause inviability of stilt eggs at exposure levels less severe than those required for selenium-induced overt teratogenesis. Analysis of egg viability (hatchability) should therefore provide a sensitive estimate of the exposure threshold for reproductive toxicity. By examining the clutchwise incidence of inviable stilt eggs as a function of selenium concentrations measured in sibling sample eggs, it was determined that the threshold exposure for impaired hatchability occurred between 4 and 9 mg/kg selenium in the egg (Ohlendorf et al., 1993). Enough additional data have accumulated since 1993 to warrant a reexamination of that threshold region by single milligram-per kilogram increments. These additional data produce an estimate of the stilt embryotoxicity threshold that is narrowed to the region between 6 and 7 mg/kg selenium in eggs (Fig. 3). Thus, the upper boundary of safe exposure levels for stilt eggs (embryos) is about 6 mg/kg, or roughly three times the normal background exposure of about 2 mg/kg.

A strong relationship between the mean concentration of selenium in impounded water and in stilt eggs at the Tulare Basin sites ($r^2 = 0.81$) is described by the equation shown in Figure 4, where egg and water selenium are expressed in milligrams per kilogram and micrograms per liter, respectively. Based on that regression equation, the average concentration of selenium in a population sample of stilt eggs would be expected to exceed 6 mg/kg when the mean selenium content of impounded drainage water exceeds 6 μg/L. If the average exposure of embryos is 6 mg/kg selenium when water contains 6 μg/L, then roughly 50%

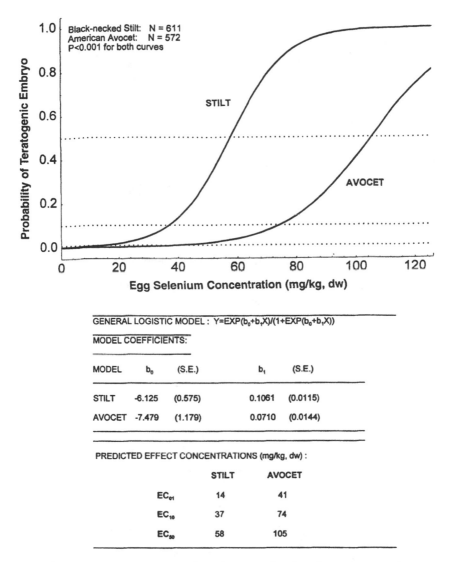

GENERAL LOGISTIC MODEL : Y=EXP(b₀+b₁X)/(1+EXP(b₀+b₁X))

$$Y = \text{EXP}(b_0 + b_1 X)/(1 + \text{EXP}(b_0 + b_1 X))$$

MODEL COEFFICIENTS:

MODEL	b_0	(S.E.)	b_1	(S.E.)
STILT	-6.125	(0.575)	0.1061	(0.0115)
AVOCET	-7.479	(1.179)	0.0710	(0.0144)

PREDICTED EFFECT CONCENTRATIONS (mg/kg, dw) :

	STILT	AVOCET
EC_{01}	14	41
EC_{10}	37	74
EC_{50}	58	105

FIGURE 2 Logistic response curves for selenium-induced embryo teratogenesis among black-necked stilt and American avocet populations exposed to agricultural drainage water.

of individual eggs still exceed the toxic risk threshold (see Salton Sea data presented in Section II.B.3). Thus, the zero-exceedance threshold point must be associated with water containing less than 6 μg/L total recoverable selenium. During the 1987–89 studies, eggs of stilts were collected at two sites with

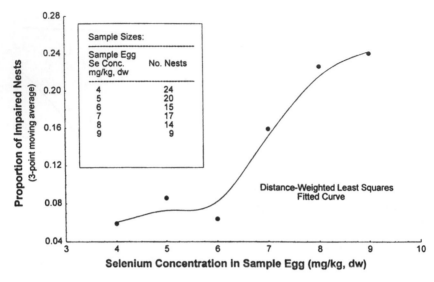

FIGURE 3 Response threshold for selenium-induced inviability of black-necked stilt eggs among populations exposed to agricultural drainage water.

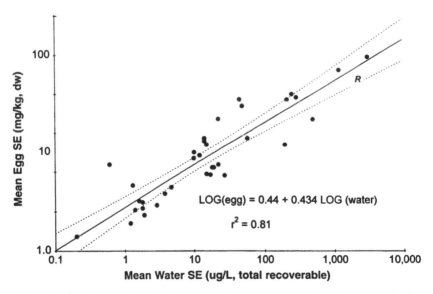

FIGURE 4 Regression relationship for mean selenium concentrations in black-necked stilt eggs as a function of mean selenium concentrations in impounded drainage water at study sites in the Tulare Basin, California. A data point for the Red Rock Ranch Agroforestry Demonstration Site is plotted as the letter R.

waterborne selenium in the 3 to 6 μg/L range (Skorupa et al., unpublished data). At a site that averaged 3.8 μg/L selenium in impounded water, the selenium content of five randomly sampled stilt eggs ranged from 2.2 to 5.6 mg/kg (zero exceedance, and a maximum value close to the 6 ppm safe limit). At a site that averaged 4.7 μg/L selenium in impounded water, two of eight randomly sampled stilt eggs contained more than 6 mg/kg selenium (25% exceedance rate). Consequently, the reproductive effects threshold for black-necked stilts nesting at drainage evaporation facilities in the Tulare Basin appears to be about 4 μg/L selenium in impounded water (about 10 times normal background for saline sinks).

This estimate of a toxicity threshold point at 4 μg/L selenium in impounded drainage water is species-specific. As illustrated in Figure 2, even closely related species can substantially vary in their sensitivity to equivalent selenium exposures. The reproductive toxicity threshold point for stilts would be overly protective for avocets. Conversely, it might not be fully protective for other waterbird species. For example, based on a sample of 126 eggs from several species of dabbling ducks, the EC_{50} for overt teratogenesis is 31 mg/kg egg selenium (Skorupa et al., unpublished data) compared to 58 and 105 mg/kg for stilts and avocets (Fig. 2). These comparative results suggest that ducks may be roughly twice as sensitive as stilts to selenium-induced embryotoxicity and that depressed egg viability might be possible at sites with less than 4 μg/L selenium in impounded drainage water.

About half the drainage water evaporation facilities in the Tulare Basin are no longer in operation. Those still running are regulated by means of Waste Discharge Requirements (WDR permits, as provided for pursuant to the California Water Code) that require creation of clean wetlands to mitigate unavoidable toxic impacts to breeding waterbirds. The mitigation wetlands are not allowed to average more than 2.7 μg/L total recoverable selenium in impounded water (e.g., Central Valley Regional Water Quality Control Board, 1993).

3. Salton Sea, California: Regional Drainage Water Terminal Sink

The Salton Sea (B3 in Fig. 1), with a surface area of about 93,000 ha, is California's largest inland body of water. The present sea was created in 1905–07 when the entire flow of the Colorado River was accidentally diverted for about 16 months into the Salton Trough. Consistent with a desert climate, where evaporation greatly exceeds precipitation, the elevation of the sea fell by 18 m, and the salinity increased to 40,000 mg/L during 1907–25. Since the early 1920s the sea has served as a terminal sink for irrigation drainage water. By 1925 the elevation of the sea stopped declining because of counterbalancing inflows of drainage water. As agricultural development proceeded, and drainage water inflows increased, the sea has regained 10 m of lost elevation. Currently, more than 125,000 ha · m/yr (1 million acre · ft/yr) of agricultural drainage water is diverted to three major river channels for conveyance to the Salton Sea (Saiki, 1990; Setmire et al., 1990; 1993).

Irrigation in the Coachella, Imperial, and Mexicali Valleys occurs with water diverted from the Colorado River. This source water contains selenium concentrations of about 2 μg/L. Evaporative concentration of selenium in the shallow groundwater results in subsurface drainage containing up to 360 μg/L selenium. The final mixture of surface and subsurface drainage conveyed to the Salton Sea contains 2 to 10 μg/L selenium. Mass loading to the Salton Sea amounts to more than 8000 kg (>9 tons) of selenium per year. Although the water column of the sea contains only 1.5 μg/L total recoverable selenium (Westcot et al., 1990), food chain fauna and tissues of fish and birds exhibit substantively elevated concentrations of selenium (Table 1). The levels of avian exposure to selenium revealed by the systematic surveys of Setmire et al. (1990, 1993) were sufficiently elevated to prompt follow-up studies of reproductive performance among black-necked stilts during 1992–93 (Bennett, 1997).

In 1992, 38 black-necked stilt eggs were sampled and averaged 6.6 mg/kg selenium. Twenty-two, or 58%, of the eggs exceeded the toxic risk threshold of 6 mg/kg associated with impounded drainage water in the Tulare Basin. The only reproductive performance variable measured in 1992 was the incidence of embryo teratogenesis. No selenium-induced terata were found among the 20 eggs that contained assessable embryos. However, because almost all the eggs contained selenium concentrations well below the 1% effect concentration (EC_{01}) of 14 mg/kg for embryo teratogenesis in stilts (Fig. 2), a failure to find selenium-induced terata would be the predicted outcome. In 1993, 44 stilt eggs were sampled and averaged 5.8 mg/kg selenium. Twenty-one, or 48%, of the eggs exceeded the 6 mg/kg toxic risk threshold. Once again, as expected, no selenium-induced terata were found among 28 eggs that contained assessable embryos.

During 1993, however, the nests from which eggs were sampled also were monitored and the viability of sibling eggs was recorded. Among the 23 nests that survived to full term, 13% contained one or more inviable sibling eggs. Normally 8.9% of stilt nests contain one or more inviable eggs due to infertility and other natural causes. If the slightly elevated incidence of "affected" nests were due to selenium-induced inviability of eggs, it would represent about a 5% depression in nesting proficiency, where proficiency is defined as the proportion of nests in which all sibling eggs are viable [(0.911 − 0.87)/0.911 = 0.045 = ca. 5%]. This putative magnitude of reproductive depression very closely matches predictions based on 410 sample eggs and sibling fates from full-term nests of stilts that were monitored at Kesterson Reservoir and in the Tulare Basin. Based on the Kesterson–Tulare logistic regression curve for nesting proficiency of stilts, the exposure-specific ($N = 23$) predicted probability of affected nests for the Salton Sea sample is 11.9% (Skorupa et al., unpublished data). The close match between the 11.9% prediction and the 13% observed value supports the view that the apparent 5% reproductive depression is biologically real. For such a small putative effect, about 225 full-term nests would have to be monitored to statistically distinguish between a true effect and random sampling error. Additional nest monitoring has been recommended for the Salton Sea (Bennett, 1997).

Because stilt eggs collected during the 1993 study came from numerous locations, only a few of which constituted Salton Sea "shoreline" sites, it is not precisely known what concentration of selenium in water can be associated with this case study. It is highly probable that the birds in this study were predominantly using wetlands with selenium concentrations in water of 10 μg/L or less.

4. Kendrick Reclamation Project, Wyoming: Seepage Wetlands

The Kendrick Reclamation Project (B4 in Fig. 1), near Casper, Wyoming, was constructed to store and divert water from the North Platte River for use within the Casper–Alcova Irrigation District. The project has been in operation since 1946, and by the mid-1960s it had expanded from an initial 425 ha of irrigated land to the present service area of 8000 to 10,000 ha. Soil in the service area is derived principally from seleniferous Cretaceous age geologic formations. Surveys of shallow groundwater revealed localized selenium concentrations as high as 23,000 μg/L. Although the high bioavailability of selenium to plants within the Kendrick Project area had been reported much earlier (Rosenfeld and Beath, 1964), the potential for fish and wildlife exposure to selenium was not extensively investigated until 1986–90 (Peterson et al., 1988; See et al., 1992b).

Highly elevated selenium concentrations in bird eggs ($\leq$135–160 mg/kg), along with elevated frequencies of overt embryonic terata consistent with the Kesterson syndrome, were documented at two wetlands within the project area: Rasmus Lee Lake and Goose Lake (Table 1). Both lakes functioned as terminal sinks during 1986–90. Both sites are considered seepage wetlands that are at least partially dependent on seepage from Casper Canal, the primary conveyance ditch for delivery of irrigation water (Peterson et al., 1988). Rasmus Lee and Goose Lakes also receive direct and indirect irrigation drainage and natural runoff, including snowmelt. The selenium content of those diverse inflows is mostly uncharacterized. Selenium concentrations in impounded water averaged 38 μg/L in Rasmus Lee Lake and 54 μg/L in Goose Lake. Based on ratios of selenium and chloride, it was concluded that evaporative concentration of Casper Canal source water (averaging 1 μg/L selenium) was the only mechanism necessary to explain the highly elevated concentrations of selenium in water impounded at Rasmus Lee and Goose Lakes (See et al., 1992a). Consistent with highly elevated selenium in impounded water, sediments and food chain fauna at Rasmus Lee and Goose Lakes also contained very high concentrations of selenium (Table 1).

The most unambiguous selenium poisoning occurred among American avocets nesting at Rasmus Lee Lake in 1989. At least 6% of all embryos were teratogenic (ca. 0.15% is normal for avocets), including some embryos exhibiting the suite of multiple malformations (eyes, limbs, bill) that is particularly character-istic of avian selenosis. More than 50% of the full-term avocet nests ($N = 47$) contained one or more inviable eggs (ca. 12% is normal for avocets). These high rates of embryo deformity and inviability indicate a system that is far beyond the threshold for toxic effects, a conclusion consistent with the exposure data. A sample of 86 eggs had a median selenium concentration of 79 mg/kg (See et al., 1992a). The true median selenium content of eggs was probably somewhat lower because many nonrandomly selected "effect" eggs were included in the sample (See et al., 1992b).

The American avocet is relatively insensitive to selenium poisoning (Fig. 2), yet avocets experienced severe reproductive failure at Rasmus Lee Lake in 1989. Thus, the level of selenium contamination prevalent at Rasmus Lee Lake (Table 1) may be an order of magnitude or more above toxic thresholds for sensitive species. However, among randomly sampled avocet embryos at Kendrick ($N = 28$), the incidence of teratogenesis was statistically consistent with the response curve for Tulare Basin avocets ($\chi^2 = 1.29$, $P > 0.05$; Skorupa and Ramirez, unpublished data) suggesting that Tulare Basin threshold estimates may also apply at Kendrick.

Poor reproductive performance also was documented for eared grebes (Podi-ceps nigricollis) and Canada geese (Branta canadensis). Grebe eggs exhibited selenium concentrations comparable to the avocet eggs, but interpretation of exposure–response data for grebes is complicated because grebes build floating nests. Water often seeps into grebe nests, and salts in the water can pass through the eggshells. Thus, potential embryotoxic effects from the ionic composition of the water (e.g., sulfates) would be difficult to partition from the effects of mater-nally deposited selenium (Skorupa et al., unpublished data). Severe deformities, such as anophthalmia, exhibited by eared grebe embryos at Goose Lake (See et al., 1992b), were very likely attributable to selenium poisoning, but the overall rates of embryonic inviability in eared grebe eggs could have been due to more than just selenium. The data for Canada geese do not present a strong case for selenium poisoning because embryo exposure to selenium was not consistent with the severity of observed effects. Unless Canada geese are extremely sensitive to selenium compared to known response curves (Fig. 2), or are exposed to a different form of selenium in their herbivorous diet (known response curves are from species with predominantly animal diets), it would be unreasonable to conclude that the severe reproductive depression documented for geese at Rasmus Lee Lake was solely the result of selenium poisoning.

The level of selenium in impounded water at Rasmus Lee and Goose Lakes (roughly 35–60 μg/L) was far above threshold levels for wildlife poisoning. Other than that conclusion, the data from Kendrick do not provide much insight on toxic threshold points. The U.S. Bureau of Reclamation and the Casper–Alcova

Irrigation District are planning a remediation strategy for reducing selenium levels in Rasmus Lee and Goose Lakes. Continued monitoring of these sites as remediation actions are implemented and concentrations of selenium are reduced could have great potential for precisely delineating toxic thresholds.

5. Ouray National Wildlife Refuge, Utah: Seepage Wetlands

Ouray National Wildlife Refuge (Ouray NWR: B5 in Fig. 1) lies adjacent to the Green River near the town of Ouray, Utah. The refuge was established in 1960 as part of the mitigation for Flaming Gorge Reservoir and is managed primarily as waterfowl habitat. Agricultural land just northwest of the refuge is irrigated with water delivered from Pelican Lake by the Ouray Park Irrigation Company. The natural flow of groundwater toward the Green River results in subsurface seepage of irrigation drainage into wetlands near the western boundary of Ouray National Wildlife Refuge. The seepage contains elevated concentrations of selenium, probably from a combination of evaporative concentration of the source water (Pelican Lake water contains ≤ 1 μg/L selenium) and leaching of selenium-enriched geologic formations (Stephens et al., 1992).

A pair of hydrologically connected ponds, the North and South Roadside Ponds, are maintained by seepage inflows and also periodically receive surface inflow of irrigation drainage water. The overall average selenium content of seepage and surface inflows to the Roadside Ponds was not determined, but upgradient shallow groundwater contained < 1–830 μg/L selenium. Impounded water in these flow-through ponds averaged about 40 μg/L selenium during 1988–89. This level of contamination was sufficient to cause highly elevated selenium concentrations in food chain fauna, fish tissues, and bird tissues compared to normal reference values (Table 1). The most abundant species of waterbird nesting at the Roadside Ponds was the American coot (Stephens et al., 1992).

A sample of 21 coot eggs collected at random from the Roadside Ponds contained a geometric mean concentration of 50 mg/kg selenium (ca. 25–50 times normal). Embryo teratogenesis was documented for about 10% of the nests that survived to full term, but many inviable eggs were not assessable for embryo condition and therefore the percentage of teratogenic nests is probably underestimated. More than 85% of all full-term eggs failed to hatch, a severe level of reproductive depression comparable to the observations for American coots at Kesterson Reservoir in California. Although fewer than 10 duck nests were monitored, embryo teratogenesis also was documented for two species of ducks in eggs that contained about 20 to 45 mg/kg selenium. This case study provides another example of environmental contamination and avian exposure that is well beyond toxic thresholds. At another Ouray NWR impoundment (Sheppard Bottom-5), however, two eggs of black-necked stilts sampled in 1989 contained 5.3 and 5.4 mg/kg selenium (Peltz and Waddell, 1991), which is just below the embryotoxicity threshold of greater than 6 mg/kg for stilt eggs (see Tulare Basin, Section II.B.2). Water at Sheppard Bottom-5 sampled in 1987 and 1988 contained 2 to 4 μg/L selenium ($N = 4$, mean = 3.25 μg/L; Stephens et al., 1988; 1992), which appears very consistent with the toxic threshold estimate of about 4 μg/L waterborne selenium determined for stilts in the Tulare Basin, California (see above).

6.   Red Rock Ranch, California: Agroforestry Demonstration Site

The California Department of Food and Agriculture and several cooperating agencies devised an integrated system of irrigation, drainage, and salt management that employs a series of increasingly salt-tolerant crops (including trees) and reuse of drainage water to greatly reduce the flow of saline wastewater from irrigated croplands (Cervinka, 1990). This "agroforestry" system was developed primarily to meet the need for managing irrigation salt loads in the San Joaquin Valley. A reduced flow of more highly concentrated saline drainage water facilitates salt recovery and maintenance of a systemwide salt balance. An additional anticipated benefit of the proposed system was the opportunity to replace traditional evaporation basins that create unavoidable selenium hazards for wildlife (see Kesterson Reservoir and Tulare Basin case studies above) with much smaller, plastic-lined solar evaporators that could be managed to prevent wildlife exposure to selenium. With the saline drainage water discharged to the solar evaporator at a rate equivalent to daily evaporative loss, no more than a thin film of water would ever cover the solar evaporator's plastic liner (or salt crust) and no aquatic habitat suitable for waterbirds would be created.

Red Rock Ranch (B6 in Fig. 1), near Five Points, California, includes one section of land (259 ha) set aside as an agroforestry demonstration site. The demonstration site includes 251 ha of traditional cropland, 5 ha of tree plantation, 1.8 ha for halophyte crops (such as Salicornia), and a small (0.86 ha) solar evaporator basin (Westside Resource Conservation District, 1995). Subsurface drainage systems were installed and trees planted in 1994. The solar evaporator basin was completed and halophytes planted in 1995. The full system of traditional irrigation integrated with sequential recovery and blending of drainage water to irrigate increasingly salt-tolerant crops, trees, and halophytes prior to terminal drainage discharge to the solar evaporator was operational in July 1995.

In May 1996 staff from the Central Valley Regional Water Board discovered that nonuniform distribution of drainage water inflow to the solar evaporator was great enough to cause patches of ponding (puddles) sufficient to attract breeding shorebirds. A sample of the water impounded in the solar evaporator contained more than 11,000 μg/L selenium (Table 1). The irrigation furrows of the halophyte plot were also flooded with standing water, and a survey of nesting shorebirds conducted by U.S. Fish and Wildlife Service and Regional Water Board staff in early June 1996 revealed that 12 of 13 nests at the demonstration site were located in the halophyte plot. Water entering the halophyte plot during the spring of 1996 averaged about 1600 μg/L selenium (Westside Resource Conservation District, 1996). Representative sample eggs from 7 stilt nests contained a geometric mean selenium concentration of 58 mg/kg. Assuming that these stilts fed mostly in the halophyte plot, a mean egg selenium of 58 mg/kg is consistent with the water-to-egg regression for Tulare Basin stilt eggs (see point R plotted on Fig. 4). Overall, the status of 30 stilt embryos was determined, of which 17 (56.7%) were teratogenic (Skorupa et al., unpublished data). That is the highest incidence of selenium-induced avian teratogenesis reported by any field study to date, and it is consistent with the predicted rate of 50.4% based on the selenium content of the 7 sample eggs and the response curve for stilts presented in Figure 2. The status of 7 killdeer (Charadrius vociferus) embryos was also determined of which 1 (14%) contained a teratogenic embryo. Three representative sample eggs for killdeer contained a geometric mean selenium concentration of 19 mg/kg, but the teratogenic egg contained 57 mg/kg. Killdeer feed in upland as well as aquatic habitats and typically exhibit lower mean exposure to selenium than stilts when the two species co-occur at the same site (Skorupa et al., unpublished data).

During the summer of 1996, the method of water delivery to the solar evaporator was modified to provide a more even distribution of water, and this measure is expected to eliminate the ponding of water. Additionally, a wildlife monitoring and management program was devised in an attempt to prevent future avian nesting in the halophyte plot or elsewhere at the demonstration site (Westside Resource Conservation District, 1996).

C. Other Case Studies

I. Sweitzer Lake, Colorado: Mining Drainage?/Ambient Seleniferous Geology?/ Irrigation Drainage

Sweitzer Lake (aka Garnet Mesa Reservoir; C1 in Fig. 1) near Delta, Colorado, was built in 1954 for recreational purposes. Sweitzer Lake occurs in an area with naturally seleniferous geological formations, but there is also a great deal of mining activity and irrigation drainage in the regions surrounding Sweitzer Lake (Barnhart, 1957). It is unestablished, however, how much of the cumulative selenium loading into Sweitzer Lake is of natural versus anthropogenic origins. Initial (1950s) water sampling revealed concentrations of selenium exceeding 100 μg/L. Biotic selenium concentrations as high as about 20 mg/kg in benthic food chain fauna, and 40 mg/kg in fish hepatic tissue, were reported (Table 1). This level of exposure was associated with progressive mortality of stocked game fishes (7 species) attributed to excessive dietary intake of selenium (Barnhart, 1957; Lemly, 1985a; Butler et al., 1991).

In 1974 and 1977, the Colorado Division of Wildlife measured more than 100 mg/kg selenium in muscle tissue of fish and decided to stop stocking the lake, although catfish (Ictalurus spp.) were restocked in 1984. In 1978 water with selenium concentrations as high as 25 μg/L was being discharged to Sweitzer Lake via agricultural drainage ditches, and impounded water contained up to 45 μg/L selenium. Samples of impounded water in the late 1980s contained about 10 to 25 μg/L selenium except for a deep-water sample containing 170 μg/L; muscle and eggs of catfish averaged about 30 mg/kg selenium, and a multispecies collection of six waterbird eggs contained 5.6 to 18 mg/kg selenium (Butler et al., 1991).

At that level of exposure, there was no evidence of successful reproduction among catfish. It was not determined, however, whether habitat conditions were suitable for catfish reproduction. It is not unusual for man-made reservoirs to be devoid of suitable spawning substrate for catfish (M. K. Saiki, personal communication). Large populations of green sunfish and carp (Cyprinus carpio) that included various age classes were reported by Butler et al. (1991), but selenium concentrations in eggs from those species were not measured. About 85% of waterbird eggs exceeded the 6 mg/kg toxic threshold point for black-necked stilts; however, stilts were not one of the species whose eggs were sampled at Sweitzer Lake. Insufficient nesting activity occurred at Sweitzer Lake to permit a systematic examination of avian reproductive proficiency. Based on the catfish and avian egg residue data, 10 to 25 μg/L waterborne selenium at Sweitzer Lake is distinctly above toxic threshold points.

The most recent environmental monitoring conducted at Sweitzer Lake occurred in 1995 (R. Krueger and B. Osmundson, U.S. Fish and Wildlife Service, unpublished data). A water sample collected in April contained 24 μg/L selenium. Sediment, aquatic invertebrates, fish, and avian eggs all contained highly elevated concentrations of selenium (Table 1). The eight American avocet eggs collected in 1995 are of particular significance because of the interpretive data available for avocets (Fig. 2). Additionally, a "stilt response standard" can be inferred directly from avocet exposure data because stilt and avocet eggs usually contain similar concentrations of selenium when these species co-occur at a study site (e.g., Ohlendorf and Skorupa, 1989; Skorupa and Ohlendorf, 1991; Ohlendorf and Hothem, 1994). The avocet eggs contained 23 to 32 mg/kg selenium with a geometric mean of 27 mg/kg. For a selenium-tolerant species like the avocet, that level of exposure would not be expected to have teratogenic impacts (Fig. 2). Based on a more sensitive standard, such as a "stilt response standard," that level of exposure would be expected to cause about a 4% incidence of embryo teratogenesis and an overall 21% depression in nesting proficiency (Fig. 2; Skorupa et al., unpublished data). Therefore, at Sweitzer Lake, a clear potential for moderate reproductive impacts among sensitive species of waterbirds is associated with impounded water containing less than 25 μg/L selenium.

2. Swedish Lakes Project, Sweden: Mercury Remediation Treatments

Following initial experimentation at Lake Oltertjarn in 1985–86 (Paulsson and Lundbergh, 1989), 11 additional lakes widely distributed across Sweden (C2 in Fig. 1) were treated with selenite in an attempt to mitigate high levels of mercury in edible fish (Paulsson and Lundbergh, 1991, 1994). Treatments consisted of a leachable rubber matrix containing sodium selenite suspended in a sack 1 to 2 m below the lake surface for 2 years. The selenium-depleted rubber skeletons, remaining after continuous leaching of sodium selenite, were removed at intervals of several months. During the first year of treatments (beginning in September 1987), the doses were calibrated for a target lake concentration of 3 to 5 μg/L selenium (lakes initially contained about 0.1 μg/L selenium). On average, the target concentration was achieved at most of the lakes, although up to 25 to 35 μg/L was measured within 100 m of the leach sacks. Four lakes, however, never exceeded about 2.6 μg/L average waterborne selenium. Because mitigation of mercury residues was equally successful in the four low-selenium lakes just cited and in the target concentration (3–5 μg/L) lakes, the dosing was adjusted in the second year of treatment for a target lake concentration of 1 to 2 μg/L selenium.

Prior to treatment, concentrations of selenium in pike (Esox lucius) muscle tissue averaged 0.7 to 2.4 mg/kg (1.3 mg/kg grand mean) in the 11 lakes. After the first year of treatment, muscle concentrations averaged 0.9 to 2.3 mg/kg selenium (1.6 mg/kg grand mean), and after 2 years of treatment they averaged 2.8 to 7.4 mg/kg (4.6 mg/kg grand mean). There was no evidence of catastrophic declines of pike populations in any of the lakes.

Prior to treatment, concentrations of selenium in perch (Perca fluviatilis) muscle tissue averaged 0.8 to 2.0 mg/kg. After the first year of treatment these tissues averaged 6 to 36 mg/kg selenium. By the end of the second year of treatment, researchers were unable to find any perch in four lakes and had a severely reduced catch from a fifth lake. Among the five lakes with an apparent collapse of perch populations, muscle tissues had averaged 6.9 to 36 mg/kg selenium (23 mg/kg grand mean) at the 1-year sampling point; by comparison, among the other six lakes muscle tissue of perch had averaged only 6 to 18 mg/kg selenium (12 mg/kg grand mean) at the 1-year sampling point, and in some cases had declined by the 2-year sampling point (consistent with lower dosing of lakes in the second year). At the end of 2 years, muscle concentrations of selenium in the six lakes where perch populations persisted averaged 6.9 to 26 mg/kg (15 mg/kg grand mean). The authors could not clearly establish the cause of the collapse of some perch populations, but they concluded that the possibility of selenium poisoning could not be excluded (Paulsson and Lundbergh, 1994). They also concluded that one of their "most important findings" was the need to keep waterborne selenium levels below 2 μg/L to avoid undesirable levels of selenium bioaccumulation in fish and unintentional side effects. Considering the differential persistence of pike (which coincided with substantively lower selenium exposure than perch), the absolute magnitude of the perch tissue data (well into the toxic range for other species of fish), and the stability of perch populations after selenium treatments of lakes were reduced to 2 μg/L or less, the circumstantial case for selenium poisoning of perch in this field study seems very strong.

III. MAJOR CASE STUDY GAPS

A. Mining Associated Case Studies

The mining of sulfide ores is commonly identified as a major source of anthropogenically mobilized selenium (see, e.g., Eisler, 1985). Despite the recognition that deep pit mining is producing numerous "pit lakes" with elevated levels of waterborne selenium (Miller et al., 1996), no case studies of wildlife at pit lakes nor biotic exposure assessments for selenium have yet been reported. A recent risk assessment for wetlands affected by copper smelting and refining on the south shore of the Great Salt Lake, Utah, reported selenium concentrations as high as 69 mg/kg in eggs of black-necked stilts (Fairbrother et al., 1997). In the San Joaquin Valley, stilt eggs containing that much selenium would have about a 75% probability of embryo teratogenesis (Fig. 2). Unfortunately, the few high-exposure stilt eggs in the Great Salt Lake sample were cracked and did not contain embryos that were assessable for terata. Future case studies of wildlife exposure to seleniferous mining drainage should have substantial potential to yield insights regarding toxic threshold points and the generality of currently delineated thresholds.

B. Feedlot and Feedbarn Associated Case Studies

As a result of the supplementation of livestock diets with selenium, it is not uncommon for the liquid manure in pits beneath feedlots or feedbarns to contain highly elevated concentrations of selenium (50–150 μg/L; see, e.g., Oldfield, 1994). Some livestock production schemes route this liquid manure to outdoor holding ponds where the potential for wildlife exposure exists (personal observation). To date, no environmental hazards associated with selenium supplementation of livestock feeds are known, but no systematic examination of wildlife use or selenium exposure at outdoor manure ponds has been reported either. The biochemistry of selenium in liquid manure could be quite unique compared to other sources of environmental selenium (Council for Agricultural Science and Technology, 1994), suggesting that future case studies in this arena could provide important new insights.

IV. LESSONS FROM NATURE

A. Uniformity of Toxic Thresholds

Research at Belews Lake identified a toxicity threshold for fish of 2 to 5 μg/L waterborne selenium (predominantly as selenite). It is now clear that there is nothing unique about the results from Belews Lake. The Martin Reservoir (fish and selenite), Tulare Basin (birds and selenate), and Swedish lakes (fish and selenite) case studies also provide strong documentation of a toxicity threshold at 5 μg/L waterborne selenium or less. The Salton Sea (birds and selenate) and Ouray NWR (Sheppard Bottoms; birds and selenate) case studies provide circumstantial evidence for a threshold of 5 μg/L or less. Thus, all six of the real-world toxic episodes that included biotic exposures near the Belews Lake threshold region uniformly support a criterion of $\leq$5 μg/L for selenium in water. Equally noteworthy, no case studies of selenium exposure and response among biota in nature have *affirmatively* documented discordant estimates of toxicity thresholds.

The most extensive and detailed set of real-world toxic threshold data has been produced by long-term studies of birds exposed to irrigation drainage water in the San Joaquin Valley of California. For a focal species of intermediate sensitivity, the black-necked stilt, the toxic threshold for selenium in eggs (>6 mg/kg) was associated with 3 to 4 μg/L selenium (predominantly as selenate) in impounded water. More importantly, the exposure–response relationships and bioaccumulation curves established for the Tulare Basin have proven reliable when applied to data from the other five irrigation drainage water case studies. Typical of this between-site convergence of data are the nearly identical teratogenesis response curves for stilt eggs from the Kesterson and Tulare Basin case studies (Fig. 5). Although insufficient data have been accumulated elsewhere to provide comparable site-specific response curves, additional examples of the interpretive consistency of data between sites were presented above for every irrigation drainage water case study. Thus, even the irrigation drainage water case studies with levels of biotic exposure and response too severe to yield direct estimates of threshold points (e.g., Kesterson, Kendrick, Ouray NWR—Roadside Ponds) indirectly support the threshold estimates derived from the Tulare Basin research.

B. Selenite Versus Selenate

In nature, selenite-dominated waters appear to have much steeper environmental response curves than selenate-dominated waters. Although response thresholds below 5 μg/L have been identified for both types of water, the biotic consequences of threshold exceedance appear much more severe for selenite-dominated waters. For example, although a toxicity threshold point of about 3 to 4 μg/L waterborne selenium was identified earlier in this chapter for black-necked stilts exposed to selenate-dominated irrigation drainage water in the Tulare Basin, exposures of

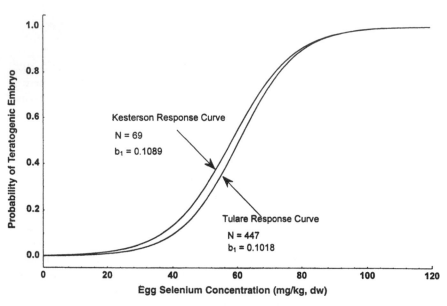

FIGURE 5   Comparison of teratogenesis response curves generated from black-necked stilt data at Kesterson Reservoir versus Tulare Basin evaporation basins. The selenium response logistic regression coefficients, b_1, are not significantly different.

stilts to irrigation drainage containing up to 10 μg/L selenium result in reproductive depression of only a few percentage points (e.g., Salton Sea case study and Tulare Basin regression relationships). It was only because very large data sets (with the statistical power to detect small effects) were accumulated for stilts exposed to irrigation drainage that a precise threshold point below 10 μg/L was ultimately identified.

In clear contrast to the situation for selenate, by the time selenite-dominated waters reach concentrations of 10 μg/L total selenium, fish populations are completely collapsing (e.g., Belews Lake, Hyco Reservoir, Martin Reservoir, and Swedish lakes), and avian populations are exhibiting the severe reproductive impacts more typical of selenate-dominated waters with total selenium exceeding 100 μg/L (e.g., Martin Reservoir, Chevron Marsh). The bioaccumulation study at Chevron Marsh clarified the basis for this trend by showing that bioaccumulation of selenium into the food chain is much more proficient when the source water is selenite-dominated than when it is selenate-dominated. Consequently, selenite-dominated aquatic systems have "supercharged" food chains compared to selenate-dominated systems.

C. Offstream Versus Instream Aquatic Systems

All 12 real-world toxic episodes of selenium exposure presented in this chapter are from off-stream aquatic environments. This pattern in nature probably explains why advocates for strengthening or relaxing regulatory criteria for aquatic selenium generally are polarized along a boundary that separates scientists whose primary experience is with off-stream systems from those whose primary experience is with in-stream systems. This axis of contention persists, at least in part, because it has not been demonstrated to any degree of scientific certainty whether the lack of well-documented in-stream toxic episodes is the result of categorical differences in toxic risk or the result of categorical bias against documenting in-stream risks (due to methodological constraints associated with studies of in-stream biota).

Two primary factors have greatly facilitated documentation of off-stream toxic episodes. First, off-stream case studies of fish have dealt with demographically closed populations. For demographically closed populations, even moderate selenium-induced reproductive deficiency leads to population collapse, an easily detected secondary response. Second, off-stream case studies of demographically open bird populations have relied on sampling eggs (embryo viability). Selenium-impaired embryo viability is a primary response not dependent on demographically closed populations for valid assessment of population response. Equally important, samples of bird eggs provide unbiased measures of biotic response to selenium exposure because the health of the embryo inside the egg does not influence a scientist's probability of sampling the egg, whereas samples of free-living birds or fish are "self-selected" to be insensitive (biased) measures of biotic response because only survivors (live specimens) are sampled. Clearly, only eggs provide unbiased exposure–response data in field studies.

Unfortunately, most in-stream studies of biotic response to selenium exposure are based on samples of free-living fish (biased survivors) from demographically open populations (i.e., populations that are regularly replenished by immigration of individuals from outside the segment of water being studied). Thus, it is common for in-stream studies to report the counterintuitive combination of abnormally elevated levels of selenium in fish tissue associated with what is viewed as a normally abundant and diverse fish fauna. It is then usually concluded that *conventional wisdom* regarding toxic thresholds for selenium, as derived from clinical data and off-stream field data, does not apply to the in-stream case at hand. Such studies meet neither the closed-population nor the unbiased exposure–response sampling criterion of a typical off-stream study and therefore are categorically very low power approaches for detecting toxic episodes.

To further illustrate this point, consider the six irrigation drainwater case studies summarized in this chapter. In all cases, normally abundant and diverse bird faunas, or even superabundant bird populations, were present at the study sites despite selenium concentrations in avian liver or muscle tissues that were abnormally elevated. Without the egg data, demonstrating embryotoxic effects, researchers could easily have concluded in at least five of the six cases that the high levels of selenium in tissues of free-ranging birds were not associated with any evidence of harmful effects (Kesterson being the exception because adult mortality was widespread and led to the sampling of dead as well as live specimens of free-ranging birds). Of course in all five cases we know that the researchers would have been committing type II error (acceptance of a false negative). The crucial flaw of the dominant paradigm for in-stream studies is the interpretation of what are effectively only exposure surveys, because of their low power to detect a biotic response, as exposure–response studies. Such interpretations are highly vulnerable to type II error.

The preceding observations do not imply an absence of scientifically valid categorical reasons to question the applicability of off-stream toxicity results to in-stream risk management. For example, the lower primary productivity categorically associated with in-stream aquatic environments compared to off-stream systems should have major implications for the proportional within-system fluxes of selenium partitioned into biotic versus abiotic compartments. Similarly the categorically higher input of allochthonous energy into in-stream systems as opposed to off-stream systems could cause in-stream biota to be less tightly linked to waterborne levels of selenium (M. K. Saiki, personal communication). If it can be demonstrated that such differences result in categorically different bioaccumulation dynamics in-stream as opposed to off-stream, or that species-specific exposure–response curves for selenium in fish eggs are unequivocally different for off-stream and in-stream data sets, then a strong basis for context-dependent regulatory discrimination would indeed be established.

Since, however, all in-stream water can eventually become off-stream water, the off-stream–in-stream issue is moot. Even if valid site-specific arguments could be made for setting in-stream selenium criteria above off-stream toxic threshold values, from a systems level perspective the cumulative effects of maintaining in-stream water above off-stream toxic thresholds could be severe for fish and wildlife populations. For example, it is very conceivable that the millions of fish and birds that utilize California's Salton Sea would be in serious jeopardy if the cumulative effects of implementing site-specific selenium criteria within the Colorado River and its major tributaries caused even a 1 to 2 μg/L increase downstream in the water diverted for irrigating the Coachella, Imperial, and Mexicali Valleys. That seemingly trivial increase would double the selenium load delivered to a Salton Sea system already exhibiting threshold toxicity. One of the lessons nature has provided at the Salton Sea is that the off-stream ecotoxicology of selenium should be considered as a limiting constraint on in-stream regulatory largesse.

D. Sulfate Dependency

It has been demonstrated clinically that environmentally relevant concentrations of sulfate inhibit the uptake of selenate by algae (e.g., Williams et al., 1994). Sulfate has also been clinically demonstrated to inhibit the bioconcentration of selenate by aquatic invertebrates (Hansen et al., 1993). Consequently, it has been suggested that aquatic ecosystems with relatively low sulfate levels may be particularly susceptible to selenium toxicity (i.e., may have lower toxic threshold points). Further, it has been asserted that EPA's chronic criterion of 5 μg/L for selenium is inapplicable to high-sulfate waters because the criterion was derived from research at Belews Lake, which is a low-sulfate aquatic ecosystem (Mongan and Miller, 1995).

Real-world toxic episodes of fish and wildlife exposure to aquatic selenium, however, have not been restricted to low-sulfate systems. At least 7 of the 12 case studies reviewed for this chapter were high-sulfate systems. Moreover, the toxic threshold point (3–4 μg/L selenium) for a waterbird utilizing high-sulfate irrigation drainage water is consistent with the toxic threshold point (2–5 μg/L selenium) originally delineated by researchers at Belews Lake for fish in a low-sulfate system. The apparent absence of sulfate dependency in real-world episodes of selenium poisoning is consistent with Birkner's (1978) survey of 30 field sites in Colorado and Wyoming, which led him to conclude that levels of dissolved sulfate (5–9611 mg/l) did not influence the level to which selenium is accumulated by aquatic organisms. In the Tulare Basin (Fig. 4), sulfate concentrations from 2,000 to 100,000 mg/L did not significantly confound rates of selenium accumulation in avian eggs (via maternal dietary exposure to aquatic organisms).

Thus, the sulfate issue appears to be a good example of the lab-to-field dilemma (Landis and Yu, 1995). Both Hansen et al. (1993) and Williams et al. (1994) explicitly discussed the possibility that their lab results (based on only 48–96 h exposures in grossly simplified systems) might not be relevant to the field. The lesson from nature is that toxic threshold points for selenium are not sulfate-dependent.

V. CONCLUSION: RISK MANAGEMENT

Conceptually, a *national* water-based criterion for selenium is a safety net set at the lowest height necessary to be reasonably certain that across a broad (national) array of environmental permutations, no potentially injurious falls will occur below the level of the safety net. Consider the safety net analogy further: even though the trapeze artist may work at a height of 35 m above the ground, the safety net is set at 2 m above the ground, not 30 m, because the artist *potentially* could fall at any point along the climb up to 35 m or along the climb down from it, and from 2 m upward that fall could *potentially* be injurious. Risk management must focus on the *constraining potentials for risk*. The fact that in many individual cases trapeze artists routinely climb up to and down from a height of 35 m without falling does not justify raising the net from 2 m to 30 m (even though the greatest risk of falling is from 35 m, the height at which the trapeze artist performs). However, even one injurious fall from 2 m would be sufficient to justify lowering the safety net to below 2 m.

If we consider short-term catastrophic demographic impacts to fish or wildlife as the "fall" we want our selenium criterion "safety net" to protect against (a minimal level of protection), where "catastrophic" might be defined as a population collapse of 50% or more among sensitive species, then the real-world data from nature support the viewpoint that EPA's current chronic criterion for selenium of 5 μg/L is set too high. Short-term catastrophic demographic impacts on birds at Martin Reservoir and on fish in Swedish Lakes were evident at below 5 μg/L. If we want our safety net to be just under known toxic thresholds (a reasonable level of protection), even if those thresholds are associated with modest demographic impacts, then nature's lessons at Belews Lake (second-generation studies, i.e., Lemly, 1993b; 1997), Hyco Reservoir, Chevron Marsh, Tulare Basin, Salton Sea, and Ouray NWR (Sheppard Bottom) also dictate a national water-based criterion for selenium of <5 μg/L.

Managing the potential for risk across a broad (national) array of environmental permutations, where the riskiest permutations are not fully predictable solely from concentrations of selenium in a water column, means that there will be site-specific cases where *potential* risk is unlikely ever to be *realized*. Thus arises the question of how to identify and appropriately manage site-specific risk, including establishment of site-specific criteria. To delve into that topic in detail is beyond the scope of this chapter, but nature's relevant lesson is that avian and fish eggs are the appropriate metric for precise and reliable site-specific risk management. So far, all the real-world data suggest that selenium concentrations in bird and fish eggs incorporate and boil down all the confounding between-site variability in environmental permutations to a universal currency that is very reliable for risk assessment and management.

Consequently, nature's two bottom lines are that it is not unusual for toxic risk to fish and wildlife populations to be associated with less than 5 μg/L selenium in impounded water (thus the current national chronic criterion of 5 μg/L is not an adequate safety net), and that only bird and fish eggs provide a risk metric reliable enough to justify fiddling with the safety net on a site-specific basis (and even then, downstream effects must be given full consideration).

ACKNOWLEDGMENTS

I am indebted to many colleagues whose assistance made large-scale data collection in the Tulare Basin possible, especially D. A. Barnum, T. M. Charmley, S. J. Detwiler, J.

Henderson, W. L. Hohman, R. L. Hothem, P. J. Leonard, C. M. Marn, K. Marois, T. Mauer, G. Montoya, M. Morse, H. M. Ohlendorf, D. U. Palawski, D. L. Roster, S. E. Schwarzbach, R. G. Stein, D. Welsh, and J. Winckel. I also owe a debt of gratitude to a multitude of colleagues in government, academia, and private consulting whose field work was incorporated directly or indirectly into this review. Technical review comments on a draft version of this chapter were provided by H. M. Ohlendorf , M. K. Saiki, and S. E. Schwarzbach. Substantial financial support making this contribution possible was provided by the U.S. Bureau of Reclamation, the California Department of Water Resources, the U.S. Department of Interior's National Irrigation Water Quality Program, and the U.S. Fish and Wildlife Service.

REFERENCES

Afifi, A. A., and V. Clark. 1996. *Computer-Aided Multivariate Analysis*, 3rd ed., p. 286. Chapman & Hall, New York.

Barnhart, R. A. 1957. Chemical factors affecting the survival of game fish in a western Colorado reservoir. M.S. thesis. Colorado State University, Fort Collins.

Barnum, D. A. 1994. Low selenium in waterfowl wintering at Kern National Wildlife Refuge. National Biological Survey, Research Information Bull. No. 25. U.S. Department of Interior, Fort Collins, CO.

Bennett, J. 1997. Biological effects of selenium and other contaminants associated with irrigation drainage in the Salton Sea area, California, 1992–94. Report to the National Irrigation Water Quality Program. U.S. Department of Interior, Washington, DC.

Beyer, W. N. 1990. Evaluating soil contamination. Biol. Report No. 90(2):1–25. U.S. Fish and Wildlife Service, Washington, DC.

Birkner, J. H. 1978. Selenium in aquatic organisms from seleniferous habitats. Ph.D. thesis. Colorado State University, Fort Collins.

Butler, D. L., R. P. Krueger, B. C. Osmundson, A. L. Thompson, and S. K. McCall. 1991. Reconnaissance investigation of water quality, bottom sediment, and biota associated with irrigation drainage in the Gunnison and Uncompahgre River Basins and at Sweitzer Lake, West-Central Colorado, 1988–89. U.S. Geological Survey, Water Resources Investi. Report No. 91-4103. USGS, Denver.

Central Valley Regional Water Quality Control Board. 1993. Waste Discharge Requirements for Tulare Lake Drainage District. Order No. 93-136, CVRWQCB, Fresno, CA.

Cervinka, V. 1990. A farming system for the management of salt and selenium on irrigated land (Agroforestry). Agricultural Resources Branch, California Department of Food and Agriculture, Sacramento.

Chevron USA. 1996. Richmond Refinery Water Enhancement Wetland, 5 year wetland management plan. Amended proposal submitted to the California Regional Water Quality Control Board, San Francisco Bay Region, Oakland.

CH2M HILL. 1994. Bird use and reproduction at the Richmond Refinery Water Enhancement Wetland. Final Report to Chevron USA, Richmond, CA, and to the California Regional Water Quality Control Board, San Francisco Bay Region, Oakland.

CH2M HILL. 1995. Selenium bioaccumulation study at Chevron's Richmond Refinery Water Enhancement Wetland. Final Report to Chevron USA, Richmond, CA, and to the California Regional Water Quality Control Board, San Francisco Bay Region, Oakland.

CH2M HILL, H. T. Harvey and Associates, and G. L. Horner. 1993. Cumulative impacts of agricultural evaporation basins on wildlife. Final Report to the California Department of Water Resources, Sacramento.

Chilcott, J. E., D. W. Westcot, A. L. Toto, and C. A. Enos. 1990a. Water quality in evaporation basins used for the disposal of agricultural subsurface drainage water in the San Joaquin Valley, California, 1988 and 1989. California Regional Water Quality Control Board, Central Valley Region, Sacramento.

Chilcott, J. E., C. A. Enos, D. W. Westcot, and A. L. Toto. 1990b. Sediment quality in evaporation basins used for the disposal of agricultural subsurface drainage water in the San Joaquin Valley, California 1988 and 1989. California Regional Water Quality Control Board, Central Valley Region, Sacramento.

Cleveland, L., E. E. Little, D. R. Buckler, and R. H. Wiedmeyer. 1993. Toxicity and bioaccumulation of waterborne and dietary selenium in juvenile bluegill (Lepomis macrochirus). Aquat. Toxicol. 27:265–280.

Coughlan, D. J., and J. S. Velte. 1989. Dietary toxicity of selenium-contaminated red shiners to striped bass. Trans. Am. Fish. Soc. 118:400–408.

Council for Agricultural Science and Technology. 1994. Risk and benefits of selenium in agriculture. Issue Paper No. 3, CAST, Ames. IA

Coyle, J. J., D. R. Buckler, C. G. Ingersoll, J. F. Fairchild, and T. W. May. 1993. Effect of dietary selenium on the reproductive success of bluegills (Lepomis macrochirus). Environ. Toxicol. Chem. 12:551–565.

Crane, M., T. Flower, D. Holmes, and S. Watson. 1992. The toxicity of selenium in experimental freshwater ponds. Arch. Environ. Contam. Toxicol. 23:440–452.

Cumbie, P. M., and S. L. Van Horn. 1978. Selenium accumulation associated with fish mortality and reproductive failure. Proc. Annu. Conf. Southeast. Assoc. Fish Wildl. Agencies 32:612–624.

Cutter, G. A. 1986. Speciation of selenium and arsenic in natural waters and sediments, Vol. 1: Selenium speciation. Final Report to the Electric Power Research Institute, Palo Alto, CA.

DuBowy, P. 1989. Effects of diet on selenium bioaccumulation in marsh birds. J. Wildl. Manage. 53:776–781.

Eisler, R. 1985. Selenium hazards to fish, wildlife, and invertebrates: A synoptic review. Patuxent Wildlife Research Center Contaminant Hazards Reviews Report No. 5. U.S. Fish and Wildlife Service, Laurel, MD.

Fairbrother, A., R. S. Bennett, L. A. Kapustka, E. J. Dorward-King, and W. J. Adams. 1997. Risk to birds from selenium in the southshore wetlands of Great Salt Lake, Utah, In W. J. Adams (ed.), Understanding Selenium in the Aquatic Environment. Proceedings of Kennecott Salt Lake City Symposium. Kennecott Utah Copper, Magna, UT, March 6–7, 1997.

Garrett, G. P., and C. R. Inman. 1984. Selenium-induced changes in fish populations of a heated reservoir. Proc. Annu. Conf. Southeast. Assoc. Fish Wildl. Agencies 38:291–301.

Gillespie, R. B., and P. C. Baumann. 1986. Effects of high tissue concentrations of selenium on reproduction by bluegills. Trans. Am. Fish. Soc. 115:208–213.

Gober, J. 1994. The relative importance of factors influencing selenium toxicity, In R. A. Marston and V. R. Hasfurther (eds.), *Effects of Human-Induced Changes on Hydrologic Systems. Proceedings of Annual Summer Symposium of the American Water Resources Association, June 26–29, 1994, Jackson Hole, Wyoming*, pp. 1021–1031. American Water Resources Association, Bethesda, MD.

Great Lakes Science Advisory Board. 1981. Report of the Aquatic Ecosystem Objectives Committee. International Joint Commission, Washington, DC, and Ottawa.

Hamilton, S. J., and B. Waddell. 1994. Selenium in eggs and milt of razorback sucker (*Xyrauchen texanus*) in the middle Green River, Utah. *Arch. Environ. Contam. Toxicol.* 27:195–201.

Hamilton, S. J., K. J. Buhl, N. L. Faerber, R. H. Wiedmeyer, and F. A. Bullard. 1990. Toxicity of organic selenium in the diet to Chinook salmon. *Environ. Toxicol. Chem.* 9:347–358.

Hansen, L. D., K. J. Maier, and A. W. Knight. 1993. The effect of sulfate on the bioconcentration of selenate by *Chironomus decorus* and *Daphnia magna. Arch. Environ. Contam. Toxicol.* 25:72–78.

Heinz, G. H. 1996. Selenium in birds. In W. N. Beyer, G. H. Heinz, and A. W. Redmon (eds.), *Interpreting Environmental Contaminants in Animal Tissues*, pp. 453–464. Lewis Publishers, Boca Raton, FL.

Hermanutz, R. O., K. N. Allen, T. H. Roush, and S. F. Hedtke. 1992. Effects of elevated selenium concentrations on bluegills (*Lepomis macrochirus*) in outdoor experimental streams. *Environ. Toxicol. Chem.* 11:217–224.

Hothem, R. L., and H. M. Ohlendorf. 1989. Contaminants in foods of aquatic birds at Kesterson Reservoir, California, 1985. *Arch. Environ. Contam. Toxicol.* 18:773–786.

Irwin, R. J. 1996. Draft selenium profile. *Prototype Contaminants Encyclopedia.* U.S. National Park Service, Fort Collins, CO.

Jenkins, D. W. 1980. Biological monitoring of toxic trace metals, Vol. 2: Toxic trace metals in plants and animals of the world (Parts I, II, and III). Environmental Monitoring Systems Laboratory, EPA-660/3-80-090, EPA-660/3-80-091, and EPA-660/3-80-092. U.S. Environmental Protection Agency, Las Vegas, NV.

King, K. A. 1988. Elevated selenium concentrations are detected in wildlife near a power plant. Research Information Bull. No. 88-31, U.S. Fish and Wildlife Service, Fort Collins, CO.

King, K. A., T. W. Custer, and D. A. Weaver. 1994. Reproductive success of barn swallows nesting near a selenium-contaminated lake in east Texas, USA. *Environ. Pollut.* 84:53–58.

Landis, W. G., and M. Yu. 1995. *Introduction to Environmental Toxicology*, p. 28. Lewis Publishers/CRC Press, Boca Raton, FL.

Lemly, A. D. 1985a. Ecological basis for regulating aquatic emissions from the power industry: The case with selenium. *Regul. Toxicol. Pharmacol.* 5:465–486.

Lemly, A. D. 1985b. Toxicology of selenium in a freshwater reservoir: Implications for environmental hazard evaluation and safety. *Ecotoxicol. Environ. Saf.* 10:314–338.

Lemly, A. D. 1993a. Guidelines for evaluating selenium data from aquatic monitoring and assessment studies. *Environ. Monit. Assess.* 28:83–100.

Lemly, A. D. 1993b. Teratogenic effects of selenium in natural populations of freshwater fish. *Ecotoxicol. Environ. Saf.* 26:181–204.

Lemly, A. D. 1993c. Metabolic stress during winter increases the toxicity of selenium to fish. *Aquat. Toxicol.* 27:133–158.

Lemly, A. D. 1995. A protocol for aquatic hazard assessment of selenium. *Ecotoxicol. Environ. Saf.* 32:280–288.

Lemly, A. D. 1997. Ecosystem recovery following selenium contamination in a freshwater reservoir. *Ecotoxicol. Environ. Saf.* 36:275–281.

Lillebo, H. P., S. Shaner, D. Carlson, N. Richard, and P. DuBowy. 1988. Water quality criteria for selenium and other trace elements for protection of aquatic life and its uses in the San Joaquin Valley. California State Water Resources Control Board Order No. W.Q. 85-1 Technical Committee Report, Appendix D. CSWRCB, Sacramento.

Lorentzen, M., A. Maage, and K. Julshamm. 1994. Effects of dietary selenite or selenomethionine on tissue selenium levels of Atlantic salmon (*Salmo salar*). *Aquaculture* 121:359–367.

Maier, K. J., and A. W. Knight. 1994. Ecotoxicology of selenium in freshwater systems. *Rev. Environ. Contam. Toxicol.* 134:31–48.

Martin, D. B., and W. A. Hartman. 1984. Arsenic, cadmium, lead, mercury, and selenium in sediments of riverine and pothole wetlands of the north central United States. *J. Assoc. Off. Anal. Chem.* 6:1141–1146.

Miller, G. C., W. B. Lyons, and A. Davis. 1996. Understanding the water quality of pit lakes. *Environ. Sci. Technol.* 30:118A–123A.

Mongan, T. R., and W. J. Miller. 1995. Comments on the applicability of EPA's 5 ppb national water quality criterion for selenium to the San Joaquin River. Technical review prepared for Workshop on Beneficial Uses and Water Quality Criteria for Grasslands Water Bodies, June 23, 1995. California Regional Water Quality Control Board, Central Valley Region, Sacramento.

Moore, S. B., S. J. Detwiler, J. Winckel, and M. D. Weegar. 1989. Biological residue data for evaporation ponds in the San Joaquin Valley, California. San Joaquin Valley Drainage Program. U.S. Bureau of Reclamation and California Department of Water Resources, Sacramento.

Moore, S. B., J. Winckel, S. J. Detwiler, S. A. Klasing, P. A. Gaul, N. R. Kanim, B. E. Kesser, A. B. DeBevec, K. Beardsley, and L. K. Puckett. 1990. Fish and wildlife resources and agricultural drainage in the San Joaquin Valley, California (two vols.). San Joaquin Valley Drainage Program. U.S. Bureau of Reclamation and California Department of Water Resources, Sacramento.

Ogle, R. S., and A. W. Knight. 1989. Effects of elevated foodborne selenium on growth and reproduction of the fathead minnow (*Pimephales promelas*). *Arch. Environ. Contam. Toxicol.* 18:795–803.

Ohlendorf, H. M. 1989. Bioaccumulation and effects of selenium in wildlife. In L. W. Jacobs (ed.), *Selenium in Agriculture and the Environment*, pp.133–177. American Society of Agronomy and Soil Science Society of America, Madison, WI.

Ohlendorf, H. M., and R. L. Hothem. 1994. Agricultural drainwater effects on wildlife in central California. In D. J. Hoffman, B. A. Rattner, G. A. Burton, Jr., and J. Carins (eds.), *Handbook of Ecotoxicology*, pp. 577–595. Lewis Publishers, Boca Raton, FL.

Ohlendorf, H. M., and G. M. Santolo. 1994. Kesterson Reservoir—Past, present, and future: an ecological risk assessment. In W. T. Frankenberger, Jr., and S. Benson (eds.), *Selenium in the Environment*, pp. 69–117. Dekker, New York.

Ohlendorf, H. M., and J. P. Skorupa. 1989. Selenium in relation to wildlife and agricultural drainage water. In S. C. Carapella, Jr. (ed.), *Proceedings of the Fourth International Symposium on Uses of Selenium and Tellurium*, pp. 314–338. Selenium–Tellurium Development Association, Darien, CT.

Ohlendorf, H. M., D. J. Hoffman, M. K. Saiki, and T. W. Aldrich. 1986a. Embryonic mortality and abnormalities of aquatic birds: Apparent impacts of selenium from irrigation drainwater. *Sci. Total Environ.* 52:49–63.

Ohlendorf, H. M., R. L. Hothem, C. M. Bunck, T. W. Aldrich, and J. F. Moore. 1986b. Relationships between selenium concentrations and avian reproduction. *Trans. North Am. Wildl. Nat. Resourc. Conf.* 51:330–342.

Ohlendorf, H. M., A. W. Kilness, J. L. Simmons, R. K. Stroud, D. J. Hoffman, and J. F. Moore. 1988. Selenium toxicosis in wild aquatic birds. *J. Toxicol. Environ. Health* 24:67–92.

Ohlendorf, H. M., R. L. Hothem, C. M. Bunck, and K. C. Marois. 1990. Bioaccumulation of selenium in birds at Kesterson Reservoir, California. *Arch. Environ. Contam. Toxicol.* 19:495–507.

Ohlendorf, H. M., J. P. Skorupa, M. K. Saiki, and D. A. Barnum. 1993. Food-chain transfer of trace elements to wildlife. In R. G. Allen and C. M. U. Neale (eds.), *Management of Irrigation and Drainage Systems: Integrated Perspectives.* pp. 596–603. American Society of Civil Engineers, New York.

Oldfield, J. E. 1994. Impacts of agricultural uses of selenium on the environment, *Proceedings of the Fifth International Symposium on Uses of Selenium and Tellurium.* Selenium–Tellurium Development Association, Darien, CT.

O'Toole, D., M. Raisbeck, J. C. Case, and T. D. Whitson. 1996. Selenium-induced "blind staggers" and related myths: A commentary on the extent of historical livestock losses attributed to selenosis on western US rangelands. *Vet. Pathol.* 33:104–116.

Paulsson, K., and K. Lundbergh. 1989. The Se method for treatment of lakes for elevated levels of mercury in fish. *Sci. Total Environ.* 87/88:495–507.

Paulsson, K., and K. Lundbergh. 1991. Treatment of mercury contaminated fish by selenium addition. *Water Air Soil Pollut.* 56:833–841.

Paulsson, K., and K. Lundbergh. 1994. Selenium treatment of mercury-contaminated water systems. *Proceedings of the Fifth International Symposium on Uses of Selenium and Tellurium*, pp. 287–290. Selenium–Tellurium Development Association, Darien, CT.

Peltz, L. A., and B. Waddell. 1991. Physical, chemical, and biological data for detailed study of irrigation drainage in the Middle Green River Basin, Utah, 1988–89, with selected data for 1982–87. U.S. Geological Survey Open File Report No. 91-530. USGS, Salt Lake City, UT.

Peterson, D. A., W. E. Jones, and A. G. Morton. 1988. Reconnaissance investigation of water quality, bottom sediment, and biota associated with irrigation drainage in the Kendrick Reclamation Project area, Wyoming, 1986–87. U.S. Geological Survey, Water Resources Investi. Report No. 87-4255. USGS, Cheyenne, WY.

Peterson, J. A., and A. V. Nebeker. 1992. Estimation of waterborne selenium concentrations that are toxicity thresholds for wildlife. *Arch. Environ. Contam. Toxicol.* 23:154–162.

Presser, T. S. 1995. Selenium perspectives. Memo to Technical Advisory Committee on Re-use of the San Luis Drain, Jan. 27, 1995. U.S. Geological Survey, Menlo Park, CA.

Presser, T. S., and H. M. Ohlendorf. 1987. Biogeochemical cycling of selenium in the San Joaquin Valley, California, USA. *Environ. Manage.* 11:805–821.

Robinson, J. A., L. W. Oring, J. P. Skorupa, and R. Boettcher. 1997. American avocet (*Recurvirostra americana*), In A. Poole and F. Gill (eds.), *Birds of North America*, No. 275. Academy of Natural Sciences, Philadelphia, and American Ornithologists' Union, Washington, DC.

Rosenfeld, I., and O. A. Beath. 1964. *Selenium Geobotany, Biochemistry, Toxicity, and Nutrition.* Academic Press, New York.

Saiki, M. K. 1989. Selenium and other trace elements in fish from the San Joaquin Valley and Suisun Bay, 1985. In A. Q. Howard (ed.), *Selenium and Agricultural Drainage: Implications for San Francisco Bay and the California environment*, pp. 35–49. Bay Institute of San Francisco, Sausalito, CA.

Saiki, M. K. 1990. Elemental concentrations in fishes from the Salton Sea, southeastern California. *Water Air Soil Pollut.* 52:41–56.

Saiki, M. K., and T. P. Lowe. 1987. Selenium in aquatic organisms from subsurface agricultural drainage water, San Joaquin Valley, California. *Arch. Environ. Contam. Toxicol.* 16:657–670.

Saiki, M. K., and R. S. Ogle. 1995. Evidence of impaired reproduction by western mosquito-fish inhabiting seleniferous agricultural drainwater. *Trans. Am. Fish. Soc.* 124:578–587.

Saiki, M. K., M. R. Jennings, and W. G. Brumbaugh. 1993. Boron, molybdenum, and selenium in aquatic food chains from the lower San Joaquin River and its tributaries, California. *Arch. Environ. Contam. Toxicol.* 24:307–319.

Schuler, C. A., R. G. Anthony, and H. M. Ohlendorf. 1990. Selenium in wetlands and waterfowl foods at Kesterson Reservoir, California, 1984. *Arch. Environ. Contam. Toxicol.* 19:845–853.

Schmitt, C. J., and W. J. Brumbaugh. 1990. National contaminant biomonitoring program: Concentrations of arsenic, cadmium, copper, lead, mercury, selenium, and zinc in U.S. freshwater fish, 1976–1984. *Arch. Environ. Contam. Toxicol.* 19:731–747.

See, R. B., D. A. Peterson, and P. Ramirez, Jr. 1992a. Physical, chemical and biological data for detailed study of irrigation drainage in the Kendrick Reclamation Project area, Wyoming, 1988–90. U.S. Geological Survey, Open File Report No. 91-533. USGS, Cheyenne, WY.

See, R. B., D. L. Naftz, D. A. Peterson, J. G. Crock, J. A. Erdman, R. C. Severson, P. Ramirez, Jr., and J. A. Armstrong. 1992b. Detailed study of selenium in soil, representative plants, water, bottom sediment, and biota in the Kendrick Reclamation Project area, Wyoming, 1988–90. U.S. Geological Survey Water Resources Investi. Report No. 91-4131, Cheyenne, WY.

Setmire, J. G., J. C. Wolfe, and R. K. Stroud. 1990. Reconnaissance investigation of water quality, bottom sediment, and biota associated with irrigation drainage in the Salton Sea area, California, 1986–87. U.S. Geological Survey Water Resources Invest. Report No. 89-4102. USGS, Sacramento.

Setmire, J. G., R. A. Schroeder, J. N. Densmore, S. L. Goodbred, D. J. Audet, and W. R. Radke. 1993. Detailed study of water quality, bottom sediment, and biota associated with irrigation drainage in the Salton Sea area, California, 1988–90. U.S. Geological Survey Water Resources Invest. Report No. 93-4014. USGS, Sacramento.

Sibley, C. G., J. E. Ahlquist, and B. L. Monroe, Jr. 1988. A classification of the living birds of the world based on DNA–DNA hybridization studies. *Auk* 105:409–423.

Skorupa, J. P., and H. M. Ohlendorf. 1991. Contaminants in drainage water and avian risk thresholds. In A. Dinar and D. Zilberman (eds.), *The Economics and Management of Water and Drainage in Agriculture*, pp. 345–368. Kluwer Academic Publishers, Boston.

Sorensen, E. M. B. 1986. The effects of selenium on freshwater teleosts. In E. Hodgson (ed.), *Reviews in Environmental Toxicology*, Vol. 2, pp. 59–116. Elsevier Science Publishers, New York.

Sorensen, E. M. B. 1988. Selenium accumulation, reproductive status, and histopathological changes in environmentally exposed redear sunfish. *Arch. Toxicol.* 61:324–329.

Sorensen, E. M. B. 1991. *Metal Poisoning in Fish*. Lewis Publishers/CRC Press, Boca Raton, FL.

Sorensen, E. M. B., P. M. Cumbie, T. L. Bauer, J. S. Bell, and C. W. Harlan. 1984. Histopathological, hematological, condition factor, and organ weight changes associated with selenium accumulation in fish from Belews Lake, North Carolina. *Arch. Environ. Contam. Toxicol.* 13:153–162.

S. R. Hansen and Associates. 1994. The fate of selenium in the bio-oxidation pond at the Tosco Avon refinery. Final Report to California Regional Water Quality Control Board, San Francisco Bay Region, Oakland.

Stephens, D. W., B. Waddell, and J. B. Miller. 1988. Reconnaissance investigation of water quality, bottom sediment, and biota associated with irrigation drainage in the Middle Green River Basin, Utah, 1986–87. U. S. Geological Survey, Water Resources Investi. Report No. 88-4011. USGS, Salt Lake City, UT.

Stephens, D. W., B. Waddell, L. A. Peltz, and J. B. Miller. 1992. Detailed study of selenium and selected elements in water, bottom sediment, and biota associated with irrigation drainage in the Middle Green River Basin, Utah, 1988–90. U.S. Geological Survey, Water Resources Investi. Report No. 92-4084. USGS, Salt Lake City, UT.

Texas Parks and Wildlife Department. 1990. Draft interim report: Selenium in fish tissues from Martin Lake, Texas, 1986–1989. Texas Parks and Wildlife Department, Austin.

UC Committee. 1988. The evaluation of water quality criteria for selenium, boron, and molybdenum in the San Joaquin River basin. University of California Committee on San Joaquin River Water Quality Objectives. Drainage, Salinity, and Toxic Constituent Series, Report No. 4. UC Salinity/Drainage Task Force and UC Water Resources Center, Davis.

U.S. Bureau of Reclamation. 1986. Kesterson Program EIS: Technical appendices. USBR, Mid-Pacific Region, Sacramento.

U.S. Environmental Protection Agency. 1987. Ambient water quality criteria for selenium—1987. USEPA, Office of Water Regulations and Standards, Washington, DC.

U.S. Fish and Wildlife Service. 1990a. Effects of irrigation drainwater contaminants on wildlife. Final report to the San Joaquin Valley Drainage Program. USFWS, Patuxent Wildlife Research Center, Laurel, MD.

U.S. Fish and Wildlife Service. 1990b. Agricultural irrigation drainwater studies. Final report to the San Joaquin Valley Drainage Program. USFWS, National Fisheries Contaminant Research Center, Columbia, MO.

Walsh, D. F., B. L. Berger, and J. R. Bean. 1977. Mercury, arsenic, lead, cadmium, and selenium residues in fish, 1971–73: National pesticide monitoring program. *Pestic. Monit. J.* 11:5–34.

Wells, F. C., G. A. Jackson, and W. J. Rogers. 1988. Reconnaissance investigation of water-quality, bottom sediment, and biota associated with irrigation drainage in the lower Rio Grande Valley and Laguna Atascosa National Wildlife Refuge, Texas, 1986–87. U.S. Geological Survey, Water Resources Investi. Report No. 87-4277. USGS, Austin, TX.

Welsh, D., and O. E. Maughan. 1994. Concentrations of selenium in biota, sediments, and water at Cibola National Wildlife Refuge. *Arch. Environ. Contam. Toxicol.* 26:452–458.

Westcot, D. W., S. Rosenbaum, B. Grewell, and K. K. Belden. 1988. Water and sediment quality in evaporation basins used for the disposal of agricultural subsurface drainage water in the San Joaquin Valley, California. California Regional Water Quality Control Board, Central Valley Region, Sacramento.

Westcot, D. W., C.A. Enos, J. E. Chilcott, and K. K. Belden. 1990. Water and sediment quality survey of selected inland saline lakes. California Regional Water Quality Control Board, Central Valley Region, Sacramento.

Westside Resource Conservation District. 1995. Integrated system for agricultural drainage water management on irrigated farmland. Semiannual Report to U.S. Bureau of Reclamation (August 1995), Red Rock Ranch Agroforestry Grant Number 4-FG-20-11920. Westside Resource Conservation District, Five Points, CA.

Westside Resource Conservation District. 1996. Integrated system for agricultural drainage water management on irrigated farmland. Semiannual Report to U.S. Bureau of Reclamation (September 1996), Red Rock Ranch Agroforestry Grant Number 4-FG-20-11920. Westside Resource Conservation District, Five Points, CA.

White, D. H., J. R. Bean, and J. R. Longcore. 1977. Nationwide residues of mercury, lead, cadmium, arsenic, and selenium in starlings, 1973. *Pestic. Monit. J.* 11:35–39.

White, J. R., P. S. Hofmann, D. Hammond, and S. Baumgartner. 1987. Selenium verification study, 1986. Final Report to California State Water Resources Control Board, Bay–Delta Project and Water Pollution Control Laboratory, California Department of Fish and Game, Sacramento.

Wilber, C. G. 1980. Toxicology of selenium: A review. *Clin. Toxicol.* 17:171–230.

Williams, M. J., R. S. Ogle, A. W. Knight, and R. G. Burau. 1994. Effects of sulfate on selenate uptake and toxicity in the green alga *Selenastrum capricornutum*. *Arch. Environ. Contam. Toxicol.* 27:449–453.

Woock, S. E., and P. B. Summers, Jr. 1984. Selenium monitoring in Hyco Reservoir (NC) waters (1977–1981) and biota (1977–1980). *The Effects of Trace Elements on Aquatic Ecosystems. Proceedings from a Workshop, March 23–24, 1982, Raleigh, North Carolina*, pp. 6.1–6.27. Carolina Power and Light Company, New Hill, NC.

Zahm, G. R. 1986. Kesterson Reservoir and Kesterson National Wildlife Refuge: History, current problems and management alternatives. *Trans. North Am. Wildl. Nat. Resourc. Conf.* 51:324–329.

19

Magic Numbers, Elusive Lesions: Comparative Pathology and Toxicology of Selenosis in Waterfowl and Mammalian Species

Dónal O'Toole and Merl F. Raisbeck
University of Wyoming, Laramie, Wyoming

> I gave careful consideration to the possibility that anatomical dissection might be used to check speculation.
> Andreas Vesalius (1539), *Epistola, docens venam axillarem dextri cubiti in dolore laterali secandem: & melancholicum succum ex venae portae remis dedem pertinaentibus, purgari.* (O'Malley, 1965)

Characterization of disease syndromes caused by excess dietary selenium (selenosis) represents one of the historic research achievements of the U.S. agricultural experiment station system. In the late 1920s and 1930s this success was exemplified by the interdisciplinary work of Kurt Franke and his colleagues at South Dakota State University (Franke et al., 1934; Moxon, 1937; Moxon and Rhian, 1943).

Research into selenosis also had its failures, particularly with regard to disease syndromes incorrectly attributed to the element. For some years, based on experimental studies using rats fed seleniferous grain or inorganic selenium, selenium was considered a cause of hepatic tumors (Nelson et al., 1943). Since selenium is an essential micronutrient in animal rations, this led to continuing regulation of selenium supplements by the FDA under the Delaney clause, and recurrent debates about benefits and risks of such supplementation (Oldfield et al., 1994). Subsequently it was established that original studies were flawed by poor experimental design and the inability of investigators to distinguish age-related changes from hepatic carcinomata (Shamberger, 1985). Excess dietary concentrations of selenate and selenite in semipurified diets did not cause an increased incidence of hepatic tumors in experimental studies using rats (Harr et al., 1967). Selenium is now recognized as a mild anticarcinogen for some tumor classes (Shamberger, 1985; Clark et al., 1996). A second example of the difficulties inherent in linking excess dietary selenium to spontaneous disease involved field studies conducted by chemists between the 1920s and 1950s on seleniferous rangelands in the western United States. Investigators concluded that selenosis in herbivores was widespread (Draize and Beath, 1935; Rosenfeld and Beath, 1946). The myth of vast seleniferous "poison strips" is now a fixture of popular western folklore (Harris, 1985, 1991; McPhee, 1986; Stashak, 1987; Boon, 1989; Anonymous, 1991). In retrospect it is likely that the identified syndromes—"blind staggers," "dishrag heart," and heavy losses in trailed sheep— were attributable to other factors (Raisbeck et al., 1993, 1995; O'Toole et al., 1996). A third instance involves deg-nala disease, a seasonally endemic form of dry gangrene of hooved stock in riverine rice-growing areas of the Indian subcontinent (Shirlaw, 1939). Some workers attribute the syndrome to selenium intoxication (Arora et al., 1975, 1987; Prasad and Arora, 1991). This is a surprising conclusion since dry gangrene, the principal clinical feature of deg-nala, does not occur as a result of experimentally induced selenosis in any species.

Selenosis is now the suspected cause of emaciation, reproductive failure, and teratogenesis in free-ranging populations of waterfowl in some North American wetlands. There is little doubt that avian selenosis was a significant disease factor in the Kesterson Reservoir site in California in the early 1980s (Ohlendorf et al., 1986, 1988) and that it remains a problem in the Kendrick Reclamation District in Wyoming (Ramirez and Armstrong, 1992; Ramirez et al., 1994). The key question is the extent and severity of avian selenosis in wetlands elsewhere in the western United States. The irrigation industry argues the problem is minor, but concedes it is of significant magnitude in "localized" areas (Deason, 1989). Wildlife biologists, hydrologists, and environmentalists remain concerned about unregulated flows of selenium-contaminated agricultural drainage water, and possible widespread impacts on fish and waterfowl (Bobker, 1993; Lemly, 1993). Diagnosis of avian selenosis is currently based on chemical analyses. Quantitative thresholds, combined in some instances with observations of congenital deformities, emaciation, or declines in bird numbers, are presumed to be diagnostic (Skorupa and Ohlendorf, 1991; Peterson and Nebeker, 1992; Maier and Knight, 1994; Presser et al., 1994; Seiler, 1995; Heinz, 1996; Lemly, 1996; Ohlendorf, 1996; Skorupa et al., 1996).

A convention of diagnostic medicine is that "the study of things caused precedes study of the causes of things" (Innes and Saunders, 1962). In the present context, this implies that residues analysis should be corroborated by conventional histopathology. The primary purpose of this chapter is to review the lesions of selenium toxicosis in mammalian and avian species. Particular attention is drawn to those of diagnostic application with which we have some direct experience. As the chapter title suggests, we do not consider that either numbers (concentrations of selenium in tissues) or lesions (morphologic analysis) alone generate an assured diagnosis of selenosis. Interpretation of selenium concentrations is complicated by dietary interactions (Moxon and Rhian, 1943; Levander et al., 1970; Arnold et al., 1973; Palmer et al., 1980; Kezhou et al., 1987), the narrow margin between "safe" and "toxic" concentrations in tissue (Heinz, 1996), interspecies differences in accumulation and excretion rates (Goede and Wolterbeek, 1994), inter- and intraspecies sensitivity to intoxication (M. I. Smith et al., 1937; Miller and Schoening, 1938; Herigstad et al., 1973; Goede, 1985; Goede and DeBruin, 1985; Heinz, 1996; Weimeyer and Hoffman, 1996), seasonal effects (Albers et al., 1996), conditioned aversion (Heinz and Sanderson, 1990), the many chemical forms of selenium and their respective capacity to accumulate in eggs and tissue (Whanger, 1989; Heinz, 1996), and uncertainty about proximate selenium metabolite(s) responsible for cell damage. Morphologic analysis is similarly constrained by, among other factors, its inherently qualitative character, the nonspecific nature of some selenium-induced lesions, and the absence of histological techniques for demonstrating selenium in tissue (O'Toole et al., 1995).

It is perhaps worth remembering that the current reference dose for toxic human exposure may overlap the recommended daily allowance for selenium, based on epidemiological studies in China (Canady and Hodes, 1994). In spite of this, human selenosis remains an extremely rare condition. From a regulatory and diagnostic standpoint, selenosis in mammals and avian species can be difficult to diagnose with confidence, particularly when the number of animals available for analysis is small. In the absence of specific biomarkers for selenosis, combining residue analysis with morphology is the most reliable diagnostic approach at this time.

I. INTEGUMENT

A common theme of subchronic and chronic selenium toxicosis in various species is the involvement of keratin-forming cells (keratinocytes), particularly those that produce hard (sulfur-rich) keratins in hooves, horns, hair, feathers, beak, and digital nails (Tables 1 and 2). Appreciable quantities of selenium accumulate in these structures, and in some instances they approach the high concentrations found in liver and kidney (Dudley, 1936; Arnold et al., 1973; Morris et al., 1983; Goede and Wolterbeek, 1994). Abnormal terminal maturation of keratinocytes in stratum spinosum occurs in the mild or early stages of selenium intoxication (Fig. 1a). More advanced lesions are associated with ballooning degeneration and death of keratinocytes, also predominantly in stratum spinosum. This results in poor quality keratin in the cornified layer and, at some sites, full-thickness

Subchronic Lesions of Selenosis in Mammalian and Avian Species

	Hard keratin-containing tissue							
	Hoof or nail	Horn	Hair or feather follicles	Liver	Heart	Emaciation	Spinal cord	Anemia
	−	−	−	−	+	−	−	−
	nd	nd	nd	±	nd	nd	nd	nd
	−	−	−	+	+	−	−	−
	nd	nd	nd	nd	nd	nd	nd	nd
	±	na	−	−	+	−	−	±
	+	na	+	−	±	+	+	−
	nd	na	−	+	−	+	−	−
	nd	na	−	+	−	+	−	−
	nd	na	−	+	nd	+	−	+
	−	na	−	nd	−	−	−	±
nate	−	na	−	−	−	−	−	−
	−	na	−	+	−	+	−	+

'severe lesion; +, mild; ±, inconsistent data or lesion; −, no lesion; nd, no data; na, not applicable.

...ic Lesions of Selenosis in Mammalian and Avian Species

	Hard keratin-containing tissue							
	Hoof or nail	Horn	Hair or feather follicles	Liver	Heart	Emaciation	Spinal cord	Anemia
	+	+	+	±	-	+	-	±
	+	+	nd	nd	nd	nd	nd	nd
	-	-	-	+	+	+	-	-
	+	+	-	nd	nd	nd	nd	nd
	+	na	+	±	-	+	+	±
	-	na	+	-	-	-	nd	±
	-	na	-	+	-	+	nd	+
	-	na	-	+	+	+	nd	nd
	+	na	+	-	-	-	-	+
mate	+	na	-	-	-	-	-	-
	+	na	+	-	-	+	-	-

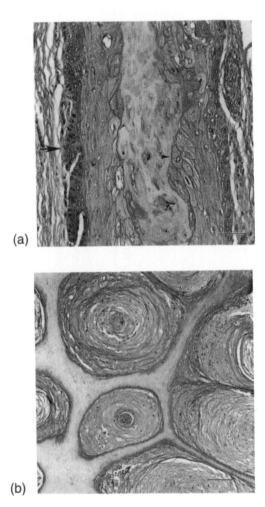

(a)

(b)

FIGURE I Microscopic appearance of hard keratin-forming epithelium in equine seleno-sis illustrating ballooning degeneration in hematoxylin- and eosin-stained (HE) sections. (a) Primary laminae of hoof. The intralaminar *stratum corneum* is irregular. There is ballooning degeneration of cells in *stratum spinosum* (arrowhead). The absence of significant lesions in the germinal layer (arrow) accounts for the ability of hoof epithelium to form healthy new hoof matrix when a seleniferous diet is withdrawn. Bar: 25 μm. (b) Horn tubules in zone of dystrophic hoof growth. Horn tubules in *stratum medium* are abnormally dilated, and intertubular horn matrix is correspondingly reduced. Tubules are filled with laminated accretions of ballooned keratinocytes. Bar: 100 μm.

epithelial necrosis (Figs. 1b, and 2d). Few microscopic studies have been performed on the nail and hoof lesions of selenosis, partly because of the difficulty of preparing satisfactory histological sections from these hard tissues.

A. Avian Species

A memorable image emerging from Kesterson was of American coots with abnormally thin plumage (Ohlendorf, 1996). This lesion is striking since well-defined causes of bilaterally symmetrical alopecia are rare in free-ranging waterfowl (Robinson, 1996). Alopecia of the head and neck was reproduced in mallard ducks on diets containing more than 25 mg/kg selenium fed as selenomethionine for 2 or more months (Albers et al., 1996; Green and Albers, 1997; O'Toole and Raisbeck, 1997) (Fig. 2a). Alopecia was preceded by decreased water repellence and generally disheveled plumage, suggesting that excess dietary selenium alters the microarchitectural structure of feathers that renders them waterproof. The basis for alopecia was not established. A reasonable assumption is that selenium accumulated in the epithelial collar of the follicle, where it selectively damaged keratinocytes, but this remains to be tested. The other distinctive gross lesion in experimentally intoxicated mallards is fracture and loss of digital nails, and necrosis of the maxillary nail (Fig. 2b,c). These lesions have not yet been reported in naturally intoxicated birds. We are not aware of other diseases in waterfowl that result in bilaterally symmetrical alopecia and necrosis of digital and maxillary nails. Fungi, frostbite, chemicals, and plants such as bishop's weed (*Ammi majus*) cause lesions that might be confused with selenosis. Various viral, parasitic, and endocrine causes of alopecia are also recognized in nonwaterfowl species (Peckham, 1972; Riddell, 1991; Lawton, 1993).

B. Herbivores

Chronic selenosis in herbivores ("alkali disease") is a sporadic disease that is most commonly seen in arid seleniferous areas such as Wyoming and South Dakota. Equine cases have occurred recently in selenium-deficient parts of Idaho in pastures contaminated by runoff from mine wastes, and in California pastures that were irrigated using water from seleniferous aquifers (F. D. Galey, personal communication). Alkali disease can also occur in areas with high precipitation rates, such as low-lying, poorly drained areas with alkaline, organic-rich soils in Ireland (Fleming and Walsh, 1957) and dense rain- and eucalypt-forested areas of Queensland in Australia (Knott et al., 1958). There are no well-documented descriptions of selenium-induced lesions of the integument of poisioned sheep (Glenn et al., 1964a, 1964b). Lesions consistent with selenosis were described in goats in a seleniferous area of northeastern India (Gupta et al., 1982). Selenosis was claimed to occur in free-ranging pronghorn (*Antilocapra americana*) in selenif-

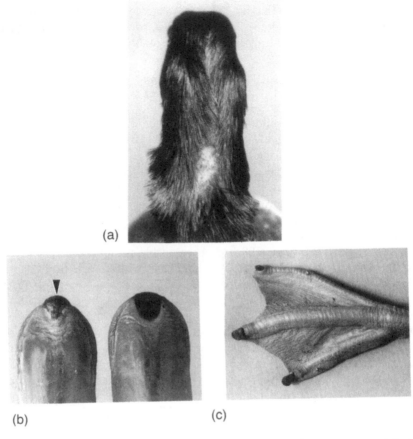

(a)

(b)   (c)

FIGURE 2 Head, beaks, and digits of adult male mallards fed an experimental high-selenium diet containing selenomethionine. (a) Posterior view of head and dorsal midline of neck. Note bilaterally symmetrical feather loss at the nape. The skin itself is unremarkable. (b) Dorsal view of distal maxillary breaks from a control (right) and an affected duck (left). Degeneration of the beak is restricted to the black-pigmented maxillary nail (arrowhead). (c) Foot of a duck with sloughed digital nails. Nail beds are covered by serocellular crusts. (d) Histological appearance of necrotic pigmented maxillary nail from affected duck shown in (b). Note necrosis of epidermis, separation of the abnormally nucleated (parakeratotic) stratum corneum, and minimal inflammatory change in dermis. Bar: 100 μm. HE stain. (Figures 2a and 2c reprinted with permission of *Veterinary Pathology*, O'Toole and Raisbeck, 1997.)

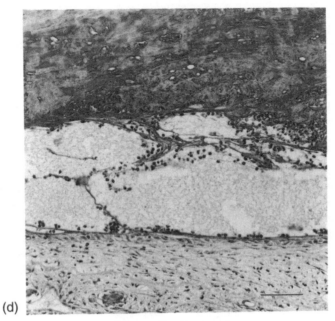

(d)

FIGURE 2   Continued

erous areas of the American West (Post, 1958, 1959). A recent study using seleniferous hay fed to pronghorn for an extended period did not result in clinical signs or lesions of selenosis after 5 months (Raisbeck et al., 1996). This suggests that herbivorous species native to seleniferous areas may tolerate higher dietary concentrations of the element than domesticated species. Additional studies are required to corroborate this.

The most common cause of chronic selenosis is consumption of seleniferous grass or hay containing more than 5 mg/kg selenium (Moxon and Rhian, 1943; Raisbeck et al., 1993; Witte et al., 1993). Herbivores will consume forages such as bluegrass and western wheatgrass that contain less than 50 mg/kg selenium under natural conditions. In Queensland, a secondary selenium accumulating plant containing more than 200 mg/kg selenium was reponsible for intoxication (Knott et al., 1958; Knott and McCray, 1959). Obligate (primary) selenium-accumulating plants contain considerably higher concentrations of selenium (up to 50,000 mg/kg) but are so unpalatable (Marsh, 1924) that they rarely cause disease.

Lesions in cattle, horses, goats, and domesticated buffalo (*Bubalus bubalis*) involve hooves and hair of the mane and tail (tail switch), which is the basis for the colloquial names "change hoof disease" and "bobtail" (Fig. 3a–e). Lesions in hooves may develop within 2 weeks, although 1 to many months of dietary exposure is more common (Knott and McCray, 1959; Olson and Embry, 1973; Twomey et al., 1977; Crinion and O'Connor, 1978; Hultine et al., 1979; Gupta et al., 1982). The shortest reported interval between exposure and lameness was 5 days (following iatrogenic intoxication); hoof cracks developed 3 weeks later (Dewes and Lowe, 1987). Loss of the mane and tail switch may be the earliest (or only) clinical sign, followed by corrugations or cracks in the hoof wall (Moxon,

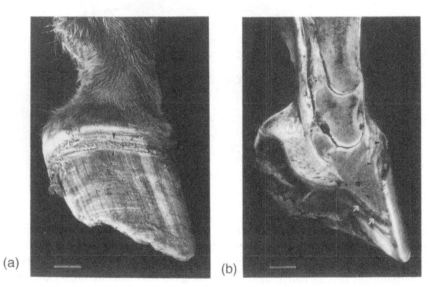

(a)   (b)

FIGURE 3 Hoof from a horse with spontaneous alkali disease. (a) Lateral view of hoof. Note 6 mm wide ridge (between arrowheads) in wall of hoof some 10 mm distal to the periople. The circumferential crack in the wall is covered by a dried serocellular crust of exudate. Bar: 2.5 cm. (b) Midsagittal view of same hoof. A 3 cm crack extends obliquely across the wall of the hoof (arrowheads). Distal to the crack, the hoof wall is darkly discolored by seepage of blood and exudate into horn tubules (1). There is horizontal cleavage in tubular horn of the sole (arrows). Highest concentrations of selenium in the hooves of this horse were in fissured horn from hoof wall (15.8 mg/kg Se) and sole (19.8 mg/kg Se). Old, nonfissured tubular horn had considerably lower concentrations of selenium (1.1 mg/kg). Bar: 2.5 cm. (c) Higher magnification of defect in hoof wall. Note parallel arrangement of horn tubules in old, normal horn (arrows in 1). Horn tubules in poor quality horn (2) between a proximal and a distal crack have a wavy configuration. Newly formed horn (3), produced after the horse was removed from seleniferous forage, is essentially normal. Bar: 1 cm. (d) Horizontal cleavage in tubular horn of sole (between arrowheads). Bar: 1 cm. (e) Sole of hoof demonstrating partial separation between the sole and hoof wall (arrow). Deep cracks in the sole radiate from the frog. Bar: 2.5 cm.

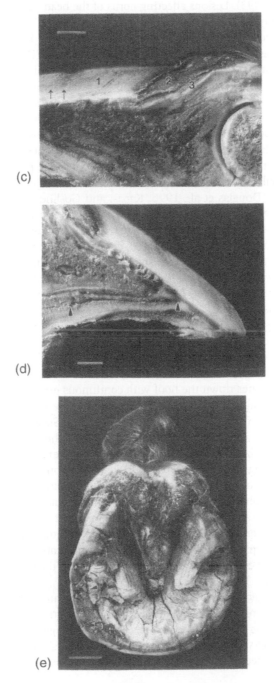

(c)

(d)

(e)

FIGURE 3   Continued

1937, Witte et al., 1993). Lesions affecting horns of the head may occur in cattle and buffalo (Moxon, 1937; Gupta et al., 1982).

The keratinized wall and sole of the hoof in cattle and horses consists of solid horn matrix perforated by regularly spaced horn tubules (Dyce et al., 1996). Keratinocytes are oriented in a steep spiral around the axis of tubules (Bertram and Gosline, 1986). Tubules are continually formed by epidermal papillae. Newly formed hoof migrates at a set rate (about 1 cm/month) by a sliding mechanism and is worn away at bearing surfaces (Budras et al., 1989). A small proportion of the hoof wall (about 20%), called laminar horn, is formed by primary and (in horses) secondary laminae. Hoof material is one of the toughest known natural multidirectional composite materials, and this is due in large part to inter- and intramolecular disulfide cross-links between keratin fibers and other components in the hoof matrix (Douglas et al., 1996; Kapasi and Gosline, 1996).

The distinctive crack that develops in hooves of herbivores with selenosis appears to be due primarily to degeneration and loss of keratinocytes at the tips of papillae. This results in abnormally dilated, misshapen horn tubules filled with cellular debris and old blood (Fig. 1b). There is a corresponding decrease in the volume of the intertubular horn matrix. The normal spiral orientation of peritubular keratinocytes is probably lost. Production of this zone of weak, poor quality horn results in coalescing defects and one or more grossly evident circumferential cracks immediately distal to the periople. The highest concentration of selenium in the hoof is associated with dystrophic hoof matrix in the innermost part of the wall, which is where samples for chemical analysis should be collected from at necropsy (Raisbeck et al., 1997).

The crack migrates down the hoof with continuous growth of the hoof. In severe cases, the papillae of the sole also are affected, resulting in horizontal and vertical cleavages (Fig. 3d,e) and subsolar abscesses (Stashak, 1987). Other microscopic lesions in affected hoofs involve laminar epithelium. These consist of irregularly keratinized laminar plates, caps of parakeratotic or ballooned keratinocytes overlying laminar tips, premature keratinization in primary laminae, and scattered multinucleated keratinocytes. In severe intoxication, animals shed their hooves. In milder cases with a prolonged clinical course, hooves become long and misshapen as animals attempt to reduce discomfort either by not walking or by bearing weight on their heels (Stashak, 1987). Pain and reluctance to walk for food is the principal reason for weight loss in range cattle and horses with chronic selenosis. There is no effective treatment other than supportive care (Traub-Dargatz et al., 1986; Raisbeck et al., 1993).

Any process that damages epidermis of the periople and coronary area can mimic the hoof lesions of selenosis. Intense vasoconstriction due to frostbite, ergotism, or tall fescue grass toxicosis causes necrosis (dry gangrene) of tail tips, digits, and pinnae. Dry gangrene is not a feature of selenosis in our experience. "Frozen feet," attributed to selenosis combined with extremely cold temperatures

(Davidson, 1940; Williams et al., 1941), is usually uncomplicated frostbite. Epitheliotropic infectious agents such as vesicular stomatitis virus may destroy sufficient epithelium of the hoof to cause circumferentially cracked hooves.

Hoof lesions in selenosis are sometimes referred to as laminitis (Fleming and Walsh, 1957; Knott et al., 1958), which literally means inflammation of the laminar part of the hoof. Histological lesions and radiography should serve to distinguish true laminitis from selenium-induced degeneration. Unfortunately, few histological studies have been published on the acute and chronic changes of laminitis (Nilsson, 1963; Kameya et al., 1980; Roberts et al., 1980; Galey et al., 1991; Ekfalck et al., 1992) and the coronary part of the hoof is rarely examined in foundered horses (Obel, 1948). The target tissues in laminitis are presumed to be the primary and secondary epithelial laminae following vascular changes. In selenosis, by contrast, it appears to be the horn-tubule-forming papillae of the coronet and sole that develop the most significant lesions. Cutaneous lesions of *mal do eucalipto*, a disease of cattle and sheep due to poisoning by the mushroom *Ramaria flavo-brunnescens*, are clinically and morphologically indistinguishable from selenosis (Sallis et al., 1993). The mushroom does not contain toxic amounts of selenium (Kommers and Santos, 1995). A legume, *Leucana leucocephala*, causes alopecia and swelling of the coronet band, leading to hoof damage similar to that of alkali disease (Cheeke and Schull, 1985).

Deg-nala (degnala) disease in Pakistan and northwestern India is a seasonally endemic form of dry gangrene that affects buffalo and, to a lesser extent, cattle fed rice straw (Irfan, 1971; Basak et al., 1994). Some authors consider it synonymous with chronic selenosis, since both diseases affect the digits and tail. Clinical signs include cutaneous edema, dry gangrene with sloughing of extremities, and aural, lingual, and ocular lesions. The few histological accounts of the lesions of deg-nala describe an obliterative arteriopathy in affected tissues (Shirlaw, 1939; Irfan, 1971). A cutaneous disease that investigators considered identical to deg-nala was produced experimentally in domestic buffalo using selenium (Arora et al., 1987; Dhillon et al., 1990; Prasad and Arora, 1991). Gupta et al. (1982) suggest that deg-nala is probably not selenium toxicosis, and on the basis of what has been published to date we agree.

C. Other Mammalian Species

Swine with naturally acquired and experimentally induced selenium toxicosis develop lesions of the hair and hooves (Moxon, 1937; Miller and Schoening, 1938; Wahlstrom and Olson, 1959; Harrison et al., 1983; Goehring et al., 1984; Mihailovic et al., 1992). Within 1 month of exposure to selenium, some swine develop bilaterally symmetrical alopecia along the dorsal midline extending from poll to tail. Swelling of the periople leads to lameness and cracked claws, including the dewclaws (Miller and Schoening, 1938). Parakeratosis and abnormal keratin are present in fissured parts of the hoof (Mihailovic et al., 1992). Integumentary lesions are overshadowed in importance by paresis due to poliomyelomalacia, the principal cause of death in intoxicated pigs that do not die acutely (see below).

Lesions of the nails may develop in people and long-tailed macaque monkeys, although histological descriptions are lacking (Yang et al., 1983; Jensen et al., 1984; Yang and Zhou, 1994). In one short-term oral toxicity study, some long-tailed macaques developed widespread xerosis, hyperkeratosis, and dermatitis of the limbs and neck, in addition to mucocutaneous lesions (Cukierski et al., 1989). There are anecdotal historical reports of widespread alopecia and onychodystrophy/onycholysis in people in a seleniferous area of Ireland (Fleming and Walsh, 1957). There are no reports of abnormal nail growth in dogs or rodents in spite the widespread use of these experimental species for selenium toxicity trials.

II. HEPATOPATHY

A. Avian Species

The number of studies of the hepatic lesions in birds is limited. There is a disparity between the severity of the hepatic lesions in free-ranging birds and those in experimentally intoxicated mallards, possibly as a result of the differences in chemical forms of ingested selenium, day-to-day variations in intake, and other dietary interactions. At Kesterson, some birds had chronic hepatic disease with cirrhotic-type change (diffuse nodular hyperplasia, fibrosis, and hepatocellular necrosis), intra-abdominal effusions and fibrinous perihepatitis (Ohlendorf et al., 1986, 1988). Experimentally induced lesions are less severe. Terminally intoxicated birds can have high hepatic selenium concentrations with minimal gross or microscopic hepatic changes (Table 3) (Heinz, 1996; O'Toole and Raisbeck, 1997). Mallards fed more than 80 mg/kg selenium as seleno-L-methionine may die within 1 to 2 weeks of exposure with massive (panacinar) acute multifocal hepatocellular necrosis (Fig. 4a). These lesions may be apparent grossly as irregular areas of hepatic discoloration (Albers et al., 1996). In subchronic toxicosis, hepatic lesions develop after 20 days. Lesions include anisokaryosis, single-cell necrosis in hepatocytes, bile duct hyperplasia, and accumulation of iron pigment in Kupffer cells and periportal macrophages (Fig. 4b,c). There are also associated serum chemical changes (Hoffman et al., 1991). Intrahepatic iron pigment accumulation probably accounts for diffuse yellow discoloration seen in some livers, resulting in their distinct "bronzed" appearance. Iron accumulation may be due to the effects of self-starvation combined with mild hemolytic anemia and intrahepatic uptake of heme breakdown products. Strongly autofluorescent ceroid lipofuscin pigment accumulates in Kupffer cells (Fig. 4d).

We are uncertain whether sudden death of birds on high-selenium diets is due exclusively to hepatic necrosis. Self-starvation occurs in avian and mammalian

...sue Selenium and Lesions in Ducks Fed 60 ppm Selenium as Seleno-L-Methionine

Weight loss (%)	Days on diet	Liver score	Tissue Se (ppm)		Hepatic lesions[b,c]			Bile du...
			Liver	Blood	Necrosis	Iron	Karyomegaly	
35	22	0	23.7	3.2	0	1+	0	
36	22	0	57.4	4.3	0	0	0	
44	39	2+	66.9	4.3	0	2+	1+	
23	50	0	36.0	15.0	0	0	0	
19	50	0	65.0	12.1	0	1+	0	
39	48	3+	30.8	—[d]	2+	3+	2+	
33	50	3+	151.8	15.2	1+	2+	3+	
37	50	3+	66.0	23.9	0	2+	1+	
25	50	0	48.6	18.9	0	1+	0	
25	50	0	37.9	11.5	0	0	0	
45	50	3+	67.7	13.3	2+	2+	1+	
25	50	2+	75.0	18.2	0	2+	0	

nd 7–12 were euthanized due to weight loss, duck 6 died spontaneously.
ion scores: O, no lesion; +, mild lesion; 2+, mild moderate lesion; 3+, moderate lesion.

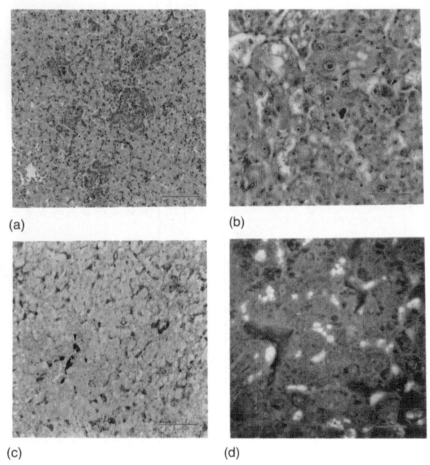

(a)   (b)

(c)   (d)

FIGURE 4 Microscopic lesions in liver of adult mallards with acute (a) and subchronic (b–d) hepatopathy. (a) Severe acute bridging hepatocellular necrosis. (b) Single-cell necrosis (arrowhead) and karyomegaly. There is intracytoplasmic accumulation of dark-staining iron pigment in macrophages. (c) Iron accumulation in Kupffer cells (arrowhead) and periportal macrophages (arrow). This probably accounts for the distinctive gross yellow coloration of livers in some waterfowl with subchronic selenois. (d) Moderate accumulation of autofluorescent ceroid lipofuscin pigment in Kupffer cells and macrophages. Bar: 25 μm. Stains: HE (a, b) and Perls' prussian blue (c); unstained section (d) examined in ultraviolet light.

species offered diets containing excessive amounts of selenium. Conditioned aversion is evident in mallards offered feed containing more than 10 mg/kg (Heinz and Sanderson, 1990). The morphological effects of terminal starvation (weight loss of 50%) (Jordan, 1953; Loesch and Kaminski, 1989) may exacerbate some of the direct cytotoxic effects of selenium in liver and other organs.

B. Herbivores

Few detailed morphological studies of the changes in the liver of selenium-intoxicated herbivores are available. In our limited experience such lesions are minimal, and others report similar negative findings (Knott et al., 1958). Available accounts of severe hepatic changes in range animals with selenosis are unreliable, since prosectors were unfamiliar with histopathology and the effects of autolysis (O'Toole et al., 1996). Mild increases in serum bilirubin were noted in a horse fed a high-selenium native plant (Knott and McCray, 1959). There are several reports of abnormal serum chemistries in horses with selenosis, such as elevated sorbitol dehydrogenase activity, elevated serum bilirubin, and prolonged bromo-sulfophthalein times, all of which indicate hepatic insult (Knott et al., 1958; Twomey et al., 1977; Crinion and O'Connor, 1978). Cattle that died acutely of selenosis were said to have multifocal necrosis, but details were not supplied (Shortridge et al., 1971). Mild hepatic changes were reported in sheep with acute and chronic selenosis (Gardiner, 1966; Glenn et al., 1964a; Smyth et al., 1990a).

C. Other Mammalian Species

Hepatic lesions and ascites were one of the first significant lesions identified in rodents and dogs fed seleniferous grains (Franke, 1934; Munsell et al., 1936; Lillie and Smith, 1940; Moxon and Rhian, 1943). Rats that died early in the course of exposure had congested livers and jaundiced carcasses (Franke, 1934). Surviving rats and dogs had chronic diffuse hepatic lesions suggestive of cirrhosis, either with or without jaundice and ascites. Similar lesions of acute hepatocellular necrosis followed by periportal cirrhosis were induced in rats, rabbits, and cats fed sodium selenite and selenate (Munsell et al., 1936; M. I. Smith et al., 1937). A large-scale study using rats described three lesion patterns in the liver: acute toxic hepatitis with ascites and pleural effusion; chronic toxic hepatopathy with fibroplasia, karyomegaly, and bile duct hyperplasia; and nodular hyperplasia (Harr et al., 1967). Atrophy and/or nodular hyperplasia of the liver is used as an end point in dietary studies involving excessive selenium intake, particularly for defining interactions with other dietary constituents (Levander et al., 1970). Pair-fed control studies indicate that hepatic lesions of selenosis in rats are unrelated to reduced dietary intake (Munsell et al., 1936). Various mild hepatic lesions were described in pigs (Van Vleet et al., 1974; Baker et al., 1989; Mihailovic et al., 1992).

III. ANEMIA

A. Avian Species

Mild anemia, such as a decrease in hematocrit of 4 to 7%, was reported in some studies (Hoffman et al., 1991, 1996). Increased fragility of erythrocytes or "tying up" of iron was considered responsible. In other studies there was no evidence of anemia (Albers et al., 1996).

B. Herbivores

Mild anemia may occur in some cases of spontaneous alkali disease in horses and in cattle (Knott et al., 1958; Twomey et al., 1977). It is more often absent (Witte et al., 1993; Raisbeck et al., 1993).

C. Other Mammalian Species

Anemia is consistently produced in rats fed more than 15 mg/kg dietary selenium, provided they survive the initial 60 days on the diet (Franke, 1934; M. I. Smith et al., 1937). Hemolysis was the basis for anemia in rats fed an artificial diet (Halverson et al., 1970). There is considerable individual variation among rats in the severity of the anemia. The anemia is microcytic/hypochromic and regenerative in character. One early report described "hypoplasia in greater or lesser degree" of myeloerthroid elements in bone marrow, extramedullary hematopoiesis ("myeloid metaplasia"), and reticulocytosis (M. I. Smith et al., 1937). Guinea pigs also developed a microcytic hypochromic anemia, particularly when fed organic selenium (Das et al., 1989). Some cats developed mild anemia in response to chronic selenosis (M. I. Smith et al., 1937). Rabbits fed naturally seleniferous vegetation that was collected from a farm where chronic selenosis was diagnosed in horses exhibited gradual but unspecified decreases in hemoglobin over a 2-month period (Fleming and Walsh, 1957). Swine developed bone marrow depression and, in one episode of iatrogenic selenosis, mild anemia was reported (Wilson et al., 1983). Anemia was not found in other studies of pigs with selenosis (Harrison et al., 1983; Goehring et al., 1984; Baker et al., 1989).

IV. HATCHING SUCCESS, TERATA, AND EMBRYOLETHALITY

A. Avian Species

Poor hatchability of chicken eggs was the one of original complaints that led investigators in South Dakota to identify selenium as a toxicant. Reduced hatchability was due in large part to the development of deformed chicks ("monstrosities") that were unable to leave the egg (Moxon, 1937; Arnold et al., 1973). Chickens and mallard ducks developed a range of teratogenic lesions following maternal exposure to selenium, in addition to reduced survival of early (1–7 day old) embryos and newly hatched chicks. This effect developed within a week of dietary exposure and disappeared within a week of withdrawing the ration (Moxon, 1937; Poley et al., 1937; Heinz, 1993).

Hatching failure, due in part to the development of terata, is a sensitive index of excess dietary selenium exposure in birds. There is some variation in the toxicity of various selenium compounds for fertile eggs, since inorganic forms of selenium such as selenite and selenate tend to be embryotoxic, whereas selenomethionine tends to cause terata (Heinz et al., 1987). The principal lesions involve craniofacial and musculoskeletal deformation, particularly of beak, wings, and legs, as well as anomalous development of brain and eyes (Ohlendorf, 1996). A distinctive feature is hypoplasia of the maxillary or mandibular beak, combined with rotation, deviation from the midline, and spatulate narrowing (Poley et al., 1937; Hoffman et al., 1988; Raisbeck and O'Toole, unpublished) (Fig. 5a,b). "Anophthalmia" is reported frequently, but it is not clear whether the contents of orbits are examined routinely using histology to confirm this diagnosis. Abnormalities of the limbs include fusion of distal long bones (symmelia), aplasia, rotation, hypoplasia, and increased or decreased numbers of digits (Franke and Tully, 1935, 1936; Poley et al., 1937; Hoffman and Heinz, 1988; Hoffman et al., 1988). A common abnormality was fusion and webbing of the two outside digits (digits III and IV) in one study using various forms of selenium injected into air cells on the fourth day of incubation (Palmer, 1973).

Edema of the neck and head combined with abnormal plumage ("greased" or wiry down) was a consistent lesion identified in chickens (Franke and Tully, 1936; Moxon, 1937; Poley et al., 1937; Arnold et al., 1973). Other lesions included abnormal location of the yolk sac (Franke and Tully, 1936) and gastroschisis (Hoffman et al., 1988). "Soft" beaks were noted in the original report of spontaneous avian selenosis (Franke and Tully, 1935) and constituted a major gross abnormality noted in one experimental study (Arnold et al., 1973). Other reproductive effects of selenium, such as reduced egg weight, are probably the result of decreased food consumption by hens (Poley et al., 1937). Exposure of chick embryos later in gestation (14 days in ovo) caused edema of the neck and death, but not teratogenesis (Halverson et al., 1965).

It is sometimes assumed that craniofacial abnormalities and micromelia are unique morphological markers of selenosis. A variety of chemicals, physical factors, infectious agents, and inherited defects cause similar or identical lesions in domestic chickens and turkeys (Landauer, 1967; Riddell, 1975a, 1975b, 1991). The sensitivity of embryos of various avian species to the embryotoxic effects of selenium is as follows: domestic chicken > quail > mallard > black-crowned night heron = screech-owl (G. J. Smith et al., 1988; Weimeyer and Hoffman, 1996). Although the mallard is assumed to be a satisfactory avian model for at-risk free-ranging avian species, we are not aware of controlled experimental studies that directly compare the sensitivity of mallards to avian species that are most at risk (avocet, American coot, eared grebe, black-necked stilt, gadwall, and

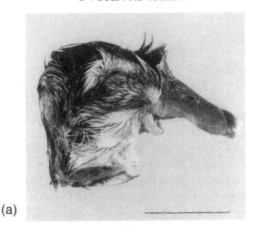

(a)

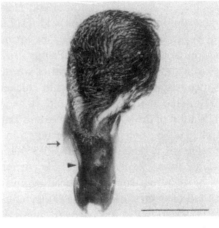

(b)

FIGURE 5 Heads of embryonic mallards; dams had been exposed to 15 mg/kg selenium as seleno-L-methionine. (a) The principal abnormality is hypoplasia of the mandible (brachygnathia). (b) There is spoonbill narrowing (arrowhead) with lateral deviation of the maxillary beak. The mandible in this embryo is also hypoplastic (arrow).

Canada goose). Current studies by the U.S. Department of the Interior (DOI) are guided in part by field studies that indicated the rate of spontaneous malformations in waterfowl is 0.5 to 0.3% (Skorupa et al., 1996). A rate of 0.2% is estimated for American avocets and black-necked stilts (Skorupa, personal communication). These values are low compared to rates of spontaneous abnormalities, including terata, stunting, and generalized edema, in domestic chickens

(5.0–9.5%), captive mallards (0.6% or 4.2%), and mammalian species (Romanoff and Romanoff, 1972; Heinz et al., 1987; 1989; Hoffman and Heinz, 1988; Schardein, 1993).

The only published histological study of the basis for these lesions concluded that they developed as early as 3 to 5 days of gestation (Gruenwald, 1958). Terata were attributed to localized necrosis and hemorrhage in the developing nervous system, particularly optic and otic vesicles, diencephalon, and spinal cord, as well as wing and leg buds, and somites of the tail region. Defects in the beak and face were assumed to be secondary to failure of a normal brain to develop. The gross defects are suggestive of direct effects on chondroid anlangen. The generation of selenide and activated precursors of methylated metabolites may be a key step for all forms of selenium-induced cellular injury, including embryotoxicity (Burk, 1991; Spallholz, 1994). Studies using avian and mammalian in vitro embryos, limb explants, or cultured limb bud mesenchyme exposed to selenium resulted in abnormalities such as inhibition of fibroblasts undergoing chondroid metaplasia (Danielson et al., 1990; Usami and Ohno, 1996). Arrested growth of cartilage in long bones was found when limb explants were exposed in vitro to supraphysiologic concentrations of inorganic but not organic selenium (Rousseaux et al., 1993). The pertinence of such experiments to intoxicated in vivo embryos is open to question.

B. Herbivores

Mammalian species, including herbivores, rarely if ever develop congenital deformations as a result of natural selenium toxicosis (Willhite, 1993; Raisbeck and O'Toole, unpublished). Original accounts of "alkalied" livestock in seleniferous parts of South Dakota in the 1920s and 1930s described infertility, not terata. When alkali disease occurs, conception rates are normal provided the nutritional status of the dam is in other respects adequate (Witte et al., 1993). Negative results were obtained in a recent experimental attempt to induce teratogenic effects in cattle with subclinical selenosis (M. J. Yaeger, personal communication).

There is one account of teratatogensis in herbivores attributed to selenium intoxication (Rosenfeld and Beath, 1947). In this large-flock outbreak, lambs had retinal dysplasia, colobomatous microphthalmia, "dwarfed" extremities and underdeveloped reproductive organs. A high proportion of lambs died at birth. Musculoskeletal defects in two live lambs were illustrated in a later publication but not described in detail (Beath et al., 1953). Tissue concentrations of selenium in affected lambs were elevated, but the similarity of the described lesions to those of hypovitaminosis A (Palludan, 1961; Van der Lugt and Prozesky, 1989) suggests that selenosis was not the only factor involved (Willhite, 1993).

A large-scale experimental study using inorganic selenium fed to pregnant sheep did not result in developmental anomalies in fetal lambs (Glenn et al.,

1964a). The principal effect of selenosis on mammalian embryos is probably mediated via weight loss in the dam. There are several reports of abnormal *in utero* growth of hooves in foals (Moxon and Rhian, 1943; F. D. Galey, *personal communication*). Gupta et al. (1982) stated that failure to conceive and abortion were "frequent" in herbivores with selenosis, but supplied no additional details.

C. Other Mammalian Species

Congenital abnormalities due to selenium occur in laboratory species such as hamsters, rabbits, and long-tailed macaque monkeys at near-lethal dietary concentrations for the dam (Schardein, 1993; Canady and Hodes, 1994). Defects include encephalocele, exencephaly, and rib fusion. Female rats fed a selenized diet either died of liver failure or were infertile. The main effect of selenium toxicity in primates is reproductive wastage due to increased nonviable fetuses and ovarian toxicity (Tarantal et al., 1991). Piglets exposed to excess dietary selenium in the second half of gestation exhibited epithelial degeneration in digits at birth (Mensink et al., 1990).

V. MYOCARDIAL DEGENERATION

A. Avian Species

There are no detailed accounts of myocardial lesions in selenosis in birds. Two recent experimental studies using adult mallards fed up to 80 mg/kg selenium did not find abnormalities in the heart (Albers et al., 1996; O' Toole and Raisbeck, 1997). Cardiac lesions were not reported in carcasses collected at Kesterson Reservoir in California (Ohlendorf, 1996).

B. Herbivores

Multifocal myocardial necrosis occurs due to excessive dietary exposure to selenite and selenate of sheep, cattle and pigs, and possibly horses (Miller and Williams, 1940; Gardiner, 1966; Morrow, 1968; Smyth et al., 1990a, 1990b; Stowe et al., 1992; Taylor and Mullaney, 1984). The most detailed studies of this lesion were in experiments where lambs or pregnant ewes were fed sodium selenate (Glenn et al., 1964a; 1964b; Smyth et al., 1990a, 1990b). A high proportion of ewes that died had acute myocardial necrosis with postnecrotic fibrosis, resulting in multifocal areas of pallor in the left ventricular free wall and interventricular septum (Glenn et al., 1964a). Abnormal electrocardiographic findings suggested ventricular enlargement in naturally intoxicated cattle (Twomey et al., 1977).

A pregnant heifer fed sodium selenate became dull after 81 days and died of heart failure due to multifocal myocardial necrosis (O'Toole and Raisbeck, unpublished) (Fig. 6). Lesions did not involve the right ventricle and so were similar to those of nutritional myodegeneration due to combined selenium–vitamin E deficiency ("white muscle disease") (Kennedy and Rice, 1988, 1992). Following one-time exposure to toxic concentrations of selenium, heart failure due to myocardial injury may account for lingering clinical effects (anorexia, dyspnea, weakness, sudden death) and major findings at necropsy (myocardial pallor, hepatic swelling, pleural, abdominal and pericardial effusions).

In the largest reported episode of iatrogenic selenosis in cattle, 376 of 557 3-month-old calves died within 5 weeks of receiving a single injection of sodium selenite (Shortridge et al., 1971). The number of animals examined postmortem was small, but the combination of histological findings and clinical signs suggests that heart failure was a proximate cause of death. In a recent experimental study of subclinical selenosis of pregnant beef cattle, a calf exposed *in utero* had multifocal myocardial necrosis attributed to selenosis (M. J. Yaeger, personal communication). It is not clear whether the principal chemical form of selenium in vegetation, selenomethionine, can also cause myocardial necrosis.

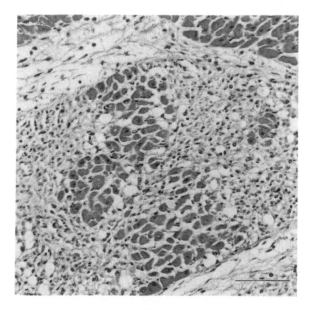

FIGURE 6 Myocardial degeneration and fibrosis in a heifer fed sodium selenite for 83 days. There is replacement fibrosis in areas where cardiocytes were lost. Bar: 100 μm. HE stain.

C. Other Mammalian Species

There are few detailed accounts of myocardial injury in nonherbivorous species. Some pigs with iatrogenic poliomyelomalacia due to excessive selenium supplementation also had mild cardiac and/or skeletal muscle lesions (Orstadius, 1960; Diehl et al., 1975; Hill et al., 1985; Baker et al., 1989; Stowe et al., 1992). Microangiopathy was claimed to be the basis for myonecrosis (Herigstad et al., 1973). Rats fed high-selenium diets had pericardial, pleural, and abdominal effusions, as well as "ischemic infarction," "parenchymal degeneration," replacement fibrosis, and chronic myocarditis–endocarditis (M. I. Smith et al., 1937; Harr et al., 1967).

VI. CENTRAL NERVOUS SYSTEM

A. Birds

Central nervous system lesions have not been reported in birds, other than those intoxicated in ovo.

B. Herbivores

A vexing question is the association of selenosis with a neurological disease in cattle called "blind staggers." This disease was originally described in 1935, with additional details published in 1946 (Draize and Beath, 1935; Rosenfeld and Beath, 1946). "Blind staggers" is a nonspecific, catch-all stockman's term for diseases of cattle that result in a combination of blindness, lassitude, and incoordination (Jensen et al., 1956). A small number of cattle in one selenium study developed polioencephalomalacia, but it was not established whether this lesion was due to selenium or to intercurrent disease (Maag and Glenn, 1967).

The hypothesis that intoxication by unspecified chemical form(s) of selenium manifests itself as an encephalopathy rather than alkali disease was tested recently with negative results (O'Toole and Raisbeck, 1995). Range cattle in Wyoming that develop with what stockmen and veterinarians call "blind staggers" most often have lesions attributable to sulfate-associated polioencephalomalacia, probably due to absorption of excessive intraruminal hydrogen sulfide from sulfate in water or feed (Gould et al., 1997). We have never seen a "blind staggers"–like syndrome or polioencephalomalacia in cattle with experimentally induced or spontaneous selenosis. Selenium-induced "blind staggers" should be assumed to be a misnomer until a neurological syndrome has been reliably documented in cattle as a direct consequence of a seleniferous diet.

C. Other Mammalian Species

Paresis in swine was recognized in the 1930s in intoxicated swine (Miller and Schoening, 1938). The term "blind staggers" is still applied to this syndrome in pigs, which is unfortunate because the term lacks clinical precision and perpetuates the association with selenosis (Diehl et al., 1975). It was not until 1982 that the basis for paresis was identified as poliomyelomalacia (Wilson and Drake, 1982; Harrison et al., 1983; Wilson et al., 1983). This porcine syndrome is the only disease of the central nervous system that has been consistently produced by excess dietary selenium. Clinical signs, chiefly acute-onset flaccid tetraparesis, generally develop within 30 days of exposure. Severe bilaterally symmetrical malacic lesions of ventral gray columns of the spinal cord are confined to cervical and lumbar intumescences. The initial morphological lesion probably involves neuroglia and endothelium, suggesting that selenium is gliotoxic (Summers et al., 1995). In some pigs, lesions also occur in the brainstem, localized to motor nuclei of the fifth and seventh cranial nerves, as well as in cuneate and gracile nuclei, reticular formation, caudal colliculi and thalamus (Panter et al., 1996). A similar porcine poliomyelomalacia is produced by a niacin antagonist (O'Sullivan and Blakemore, 1980). The hypothesis that selenium causes the lesion through niacin antagonism was tested in one elegant study but not conclusively proven (Wilson et al., 1989).

Poliomyelomalacia is seen in small ruminants in Africa and the United States, but an association with selenosis has not been established (Summers et al., 1995). A reported series of cases of poliomyelomalacia in sea lions was attributed to selenium (Edwards et al., 1989). Polioencephalomalacia developed in a young dog injected for 4 months with a selenium-tocopherol preparation (Turk, 1980). No analysis of tissues for selenium was performed, so the possibility that the neurological lesions were due to hypoxia secondary to idiopathic epilepsy rather than selenosis was not excluded. A study using dogs resulted in clinical signs interpreted as blindness and encephalopathy (Moxon and Rhian, 1943), but appropriate neurologic or ophthalmoscopic examination was not attempted and brains were not assessed histologically. Early experimental studies of the effects of selenized grain on rats resulted in posterior paralysis (Franke, 1934). Again, the central nervous system of intoxicated rats has not been examined to establish the basis for this syndrome.

VII. RELATIONSHIP OF CLINICAL EFFECTS TO POSTULATED MECHANISMS OF SELENIUM TOXICITY

Several mechanisms are postulated to explain the manifestations of selenium intoxication. As noted previously, the suggestion that selenium replaces sulfur in keratin or keratin-associated proteins, thereby compromising structural integrity, dates back to the 1930s (Dudley, 1936; Moxon, 1937). This hypothesis has never been rigorously tested, in part because of difficulties in analyzing hard keratins from horn, hair, and nail. It remains a plausible suggestion, given the higher content of sulfur in hard versus soft keratin (5% vs. 1% of dry matter, respectively) (Ekfalck, 1990; Powell and Rogers, 1990). Ganther (1968) demonstrated that selenite reacts with protein thiols and suggested that selenium toxicity is due to denaturation of critical subcellular macromolecules, including enzymes. The niacin antagonism hypothesis has been neither conclusively proved nor disproved (Wilson et al., 1989).

The currently prevailing theory indicts oxidative stress as the pivotal biochemical lesion in selenosis. Spallholz (1994) summarized a series of in vitro experiments and suggested that some chemical forms of selenium produce reactive oxygen species. For example, selenite reacts with reduced glutathione in vitro to produce superoxide anion and elemental selenium (Seko et al., 1989). Acute selenosis in mammals produces several indicators of oxidative injury (Dougherty and Hoekstra, 1982; Csallany et al., 1984; LeBoeuf et al., 1985). Veterinarians have long recognized that vitamin E–deficient animals are more susceptable to acute selenium toxicity (Van Vleet et al., 1974). In birds, chronic dietary selenium exposure increased the ratio of oxidized to reduced hepatic glutathione (Hoffman et al., 1991, 1996) and increased thiobarbituric acid reactive substances (Hoffman et al., 1989, 1996). Ceroid lipofuscin, a lipid peroxidation product, is increased in the livers of some birds with subchronic selenosis (O'Toole and Raisbeck, 1997; Fig. 4d).

The link between oxidant stress and epithelial lesions remains to be demonstrated. We were unable to demonstrate pro-oxidant effects in experimentally intoxicated calves (Raisbeck, unpublished), but the indicator used (circulating vitamin E) is relatively insensitive. Nevertheless, severe decreases in glutathione peroxidase activity were noted in some horses and cattle shortly before or immediately after onset of alkali disease (Raisbeck and Hamar, unpublished). It seems reasonable to assume that selenium-catalyzed oxidant damage will be most severe in cells with weak antioxidant defenses, high rates of selenium or sulfur metabolism, high endogenous selenium concentrations, and relatively poor methylation (detoxification) potential. The localization of alkali disease lesions in stratified squamous epithelium of hair/feather follicles, hoof, and horn is consistent with two and possibly all four of these conditions.

VIII. TOXICOKINETICS OF SELENIUM

The toxicokinetics of selenium are complex and poorly delineated. Uptake, distribution, metabolism, and elimination are influenced not only by the chemical form of the element but also by dose, nutritional, and physiological status of the animal, and interactions with other trace and macro elements. Most selenium toxicokinetic data are derived from studies using diets deficient or adequate in selenium. Such data may not extrapolate well to toxic exposure. Nevertheless some tentative generalities can be drawn.

Selenium from selenite or selenate is distributed more to liver and erythrocytes and is more rapidly eliminated than selenium from selenomethionine (Behne et al., 1990; Willhite et al., 1990; Thompson et al., 1993). Blood and hepatic concentrations are somewhat higher after acute (single dose) exposure to selenate or selenite than selenomethionine (Willhite et al., 1990). The elimination of parenteral selenium from selenite in nutritionally adequate mammals is dose-dependent and exhibits bi- or triphasic elimination, with higher dosages eliminated more slowly (Blodgett and Bevell, 1987; McMurray et al., 1987). This probably reflects the simultaneous operation of both zero- and first-order processes, since elimination kinetics also vary between organ systems. Although we are unaware of comparable kinetic studies in birds, data from subchronic oral exposure studies using selenomethionine in mallards suggest a similar situation (Heinz et al., 1990).

With subchronic or chornic exposures, selenomethionine achieves higher tissue concentrations than selenite or selenate in mammalian and avian species (Heinz et al., 1988; Raisbeck et al., 1995) probably as a result of nonspecific substitution for methionine in protein (Burk, 1991). The biological availability of this increased body burden for either nutrition or intoxication may be less than an equivalent concentration from selenite, since protein incorporation takes precedence over other pathways in methionine-deficient animals (Waschulewski and Sunde, 1988; Burk, 1991).

Selenium from selenomethionine accumulates to higher concentrations and is eliminated more quickly in avian than in mammalian species. For example, selenium in whole blood reached plateau concentrations at 20 days and in less than 41 days in mallard ducks fed 15 and 25 mg/kg selenium as seleno-L-methionine, respectively (Fig. 7). By contrast, selenium in whole blood of steers fed 25 mg/kg selenium as seleno-L-methionine did not reach a plateau until 50 to 60 days (Fig. 7) (Raisbeck et al., 1995; O'Toole and Raisbeck, 1997; Raisbeck, unpublished). Selenium concentrations in eggs from hens fed either selenate or selenomethionine reached plateau concentrations in 2 weeks, and selenium in eggs returned to nonteratogenic concentrations within 1 or 2 weeks of discontinuing treated diets (Poley et al., 1937; Arnold et al., 1973; Heinz, 1993). Heinz et al. (1990) reported 9.8 days as the "half-time" of selenium in whole blood of mallard hens fed increasing concentrations of selenomethionine. In our experience it requires 2 to 3 months for selenium in whole blood of cattle or horses to return to a normal range (i.e., $< 0.3 \mu m/kg$) following intoxication by seleniferous hay or pasture.

Ruminants reportedly excrete more dietary selenium in feces than monogastric species as a result of conversion of selenium to insoluble forms by ruminal microrganisms (Butler and Peterson, 1961). Indeed, we have observed that horses

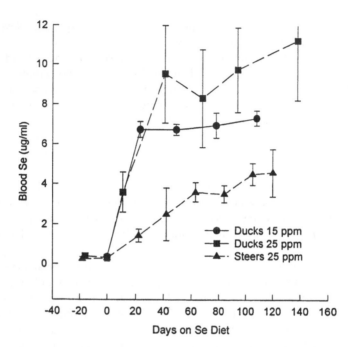

Figure 7 Comparative blood concentrations of adult mallard ducks and yearling steers fed varying concentrations of selenium as seleno-L-methionine for more than 100 days (ducks: 15 and 25 mg/kg Se; steers: 25 mg/kg). Note the considerably higher blood concentrations attained by ducks. Blood selenium concentrations plateau earlier in ducks than in steers.

seem to be more frequently affected by alkali disease than cattle under field conditions. The explanation for this observation may be more complex than the absence of a rumen in monogastric species. For example, elimination of parenteral selenomethionine was promoted by increased dietary organic matter, presumably as a result of greater ruminal activity (Langlands et al., 1986). Limited experience by our group indicates that horses and cattle pastured together on moderately seleniferous forages achieve similar blood concentrations of selenium (Raisbeck, unpublished).

IX. DIAGNOSTIC CRITERIA FOR SELENOSIS IN HERBIVORES AND WATERFOWL

At our laboratory we use the following criteria for subchronic/chronic selenosis (values expressed on an as-received, wet weight basis). We consider a firm diagnosis to be established in cattle and horses that have blood concentrations exceeding 1.0 μg/kg (or hepatic or renal concentrations > 2.0 mg/kg) combined with typical clinical signs or histological lesions of dystrophic growth of the hoof and/or hair loss. Selenium concentrations exceeding 5.0 mg/kg in hair and dystrophic hoof wall are also useful, provided samples are taken from sites that were growing and actively accumulating selenium during toxic exposure. For adult waterfowl, whole-blood concentrations exceeding 10 μg/kg (or hepatic concentrations > 10 mg/kg) are suspicious for selenosis, particularly when one or more of the following conditions is present: emaciation, poor quality or shed nails, bilaterally symmetrical alopecia of the head and neck, toxic hepatic lesions, and necrosis of maxillary nails. We have not found altered mean organ weights of kidney, spleen, pancreas, heart, and thymus to be of diagnostic value, although others suggest their use (Albers et al., 1996). Affected birds may have no lesions other than emaciation, a common nonspecific feature of many diseases in free-ranging waterfowl (Wobeser, 1997). Examination of a representative number of birds (about 10 per species) increases the probability of a conclusive positive or negative diagnosis. Teratogenic lesions consistent with selenosis are deformed or hypoplastic beaks, micromelia, exencephaly, and gastroschisis, combined with selenium concentrations in whole egg that are in excess of the 3 mg/kg threshold suggested by Heinz (1996). In field situations where selenosis is suspected, the "focused" necropsy approach (i.e., just looking for evidence of selenium toxicosis) is likely to result in the overlooking of concurrent diseases. Detailed necropsy examinations are the foundation of accurate diagnosis in waterfowl disease investigation (Wobeser, 1997). Factors capable of causing a selenium-like syndrome should be excluded by appropriate samples and analysis, including microbiology.

X. DEPARTMENT OF INTERIOR STUDIES OF SELENIUM CONTAMINATED WETLANDS: ARE THEY TELLING US WHAT WE NEED TO KNOW?

It is now clear on the basis of the water resources investigations by DOI personnel that selenium contaminates wetlands in several parts of the western United States. Such wetlands suffer from other problems, particularly salinity (>2% salt and/or > 4000 μS/cm and atypical ionic concentrations that can affect the survival of waterfowl (R. J. Hoffman, 1994; Lambing et al., 1994; Leighton and Wobeser, 1994). Tissues from waterfowl and fish at these sites may have concentrations of selenium that exceed the DOI's estimates for "levels of concern" and the "toxicity thresholds" (see Table 12.1 in Ohlendorf, 1996). What is missing in many recent field studies is a correlation between tissue concentrations and disease syndromes and/or lesions that match credible descriptions of selenosis in waterfowl. For example, selenium-induced problems in waterfowl were considered likely in the Stillwater Wildlife Management Area in Nevada (Lemly, 1993; Presser et al., 1994; Seiler, 1995), yet terata consistent with selenosis were not seen in a DOI study conducted between 1987 and 1990 (R. J. Hoffman, 1994). In other areas (Sun River, MT; Grasslands, CA), predictions based on tissue residues derived from the Kesterson model were not borne out in field studies (Palawski et al., 1991; Paveglio et al., 1992; Hothem and Welch, 1994; Lambing et al., 1994; Ohlendorf and Hothem, 1995). In the selenium-contaminated Kendrick site in Wyoming, species such as the Canada goose—which had the highest frequency of dead embryos—also had the lowest mean selenium concentrations in eggs, although still above the 3 mg/kg reproductive impairment threshold (Ramirez and Armstrong, 1992).

This does not mean that thresholds should not be used. Instead they should be interpreted flexibly, as an index of exposure rather than "proof" of disease. The exclusive use of DOI thresholds to address the extent of selenosis in the absence of morphological corroboration reminds us of an old definition of a statistican: someone who draws mathematically straight lines between unwarranted assumptions and foregone conclusions.

The complexity of selenium's effects on biota indicates that we do not need ever more intricate numerical toxicity models or an unending series of large-scale experimental studies to examine the effects of various chemical forms of selenium in multiple avian species. The focus should be on the one real-world experiment that matters: natural wetlands containing high concentrations of total recoverable selenium. An investigative approach that addresses two basic questions should serve us better. First, do lesions in sick, deformed, and dead birds tally with those in experimentally induced studies or well-documented episodes of selenosis? Second, what are the cumulative impacts of selenosis on disease patterns—including infectious diseases—among avian populations at selenium-contaminated wetlands? Such investigations are warranted in advance of costly attempts to remediate environments contaminated by selenium. It remains to be seen whether it is practicable to combine diagnostic precision with the political realities that attend debates about the use of water in remaining wetlands in the western United States.

REFERENCES

Anonymous. 1991. Selenium poisoning. In C. M. Fraser (ed.), *The Merck Veterinary Manual,* pp. 1727–1728. Merck & Co., Rahway, NJ.

Albers, P. H., D. E. Green, and C. J. Sanderson. 1996. Diagnostic criteria for selenium toxicosis in birds. I. Dietary exposure, tissue concentrations, and macroscopic effects. *J. Wildl. Dis.* 32:468–485.

Arnold, R. L., O. E. Olsen, and C. W. Carlson. 1973. Dietary selenium and arsenic additions and their effects on tissue and egg selenium. *Poult. Sci.* 52:847–854.

Arora, S. P., P. Kaur, S. S. Khiwar, R. C. Chopra, and R. S. Ludri. 1975. Selenium levels in fodders and its relationship with Degnala disease. *Indian J. Dairy Sci.* 28:249–253.

Arora, S. P., T. Prasad, and R. C. Chopra. 1987. Selenomethionine in the aetiology of Degnala disease in buffaloes. *Indian J. Anim. Nutr.* 4:48–51.

Baker, D. C., L. F. James, W. J. Hartley, K. E. Panter, H. F. Mayland, and J. Pfister. 1989. Toxicosis in pigs fed selenium-accumulating *Astragalus* plant species or sodium selenite. *Am. J. Vet. Res.* 50:1396–1399.

Basak, D. N., L. N. Banerjee, M. Mitra, P. Roy, and A. Chakrabart. 1994. Incidence of spontaneous Degnala-like disease in buffaloes in and around Nakasipara P. S. of Nadia district, West Bengal. *Indian Vet. J.* 71:1225–1228.

Beath, O. H., C. S. Gilbert, H. F. Eppson, and I. Rosenfeld. 1953. Poisonous plants and livestock poisoning. Univ. Wyoming Agric. Exp. Stn. Bull. 324:1–94.

Behne, D., A. Kryiakopoulos, H. Gessner, S. Schneid, and H. Gessner. 1990. Effects of chemical form and dosage on incorporation of selenium into tissue proteins in rats. *J. Nutr.* 121:806–814.

Bertram, J. E. A., and J. M. Gosline. 1986. Fracture toughness design in horse hoof keratin. *J. Exp. Biol.* 125:29–47.

Blodgett, D. J., and R. F. Bevill. 1987. Pharmacokinetics of selenium administered parenterally at toxic doses in sheep. *Am. J. Vet. Res.* 22:422–428.

Bobker, G. 1993. *Death in the Ponds. Selenium-Induced Waterbird Deaths and Deformities at Agricultural Evaporation Ponds*, pp. 1–52. Bay Institute of San Francisco, Sausalito, CA.

Boon, D. Y. 1989. Potential selenium problems in Great Plains soils. In *Selenium in Agriculture and the Environment.* Soil Science Society of America, Special Publication No. 23:107–121. SSSA, Madison, WI.

Budras, K.-D., R. L. Hullinger, and W. O. Sack. 1989. Light and electron microscopy of keratinization in the laminar epidermis of the equine hoof with reference to laminitis. *Am. J. Vet. Res.* 50:1150–1160.

Burk, R. F. 1991. Molecular biology of selenium and implications for its metabolism. *FASEB J.* 5:2274–2279.

Butler, G. W., and P. J. Peterson. 1961. Aspects of the fecal excretion of selenium in sheep. *N. Z. J. Agric. Res.* 4:484–491.

Canady, R. A., and C. S. Hodes. 1994. Toxicological profile for selenium, pp. 1–243. Draft for public comment, U.S. Department of Health and Human Services, Agency for Toxic Substances and Disease Registry, Atlanta.

Cheeke, P. R., and L. R. Schull. 1985. *Natural Toxicants in Feeds and Poisonous Plants*, pp. 257–260. AVI Publishers, Westport, CT.

Clark, L. C., G. F. Combs, B. W. Turnbull, E. H. Slate, D. K. Chalker, J. Chow, L. S. Davis, R. A. Glover, G. F. Graham, E. G. Gross, A. Krongrad, J. L. Lesher, H. K. Park, B. B. Sanders, C. L. Smith, and J. R. Taylor (1996). Effects of selenium supplementation for cancer prevention in patients with carcinoma of the skin. A randomized controlled trial. *J. Am. Med. Assoc.* 276:1957–1963.

Crinion, R. A. P., and J. P. O'Connor. 1978. Selenium intoxication in horses. *Ir. Vet. J.* 32:81–86.

Csallany, A. S., L. Su, and B. Menken. 1984. Effect of selenite, vitamin E and *N,N'*-diphenyl-*p*-phenylenediamine on liver organic solvent soluble lipofuscin pigments in mice. *J. Nutr.* 114:1482–1587.

Cukierski, M. J., C. C. Willhite, B. L. Lasley, T. A. Hendrie, S. A. Book, D. N. Cox, and A. G. Hendrickx. 1989. 30-Day oral toxicity study of L-selenomethionine in female long-tailed macaques (*Macaca fascicularis*). *Fundam. Appl. Toxicol.* 13:26–39.

Danielson, B. R. G., M. Danielson, A. Khayat, and M. Wide. 1990. Comparative embryotoxicity of selenite and selenate: Uptake in murine embryonal and fetal tissues and effects on blastocysts and embryonic cells *in vitro*. *Toxicology* 63:123–136.

Das, P. M., J. R. Sadana, R. K. Gupta, and R. P. Gupta. 1989. Experimental selenium toxicity in guinea pigs: Haematological studies. *Ann. Nutr. Metab.* 33:347–353.

Davidson, W. B. 1940. Selenium poisoning. *Can. J. Comp. Med.* 4:19–25.

Deason, J. P. 1989. Irrigation-induced contamination: How real a problem? *J. Irrigat. Drain. Eng.* 115:9–20.

Dewes, H. F., and M. D. Lowe. 1987. Suspected selenium poisoning in a horse. *N. Z. Vet. J.* 35:53–54.

Dhillon, K. S., K. S. Dhillon, A. K. Srivastava, B. S. Gill, and Jasmer-Singh. 1990. Experimental chronic selenosis in buffalo calves. *Indian J. Anim. Sci.* 60:532–535.

Diehl, J. S., D. C. Mahan, and A. L. Moxon. 1975. Effects of single intramuscular injections of selenium at various levels to young swine. *J. Anim. Sci.* 40:844–850.

Dougherty, J. J., and W. G. Hoekstra. 1982. Stimulation of lipid peroxidation in vivo by injected selenite and lack of stimulation by selenate. *Proc. Soc. Exp. Biol. Med.* 169:209–215.

Douglas, J. E., C. Mittal, J. J. Thomason, and J. C. Jofriet. 1996. The modulus of elasticity of equine hoff wall: Implications for the mechanical function of the hoof. *J. Exp. Biol.* 199:1829–1836.

Draize, J. H., and O. A. Beath. 1935. Observations on the pathology of blind staggers and alkali disease. *J. Am. Vet. Med. Assoc.* 39:753–763.

Dudley, H. C. 1936. Toxicology of selenium: I. A study of the distribution of the distribution of selenium in acute and chronic cases of selenium poisoning. *Am. J. Hyg.* 23:169–180.

Dyce, K. M., W. O. Sack, and C. J. G. Wensing. 1996. *Textbook of Veterinary Anatomy*, pp. 358–360, 595–600. Saunders, Philadelphia.

Edwards, W. C., D. L. Whitenack, J. W. Alexander, and M. A. Solangi. 1989. Selenium toxicosis in three California sea lions. (*Zaophus californianus*). *Vet. Hum. Toxicol.* 31:568–570.

Ekfalck, A. 1990. Amino acids in different layers of the matrix of the normal equine hoof. Possible importance of the amino acid pattern for research on laminitis. *J. Vet. Med. Ser. B* 37:1–8.

Ekfalck, A., H. Rodriguez, and N. Obel. 1992. Histopathology in post-surgical laminitis with a peracute course in a horse. *Equine Vet. J.* 24:321–324.

Fleming, G. A., and T. Walsh. 1957. Selenium occurrence in certain Irish soils and its toxic effects on animals. *Proc. R. Ir. Acad. Sect. B. Biol. Geol. Chem. Sci.*, pp. 151–166.

Franke, K. W. 1934. A new toxicant occurring naturally in certain samples of foodstuffs: I. Results obtained in preliminary feeding trials. *J. Nutr.* 8:597–607.

Franke, K. W., and W. C. Tully, 1935. A new toxicant occurring naturally in certain samples of plant foodstuffs: V. Low hatchability due to deformities in chicks. *Poult. Sci.* 14:273–279.

Franke, K. W., and W. C. Tully. 1936. A new toxicant occurring naturally in certain samples of plant foodstuffs: VII. Low hatchability due to deformities in chicks produced from eggs obtained from chickens of known history. *Poult. Sci.* 15:316–318.

Franke, K. W., T. D. Rice, A. G. Johnson, and H. W. Schoening. 1934. Report on a preliminary field survey of the so-called "alkali disease" of livestock. U.S. Department of Agriculture Circ. no. 320:1–10.

Galey, F. D., H. E. Whiteley, T. E. Goetz, A. R. Kuenstler, C. A. Davis, and V. R. Beasley. 1991. Black walnut (*Jugulans nigra*) toxicosis: A model for equine laminitis. *J. Comp. Pathol.* 104:313–326.

Ganther, H. E. 1968. Selenotrisulfides. Formation by reaction by thiols with selenous acid. *Biochemistry* 7:2898–2905.

Gardiner, M. R. 1966. Chronic selenium toxicity studies in sheep. *Aust. Vet. J.* 42:442–448.

Glenn, M. W., R. Jensen, and L. A. Griner, 1964a. Sodium selenate toxicosis: The effects of extended oral administration of sodium selenate on mortality, clinical signs, fertility, and early embryonic development in sheep. *Am. J. Vet. Res.* 25:1479–1485.

Glenn, M. W., R. Jensen, and L. A. Griner. 1964b. Sodium selenate toxicosis: Pathology and pathogenesis of sodium selenate toxicosis in sheep. *Am. J. Vet. Res.* 25:1486–1494.

Goede, A. A. 1985. Mercury, selenium, arsenic and zinc from waders from the Dutch Wadden Sea. *Environ. Pollut. Ser. A Ecol. Biol.* 37:287–309.

Goede, A. A., and M. DeBruin. 1985. Selenium in a shore bird, the dunlin, from the Dutch Waddenzee. *Mar. Pollut. Bull.* 16:115–117.

Goede, A. A., and H. T. Wolterbeek. 1994. Have high selenium concentrations in wading birds their origin in mercury? *Sci. Total Environ.* 144:247–253.

Goehring, T. B., I. S. Palmer, O. E. Olson, G. W. Libal, and R. C. Wahlstrom. 1984. Toxic effects of selenium on growing swine fed corn–soybean meal diets. *J. Anim. Sci.* 59:733–737.

Gould, D. H., B. A. Cummings, and D. W. Hamar. 1997. In vivo indicators of pathologic ruminal sulfide production in steers with diet-induced polioencephalomalacia. *J. Vet. Diagn. Invest.* 9:72–76.

Green, D. E., and P. H. Albers. 1997. Diagnostic criteria for selenium toxicosis in aquatic birds: Histologic lesions. *Wildl. Dis.* 33:385–404.

Gruenwald, P. 1958. Malformations caused by necrosis in the embryo illustrated by the effect of selenium compounds on chick embryos. *Am. J. Pathol.* 34:77–103.

Gupta, R. C., M. S. Kwatra, and N. Singh. 1982. Chronic selenium toxicity as a cause of hoof and horn deformities in buffalo, cattle and goat. *Indian Vet. J.* 59:738–740.

Halverson, A. W., L. G. Jerde, and C. L. Hills. 1965. Toxicity of inorganic selenium salts to chick embryos. *Toxicol. Appl. Pharmacol.* 7:675–697.

Halverson, A. W., D. Ding-Tsay, K. C. Treibwasser, and E. I. Whitehead. 1970. Development of hemolytic anemia in rats fed selenite. *Toxicol. Appl. Pharmacol.* 17:151–159.

Harr, J. R., J. F. Bone, I. J. Tinsley, P. H. Weswig, and R. S. Yamamoto. 1967. Selenium toxicity in rats: II. Histopathology. In O. H. Muth (ed.), *Selenium in Biomedicine*, pp. 153–178. AVI Publishing, Westport CT.

Harris, T. 1985. Selenium. Toxic trace element threatens the West; *The Bee* uncovers conspiracy of silence. *Sacramento Bee*, Sept 8–10, 1985, pp. 1, 16.

Harris, T. 1991. *Death in the Marsh*, pp. 125–147. Island Press, Washington, DC.

Harrison, L. H., B. M. Colvin, B. P. Stuart, L. T. Sangster, E. J. Gorgacz, and H. S. Gosser. 1983. Paralysis in swine due to focal symmetrical poliomalacia: Possible selenium toxicosis. *Vet. Pathol.* 20:265–273.

Heinz, G. H. 1993. Selenium accumulation and loss in mallard eggs. *Environ. Toxicol. Chem.* 12:775–778.

Heinz, G. H. 1996. Selenium in birds. In W. N. Beyer, G. H. Heinz, and A. W. Redmon-Norwood (eds.), *Environmental Contaminants in Wildife. Interpreting Tissue Concentrations*, pp. 447–458. CRC/Lewis Publishers, Boca Raton, FL.

Heinz, G. H., and C. J. Sanderson. 1990. Avoidance of selenium-treated food by mallards. *Environ. Toxicol. Chem.* 9:1155–1158.

Heinz, G. H., D. J. Hoffman, A. J. Krynitsky, and D. M. Weller. 1987. Reproduction in mallards fed selenium. *Environ. Toxicol. Chem.* 6:423–433.

Heinz, G. H., D. J. Hoffman, and L. G. Gold. 1989. Impaired reproduction of mallards fed an organic form of selenium. *J. Wildl. Manage.* 53:418–428.

Heinz, G. H., G. W. Pendleton, A. J. Kreynitsky, and L. G. Gold. 1990. Selenium accumulation and elimination in mallards. *Arch. Environ. Contam. Toxicol.* 19:374–379.

Herigstad, R. R., C. K. Whitehair, and O. E. Olson. 1973. Inorganic and organic selenium toxicosis in young swine. Comparison of pathologic changes with those of swine with vitamin E–selenium deficiency. *Am. J. Vet. Res.* 34:1227–1238.

Hill, J., F. Alliston, and C. Halpin. 1985. An episode of acute selenium toxicity in a commercial piggery. *Aust. Vet. J.* 62:207–209.

Hoffman, D. J., and G. H. Heinz. 1988. Embryotoxic and teratogenic effects of selenium in the diet of mallards. *J. Toxicol. Environ. Health* 32:449–490.

Hoffman, D. J., H. M. Ohlendorf, and T. W. Aldrich. 1988. Selenium teratogenesis in natural populations of aquatic birds in central California. *Arch. Environ. Contam. Toxicol.* 17:519–525.

Hoffman, D. J., G. H. Heinz, and A. J. Krynitsky. 1989. Hepatic glutathione metabolism and lipid peroxidation in response to excess dietary selenomethionine and selenite in ducklings. *J. Toxicol. Environ. Health* 27:263–271.

Hoffman, D. J., G. H. Heinz, L. J. LeCaptain, C. M. Bunck, and D. E. Green. 1991. Subchronic hepatotoxicity of selenomethionine in mallard ducks. *J. Toxicol. Environ. Health* 32:449–464.

Hoffman, D. J., G. H. Heinz, L. J. LeCaptain, J. D. Eisemann, and G. W. Pendleton. 1996. Toxicity and oxidative stress of different forms of organic selenium and dietary protein in mallard ducklings. *Arch. Environ. Contam. Toxicol.* 31:120–127.

Hoffman, R. J. 1994. Detailed study of irrigation drainage in and near wildlife management areas, west-central Nevada, 1987–90: Part C. Summary of irrigation–drainage effects on water quality, bottom sediment and biota, pp. 1–32. U.S. Geological Survey Water Resources Invest. Report No. 92-4024C.

Hothem, R. L., and D. Welch. 1994. Contaminants in eggs of aquatic birds from the Grasslands of central California. *Arch. Environ. Contam. Toxicol.* 27:180–185.

Hultine, J. D., M. E. Mount, K. J. Easley, and F. W. Oehme. 1979. Selenium toxicosis in the horse. *Equine Pract.* 1:57–60.

Innes, J. R. M., and L. Z. Saunders. 1962. Preface, p. xi. *Comparative Neuropathology*. Academic Press, New York.

Irfan, M. 1971. The clinical picture and pathology of "Deg Nala disease" in buffaloes and cattle in West Pakistan. *Vet. Rec.* 88:422–424.

Jensen, R., L. A. Griner, and O. R. Adams. 1956. Polioencephalomalacia of cattle and sheep. *J. Am. Vet. Med. Assoc.* 129:311–321.

Jensen, R., W. Closson, and R. Rothenberg. 1984. Selenium intoxication. *New York Morbid. Mortal. Wkly. Report* 33:157–158.

Jordan, J. S. 1953. Effects of starvation on wild mallards. *J. Wildl. Manage.* 17:304–305.

Kameya, T., K. Kiryu, and M. Kaneko. 1980. Histopathogenesis of thickening of the hoof wall laminae in equine laminitis. *Jpn. J. Vet. Sci.* 42:361–371.

Kapasi, M. A., and J. M. Gosline. 1996. Strain-rate-dependent mechanical properties of the equine hoof wall. *J. Exp. Biol.* 199:1133–1146.

Kennedy, S., and D. A. Rice. 1988. Selective morphologic alterations of the cardiac conduction system in calves deficient in vitamin E and selenium. *Am. J. Pathol.* 130:315–325.

Kennedy, S., and D. A. Rice. 1992. Histopathologic and ultrastructural myocardial alterations in calves deficient in vitamin E and selenium and fed polyunsaturated fatty acids. *Vet. Pathol.* 29:129–138.

Kezhou, W., H. D. Stowe, A. M. House, K. Chou, and T. Thiel. 1987. Comparison of cupric and sulfate ion effects on chronic selenosis in rats. *J. Anim. Sci.* 64:1467–1475.

Knott, S. G., and C. W. R. McCray. 1959. Two naturally occurring outbreaks of selenosis in Queensland. *Aust. Vet. J.* 35:161–165.

Knott, S. G., C. W. R. McCray, and W. T. K. Hall. 1958. Selenium poisoning in horses in north Queensland. Queensland Department of Agriculture and Stock, Div. An. Ind. Bull. 41:1–16.

Kommers, G. D., and M. N. Santos. 1995. Experimental poisoning of cattle by the mushroom *Ramaria flavobrunnescens* (Clavariaceae): A study of the morphology and pathogenesis of lesions in hooves, tail, horns, and tongue. *Vet. Hum. Toxicol.* 37:297–302.

Lambing, J. H., D. A. Nimick, J. R. Knapton, and D. U. Palawski. 1994. Physical, chemical, and biological data for detailed study of the Sun River Irrigation Project, Freezout Lake Wildlife Management Area and Benton Lake National Wildlife Refuge, west-central Montana, 1990–92, with selected data for 1987–89, pp. 1–171. U.S. Geological Survey Water Resources Invest. Report No. 94-120: USGS, Denver.

Landauer, W. 1967. The hatchability of chicken eggs as influenced by environment and heredity, pp. 92–101, 115–117, 143–177. Storrs Agric. Exp. Stn. Monogr. 1 (rev.), Storrs, CT.

Langlands, J. P., J. E. Bowles, G. E. Donald, and A. J. Smith. 1986. Selenium excretion in sheep. *Aust. J. Agric. Res.* 37:201–209.

Lawton, M. P. C. 1993. Feather loss in birds. In P. H. Locke, R. G. Harvey, and I. S. Mason (eds.), *Manual of Small Animal Dermatology*, pp. 171–176. British Small Animal Veterinary Association, Cheltenham, Gloucestershire, U.K.

LeBoeuf, R. A., K. L. Zentner, and W. G. Hoekstra. 1985. Effect of dietary selenium concentration and duration of selenium feeding on hepatic glutathione and concentrations in rats. *Proc. Soc. Exp. Biol. Med.* 180:348–352.

Leighton, F. A., and G. Wobeser. 1994. Salinity and selenium content in western Canadian wetlands. *Wildl. Soc. Bull.* 22:111–116.

Lemly, A. D. 1993. Subsurface agricultural irrigation drainage: The need for regulation. *Regul. Toxicol. Pharmacol.* 17:157–180.

Lemly, A. D. 1996. Assessing the toxic threat of selenium to fish and aquatic birds. *Environ. Monit. Assess.* 43:19–35.

Levander, O. A., M. L. Young, and S. A. Meeks. 1970. Studies on the binding of selenium by liver homogenates from rats fed diets containing either casein or casein plus linseed oil meal. *Toxicol. Appl. Pharmacol.* 16:79–87.

Lillie, R. D., and M. I. Smith. 1940. Histogenesis of hepatic cirrhosis in chronic food selenosis. *Am. J. Pathol.* 16:223–228.

Loesch, C. R., and R. M. Kaminski. 1989. Winter body-weight patterns of female mallards fed agricultural seeds. *J. Wildl. Manage.* 53:1081–1087.

Maag, D. D., and M. W. Glenn. 1967. Toxicity of selenium: Farm animals. In O. H. Muth (ed.), *Selenium in Biomedicine,* pp. 127–140. AVI Publishing, Westport, CT.

Maier, K. J., and A. W. Knight. 1994. Ecotoxicology of selenium in freshwater systems. *Rev. Environ. Contam. Toxicol.* 134:31–48.

Marsh, C. D. 1924. Stock-poisoning plants of the range. U.S. Department of Agriculture, Bull. No. 1245, pp. 1–36.

McMurray, C. H., W. B. Davidson, and W. J. Branchflower. 1987. The distribution of selenium in the tissues of lambs following intramuscular administration of different levels of sodium selenite. *Br. Vet. J.* 143:51–58.

McPhee, J. 1986. *Rising from the Plains,* pp. 7–9. Noonday Press, New York.

Mensink, C. G., J. P. Koeman, J. Veling, and E. Gruys. 1990. Hemorrhagic claw lesions in newborn piglets due to selenium toxicosis during pregnancy. *Vet. Rec.* 126:620–622.

Mihailovic, M., G. Matic, P. Lindberg, and B. Rigic. 1992. Accidental selenium poisoning of growing pigs. *Biol. Trace Elem. Res.* 33:63–69.

Miller, W. T., and H. W. Schoening. 1938. Toxicity of selenium fed to swine in the form of sodium selenite. *J. Agric. Res.* 56:831–842.

Miller, W. T., and K. T. Williams. 1940. Effects of feeding repeated small doses of selenium as sodium selenite to equines. *J. Agric. Res.* 61:353–368.

Morris, J. S., and M. J. Stampfer, and W. Willett. 1983. Dietary selenium in humans. Toenails as an indicator. *Biol. Trace. Elem. Res.* 5:529–537.

Morrow, D. A. 1968. Acute selenite toxicity in lambs. *J. Am. Vet. Med. Assoc.* 152:1625–1629.

Moxon, A. L. 1937. Alkali disease or selenium poisoning. South Dakota State College of Agriculture and Mechanic Arts, Agric. Exp. Stn. Bull. No. 311:1–91.

Moxon, A. L., and M. Rhian. 1943. Selenium poisoning. *Physiol. Rev.* 23:305–337.

Munsell, H. E., G. M. DeVaney, and M. H. Kennedy. 1936. Toxicity of food containing selenium as shown by its effect on the rat. U.S. Department of Agriculture Tech. Bull. No. 534:1–25.

Nelson, A. A., O. G. FitzHugh, and H. O. Calvery. 1943. Liver tumors following cirrhosis caused by selenium in rats. *Cancer Res.* 3:230–236.

Nilsson, S. A. 1963. Clinical, morphological, and experimental studies of laminitis in cattle. *Acta Vet. Scand.* 4 (suppl. 1):1–304.

O'Malley, C. D. 1965. *Andreas Vesalius of Brussels, 1514–1564,* p. 82. University of California Press, Los Angeles.

O'Sullivan, B. M., and W. F. Blakemore. 1980. Acute nicotinamide deficiency in the pig induced by 6-aminonicotinamide. *Vet. Pathol.* 103:748–756.

O'Toole, D., and M. F. Raisbeck. 1995. Pathology of experimentally induced chronic selenosis (alkali disease) in yearling cattle. *J. Vet. Diagn. Invest.* 7:364–373.

O'Toole, D., and M. F. Raisbeck. 1997. Experimentally-induced selenosis of adult mallard ducks: Clinical signs, lesions and toxicology. *Vet. Pathol.* 34:330–340.

O'Toole, D., L. E. Castle, and M. F. Raisbeck. 1995. Comparison of histochemical autometallography (Danscher's stain) to chemical analysis for detection of selenium in tissues. *J. Vet. Diagn. Invest.* 7:281–284.

O'Toole, D., M. F. Raisbeck, J. C. Case, and T. D. Whitson. 1996. Selenium-induced blind staggers and related myths. A commentary on the extent of historical livestock losses attributed to selenosis on western U.S. rangelands. *Vet. Pathol.* 33:104–116.

Obel, N. 1948. Studies on the histopathology of acute laminitis. Dissertation. Almqvist & Wiksell, Uppsala.

Ohlendorf, H. M. 1996. Selenium. In A. Fairbrother, L. N. Locke, and G. L. Hoff (eds.), *Noninfectious Disease of Wildlife,* 2nd ed., pp. 128–140. Iowa State University Press, Ames.

Ohlendorf, H. M., and R. L. Hothem. 1995. Agricultural drainwater effects on wildlife in central California. In D. J. Hoffman, B. A. Rattner, G. A. Burton, and J. Cairns (eds.), *Handbook of Ecotoxicology,* pp. 577–575. Lewis Publishers, Boca Raton, FL.

Ohlendorf, H. M., D. J. Hoffman, M. K. Saili, and T. W. Aldrich. 1986. Embryonic mortality and abnormalities of aquatic birds: Apparent impacts of selenium from irrigation water. *Sci. Total Environ.* 52:49–63.

Ohlendorf, H. M., A. W. Kilness, J. L. Simmons, R. K. Stroud, D. J. Hoffman, and J. F. Moore. 1988. Selenium toxicosis in wild aquatic birds. *J. Toxicol. Environ. Health* 24:67–92.

Oldfield, J. E., R. Burau, G. Moller, H. M. Ohlendorf, and D. Ullrey. 1994. Risks and benefits of selenium in agriculture. Council for Agricultural Science and Technology, Issue Paper No. 3:1–6. CAST, Ames, IA.

Olson, O. E., and L. B. Embry. 1973. Chronic selenite toxicity in cattle. *Proc. South Dakota Acad. Sci.* 52:50–58.

Orstadius, K. 1960. Toxicity of a single subcutaneous dose of sodium selenite in pigs. *Nature* 188:1117.

Palawski, D. U., W. E. Jones, and K. DuBois. 1991. Contaminant biomonitoring at the Benton Lake National Wildife Refuge in 1988, pp. 1–35. U.S. Fish and Wildlife Service, Helena, MT.

Palludan, B. 1961. The teratogenic effect of vitamin-A deficiency in pigs. *Acta. Vet. Scand.* 2:32–59.

Palmer, I. S., R. L. Arnold, and C. W. Carlson. 1973. Toxicity of various selenium salts to chick embryos. *Poult. Sci.* 52:1841–1846.

Palmer, I. S., O. E. Olson, A. W. Halverson, R. Miller, and C. Smith. 1980. Isolation of factors in lineseed oil meal protective against chronic selenosis in rats. *J. Nutr.* 110:145–150.

Panter, K. E., W. J. Hartley, L. F. James, H. F. Mayland, B. L. Stegelmeier, and P. O. Kechele. 1996. Comparative toxicity of selenium from seleno-DL-methionine, sodium selenate, and *Astragalus bisulcatus* in pigs. *Fundam. Appl. Toxicol.* 32:217–223.

Paveglio, F. L., C. M. Bunck, and G. H. Heinz. 1992. Selenium and boron in aquatic birds from central California. *J. Wildl. Manage.* 56:31–42.

Peckham, M. C. 1972. Vices and miscellaneous diseases. In M. S. Hofstad, B. W. Calnek, C. F. Helmboldt, W. M. Reid, and H. W. Yoder (eds.), *Diseases of Poultry*, 6th ed., pp. 1079–1081. Iowa State University Press, Ames.

Peterson, J. A., and A. V. Nebeker. 1992. Estimation of waterborne selenium concentrations that are toxicity thresholds for wildlife. *Arch. Environ. Contam. Toxicol.* 23:154–162.

Poley, W. E., A. L. Moxon, and K. D. Franke. 1937. Further studies on the effects of selenium poisoning on hatchability. *Poult. Sci.* 16:219–225.

Post, G. 1958. Diagnosis in mammals and birds. Pittman-Robinson Job Completion Report, pp. 5–13. Project FW-3-R-5, Work Plan No. 1, Job No. 1W. Wyoming Game and Fish Commission, Cheyenne.

Post G. 1959. Diagnosis in mammals and birds. Pittman-Robinson Job Completion Report, pp. 1–22. Project FW-3-R-7, Work Plan No 1, Job No. 1W. Wyoming Game and Fish Commission, Cheyenne.

Powell, B. C., and G. E. Rogers. 1990. Hard keratin IF and associated proteins. In R. D. Golman and P. M. Steinert (eds.), *Molecular Biology of Intermediate Filaments*, pp. 267–300. Plenum Publishing, New York.

Prasad, T. and S. P. Arora. 1991. Influence of different sources of injected selenium on certain enzymes, glutathione and adenosylmethionine concentration in buffalo (*Bubalus bubalus*) calves. *Br. J. Nutr.* 66:261–267.

Presser, T., M. A. Sylvester, and W. F Low. 1994. Bioaccumulation of selenium from natural geologic sources in western states and its potential consequences. *Environ. Manage.* 18:423–436.

Raisbeck, M. F., E. R. Dahl, D. A. Sanchez, E. L. Belden, and D. O'Toole. 1993. Naturally occurring selenosis in Wyoming. *J. Vet. Diagn. Invest.* 5:84–87.

Raisbeck, M. F., D. T. O'Toole, E. L. Belden, D. A. Sanchez, and R. A. Siemion. 1995. Selenium or sulfur? A comparison of toxic effects in mammals on western rangelands. In *Planning, rehabilitation and treatment of disturbed lands.* R. C. Severson, S. E. Fisher, and L. P. Gough, (eds.), *Sixth Billings Symposium,* Vol. 1, pp. 139–151. Reclamation Research Unit Publ. No. 9301, Billings, MT.

Raisbeck, M. F., D. O'Toole, R. A. Schamber, E. L. Belden, and T. L. Robinson. 1996. toxicologic effects of a high-selenium hay diet in captive pronghorn antelope (*Antilocapra americana*). *J. Wildl. Dis.* 32:9–16.

Raisbeck, M. F., D. O'Toole, and E. L. Belden. 1997. Selenium. In *Current Veterinary Therapy,* Vol. 4, J. L. Howard (ed.), *Food Animal Practice.* Saunders, Philadelphia (in press).

Ramirez, P., and J. A. Armstrong. 1992. Biota. In Detailed study of selenium in soil, representative plants, water, bottom sediment, and biota in the Kendrick reclamation project area, Wyoming 1988–90, pp. 69–114. U.S. Geological Survey, Water Resources Invest. Report No. 91-4131. USGS, Cheyenne.

Ramirez, P. M., Jennings, and K. Dickerson. 1994. Selenium in fish and aquatic bird food chain. Kendrick reclamation project, Natrona County, Wyoming. In *Effects of Human-Induced Changes on Hydrologic Systems*, pp. 1043–1053. Annual Symposium, American Water Resources Association.

Riddell, C. 1975a. Pathology of developmental and metabolic disorders of the skeleton of domestic chickens and turkeys: I. Abnormalities of genetic or unknown origin. *Vet. Bull.* 45:629–640.

Riddell, C. 1975b. Pathology of development and metabolic disorders of the skeleton of domestic chickens and turkeys: II. Abnormalities due to nutritional or toxic factors. *Vet. Bull.* 45:705–718.

Riddell, C. 1991. Developmental, metabolic, and miscellaneous disorders. In B. W. Calnek, H. J. Barnes, C. W. Beard, W. M. Reid, and H. W. Yoder (eds.), *Diseases of Poultry*, 9th ed., p. 854. Iowa State University Press, Ames.

Roberts, E. D., R. Ochoa, and P. F. Haynes. 1980. Correlation of dermal–epidermal laminar lesions of equine hoof with various disease conditions. *Vet. Pathol.* 17:656–666.

Robinson, I. 1996. Feathers and skin. In P. H. Beynon, N. A. Forbes, and N. H. Harcourt-Brown (eds.), *Manual of Raptors, Pigeons and Waterfowl*, pp. 305–310. British Small Animal Veterinary Association, Cheltenham, Gloucestershire, U.K.

Romanoff, A. L., and A. J. Romanoff. 1972. *Pathogenesis of the Avian Embryo. An Analysis of Causes of Malformation and Prenatal Death*, p. 30. Wiley-Interscience, New York.

Rosenfeld, I., and O. A. Beath. 1946. Pathology of selenium poisoning. University of Wyoming Agric. Exp. Stn. Bull. 275:1–27.

Rosenfeld, I., and O. A. Beath. 1947. Congential malformations of eyes of sheep. *J. Agric. Res.* 75:93–103.

Rousseaux, C. G., M. J. Politis, and J. Keiner. 1993. The effects of sodium selenite and selenomethionine on murine limb development in culture. *Environ. Toxicol. Chem.* 12:1283–1290.

Sallis, E. S., F. Riet-Correa, and M. C. Mendez. 1993. Experimental intoxication by *Ramaria flavobrunnescens* in sheep. *N.Z. Vet. J.* 41:224.

Schardein, J. L. 1993. *Chemically Induced Birth Defects*, 2nd ed., pp. 15, 738–730. Dekker, New York.

Seko, Y., Y. Saito, J. Kitahara, and N. Imura. 1989. Active oxygen generation by the reaction of selenite with reduced glutathione in vitro. In A. Wendel (ed.), *Selenium in Biology and Medicine*, pp. 70–73. Springer-Verlag, Berlin.

Seiler, R. L. 1995. Prediction of areas where irrigation drainage may induce selenium contamination of water. *J. Environ. Qual.* 24:973–979.

Shamberger, R. J. 1985. The genotoxicity of selenium. *Mutat. Res.* 154:29–48.

Shirlaw, J. F. 1939. Deg-nala disease of buffaloes: An account of the lesions and essential pathology. *Indian J. Vet. Sci. Anim. Husb.* 9:173–177.

Shortridge, E. H., P. J. O'Hara, and P. M. Marshall. 1971. Acute selenium poisoning in cattle. *N. Z. Vet. J.* 19:47–50.

Skorupa, J. P., and H. M. Ohlendorf. 1991. Contaminants in drainage water and avian risk thresholds. In A. Dinar and D. Zilberman (eds.), *The Economics and Management of Water and Drainage in Agriculture*, pp. 345–368. Kluwer Academic Publishers, Boston.

Skorupa, J. P., S. P. Mormon, and J. S. Sefchik-Edwards. 1996. Guidelines for interpreting selenium exposures of biota associated with nonmarine acquatic habitats, pp. 1–74. Prepared for the National Water Quality Program, Sacramento, CA.

Smith, G. J., G. H. Heinz, D. J. Hoffman, J. W. Spann, and A. J. Krynitsky. 1988. Reproduction in black-crowned night herons fed selenium. *Lake Reservoir Manage.* 4:175–180.

Smith, M. I., E. F. Stohlman, and R. D. Lillie. 1937. The toxicity and pathology of selenium. *J. Pharm. Exp. Ther.* 60:449–471.

Smyth, J. B. A., J. H. Wang, R. M. Barlow, D. J. Humphreys, M. Robins, and J. B. Stodulski. 1990a. Experimental acute selenium intoxication in lambs. *J. Comp. Pathol.* 102:197–209.

Smyth, J. B. A., J. H. Wang, R. M. Barlow, D. J. Humphreys, M. Robins, and J. B. Stodulski. 1990b. Effects of concurrent oral administration of monensin of the toxicity of increasing doses of selenium in lambs. *J. Comp. Pathol.* 102:443–455.

Spallholz, J. E. 1994. On the nature of selenium toxicity and carcinostatic activity. *Free Radical Biol. Med.* 17:45–64.

Stashak, T. S. 1987. *Adam's Lameness in Horses,* pp. 277–289, 541–543. Lea & Febiger, Philadelphia.

Stowe, H. D., A. J. Eavey, L. Granger, S. Halstead, and B. Yamini. 1992. Selenium toxicosis in feeder pigs. *J. Am. Vet. Med. Assoc.* 201:292–295.

Summers, B. A., J. F. Cummings, and A. deLahunta. 1995. Selenium poisoning and focal symmetrical poliomyemomalacia, pp. 258–261. In *Veterinary Neuropathology.* Mosby, St. Louis, MO.

Tarantal, A. F., C. C. Willhite, B. L. Lasley, C. J. Murphy, C. J. Miller, M. J. Cukierski, S. A. Book, and A. G. Hendrickx. 1991. Developmental toxicity of L-selenomethionine in *Macaca fascicularis. Fundam. Appl. Toxicol.* 16:147–160.

Taylor, R. F., and T. P. Mullaney. 1984. Selenium toxicosis in neonatal lambs. *Proc. Am. Assoc. Vet. Lab. Diagn.* 27:369–378.

Thompson, C. D., M. F. Robinson, J. A. Butler, and P. D. Whanger. 1993. Long-term supplementation with selenate and selenomethionine: Selenium and glutathione peroxidase in blood components of New Zealand women. *Br. J. Nutr.* 69:577–588.

Traub-Dargatz, J. L., A. P. Knight, and D. W. Hamar. 1986. Selenium toxicity in horses. *Comp. Cont. Educ.* 8:771–776.

Turk, J. R. 1980. Chronic parenteral selenium administration in a dog. *Vet. Pathol.* 17:493–496.

Twomey, T., R. A. P. Crinion, and D. B. Glazier. 1977. Selenium toxicity in cattle in Co. Meath. *Ir. Vet. J.* 31:41–46.

Usami, M., and Y. Ohno. 1996. Teratogenic effects of selenium compounds on cultured postimplantation rat embryos. *Teratog. Carcinog. Mutatgen.* 16:27–36.

Van der Lugt, J. J., and L. Prozesky. 1989. The pathology of blindness in newborn calves caused by hypovitaminois A. *Onderstepoort J. Vet. Res.* 56:99–109.

Van Vleet, J. F., K. B. Meyer, and H. J. Olander. 1974. Acute selenium toxicosis induced in baby pigs by parenteral administration of selenium–vitamin E preparations. *J. Am. Vet. Med. Assoc.* 165:543–547.

Wahlstrom, R. C., and O. E. Olson. 1959. The effect of selenium on reproduction in swine. *J. Anim. Sci.* 18:141–145.

Waschulewski, I. H., and R. A. Sunde. 1988. Effect of dietary methionine on tissue selenium and glutathione peroxidase activity in rats given selenomethionine. *Br. J. Nutr.* 60:57–68.

Weimeyer, S. N., and D. J. Hoffman. 1996. Reproduction in eastern screech-owls fed selenium. *J. Wildl. Manage.* 60:332–341.

Whanger, P. D. 1989. Selenocompounds in plants and their effects on animals. In P. R. Cheeke (ed.), *Toxicants of Plant Origin*, Vol. III: *Proteins and Amino Acids*, pp. 141–167. CRC Press, Boca Raton, FL.

Williams, K. T., H. W. Lakin, and H. C. Byers. 1941. Selenium occurrence in certain soils in the United States with a discussion on related topics: Fifth report. U.S. Department of Agriculture, Tech. Bull. No. 758:1–69.

Willhite, C. C. 1993. Selenium teratogenesis. Species-dependent response and influence on reproduction. *Ann. New York Acad. Sci.* 678:169–177.

Willhite, C. C., V. H. Ferm, and L. Zeise. 1990. Route dependent pharmacokinetics, distribution and placental permeability of organic and inorganic selenium in hamsters. *Teratology* 42:359–371.

Wilson, T. M., and T. R. Drake. 1982. Porcine focal symmetrical poliomyelomalacia. *Can. J. Comp. Med.* 46:218–220.

Wilson, T. M., R. W. Scholz, and T. R. Drake. 1983. Selenium toxicity and porcine focal symmetrical encephalomalacia: Description of a field ourbreak and experimental reproduction. *Can J. Comp. Med.* 47:412–421.

Wilson, T. M., P. G. Cramer, R. L. Owen, C. R. Knepp, I. S. Palmer, A. deLahunta, J. L. Rosenberger, and R. H. Hammerstedt. 1989. Porcine focal symmetrical poliomyelomalacia: Test for interaction between dietary selenium and niacin. *Can. J. Vet. Res.* 53:454–461.

Witte, S. T., L. A. Will, C. R. Olsen, J. A. Kinker, and P. Miller-Graber. 1993. Chronic selenosis in horses fed locally produced alfalfa hay. *J. Am. Vet. Med. Assoc.* 202:406–409.

Wobeser, G. A. 1997. *Diseases of Waterfowl*, 2nd ed., pp. 211–213, 237–238. Plenum Press, New York.

Yang, G., and Zhou, R. 1994. Further observation on the human maximum safe dietary selenium intake in a seleniferous area of China. *J. Trace Elem. Electrolytes Health Dis.* 8:159–165.

Yang, G., S. Wang, R. Zhou, and S. Sun. 1983. Endemic selenium intoxication of humans in China. *Am. J. Clin. Nutr.* 37:872–881.

20

Prediction of Lands Susceptible to Irrigation-Induced Selenium Contamination of Water

RALPH L. SEILER

United States Geological Survey, Carson City, Nevada

I. INTRODUCTION

At a conference in 1985, shortly after selenium was identified as the cause of collapse of the warm-water fishery and of death and deformities among aquatic birds at California's Kesterson Reservoir, U.S. Geological Survey (USGS) geochemist Ivan Barnes predicted where selenium problems would occur. Responding to a question from the audience, he said "If you want some generalizations, I'll give them to you. If there is a continental climate, with marine pyrite, and less than 20 inches (50.8 cm) of precipitation a year, there *will* be a selenium problem. If there is a Mediterranean climate with marine pyrite and less than 12 inches (30.5 cm) of precipitation a year, there *will* be a selenium problem" [emphasis his]. Although the events at Kesterson Reservoir were a surprise to the public and to government officials, Barnes's (1985) response drew on nearly 50 years of selenium research in the western United States by university and government scientists.

The sources and distribution of seleniferous soils in the western United States were intensively researched during the 1930s and 1940s after the discovery that selenium in pasturage caused a fatal disease of cattle and horses. However, by the time the Bureau of Reclamation built the upper 85-mile reach of the San Luis Drain, which terminated at Kesterson Reservoir, awareness of selenium as a toxicant had faded. The potential toxic effects of selenium were not considered when the decision was made in the mid-1970s to use Kesterson Reservoir as the terminus of the drain. Early water quality concerns focused on salinity, boron, nitrates, and pesticides (Tanji et al., 1986). The events in 1982–1983 at Kesterson Reservoir led to the rediscovery of selenium as an important environmental toxicant.

The Department of the Interior initiated the National Irrigation Water Quality Program (NIWQP) to determine whether what happened at Kesterson Reservoir could happen elsewhere in the United States. Between 1986 and 1993, the NIWQP investigated 26 areas in 14 states (Fig. 1) for irrigation-induced contamination. Data from the 26 study areas were compiled and evaluated to identify factors common to seleniferous areas and to enhance the ability to predict where irrigation-induced selenium contamination of water and biota will occur.

Previous investigators have identified certain combinations of climatic and geologic settings that are likely to result in selenium problems. These conclusions

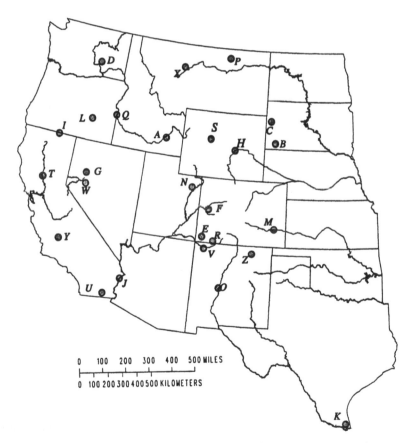

FIGURE I   Locations of areas investigated by the National Irrigation Water Quality Program for irrigation-induced contamination. (Letters are keyed to Table 1.)

are still largely valid. Lakin and Byers (1941) summarized some of the research done during the 1930s and 1940s and concluded that "All areas of soils derived from material of Cretaceous age are then open to suspicion of the presence of harmful quantities of selenium. . . ." Trelease and Beath (1949) noted that even in areas of seleniferous sediments, seleniferous vegetation has not been found when annual rainfall exceeds 50.8 cm. Barnes (1985) identified a specific combination of minerals and precipitation rates that is likely to result in selenium problems.

This chapter uses data from the 26 NIWQP investigations to test the predictions of previous investigators. A new variable is proposed to replace precipitation as an explanatory variable, and a decision tree and map are described that can be used to predict where irrigation drainage is likely to result in selenium problems.

II. SOURCES OF SELENIUM TO IRRIGATED SOILS

A. Seleniferous Rocks

The importance of reduced sulfur compounds and marine sedimentary rocks of Cretaceous age as sources of selenium in the western United States has been known since the 1930s. Byers (1935) stated "The source of selenium in soils has been shown to be sulfide minerals occurring in the soil parent materials. So far as yet known the seleniferous soil-forming material is, for the most part, shales of the Cretaceous period." The regional importance of Upper Cretaceous marine sedimentary rocks as sources of selenium is indicated by their large areal extent (Fig. 2); they form the near surface bedrock beneath about 805,000 km^2 of land in the 17 conterminous western states. Much of the selenium in the crust of the earth occurs in pyrite and other sulfide minerals (Berrow and Ure, 1989). These occurrences demonstrate the important association between selenium and reduced sulfur in geologic material.

Understanding general mechanisms by which selenium is incorporated into rocks is important so that seleniferous rocks of all ages and areas can be identified. Selenium is associated with volcanic sulfur, and the amount of selenium in volcanic sulfur can range from traces up to 5% (Berrow and Ure, 1989). Trelease and Beath (1949) and Berrow and Ure (1989) describe general geologic mechanisms by which marine sedimentary rocks became seleniferous through erosion of seleniferous volcanic materials and atmospheric deposition of selenium from volcanic gases and dusts. Presser (1994a, 1994b) suggests that in addition to these geologic mechanisms, bioaccumulation of selenium in ancient seas, followed by deposition and diagenesis of the seleniferous organic matter, may constitute a primary mechanism of selenium enrichment in ancient sedimentary deposits.

Tertiary marine sedimentary deposits are likely to be seleniferous if they were laid down in the same depositional environment as the Upper Cretaceous marine sedimentary rocks. Marine sedimentary deposits of early Tertiary age in

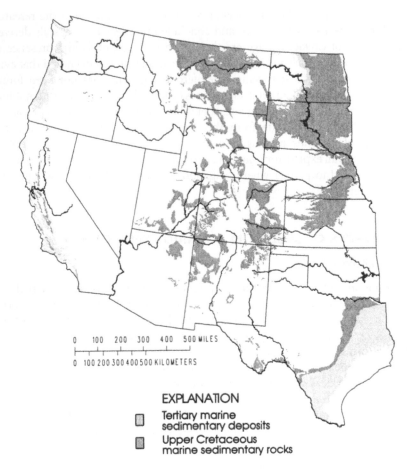

EXPLANATION

☐ Tertiary marine
   sedimentary deposits

☒ Upper Cretaceous
   marine sedimentary rocks

FIGURE 2 Locations of areas in the western United States where potentially seleniferous rocks form the bedrock.

the western United States may be generally seleniferous; Presser (1994b) identifies Upper Cretaceous–Paleocene, Eocene–Oligocene, and Miocene marine formations from the Coast Ranges of California that are seleniferous. Marine sedimentary rocks of Tertiary age form the bedrock beneath almost 219,000 km^2 of land in the western United States (Fig. 2).

Some, but not all, Tertiary continental sedimentary deposits are seleniferous. Continental sedimentary deposits of Tertiary age derived by reworking of Upper Cretaceous marine sedimentary rocks also may be seleniferous. Trelease and Beath (1949) note that because of the diverse conditions that controlled their deposition, Tertiary age rocks differ greatly in selenium content. Whether a Tertiary continental sedimentary deposit is seleniferous depends on several factors. Such factors include whether the parent rock was seleniferous, whether reduced selenium in the rock was exposed to the strongly oxidizing conditions required to mobilize the selenium, and whether the rock was leached enough to remove any selenium. Continental sedimentary rocks of Tertiary age form the bedrock in about 948,000 km² of land in the western United States.

B.   Transfer of Selenium from Seleniferous Rocks to Irrigated Soils and Groundwater

Irrigation water applied to soils can dissolve and mobilize selenium in the soils and create hydraulic gradients that cause the discharge of seleniferous groundwater into drains. Since drainage from agricultural areas is a primary source of selenium to wildlife areas, it also is important to understand processes that allow selenium derived from seleniferous rocks to accumulate in soils and groundwater in agricultural areas. Three general mechanisms by which soils and ground water become seleniferous are discussed here.

In the simplest case, which occurs in most of the areas with selenium problems that were investigated by NIWQP, the selenium in the soil and groundwater originates in a seleniferous rock beneath the soils. Soils are seleniferous because of selenium-containing minerals that remain in the soil following pedochemical weathering of the parent rock. Groundwater becomes seleniferous when water in contact with the soil and rock reacts with the selenium containing minerals and solubilizes the selenium. Areas with soils and groundwater that may become seleniferous by the "beneath" mechanism can be identified by geologic maps showing the bedrock distribution of seleniferous sediments and deposits.

The most important process by which soils and groundwater becomes seleniferous in the San Joaquin Valley of California is the transport of selenium from upland areas in the mountains surrounding the valley. Presser (1994a, 1994b) describes how seleniferous sediments tens of miles upland of irrigated lands can contribute selenium to lands downslope through processes of active weathering, alluvial fan building, and local drainage. Areas with soils that may become seleniferous by the "upland" mechanism can be identified by geologic and topographic maps because they will be adjacent and downslope of seleniferous deposits.

Soils and groundwater can become seleniferous if dissolved or suspended selenium is imported into an area in surface water. Importation is similar to the upland mechanism, except that the source of the selenium can be discharges to a river or lake of selenium resulting from human industrial or agricultural activities or present in naturally occurring seleniferous rocks from hundreds of miles upstream. Fish kills in Belews Lake, North Carolina, caused by seleniferous ash from a nearby coal-fired power plant (Lemly, 1985), present an example of importation from an industrial source. An example of importation of natural and agriculturally derived selenium occurs in the Imperial Valley in the Salton Sea area (U, see Fig. 1). Hundreds of miles upstream from the Salton Sea, the mainstem of the Colorado River and several of its major tributaries traverse Upper Cretaceous marine sedimentary rocks. Those rivers transport selenium mobilized by the mechanisms described previously. Additionally, selenium in drain-water from irrigated agriculture in basins tributary to the Colorado River, including the Green River, Gunnison River, and the San Juan River, may eventually end up in the Salton Sea area

III. CHARACTERIZATION OF NIWQP STUDY AREAS

A. Selenium Concentrations in Surface Water

The 26 study areas were classified as seleniferous or not depending on the selenium concentrations in surface water in and downstream of irrigated areas. Selenium concentrations in surface water from the NIWQP areas were compared with 5 μg/L, the chronic criterion for selenium for the protection of freshwater aquatic life (U.S. Environmental Protection Agency, 1987), and 3 μg/L, the low end of a range (3–20 μg/L total recoverable selenium) described as hazardous to some species of aquatic birds under some environmental conditions (Skorupa and Ohlendorf, 1991). The NIWQP investigations typically measured filtered, not total recoverable selenium, in water samples. However, Seiler (1996) presents evidence that in NIWQP samples, filtered and total selenium concentrations are nearly the same over a wide range of concentrations.

Selenium concentrations in water only from surface water sites in and downstream from irrigated lands were used in summarizing selenium data. Data for flowing and impounded bodies of water were combined in this analysis, because no samples or very few samples were collected from lakes and ponds in some areas. Summary statistics of data that include censored values were estimated using either probability plotting methods (Helsel and Hirsch, 1992) or applied lognormal maximum likelihood estimation methods (Helsel and Cohn, 1988).

For comparison with criteria, the 75th percentile of the selenium concentrations was chosen to represent the degree of selenium hazard in water from the area. The 75th percentile is the concentration that is exceeded by 25% of the samples from the area. Classifying areas as seleniferous if 25% of the samples exceed water quality criteria seems to be a reasonable standard; fish and aquatic birds probably are exposed to seleniferous water if that high a percentage of samples from an area exceed criteria.

Fourteen of the 26 areas were classified as seleniferous because the selenium concentration in 25% or more of the surface water samples equaled or exceeded 3 $\mu g/L$ (Table 1, Fig. 3). In 12 of the 14 areas, the selenium concentration in 25% or more of the surface water samples also equaled or exceeded the chronic criterion for selenium for the protection of freshwater aquatic life. In 7 of the 12 nonseleniferous areas, selenium concentrations in 75% or more of the surface water samples were less than the analytical reporting limit (1 $\mu g/L$), and in 3 areas, selenium was not detected in any surface water sample.

In two areas, sample bias greatly affects the 75th percentile selenium concentration (Seiler, 1995). In the Kendrick area (H) almost 17% of the samples were repeat samples from a lake that was known to be selenium-contaminated. Sampling primarily at the most seleniferous sites in an area results in a higher 75th percentile of the selenium concentrations than if an unbiased data set is used. In the San Juan River area (V), there is another type of bias. Almost 25% of the samples in that area were from the main channels of large rivers. Sampling primarily at main channel sites results in a lower 75th percentile of the selenium concentrations because sites on main channels of large rivers tend to have lower selenium concentrations than do drains or ponds.

B.   Physical Classification of Areas

1.   Geology

The King and Beikman (1974) map of the geology of the United States was used as the source of geological information about the study areas. The 26 study areas were classified into four groups primarily on the basis of their association with marine sedimentary rock and deposits:

1. Areas where the bedrock beneath irrigated land is mainly Upper Cretaceous marine sedimentary rocks [(uK) in Table 1, note c)]
2. Areas where the bedrock in mountains upland of irrigated land includes Upper Cretaceous marine sedimentary rocks or, is a combination of Upper Cretaceous and Tertiary marine sedimentary deposits [(uK)−M or (uK + T)−M in Table 1, note c)]
3. Areas where rivers supplying water for irrigation traverse Upper Cretaceous or Tertiary marine sedimentary deposits upstream of irrigated lands [(uK)−U or (uK + T)−U in Table 1, note c]
4. Areas that are not associated with Upper Cretaceous marine sedimentary rocks (— in Table 1)

Upper Cretaceous marine sedimentary rocks are by far the most important geologic sources of selenium in the NIWQP data set. These marine rocks compose all of the near-surface bedrock beneath irrigated land in eight NIWQP areas and

Study area	Selenium concentration $(\mu g/L)$[b]	Geology: Marine sedimentary rocks[c]	Hydrology: Terminal lakes or ponds[d]	Mean annual precipitation (cm)	Free-water surface evaporation (cm)[e]
merican Falls Reservoir, ID	1.0	—	No	27.7	103
ngostura Reclamation Unit, SD	4.5	(uK)	Yes	41.7	116
elle Fourche Reclamation Project, SD	5	(uK)	Yes	36.6	102
olumbia River Basin, WA	<1	—	Yes	203	101
olores–Ute Mountain Area, CO	7.0	(uK)	No	30.5	134
iunnison River Basin–Grand Valley Project, CO	35	(uK)	No	23.5	126
lumboldt River Area, NV	2.0	—	Yes	14.0	113
endrick Reclamation Project, WY	64	(uK)	Yes	30.5	112
lamath Basin Refuge Complex, CA–OR	<1	—	Yes	33.0	100
ower Colorado River Valley, CA–AZ	2.0	(uK) – U	No	114	216
ower Rio Grande Valley, TX	1.0	(uK + T) – U	No	65.8	145
lalheur National Wildlife Refuge, OR	<1	—	Yes	25.4	109
liddle Arkansas River Basin, CO–KS	10	(uK)	No	37.1	147
liddle Green River Basin, UT	73	(uK)	Yes	19.3	110

Middle Rio Grande, NM	<1	—	Yes	23.9	162
Milk River Basin, MT	<1	(uK)	Yes	54.4[g]	101
Owyhee–Vale Reclamation Project Areas, OR–ID	2.0	—	No	24.1	109
Pine River Area, CO	6.0	(uK)	No	35.6	124
Riverton Reclamation Project, WY	5.0	(uK) – M	No	20.6	101
Sacramento Refuge Complex, CA	<1	(uK) – M	No	47.0	123
Salton Sea Area, CA	8.0	(uK) – U	Yes	7.6	186
San Juan River Area, NM	3.0	(uK)	Yes	19.3	143
Stillwater Wildlife Management Area, NV	<1	—	Yes	13.3	134
Sun River Area, MT	7.5	(uK)	Yes	30.5	91
Tulare Lake Bed Area, CA	265	(uK + T) – M	Yes	14.0	155
Vermejo Project, NM	6.0	(uK)	Yes	35.1	138

r used in Figure 1.

at the 75th percentile concentration from analyses of surface water in irrigated areas and downstream from them.

retaceous marine sedimentary rock; T. Tertiary marine sedimentary deposits; (uK), bedrock beneath irrigated land is primarily Up

ntary rock; (uK) – M or (uK + T) – M, bedrock in mountains upland from irrigated land includes rocks of the referenced age; (uK) – U d

ng water for irrigation traverse deposits of the referenced age upstream of irrigating land; —, irrigated lands are not associated with mari

logy after King and Beikman (1974).

lakes/ponds exist during nonflood years; No, terminal lakes/ponds do not exist during nonflood years.

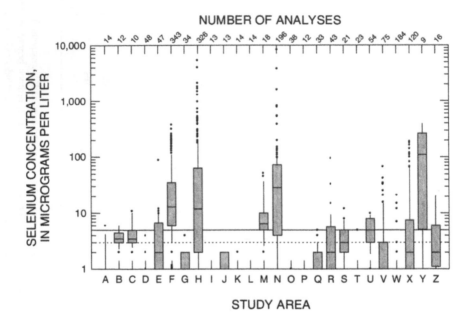

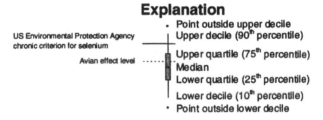

FIGURE 3 Statistical summary of selenium concentrations in filtered surface water samples from NIWQP study areas. See Table 1 for complete names of study areas. Chronic criterion from U.S. Environmental Protection Agency (1987) and avian effect level from Skorupa and Ohlendorf (1991).

compose some or most of the near-surface bedrock beneath irrigated land in four other areas. Tertiary marine sedimentary deposits are of minor importance in the NIWQP data set. They occur upland of one area (Y) in the San Joaquin Valley of California, and the Rio Grande traverses them upstream of another area in Texas (K). Areas with Tertiary continental sedimentary deposits were not classified separately because, in most instances, in the NIWQP data set they also are associated with Upper Cretaceous marine sedimentary rocks. Their significance as sources of selenium is discussed in Section IV.A.

2. Hydrology

The 26 areas were classified hydrologically (Table 1) as open basin or closed basin, depending on the existence of lakes or ponds that receive irrigation drainwater and are terminal (i.e., have no outlets) during nonflood years. In terminal lakes, evaporation can result in very high concentrations of selenium because the solutes are retained and accumulated rather being than flushed out during the spring runoff or other storm events.

3. Climate

The mean annual precipitation and evaporation (free-water surface) rates for the 26 areas are shown in Table 1. Precipitation data for the study areas were obtained from USGS series reports describing the results of investigations in the NIWQP study areas. If a range of values was given in these reports, the midpoint of the range was selected. Free-water surface evaporation (FWSE) was selected to represent the potential for evaporation from the study area. FWSE was used instead of class A pan evaporation because a national map showing FWSE (Farnsworth et al., 1982) was available at a larger scale.

In the Milk River Basin (P), all data were collected during August 1988. It is noted that 1988 was an unusually wet year in that area, with precipitation almost twice the normal amount (Seiler, 1995). In the spring and summer prior to data collection, the area received much greater amounts of water than normal, which reduced selenium concentrations because of dilution. Because of the effects of the above-normal precipitation on selenium concentrations in the Milk River Basin, the precipitation during the year of data collection rather than the mean annual precipitation is used in that area.

IV. RELATION BETWEEN SELENIUM, GEOLOGY, AND CLIMATE

A. Relation Between Selenium Concentration and Geology

All 14 of the NIWQP study areas that are seleniferous are associated, either directly or indirectly, with marine sedimentary rocks or deposits (Table 1). In 11 of the seleniferous areas, Upper Cretaceous marine sedimentary rocks form the near-surface bedrock. In the three remaining seleniferous areas, rivers that supply water for irrigation traverse marine sedimentary rocks and deposits of Late Cretaceous and Tertiary age upstream of irrigated land, or these geologic materials are found in mountains upland of irrigated land.

The importance of Tertiary continental sedimentary deposits as sources of selenium is difficult to assess using the NIWQP data set. Four study areas (N, R, S, and V) contain some land where Lower Tertiary (Paleocene–Eocene) continental sedimentary rocks form the bedrock. Three areas (M, O, and Q) contain some land where Upper Tertiary (Pliocene) continental sedimentary rocks form the bedrock. Elevated levels of selenium occurs in five of the seven areas. The two nonseleniferous areas (O and Q) are both associated with Pliocene continental sedimentary deposits and there are no nearby Upper Cretaceous marine sedimentary rocks. Of the five seleniferous areas, all are associated with Upper Cretaceous marine sedimentary rocks. Four (M, N, R, and V) also contain some lands where Upper Cretaceous marine sedimentary rocks form the bedrock and, in the fifth area (S), the mountains upland of irrigated lands include Upper Cretaceous marine sedimentary rocks.

The best evidence for the importance of rocks of Tertiary age in the NIWQP data set is in the Ouray subarea of the Middle Green River study area in Utah (N). Ponds used by wildlife in the subarea commonly contain selenium at concentrations exceeding 20 μg/L, and deformed bird embryos have been found at these ponds (Stephens et al., 1992). All the data collection sites in the Ouray subarea are associated with continental sedimentary rocks of Eocene age, and the nearest exposure of Upper Cretaceous marine sedimentary rocks is about 14 miles upstream. Trelease and Beath (1949) note that seleniferous plants grow in parts of the Eocene age Uinta Formation, which is exposed in the area (Stephens et al., 1992).

B. Relation Between Selenium Concentration and Climate

The relation between selenium concentrations and climate was explored to test the conclusions of earlier investigators. Figure 4A shows the relation between mean annual precipitation and the 75th percentile of the selenium concentrations for those areas where Upper Cretaceous marine sedimentary rocks form the near-surface bedrock in irrigated areas. The trend for increasing concentrations of selenium with decreasing precipitation is clearly shown. As suggested by Barnes (1985), annual precipitation between 30.5 and 50.8 cm appears to separate seleniferous areas from nonseleniferous areas.

A statistically significant regression exists between the log of annual precipitation and the log of selenium concentration ($r^2 = 0.83$, $p < 0.001$). Two areas (H and V) were not included in the regression because of known sample bias as discussed earlier. One area (P) was excluded because all the samples from the area contained selenium concentrations below the method detection limit of 1 μg/L. To test its predictive capability, the regression was recomputed using only data from areas where the precipitation was between 30.5 and 50.8 cm (B, C, E, M, X, and Z). No significant relation exists ($r^2 = 0.06$, $p > 0.29$) when outliers F and N are removed.

It seemed intuitively that the climatic variable should include information about evaporation rates in addition to precipitation. To test whether adding

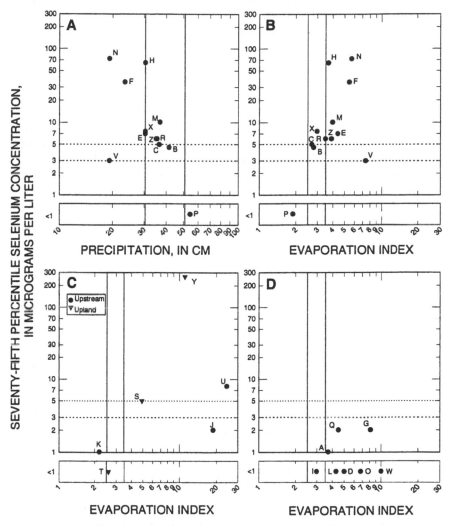

FIGURE 4   Relation between selenium concentrations, climatic variables, and geohydrologic setting. (A) and (B) Bedrock beneath irrigated land is mainly Upper Cretaceous marine sedimentary rock. (C) Bedrock in rivers upstream or mountains upland from irrigated land includes Upper Cretaceous or Tertiary marine sedimentary deposits. (D) No association with Upper Cretaceous or Tertiary marine sedimentary deposits. Dashed and solid vertical and horizontal lines are climatic and selenium criteria discussed in the text.

evaporation rate data to the climatic variable provided a better relation, precipitation and evaporation rates were combined into a single number to characterize the aridity of the study areas. This number, herein called the evaporation index (EI) (Table 1), is the annual FWSE divided by the mean annual precipitation.

Figure 4B shows the relation between the EI and the 75th percentile of the selenium concentrations for the same areas shown in Figure 4A. The trend for increasing selenium concentration with increasing EI is clearly shown. An EI value of about 2.5 is considered important because it lies between a nonseleniferous area (P, EI = 1.9) and the first areas where elevated concentrations of selenium occur (B and C, EI = 2.8). An EI value of about 3.5 is considered important because, in all but one study area, if the EI exceeds 3.5, more than 25% of the selenium concentrations exceed the chronic criterion for selenium (Fig. 4B).

A statistically significant positive relation exists between the log of EI and the log of selenium concentration ($r^2 = 0.70$, $p = 0.003$), and this relation is statistically significant even when the outliers F and N are removed ($r^2 = 0.26$, $p = 0.14$). For this reason, EI, which incorporates evaporation information, was selected as the climatic variable rather than precipitation alone.

In three of the study areas, Upper Cretaceous marine sedimentary rocks form some of the bedrock in mountains upland of irrigated land, and two of those areas (S and Y) are seleniferous (Fig. 4C). Selenium could be imported into three areas because rivers that supply the water used for irrigation traverse marine sedimentary deposits upstream of irrigated areas. One of those areas, the Salton Sea (U), is seleniferous (Fig. 4C).

Knowing where selenium is not a hazard is nearly as important as knowing where it is a hazard. In the western United States, areas that are not associated with marine sedimentary rocks and deposits are unlikely to have selenium problems. Eight of the 26 areas (A, D, G, I, L, O, Q, and W) fall within this classification (Table 1). The selenium concentration at the 75th percentile in most of these areas is less than the analytical reporting limit (1 μg/L) and is less than 3 μg/L in all of them (Fig. 4D).

Areas with low EI are unlikely to have selenium problems even if there is a source for selenium (Fig. 4B). This is illustrated by data from the Milk River Basin (P). Data were collected in the area during a flood year when sufficient water was available to dilute the contaminants; no selenium was detected in any of the NIWQP surface water samples. The EI calculated for the flood year was 1.9, which was the lowest of any of the 26 areas. The EI for average conditions would be 3.3, and data collected under normal conditions indicate that the area may be contaminated when there is less precipitation.

V. PREDICTION

Two methods were devised to allow managers to identify lands susceptible to selenium problems caused by irrigation. The methods are intended to be used as screening and ranking tools by managers. After the likelihood of selenium problems in several areas has been ranked, resources can be directed toward assessing selenium hazards in areas where selenium problems are most likely to occur. Strategies to minimize impacts of selenium from irrigated lands on wetlands can be most effective if vulnerable areas are targeted, rather than general measures being implemented over a broad area.

A. Criteria for Being Considered Susceptible to Irrigation-Induced Selenium Problems

Certain combinations of geologic, climatic, and hydrologic conditions are likely to result in selenium contamination if an area is irrigated. For the predictive tools, areas were considered geologically susceptible to irrigation-induced selenium problems if the near-surface bedrock in the area consists of Upper Cretaceous or Tertiary marine sedimentary rocks. Tertiary continental sedimentary deposits were not included because many of these deposits are not seleniferous and their inclusion could have resulted in a large amount of land being falsely identified as susceptible to selenium problems.

Areas were considered climatically susceptible to irrigation-induced selenium problems if the EI exceeds 2.5. Areas were considered hydrologically susceptible if selenium is imported into the area in water used for irrigation or if there are terminal lakes or ponds.

B. Decision Tree

Figure 5 presents a decision tree to predict the likelihood of selenium problems if an area in the western United States is irrigated. The decision tree ranks a target area into one of four classes based on the likelihood of significant problems resulting from elevated selenium concentrations in the water:

1. Selenium problem is unlikely.
2. Selenium problem is possible.
3. Selenium problem is likely.
4. Selenium problem is very probable.

From these rankings, a manager can assess the need to collect additional information.

The decision tree poses four basic questions about the geology, climate, and hydrology of an area to provide an estimate of the likelihood that selenium problems will occur. The questions are as follows:

1. Do irrigated lands lie on, or near, an area where Upper Cretaceous or Tertiary marine sedimentary rocks form the near-surface bedrock?
2. What is the evaporation index of the area?

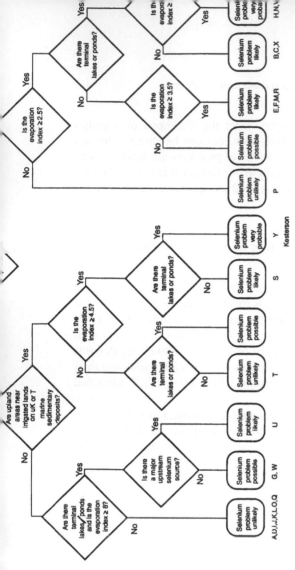

FIGURE 5 Decision tree for predicting the likelihood of selenium contamination and classification of Kesterson Re and the 26 National Irrigation Water Quality Program study areas used to derive the decision tree. (Modified and re with permission of the American Society of Agronomy.)

3. Are there terminal ponds or lakes in the area?
4. Is there an upstream source of selenium?

Answers to question 1 can be obtained by plotting the location of the target area on a geologic map. Calculating the EI for question 2 requires obtaining the annual free-water surface evaporation rate (from the evaporation map by Farnsworth et al., 1982 and the average annual precipitation (from records of individual weather stations or from state maps). Answers to questions 3 and 4 can be obtained by examining topographic and geologic maps of the area and its watershed.

C. Map of Susceptible Areas

Using a geographic information system (GIS), maps were created showing the distribution of Upper Cretaceous and Tertiary marine sedimentary deposits and EI in the western United States. Figure 6 identifies the areas that, if irrigated, are most susceptible to having selenium problems. This map was created by intersecting the geologic coverage and the coverage of areas where the EI exceeds 2.5. On this basis, about 414,000 km² of land in the western United States is identified as susceptible to having irrigation-induced selenium problems. The greater the EI, the more likely that there will be selenium problems. The dark areas on Figure 6 are the most arid (EI > 3.5) and are the most likely to have elevated concentrations of selenium. About 136,000 km² in the western United States is in this category.

When using Figure 6, it is important that land adjacent to areas mapped as susceptible be considered provisionally susceptible because selenium can be transported from source areas in mountains to irrigated areas in nearby valleys. In California, all the areas mapped as susceptible are in mountain ranges where there is no irrigation. The actual areas that are susceptible, which do not show on the map, are the irrigated areas on alluvial fans and valleys at the base of the mountains. Another reason to treat as susceptible land adjacent to areas mapped as susceptible is that the parent material for the soils may be a seleniferous Tertiary continental sedimentary deposit. The NIWQP data clearly indicate that selenium problems can develop in areas where Tertiary continental sedimentary deposits form the bedrock if there are nearby Upper Cretaceous marine sedimentary rocks.

Twelve of the 14 areas where the 75th percentile of the selenium concentrations exceeds 3 μg/L are correctly identified as susceptible because they fall on or adjacent to areas mapped as susceptible. One of the two areas with elevated selenium concentrations that was not identified is the Sun River area (X). The reason it was not identified is discussed later. The other such area is the Salton Sea area (U), which is not directly identified by the map as susceptible because selenium is imported into the area. However, Figures 2 and 6 indicate that the

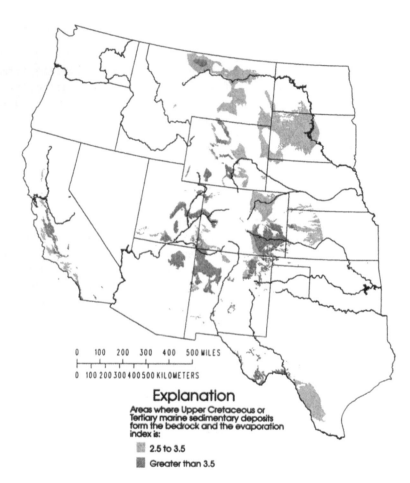

FIGURE 6   Locations of areas in the western United States susceptible to irrigation-induced selenium contamination.

Colorado River traverses seleniferous rocks, and this information could be used to predict the occurrence of selenium problems in arid downstream reaches of the river.

Two areas were identified as being susceptible where reconnaissance investigations did not turn up selenium problems. In one area, the Milk River Basin (P), all the selenium concentrations in surface water measured were less than 1 μg/L. As discussed previously, all samples from the Milk River Basin were collected during a flood year, but this area is probably seleniferous under normal circumstances (Seiler, 1995). The other area is the Sacramento Refuge Complex area (T), which lies adjacent to an area mapped as susceptible. This area had one of the lowest EI values (2.6) of the NIWQP areas; being near seleniferous rocks may not be a problem if the EI is low.

D.   Reliability of the Predictions

The fundamental assumptions for the predictions are that application of irrigation water will mobilize selenium only from soils derived from Upper Cretaceous or Tertiary marine sedimentary deposits and that subsequent evaporative enrichment is the principal factor that determines whether toxic concentrations of selenium will occur. There are known circumstances in which these predictive methods fail to identify susceptible areas because the foregoing assumptions are not met.

Both predictive methods will fail to identify areas where soils are derived from seleniferous rocks that are not Upper Cretaceous or Tertiary marine sedimentary deposits. The predictive methods may fail, as well, when selenium concentrations exceeding criteria can be reached with no, or very little, subsequent evaporative enrichment. This may occur in areas where the soils are much more seleniferous than those in areas studied by the NIWQP. In such areas, application of irrigation water may mobilize so much selenium that the chronic criterion is exceeded even though the EI is less than 2.5, which is the minimum value used in the predictive tools to identify susceptible areas.

Inaccuracies in the GIS coverage for the evaporation index reduce the reliability of the map of susceptible areas in central Montana. The configuration of the contours on the precipitation map results in underestimating values for EI in central Montana. As a result, areas in Montana that probably would have selenium problems, such as the Sun River area (X), are not identified. Somewhat similar problems occur in parts of California, Arizona, and Nevada, where the EI is computed in the entire area as if the precipitation were 25.4 cm because the lowest contour on the precipitation map is 25.4 cm and values less than that are not shown. Fortunately, in those areas of California, Arizona, and Nevada where the EI is inaccurate, the reliability of the map of susceptible areas is not greatly affected because usually the EI is much greater than 2.5 and seleniferous rocks do not form the bedrock. Where adequate data are available, development of GIS coverages for geology and EI at a localized scale could provide more accuracy.

E.   Biological Significance of Predictive Methods.

Biological data collected as part of the NIWQP were used to assess the reliability of the predictive methods. Kesterson-like deformities of bird embryos were found in four areas (F, H, N, and Y). The decision tree and map identify the potential for selenium problems in all four of those areas. The map cannot be tested by comparison with areas where no deformities were found. Thorough biological surveys were not done in all areas, and the lack of observed deformities in an area may not mean that deformities are not occurring. In some areas, not enough or no late-stage embryos were examined for deformities and, in other areas, the embryos that were examined were not from the areas where high selenium concentrations were measured in eggs or water.

The biological significance of the decision tree and map also was assessed using selenium concentrations in bird eggs collected during the NIWQP studies. Seiler and Skorupa (1995) classified the eggs from 23 areas into one of three categories (normal, elevated, and embryotoxic) on the basis of their selenium content. In 12 of 23 NIWQP areas where eggs were collected, the selenium concentration of eggs from at least one population of breeding birds was classified as embryotoxic. The map identified 9 of those 12 areas as susceptible to irrigation-induced selenium contamination.

VI. CONCLUDING PERSPECTIVES

Because selenium is toxic at low concentrations, knowing where elevated concentrations of selenium will or will not occur is important so that appropriate measures can be taken to protect wildlife. The decision tree and map presented here are intended to be used as screening and ranking tools by managers. After several areas have been ranked for likelihood of selenium problems, resources can be directed toward assessing selenium concentrations in areas where problems are most likely to occur.

The values of EI used in the decision tree and map are empirical and are derived solely from an examination of the NIWQP data set. Although data available from other areas (e.g., Kesterson Reservoir, Fig. 5) presently support the choice of these values (Seiler, 1995), additional data may result in future revision of the decision tree and map. In particular, the EI values of 4.5 and 8 used in the decision tree are based on only few data points.

Conclusions based on the decision tree and map that an area is not at risk, should be examined critically. Both the decision tree and the map are based on average climatic conditions, and although problems may not occur during normal years or wet years, they could occur during drought years and other periods of reduced water availability.

ACKNOWLEDGMENTS

I thank the many members of the U.S. Department of the Interior who participated in the NIWQP investigations and provided me with detailed information about the study areas. I also acknowledge my debt to past and present researchers for their contributions in the literature and to colleagues for suggestions and stimulating discussions that greatly improved this chapter.

REFERENCES

Barnes, I. 1985. Sources of selenium. *Selenium and Agricultural Drainage: Implications for San Francisco Bay and the California Environment*, pp. 41–51. Proceedings of the Second Selenium Symposium, Berkeley, CA, March 23, 1985.

Berrow, M. L., and A. M. Ure. 1989. Geological materials and soils. In M. Ihnat (ed.), *Occurrence and Distribution of Selenium*, pp. 213–242. CRC Press, Boca Raton, FL.

Byers, H. G. 1935. Selenium occurrence in certain soils in the United States, with a discussion of related topics. U.S. Department of Agriculture, Tech. Bull. No. 482.

Farnsworth, R. K., E. S. Thompson, and E. L. Peck. 1982. Evaporation atlas for the contiguous 48 United States. Map 3: Free water surface exchange, 1956–70: National Oceanographic and Atmospheric Administration Tech. Report No. NWS 33.

Helsel, D. R., and T. A. Cohn. 1988. Estimation of descriptive statistics for multiply censored water quality data. *Water Resour. Res.* 24:1997–2004.

Helsel, D. R., and R. M. Hirsch. 1992. *Statistical Methods in Water Resources*. Elsevier Science Publishing, New York.

King, P. B., and H. M. Beikman. 1974. Geologic map of the United States (exclusive of Alaska and Hawaii). U.S. Geological Survey Special Geologic Map. Scale 1 : 2,500,000.

Lakin, H. W., and H. G. Byers. 1941. Selenium occurrence in certain soils in the United States, with a discussion of related topics: Sixth report. U.S. Department of Agriculture, Tech. Bull. No. 783.

Lemly, A. D. 1985. Toxicology of selenium in a freshwater reservoir: Implications for environmental hazard evaluation and safety. *Ecotoxicol. Environ. Saf.* 10:314–338.

Presser, T. S. 1994a. The Kesterson effect. *Environ. Manage.* 18(3):437–454.

Presser, T. S. 1994b. Geologic origin and pathways of selenium from the California Coast Ranges to the West-Central San Joaquin Valley, In W. T. Frankenberger, Jr., and S. Benson (eds.), *Selenium in the Environment*, pp. 139–155. Dekker, New York.

Seiler, R. L. 1995. Prediction of areas where irrigation drainage may induce selenium contamination of water. *J. Environ. Qual.* 24(5):973–979.

Seiler, R. L. 1996. Synthesis of data from studies by the National Irrigation Water Quality Program. *Water Resourc. Bull.* 32(6):1233–1245.

Seiler, R. L., and J. P. Skorupa. 1995. Identification of areas at risk for selenium contamination in the western United States. In W. R. Hotchkiss, J. S. Downey, E. D. Gutentag, and J. E. Moore (eds.), *Water Resources at Risk*, pp. LL85–LL94. American Institute of Hydrology, Minneapolis, MN.

Skorupa, J. P., and H. M. Ohlendorf. 1991. Contaminants in drainage water and avian risk thresholds. In A. Dinar, and D. Zilberman (eds.), *The Economics and Management of Water and Drainage in Agriculture*, pp. 345–368. Kluwer Academic Publishers, Boston.

Stephens, D. W., B. Waddell, L. A. Peltz, and J. B. Miller. 1992. Detailed study of selenium and selected elements in water, bottom sediment, and biota associated with irrigation drainage in the Middle Green River Basin, Utah, 1988–90. U.S. Geological Survey Water Resources Invest. Report No. 92-4084.

Tanji, K., A. Lauchli, and J. Meyer. 1986. Selenium in the San Joaquin Valley. *Environment* 28(6):6–39.

Trelease, S. F., and O. A. Beath. 1949. Selenium, its geological occurrence and its biological effects in relation to botany, chemistry, agriculture, nutrition, and medicine. Published by the authors, Champlain Printers, Burlington VT.

U.S. Environmental Protection Agency. 1987. Ambient water quality criteria for selenium—1987. Report EPA-440/5-87-006. U.S. Office of Water Regulations and Standards, Washington, DC.

21

A Classification Model That Identifies Surface Waters Containing Selenium Concentrations Harmful to Waterfowl and Fish

DAVID L. NAFTZ and MATTHEW M. JARMAN
United States Geological Survey, Salt Lake City, Utah

I. INTRODUCTION

Well-established pattern recognition and classification modeling techniques applied to large, multivariate databases generated from national- or regional-scale environmental data synthesis programs can be useful in addressing environmental problems. For example, these methods may be combined with existing geochemical modeling techniques to identify, by means of readily available major ion chemical data, areas in which the concentrations of selenium in surface water may be harmful to waterfowl and fish. Classification models based on these regional- or national-scale data synthesis programs can then be used by regulatory and management personnel to assess selenium toxicity potential in areas for which data are available for major ion concentrations but not for selenium concentrations.

Selenium is a good candidate for classification modeling applications. Numerous circumstances may contribute to the lack of selenium data for surface water samples; this lack of data creates a need for a selenium hazard classification model that can identify water samples that may be harmful to waterfowl and fish. Possible applications of this model include (1) samples for which analytical techniques have large detection limits for selenium, preventing the required analytical sensitivity to detect hazardous selenium concentrations, (2) use by developing countries that are expanding their agricultural lands and have a limited selenium database from which to make environmentally sound decisions, and (3) use when the analytical reliability of selenium concentrations from an area may be in question, even though the major ion data are of good quality.

The National Irrigation Water Quality Program (NIWQP) provides an extensive database for the construction of a classification model that identifies water that may pose a selenium hazard to waterfowl and fish. The NIWQP was implemented in 1985 by the U.S. Department of the Interior (USDOI) because of concern about possible adverse effects from irrigation drainage in the United States. In 1986 reconnaissance phase investigations of irrigation drainage at nine areas in the western United States began under the NIWQP. Twenty-six studies had been completed as of 1997 (Fig. 1). A relational database of all the NIWQP data (Seiler, 1994) contains results of more than 10,000 chemical analyses of water, bottom sediment, and biota. Water quality and bottom sediment data from the NIWQP are available via the World Wide Web at:

http://www.fws.gov/~r9dec/identify/niwqp/irrgwat2.html

Preliminary evaluation of data from the NIWQP study areas indicated that selenium was the constituent most often found at elevated concentrations in water, bottom sediment, and biota (Sylvester et al., 1988; Feltz et al., 1991; Presser, 1994; Presser et al., 1994). Therefore, additional study of the physical and geochemical processes controlling selenium concentrations in water on a regional scale is justified (Seiler, 1995; Naftz, 1996a; Nolan and Clark, 1997). Naftz (1996a, 1996b) combined geochemical and statistical modeling techniques with the major ion water quality data from the NIWQP to develop a classification model that uses only major ion water quality data to identify surface waters that contain selenium concentrations that may pose a hazard to waterfowl and fish.

This chapter describes the geochemical and statistical methods used to develop the selenium-hazard classification model and demonstrates the success of the classification model in identifying Wyoming drainage basins where surface waters may pose a selenium hazard to waterfowl and fish.

II. METHODOLOGY

A. Normative Salts

Interpretation of normative salts in water samples provided a different approach to identifying hydrochemical facies instead of using dominant, unassociated cations and anions (Back, 1966). A normative salt composition of a water sample is the quantitative ideal equilibrium assemblage that would crystallize if water evaporated completely at 25°C and 1 bar pressure with atmospheric pCO_2 (Bodine and Jones, 1986). Characterization of water composition by an assemblage of

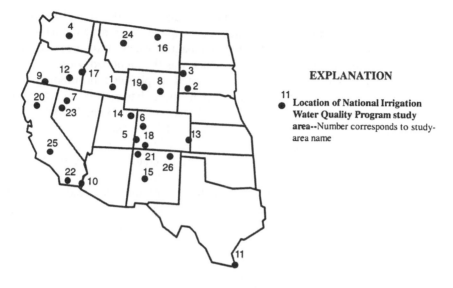

EXPLANATION

11
● Location of National Irrigation
Water Quality Program study
area--Number corresponds to study-
area name

STUDY-AREA NAMES

1. American Falls Reservoir, Idaho
2. Angostura Reclamation Unit, South Dakota
3. Belle Fourche Reclamation Project, South Dakota
4. Columbia River Basin, Washington
5. Dolores Ute Mountain Area, Colorado
6. Gunnison River Basin/Grand Valley Project, Colorado
7. Humboldt River Area, Nevada
8. Kendrick Reclamation Project Area, Wyoming
9. Klamath Basin Refuge Complex, California and Oregon
10. Lower Colorado River Valley, California and Arizona
11. Lower Rio Grande and Laguna Atascosa
    National Wildlife Refuge, Texas
12. Malheur National Wildlife Refuge, Oregon
13. Middle Arkansas River Basin, Colorado and Kansas

14. Middle Green River Basin, Utah
15. Middle Rio Grande and Bosque del Apache
    National Wildlife Refuge, New Mexico
16. Milk River Basin, Montana
17. Owyhee-Vale Reclamation Project Area,
    Oregon and Idaho
18. Pine River Area, Colorado
19. Riverton Reclamation Project, Wyoming
20. Sacramento Refuge Complex, California
21. San Juan River Area, New Mexico
22. Salton Sea Area, California
23. Stillwater Wildlife Management Area, Nevada
24. Sun River Area, Montana
25. Tulare Lake Bed Area, California
26. Vermejo Project, New Mexico

FIGURE I   Location of National Irrigation Water Quality Program study areas, 1986–1997.

salts provides information on solute origin and subsequent interaction to a greater extent than major cation–anion predominance graphs. Salt norm data are useful in evaluating the similarities and differences in water from different geochemical landscapes. Because selenium initially is mobilized during the application of water on irrigated lands in areas underlain by seleniferous materials, salt norm data may be very useful in identifying selenium-producing landscapes based on "*characteristic*" normative salt compositions. For example, salt norms calculated from water samples with elevated selenium concentrations in the San Joaquin Valley, California, were used to confirm solute sources and weathering cycles indicated by detailed mineralogical data (Presser and Swain, 1990). The normative salt assemblages [thenardite (Na_2SO_4), bloedite [$Na_2Mg(SO_4)_2 \cdot 4H_2O$], and glauberite [$Na_2Ca(SO_4)_2$]] calculated by the geochemical computer program SNORM (Bodine and Jones, 1986) were identified in the San Joaquin Valley study area and were indicative of sulfide mineral oxidation, which was the identified source of selenium. Normative salt assemblages and associated simple salt concentrations also were used to identify the weathering reactions most characteristic in selenium-producing landscapes in the western United States (Naftz, 1994, 1996a).

During construction of the classification model, SNORM was used to calculate the salt norm from the major ion chemical analysis of selected surface water samples from NIWQP study sites. The SNORM program distributes solutes into normative salts assigned from 63 possibilities. SNORM is written in FORTRAN IV and performs three major tasks: (1) solute concentrations are read, converted to appropriate concentration units, and adjusted to yield a cation–anion charge balance; (2) an equilibrium normative salt assemblage is determined according to the principles of phase equilibria, and the solutes are assigned into the salts of the assemblage; and (3) the major solute normative salts are converted into 12 single-cation–single-anion simple salts, namely, the alkaline earth and alkali salts of carbonate ($CaCO_3$, $MgCO_3$, Na_2CO_3, K_2CO_3), sulfate ($CaSO_4$, $MgSO_4$, Na_2SO_4, K_2SO_4), and chloride ($CaCl_2$, $MgCl_2$, $NaCl$, KCl). Hydrochemical facies are then described in terms of the principal simple salts they contain.

B. Statistical Techniques

Statistical techniques collectively referred to as pattern recognition analysis are useful in extracting chemical information from large, multivariate databases that may otherwise be difficult or impossible to interpret. For example, these techniques have been used to discriminate marble sources (Mello et al., 1988), to classify ancient ceramics (Heydorn and Thuesen, 1989), to differentiate sources of smoke aerosols (Voorhees and Tsao, 1985), and to identify sources of oil spills (Duewer et al., 1975).

Pattern recognition analysis techniques were applied to the simple salt data matrix calculated from the SNORM program to identify the potential "selenium" and "nonselenium" hydrochemical facies represented in the NIWQP database. The pattern recognition analysis of the NIWQP database consisted of two phases: exploratory data analysis, and applied pattern recognition analysis (Meglen, 1988) or classification modeling. During exploratory data analysis, a combination of statistical techniques (principal component analysis [PCA], cluster analysis, and individual statistical correlations) were combined with data visualization techniques using the software package PIROUETTE (Infometrix, 1992) to identify hydrochemical facies characteristic of surface waters containing dissolved (0.45 μm filter size) selenium concentrations (selenium ≥ 3.0 μg/L) found to be hazardous to some species of aquatic birds (Skorupa and Ohlendorf, 1991).

Furthermore, Lemly (1993) found that dissolved (0.45 μm filter size) selenium concentrations of 2 μg/L or greater should be considered hazardous to the long-term survival of fish and wildlife populations. Additional details on the application of exploratory data analysis to multivariate databases can be found in publications of Meglen (1988) and Naftz (1996a).

In the applied pattern recognition analysis of the data (phase 2), a "*formal*" classification model is constructed that allows for the classification of unknown water samples into the "selenium" or "nonselenium" facies identified during the exploratory data analysis phase (phase 1). The classification algorithm, soft independent modeling by class analogy (SIMCA), was used to construct the classification model. The SIMCA algorithm uses PCA to construct a separate principal component model that describes each of the hydrochemical facies identified during the exploratory data analysis phase. The NIWQP database served as the training data set during construction of the classification model (Naftz, 1996a). After construction, the classification model was used to identify water samples that may pose a hazard to waterfowl and fish when the selenium concentration is unknown. Additional details on the application of classification modeling techniques to multivariate data sets can be found in the work of Wold and Sjostrom (1977), Meglen (1988), and Infometrix (1992).

III.   RESULTS AND DISCUSSION

A.   Pattern Recognition Analysis of the NIWQP Database

Simple salt concentrations were calculated for 1962 surface water samples collected from 23 of the 26 NIWQP study areas. The study areas that were not included (Angostura Reclamation Unit, SD; Milk River Basin, MT; and Lower Rio Grande and Laguna Atascosa National Wildlife Refuge, TX) did not have the appropriate major ion data for calculation of simple salt concentrations. Results of the PCA applied to the simple salt data indicated three principal components (PCs) best explain the data set. The first three PCs accounted for 95% of the total variance of the simple salt data set. The PC scores for each of the three PCs are plotted (Fig. 2) to evaluate the occurrence of distinct clusters in the data that may indicate common geochemical processes controlling surface water chemistry among and within NIWQP study areas. The PC scores of the 1962 water samples were grouped into three distinct clusters when viewed in three dimensions and are classified as facies 1, 2, and 3 (Fig. 2). The boundaries drawn around the clusters of PC scores are not definitive, but they aid in the visualization of the data and confirm possible commonalities in geochemical processes indicated by the variations in simple salt concentrations for each of the three facies. Facies 1 samples have large scores on PC 1 relative to PCs 2 and 3, whereas facies 2 and 3 samples have small scores on PC 1 relative to PCs 2 and 3 (Fig. 2).

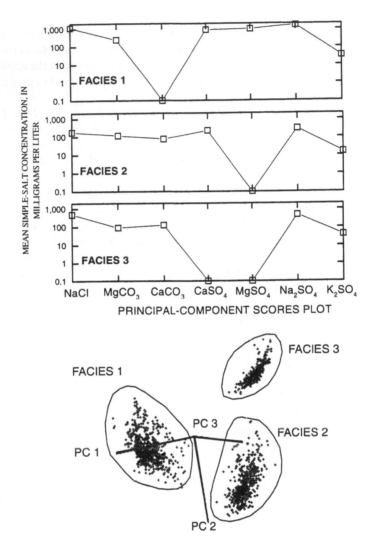

FIGURE 2   Mean simple salt concentrations for facies 1, 2, and 3 and three-dimensional plot of principal component scores for each water sample in the NIWQP database.

Variation in mean simple salt concentration in each of the first three PCs is shown in Figure 2. Facies 1 samples are distinguished by water samples without $CaCO_3$ and elevated concentrations of NaCl, $CaSO_4$, $MgSO_4$, and Na_2SO_4 relative to water samples in facies 2 and 3. The large mean concentration of sulfate-bearing simple salts in facies 1 water samples probably reflects dominant sulfuric acid weathering derived from the oxidation of sulfides and/or resolution of sulfate minerals deposited previously. The dominance of sulfuric acid weathering in facies 1 water samples is reinforced by the absence of the $CaCO_3$. The elevated concentration of NaCl in facies 1 water samples probably reflects a marine or evaporative influence. In summary, the simple salt association in facies 1 water samples is most characteristic of meteoric water derived from the weathering of marine shales with both reduced and oxidized sulfur mineral phases.

Water samples in facies 2 are distinguished from facies 1 samples by the absence of $MgSO_4$ and the presence of $CaCO_3$ (Fig. 2). Water samples in facies 3 are similar to samples in facies 2, except for the absence of both $MgSO_4$ and $CaSO_4$. The lower mean concentration of sulfate simple salts in facies 2 water samples coupled with the presence of $MgCO_3$ and $CaCO_3$ reflects a mixed system of both sulfuric and carbonic acid hydrolysis of rock-forming minerals. The increased mean concentration of $CaCO_3$ coupled with the absence of both $CaSO_4$ and $MgSO_4$ in facies 3 water samples (Fig. 2) reflects the dominance of a carbonic acid weathering regime. For all 23 study areas, the median selenium concentration for water samples in facies 1 is 10 $\mu g/L$, which exceeds the concentration that may pose a hazard to waterfowl and fish. Median selenium concentration for samples in facies 2 (2.0 $\mu g/L$) is much less than for facies 1; however, the 2.0 $\mu g/L$ median selenium concentration may still be hazardous to waterfowl and fish using the limit suggested by Lemly (1993). Median selenium concentration for samples in facies 3 (<1.0 $\mu g/L$) is less than the concentration considered hazardous to waterfowl and fish.

A bivariate plot comparing the percentage of facies 1 water samples from each of the 23 study areas to the percentage of water samples with selenium concentrations greater than or equal to 3.0 $\mu g/L$ (Fig. 3) was used to illustrate that facies 1 water samples typically contain selenium concentrations that pose a hazard to waterfowl and fish. A positive correlation ($r = 0.775$, $p < 0.0001$, $n = 23$) between the percentage of facies 1 samples in each study area and the percentage of samples with elevated selenium concentrations (≥ 3.0 $\mu g/L$) is indicated (Fig. 3). The simple salt assemblage characteristic of facies 1 water samples indicates the occurrence of conditions that could allow selenium to be oxidized from sulfide minerals such as pyrite and ferroselite to the selenate form. Once oxidized, the selenate ion could then be substituted for sulfate in the open-lattice structure of soluble sulfate salts. Precipitation of these salts could act as a temporary selenium sink during wet/dry cycles in the irrigated areas. Presser and Swain (1990) provided a detailed mineralogical analysis in a seleniferous area of San Joaquin Valley, California, and found evidence of release of the selenium by sulfuric acid weathering. This mineralogical evidence included selenium-bearing sulfide and sulfate minerals in an area that contains outcrops of marine sediments of Upper Cretaceous–Paleocene age.

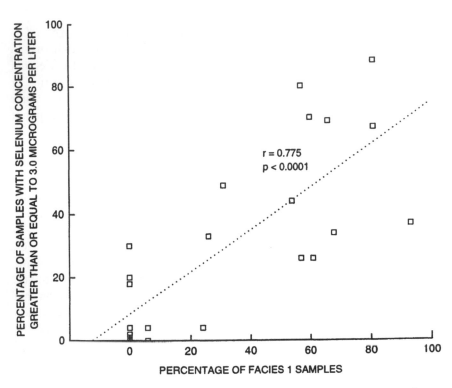

FIGURE 3 Percentage of facies 1 water samples in each study area and percentage of water samples from each study area with a selenium concentration that poses a hazard to waterfowl and fish (≥3.0 μg/L).

B. Construction of the Classification Model

Exploratory data analysis of the SNORM simple salt data from the NIWQP study areas indicated that water samples with hazardous selenium concentrations have a unique simple salt geochemical fingerprint (Naftz, 1996a). On the basis of these results, the classification algorithm, SIMCA, was applied to the simple salt data set to formalize water sample classification through construction of a classification model. The SIMCA algorithm was selected for application to the NIWQP data for two reasons. First, the SIMCA algorithm is best suited to sample-rich classes (Meglen, 1988) typified by the NIWQP database. Second, the SIMCA algorithm does not force a class assignment to an "outlier." This feature is beneficial because it preserves the option that outlier samples may belong to an unanticipated class not initially hypothesized in the training set, as opposed to forcing membership into a currently hypothesized class.

The training data set was log-transformed and mean-centered prior to application of the SIMCA alogrithm. A probability level of 0.95 was used to define class inclusion. Specific parameters were evaluated during the SIMCA classification modeling of the training data set to ensure an optimal model for classification of the test data. Two principal components were retained in each of the three classes, and the homogeneity of each class was ensured by visual inspection of the PC scores for each class. Class separation and overall modeling power of the classification model were found to be acceptable, and the resulting classification model was successful in differentiating hazardous from nonhazardous selenium samples in a test data set (Naftz, 1996a). The water samples assigned to class 1 had a high potential selenium hazard associated with elevated selenium concentrations that could be hazardous to waterfowl and fish, and water samples in class 2 had a low to moderate potential degree of selenium hazard associated with slightly elevated selenium concentrations that were derived from a mixed system of both sulfuric and carbonic acid hydrolysis of rock-forming minerals. Water samples assigned to class 3 were associated with nonhazardous selenium concentrations.

C. Application of the Classification Model

Although Naftz (1996a) showed that the classification model is successful in differentiating hazardous and nonhazardous water samples, the model has not been used to identify individual drainage basins that contain water that could pose a selenium hazard to waterfowl and fish. The classification model was used in combination with geographic information system (GIS) mapping of seleniferous geologic formations on two drainage basins (Powder River and North Platte River) in Wyoming (Figs. 4 and 5) to demonstrate how the model might be used to identify smaller drainage basins (referred to as subbasins in this chapter) that probably contain hazardous selenium concentrations in water. Both the principal drainage basins contain subbasins that have elevated selenium concentrations (Lowry and Wilson, 1986; Naftz and Rice, 1989; See et al., 1992; Naftz et al., 1993) and potentially seleniferous geologic units (Figs. 4 and 5) mapped by Case and Cannia (1988). Drainage subbasins that have a large proportion of seleniferous geologic units would most likely contain surface waters that have elevated selenium concentrations. Major ion water quality data collected after December 1969 from all surface water sampling sites in both principal drainage basins were compiled from the National Water Data Storage and Retrieval (WATSTORE) System of the U.S. Geological Survey. The classification model was applied to the most recent major ion analysis from each sampling site, and the results were compared to mapped seleniferous areas for each drainage subbasin within the Powder River and North Platte River drainage basins in Wyoming (Figs. 4 and 5).

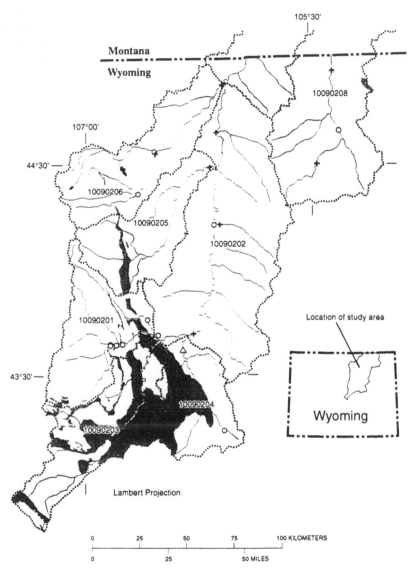

FIGURE 4 Extent of seleniferous geologic units in relation to classification modeling results from 21 surface water sites in the Powder River drainage basin, Wyoming.

EXPLANATION

+ **Class 1**—High potential selenium hazard to waterfowl and fish

○ **Class 2**—Low to moderate potential selenium hazard to waterfowl and fish

△ **Class 3**—No potential selenium hazard to waterfowl and fish

————— **Stream**

·············· **Drainage subbasin boundary**

10090202 **Hydrologic unit code**

Geologic formations, equivalents of geologic formations, or locally derived soils that have the potential to support seleniferous vegetation that (Adapted from Case and Cannia, 1988):

Is mildly toxic or unclassified

May be moderately toxic to animals in localized areas

May be highly toxic to animals in localized areas

FIGURE 4 Continued

Visual inspection of the classification modeling results for the 21 sites in the Powder River drainage basin produces general agreement with the mapped areas of geologic formations supporting highly and moderately seleniferous vegetation (Fig. 4). For example, the sampling sites at the mouth of drainage subbasin 10090203 and near the state line in subbasin 10090208 are in class 1 (high potential selenium hazard to waterfowl and fish). More than 78% of the surface area in drainage subbasin 10090203 and 50% of the surface area in drainage subbasin 10090208 (Wyoming only) contain geologic units that support highly or moderately seleniferous vegetation (Fig. 4). The site in subbasin 10090203 also was noted by Lowry and Wilson (1986) as a source of water samples with selenium concentrations consistently higher than 10 $\mu g/L$.

In contrast, sample sites in the upper parts of drainage subbasins 10090201, 10090204, and 10090206 (Fig. 4) are in class 2 (low to moderate selenium hazard to waterfowl and fish). The upper parts of all three drainage subbasins contain no areas, or only minor areas, where geologic units support highly or moderately seleniferous vegetation (Fig. 4). For example, only 17% of the area in drainage subbasin 10090201 contains seleniferous geologic units. All four sample sites in this subbasin are in class 2 (low to moderate potential selenium hazard to waterfowl and fish).

The site at the mouth of drainage subbasin 10090204 appears to be misclassified in class 3 (no selenium hazard to waterfowl and fish). This subbasin contains

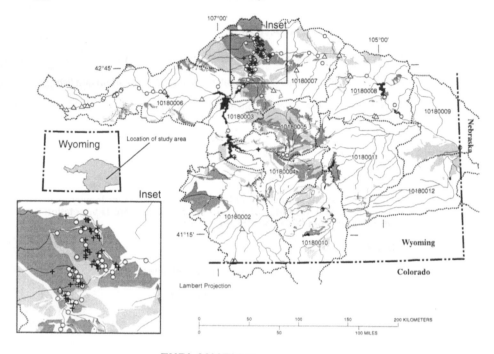

EXPLANATION

+   **Class 1**—High potential selenium hazard to waterfowl and fish

○   **Class 2**—Low to moderate potential selenium hazard to waterfowl and fish

△   **Class 3**—No potential selenium hazard to waterfowl and fish

———— **Stream**

············· **Drainage subbasin boundary**

10180012   **Hydrologic unit code**

■   **Lake or reservoir**

Geologic formations, equivalents of geologic formations, or locally derived soils that have the potential to support seleniferous vegetation that (Adapted from Case and Cannia, 1988):

☐   Is mildly toxic or unclassified

▨   May be moderately toxic to animals in localized areas

▨   May be highly toxic to animals in localized areas

FIGURE 5   Extent of seleniferous geologic units in relation to classification modeling results from 151 surface water sites in the North Platte River drainage basin, Wyoming.

large areas of geologic units that support highly or moderately seleniferous vegetation and should be classified as a subbasin with a high potential selenium hazard to waterfowl and fish. The site-specific reason for this misclassification is unknown; however, it could be related to variables not considered in the classification model. Such additional variables may include (1) specific amount of runoff that occurs in seleniferous and nonseleniferous areas of the drainage subbasin, (2) variable distribution of precipitation in the drainage subbasin, (3) average topographic slope, (4) groundwater versus surface water component within streams in the drainage subbasin, and (5) land use.

The North Platte River drainage basin in Wyoming (Fig. 5) was used as the second example for the application of the classification model. In agreement with the Powder River basin example, the classification modeling results of the 151 sampling sites in the North Platte River basin agree with the distribution of mapped areas of geologic units supporting seleniferous vegetation. Class 1 sample sites generally are located downstream of areas that have a large proportion of geologic formations supporting highly or moderately seleniferous vegetation. Examples include the northwestern part of subbasin 10180002, subbasin 10180004, and the northwestern part of subbasin 10180007, which contains the Kendrick Reclamation Project Area, an NIWQP study site where selenium hazards to waterfowl have been documented (See et al., 1992).

The two sample sites at the mouth of subbasin 10180005 appear to be misclassified as class 2 (Fig. 5). More than 58% of the surface area in drainage subbasin 10180005 contains geologic units that support highly or moderately seleniferous vegetation (Fig. 5), and therefore it is unclear why sites at the mouth of this subbasin are not in class 1. The 30 major-ion analyses available from both sites were individually classified to determine whether temporal variation at the sites was partly responsible for the apparent misclassification. Classification of the 30 analyses available from both sites indicated that individual water samples collected from both sites were in class 1 during selected time periods, consistent with the dominance of seleniferous geologic units in the drainage subbasin. The mechanism responsible for the apparent oscillation between classes 1 and 2 is unclear; however, discharge from Boles Spring and diversion of significant volumes of water for irrigation upstream of the sampling sites (Druse and Rucker, 1986) could be partly responsible.

The classification model also is successful in identifying areas that have a low to moderate or no potential selenium hazard to waterfowl and fish. For example, drainage subbasins 10180006, 10180008, 1018009, and 10180011 do not contain large areas of geologic units that support seleniferous vegetation (<9% of each subbasin drainage area). Samples sites from these subbasins are generally in class 2 (low to moderate selenium hazard to waterfowl and fish) or 3 (no selenium hazard to waterfowl and fish).

A bivariate plot showing the relation between the proportion of seleniferous geologic units in each subbasin to the proportion of class 1 (high potential selenium hazard to waterfowl and fish) sampling sites in each subbasin was used to illustrate how the model can consistently identify high potential selenium hazard subbasins (Fig. 6). For the purpose of the plot, both sites in subbasin 10180005 were designated as class 1 using the justification discussed previously. A positive correlation ($r = 0.801$, $p = 0.003$, $n = 10$) between proportions of seleniferous geologic units and class 1 sites in each drainage subbasin within the North Platte River drainage basin in Wyoming is indicated (Fig. 6).

The combination of the classification model with a GIS was successful in differentiating between drainage subbasins containing high, low to moderate, and nonhazardous concentrations of selenium. The modeling results from the two

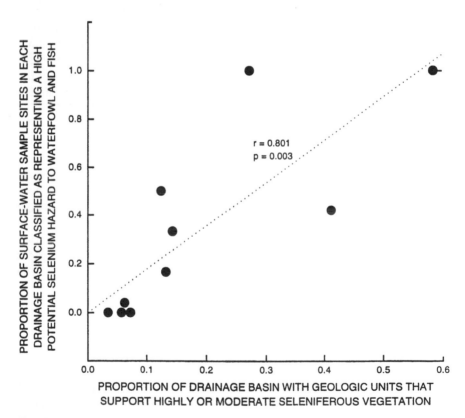

FIGURE 6 Proportion of seleniferous units in each North Platte River drainage subbasin compared to proportion of surface water sites classified as presenting a selenium hazard to waterfowl and fish.

principal drainage basins in Wyoming indicate that the same technique could be applied to other drainage basins throughout the western United States and other parts of the world to produce selenium hazard maps. These hazard maps could be used by land-use managers and regulatory personnel for numerous applications, including the placement of irrigation and wetland enhancement projects in areas where detailed selenium data are not available or are unreliable.

IV. SUMMARY

Established chemometric and geochemical techniques were applied to water quality data from 23 NIWQP study areas in the western United States. These techniques were applied to a data set to identify common geochemical processes responsible for mobilization of selenium and to develop a classification model that uses major ion concentrations to identify areas that contain elevated selenium concentrations in water that may pose a hazard to waterfowl and fish.

Pattern recognition modeling of the simple salt data computed with the SNORM geochemical program indicates that three principal components explain 95% of the total variance of the simple salt database. A three-dimensional plot of PC 1, 2, and 3 scores shows three distinct clusters that correspond to distinct hydrochemical facies denoted as facies 1, 2, and 3. Facies 1 water samples are distinguished by water without $CaCO_3$ and elevated concentrations of NaCl, $CaSO_4$, $MgSO_4$, and Na_2SO_4 relative to water samples in facies 2 and 3. Water samples in facies 1 present a high potential selenium hazard to waterfowl and fish, with a median selenium concentration of 10 $\mu g/L$. Water samples in facies 2 present a low to moderate potential selenium hazard (median selenium concentration 2.0 $\mu g/L$), and water samples in facies 3 present no selenium hazard (median selenium concentration < 1.0 $\mu g/L$).

A classification model using the SIMCA algorithm was developed using data from the NIWQP study areas. The combination of the SIMCA classification model with the mapped area of seleniferous geologic units within drainage subbasins was successful in differentiating between drainage subbasins containing hazardous and nonhazardous concentrations of selenium. The results indicate the same technique could be applied to other drainage basins throughout the western United States and other parts of the world to produce selenium-hazard maps. These hazard maps can be used by land-use managers and regulatory personnel in areas lacking detailed, reliable selenium data.

ACKNOWLEDGMENTS

Funding for this study was provided by the U.S. Department of the Interior, National Irrigation Water Quality Program. Technical review of the chapter by Melanie Clark and

Doyle Stephens improved the manuscript and is appreciated. Product names referenced in this chapter are for identification only and do not indicate endorsement by the U.S. Geological Survey.

REFERENCES

Back, W. 1966. Hydrochemical facies and ground-water flow patterns in the northern part of the Atlantic Coastal Plain. U.S. Geological Survey Prof. Paper No. 448-A.

Bodine, M. W., and B. F. Jones. 1986. THE SALT NORM: A quantitative chemical-mineralogical characterization of natural waters. U.S. Geological Survey Water-Resources Invest. Report No. 86-4086.

Case, J. C., and J. C. Cannia. 1988. Guide to potentially seleniferous areas in Wyoming. Geological Survey of Wyoming Open-File Report No. 88-1.

Druse, S. A., and S. J. Rucker. 1986. Water resources data for Wyoming, water year 1985. U.S. Geological Survey Water Data Report No. WY-85-1.

Duewer, D. L., B. R. Kowalski, and T. F. Schatzki. 1975. Source identification of oil spills by pattern recognition analysis of natural elemental composition. *Anal. Chem.* 47:1573–1578.

Feltz, H. R., M. A. Sylvester, and R. A. Engberg. 1991. Reconnaissance investigations of the effects of irrigation drainage on water quality, bottom sediment, and biota in the western United States. In G. E. Mallard and D. A. Aronson (eds.), U.S. Geological Survey Water-Resources Invest. Report No. 91-4034, pp. 319–323.

Heydorn, K., and I. Thuesen. 1989. Classification of ancient Mesopotamian ceramics and clay using SIMCA for supervised pattern recognition. *Chemometric Intell. Lab. Syst.* 7:181–188.

Infometrix. 1992. Multivariate data analysis for IBM PC systems, Version 1.1. Infometrix Inc., Seattle, WA.

Lemly, D. A. 1993. Guidelines for evaluating selenium data from aquatic monitoring and assessment studies. *Environ. Monitor. Assess.* 28:83–100.

Lowry, M. E., and J. F. Wilson, Jr. 1986. Hydrology of Area 50, Northern Great Plains and Rocky Mountain coal provinces, Wyoming and Montana. U.S. Geological Survey Water-Resources Invest. Open-File Report No. 83–545.

Meglen, R. R. 1988. Chemometrics: Its role in chemistry and measurement sciences. *Chemometric Intell. Lab. Syst.* 3:17–29.

Mello, E., D. Monna, and M. Oddone. 1988. Discriminating sources of Mediterranean marbles: A pattern recognition approach. *Archaeometry* 30:102–108.

Naftz, D. L. 1994. Using salt norms to identify selenium-producing landscapes in the western United States. In R. A. Marston and V. R. Hasfurther (eds.), *Effects of Human-Induced Changes on Hydrologic Systems*, pp. 1033–1042. American Water Resources Assocation Summer Symposium, Jackson, WY, 1994.

Naftz, D. L. 1996a. Pattern-recognition analysis and classification modeling of selenium-producing areas. *J. Chemometrics* 10:309–324.

Naftz, D. L. 1996b. Using geochemical and statistical tools to identify irrigated areas that might contain high selenium concentrations in surface water. U.S. Geological Survey Fact Sheet No. FS-077-96.

Naftz, D. L., and J. A. Rice. 1989. Geochemical processes controlling selenium in ground water after mining, Powder River Basin, Wyoming, U.S.A. *Appl. Geochem.* 4:565–576.

Naftz, D. L., R. B. See, and P. Ramirez. 1993. Selenium source identification and biogeochemical processes controlling selenium in surface water and biota, Kendrick Reclamation Project, Wyoming, U.S.A. *Appl. Geochem.* 8:115–126.

Nolan, B. T., and M. L. Clark. 1997. Selenium in irrigated agricultural areas of the western United States. *J. Environ. Qual.* 26:849–857.

Presser, T. S. 1994. The Kesterson effect. *Environ. Manage.* 18:437–455.

Presser, T. S., and W. C. Swain. 1990. Geochemical evidence for Se mobilization by the weathering of pyritic shale, San Joaquin Valley, California, U.S.A. *Appl. Geochem.* 5:703–718.

Presser, T. S., M. A. Sylvester, and W. H. Low. 1994. Bioaccumulation of selenium from natural geologic sources in western states and its potential consequences. *Environ. Manage.* 18:423–436.

See, R. B., D. L. Naftz, D. A. Peterson, J. G. Crock, J. A. Erdman, R. C. Severson, P. Ramirez, and J. A. Armstrong. 1992. Detailed study of selenium in soil, representative plants, water, bottom sediment, and biota in the Kendrick Reclamation Project Area, Wyoming, 1988–90. U.S. Geological Survey Water-Resources Invest. Report No. 91-4131.

Seiler, R. L. 1994. Synthesis of data from the U.S. Department of Interior irrigation drainage studies, western United States, In R. A. Marston and V. R. Hasfurther (eds.), *Effects of Human-Induced Changes on Hydrologic Systems*, p. 1167. American Water Resources Association Summer Symposium, Jackson, WY, 1994.

Seiler, R. L. 1995. Prediction of areas where irrigation drainage may induce selenium contamination of water. *J. Environ. Qual.* 24:973–979.

Skorupa, J. P., and H. M. Ohlendorf. 1991. Contaminants in drainage water and avian risk thresholds. In A. Dinar and D. Zilberman (eds.), *The Economics and Management of Water and Drainage in Agriculture*, p. 345–368. Kluwer Academic Publishers, Norwood, MA.

Sylvester, M. A., J. P. Deason, H. R. Feltz, and R. A. Engberg. 1988. Preliminary results of the Department of Interior's irrigation drainage studies. *Planning Now for Irrigation and Drainage*, pp. 665–667. American Society of Civil Engineers, Lincoln, NE.

Voorhees, K. J., and R. Tsao. 1985. Smoke aerosol analysis by pyrolysis–mass spectrometry/pattern recognition for assessment of fuels involved in flaming combustion. *Anal. Chem.* 57:1630–1636.

Wold, S., and M. Sjostrom. 1977. SIMCA: A method for analyzing chemical data in terms of similarity and analogy. In B. R. Kowalski (ed.), *Chemometrics, Theory and Application*, pp. 243–282. ACS Symposium Series No. 52. American Chemical Society, Washington, DC.

22

Influence of Nitrate on the Mobility and Reduction Kinetics of Selenium in Groundwater Systems

SALLY M. BENSON
Ernest Orlando Lawrence Berkeley National Laboratory, Berkeley, California

I. INTRODUCTION

In the mid-1980s, great concern arose regarding the ecological consequences and potential human health impact associated with selenium derived from agricultural drainage on the west side of the San Joaquin Valley, California. Groundwater contamination was one of the key issues, both because humans and livestock might drink the water, and because the water is an important pathway for contaminant transport within the San Joaquin Valley ecosystem. Over the past decade, much has been learned about the behavior of selenium in the San Joaquin Valley (also known as the Central Valley), including its geologic origins and transport pathways (Presser, 1994), the distribution and mobility of selenium in groundwater (Deverel and Fujii, 1988; Deverel and Millard, 1988), the important oxidation–reduction controls on selenium mobility (Dubrovsky et al., 1991; White and Dubrovsky, 1994), microbial transformations of selenium (Karlson and Frankenberger, 1988; Oremland, 1994; Zawislanski et al., 1996), and its behavior in soils and groundwater at Kesterson Reservoir (Long et al., 1990; White et al., 1991; Tokunaga et al., 1991).

Taken together, these studies indicate that selenium mobility in the groundwater system of the Central Valley depends on a combination of natural and anthropogenic factors, including the nature of the aquifer sediments, the rate of groundwater flow, and the use of nitrogen fertilizers on agricultural fields. It was found that aquifer sediments derived from the Sierra Nevada to the east maintain moderately reducing groundwater conditions, which favor selenate and selenite reduction and immobilization. On the other hand, aquifer sediments derived from the Coast Range (to the west) tend to be more oxidizing, and consequently, selenate reduction is not favored and selenate remains mobile in the groundwater. Exceptions to this general observation about selenium mobility occur where high concentrations of nitrate (>10 mg/L) are observed in Sierran sediments (White et al., 1991). One plausible explanation for this high concentration is that the presence of nitrate inhibits selenium reduction/immobilization (Weres et al., 1990; White et al., 1991; Oremland, 1994). Both field observations and laboratory experiments support this hypothesis. In addition, field and laboratory evidence suggest that denitrification occurs in the reducing Sierran sediments and that following this, selenate reduction can occur (Weres et al., 1990; White et al., 1991). This chapter describes an experiment designed to test the hypothesis that nitrate inhibits selenium reduction/immobilization under in situ conditions, to quantify denitrification rates, and to test whether selenate reduction will follow denitrification.

II. COMMON SPECIES OF SELENIUM IN GROUNDWATER SYSTEMS

Numerous studies, including those mentioned above, have built on the experimental and theoretical foundation for understanding the behavior of selenium in environmental systems provided by Cary et al. (1967), Geering et al. (1968), Cutter (1982), Doran (1982), and Elrashidi et al. (1987) among others. These theoretical and laboratory studies indicate, and field investigations confirm, that selenate [Se(VI)] and selenite [Se(IV)] are the dominant mobile forms of selenium likely to be found in groundwater systems. Both species, which occur as the oxyanions SeO_4^{2-} and SeO_3^{2-}, have been observed under oxidizing to moderately oxidizing conditions. Se(VI) is the more mobile of the two species because, unlike Se(IV), it does not adsorb strongly to surfaces of soil minerals or organic matter under near-neutral pH (Balistrieri and Chao, 1987; Neal and Sposito, 1989). Dissolved (mobile) selenium is not typically found under reducing conditions, because less soluble forms such as elemental selenium [Se(0)] are thermodynamically favored. Laboratory and field studies have demonstrated convincingly that when Se(VI) and Se(IV) are introduced into a moderately reducing environment, they are quickly transformed through microbial processes to Se(0) and/or organic selenium compounds (Long et al., 1990; Weres et al., 1990; White et al., 1991; Oremland, 1994). As mentioned above, the presence of nitrate has been shown to inhibit these transformations in the laboratory, presumably because it is a more favorable electron acceptor (Weres et al., 1990; Oremland, 1994). This observation is consistent with the sequence of reduction reaction expected in near-neutral environments (e.g., O_2, nitrate-N, SeO_4^{2-}, SeO_3^{2-}, SO_4^{2-}) (Lundquist et al., 1994).

In the field experiment presented here, nitrate and Se(VI), the most abundant mobile form of selenium at Kesterson Reservoir, were introduced as a mixture into a mildly reducing (suboxic) aquifer under controlled conditions to evaluate their interaction under in situ conditions. Results from this experiment were then compared to two earlier experiments (Long et al., 1990; White et al., 1991), in which Se(VI) and Se(IV) were introduced into the same aquifer without detectable concentrations of nitrate.

III. AQUIFER DESCRIPTION

Kesterson Reservoir is located on the west side of the Central Valley, about 150 km south of Sacramento, California. The shallow aquifer under Kesterson consists of alternating layers of sand, silt and clay of Sierran origin (Benson, 1988; Benson et al., 1991). The test described here was conducted at a depth interval of 6 to 12 m under the reservoir. The hydraulic conductivity of the sediments at the test site averaged about 3×10^{-4} m/s. The composition of the groundwater with the exception of Se concentration was typical of the drainage water that seeped through the bottom of Kesterson Reservoir into the underlying aquifer; that is, having total dissolved solids concentrations of approximately 10,000 mg/L, pH of approximately 6.8 to 7, oxidation–reduction potential Eh of 100 to 200 mV, dissolved oxygen of less that 100 μg/L, and sulfate concentrations in the range of 3000 to 6000 mg/L (Benson, 1988; White et al., 1991). Prior to the experiment, Se(VI) and Se(IV) concentrations were near the detection level of approximately 2 and 1 μg/L, respectively. Nitrate-N concentrations were less than the detection level of 0.2 mg/L.

IV. TEST SITE DESCRIPTION AND EXPERIMENTAL DESIGN

The test site consisted of an array of 10 wells listed in Table 1 and shown in Figures 1 and 2. During the experiment, a steady stream of water was pumped from well LBL-18A and reinjected back into well LBL-14. This created a steady flow field into which a pulse of a conservative tracer (fluorescein dye), and later a mixture of Se(VI) and nitrate, would be introduced. The movement of the chemicals and their interaction with the sediments could then be monitored from the sampling wells located midway between the extraction and injection wells (see Figs. 1 and 2). The injection and pumping wells were screened over a 6.1 m thick interval of the aquifer. The majority of sampling wells were screened pth, Casing Size, Slotted Interval, and Local Coordinates for the Extraction, Injection, and Monitoring Boreholes

Use	Depth (m)	Slotted interval (m)	Casing diameter (cm)	x Coordinate (m)	y C
Injection	12.19	6.10–12.19	0.10	14.21	
Extraction	12.19	6.10–12.19	0.10	−14.21	
Observation	7.62	6.10–7.62	0.05	−3.38	
Observation	9.14	7.62–9.14	0.05	−3.44	
Observation	10.67	9.14–10.67	0.05	−0.65	
Observation	12.19	10.67–12.19	0.05	0.09	
Observation	13.72	12.19–13.72	0.05	−1.36	
Observation	6.10	4.57–6.10	0.05	−2.17	
Observation	9.14	7.62–9.14	0.05	0.91	
Observation	12.19	10.67–12.19	0.05	−1.83	

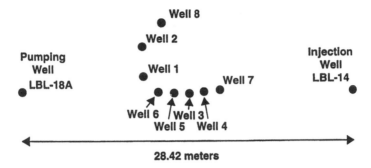

FIGURE 1 Plan view of the location of the injection, extraction, and monitoring wells.

over a 1.5 m interval to enable sampling from discrete depth intervals. Specific information for each of these wells is provided in Table 1.

The entire test took place over a 3-week interval and consisted of two phases. Table 2 summarizes the test parameters. In the first phase, a one-hour pulse of fluorescein dye (average concentration of 140 mg/L) was injected into the aquifer to determine the hydraulic properties of the test site. Samples were then collected over a one-week period until the fluorescein pulse had moved past the sampling wells. During the second phase Se(VI), as sodium selenate,

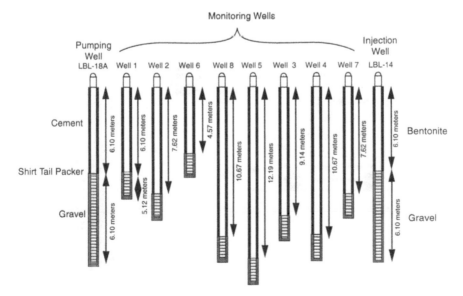

FIGURE 2 Cross-sectional view of the injection, extraction, and monitoring wells.

TABLE 2   Summary of Parameters for the Experiment

Extraction well (screened interval)	LBL-18A (6.1–12.2 m)
Injection well (screened interval)	LBL-14 (6.1–12.2 m)
Distance between injection and extraction well	28.4 m
Flow rate (Q)	5.05×10^{-3} m³/s
Tracer	
Phase 1	140 mg/L fluorescein
Pulse duration	3825 s
Phase 2	300 μg/L Se(VI)
	42 mg/L nitrate-N
Pulse duration	20,160 s

and nitrate-N, as sodium nitrate, were injected into the flow stream at a nearly constant rate over 6-hour period at concentrations of 300 μg/L and 42 mg/L, respectively. Samples were collected for a week after the mixture was injected.

   Groundwater samples were collected automatically at regular intervals throughout phases 1 and 2 using the system illustrated in Figure 3. To ensure that a representative sample of the fluid in the screened and packed-off interval was collected, the groundwater within the packed-off interval was mixed with a

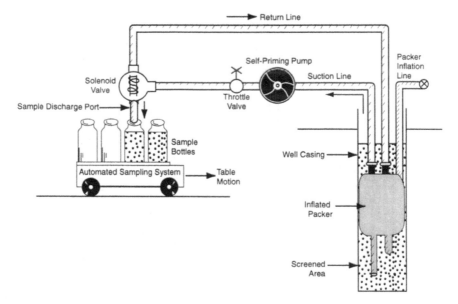

FIGURE 3   Schematic of the borehole and surface fluid sampling unit.

small circulation pump for several minutes before the sample was collected. Samples were collected at intervals ranging from 1 to 4 hours.

V. CHEMICAL ANALYSIS METHODS

All fluorescein samples were analyzed at the field site using a Kontron Spectrofluorometer (Model SFM-23) with an excitation wavelength of 491 nm and an emission wavelength of 513 nm. A detection limit of approximately 1 mg/L was achieved using this method. Se(VI) and Se(IV) concentrations were measured at an in-house laboratory using hydride generation atomic absorption spectrometry. Se(IV) was measured directly, while Se(VI) was determined from the difference between a sample digested to determine total selenium and Se(IV). The digestion procedure consisted of boiling equal amounts of sample with 5 mL of concentrated HCl for 10 minutes. A detection limit for Se(VI) of 2 μg/L was achieved using this procedure. Se(IV) levels remained at or below the detection level of approximately 1 μg/L during the entire test, and therefore, are not discussed further. Nitrate-N concentrations were measured at a commercial laboratory using cadmium reduction method (Clesceri et al., 1989). Detection limits were either 0.2 mg/L (for well 1) or 0.02 mg/L (for all other wells).

VI. INTERPRETIVE METHOD

As shown in Figure 4, a schematic of the transport of a solute through an extraction/injection well doublet intersecting a purely stratified aquifer, the solute moves most rapidly within the layers of high hydraulic conductivity connecting the extraction and injection wells. For a sampling well (located midway between the extraction and injection wells) that intersects more than one layer, a multi-peaked breakthrough curve (lower diagram of Fig. 4) is expected. It is exactly this type of behavior that was observed in this experiment. While this behavior complicates the analysis, because we must include the responses of each layer in the interpretation, it also enables us to evaluate the influence of hydrologic and biogeochemical heterogeneity on the interaction between Se(VI) and nitrate-N in the aquifer.

A. Assumptions

The key assumptions employed to develop a method to interpret the experimental data are as follows.

1. The fluids were completely mixed in the packed-off interval of the well at all times during the tracer experiments. This assumption requires essentially instantaneous mixing of fluids that enter the well.

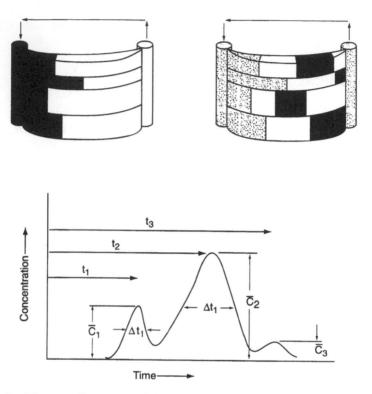

FIGURE 4 Schematic illustrations of the migration of a tracer pulse through a purely stratified aquifer, including multipeaked breakthrough curve corresponding to solute movement in this layered system.

2. Flow into the well occurs in well-defined, discrete layers, with thickness h_n, velocity v_n, and concentration of $C_n(t)$. The concentration $C_n(t)$ is assumed to be uniform over the thickness (h_n) and the width $(2r_w)$ of the flow path. The number of flow paths, thickness of each flow path, and velocity are not restricted by the model.

3. Flow out of the well occurs in well-defined, discrete flow paths, each with the same thickness and flow velocity as the flow paths that enter the well. Flow out of the well occurs at the mixed concentration $\overline{C}(t)$ in the well.

4. The length of the flow path is independent of the heterogeneity of the aquifer and is determined only by the geometry of the doublet flow system.

5. Transverse dispersion across adjacent flow paths is neglected.

6. Within a flow path, flow of nonreactive solutes is governed by the one-dimensional advection–dispersion equation (Ogata, 1977)

$$\frac{\partial C}{\partial t} = D_1 \frac{\partial^2 C}{\partial x^2} - v \frac{\partial C}{\partial x} \tag{1}$$

where D_1 is the longitudinal dispersion coefficient and the other variables are as defined before. (See Benson, 1988, for a detailed discussion of the rationale for using the one-dimensional advection–dispersion equation.)

7. Flow of reactive solutes (e.g., nitrate and selenate) is governed by the one-dimensional advection–dispersion equation with first-order decay of the solutes given by (van Genuchten and Alves, 1982)

$$\frac{\partial C}{\partial t} = D_1 \frac{\partial^2 C}{\partial x^2} - v \frac{\partial C}{\partial x} - \lambda C \tag{2}$$

where λ is the first-order decay coefficient.

B. Mathematical Model

The mass balance for transport in the aquifer–wellbore system shown in Figure 5 and subject to the foregoing assumptions is developed as follows:

$$\Delta M_t = \int_0^h \left[\frac{\partial}{\partial Z} (M_{t,i} - M_{t,o}) \right] dz \tag{3}$$

where ΔM_t is the change in the *mass* of the solute in the well in the time interval Δt, $M_{t,i}$ is the mass of solute that flows into the well in the interval dz during the time interval Δt, $M_{t,o}$ is the mass of solute that flows out of the well in the interval dz during Δt, and h is the total thickness of the slotted interval. Assuming that the interval h is divided into a set of discrete layers, each with thickness h_n, Equation (3) can be written as

$$\Delta M_t = \sum_{n=1}^{\text{No. layers}} M_{(t,i)_n} - M_{(t,o)_n} \tag{4}$$

The mass of solute entering from each layer during a short interval (Δt) can be written

$$M_{(t,i)_n} = v_n A_n C_n(t) \Delta t \tag{5}$$

where $A_n = h_n 2 r_w$, v_n is the average fluid velocity in layer n, and $C_n(t)$ is the average concentration of solute in layer n just upstream of the monitoring well during the time period Δt. The mass of solute that flows out of each layer is given by

$$M_{(t.o)_n} = v_n A_n \overline{C}(t) \Delta t \tag{6}$$

where $\overline{C}(t)$ is the concentration of the solute in the well. Combining Equations 4 and 6, and recognizing that $\Delta M_t = V \Delta C_t$ (where V is the volume of the packed-off volume of the monitoring well), the mass balance equation can be written

$$\frac{\Delta C_t}{\Delta t} = \frac{2r_w}{V} \sum_{n=1}^{\text{No. layers}} v_n h_n [C_n(t) - \overline{C}(t)] \tag{7}$$

To evaluate $C_n(t)$, the one-dimensional solution to the advective–dispersion (Eq. 1) was used. Subject to the initial condition

$$C(x,t) = 0 \tag{8a}$$

and the boundary conditions (pulse input)

$$C(0,t) = C_o \quad 0 < t \le t_o \tag{8b}$$

$$C(0,t) = 0 \quad t > t_o \tag{8c}$$

$$C(\infty,t) = 0 \tag{8d}$$

the concentration of the solute at any point along the flow path can be evaluated from

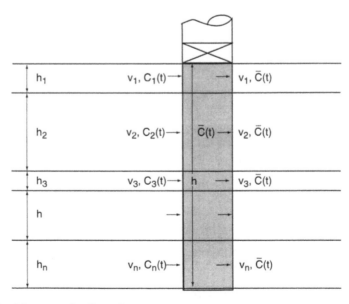

FIGURE 5   Schematic of well aquifer–wellbore interaction, showing the flow velocity entering the borehole, mixing within the borehole, and outflow at a uniform concentration, but at variable velocities.

$$C_n(x,t) = \frac{C_0}{2}\left\{ erfc\left[\frac{x - vt}{2(D_l t)^{1/2}}\right] + \exp\left(\frac{vx}{D_l}\right) erfc\left[\frac{x + vt}{2(D_l t)^{1/2}}\right]\right\}$$
$$- \frac{C_0}{2}\left\{ erfc\left[\frac{x - v(t - t_0)}{2(D_l(t - t_0))^{1/2}}\right] + \exp\left(\frac{vx}{D_l}\right) erfc\left[\frac{x + v(t - t_0)}{2(D_l(t - t_0))^{1/2}}\right]\right\} \quad (9)$$

for $t > t_0$ (Ogata, 1970).

For the case of a solute that is removed from solution by a first-order chemical reaction (e.g., denitrification or selenate reduction), the governing advection–dispersion equation is given by Equation 2. Subject to the boundary and initial conditions given in Equations 8, the solution is given by (van Genuchten and Alves, 1982)

$$C(x,t) = \frac{C_0}{2}\exp\left[\frac{(v - u)x}{2D_l}\right] erfc\left[\frac{x - ut}{2(D_l t)^{1/2}}\right]$$
$$+ \frac{C_0}{2}\exp\left[\frac{(v + u)x}{2D_l}\right] erfc\left[\frac{x + ut}{2(D_l t)^{1/2}}\right]$$
$$- \frac{C_0}{2}\exp\left[\frac{(v - u)x}{2D_l}\right] erfc\left\{\frac{x - u(t - t_0)}{2[D_l(t - t_0)]^{1/2}}\right\} \quad (10)$$
$$- \frac{C_0}{2}\exp\left[\frac{(v + u)x}{2D_l}\right] erfc\left[\frac{x + u(t - t_0)}{2[D_l(t - t_0)]^{1/2}}\right]$$

where

$$u = v\left[1 + \frac{4\lambda D_l}{v^2}\right]^{1/2}$$

A computer program was written to solve equations for the purpose of calculating the solute concentrations at the monitoring wells. An explicit time-stepping procedure was used to calculate solute concentrations at the monitoring wells by

$$\overline{C}(t_{i+1}, x) = \overline{C}(t_i, x) + \Delta\overline{C}(t_i, x) \quad (11)$$

where $\Delta\overline{C}(t_i,x)$ is calculated from Equations 7 and 9 or 10.

C. History Matching

To evaluate the transport and decay rate constants at the test site, data from the experiments were matched to the theoretical solution by a trial-and-error procedure. Adjustable parameters included the number of layers that intersected the sampling well and, for each of these layers, the transport velocity, the dispersion coefficient, and the decay coefficient. Fortunately, the shape of the breakthrough curves allowed unique values for most of these parameters to be determined with a reasonable degree of confidence (Benson, 1988).

VII. DATA INTERPRETATION

Before quantitatively evaluating the transport and reaction rates from a detailed evaluation of the breakthrough curves, it is valuable to look at some overall characteristics of the data collected during the experiment. For example, one of the simplest ways to evaluate the data obtained from this experiment is to calculate the amount of fluorescein, Se(VI), and nitrate-N recovered from the sampling wells. In theory, if no transformation or degradation takes place in the aquifer, each of the wells should recover an amount equal to $C_o \times T_o$, where C_o is the average concentration of a constituent in the injection pulse and T_o is the time interval over which the pulse is injected. A recovery less than this amount indicates that some fraction of the chemical has degraded or transformed while it traveled through the aquifer. Table 3 lists the ratio between the actual recovery and the theoretical recovery (expressed as a percent) for fluorescein, Se(VI), and nitrate-N for each of the sampling wells. Note that the average recovery for fluorescein and Se(VI) is nearly 100%, indicating that on average no significant degradation or transformation occurred. However, the average recovery for nitrate-N is only 64%, indicating that a significant amount of the nitrate has been transformed. This observation is consistent with the existence of thermodynamic conditions that favor denitrification.

Another method of interpreting these data is useful in detecting differences between the adsorption of fluorescein, nitrate-N and Se(VI) onto the aquifer sediments. By noting when each breakthrough curve's center of mass passes by a sampling well, we can infer whether one solute is traveling significantly faster or slower than the others. Table 4 lists the time when the center of mass passes by each well for the chemical of interest. As we can see, in general, the fluorescein, Se(VI), and nitrate-N travel at almost identical speeds, indicating that adsorption

TABLE 3   Summary of the Recovery Values for Fluorescein, Se(VI), and Nitrate-N

| Observation well | Recovery (%) | | |
	Fluorescein	Se(VI)	Nitrate-N
Well 1	83	110	74
Well 2	110	93	84
Well 3	73	80	59
Well 4	99	108	62
Well 5	125	112	34
Well 7	99	93	62
Well 8	83	91	76
Average	99%	98%	64%

TABLE 4   Values for the Time at Which the Center of Mass (cm) of the Tracer
Cloud Passed by the Observation Wells for Fluorescein, Se(VI), and Nitrate-N

Observation well	t_{cm} (s $\times$ 10^{-5})		
	Fluorescein	Se(VI)	Nitrate-N
Well 1	3.1	3.0	2.9
Well 2	3.9	3.8	3.9
Well 3	3.2	4.4	3.3
Well 4	3.8	4.0	4.0
Well 5	3.9	4.1	4.0
Well 7	3.1	2.9	2.9
Well 8	4.8	4.7	4.7

is insignificant. This observation confirms predictions based on laboratory experiments that selenate adsorption is small in this environment (Balistrieri and Chao, 1987; Neal and Sposito, 1989).

The most comprehensive interpretation of the experimental data is obtained by applying the model described above to determine the number of layers intersecting each well, along with the corresponding transport properties and reaction rates. Matches between the data and the theoretical solution are provided for four of the eight observation wells in Figures 6 through 9 (equally good matches were obtained with all the data). As seen from these matches, the model provides an adequate means to reproduce the observed transport and reaction rates. The parameter values determined from these matches are provided in Table 5.

Of particular significance for this study is the conclusion that measurable decay coefficients for nitrate are calculated for each of the layers intersecting the observation wells. First-order decay coefficients for nitrate-N range from 0.3 to 5×10^{-6} s^{-1}. At the same time, the decay coefficients provided in Table 5 for Se(VI) indicate that for the majority of flow paths, the decay coefficients are negligible. This suggests that little or no reduction of Se(VI) is occurring during the experiment. There are, however, several exceptions to this general conclusion. In sampling wells 2, 3, and 5, measurable decay [Se(VI) reduction] coefficients were observed in several of the layers. In all but one of these cases, the decay coefficients are significantly smaller than the corresponding decay coefficients for nitrate-N, but coincide with the layers where the nitrate coefficients are greatest.

VIII.   SUMMARY AND CONCLUSIONS

This study had a threefold objective: (1) to confirm that nitrate inhibits selenium reduction/immobilization under in situ conditions, (2) to quantify denitrification

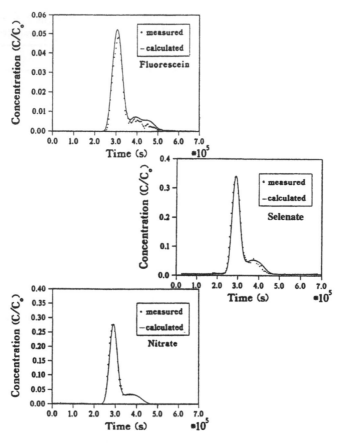

FIGURE 6 History matches for well 1 between the measured and calculated solute concentrations for fluorescein, Se(VI), and nitrate-N. See Table 5 for the parameters used to generate these curves.

rates, and (3) to establish that selenate reduction will follow nitrate removal. While this experiment certainly supports the conclusion that nitrate is removed from the groundwater by reaction with aquifer sediments and provides a range of in situ decay coefficients, it does not in itself accomplish another important objective of this study: to show that Se(VI) reduction is inhibited by the presence of nitrate. To arrive at this conclusion, we must look to other studies conducted in the aquifer under Kesterson Reservoir.

In particular, two studies have demonstrated and at least partially quantified the rate at which Se(VI) is reduced in the absence of nitrate. White et al. [9] presented data from an injection/extraction tracer experiment showing almost

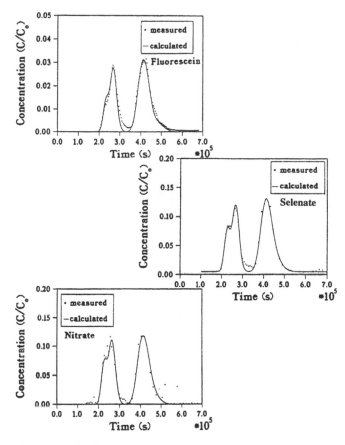

FIGURE 7 History matches for well 2 between the measured and calculated solute concentrations for fluorescein, Se(VI), and nitrate-N. See Table 5 for the parameters used to generate these curves.

complete Se(VI) reduction over a period of less than 25 days, with a half-life of about 10 days (decay coefficient of approximately 10^{-6} s^{-1}, which is similar to the rate of nitrate reduction observed here). Long et al. [10] provided an extensive data set regarding Se(VI) reduction in the uppermost portion of the aquifer with rate constants in the range of 10^{-7} to 10^{-6} s^{-1}. Taken together, the previous experiments, combined with the interpretation presented here, support the hypothesis that Se(VI) reduction is inhibited by the presence of nitrate in the Sierran aquifer sediments under Kesterson Reservoir. The final objective of this study—to demonstrate that Se(VI) reduction would proceed once denitrification had decreased nitrate-N concentrations to some appropriate level—is anecdotally sup-

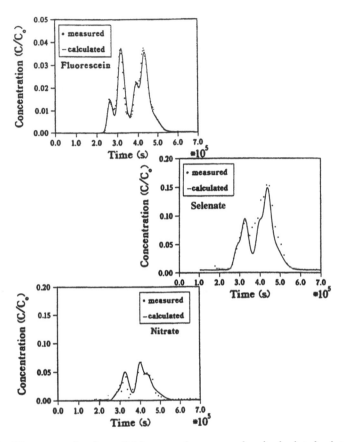

FIGURE 8 History matches for well 5 between the measured and calculated solute concentrations for fluorescein, Se(VI), and nitrate-N. See Table 5 for the parameters used to generate these curves.

ported by these experiments. As mentioned, Se(VI) reduction was observed only in several wells, and therefore it is difficult to support definitive conclusions regarding the sequential reduction of Se(VI) following denitrification. Qualitatively, however, we can point out that locations where Se(VI) reduction was observed coincided with the wells (and layers) where nitrate decay coefficients were the greatest. This lends credence to the hypothesis that sequential reduction is likely, but further controlled experiments are needed to provide quantitative support.

A final and significant observation is that the decay coefficients for nitrate-N and Se(VI) range over a wide set of values, even on the relatively small length

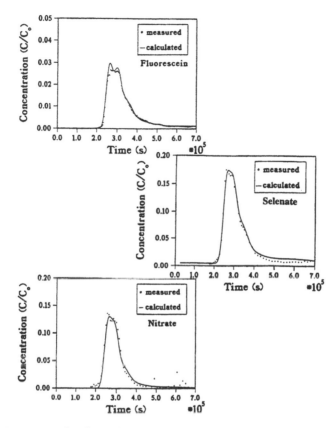

FIGURE 9 History matches for well 7 between the measured and calculated solute concentrations for fluorescein, Se(VI), and nitrate-N. See Table 5 for the parameters used to generate these curves.

scale of this experiment (30 m). Systematic relationships between the physical properties (e.g., flow velocity, dispersion, dispersion coefficient, and decay rates) were not detected, suggesting that the observed variability can be attributed instead to the geochemical and microbial heterogeneity of the aquifer system.

ACKNOWLEDGMENTS

The hard work of Ray Solbau, John Daggett, and Robert Long during this experiment is gratefully acknowledged. Thanks are also due to my coworkers Peter Zawislanski and

Flow path	Thickness h (m)	Longitudinal dispersivity α_l (m)	Fluorescein: velocity v_n (m/s $\times 10^{-5}$)	Velocity v_n (m/s $\times 10^{-5}$)	Decay coefficient λ (s^{-1} $\times 10^6$)	Velocity v_n (m/s $\times 10^{-5}$)
1-1	0.3	0.035	6.19	6.39	—	6.34
1-2	0.8	0.03	5.68	6.26	—	6.26
1-3	0.22	0.04	4.52	5.03	—	5.3
1-4	0.20	0.035	3.87	4.45	—	4.45
2-1	0.12	0.03	7.87	8.39	—	8.39
2-2	0.31	0.03	6.83	7.09	0.75	7.15
2-3	0.51	0.03	4.55	4.46	0.6	4.64
2-4	0.46	0.03	4.29	4.36	0.6	4.36
2-5	0.12	0.03	3.77	3.97	0.6	3.97
3-1	0.06	0.035	8.05	8.92	—	9.92
3-2	0.04	0.035	6.89	8.92	—	8.92
3-3	0.18	0.08	5.34	5.95	2.75	5.95
3-4	0.10	0.03	3.19	3.19	0.5	3.19
3-5	0.24	0.025	2.81	3.19	0.5	3.19
3-6	0.20	0.05	2.47	2.76	—	2.76
3-7	0.70	0.2[a]	1.74	2.03	—	2.03
4.1	0.32	0.035	6.22	6.22	—	6.22

4–2	0.15	0.03	3.87	3.87	—	3.87
4–3	0.36	0.027	3.58	3.65	—	3.58
4–4	0.33	0.03	3.13	2.76	—	2.83
4–5	0.36	0.04	2.68	2.76	1.0	2.83
5–1	0.11	0.02	5.92	5.62	1.0	5.60
5–2	0.38	0.018	4.92	4.92	—	4.92
5–3	0.08	0.03	4.30	4.30	—	4.30
5–4	0.19	0.01	4.02	4.02	—	4.02
5–5	0.39	0.01	3.62	3.65	—	3.65
5–6	0.26	0.06	3.53	3.44	—	3.44
5–7	0.11	0.03	3.24	3.31	—	3.31
7–1	0.33	0.025	5.13	5.35	—	5.35
7–2	0.37	0.025	4.44	4.7	—	4.7
7–3	0.25	0.04	3.83	4.05	—	4.05
7–4	0.29	0.2[a]	3.21[a]	3.53	—	3.53
7–5	0.28	0.2[a]	1.99[a]	2.19	—	2.19
8–1	0.03	0.028	4.71	4.71	—	4.71
8–2	0.10	0.028	3.86	4.71	—	4.71
8–3	0.63	0.04	3.96	3.92	—	3.92
8–4	0.50	0.04	3.37	3.50	—	3.50

Nari Narasimhan for their helpful review comments. This work was supported by the U.S. Bureau of Reclamation under U.S. Department of Interior Interagency Agreement 9-AA-20-20, through the U.S. Department of Energy Contract No. DE-AC03-76F00098.

REFERENCES

Balistrieri, L. S., and T. T. Chao. 1987. Selenium adsorption by goethite. *J. Soil Sci. Soc. Am.* 51:1145–1151.

Benson, S. M. 1988. Characterization of the hydrogeologic and transport properties of the shallow aquifer under Kesterson Reservoir, Merced County, California, Ph.D. thesis, University of California, Berkeley.

Benson, S. M., A. F. White, S. Halfman, S. Flexser, and M. Alavi. 1991. Groundwater contamination at the Kesterson Reservoir, California: 1. Hydrogeologic setting and conservative solute transport. *Water Resour. Res.* 27:1071–1084.

Cary, E. E., G. A. Wieczorek, and W. H. Allaway. 1967. Reactions of selenite-selenium added to soils that produce low-selenium forages. *Proc. Soil Sci. Soc. Am.* 31:21–26.

Clesceri, L. S., E. A. E. Greenberg, and R. R. Trussell (eds.). 1989. *Standard Methods for the Examination of Water and Waste Water*, 17th ed.

Cutter, G. A. 1982. Selenium in reducing water. *Science* 217:829.

Deverel, S. J., and R. Fujii. 1988. Processes affecting the distribution of selenium in shallow groundwater of agricultural areas, western San Joaquin Valley, California. *Water Resourc. Res.* 24:516–524.

Deverel, S. J., and S. P. Millard. 1988. Distribution and mobility of selenium and other trace elements in shallow groundwater of the western San Joaquin Valley, California. *Environ. Sci. Technol.* 22:697.

Doran, J. W. 1982. Microorganisms and the biological cycling of selenium. *Adv. Microbial Ecol.* 6:1.

Dubrovsky, N. M., J. M. Neil, R. Fujii, R. S. Oremland, and J. T. Hollobaugh. 1991. Influences of redox potentials on selenium distribution in groundwater. U.S. Geological Survey, Open File Report No. 90-138.

Elrashidi, M. A., D. A. Adriano, S. M. Workman, and W. L. Lindsay. 1987. Chemical equilibria of selenium in soils: A theoretical development. *Soil Sci.* 144:141–152.

Geering, H. R., E. E. Cary, L. H. P. Jones, and W. H. Allaway. 1968. Solubility and redox criteria for the possible form of selenium in soils. *Proc. Soil Sci. Soc. Am.* 32:35–40.

Karlson, U., and W. T. Frankenberger, Jr. 1988. Effects of carbon and trace element addition on alkylselenide production from soil. *Soil Sci. Soc. Am.* 52:1940.

Long, R. H., S. M. Benson, T. K. Tokunaga, and A. Yee. 1990. Selenium immobilization in a pond bottom sediment at Kesterson Reservoir. *J. Environ. Qual.* 19:302–311.

Lundquist, T. J., M. B. Gerhardt, F. B. Green, R. B. Tresan, R. D. Newman, and W. J. Oswald. 1994. The algal–bacterial selenium removal system: Mechanisms and field study. In W. T. Frankenberger, Jr., and S. Benson (ed.), *Selenium in the Environment*, pp. 251–278. Dekker, New York.

Neal, R. H., and G. Sposito. 1989. Selenate adsorption on alluvial soils. *J. Soil Sci. Soc. Am.* 53:70.

Ogata, A. 1970. Theory of dispersion in a granular medium. U.S. Geological Survey Prof. Paper No. 411-I. USGS, Washington, DC.

Oremland, R. S. 1994. Biogeochemical transformations of selenium in anoxic environments. In W. T. Frankenberger, Jr., and S. Benson (ed.), *Selenium in the Environment*, pp. 389–419. Dekker, New York.

Presser, T. S. 1994. Geologic origin and pathways of selenium from the California Coast Ranges to the West Central San Joaquin Valley. In W. T. Frankenberger, Jr., and S. Benson (ed.), *Selenium in the Environment*, pp. 139–156. Dekker, New York.

Tokunaga, T. K., D. S. Lipton, S. M. Benson, A. Yee, J. Oldfather, E. C. Duckart, P. W. Johannis, and K. E. Halvorsen. 1991. Soil selenium fractionation, depth profiles and time trends in a vegetated upland at Kesterson Reservoir. *Land, Air Soil Pollut.* 57/58:3141.

van Genuchten, M. T., and W. J. Alves. 1982. Analytical solutions to the one-dimensional convective dispersion equation. U.S. Department of Agriculture Tech. Bull. No. 1661. USDA, Washington, DC.

Weres, O., H. R. Bowman, A. Goldstein, E. C. Smith, and L. Tsao. 1990. Air, the effect of nitrate and organic matter upon the mobility of selenium in groundwater and in a water treatment process. *Water Soil Pollut.* 49:251.

White, A. F., and N. M. Dubrovsky. 1994. Chemical oxidation–reduction controls on selenium mobility in groundwater systems. In W. T. Frankenberger, Jr., and S. Benson (ed.), *Selenium in the Environment*, pp. 185–221. Dekker, New York.

White, A. F., S. M. Benson, A. W. Yee, H. W. Wollenberg, and S. Flexser. 1991. Groundwater contamination at the Kesterson Reservoir, California, 2. Geochemical parameters influencing selenium mobility. *Water Resourc. Res.* 27:1085–1098.

Zawislanski, P. T., et al. 1996. The pond 2 selenium volatilization study: A synthesis of five years of experimental results 1990–1995. Lawrence Berkeley National Laboratory Report No. 39516. LBNL, Berkeley, CA.

23

The Uptake and Metabolism of Inorganic Selenium Species

JOHN B. MILNE
University of Ottawa, Ottawa, Ontario, Canada

I. INTRODUCTION

Before the middle of this century, the public perception of selenium was similar to the present-day view of arsenic. That is, selenium was regarded as highly toxic (Moxon and Rhian, 1943) and carcinogenic (Nelson et al., 1943). The first report to note any nutritional benefit from selenium described the prevention of liver necrosis in rats when selenium was included in the diet (Schwarz and Foltz, 1957). Since then, many beneficial aspects of selenium to animal metabolism have been catalogued, among them improved animal husbandry (Ullrey, 1978), and a negative correlation has been found for some forms of cancer in humans (Jansson, 1980). Selenium appears nowadays as a supplement in many vitamin preparations. The late recognition of the importance of selenium in metabolism arises from its low abundance and the narrow range of concentration between beneficial and toxic effects. Selenium is ranked 69th in abundance in the earth's crust, with an average concentration of 0.09 mg/kg, and 29th in seawater at 0.004 mg/kg (Weast, 1969). Concentrations of the order of hundreds of milligrams per kilogram have however been found in some shales and accumulator plants (Mayland et al., 1989). Toxic effects in animals become apparent with diets containing above 5 to 15 mg/kg Se and deficiencies with diets containing below 0.05 to 0.10 mg/kg.

Selenium exhibits a broad range of oxidation states: $+6$ in selenates ($HSeO_4^-$, SeO_4^{2-}) and selenic acid (H_2SeO_4), $+4$ in selenites ($HSeO_3^-$, SeO_3^{2-}) and selenous acid (H_2SeO_3), 0 in elemental selenium, and -2 in selenides (Se^{2-}, HSe^-), hydrogen selenide (H_2Se), and organic selenides (R_2Se). Selenium also

459

shows some tendency to form catenated species like organic diselenides (RSeSeR). The Pourbaix diagram for selenium, (Fig. 1) shows the reduction potential/pH existence range in water of the common inorganic selenium species. Within the normal physiological pH range and the reduction potential range permitted by water, only Se, SeO_3^{2-}, $HSeO_3^-$, and SeO_4^{2-} can exist at thermodynamic equilibrium. While ionic reactions are expected to be rapid in water, oxidation–reduction reactions may be slow and Figure 1 tells nothing about the possible existence of HSe^- in living systems and some environments where anoxic conditions arise. The large kinetic barrier to the oxidation of HSe^- by water permits this species to exist in aqueous solution, but special care must be taken to protect against oxidation by oxygen in experiments designed to explore the reducing region of the existence range of selenium (Bjoernstedt et al., 1996). The parallel behavior of comparable species of sulfur and selenium in living systems has often been stressed, but it must be recalled that their chemistries show many differences. For instance, selenate is comparable to chromate in oxidizing strength and far stronger than sulfate [$E^0(SeO_4^{2-}/H_2SeO_3) = 1.15$ V; $E^0(Cr_2O_7^{2-}/Cr^{3+}) = 1.33$ V;

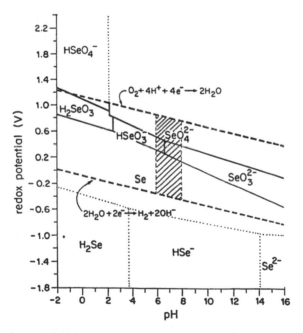

FIGURE I Pourbaix diagram for selenium: hatched region delineates normal physiological conditions for the living cell interior, dashed lines show the equilibrium potentials for water decomposition to hydrogen and oxygen. (Adapted with permission from W. Kaim and B. Schwederski, 1994.)

$E^0(SO_4^{2-}/H_2SO_3) = 0.200$ V (standard potentials in acid solution: Weast, 1969)], while selenide is a much stronger reducing agent than sulfide [$E^0(Se/H_2Se) = -0.36$ V; $E^0[S/H_2S] = 0.14$ V).

Many excellent reviews pertinent to the environmental chemistry of selenium have been published; (Mayland et al., 1989; McNeal and Balisteri, 1989; Ohlendorf, 1989; Stadtman, 1990; Thompson-Eagle and Frankenberger, 1992; Heider and Boeck, 1993) and a book has recently appeared (Frankenberger and Benson, 1994). These publications should be referred to for a more comprehensive coverage of the earlier literature on the subject.

II. UPTAKE OF SELENIUM SPECIES

The first step in the interaction of living systems with SeO_4^{2-}, $HSeO_3^-$ and Se^0, the common selenium species in the environment, is uptake through the cell wall. No special scavenging and conservation systems like those for iron and other elements are yet known for selenium.

A. Selenate

In spite of its oxidizing strength, SeO_4^{2-} exhibits considerable kinetic stability in the presence of reducing agents (Cotton and Wilkinson, 1988). The radius of SeO_4^{2-} is comparable to that of SO_4^{2-} (Frausto da Silva and Williams, 1991), and uptake by the cell is expected to take place via the same ion channels or permeases for both anions. Competition between sulfate and selenate uptake has been observed in many species: algae (Kumar and Prakash, 1971; Williams et al., 1994; Riedel and Sanders, 1996), aquatic plants (Bailey et al., 1995), crustacea (Ogle and Knight, 1996), fungi (Gharieb et al., 1995), HeLa cells (Yan and Frenkel, 1994), and wheat (Richter and Bergmann, 1993). Although selenate and sulfate uptake are related in *E. coli* (Lindblow-Kull et al., 1985), interestingly, the active bacteria in sediments and drainage water that were studied for selenate reduction function independently of sulfate concentration (Oremland et al., 1989; Gerhardt et al., 1991). Competition with selenate has also been observed for phosphate in green algae (Riedel and Sanders, 1996) and chromate and tungstate in anaerobic bacteria (Oremland et al., 1989). This may also be due to competitive uptake.

B. Selenite

Relative to selenate, selenite is more reactive because of its polar character and its basicity. No evidence has yet been presented to show that $HSeO_3^-$ or SeO_3^{2-} is taken up intact into the cell interior. Evidence discussed below indicates that selenite is reduced rapidly, even before uptake in some cases. It is therefore difficult to distinguish between uptake and metabolic processes. Nevertheless, sulfate ion interferes with selenite uptake in wheat plants (Riedel and Sanders, 1996), in *Vigna radiata* (Lalitha and Easwari, 1995), and in *Fusarium* sp. (Gharieb et al., 1995), although the effect is not as pronounced as for selenate.

However, while nonspecific uptake of selenite appears to occur in some organisms, specific transport systems exist in others. Sulfate competition is insignificant in *Ruppia maritima* (Bailey et al., 1995), and specific uptake systems have been demonstrated in some microorganisms (Heider and Boeck, 1993). Sulfate, selenate, and selenite are considered by some to use the same transporter in *E. coli* (Lindblow-Kull et al., 1985), although not all investigators reach this conclusion (Brown and Shrift, 1982). In the plant, *Vigna radiata*, a dicarboxylate port is thought to be involved, although thiol antagonists also decrease selenite uptake significantly (Lalitha and Easwari, 1995) and two different permeases may be involved. Selenite uptake in green algae, unlike selenate, is increased substantially at lower pH values, a property that represents another difference between these anions (Riedel and Sanders, 1996).

C. Selenium

Elemental selenium is not measurably soluble in water. The amount of elemental selenium present in soils is very small (Elrashidi et al., 1989) and arises principally from the action of microorganisms. A recent estimate of concentration for soils in Japan is 1 to 2 μg/kg (Yamada et al., 1996). The bioavailability of elemental selenium in soils to forage crops is estimated to be two orders of magnitude less than that of selenite (Mayland et al., 1989). A comparable relative bioavailability of elemental selenium to selenite is found in chick feeds (Coombs et al., 1996). It has been reported that elemental selenium is slowly metabolized by several bacteria (Doran and Alexander, 1977; Sarathchandra and Watkinson, 1981; Bacon and Ingledew, 1989), and the translocation of elemental selenium into the soft tissue of *Macoma balthica* has been reported (Luomo et al. 1992). In view of the insolubility of elemental selenium, uptake may be preceded by air oxidation or, in reducing environments, thiols may facilitate the solubilization (Amaratunga and Milne, 1994).

III. METABOLISM

The metabolic reduction of selenate and selenite by living systems produces elemental selenium, selenoproteins, and methylated products such as dimethyl-selenide (DMSe), depending on growth conditions and organism. The metabolic oxidation of elemental selenium has received little attention. While much of the research on the environmental chemistry of selenium has stressed the involvement of specialized enzymes in living systems, nonspecific interactions may play an important, if not dominant role (Fenchel and Blackburn, 1979). The importance of specific versus nonspecific interactions will be determined in large part by the concentration of selenium present in the local environment, and Paracelsus's observation that "the dosage determines the poison" (Pachter, 1961) is especially important for selenium. In general, the interior of a living cell is more reducing than its external environment, and bioreduction of selenate and selenite are the principal chemical processes occurring.

A. Selenate

Microorganisms exhibit greater difficulty in reducing selenate than in reducing selenite (Doran, 1982; Maiers et al., 1988), and thus selenate is often a spectator in living systems. However, growth is inhibited more by selenate than selenite in some species, such as algae (Wheeler et al., 1982; Bennett, 1988), *Fusarium* sp. (Gharieb et al., 1995), and wheat plants (Richter and Bergmann, 1993). Usually, selenite is more toxic than selenate as, for instance, in *Cricosphaera elongata* (Boisson et al., 1995), *Mortierella* sp. (Zieve et al., 1985), and *E. coli* (Milne, unpublished). Table 1 lists the microorganisms that reduce selenate and selenite.

Slow and limited reduction of selenate has been observed for *E. coli* (Tiels and Cheldelin, 1949), but growth is unaffected (Milne, unpublished). *Desulfovibrio* (Postgate, 1952; Tomei et al., 1995) and *Wolinella succinogens* (Tomei et al., 1992) can be adapted to grow in the presence of selenate and produce elemental selenium. Interestingly, the selenium is observed only after cessation of growth. However, methylated products result from the bioreduction of selenate by bacteria isolated from seleniferous environments (*Corynebacterium* sp. and *Pseudomonas fluorescens*) and some fungi (*Fusarium* sp., *Scopulariopsis brevicaulis*, *Schizopyllum commune*, *Aspergillus niger*, *Candida humicola*, *Acremonium falciforme*, *Alternaria alternata*) (Thompson-Eagle and Frankenberger, 1992). Sediments from anoxic marine and evaporation pond environments contain organisms capable of reducing selenate to elemental selenium. In media containing acetate, a bacterium isolated from these slurries exhibited selenate-dependent growth accompanied by selenium precipitation (Oremland et al., 1989). An organism initially identified as a pseudomonad respires anaerobically using selenate and acetate (Macy et al., 1989). This organism was subsequently shown to represent a new genus of gram-negative bacterium and was named *Thaurera selenatis* (Macy et al., 1993). *T. selenatis* reduces selenate to selenite and is not inhibited by the presence of nitrate. However, reduction of selenite to elemental selenium takes place only if denitrification is also occurring, implying that nitrate reductase is necessary for this stage. A pilot plant for the bioremediation of drainage water from the San

TABLE I   Microorganisms Known to Reduce Selenium

Microorganism	Substrate	Product[a]	Ref.
Bacteria			
Aeromonas sp.	SeO_3^{2-}	DMSe	Chau et al., 1976
	SeO_3^{2-}	Se	Silverberg et al., 1976
Arthrobacter protopharmeae	SeO_3^{2-}	Se	Riadi and Barford, 1996
Bacillus pasteurii	SeO_3^{2-}	Se	Riadi and Barford, 1996
Bacillus subtilis	SeO_3^{2-}	Se (w)	Riadi and Barford, 1996
Bradyrhizobium japonicum	SeO_3^{2-}	SeDC	Hsu et al., 1990
Clostridium pasteurianum	SeO_3^{2-}	Se (m)[b]	Harrison et al., 1980
Cornyebacterium diphtheriae	SeO_3^{2-}	Se (c)	Levine, 1925
Cornyebacterium sp.	SeO_4^{2-}	DMSe	Doran and Alexander, 1977
	SeO_3^{2-}	DMSe	
	Se	DMSe	
Desulfovibrio vulgaris	SeO_4^{2-}	Se[b]	Postgate, 1949
Desulfovibrio desulfuricans	SeO_4^{2-}	Se (w)[b]	Tomei et al., 1995
	SeO_3^{2-}	Se (c)[b, c]	
	SeO_4^{2-}	H_2Se[d]	Woolfolk and Whiteley, 1962
	SeO_3^{2-}	H_2Se	Zehr and Oremland, 1987
Escherichia coli	SeO_4^{2-}	Se	Gerrard et al., 1974
	SeO_3^{2-}	Se (w)	Fels and Cheldelin, 1949
	SeO_3^{2-}	Se (c)	Silverberg et al., 1976
Flavobacterium sp.	SeO_3^{2-}	DMDSe	Chau et al., 1976
	SeO_3^{2-}	Se	Silverberg et al., 1976
Lactobacillus sp.	SeO_3^{2-}	Se	Calomme et al., 1995
Microbacterium arborescens	SeO_3^{2-}	Se	Combs et al., 1996
Micrococcus lactilyticus	SeO_3^{2-}	Se, H_2Se[d]	Woolfolk and Whiteley, 1962
Pseudomonas fluorescens	SeO_4^{2-}	DMSe, DMDSe	Chasteen et al., 1990
Pseudomonas mesophilica	SeO_4^{2-}	Se	Barton et al., 1992
	SeO_3^{2-}	Se(c)	
Pseudomonas stutzeri	SeO_4^{2-}	Se	Lortie et al., 1992
	SeO_3^{2-}	Se	
Pseudomonas picketti	SeO_3^{2-}	Se	Riadi and Barford, 1996
Pseudomonas sp.	SeO_3^{2-}	Volatile Se	Chau et al., 1976
Rhodobacter sphaeroides	SeO_4^{2-}	Se	Moore and Kaplan, 1992
	SeO_3^{2-}	Se (c,w)	
Salmonella heidelberg	SeO_3^{2-}	Se (c)	McCready et al., 1966
Staphylococcus hemolytica	SeO_3^{2-}	Se	Riadi and Barford, 1996
Streptococcus faecalis	SeO_3^{2-}	Se	Tilton et al., 1967
Thauera selenatis	SeO_4^{2-}	Se	Macy et al., 1993
	SeO_3^{2-}	Se	

TABLE I Continued

Microorganism	Substrate	Product[a]	Ref.
Wolinella succinogenes	SeO_4^{2-}	Se	Tomei et al., 1992
	SeO_3^{2-}	Se (c)	
Fungi			
Acremonium falciforme	SeO_3^{2-}	DMSe	Frankenberger and Karlson, 1994
Alternaria alternata	SeO_4^{2-}	DMSe	Thompson-Eagle and Frankenberger, 1990
	SeO_3^{2-}	DMSe	
Aspergillus funiculosis	SeO_3^{2-}	Se (c,w)	Gharieb et al., 1995
Aspergillus parasiticus	SeO_3^{2-}	Se	Moss et al., 1987
Cephalosporium sp.	SeO_3^{2-}	DMSe	Barkes and Fleming, 1974
Fusarium sp.	SeO_4^{2-}	DMSe	Barkes and Fleming, 1974
	SeO_3^{2-}	Se	Gharieb et al., 1995
Mortierella sp.	SeO_3^{2-}	Se, DMSe	Zieve et al., 1985
Mucor hiemalis	SeO_3^{2-}	Se	Gharieb et al., 1995
Mucor SK	SeO_3^{2-}	Se (w)	Gharieb et al., 1995
Neurospora crassa	SeO_3^{2-}	Se (c)	Zalokar, 1953
Penicillium chrysogenum	SeO_3^{2-}	Se	Gharieb et al., 1995
Penicillium citrinium	SeO_3^{2-}	DMSe, DMDSe	Chasteen et al., 1990
Penicillium funiculosum	SeO_3^{2-}	Se	Gharieb et al., 1995
Penicillium sp.	SeO_4^{2-}	DMSe	Fleming and Alexander, 1972
	SeO_3^{2-}	DMSe	
Penicillium sp.	SeO_4^{2-}	DMSe	Barkes and Fleming, 1974
Penicillium sp.	SeO_3^{2-}	Se, DMSe	Brady et al., 1996
Schizophyllum commune	SeO_4^{2-}	DMSe	Challenger and Charlton, 1947
Scopulariopsis brevicaulis	SeO_4^{2-}	DMSe	Challenger and North, 1934
	SeO_3^{2-}	DMSe	
Scopulariopsis sp.	SeO_4^{2-}	DMSe	Barkes and Fleming, 1974
Trichoderma reesei	SeO_3^{2-}	Se	Gharieb et al., 1995
Ulocladium tuberculatum	SeO_3^{2-}	DMSe	Frankenberger and Karlson, 1994
Yeast			
Candida albicans	SeO_3^{2-}	Se (c)	Falcone and Nickerson, 1963
Candida glabrata	SeO_3^{2-}	Se	Gharieb et al., 1995
Candida humicola	SeO_3^{2-}	Volatile Se	Zieve and Peterson, 1981
Candida lypolitica	SeO_3^{2-}	Se	Gharieb et al., 1995
Rhodotorula rubra	SeO_3^{2-}	Se	
Saccharomyces cerevisiae	SeO_3^{2-}	Se	

[a] Sites of Se deposit given in parentheses: DMSe, dimethyl selenide; DMDSe, dimethyl diselenide; SeDC, selenodicysteine; c, cytoplasm; m, medium; w, cell wall.
[b] Possible precipitation by H_2S.
[c] No DMSe detected.
[d] In the presence of hydrogen.

Joaquin Valley is currently being tested using this organism (Cantafio et al., 1996). Selenium recoveries up to 96% have been demonstrated with this process, and the product consists of 97.9% elemental selenium. A *Pseudomonas stutzeri* isolate also is able to reduce selenate to elemental selenium in trypticase soy broth (Lortie et al., 1992).

The relatively efficient and rapid reduction of selenate by *T. selenatis* and *Ps. stutzeri* is unique among the microorganisms capable of this reduction and implies that special processes have been developed by these bacteria which have been isolated from selenium-rich environments. Selenates are not easy to reduce in vitro for kinetic reasons, and the reaction normally requires acidic conditions and high temperature. Reduction by H_2S (Vogel, 1952) or by 2-mercaptoethanol (Milne, unpublished) is not observed even over a period of days at 100°C. Thiols, in particular glutathione, are regarded as a principal defense in living systems against oxidative stress (Mannervik et al., 1989) and, if the reduction of selenate is slow in vitro and in most microorganisms, then the ability of *T. selenatis* and *Ps. stutzeri* to carry out the process is exceptional. The mechanism is apparently distinct from sulfate reduction inasmuch as none of the organisms capable of selenate respiration are able to use sulfate in this way (Heider and Boeck, 1993, and references therein). Understanding of the chemistry of the process of detoxification of selenate by living systems is poor at present.

B.   Selenite

I.   Microorganisms

Selenite-containing media have been used for many years in the selection of bacteria (Guth, 1916). A broad range of bacteria, fungi, and yeast are able to grow in the presence of selenite or can adapt to its presence in culture media; microorganisms known to reduce selenite are listed in Table 1. The end products of the bioreduction of selenite parallel those formed by selenate metabolism. In addition to elemental selenium or a methylated form of selenium, such as DMSe, there is also a wide range of selenium-containing enzymes (Boeck et al., 1990) and proteins (Stadtman, 1990; Heider and Boeck, 1993). The selenite uptake and metabolic stages are intimately related, but it is clear that the first chemical reaction to take place is that between selenite and thiol compounds that are excreted by the cell (Tomei et al., 1995), contained within the cell wall (Gerrard et al., 1974), or contained in the cell interior (McCready et al., 1966). Several lines of evidence support this. The addition of Hg^{2+} and Pb^{2+} to seleniferous soils inhibits the biomethylation of selenium (Karlson and Frankenberger, 1988). Thiols are known to be strongly bound by heavy metal cations. Selenite uptake by *Salmonella heidelberg* was shown to proceed by an intermediate Se(2+) oxidation state that is rare in the chemistry of selenium (McCready et al., 1966) but is readily accounted for by the formation of a selenotrisulfide via a reaction shown to occur when selenous acid reacts with thiols (RSH) (Ganther, 1968).

$$H_2SeO_3 + 4RSH \longrightarrow RSSR + RSSeSR + 3H_2O \tag{1}$$

This reaction has now been shown to be quite general, and details of the mechanism have been worked out (Kice, 1981; Milne and Milne, 1996, and references therein). Several thiols of biological importance, such as glutathione, acetyl–coenzyme A and cysteine, have been shown to react in this way, and nuclear magnetic resonance (NMR) spectroscopy has been used to characterize and identify several of these (Rabenstein and Tan, 1988). Dicysteylselenotrisulfide has been detected in the hydrolysate from the hydrogenase produced by *Bradyrhizobium japonicum* grown in selenite-containing media (Hsu et al., 1990). An in vivo NMR study of the reduction $HSeO_3^-$ (94% ^{77}Se enriched) by *E. coli*, previously adapted to grow in the presence of selenite, showed the growth of a ^{77}Se NMR signal at 645 ppm accompanied by a decrease in the selenite signal at 1310 ppm over a period of 30 minutes at 19°C (Milne et al., 1994). The new chemical shift appears in the expected range for selenotrisulfides, 550 to 700 ppm (Rabenstein and Tan, 1988). After this initial growth, the new signal weakens in intensity over the next 5 hours, and red selenium precipitates out of the solution. Selenotrisulfides are known to decompose slowly to elemental selenium and the disulfide (reaction 2) at basic pH (Ganther, 1968; Afsar et al., 1989) and in the presence of thiol in excess of that required for reaction 1, (Amaratunga and Milne, 1994).

$$RSSeSR \longrightarrow Se + RSSR \tag{2}$$

Decomposition is accelerated by glutathione reductase (Ganther, 1971). Thiol groups are found in many biologically important molecules—cysteine, glutathione, acetyl–coenzyme A, metallothioneins, thioredoxin, and dihydrolipoic acid, to mention a few. Glutathione is probably the most widely distributed, and the most abundant thiol in living systems. It is found in particularly high concentrations in gram-negative bacteria (Fahey et al., 1978). In *E. coli* it is thought to act as a detoxifying agent for selenium, and its production is stimulated by the presence of selenite (Schmidt and Konetzka,1986). Cell-free extracts of *Candida albicans* (Falcone and Nickerson, 1963), *Streptococcus faecalis* (Tilton et al., 1967), and *Clostridium pasteurianum* (Harrison et al., 1980) reduce selenite to elemental selenium. A broad range of thiols are probably oxidized in this process. It is interesting that cell-free extracts containing selenotrisulfide inhibit growth in *E. coli* (Milne et al., 1994) and *Salmonella heidelberg* (McCready et al., 1966). Equimolar selenium concentrations of selenite and bis(glutathione)selenium exhibit equivalent toxicity toward *E. coli* (Milne et al., 1994). The precise mode of action of this toxicity is unknown.

The site of precipitation of selenium within the cell varies with different organisms. Precipitation external to the cell has been observed for *Clostridium pasteurianum* (Harrison et al., 1980), *Wolinella succinogenes* (Tomei et al., 1992), and *Desulfovibrio desulfuricans* (Tomei et al., 1995), all organisms that produce hydrogen sulfide, which can yield insoluble selenium and selenium sulfide. For most organisms studied, the selenium was deposited in either the cell wall or cytoplasmic regions as indicated in Table 1. For *E. coli* and *W. succinogenes*, selenium was found in both regions, although this may be dependent on selenite concentration in the media used, or on the bacterial isolates present in the culture. In all cases where red selenium was observed, the selenite concentration in the medium was high, ranging from 5 to 800 mg/L. In more dilute media no precipitation was observed.

Volatile selenium products such as DMSe and dimethyl diselenide, (DMDSe), also have been detected as products of selenite microbial reduction. For most of these microorganisms, elemental selenium has also been observed. Whether elemental or volatile selenium are produced in these cases is probably determined by the selenite concentration of the medium. For the bacteria studied until now for which DMSe or DMDSe was detected as the only reduced product (Thompson-Eagle and Frankenberger, 1992), concentrations of selenium in the media have been 5 mg/L or lower. In *Lactobacillus*, red selenium was observed only with media containing 5 mg/L or more as Se, and below this concentration all the selenium was apparently incorporated into protein (Calomme et al., 1995). No test for volatile selenium was done in this work. It is interesting that no volatile selenium products were detected for *D. desulfuricans* adapted to grow in the presence of selenate or selenite (Tomei et al., 1995).

Selenoproteins and tRNAs for selenocysteine from microorganisms have been extensively studied, and several recent reviews have appeared on the subject (Stadtman, 1990, 1996; Heider and Boeck, 1993). Among the enzymes that have been characterized are formate dehydrogenases from *E. coli* and *Salmonella*, a glycine reductase selenium protein A from *Clostridia*, and hydrogenases from methanococci and *Desulfovibrio*, all of which incorporate selenocysteine. More recently, glutathione peroxidases (GSH-Px) from *Plasmodia* (Gamain et al., 1996) and *Schistosoma* (Maiorino et al., 1996) have been isolated and, like mammalian GSH-Px, probably also incorporate selenocysteine. Other selenoenzymes incorporate selenomethionine or as yet uncharacterized forms of selenium. All these enzymes exhibit increased activity if selenium is added to the growth media for the organism.

2. Other Organisms

There are parallels between the microbial metabolism of selenium and that of other organisms. For instance, selenite uptake in the mitochondria of *Vigna radiata* is hindered by the presence of thiol receptor antagonists like $CdCl_2$ and mersalyl acid (Lalitha and Easwari, 1995), and a decrease in glutathione concentration induced in rat intestinal segments by buthionine sulfoxamine inhibits selenite uptake (Vendeland et al., 1992), showing that the thiol–selenite interaction is important in these eukaryotes as well. In this connnection, it is interesting to note that the toxic effects of heavy metals are ameliorated by the presence of selenite in animal feeds (Jansson, 1980; Suzuki, 1988). Studies of selenite reduction by the thioredoxin system in lymphocytes (Spyrou et al., 1996) and by mammalian metallothionein (MT) (Chen and Whanger, 1994) have recently been reported. These reductions probably proceed via reaction 1. Subsequent metabolic steps have yet to be elucidated, but the excreted terminal selenium containing products are known for several species.

The foul odor, characteristic of volatile selenium compounds, has been recognized in mammals suffering from selenium intoxication for over 170 years (Guth, 1916, and references therein). Methylation is the usual detoxification route used by mammals to yield trimethylselenoniumion (TMSe$^+$) in urine and volatile selenium products such as DMSe in breath (Hassoun et al., 1995). No evidence of the production of elemental selenium has been reported for mammals, probably because toxic effects set in well before the necessary level of Se in diet can be reached. However, anyone who has worked with selenite solutions is familiar with the red selenium precipitate that forms on skin if the solution is not thoroughly washed away. It is interesting that in adult ruminants nearly all ingested selenite is excreted in feces in a form unavailable to plants, probably as elemental selenium, indicative of microbial action; but in young animals, which have not yet established rumen function, most of the selenium is excreted in urine, indicating detoxification by methylation to form TMSe$^+$ characteristic of mammals (Mayland et al., 1989). Several selenoenzymes, such as GSH-Px, plasma protein P, and thyroxine deiodinase (Heider and Boeck, 1993), are known in mammals, and the detection of the selenocysteyl-tRNA in several eukaryotes shows that selenocysteine-containing proteins are also end products of selenium metabolism in these cells (Heider and Boeck, 1993).

Organic selenium compounds are common as products of plant metabolism. Reduction to elemental selenium has not been reported. Nonaccumulator plants such as grains and grasses contain selenocysteine, selenocystine, and selenomethionine, incorporated into protein, as well as Se-methylselenomethionine selenonium salt (Ohlendorf, 1989). Recently selenocysteyl-tRNA has been reported in sugar beet, *Beta vulgaris* (Nève, 1992). Nonaccumulator plants rarely contain more than 50 mg/kg selenium even when grown in seleniferous environments, and selenium is generally toxic to such plants (Mayland et al., 1989). It has been suggested that the toxicity arises from the replacement of sulfur in plant protein by selenium and subsequent inactivation of these proteins. Some plants possess enzymes able to cleave the Se—C bond in Se-methylselenomethionine selenonium salts to produce DMSe and homoserine (Lewis et al., 1971; Lewis, 1976).

Selenium accumulators such as *Astralagus biculatus* have been shown to form selenocystathionine from selenate and selenite, which subsequently yields methylselenocysteine (Lewis, 1976). Recently a specific selenocysteine methyltransferase for this process has been isolated from cultured cells of *Astralagus biculatus*, and it has been proposed that this process diverts selenium out of the sulfur metabolic pathways and accounts for the high selenium tolerance of the plant (Neuhierl and Boeck, 1996). Accumulator plants, which are notoriously toxic to animals (Mayland et al., 1989), have been found to have selenium concentrations in the order of several thousand milligrams per kilogram (Ohlendorf, 1989).

3. Overview

It is probably generally true in all living systems that the initial bioreduction for both selenate and selenite leads to selenotrisulfide (Ganther, 1971). It should be noted, however, that growth inhibition in human lymphocytes by selenite and selenate is thought to take place by different mechanisms (Spyrou et al., 1996). Subsequent metabolic steps beyond selenotrisulfide have received little attention. Reduction to elemental selenium may occur by means of specific reductases or by further reaction with biological thiols. The production of volatile organic selenium products, selenonium ions, and selenoproteins requires the formation of a direct Se—C bond. The way in which this takes place in living systems is not well understood, although some progress has been made in discovering how selenocysteine is formed. In prokaryotes, selenocysteine synthesis is reported to take place by transfer of selenium to seryl-tRNA via a chain of Se donors, among them, selenophosphate (Heider and Boeck, 1993; Stadtman, 1996). Selenophosphate is formed from selenide and ATP in prokaryotes and eukaryotes (Stadtman, 1996). Selenide ion, which is kinetically stable only in water under standard conditions (Fig. 1) and is especially subject to air oxidation, has recently been shown to be the product of selenite reduction by the mammalian thioredoxin system (Bjoernstedt et al., 1996).

C. Selenium

Little is known about the metabolic pathways involving elemental selenium in living systems. Induced *Thiobacillus ferrooxidans* is reported to reduce elemental selenium to produce small amounts of selenide (Bacon and Ingledew, 1989), and *Corynebacterium* sp. generates DMSe from the element (Doran and Alexander, 1977). A strain of *Bacillus megaterium* has been shown to oxidize elemental selenium to selenite (Sarathchandra and Watkinson, 1981).

IV. CONCLUSION

In general, bioreduction dominates the chemistry of selenium in living systems. The reduction pathway for selenite is best understood, although knowledge of the details of the later stages of the process for the formation of organic and elemental selenium is limited. The reduction of selenate is generally much slower than, but probably proceeds via, selenite, resulting in the same final products. The initial stages of this pathway have not been studied. Several aspects of the bioreduction in general require further study. Is the final product—selenoprotein, elemental selenium, or some methylated form of selenium—determined by the degree of exposure of the organism to selenite or by the species or isolate studied? What factors determine the site of elemental selenium deposition in prokaryotes? Why is elemental selenium deposition commonly observed in prokaryotes but not in eukaryotes? In view of the growing realization of the importance of selenium in questions of good health and the concern about its environmental impact, a better understanding of the biological chemistry of this element is needed.

REFERENCES

Afsar, H., I. Tor, and R. Apak. 1989. Reduction of selenium(IV) and complexation of selenium(0) with mercaptoethanol. *Analyst* 114:1315–1318.

Amaratunga, W., and J. B. Milne. 1994. Studies on the interaction of selenite and selenium with sulphur donors: Part 2. A kinetic study of the reaction with 2-mercaptoethanol. *Can. J. Chem.* 72:2506–2515.

Bacon, M., and W. J. Ingledew. 1989. The reductive reactions of *Thiobacillus ferrooxidans* on sulfur and selenium. *FEMS Microbiol. Lett.* 58:189–94.

Bailey, F. C., A. W. Knight, R. S. Ogle, and S. J. Klaine. 1995. Effect of sulfate level on selenium uptake by *Ruppia maritima*. *Chemosphere* 30:579–591.

Barkes, L., and R. W. Fleming. 1974. Production of dimethylselenide gas from inorganic selenium by eleven fungi. *Bull. Environ. Contam. Toxicol.* 12:308–311.

Barton, L. L., F. A. Fekete, E. V. Marietta, H. E. Nuttall, Jr., and R. Jain. 1992. Potential for bacterial remediation of waste sites containing selenium or lead. In D. A. Sabitini and R. C. Knox (eds.), *Transport and Remediation of Subsurface Contaminants: Colloidal, Interfacial and Surfactant Phenomena*, pp. 96–106. ACS Symposium Series No. 491. American Chemical Society, Washington, DC.

Bennett, W. N. 1988. Assessment of selenium toxicity in algae using turbidostat culture. *Water Research* 22:939–942.

Bjoernstedt, M., B. Odlander, S. Kuprin, H.-E. Claesson, and A. Holmgren. 1996. Selenite incubated with NaDPH and mammalian thioredoxin reductase yields selenide, which inhibits lipooxygenase and changes the electron spin resonance spectrum of the active site iron. *Biochemistry* 35:8511–8516.

Boeck, A., C. Baron., K. Forchhammer, J. Heider, W. Leinfelder, G. Sawyers, B. Veprek, E. Zehelein, and F. Zinoni. 1990. From nonsense to sense: UGA encodes selenocysteine in formate dehydrogenase and other selenoproteins. p. 61-68. *The Molecular Basis of Bacterial Metabolism, 41st Colloquium, Mosbach*, pp. 61-68. Springer-Verlag, Berlin.

Boisson, F., M. Gnassia-Barelli, and M. Romeo. 1995. Toxicity and accumulation of selenite and selenate in the unicellular marine alga, *Cricosphaera elongata*. *Arch. Environ. Contam. Toxicol.* 28:487-493.

Brady, J. M., J. M. Tobin, and G. M. Gadd. 1996. Volatilization of selenite in aqueous medium by a *Penicillium* species. *Mycol. Res.* 100:955-961.

Brown, T. A., and A. Shrift 1982. Selective assimilation of selenite by *Escherichia coli*. *Can. J. Microbiol.* 28:307-310.

Calomme, M. R., K. Van den Branden, and D. A. Vanden Berghe, 1995. Selenium and *Lactobacillus* species. *J. Appl. Bacteriol.* 79:331-340.

Cantafio, A. W., K. D. Hagen, G. E. Lewis, T. L. Bledsoe, K. M. Nunan, and J. M. Macy. 1996. Pilot-scale selenium bioremediation of San Joaquin drainage water with *Thaura selenatis*. *Appl. Environ. Microbiol.* 62:3298-3303.

Challenger, F., and P. T. Charlton 1947. Studies on biological methylation: Part X. *J. Chem. Soc.* 1947:424-429.

Challenger, F., and H. E. North. 1934. The production of organometalloid compounds by microorganisms: Part II. Dimethyl selenide. *J. Chem. Soc.* 1934:68-71.

Chasteen, T. G., G. M. Silver, J. W. Birks, and R. Fall. 1990. Fluorine-induced chemiluminescence detection of biologically methylated tellurium, selenium and sulfur compounds. *Chromatographia* 30:81-85.

Chau, Y. K., P. T. S. Wong, B. A. Silverberg, P. L. Luxon, and G. A. Bengert. 1976. Methylation selenium in the aquatic environment. *Science* 192:1130-1131.

Chen, R. W., and P. D. Whanger. 1994. Interaction of selenium and arsenic with metallothionein: Effect of vitamin B_{12}. *J. Inorg. Biochem.* 54:267-276.

Combs, G. F., C. Garbisu, B. C. Lee, A. Yee, D. E. Carlson, N. R. Smith, A. C. Magyarosy, T. Leighton, and R. B. Buchanan. 1996. Bioavailability of selenium accumulated by selenite-reducing bacteria. *Biol. Trace Elem. Res.* 52:209-225.

Cotton, F. A., and G. Wilkinson. 1988. *Advanced Inorganic Chemistry*, 5th ed. Wiley, New York, pp. 449.

Doran, J. W. 1982. Microorganisms and the biological cycling of selenium. In K. L. Marshall (ed.), *Advances in Microbial Ecology*, pp. 1-32. Plenum Press, New York.

Doran, J. W., and M. Alexander. 1977. Microbial formation of volatile selenium compounds in soil. *Soil Sci. Soc. Am. J.* 41:70-73.

Elrashidi, M. A., D. C. Adiano, and W. L. Lindsay. 1989. Solubility, speciation and transfomrations of selenium in soils. Selenium in Agriculture and the Environment, pp. 51-63. Soil Science Society of America Special Publ. No. 23. SSSA, Madison, WI.

Fahey, R. C., W. C. Brown W. B. Adams, and M. B. Worsham. 1978. Occurrence of glutathione in bacteria. *J. Bacteriol.* 133:1126-1129.

Falcone, G., and W. J. Nickerson. 1963. Reduction of selenite by intact yeast cells and cell-free preparations. *J. Bacteriol.* 85:754-762.

Fels, I. D., and V. H. Cheldelin. 1949. Selenate inhibtion studies. III. Reversal of selenate inhibition in *E. coli*. *Arch. Biochem.* 22:323-324.

Fenchel, T., and T. H. Blackburn. 1979. *Bacteria and Mineral Cycling.* Academic Press, London.

Fleming, R. W., and M. Alexander 1972. Dimethylselenide and dimethyltelluride formation by a strain of Penicillium. *Appl. Microbiol.* 24:424–429.

Frankenberger, W. T., Jr., and S. Benson (eds.). 1994. *Selenium and the Environment.* Dekker, New York.

Frankenberger, W. T., Jr., and U. Karlson. 1994. Microbial volatilization of selenium from soils and sediments. In W. T. Frankekberger, Jr., and S. Benson (eds.), *Selenium and the Environment,* pp. 369–387. Dekker, New York.

Frausto da Silva, J. J. R., and R. J. P. Williams. 1991. *The Biololgical Chemistry of the Elements,* p. 61. Clarendon Press, Oxford.

Gamain, B., J. Arnand, A. Favier, D. Camus, D. Dive, and C. Slomianny. 1996. Increase in glutathione peroxidase activity in malaria parasites after selenium supplementation. *Free Radical Biol. Med.* 21:559–565.

Ganther, H. E. 1968. Selenotrisulfides. Formation by the reaction of thiols with selenious acid. *Biochemistry* 7:2898–2905.

Ganther, H. E. 1971. Reduction of the selenotrisulfide derivative of glutathione to a persulfide analog by glutathione reductase. *Biochemistry* 10:4089–4098.

Gerhardt, M. B., F. B. Green, R. D. Newman, T. J. Lundquist, R. B. Tresan, and W. J. Oswald. 1991. Removal of selenium using a novel algal–bacterial process. *Res. J. Water Pollut. Control Fed.* 63:799–805.

Gerrard, T. L., J. N. Telford, and H. H. Williams. 1974. Detection of selenium deposits in *E. coli* by electron microscopy. *J. Bacteriol.* 119:1057–1060.

Ghaneb, M. M., S. C. Wilkinson, and G. M. Gadd. 1995. Reduction of selenium oxyanions by unicellular polymorphic and filamentous fungi: Cellular location of reduced selenium and implications for tolerance. *J. Ind. Microbiol.* 14:300–311.

Guth, F. 1916. Selennährboden für die elektive Züchtung von Typhusbacillen. *Zentr. Bakteriol.* 1. Abt., 77:487–496.

Harrison, G. I., E. J. Laishley, and H. R. Krouse. 1980. Stable isotope fractionation by *Clostridium pasteurianum:* 3. Effect of SeO_3^{2-} on the physiology and associated sulphur isotope fractionation during SO_3^{2-} and SO_4^{2-} reductions. *Can. J. Microbiol.* 26:952–958.

Hassoun, B. S., I. S. Palmer, and C. Dwivedi. 1995. Selenium detoxification by methylation. *Res. Commun. Mol. Pathol. Pharmacol.* 90:133–142.

Heider, J., and A. Boeck. 1993. Selenium metabolism in micro-organisms. *Adv. Microbiol. Physiol.* 35:71–109.

Hsu, J.-C., M. A. Beilstein, P. D. Whanger, and H. J. Evans. 1990. Investigation of the form of selenium in the hydrogenase from chemolithotrophically cultured *Bradyrhizobium japonicum. Arch. Microbiol.* 154:215–220.

Jansson, B. 1980. The role of selenium as a cancer-protecting trace element. In H. Sigel (ed.), *Metal Ions in Biological Systems,* pp. 281–311. Dekker, New York.

Kaim, W., and B. Schwederski. 1994. *Bioinorganic Chemistry: Inorganic Elements in the Chemistry of Life.* Wiley, Chichester.

Karlson, U., and W. T. Frankenberger, Jr. 1988. Effects of carbon and trace element addition on alkylselenide production by soil. *Soil Sci. Soc. Amer.* 52:1640–1644.

Kice, J. L. 1981. The mechanism of the reaction of thiols with selenite and other Se(IV) species. In J. E. Spallholz, J. L. Martin, and H. E. Ganther (eds.), *Selenium in Biology and Medicine*, pp. 17–32. AVI Publishers, Westport, CT.

Kumar, H. D., and G. Prakash. 1971. Toxicity of selenium to blue-green algae. *Ann. Bot.* 35:697–705.

Lalitha, K., and K. Easwari. 1995. Kinetic analysis of ^{75}Se uptake by mitochondria of germinating *Vigna radiata* of different selenium status. *Biol. Trace Elem. Res.* 48:67–89.

Levine, V. E. 1925. The reducing properties of microorganisms with special reference to selenium compounds. *J. Bacteriol.* 10:217–263.

Lewis, B. G. 1976. Selenium in biological systems and pathways for its volatilization in higher plants. In J. O. Nriagu (ed.), *Environmental Biogeochemistry*, pp. 389–409. Ann Arbor Science, Ann Arbor, MI.

Lewis, B. G., C. M. Johnson, and T. C. Broyer. 1971. Cleavage of Se-methylselenomethionine selenonium salt by a cabbage leaf enzyme fraction. *Biochim. Biophys. Acta* 237:603–605.

Lindblow-Kull, C., F. J. Kull, and A. Schrift. 1985. Single transporter for sulfate, selenate and selenite in *Escherichia coli* K-12. *J. Bacteriol.* 163:1267–1269.

Lortie, L., W. D. Gould, S. Rajan, R. G. L. McCready, and K.-J. Cheng. 1992. Reduction of selenate and selenite to elemental selenium by *Pseudomonas stutzeri* isolate. CANMET Mineral Sciences Report No. MSL 92-15(J). Energy, Mines and Resources, Ottawa, Canada.

Luomo, S. N., C. Johns, N. S. Fischer, N. A. Steinberg, R.S. Oremland, and J. R. Reinfelder. 1992. Determination of selenium bioavailability to a benthic bivalve from particulate and solute pathways. *Environ. Sci. Technol.* 26:485–491.

Macy, J. M., T. A. Michel, and T. G. Kirsch. 1989. Selenate respiration by a *Pseudomonas* species: A new mode of anaerobic respiration. *FEMS Microbiol. Lett.* 61:195–198.

Macy, J. M., S. Rech, G. Auling, M. Dorsch, E. Stackebrandt, and L. Sly. 1993. *Thaurera selenatis* gen. nov. sp. nov., a member of the beta-subclass of Proteobacteria with a novel type of anaerobic respiration. *Int. J. Sys. Bacteriol.* 43:135–142.

Maiers, D. T., P. L. Wichlacz, D. L. Thompson, and D. F. Bruhn. 1988. Selenate reduction by bacteria from a selenium rich environment. *Appl. Environ. Microbiol.* 54:2591–2593.

Maiorino, M., C. Roche, M. Kiess, K. Koenig, D. Kawlik, M. Matthes, E. Naldina, R. Pierce, and L. Flohe. 1996. A selenium-containing phospholipid–hydroxide glutathione peroxidase in *Schistosoma mansoni*. *Eur. J. Biochem.* 283:838–844.

Mannervik, B., M. Widersten, and P. G. Board. 1989. Glutathione-linked enzymes in detoxification reactions. In N. Taniguchi, T. Higashi, Y. Sakamoto, and A. Miester (eds.), *Glutathione Centennial*, pp. 23–34. Academic Press, San Diego, CA.

Mayland, H. F., L. F. James, K. E. Panter, and J. L. Sonderegger. 1989. Selenium in seleniferous environments. Selenium in Agriculture and the Environment, pp. 15–49. Soil Science Society of America Special Publ. No. 23. SSSA, Madison, WI.

McCready, R. G. L., J. N. Campbell, and J. I. Payne. 1966. Selenite reduction by *Salmonella heidelberg*. *Can. J. Microbiol.* 12:703–714.

McNeal, J. M., and L. S. Balisteri. 1989. Geochemistry and occurrence of selenium: An overview. Selenium in Agriculture and the Environment, pp. 1–13. Soil Science Society of America Special Publ. No. 23. SSSA, Madison, WI.

Milne, C. J., and J. B. Milne. 1996. Studies on the interaction of selenite and selenium with sulphur donors: Part 5. Thiocyanate. *Can. J. Chem.* 74:1889–1895.

Milne. J., L. Lorusso, W. Lear, and R. Charlebois. 1994. The bioremediation of Se-containing effluents—The reduction of Se by *Escherichia coli*. In Y. Palmieri (ed.), *Proceedings of the Fifth International Symposium on Uses of Selenium and Tellurium*, Grimbergen, Belgium, pp. 63–70. Selenium–Tellurium Development Association, Darien, CT.

Moore, M. D., and S. Kaplan. 1992. Identification of intrinsic high-level resistance to rare-earth oxides and oxyanions in members of the class Proteobacteria: Sesquioxide reduction in *Rhodobacter sphaeroides*. *J. Bacteriol.* 174:1505–1514.

Moss, M. O., F. Badii, and G. Gibbs. 1987. Reduction of biselenite to elemental selenium by *Aspergillus parasiticus*. *Trans. Br. Mycol. Soc.* 89:578–580.

Moxon, A. L., and M. Rhian. 1943. Selenium poisoning. *Physiol. Rev.* 23:305–337.

Nelson, A. A., O. G. Fitzhugh, and H. O. Calvery. 1943. Liver tumors following cirrhosis caused by selenium in rats. *Cancer Res.* 3:320–326.

Nève, J. 1992. Historical perspective on the identification of type I iodothyronine deiodinase as the second mammalian selenoenzyme. *J. Trace Elem. Electrolytes Health Dis.* 6:57–61.

Neuhierl, B., and A. Boeck. 1996. On the mechanism of selenium tolerance in selenium-accumulating plants. Purification and characterisation of a specific selenocysteine methyltransferase (SecMT) from cultivated cells of *Astralagus biculatus*. *Eur. J. Biochem.* 239:235–238.

Ogle, R. S., and A. E. Knight. 1996. Selenium bioaccumulation in aquatic ecosystems: 1. Effects of sulfate on the uptake and toxicity of selenate in *Daphnia magna*. *Arch. Environ. Contam. Toxicol.* 30:274–279.

Ohlendorf, H. M. 1989. Bioaccumulation and effects of selenium on wildlife. Selenium in Agriculture and the Environment, pp. 133–177. Soil Science Society of America Special Publ. No. 23. SSSA, Madison, WI.

Oremland, R. S., J. T. Hollibaugh, A. S. Maest, T. S. Presser, L. G. Miller, and C. W. Cullbertson. 1989. Selenate reduction to elemental selenium by anaerobic bacteria in sediments and culture: Biogeochemical significance of a novel sulfate-independent respiration. *Appl. Environ. Microbiol.* 55:2333–2343.

Pachter, H. M. 1961. Magic into Science, Collier Books, New York. p. 79.

Postgate, J. 1949. Competitive inhibition of sulphate reduction by selenate. *Nature* 164:670–671.

Postgate, J. 1952. Competitive and non-competitive inhibitors of bacterial sulphate reduction. *J. Gen. Microbiol.* 6:128–142.

Rabenstein, D. L., and K.-S. Tan. 1988. ^{77}Se NMR studies of bis(alkylthio)selenides of biological thiols. *Magn. Resonance Chem.* 26:1079–1085.

Riadi, L., and J. P. Barford. 1996. Bioremediation of process waters contaminated with selenium. *Inst. Chem. Eng. Symp. Ser.* 137:209–217.

Richter, D., and H. Bergmann. 1993. Selenium uptake by wheat plants. In M. Anke (ed.), *Mengen-Spurenelem.*, *13th Arbeitstag*, 1993, pp. 149–154. Verlag MTV Hammerschmidt, Gersdorf, Germany.

Riedel, G. F., and J. G. Sanders. 1996. The influence of pH and media composition in the uptake of inorganic selenium by *Chlamydomonas reinhardtii*. *Environ. Toxicol. Chem.* 15:1577–1583.

Sarathchandra, S. U., and J. H. Watkinson. 1981. Oxidation of elemental selenium to selenite by *Bacillus megaterium*. *Science* 211:600–601.

Schmidt, M. G., and W. A. Konetzka. 1986. Glutathione overproduction by selenite-resistant *Escherichia coli*. *Can. J. Microbiol.* 32:825–827.

Schwarz, K., and C. M. Foltz. 1957. Selenium as an integral part of factor 3 against dietary necrotic liver degeneration. *J. Am. Chem. Soc.* 79:3292–3293.

Silverberg, B. A., P. T. S. Wong, and Y. K. Chau. 1976. Localization of selenium in bacterial cells using TEM and energy dispersive X-ray analysis. *Arch. Microbiol.* 107:1–6.

Spyrou, G., M. Bjornstedt, S. Skoog, and A. Homgren. 1996. Selenite and selenate inhibit human lymphocyte growth via different mechanisms. *Cancer Res.* 56:4407–4412.

Stadtman, T. C. 1990. Selenium biochemistry. *Annu. Rev. Biochem.* 59:111–157.

Stadtman, T. C. 1996. Selenocysteine. *Annu. Rev. Biochem.* 65:83–100.

Suzuki, T. 1988. Selenium: Its roles in metal–metal interaction. In *Proceedings of the Asia–Pacific Symposium on Environmental and Occupational Toxicology*, 1988, No. 8, pp. 21–30. International Center for Medical Research, Kobe University, Kobe, Japan.

Thompson-Eagle, E. T., and W. T. Frankenberger, Jr. 1990. Volatilization of selenium from agricultural evaporation pond water. *J. Environ. Qual.* 19:125–131.

Thompson-Eagle, E. T., and W. T. Frankenberger, Jr. 1992. Bioremediation of soils contaminated with selenium. In R. Lal and B. A. Stewart (eds.), *Advances in Soil Science, 1992*, pp. 261–309. Springer-Verlag, New York.

Tilton, R. C., H. B. Gunner, and W. Litsky. 1967. Physiology of selenite reduction by enterococci. *Can. J. Microbiol.* 13:1175–1182.

Tomei, F. A., L. L. Barton, C. L. Lemanski, and T. G. Zocco. 1992. Reduction of selenate and selenite to elemental selenium by *Wolinella succinogenes*. *Can. J. Microbiol.* 38:1328–1333.

Tomei, F. A., L. L. Barton, C. L. Lemanski, T. G. Zocco, N. H. Fink, and L. O. Sillerund. 1995. Transformation of selenate and selenite to elemental selenium by *Desulfovibrio desulfuricans*. *J. Ind. Microbiol.* 14:329–336.

Ullrey, D. E. 1978. Role of selenium in animal health and disease. *Third World Congress on Animal Feed*, Vol. 7, 1980, pp. 283–288.

Vendeland, S. C., J. A. Butler, and P. D. Whanger. 1992. Intestinal absorption of selenite, selenate and selenomethionine in the rat. *J. Nutr. Biochem.* 3:359–365.

Vogel, A. I. 1952. *Qualitative Chemical Analysis*, p. 438, Longmans, London.

Weast, R. C. (ed.). 1969. *Handbook of Chemistry and Physics*, 50th ed. Chemical Rubber Co., Cleveland, OH.

Woolfolk, C. A., and H. R. Whiteley. 1962. Reduction of inorganic compounds with molecular hydrogen by *Micrococcus lactilyticus*. *J. Bacteriol.* 84:647–658.

Wheeler, A. E. E., A. Zingaro, K. Irgolic, and N. R. Bottino. 1982. The effect of selenate, selenite and sulphate on the growth of six unicellular marine algae. *J. Exp. Mar. Biol. Ecol.* 57:181–194.

Williams, M. J., R. S. Ogle, A. W. Knight, and R. G. Burau. 1994. Effects of sulfate on selenate uptake and toxicity in green algae *Selenastrum capricornutum*. *Arch. Environ. Contam. Toxicol.* 27:449–453.

Yamada, H., M. Usuki, K. Hashiuchi, S. Kajiyama, and K. Yonebayashi. 1996. Selective determination of elemental selenium in soil extract. *Nippon Dojo Niryogaku Zasshi* 67:155–161 (*Chem. Abstr.* 125:32601n).

Yan, L., and G. D. Frenkel. 1994. Effect of selenite on cell surface fibronectin receptor. *Biol. Trace Elem. Res.* 46:79–89.

Zalokar, M. 1953. Reduction of selenite by *Neurospora. Arch. Biochem. Biophys.* 44:330–337.

Zehr, J. P., and R. S. Oremland. 1987. Reduction of selenate to selenide by sulfate-respiring bacteria. *Appl. Environ. Microbiol.* 53:1365–1369.

Zieve, R., and P. J. Peterson. 1981. Factors influencing the volatilization of selenium from soil. *Sci. Total Environ.* 19:277–284.

Zieve, R., P. J. Ansell, T. W. K. Young, and P. J. Peterson. 1985. Selenium volatilization by *Mortierella* species. *Trans. Br. Mycol. Soc.* 84:177–179.

24
Microbial and Cell-Free Selenium Bioreduction in Mining Waters

D. Jack Adams and Tim M. Pickett
Weber State University, Ogden, Utah

I. INTRODUCTION

This chapter discusses factors important for removal and recovery of selenium from mining waste and process waters. Selenium reduction is presented as an example of microbial metal/metalloid reduction with potential application to the treatment of high volume wastewaters. Mining and mineral processing wastewaters are generally high in volume and low in pollutant concentrations. Process solutions are usually lower volume and have a higher metal concentration, while acid drainage sites, which vary considerably in volume, have a lower pH and a high metal content. While acid drainage is not covered in this chapter, it represents a serious, widespread, complex, and ultimately long-term environmental problem.

The U.S. Bureau of Mines estimated that in the United States alone, there is approximately 50 billion tons of waste rock existing from past mining activities, with another 2 billion tons being generated each year (National Research Council, 1995). Associated with this potential source of metal contamination are billions of gallons of metal-contaminated wastewaters requiring treatment; much of this material contains selenium. Potentially adding to the problem is the increased use of biooxidation technologies, which allow for the processing of low grade, acid-producing sulfide ores. Often different from wastewaters, mining process waters can contain high concentrations of multiple metal and inorganic contaminants, such as cyanide, that are toxic to wildlife and microorganisms. Treatment technologies for the removal of dissolved metals and other inorganics are needed for mining and other high volume process and waste waters.

Biological and microbiological methods are more frequently being considered to remove and recover metals from contaminated aqueous solutions. Current applications range from large-scale removal of metals from sewage, industrial effluents, and mining waters, to smaller scale metal recovery processes. Large-scale processes are generally unsophisticated, nonspecific, and often do not remove metals to desired levels. Metal removal and recovery processes from solutions may be based on one or more microbial pathways or attributes including (1) binding of metals to cell surfaces or biomolecules within cells, (2) translocation of metals into cells via active and passive methods, (3) formation of metal-containing precipitates through reaction with cellular polymers and produced materials, such as hydrogen sulfide, phosphates, and carbonates, (4) biotransformation of metals through oxidation/reduction/volatilization methods, and (5) metal removal and recovery through use of cellular components such as ligands and enzymes. Literature and test data show that selenium biotransformation depends at least somewhat on microorganisms, nutrients, oxygen, pH, temperature, and cocontaminants and can occur under both aerobic and anaerobic conditions.

Metals and metalloids, including selenium, can be used as electron acceptors by microorganisms and are reduced in the process. In general, microbial metal/metalloid reduction may represent either a detoxification mechanism or the end step in the respiration process, resulting in selenium biotransformation. Examples of laboratory and pilot-scale tests demonstrate selenium reduction using various reactor systems with live microbes as biofilms, as immobilized live cell preparations, and in cell-free systems.

II. BACKGROUND

Selenium (Se), a naturally occurring element, atomic number 34, atomic weight 78.96, lies between sulfur and tellurium in group 16 and between arsenic and bromine in period 4 of the periodic table of the elements. Selenium is the thirtieth most abundant element and is widely dispersed in igneous rock; in hydrothermal deposits it is associated isothermally with silver, gold, antimony, and mercury; it appears in large quantities, but in low concentrations, in sulfide and porphyry copper deposits; and it is associated with various types of sedimentary rock (Hoffman and King, 1997). Selenium is a common water contaminant throughout the world and is a problem contaminant in at least six western U.S. states. Selenium is commonly found in mining wastewaters in concentrations ranging from a few micrograms per liter up to 12 mg/L, and in process solutions at concentrations exceeding 33 mg/L. While most of these operations are currently zero-discharge facilities, eventual treatment prior to closure will necessitate removing selenium to meet discharge requirements.

Inorganic selenium is most commonly found in four oxidation states (Se^{6+}, Se^{4+}, Se^0, and Se^{2-}). Selenate (SeO_4^{2-}) and selenite (SeO_3^{2-}) are highly water soluble, while elemental selenium (Se⁰) is much less soluble in water. The most reduced form, hydrogen selenide (H_2Se), occurs as a toxic gas but is readily oxidized to elemental selenium in the presence of air. Elemental selenium can be oxidized to selenite by microorganisms or converted chemically in alkaline or mildly acidic conditions; further oxidation results in the conversion to selenate. Well-aerated surface waters, especially those with alkaline conditions, represent a highly oxidized condition that contain the majority of selenium as selenate. The relative proportions of selenite and selenate depend on water redox potentials and pH. Selenite is reduced to elemental selenium under mildly reducing conditions, while selenate reduction occurs under stronger reducing conditions (Maiers, 1988; Hughes and Poole, 1989; Ehrlich, 1990).

Selenium forms covalent compounds with most other substances and is necessary in small amounts for most life-forms. Selenium is a chemical analog of sulfur and can interfere with normal cellular metabolism, particularly those aspects involving sulfur (Anderson and Scarf, 1983; Hoffman and King, 1997). Selenium is a teratogen in mammals and birds, and birth defects in several species have been attributed to it (National Research Council, 1976). Selenium poisoning, which can cause death and/or mutations in fish and waterfowl, has been documented extensively over the last decade (Bainbridge et al., 1988; Frankenberger and Benson, 1994). A highly publicized case of selenium poisoning occurred in California's Kesterson Wildlife Reservoir, resulting in the closure of this facility (Bainbridge et al., 1988).

Cost-effective, efficient, off-the-shelf technology is not available for selenium removal from large-volume waters. While reported physical–chemical methods for selenium removal will decrease selenium concentrations to acceptable levels, the treatment costs are prohibitive for large volumes. Conventional treatments include lime precipitation (Koren et al., 1992), chemical reduction (Marchant et al., 1978), activated alumina adsorption (Trussel et al., 1980), ion exchange (Maneval et al., 1985; Boegal and Clifford, 1986), and reverse osmosis (Wilmoth et al., 1978), many of which result in a mixed-metal waste product that can increase treatment/disposal costs.

III. BIOLOGICAL TREATMENT TECHNOLOGIES

Numerous biological technologies have been tested for selenium removal from water (Case et al., 1990; U.S. DOI and California Resources Agency, 1990). For example, a combined microalgal–bacterial treatment was demonstrated to effectively remove selenium from Kesterson agricultural waters (Gerhardt et al., 1991). Bench-scale processes for selenium and general metal removal from mining and other waters using *Pseudomonas stutzeri* and other microorganisms have been demonstrated (Adams et al., 1993, 1996a, 1996b). Mixed cultures of indigenous bacteria demonstrated a 96% selenium removal at bench scale (Altringer et al., 1989; Larsen et al., 1989). Pilot tests using San Joaquin agricultural waters and *Thauera selenatis* removed 98% of the selenium in an anaerobic bioreactor and produced effluent selenium concentrations of less than 5 $\mu g/L$ (Cantafio et al., 1996). A pilot plant study was conducted in which *E. coli* was used to treat a base metal smelter, weak acid effluent containing 33 mg/L Se. A rotating biological contactor (RBC) with a 9.2 m² disk surface and 30 L capacity was used as the bioreactor. Initial test results indicated that 97% of the selenium was removed from the contaminated solution with a 4-hour retention time. Tests on other mining process waters using a bench-scale RBC, *Pseudomonas stutzeri*, molasses (1.0 g/L), and a 6-hour retention time removed 97% of the selenium (Adams et al., 1993).

Microbial volatilization of selenium through biomethylation shows promise for an in situ treatment of selenium-containing waters and sediments. Biomethylation by several fungal and bacterial species has been demonstrated (Fleming and Alexander, 1972; Barkes and Fleming, 1974; Frankenberger and Karlson, 1995; Flury et al., 1997). This microbial reaction converts toxic selenium oxyanions to less toxic gaseous dimethyl selenide, which is released into the atmosphere. Rat toxicity testing demonstrated dimethyl selenide to be about one five-hundredth as toxic as selenite (McConnell and Portman, 1952). While selenium methylation does not pose the same ecological threat as methymercury compounds, synergistic toxicity of dimethyl selenide with mercury has been reported (Parizek et al., 1971).

IV. SELENIUM REDUCTION BIOCHEMISTRY

To completely understand microbial metal/metalloid reduction, it is important to have a complete picture of the biochemistry behind these transformations. However, the biochemistry of most systems for transforming microbial metals and metalloids has not been completely characterized (Stadtman, 1974). In some systems, selenium and other metal transformations appear to be coupled with the cytochrome system (Rosen and Silver, 1987; Beveridge and Doyle, 1989; Ehrlich, 1990). Also, specific selenium-active enzymes may play a role in selenium reduction. For example, the selenium-reducing organism *Thauera selenatis* appears to reduce selenate to selenite using a selenate reductase (Rech and Macy, 1992), and selenite reduction to elemental selenium appears to be catalyzed by a periplasmic nitrite reductase (Demoll-Decker and Macy, 1993). Additionally, Macy et al. (1989) report selenate reduction in a *Pseudomonas* species as part of a novel anaerobic respiration model.

Although there has been a tremendous amount of research on microbial selenium transformation, much disparity exists between transformation efficiencies observed under carefully controlled laboratory conditions and those observed in the field. Often, there appears to be no correlation between the two sets of observations. This is probably due to the broad range of environmental and site factors involved in field applications of bioremediation technologies.

V. SITE CHARACTERIZATION, BIOASSESSMENT, AND BIOTREATABILITY ANALYSIS

Whether used as a primary treatment technology, a pretreatment, or as a polishing step, successful application of biological treatments involve site characterization, bioassessment, biotreatability testing, and bioremediation monitoring. Often these important steps are over-looked or considerably shortened because of initial assessment costs. These steps are necessary to answer the following frequently asked questions: Can bioremediation work for my selenium problem? How fast will it work? What volumes and flow rates can be processed? How much will it cost?

Responsible site evaluation, system design, treatment system development, construction, and operation requires the combined expertise of many disciplines. Input from microbiologists, chemists, geologists, hydrologists, engineers, and experienced field remediation specialists are necessary for successful execution and closure of most bioremediation projects. Site characterization, assisted through site maps, historical documentation, and sampling must determine, as best possible, the horizontal and vertical extent of the contamination. In most cases this can be achieved by answering three basic questions:

1. What form do the contaminants of interest and any cocontaminants take, and what is their concentration?
2. What contamination carrier, soils or liquids, must be treated, and what characterization is required to define the soil chemistry and site hydrology?
3. Is the contamination above, below, or in the water table, and what is the degree of contamination of these areas?

Site bioassessment for a proposed remediation system involves determination of the indigenous microorganisms and bioavailability of the contaminants, as well as identification of interferences caused by cocontaminants, ions, or required treatment conditions; site environmental conditions, including pH, redox, temperature, dissolved oxygen, and available carbon, nitrogen, and phosphate concentrations; and nutrient or amendment stability. At mining sites, other essential trace elements required for successful remediation, such as sulfur, potassium, magnesium, calcium, manganese, iron, cobalt, copper, molybdenum, and zinc, sometimes are present at high concentrations.

Following site bioassessment, more detailed biotreatability studies are needed to identify factors and parameters involved in determination of whether to consider bioremediation for a particular site. This process requires slightly different approaches for in situ versus ex situ or bioreactor treatment. For in situ treatment, it must be decided whether site adjustments or amendments of nutrients, oxygen, and/or microbes can result in transformation of selenium under the environmental conditions present. This approach requires a more thorough site characterization and bioassessment, and biotreatability testing often is more difficult. For ex situ application, the objective is to determine the optimum conditions for selenium and other target compound transformations. This testing can use reactor systems of various types, sizes, and configurations to allow microbes, nutrients, and oxygen concentrations to be controlled, monitored, and optimized more readily.

Both in situ and ex situ biotreatability testing require monitoring system performance to determine the rate and extent of metal/metalloid transformation and to predict the formation of undesirable intermediate products. Biotreatability tests do not have to determine or define the ultimate end points or finite end products, but they should define general rates and possible end products. This testing should also define key factors and parameters to monitor during remediation, (i.e., microbial populations, contaminant and nutrient concentrations, site conditions of pH, redox, dissolved oxygen).

VI. BIOAUGMENTATION AND BIOSTIMULATION

Whenever bioremediation is applied to a specific site, the question arises as to whether the bioremediation design should employ microorganisms indigenous to the site or to introduce new microbes. Biostimulation employs methods to enhance and use indigenous microbes, while bioaugmentation introduces new microbes with needed biotransformation capabilities. While both approaches have merits, generally indigenous microbes have an advantage for in situ application, and introduced species often outperform native microbes in bioreactors and some land application treatments. Indigenous microorganisms often are acclimated to the contaminated site, cocontaminants, and environmental conditions, while introduced species often are not. Careful consideration should be given to which approach is best for each contaminant mix and set of site characteristics. The evaluation should assess contaminant degradation or transformation by an indigenous microbial community or by several introduced microbes acting in concert.

A promising approach, taken by these authors and others, is development of microbial cultures for a specific site through microbial coculture under actual site or bioreactor treatment conditions. Coculture, in this instance, refers to developing a new culture by culturing site microbes and known selenium-reducing microbes together in the system waters or soil slurries of interest. Often, relatively short culture times produce a microbe or a community of microorganisms with desirable properties for the site of interest. Another promising approach is use of adapted microbes or microbial populations isolated from similar selenium-contaminated sites. However, the increased use of bioremediation may possibly result in requirements for testing and registration of acclimated or enhanced microbes and microbial populations.

VII. SELENIUM-REDUCING MICROORGANISMS

Selenium-reducing microorganisms, used in experiments described herein, were isolated from selenium-contaminated mining process and waste waters and agricultural drainage(s) throughout the western United States. They were tested in their native state and subjected to a series of adaptation, acclimation, and culture enhancement methods. These microbes were again reisolated and characterized using standard microbial tests, biochemical tests, and fatty-acid-based microbial identification (Microbial ID, Inc.). Various microorganisms, including *Alcaligenes xylosoxydans*, other *Alcaligenes spp.*, *Escherichia coli*, *Enterobacter* spp., and *Pseudomonas* spp., isolates, were obtained from agricultural drainage waters. *Hydrogenophaga pseudoflava*, *Pseudomonas pseudoalcaligenes*, *Pseudomonas stutzeri*, *Pseudomonas putida*, *Pseudomonas* spp., *Bacillus* spp., *Cellulomonas flaginia*, and numerous other microbes were isolated from mining waters.

VIII. SELENIUM ANALYSIS

Selenium analysis was conducted on filter-sterilized (Nalgene 0.20 μm) culture supernatant. Samples were stored at 4°C until analyzed. Total selenium and selenite were determined using a hydride generation atomic absorption spectrophotometry procedure (Manning, 1971). Selenite was determined directly by hydride generation. Total selenium was determined by oxidizing all selenium in the sample to selenate in a potassium persulfate–nitric acid digestion, reduction to selenite with HCl, and determination of selenite as total selenium by hydride generation. Selenate was calculated as the difference between total selenium and selenite.

IX. MINING WATERS

Mining waters are generally atypical, varying greatly in types of contaminant, contaminant levels, pH, ionic content, and so on. This makes bioremediation of mining solutions a site-specific task. Table 1 shows examples of several selenium-contaminated mining waters used in some of the selenium bioreduction experiments described. Cyanide is often present in the mining of heap leach process solutions. For most experiments, residual cyanide was destroyed by treatment with bleach and/or hydrogen peroxide before treatment to remove selenium. Cyanide destruction is necessary because at concentrations above a few milligrams per liter it either kills selenium-reducing bacteria or severely inhibits bacterial growth and selenium reduction.

X. NUTRIENT SELECTION

Figure 1, a comparison of recent and past tests, shows selenium removal by different microbes, nutrients, and aeration levels in a mine water containing 0.62 mg/L selenium at pH 7.96. Tests used 250 mL Erlenmyer flasks containing 50 mL of filter-sterilized wastewater and a 10-fold dilution of log phase microbial cells (2.0×10^8 cells/mL) as an inoculum. Optimum aeration levels depend on the microorganism, the mining water, and the nutrient supplement. Under conditions of aeration (shaker flasks, 125 rpm), peptone–glycerol–yeast (PGY) and peptone supplements with Ps. stutzeri and A. xylosoxydans resulted in significantly lower selenium levels than did ammonium sulfate and dextrose supple-

TABLE I Examples of Selected Contaminant Concentrations in Actual Mine Waters, 1997

Contaminant	Mine water contaminant concentration (mg/L)[a]					
	S-1	S-4	G-4	G-5	U-2	U-3
Arsenic	3.5	<1.0	1.5	5.7	<1.0	<1.0
Copper	2.0	3.0	35.7	0.67	<1.0	1.6
Cyanide	120.0	87.0	159.7	102.0	0	0
Iron	1.31	0.25	13.0	7.90	1.37	3.23
Mercury	<1.0	<1.0	3.7	1.7	<1.0	<1.3
Nitrates	N/A	1.4	54.1	258.0	<1.0	27.8
Selenium	0.60	14.0	15.9	31.1	1.58–4.21	.04
Sulfates	155.2	35.6	75.1	540	11.2	250
Uranium	0	0	0	0	10.1	3.5
Zinc	3.6	14.0	37	29.3	5.7	<2.9
pH	8.0	9.7	11.0	10.6	7.9	3.5

[a]Sample 1–6.
Source: Confidential.

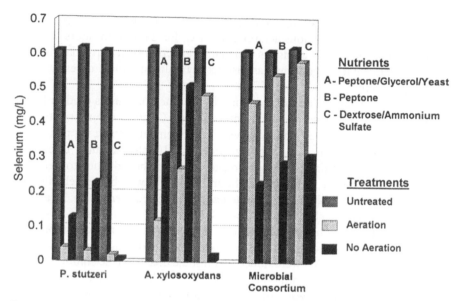

FIGURE 1 Selenium reduction by two microorganisms and a microbial consortium using different aeration treatments and nutrients. A, B, and C are the nutrient supplements tested under different aeration conditions.

ments. However, when oxygen levels were low (static flasks), the ammonium sulfate and dextrose supplement performed best. Selenium reductions to 0.02 mg/L or less were achieved with Ps. stutzeri and A. xylosoxydans. The microbial consortium, containing A. xylosoxydans, produced less dramatic results.

Additional quantitative assessments of selenate and selenite reduction by Ps. stutzeri and other isolates were performed in peptone-based media containing 10.0 mg/L selenium, followed by incubation at 24 °C with shaking at 125 rpm. Stoichiometric reductions of selenate and selenite were observed during late log and early stationary phases of cell growth. In related studies, Altringer et al. (1989), using mixed cultures of indigenous bacteria and precisely controlled mixtures of air and nitrogen, achieved 96% selenium reduction at bench scale.

Table 2 summarizes data obtained from studies comparing selenium reduction in the presence of mine waters 1 and 2, containing 1.48 and 6.2 mg/L selenate, respectively, different carbon sources, and microorganisms. A. xylosoxydans and the microbial consortium were indigenous to mining water 2, while Ps. stutzeri was not indigenous to either mining water. None of the organisms were indigenous to mining water 1. Cell growth and numbers were approximately equal when additions were made to the different carbon sources and mine water 1 at pH 7.9 and mine water 2 at pH 9.2. All microbes were used from cultures in log phase, and microbial concentrations were at 2×10^8 after dilution. Three replicate samples were tested, without aeration, at 24°C, for 24 and 36 hours with the listed carbon source and additives. Selenium reduction by a particular microbe or consortia of microorganisms has distinct optimum and often different nutrient requirements.

Figure 2 shows selenium reduction by the microorganisms used in Figure 1 and Table 2. Dextrose and ammonium sulfate (1 g/L each) nutrients were added to mine water 1. All microbes were used from cultures in log phase, and microbial concentrations were approximately 2×10^8 after dilution. In cultures started with a 10-fold dilution of cells in late log phase, the reduction to elemental selenium was complete within 6 to 18 hours. Generally, with higher cell dilutions of 50- or 100-fold, reduction to elemental selenium required 24 to 36 and 36 to 72 hours, respectively.

Screening tests, using added selenium concentrations ranging from 10 to 10,000 mg/L, were conducted with various selenium-reducing bacteria. Reduced cell growth and cell yields were noticed with some unadapted selenium-reducing isolates at selenium concentrations of 100 mg/L or higher. Inhibition at lower selenium concentrations was not observed in the adapted *Ps. stutzeri* or the *A. xylosoxydans* isolates, but was noticed in a consortium containing *A. xylosoxydans*.

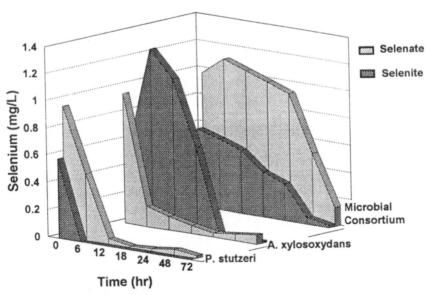

Figure 2 Selenium reduction in mine water 1 by two microorganisms and a microbial consortium using dextrose and ammonium sulfate. All tests were started using a 10-fold dilution of late log phase cells (approximately 2×10^8 cells/mL).

Most microorganisms were severely inhibited in both growth and selenium reduction at selenium concentrations of 1000 mg/L or less. One Ps. *stutzeri* isolate was only slightly inhibited in growth and selenium reduction at selenium concentrations of 1000 mg/L or less. In a similar study, Barnes et al. (1991) reported that selenium concentrations greater than 0.5 mg/L caused a significantly longer lag phase, slower growth rates, and lower cell yields compared to lower concentrations of selenate using a Ps. *stutzeri* isolate.

Testing also examined selenium reduction by various microorganisms at different temperatures. Selenium reduction by Ps. *stutzeri* was observed at temperatures of 25 to 35°C. Selenium reduction decreased approximately 60% at 20°C, compared to maximum selenium reduction observed at 30°C. However, selenium reduction by different Ps. *stutzeri* isolates varied up to 50% at 20°C. No temperature–nutrient relationship was observed. Barnes et al. (1991) suggested that the difference in selenium reduction at lower temperatures by a Ps. *stutzeri* isolate might indicate that proteins involved in selenium reduction are plasmid-encoded and/or temperature-sensitive.

XI. ANION INHIBITION OF SELENIUM REDUCTION

Table 3 shows results of the authors' recent screening tests conducted in a peptone medium (pH 7.6) to investigate inhibition of selenium reduction by nitrate, nitrite, sulfate, and sulfite. Test data for two microorganisms are presented. Various microorganisms were incubated in triplicate on a rotary shaker (125 rpm) at 25°C in flasks containing initial selenium concentrations of 50 mg/L. Selenate and selenite reduction by various microbes was inhibited to different degrees by the anions. Similar results are reported by Adams et al. (1993) and in tests with other Ps. *stutzeri* isolates (Barnes et al., 1991). Barnes et al. reported complete inhibition of selenate reduction at nitrate concentrations of 5 mM, and partial (75%) inhibition was observed at 1 mM nitrate. These researchers suggested that when nitrate levels approach that of the selenate present, cells use nitrate for respiration in preference to selenate. This conclusion is also supported by an in situ study of aquatic sediments conducted by Steinberg and Oremland (1990). However, Adams et al. (1993), observed that anion inhibition of Ps. *stutzeri* selenate or selenite reduction was isolate-dependent; only slight inhibition was observed with some isolates. Barnes et al. (1991) also reported no inhibition of selenate reduction by sulfur, an observation noted by the authors in testing Ps. *stutzeri* mining site isolates. This observation also is supported by Oremland et al. (1989) and by Steinberg and Oremland (1990). However, sulfite ions were reported to significantly reduce selenite reduction by one Ps. *stutzeri* isolate (Adams et al., 1993).

mparison of Microorganisms and Nutrients for Selenium Reduction in Two Mine Waters, 1996

	Mine water 1												Mine water 2						Co	
	Ps. stutzeri				A. xylosoxydans				Consortium				Ps. stutzeri				Consortium			
ce	A	B	C	D	A	B	C	D	A	B	C	D	A	B	C	D	A	B	C	D	A	B
	+	RR	+	R					+	+			+	RR	+	R	+	+	w	w		+
	+	RR	+	R					+	+	w	w	+	RR	R	R	+	R	w	w		R
	+	RR	+	R	+	+		w	+	R	w	w	+	RR	w	w	+	R	+	R		R
	+	RR	+	R	+	+			+	R	+		+	RR	w	w	+	+	R	R		
	+	RR		RR					+	+			+	RR	R	R	+					
	+	RR	+	RR					+	+	+		+	RR	+	RR	+	w				+
	+	R	+	+					+	+	+		w	R	+		+					+
	+	R	w	R	+	+			+	+	+		+	R	w	R	+	+	+	+		
	+	R	+	RR	+	+			+	+	R		+	R	w	R	+	+	+	+		v

											+							
vate	w	+		R	+	w			w	w			+	w	w		+	w
. succinate	+	+			+	+	+	+	+	R	+	+	w	w			+	R
	w	w	+	+	+	+		w	+	+			w	+			+	+
	+	R	+	R	+	+		w	+	+			+	R	w	w	+	w
		w		w						+				+		w	+	+
utyric acid					w	w			w	w							w	+
	+	+	+	R	+	R			+	w		w	+	+	R	RR	+	w
l	w	+	+	R									w	w	w			
id	w	w	+		+	+			w	+			w	w	+	+	+	+
	w	R	+	R	+	R	w	R	+	R		R	+	R	+	R	+	R
l	+	RR	+	R	+	+	w	R	+	+	+	R	+	RR	w	w	w	w
.d	+	RR	+	RR	+	R	+	R	+	R	+	+	+	RR	+	R	+	R
	w	R	+	R									w	R	+	RR		

ne (0.85%); B, normal saline (0.85%) plus 50 mg/L Se as sodium selenate; C, normal saline (0.85%) made with wastewater; D, normal
stewater plus 50 mg/L Se as sodium selenate. *Results key*: blank, no cell growth, no selenium reduction; w, light microbial growth/r
+, good microbial growth/no reduction to elemental Se; R, low to moderate reduction to elemental Se; RR, moderate to high reduction

TABLE 3 Microbial Selenium Reduction Inhibition Achieved by Various Anions, 1996

Microorganism	Selenium species	Anion	Concentration (M)	Decrease in Se reduction (%)
E. coli	SeO_3^{2-}	SO_4	10^{-2}	6
		NO_3	10^{-2}	37
		NO_2	10^{-2}	43
		SO_3	10^{-3}	45
E. coli	SeO_4^{2-}	SO_4	10^{-2}	8
		NO_3	10^{-2}	17
		NO_2	10^{-2}	22
		SO_3	10^{-3}	67
Ps. stutzeri	SeO_3^{2-}	SO_4	10^{-2}	1
		NO_3	10^{-2}	9
		NO_2	10^{-2}	5
		SO_3	10^{-3}	85
Ps. stutzeri	SeO_4^{2-}	SO_4	10^{-2}	3
		NO_3	10^{-2}	1
		NO_2	10^{-2}	3
		SO_3	10^{-3}	10

XII. ENHANCED/IMMOBILIZED MICROBIAL AND CELL-FREE BIOREACTORS

Without the requisite bioassessment and biotreatability studies, both in situ and ex situ selenium reduction processes can be quickly overshadowed by inadvertent stimulation of indigenous bacteria. Normal biofilms developed for selenium reduction and removal can be quickly overgrown as the bioreactor system is exposed to waters containing indigenous microbes and renewed nutrients. Overgrowth of the selenium-reducing population in a bioreactor can be delayed by optimizing the bioreactor and nutrient selection for the chosen selenium reducers. However, once nutrients have been added, time, nutrients, and indigenous microbes slowly erode the biofilm, and consequently the selenium-reducing capability. Selenium-reducing capability can be extended by several practices: encapsulation of the selenium-reducing microbes, use of selenium-reducing enzymes and/or cell-free preparations, and by more permanent immobilization of the biofilm on the support surface.

Enhanced biofilms of Ps. stutzeri have been developed on activated carbon by pretreatment of the carbon with selected nutrients and biopolymers. Biofilms established in such a manner have supported single microbes and microbial consortia for selenium reduction for periods up to 9 months. Figure 3 compares

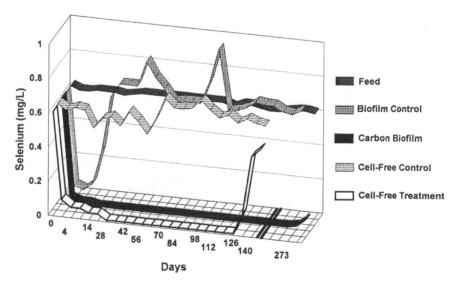

FIGURE 3 Selenium reduction by an enhanced/immobilized biofilm on activated carbon and a calcium alginate immobilized, cell-free preparation. Tests used a single-pass bioreactor with a retention time of 18 hours. Mine water feed, sterilized carbon, and empty calcium alginate bead controls are shown.

an immobilized cell-free preparation with an enhanced/immobilized carbon biofilm of *Ps. stutzeri* using a mine water containing 0.62 mg/L selenium as selenate. Tests used a single-pass bioreactor with a retention time of 18 hours. Cell-free extracts were produced by disrupting the bacterial cells and immobilizing the lysate in calcium alginate beads (Adams et al., 1993, 1996a). The immobilized enzyme preparation performed for approximately 4 months, removing selenium to below 0.01 mg/L. The enhanced carbon biofilm preparation removed selenium to below 0.01 mg/L for 9 months without supplemental nutrients.

Cell-free systems hold considerable potential for the removal of various inorganics and metal/metalloid contaminants. As mentioned earlier, cyanide is often a cocontaminant with selenium in heap leach mining process solutions and severely inhibits selenium reduction. Figure 4 shows test results from cell-free preparations of *Pseudomonas pseudoalcaligenes*, *Pseudomonas stutzeri*, and cyanide-oxidizing and selenium-reducing microorganisms combined and immobilized in calcium alginate beads. Tests were conducted in 1 in. diameter columns operated in single-pass, upflow mode with a retention time of 9 to 18 hours, at ambient temperature (25°C), using a process solution containing 102 mg/L cyanide and 31.1 mg/L selenium, at pH 10.6. Simultaneous cyanide removal and selenium reduction were demonstrated for 2 weeks (process water limitation). Cyanide

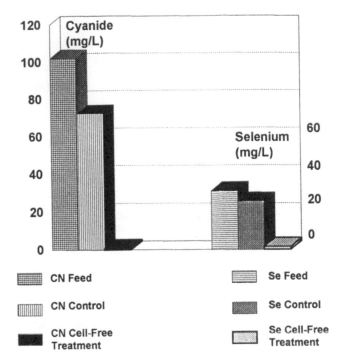

FIGURE 4 Simultaneous cyanide oxidation and selenium reduction by cell-free preparations immobilized in calcium alginate beads. Tests used a single-pass bioreactor (retention
time, 9–18 h). Process solution cyanide and selenium concentrations and empty calcium
alginate bead controls are shown.

levels decreased from 102 mg/L to below 1 mg/L, and selenium concentrations
decreased from 31.1 mg/L to 1.6 mg/L.

XIII. SCANNING ELECTRON MICROSCOPY ANALYSIS
OF SELENIUM PRECIPITATES

Depending on the mining water used and the oxygen content of the water, either
red amorphous or dark brown to black crystalline selenium precipitates are
formed. However, the dark brown to black crystalline selenium precipitates have
been noticed only in systems with low dissolved oxygen. Preliminary analysis of
the selenium precipitates by scanning electron microscopy (SEM) indicates a
relatively pure selenium by-product from bioreactors of both types (i.e., those
using live cells immobilized in calcium alginate and those using cell-free prepara-

tions). SEM scans similar to the one in Figure 5 were obtained from both immobilized live cells and immobilized cell-free preparations of *Ps. pseudoalcaligenes* and *Ps. stutzeri* used to treat complex mining waters containing uranium, copper, arsenic, and other metals. Selenium was the predominant metal/metalloid detected. Silicon is from the glass slide mount, and gold was deposited on the sample through the sputter-coating process. Calcium present in the sample is from the immobilization polymer. These preliminary data indicate the selectivity of the microbes examined for selenium reduction in complex waters. Production and recovery of a high purity selenium product from wastewaters could partially offset treatment costs.

XIV. CONCLUSIONS

Selenium contamination is a widespread environmental problem throughout the western United States and elsewhere. Although selenium is naturally occurring, anthropogenic activities (mining and agriculture) have greatly contributed to the

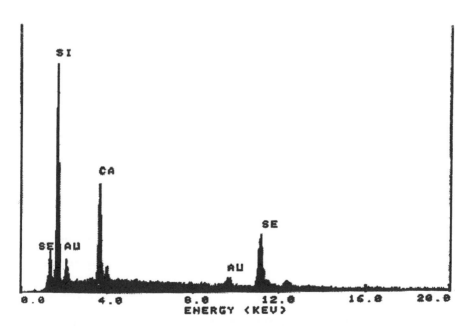

FIGURE 5 SEM analysis of microbial selenium reduction precipitate formed in calcium alginate beads when treating a mining water containing 15.7 mg/L selenium.

problem. Economical off-the-shelf technologies are not available for large-volume treatment of selenium-contaminated waters.

Conventional physical–chemical treatment technologies are generally non-specific, and too costly for the complete treatment of large-volume, complex waters, such as are produced by the mining industry. Selenium bioremediation processes have the potential for providing a low cost treatment and have been demonstrated in laboratory and field tests by the Center for Bioremediation. Based on these tests, estimated treatment costs range from $0.10/1000 gallons using only agricultural grade nutrients, to $1.50 ± $0.50/1000 gallons for immobilized or encapsulated cells for difficult-to-treat situations.

Site- and water-specific characterization, bioassessment, and biotreatability analyses are required for successful biotreatment field application. Selenium-reducing microorganisms, or microbial consortia, have distinct requirements dictated by each specific site or water to be treated. Additional considerations, such as interfering ions, redox conditions, temperature, and overgrowth by indigenous microorganisms, must also be addressed. Enhanced/Immobilized live cell and cell-free bioreactors were found to be effective for selenium removal from actual wastewaters for extended periods.

Cell-free systems show promise for the treatment of complex waters and waters with cocontaminants toxic to living systems. Immobilized, cell-free preparations were used to simultaneously remove selenium and cyanide from an actual process solution. Advantages of cell-free systems over live systems include potential for greatly increased kinetics, absence of a requirement for nutrients, and elimination of the effects of toxic process solutions. Cell-free bioreactors can be engineered to be resistant to microbial overgrowth and degradation.

These and other passive biotreatments, such as biomethylation of selenium in soil and water systems, show great potential. Field testing of these biotechnologies, to demonstrate cost-effective treatment of large water volumes to meet discharge or drinking water criteria, will validate this potential.

ACKNOWLEDGMENTS

The authors thank Kenneth R. Gardner and Douglas N. Esplin for their assistance in microbial screening and identification and in experimental setup and monitoring.

REFERENCES

Adams, D. J., P. B. Altringer, and W. D. Gould. 1993. Bioreduction of selenate and selenite. In A. E. Torma, M. L. Apel., and C. E. Brierley (eds.), *Biohydrometallurgy*, pp. 755–771. Minerals, Metals and Materials Society. Warrendale, PA.

Adams, D. J., T. M. Pickett, and J. R. Montgomery. 1996a. Biotechnologies for metal and toxic inorganic removal from mining process and waste solutions. *Randol Gold Forum Proceedings*. Olympic Valley, CA.

Adams, D. J., K. Fukushi, and S. Ghosh. 1996b. Development of enriched microbial cultures for enhanced metal removal. *Proceedings*, American Chemical Society, Emerging Technologies in Hazardous Waste Management VIII, Birmingham, AL.

Altringer, P. B., D. M. Larsen, and K. R. Gardner. 1989. Bench scale process development of selenium removal from wastewater using facultative bacteria. In *Biohydrometallurgy*, pp. 643–657. Minerals, Metals and Materials Society. Warrendale, PA.

Altringer, P. B., R. H. Lien, and K. R. Gardner. 1991. Biological and chemical selenium removal from precious metals solutions, In *Environmental Management for the 1990's*, pp. 135–142. Society of Mining Engineers, Littleton, CO.

Anderson, J. W., and A. R. Scarf. 1983. Selenium and plant metabolism. In D. A. Robb and D. S. Pierport (eds.), *Metals and Micronutrients: Uptake and Utilization by Plants*, pp. 241–275. Academic Press, New York.

Bainbridge, D., V. Wegrzyn, and N. Albasel. 1988. *Selenium in California*, Vol. 1. California State Water Resources Control Board, Contract IAA No. 5-249-300-0. CSWRCB, Sacramento.

Barkes, L., and R. Fleming. 1974. Production of dimethylselenide gas from inorganic selenium by eleven soil fungi. *Bull. Environ. Contam. Technol.* 12:308–311.

Barnes, J. M., E. McNew, J. Polman, J. McCune, and A. Torma. 1991. Selenate reduction by *Pseudomonas stutzeri*. In R. W. Smith and M. Misra (eds.), *Mineral Bioprocessing*. Minerals, Metals and Materials Society, Warrendale, PA.

Beveridge, T. J. and R. J. Doyle (eds.). 1989. *Metal Ions and Bacteria*. Wiley, New York.

Boegel, J. V., and D. A. Clifford. 1986. Selenium oxidation and removal by ion exchange. EPA 600/S2-86/031. U.S. Environmental Protection Agency, Washington, DC.

Cantafio, A. W., K. D. Hagen, G. E. Lewis, T. L. Bledsoe, K. M. Nunan, and J. M. Macy. 1996. Pilot-scale selenium bioremediation of San Joaquin drainage water with *Thauera selenatis*. *Appl. Environ. Microbiol.* 62:3298–3303.

Case, J. C., L. R. Zelmer, M. T. Harris, R. L. Anderson, and L. L. Larsen. 1990. Selected bibliography on selenium. Geological Survey of Wyoming, Bull. No. 69, Laramie.

Demoll-Decker, H., and J. Macy. 1993. The periplasmic nitrite reductase of *Thauera selenatis* may catalyze the reduction of selenite to elemental selenium. *Arch. Microbiol.* 160:241–247.

Ehrlich, H. L. 1990. *Geomicrobiology*, 2nd ed. Dekker, New York.

Fleming, R. W., and M. Alexander. 1972. Dimethylselenide and dimethyltelluride formation by a strain of *Penicillium*. *Appl. Microbiol.* 24:424–429.

Flury, M., W. T. Frankenberger, Jr., and W. A. Jury. 1997. Long-term depletion of selenium from Kesterson dewatered sediments. *Sci. Total Environ.* May 30, 1997, Vol. 198(3), p. 259–270.

Frankenberger, W. T., Jr., and S. Benson (eds.). 1994. *Selenium in the Environment*. Dekker, New York.

Frankenberger, W. T., Jr., and U. Karlson. 1995. Volatilization of selenium from a dewatered seleniferous sediment: A field study. *J. Ind. Microbiol.* 14:226–232.

Gerhardt, M. B., F. B. Green, R. D. Newman, T. J. Lundquist, R. B. Tresan, and W. J. Oswald. 1991. Removal of selenium using a novel algal–bacterial process. *Res. J. Water Pollut. Control Fed.* 63:799–805.

Hoffman, J. E., and M. G. King. 1997. Selenium and selenium compounds. In J. I. Kroschwitz and M. Howe Grant (eds.), *Kirk–Othmer Encyclopedia of Chemical Technology*, 4th ed., Vol. 21, pp. 686–719. Wiley, New York.

Hughes, H. M., and R. K. Poole. 1989. *Metals and Micro-organisms*. Chapman and Hall. London.

Koren, D. W., W. D. Gould, and L. Lortie, 1992. Selenium removal from waste water. Canadian Institute of Metallurgy meeting, Edmonton, Canada.

Larsen, D. M., K. R. Gardner, and P. B. Altringer. 1989. Biologically assisted control of selenium in process waste waters. In B. J. Sheiner, F. M. Doyle, and S. K. Kawatra (eds.), *Biotechnology in Minerals and Metals Processing*, pp. 177–185. Society of Mining Engineers, Littleton, CO.

Macy, J. M., T. A. Michel, and D. G. Kirsh. 1989. Selenate reduction by a *Pseudomonas* species: A new mode of anaerobic respiration. *FEMS Microbiol. Lett.* 61:195–198.

Maiers, D. T. 1988. Selenate reduction by bacteria from a selenium rich environment. *Appl. Environ. Microbiol.* 54:2591–2593.

Maneval, J. E., G. Klein, and J. Sinkovic. 1985. Selenium removal from drinking water by ion exchange. EPA/600/2-85/074. U.S. Environmental Protection Agency, Washington, DC.

Manning, D. C. 1971. *At. Absorpt. Newsl.* (Perkin-Elmer Corp. Norwalk, CT) 10:123.

Marchant, W. N., R. O. Dannenberg, and P. T. Brooks. 1978. Selenium removal from acidic waste water using zinc reduction and lime neutralization. U.S. Bureau of Mines Report of Investigations No. 8312.

McConnell, K. P., and O. W. Portman. 1952. Toxicity of dimethyl selenide in the rat and mouse. *Proc. Soc. Exp. Biol. Med.* 79:230–231.

National Research Council Subcommittee on Selenium. 1976. Selenium. National Academy of Sciences, Washington, DC.

National Research Council. 1995. Review of U.S. Bureau of Mines Biotechnology Program. NRC/USBM, Salt Lake City, UT.

Oremland, R. S., J. T. Hollibaugh, A. S. Maest, T. S. Presser, L. G. Miller, and C. W. Culbertson. 1989. Selenate reduction to elemental selenium by anaerobic bacteria in sediments and culture: Biogeochemical significance of a novel, sulfate independent respiration. *Appl. Environ. Microbiol.* 55:2333–2343.

Parizek, J., I. Ostadalova, J. Kalouskova, A. Babicky, and J. Benes. 1971. The detoxifying effects of selenium interrelations between compounds of selenium and certain metals. In W. Mertz and W. E. Cornatzer (eds.), *Newer Trace Elements in Nutrition*, pp. 85–122. Dekker, New York.

Rech, S. A., and J. M. Macy. 1992. The terminal reductases for selenate and nitrate respiration in *Thauera selenatis* are two distinct enzymes. *J. Bacteriol.* 174:7316–7320.

Rosen, B. P. and S. Silver (eds.). 1987. *Ion Transport in Prokaryotes*. Academic Press, San Diego, CA.

Stadtman, T. C. 1974. Selenium biochemistry. *Science.* 183:915–922.

Steinberg, N. A., and R. S. Oremland. 1990. Dissimilatory selenate reduction potentials in a diversity of sediment types. *Appl. Environ. Microbiol.* 56:3550–3557.

Trussel, R. R., A. Trussel, P. Kreft, and J. M. Montgomery. 1980. Selenium removal from ground water using activated alumina. EPA 600/2-80-153. U.S. Environmental Protection Agency, Washington, DC.

U.S. Department of Interior and California Resources Agency. 1990. San Joaquin Valley Drainage Program Draft Final Report.

Wilmoth, B. C., T. L. Baugh, and D. W. Decker. 1978. Removal of selected trace elements from acid mine drainage using existing technology. *Proceedings of the 33rd Purdue Industrial Waste Conference*, pp. 886–894. Purdue University.

25

Bioreactors in Removing Selenium from Agricultural Drainage Water

LAWRENCE P. OWENS
California State University at Fresno, Fresno, California

I. INTRODUCTION

In the last 12 to 13 years, many processes for removal of selenium from agricultural drainage water have been investigated. Hanna et al. (1990) reviewed a number of the physical, chemical, and biological processes that were so examined during the mid- to late 1980s. Applications of physical–chemical methods for selenium removal are limited because the complex nature of drainage water and its high sulfate content result in treatment interferences and/or high costs. Biological methods of selenium removal have shown the most promise and have been researched most extensively.

While aerobic biological processes have been investigated, the majority of the work, and the most fully developed processes, utilize anoxic biological treatment. The underlying principle for the biological removal of selenium is that under anoxic conditions (no molecular oxygen) with a carbon source for food, bacteria will convert the soluble selenate form of selenium present in drainage water to insoluble elemental selenium. The elemental selenium can then be settled or filtered from the water. This can be illustrated by the following pseudoreaction:

Se($+$VI) + bacteria + organic carbon $\longrightarrow$ Se($+$IV) $\longrightarrow$ Se(0)
Selenate Selenite Elemental Se
(soluble) (soluble) (particle)

Reactor configuration influences the efficiency and speed of the conversion, and so reactors of different types have been investigated.

Three investigators have been responsible for the most significant research and demonstration of selenium removal by biological processes in reactor systems:

EPOC AG at Murrieta Farms, near Mendota, California, Dr. Joan Macy at University of California, Davis (UC Davis), and the Engineering Research Institute (ERI) at California State University, Fresno (CSUF). The results from these studies are summarized in this chapter, with primary emphasis on the work directed by the author at the Engineering Research Institute. Cost estimates for a large-scale system are presented in Section V.

II. EPOC AG

EPOC AG (aka Binnie California, Inc.) conducted pilot-scale studies using biological processes from July 1985 through March 1988 (EPOC AG, 1987; Binnie California et al., 1988; Squires et al., 1989). The studies were conducted at Murrieta Farms, in the San Joaquin Valley.

Many reactor configurations were tested, but a two-stage combination of upflow anaerobic sludge blanket reactor (UASBR) followed by a fluidized-bed reactor (FBR) was the best and handled scaling problems better than earlier reactor types. The biological reactors were followed by a crossflow microfilter for removal of the particulate selenium. The effluent from the microfilter averaged less than 30 μg/L of soluble selenium. When a soil column was added as a polishing stage, soluble selenium in the effluent from the column was less than 10 μg/L.

As in the other bioreactors described in this chapter, calcium carbonate precipitation occurred within the reactors. Since the high calcium content of drainage water is expected to cause this result in any system, the selection of an appropriate reactor configuration is important. The UASBR, which generates a granular biomass, actually benefits from some precipitation as an aid to granulation and promotion of good settling characteristics of the sludge. Fluidized beds and packed beds may experience greater problems with clogging from precipitation when used as first-stage reactors. This was confirmed in the ERI studies described later.

Based on the pilot plant results, a prototype plant with a capacity of 3785 m³/d was designed and permits were obtained, but a lack of funding prevented completion of the project.

III. DR. JOAN MACY AT UC DAVIS

Dr. Joan Macy, a microbiologist at UC Davis, isolated a new selenate-respiring bacterium, *Thauera selenatis* (Macy, 1992). *T. selenatis* was inoculated into a lab-scale biological reactor system consisting of a one-liter UASBR followed by a one-liter FBR (Macy et al., 1993). The UASBR would be better described as an expanded bed reactor or FBR, since it was filled with 400 g of sand. Hydraulic retention time in each reactor was approximately 140 minutes at a flow of 6.5 milliliters per minute mL/min. Because the reactors were not operated under sterile conditions, a population of denitrifying bacteria developed in the reactor, presumably originating from bacteria present in the influent drainage water. Denitrifying bacteria are required for the selenium reduction to proceed all the way to elemental selenium in significant quantities. Nitrate-respiring bacteria were present in numbers three orders of magnitude greater than the selenate-respiring bacteria. *T. selenatis* requires acetate as carbon source and ammonium as a supplemental nitrogen source. Selenium oxide concentrations (selenate + selenite) were reduced from 350 to 450 μg/L in the influent drainage water to an average of 9.8 μg/L after treatment. Note that the concentration of selenium oxides is less than the total soluble selenium, which would probably be less than 30 μg/L in this effluent. The advantage of *T. selenatis* in a reactor system is that the reduction of selenate to selenite occurs independently of nitrate reduction, through a separate selenate reductase enzyme, thus improving the kinetics of the process. Further reduction of selenite to elemental selenium occurs via the nitrite reductase system found in the denitrifying bacteria. One disadvantage of the use of *T. selenatis* is the cost of the nutrients required for growth. Acetate is more expensive than methanol as a carbon source, and the additional requirement of ammonium is another expense.

IV. ENGINEERING RESEARCH INSTITUTE AT CALIFORNIA STATE UNIVERSITY, FRESNO

The San Joaquin Valley Drainage Program (see Chapter 10, this volume) concluded that selenium removal by biological treatment held the most promise for success and should be continued. As a result of this conclusion, the Adams Avenue Agricultural Drainage Research Center (AAADRC) was developed as a cooperative effort between the California Department of Water Resources, the U.S. Bureau of Reclamation, Westlands Water District, and the ERI at CSUF. The AAADRC is located west of Tranquillity, California, in the Westlands Water District, approximately 75 km west of Fresno.

A. Laboratory Studies

While construction of facilities at the AAADRC was proceeding, laboratory-scale projects were being conducted at CSUF. These projects were directed at looking at different reactor configurations and providing initial information to aid in start-up of larger units at the AAADRC. The reactors investigated were the sequencing batch reactor (SBR), packed-bed reactors (PBR), slow sand filters (SSF), and the UASBR.

1. Sequencing Batch Reactors

A semi-pilot-scale SBR system consisted of two 110 L reactors operated in parallel (Johnson, 1992). Two different carbon sources for the biological reaction were evaluated. The bacterial population in one reactor received acetate while methanol was fed to the other reactor. Anaerobic conditions were maintained in the closed systems.

Since a larger, 40 m³ SBR was originally planned for the AAADRC, the results from this experiment were important. Results for removal of soluble selenium are summarized in Table 1. While these removals were encouraging and showed that the biological process can effectively reduce the selenium in the drainage water, the development of the large SBR at the AAADRC was shelved because of the way in which the bacteria were growing in the system. Rather than growing in a flocculent suspension as was intended, the bacteria grew in an attached film on the walls and piping of the reactor. In the 110 L reactors, effective treatment was achieved with attached bacterial growth, but for a system built on a large scale, the ratio of surface area to reactor volume would be smaller. Decreased surface area/volume ratios would result in a lower bacterial density in the reactor, and lower treatment efficiencies would be expected. A packing material added to the reactor would provide additional surface area, so continuous flow PBRs were chosen for further study. FBRs, with bacteria growing on a fluidized bed of sand, also provide surface area for attached growth and were chosen to replace the SBR system at the AAADRC. FBRs had been demonstrated to be effective in studies conducted by EPOC AG (1987) at Murrieta Farms.

2. Packed-Bed Reactors

Two laboratory-scale PBRs, 1.75 m in height with an internal diameter of 8.6 cm, were randomly packed with 2.5 cm diameter Jaeger Tripack (Jakhar, 1993). For the first 4 months of operation, the reactors averaged 54% removal

TABLE I Removal of Soluble Se in Sequencing Batch Reactors

Carbon source	Average total soluble selenium (μg/L)		Removal (%)
	Influent	Effluent	
Acetate	455	45	90
Methanol	455	65	86

of soluble selenium. This unexpectedly low efficiency was due to the buildup of settled biomass in the bottom of the column. All the carbon was being used up in the first 30 cm of the reactor, leaving none for completion of the selenium conversion reaction. After the settled biomass had been removed, the reactors averaged 79% removal of soluble selenium. Effluent soluble selenium averaged approximately 50 μg/L, with the lowest recorded value being 28 μg/L.

3. Slow Sand Filters

The biological reactors that convert the selenium generally do not retain all the particulate elemental selenium that is formed. Usually, the majority of the elemental selenium leaves the reactor in the effluent. A complete treatment system requires some type of particle removal step. To evaluate the effectiveness of a complete treatment scheme, slow sand filters were used to filter the effluent from the PBRs described above (Dhaliwal, 1992). Two SSF columns were constructed of Plexiglas with an internal diameter of 5.7 cm and had approximately 1 m of #20 Monterey sand covered with 1 m of water above the sand bed.

The SSFs did not achieve much removal while the PBRs were not operating efficiently (as discussed above); after the settled biomass had been removed from the PBRs and their efficiency improved, however, the results from the SSFs were dramatic. Selenium removal for the combined system of PBR followed by SSF is shown in Figure 1. Average influent and effluent selenium concentrations for both SSFs are shown in Table 2 for the period after the packed beds were cleaned. The influent to the SSFs was the combined effluent from the PBRs.

The summary shown in Table 2 points out a significant treatment advantage of the SSF. The total selenium appearing in the effluent is essentially equal to filtering the influent through a 0.22 μm membrane filter. The simple sand filter can achieve this level of removal because it is not just a physical straining process. The residual soluble selenium and selenite in the influent were significantly reduced, from 58 μg/L to 22 μg/L and from 44 μg/L to 14 μg/L, respectively. This would indicate additional bioconversion of selenium in the SSF with subsequent removal of the newly formed particulate selenium. Pilot-scale sand filters were evaluated at the AAADRC.

4. Upflow Anaerobic Sludge Blanket Reactors

To facilitate start-up and provide preliminary data for the large reactor at the AAADRC, two lab-scale UASBRs were operated at the CSUF campus (Owens, 1992). Two one-liter Imhoff cones were used as conical-shaped UASBRs and were seeded with a granular anaerobic sludge from a defunct bakery wastewater treatment plant in Kansas City, Missouri. A recycle rate of 4 L/min was found to suspend and agitate the granules without washing them out of the reactor. The influent flow rate was 1 mL/min, which provided a hydraulic retention time

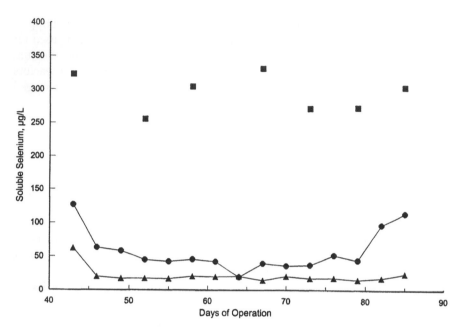

FIGURE 1   Soluble selenium removal by laboratory packed-bed reactor followed by a slow sand filter: squares, drainage influent; hexagons, PBR effluent; triangles, SSF effluent.

of 18 hours. The pH of the influent to one of the reactors was adjusted to pH 6.9 with hydrochloric acid while the other had no pH adjustment. The pH of the drainage water averaged pH 8.0.

The average removal of selenium in the reactor without pH control was 79.5% based on total selenium in the effluent and 83.4% based on soluble selenium. The average removal of selenium in the reactor with influent pH 6.9 was 71.6% based on total selenium and 86.8% based on soluble selenium. The lowest concentrations achieved in the effluent were 27 μg/L total selenium and

TABLE 2   Selenium Results from Lab-Scale Sand Filters

	Se concentration (μg/L)		
	Influent	Filter A effluent	Filter B effluent
Total Se	197	61	53
Soluble Se	58	22	22
Se(+4)	44	14	14

20 μg/L soluble selenium. These reactors were different from the SBRs and PBRs in that a majority of the particulate selenium remained in the reactor rather than coming out in the effluent.

B. Pilot-Scale Reactors at the AAADRC

Pilot-scale reactor systems were in operation at the AAADRC from September 1992 through November 1995. Seven different reactors, representing four different reactor types, were operated for varying lengths of time. The reactors were operated in both series and parallel, as shown in Figure 2. A complete description of the operation of all seven reactors, which is beyond the scope of this chapter, can be found in Owens et al. (1997) and Salamor et al. (1997). Sections IV.B 1 and IV.B.2 discuss the large UASBR and the series operation of the UASBR followed by an FBR and SSF.

1. Upflow Anaerobic Sludge Blanket Reactor: Initial Phase

The first phase of testing at the AAADRC utilized an 11 m^3 conical-bottom UASBR with internal gas–solid–liquid separator (Owens et al., 1997) as shown in Figure 3. A schematic representation is shown in Figure 4. The influent flow rate was 7.5 L/min, giving a hydraulic retention time (HRT) of about 25 hours. Effluent was recycled at 300 l/min and mixed with the influent to maintain sufficient agitation of the sludge bed. Methanol was added as the carbon source at a dosage of approximately 250 mg/L (0.25 mL of methanol per liter of drainage water).

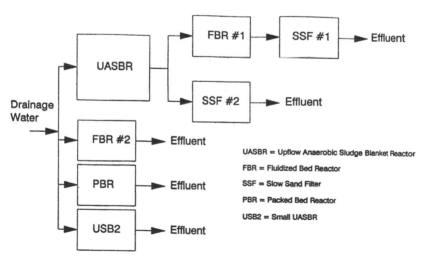

FIGURE 2 Pilot-scale reactor systems at the AAADRC.

FIGURE 3 The 11 m³ upflow anaerobic sludge blanket reactor at the AAADRC.

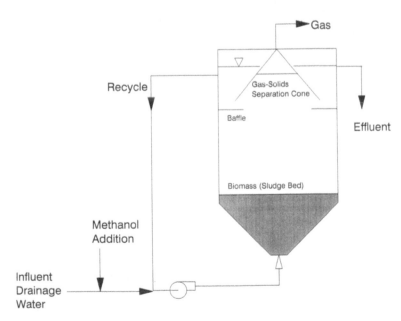

FIGURE 4 Schematic of the UASB reactor with internal gas–solids separation.

The UASBR was seeded with approximately 2 m³ of the granular sludge from the Kansas City bakery describe earlier. During a 7-month period, the agricultural drainage water had an average concentration of 45 mg/L of nitrate-N. On-site measurements of the effluent showed almost complete denitrification, with an average effluent concentration of about 3 mg/L nitrate-N. The denitrification reaction is important because most of the carbon used in the process goes to denitrification. The ratio of nitrate-N to selenium is about 90 : 1.

Selenium measurements for influent and effluent for this period are shown in Figure 5. Within 6 weeks of seeding the reactor, the UASBR achieved average selenium removals of 69% of total and 88% of soluble selenium (58 μg/L in the effluent). These results show that the fermentative granular sludge taken from the bakery reactor quickly acquired the ability to reduce selenate. At about 70 days after start-up, the effluent soluble selenium concentrations began to rise. This corresponded to the point at which reactor temperature dropped below 15°C. The lowest temperature recorded was 7°C. Even at 7°C, 35% of the soluble selenium was being converted to particulate selenium in the reactor. Also, the decrease in conversion efficiency does not appear to be due entirely to low temperatures. At about day 160, while temperatures were still low, additional mixing was provided to the reactor contents by diverting some 80 L/min of recycle flow through a perforated PVC pipe inserted from the top of the reactor into

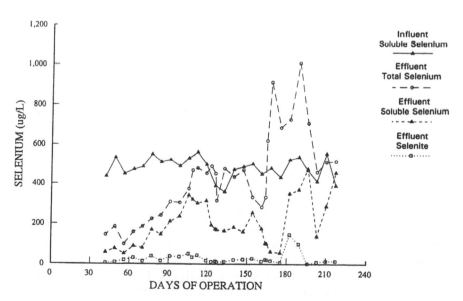

FIGURE 5   Influent and effluent selenium for the UASBR during first testing period. Additional mixing was started on day 160.

the sludge bed. The effluent total selenium increased immediately as particulate selenium that had been accumulating in the reactor was resuspended and flushed out with the effluent. At the same time, the soluble selenium dropped dramatically to 60 μg/L. Shortly after these levels were achieved, a pump failure caused a series of events that resulted in failure of the internal solids separation system, and the reactor had to be shut down for repairs. For the entire period, the influent soluble selenium averaged 482 μg/L while the effluent had an average soluble selenium of 203 μg/L, for an average reduction of 58%.

2. Upflow Anaerobic Sludge Blanket Reactor: Second and Third Phases

After repairs and modifications had been made, the reactor was reloaded with granular sludge (a second batch that had been stored on-site). The reactor was restarted, and 18 days later, effluent soluble selenium was measured at 29 μg/L, for a soluble selenium removal of 94%. However, even with modifications to the reactor, the granular sludge was being lost from the reactor. For reasons still unclear, the sludge granules were able to navigate the system of internal baffles designed to retain them and float to the top of the reactor, where they were carried out in the effluent.

In the third testing period at the AAADRC, a granular sludge was developed on-site within the UASBR using anaerobic digester effluent and waste activated sludge from local municipal wastewater treatment plants as the inoculum. It took approximately 6 months for a stable granular sludge to develop, but after this period, loss of the biomass was not a major problem, even though sludge volume did vary. The pilot-scale FBRs, SSFs, and the PBR were also added during this period in the configuration shown in Figure 2. Removals of soluble selenium of 90% or greater (from about 500 μg/L to 50 μg/L or less) were achieved by the single-stage FBR, the single-stage PBR, the UASBR followed in series by the FBR and SSF (Figure 6), and the UASBR in series with the SSF. While removals were not always this low and showed considerable variation, the ability of the process to achieve these levels of reduction was demonstrated (Owens et al., 1997; Salamor et al., 1997).

Variation of effluent selenium levels was due to planned changes in reactor operations, equipment malfunctions (e.g., problems with dosing pumps), and temperature changes (since the reactors were exposed to ambient conditions at all times of the year). It is important to note that these units were pilot-scale experimental reactors. In scaling up from a pilot-scale to a full-scale system, the problems causing effluent variation would be eliminated for the most part by having automated controls with in-line monitoring on larger equipment. Insulation of exposed reactors would also be provided. In addition, a full-scale system would be designed with a certain amount of redundancy to provide safety factors in case a malfunction did occur.

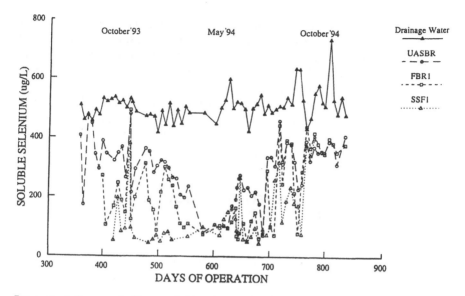

FIGURE 6 Influent and effluent soluble selenium for the treatment train of UASBR, FBR1, and SSF1 during the third testing period.

V. COSTS FOR SELENIUM TREATMENT

As part of developing plans and a budget for a prototype plant, EPOC AG worked up detailed estimates for capital and operating and maintenance costs. These estimates are given in detail in EPOC AG (1987). For a 3785 m³/d plant with two-stage biological process, crossflow filtration, and ion exchange for boron removal, a capital cost of $62/1000 m³ and an operating cost of $120/1000 m³ are given (in 1987 dollars) for a total cost of $182/1000 m³ of drainage water treated. These numbers assume that the capital cost is amortized at 4% over 20 years (bond fund rate) and that the methanol cost is $24/1000 m³, which appears low. If the ion exchange is removed from the capital cost (21.9%), and the costs are brought forward to 1997 dollars by using the ENR Construction Cost Index (Engineering News Record, 1997), the capital cost estimate grows to $80/1000 m³. Of course, selection of amortization rate and period will greatly affect the amortized capital cost.

The largest single operating cost is the use of methanol for the carbon feed source. The dosing rate at the Murrieta pilot plant was about 200 mg of methanol per liter of drainage water (EPOC AG, 1987). The dosing rate for the reactors at the AAADRC varied from 200 to 400 mg/L. Using a methanol price of $0.26/L gives methanol costs of $65 and $130/1000 m³ for the 200 and 400 mg/L dosages, respectively. Updating the operating costs to 1997 and using \$65/1000 m^3, the cost based on 200 mg/L methanol, gives an operating cost of \$275/1000 m^3. If the higher methanol dosage is used, the operating cost would be \$340/1000 m^3. The total cost for a 3785 m^3/d plant would then be estimated at \$355 to \$420/ 1000 m^3.

Calculations according to the capacity ratio exponent method, with an exponent of 0.8 as used by EPOC AG (1987), give the capital cost for a 37,850 m^3/d plant as \$50/1000 m^3. The operating costs per thousand cubic meters would still be the same, for a total cost of \$325 to \$390/1000 m^3 for a 37,850 m^3/d plant.

VI. POSSIBLE IMPROVEMENTS IN PROCESS EFFICIENCY AND COST

Total and soluble selenium in the effluent from the biological treatment process can be further reduced through physical–chemical polishing techniques. In studies performed by ERI at CSUF and at the AAADRC, the addition of ferric chloride to reactor effluent was effective in removing particulate selenium and selenite. The resulting ferric hydroxide precipitate acts as a coagulant in removing the fine colloidal elemental selenium. In addition, selenite is adsorbed to the precipitate. In a jar test of the effluent from the UASBR-FBR system, a ferric chloride dosage of 50 mg/L (as Fe) produced reductions in total selenium, soluble selenium, and selenite of 70, 50, and 79%, respectively (Salamor et al., 1997). EPOC AG (1987) used an ion-exchange resin for boron removal and selenium polishing. The resin was used as a precoat in the crossflow microfilters. Soluble selenium of less than 10 μg/L in the effluent was reported. The addition of polishing processes would increase the unit cost of treatment significantly. Also, even with polishing, it would be very difficult to reach total selenium concentrations of less than 5 μg/L.

The greatest improvement in total process cost can be made by finding an inexpensive alternative to methanol as the carbon source. A preliminary review of other carbon sources has already begun. EPOC AG (1987) used Steffens waste from sugar beet processing. If no-cost or low cost waste streams could be shown to be similar in efficiency to methanol, such sources would be ideal. Acetate or acetic acid, as required by *T. selenatis* (Macy et al., 1993), is more expensive than methanol. Another alternative is to produce the carbon feed on-site as part of the operation. Examples might include growing sugar beets, raising hogs, or anaerobically digesting agricultural waste to solubilize the carbon. The mixture of the drainage water and the carbon feed needs to have approximately 150 mg/L of total organic carbon (TOC). If the carbon feed had a TOC of 2000 mg/L, 1 L of feed would have to be added to every 12.3 L of drainage water (or 81.3 m^3 per thousand cubic meters of drainage). Another possibility is to use another commercially available feed that is simply less expensive than methanol. Ethanol, as an agriculturally based product, should be less expensive than methanol and comparable in effectiveness.

Other possible improvements would relate to process optimization. The addition of trace nutrients may improve the kinetics of the reaction, which would lead to greater throughput or smaller reactors. An example of nutrient addition is Macy's use of ammonium in lab-scale reactors to optimize the growth conditions for *T. selenatis*. Other growth factors (nutrients) may also improve reaction kinetics. As in all biological reactions, temperature plays a big part in kinetics. The metabolism of the bacteria is higher at higher temperatures (within certain limits). If on-site operations are used to produce a carbon feed source, there might also be the production of excess methane gas. This gas could be used to heat the reactors. The field tests to date have been in unheated, uninsulated reactors. Even insulating the reactors would provide some additional improvement in the process. Working on a larger scale plant, a small degree of improvement would be realized in having more reliable equipment and process control.

VII. SUMMARY

Several biological reactors have been tested for selenium removal efficiency. With the results from laboratory and pilot-scale reactors, good progress has been made in understanding the processes for efficient conversion and removal of selenium from agricultural drainage water. It has been clearly demonstrated that a two-stage biological reactor (e.g., a UASBR or a PBR in series with an FBR) followed by a sand filter or crossflow microfilter would be capable of producing an effluent total selenium of less than 50 μg/L and a soluble selenium of 20 to 30 μg/L on a continuous, consistent basis. The next step in the development of this process is to build a prototype plant of about 4000 m³/d capacity.*

REFERENCES

Binnie California, Inc., and California Department of Water Resources. 1988. Joint Operations Report: Performance Evaluation of Research Pilot Plant for Selenium Removal.
Dhaliwal, J. S. 1992. Selenium removal by slow sand filtration and chemical adsorption. M.S. project report. Department of Civil Engineering, California State University, Fresno.

---

*The research described in this chapter which was performed by the ERI at CSUF was funded by the California Department of Water Resources. Conclusions and recommendations presented in this chapter should not be construed as representing the opinion or conclusion of the Department of Water Resources. Also, mention of trade names or commercial products does not constitute endorsement or recommendation thereof.

Engineering News Record. 1997. Construction cost index. *Eng. News Record* (Jan. 6) 238:1:23.

EPOC AG. 1987. Removal of selenium from subsurface agricultural drainage by an anaerobic bacterial process. A final report on continued operation of the Murrieta Pilot Plant. Report submitted to the California Department of Water Resources.

Hanna, G. P., J. A. Kipps, and L. P. Owens. 1990. Agricultural drainage treatment technology review. Memorandum report prepared for the San Joaquin Valley Drainage Program under U.S. Bureau of Reclamation Order No. 0-PG-20-01500.

Jakhar, H. A. 1993. Selenium removal from agricultural drainage water using packed-bed, anaerobic upflow bioreactors. M.S. thesis. Department of Civil Engineering, California State University, Fresno.

Johnson, K. L. 1992. Pilot-scale anaerobic biological removal of selenium from agricultural drainage water using sequencing batch reactors. M.S. project report. Department of Civil Engineering, California State University, Fresno.

Macy, J. M. 1992. A genetic and biochemical approach to microbial purification of agricultural waste water. Report to California Department of Water Resources under Contracts B-56542 and B-57045.

Macy, J. M., S. Lawson, and H. DeMoll-Decker. 1993. Bioremediation of selenium oxyanions in San Joaquin drainage water using *Thauera selenatis* in a biological reactor system. *Appl. Microbiol. Biotechnol.* 40:588–594.

Owens, L. P. 1992. Adams Avenue Agricultural Drainage Research Center Status Report. Engineering Research Institute, California State University, Fresno.

Owens, L. P., M. Salamor, and J. E. Rex. 1997. Adams Avenue Agricultural Drainage Research Center: Final report on reactor operations for the period September 14, 1992–December 31, 1994. Submitted to California Department of Water Resources under Contract B-57806, Fresno.

Salamor, M., J. E. Rex, and L. P. Owens. 1997. Adams Avenue Agricultural Drainage Research Center: Final report on reactor operations for the period January 1, 1995–November 21, 1995. Submitted to California Department of Water Resources under Contract B-80502, Fresno.

Squires, R. C., G. R. Groves, and W. R. Johnston. 1989. Economics of selenium removal from drainage water. *J. Irrig. Drain. Eng.* 115:48–57.

26

Reduction of Selenium Oxyanions by *Enterobacter cloacae* Strain SLD1a-1

MARK E. LOSI and WILLIAM T. FRANKENBERGER, JR.
University of California at Riverside, Riverside, California

I. INTRODUCTION

Ecotoxicological impacts of selenium (Se) resulting from environmental discharge of agricultural drainage water in California's San Joaquin Valley are well documented (Ohlendorf et al., 1986; Presser and Ohlendorf, 1987). The source of the contamination is Se-rich soils, which are more extensive than originally believed and are now known to occur within several states in the western United States (Fujii et al., 1988; National Research Council, 1989; Presser et al., 1994). The major form of Se in the drainwater is selenate (SeO_4^2, Se^{61}), with lesser amounts of selenite (SeO_3^{2-}, Se^{4+}), both of which are soluble, toxic, and able to bioaccumulate (Presser and Ohlendorf, 1987; Weres et al., 1989). Accordingly, research is ongoing to study ways of removing Se oxyanions from agricultural drainage water. Although much has been learned, the goal of finding practical, cost-effective technology for treating Se-contaminated water has not yet been accomplished. Discovering a means for Se removal is fundamental to minimizing environmental contamination and ensuring wildlife protection in this region. An integrated, multiphased approach to the problem will most likely prove necessary.

Selenium is subject to a number of microbial transformations (Ehrlich, 1996), some of which may have applicability in bioremediation. These include bioreduction to the elemental form (Se^0), which is insoluble and can be physically separated from contaminated water, and methylation, which can yield volatile species (Thompson-Eagle and Frankenberger, 1992). Because of the high ratio of SeO_4^{2-} to SeO_3^{2-} in agricultural drainage water in the western United States organisms isolated and assessed for possible use in bioremediation must be able to transform SeO_4^{2-}. Several such organisms have recently been described, but much of this work focuses on bacteria that typically exhibit slow growth rates (Maiers et al., 1988), require strict anaerobicity (Tomei et al., 1992, 1995; Oremland et al., 1994), or have their reduction inhibited by alternate electron acceptors such as nitrate (NO_3^-), which is common in seleniferous waste streams (Steinberg et al., 1992). A recently isolated SeO_4^{2-}-reducing organism, *Thauera selenatis*, is reportedly a facultative anaerobe, but it reduces SeO_4^{2-} only in the absence of O_2 (Macy et al., 1993a).

A limited amount of recent work (Altringer et al., 1989; Lortie et al., 1992) has suggested that certain facultative anaerobes are able to reduce SeO_4^{2-} under aerobic or microaerophilic conditions. Working with facultative organisms may have several advantages over strict anaerobes, including increasing the flexibility of the system and reducing costs associated with maintaining strict anaerobicity and general ease in handling and processing of the organisms. In addition, faster, more prolific growth can generally be achieved with facultative anaerobes than with obligate anaerobes.

We have recently isolated a bacterium that is capable of reducing SeO_4^{2-} under microaerophilic conditions and may have potential for use in Se bioremediation processes (Losi and Frankenberger, 1997a). This chapter describes our work to date with the organism, including isolation and various aspects of SeO_4^{2-} reduction in growth and washed cell suspension experiments. The experiments focused mainly on reduction of SeO_4^{2-} because, as noted, this is the major species present in agricultural drainage water. The goal of this work is to obtain useful information regarding the potential suitability of the organism for use in a bioreactor system to treat Se-contaminated water.

II. ISOLATION AND IDENTIFICATION OF THE ORGANISM

Several waters and soils from the San Joaquin Valley were sampled and used to inoculate SeO_4^{2-}-amended agar plates, which were then incubated aerobically. One water sample from the San Luis Drain (which received subsurface agricultural drainage water along an 85-mile stretch of land ending at Kesterson Reservoir) yielded a bacterium capable of SeO_4^{2-} reduction under the conditions set forth. The Se concentration of the San Luis Drain water sample was 9.9 μM (782 μg/ L Se), 80% of which was the SeO_4^{2-} form. This organism stained gram-negative, exhibited rod-shaped morphology, and produced the characteristic brick-red color only when the medium was spiked with Se. It was identified by fatty acid analysis (two independent analyses) and Biolog Microplate as *Enterobacter cloacae* with similarity coefficients of 0.924, 0.621, and 0.836, respectively. We have deposited the strain with the American Type Culture Collection under accession number ATCC 700258, and it will henceforth be referred to as SLD1a-1.

Growth was observed on anaerobically incubated plates containing NO_3^-, SeO_4^{2-}, and both anions, but not on plates without an electron acceptor. Aerobically incubated plates (with no added NO_3^- or SeO_4^{2-}) were also positive for growth. This constitutes evidence that the organism is a facultative anaerobe, using SeO_4^{2-} and NO_3^- as terminal electron acceptors in respiratory metabolism. During anaerobic growth on solid medium, complete reduction of SeO_4^{2-} to Se^0 (as indicated by the red color) was observed only on plates where NO_3^- was present along with SeO_4^{2-}. This result suggests that reduction of SeO_4^{2-} supports growth and that NO_3^- is required for reduction of SeO_3^{2-} to Se^0 during anaerobic growth. This behavior is markedly similar to that of *T. selenatis*, which respires SeO_4^{2-}, but not SeO_3^{2-}, and reduces SeO_3^{2-} to Se^0 only in the presence of NO_3^- via a periplasmic NO_2^- reductase system (DeMoll-Decker and Macy, 1993; Macy, 1994).

III. GROWTH EXPERIMENTS

A set of growth experiments was carried out to assess the capability of SLD1a-1 to remove Se (added as SeO_4^{2-}) from solution in flasks that were open to the atmosphere, and to study the process.

A. Reduction of SeO_4^{2-} at Various Concentrations

An initial experiment involved determining the percentage of SeO_4^{2-} removed at various SeO_4^{2-} concentrations somewhat representative of typical levels found in Se-tainted drainage water. Table 1 shows removal of SeO_4^{2-} from solution (tryptic soy broth) by SLD1a-1 at initial SeO_4^{2-} concentrations ranging from 13 to 1266 μM. Results indicate that SLD1a-1 is capable of reducing SeO_4^{2-} effectively over this range, with the greatest percentage removed at 127 μM over 48 hours (Table 1). The rate increased linearly with Se concentration, indicating that the capacity of this organism to reduce and remove Se from solution was limited by the amount of SeO_4^{2-} available and the maximum rate had not yet been reached. Agricultural drainage waters in the San Joaquin Valley contain Se (mostly as SeO_4^{2-}) at levels that have ranged from 1.8 to 17.7 μM in water entering the San Luis Drain to as high as 53 μM in other locations (Sylvester, 1990). Thus, SeO_4^{2-} concentrations typical of seleniferous agricultural drainage water appear to be within the range of the organism's removal capabilities, as well as much higher concentrations.

Subsequent experiments were conducted at 127 or 633 μM, which represents a compromise between using environmentally relevant concentrations, ac-

TABLE I   Reduction of SeO_4^{2-} at Various Concentrations by *E. cloacae* Strain SLD1a-1 Growing in Tryptic Soy Broth[a]

Initial concentration of SeO_4^{2-}, μM	Live cells			Controls[b]	
	Final concentration of Se μM[c]	% Removed in 48 h		Final concentration of Se μM[c]	% Removed in 48 h
13.0	5.0 (0.6)	61.5		12.2 (1.6)	6.2
63.3	6.8 (1.1)	89.2		61.3 (2.1)	2.3
127	6.9 (1.5)	94.5		120.4 (5.2)	5.2
316	24.0 (2.6)	92.4		305.2 (5.9)	3.4
633	50.1 (4.4)	92.1		610.8 (10.2)	3.5
1266	99.8 (6.45)	92.1		1205 (30.7)	4.8

[a]Treatments were replicated in triplicate.
[b]Controls were identical except that the inoculum had been boiled for 30 minutes. Final concentrations for controls were not significantly different from original concentrations (Duncan's multiple range test, $P > 0.05$).
[c]Values in parentheses represent standard error of the mean.

commodating analytical methodology, and obtaining the best possible information regarding the metabolic capabilities of this organism. The data suggest that SLD1a-1 cannot lower the Se concentration in liquid culture below about 5 μM. Although this is well above the interim maximum mean monthly Se concentration limit of 0.025 to 0.063 μM (State of California, 1987), actual minimum achievable concentrations by this organism under various conditions have yet to be determined.

B.   X-Ray Diffraction

It was important to confirm the presence of Se^0 in the solution culture experiments. Analysis of the cells/precipitate by X-ray diffraction (XRD) yielded a diffraction pattern identified as that of elemental Se^0 (Fig. 1). Since XRD detects crystalline compounds only, we conclude that the compound produced was crystalline red Se^0. Another component was present but was not positively identified. It should be mentioned that no attempt was made to identify all reduction products in this study. Although observation of the red precipitate and X-ray diffraction data is ample evidence for the presence of Se^0, the possibility cannot be excluded that there are additional reduced products such as selenoamino acids and/or methylated selenides. Identification and quantification of all products formed seems important from a treatment standpoint, since reduced Se compounds vary

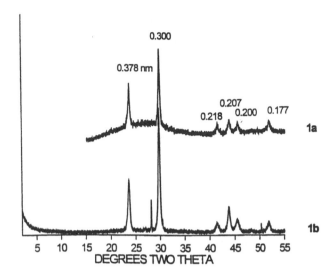

FIGURE I   X-ray diffractograms of red Se⁰ prepared abiotically by reacting glucose with Na_2SeO_3 under heat and pressure (1a) and precipitate formed by reduction of SeO_4^{2-} by *E. cloacae* strain SLD1a-1 (1b). Both compounds were identified as Se⁰ by means of Seimens Diffract At software. The biologically formed compound (1b) is observed to have another component not positively identified but evidenced by peaks that exist in addition to those of Se⁰.

markedly in terms of toxicity. This information also would be useful in determining possible secondary treatment options to further lower Se levels in treated water.

C.   Removal of SeO_4^- in the Presence of NO_3^-

As mentioned, NO_3^- is a common anion in agricultural drainage water, and interferes with reduction of SeO_4^{2-} in some organisms. Therefore, an experiment was conducted in which removal of SeO_4^{2-} at three levels, with and without equimolar amounts of NO_3^-, was quantified over time in solution culture. Nitrate had no effect at any of the three concentrations, demonstrating that under these conditions, reduction of SeO_4^{2-} occurred in the presence of, and was uninhibited by NO_3^- (Fig. 2). This is in contrast with an earlier study that found that microbial SeO_4^{2-} reduction occurred only after NO_3^- was depleted (Steinberg et al., 1992). Figure 3 plots growth (OD_{600}) and O_2 concentration in the medium over time; the growth curve (optical density at a wavelength of 600 nm) is nearly identical to growth curves measured in the previously described experiment while the organism was growing in SeO_4^{2-}-amended medium, until the OD measurement was obfuscated by precipitation and suspension of Se⁰, which occurred after 20

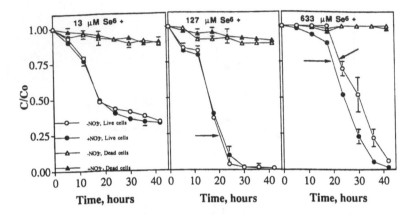

Figure 2   Removal of Se^{6+} from solution over time by *E. cloacae* strain SLD1a-1 growing aerobically in tryptic soy broth with (+) and without (−) equimolar amounts of NO$_3^-$. Arrows indicate approximate time at which a red color could be observed in the medium.

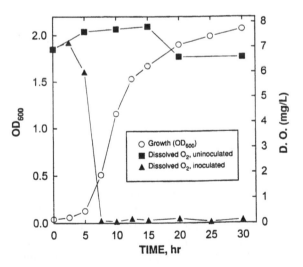

Figure 3   Growth (OD$_{600}$) of *E. cloacae* strain SLD1a-1 in tryptic soy broth and dissolved oxygen (D.O.) concentration in the medium as a function of time. The experiment was conducted in Erlenmeyer flasks that were open to the atmosphere (capped with foam plugs) and shaking on a rotary platform shaker at 150 rpm.

hours (not shown). A comparison of Figures 2 and 3 reveals a lag in SeO_4^{2-} reduction relative to growth. Figure 3 shows O_2 levels dropping rapidly in the early stages of exponential growth. This suggests that as O_2 levels fall below a critical level, SeO_4^{2-} is then used as an alternate terminal electron acceptor in respiration. Further experimentation has supported this contention (see Section IV.C).

IV. WASHED CELL SUSPENSION

A.   Evidence of Enzymatic Processes/Concomitant Reduction of SeO_3^{2-}

A washed cell technique was developed to rapidly assess a number of parameters, as well as to obtain information regarding mechanisms involved in Se reduction (Losi and Frankenberger, 1997a). Figure 4 depicts the removal of SeO_4^{2-} and production and removal of Se^{4+} and Se^{2-} compounds by the washed cell suspension. Suspensions did not reduce SeO_4^{2-} when an electron donor was not added, or when cells had been heat-killed (data not shown). In addition, parallel experiments showed that no growth occurred in the suspension for 10 hours. This suggests that the process is enzymatic and not due to reactions with by-products of growth in the medium. As was observed in the growth experiments, O_2

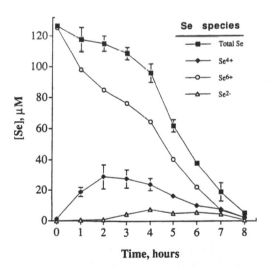

FIGURE 4   Reduction of Se^{6+} and production of Se^{4+} and Se^{2-} over time by washed cell suspensions of *E. cloacae* strain SLD1a-1 in 0.05 M sodium phosphate buffer with glucose as the electron donor. Solid and open symbols represent measured and calculated values, respectively (see text). Initial Se concentration was 127 μM as Na_2SeO_4, concentrations of biomass was 5.1 mg/mL, and error bars represent standard error of the mean, $n = 3$.

depletion was rapid, occuring within the first 0.5 hour, and levels remained below 0.2 mg/L for 8 hours, presumably when electron transport mechanisms slowed as a result of depletion of the carbon source (data not shown). These results reveal that resting cell suspensions of SLD1a-1 reduced SeO_4^{2-} and SeO_3^{2-} concomitantly, with SeO_3^{2-} accumulating to a minor degree and then being rapidly removed (Fig. 4). The analysis also indicated the presence of more reduced products (Se^{2-}), but these were negligible.

B. Concomitant Reduction of NO_3^-

Experiments were conducted to determine the fate of NO_3^- during reduction of SeO_4^{2-}. Results presented in Figure 5 show that NO_3^- and SeO_4^{2-} are reduced concurrently, with the reduction of NO_3^- being much more rapid. Figure 5 also shows that in the washed cell suspension, reduction of SeO_4^{2-} was slightly inhibited by the presence of an equimolar amount of NO_3^- which is in contrast with results from the growth experiment (Fig. 2). This suggests that the inhibitory

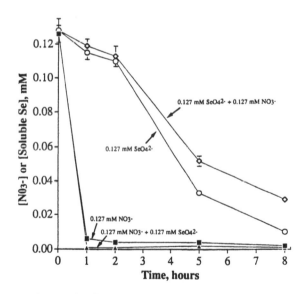

FIGURE 5 Results of washed cell suspension experiment showing time course removal of NO_3^- (solid symbols) in the presence and absence of SeO_4^{2-} and removal of Se (added as SeO_4^{2-}, open symbols) in the presence and absence of NO_3^-. Detection limits were 0.016 mM for Se and 0.014 mM for NO_3^-. The concentration of dry biomass in each flask was 6.4 mg/mL, and error bars represent standard error of the mean, $n = 3$. No significant removal of NO_3^- or Se was observed in controls with no cells. Final Se concentrations were statistically different (Duncan's multiple range test, $P > 0.05$).

effect of NO_3^- may vary according to environmental conditions. For example, growth experiments involve N uptake mechanisms for assimilation competing with NO_3^- used as an electron acceptor.

Interestingly, inhibition of SeO_4^{2-} reduction continued after NO_3^- fell to very low levels (Fig. 5). If SeO_4^{2-} and NO_3^- were reduced via the same reduction mechanism, it would be expected that once NO_3^- had been depleted, SeO_4^{2-} reduction would increase. If different mechanisms were functioning, interference should not occur at all. Since only NO_3^- and total soluble Se (SeO_4^{2-} + SeO_3^{2-}) were quantified, the nature of the interference (which may involve the step in which SeO_3^{2-} is reduced to Se^0) cannot be inferred from this experiment.

C.   Enzymes Involved in Reduction of SeO_4^{2-}

The rate of reduction and removal of SeO_4^{2-} or SeO_3^{2-} was not increased when cultures were pregrown in the presence of either oxyanion (Table 2). This experiment was designed to test the hypothesis that induction was operative, whereby exposure to the oxyanions would stimulate production of enzyme(s) active in reduction. The data in Figures 2 and 3 show a lag in SeO_4^{2-} reduction relative to the logarithmic growth phase of the organism. This lag in reduction of SeO_4^{2-} is in contrast with other studies in which SeO_4^{2-} is reduced during the logarithmic growth phase (Maiers et al., 1988; Lortie et al., 1992). Oremland et al. (1994) found that the ability of strain SES-3 to reduce SeO_4^{2-} was present when cells were pregrown in SeO_4^{2-} and absent in cells pregrown in NO_3^- and thus concluded that enzymes involved in reduction of SeO_4^{2-} were induced.

The data in Table 2 suggest that enzyme(s) responsible for SeO_4^{2-} reduction by SLD1a-1 are constitutive, indicating that pregrowing the organism in the presence of SeO_4^{2-} will provide no advantage in use for bioremediation. The hypothesis that the lag in SeO_4^{2-} reduction relative to growth (Figs. 2 and 3) was related to differences in cell physiology at various growth stages was tested by quantifying removal of SeO_4^{2-} by means of washed suspensions of cells harvested in early logarithmic phase, late logarithmic phase, or stationary phase. These experiments (data not shown) yielded SeO_4^{2-} removal rates that were indistinguishable from each other.

We have demonstrated that reduction of SeO_4^{2-} or NO_3^- as the sole electron acceptor can support growth of this organism, which constitutes evidence of reduction due to respiratory metabolism. As mentioned, data shown in Figure 3 suggest that the lag in SeO_4^{2-} reduction observed in Figure 2 may be attributed to the rapid consumption of O_2 leading to microaerophilic conditions, whereby SeO_4^{2-} then becomes an alternate electron acceptor. These and additional observations supporting this contention are summarized in Table 3.

duction of SeO_4^{2-} and SeO_3^{2-}, Both at Initial Concentrations of 127 and 633 mM in 0.05 M Phosphate Buffer by a V...
of *E. cloacae* strain SLD1a-1[a]

centration of Se in medium, mM[b]	% SeO_4^{2-} reduced				% SeO_3^{2-} reduced		
	At 2 h		At 8 h		At 2 h		A...
	127 mM	633 mM	127 mM	633 mM	127 mM	633 mM	127 mM
	26.3 (2.1)	15.2 (3.0)	93.2 (3.2)	61.1 (4.2)	46.2 (3.9)	29.4 (4.0)	93.3 (5.1)
	29.1 (2.5)	13.2 (2.1)	91.8 (3.9)	57.3 (4.1)	48.1 (2.0)	27.9 (2.9)	92.5 (4.2)
	27.5 (1.1)	12.9 (1.9)	92.0 (1.9)	59.9 (3.0)	49.5 (3.4)	26.3 (2.1)	93.1 (3.4)
	28.5 (1.9)	14.7 (2.3)	91.7 (4.6)	58.2 (3.9)	48.9 (4.5)	27.7 (3.8)	92.6 (6.2)
	27.9 (3.0)	15.0 (2.8)	93.0 (2.7)	60.2 (4.6)	47.0 (4.1)	28.6 (2.9)	93.0 (5.2)

...grown in the presence and absence of SeO_4^{2-} and SeO_3^{2-} to determine whether reduction of Se oxyanions by this organism is induced...
...re observed (Duncan's multiple range test, $P > 0.05$). Standard error of the mean is shown in parentheses.
...tes, in parentheses, as follows Se^{6+} or Se^{4+}, SeO_4^{2-} or SeO_3^{2-}, respectively.

TABLE 3 Summary of Observations/Information Supporting the Hypothesis That Reduction of SeO_4^{2-} by *E. cloacae* strain SLD1a-1 Occurs Via Terminal Reductases as a Result of Respiratory Metabolism

1. Growth was observed on anaerobically incubated agar plates containing SeO_4^{2-} as the sole terminal electron acceptor. No such growth was observed on control plates without SeO_4^{2-}.
2. Reduction of SeO_4^{2-} occurs rapidly in washed resting cell suspensions, but only when an electron donor is added.
3. Reduction of SeO_4^{2-} coincides with a decrease in O_2 content in growth and washed cell suspension experiments. Furthermore, in growth experiments sparging the flasks with room air inhibits SeO_4^{2-} reduction, but not growth, and decreasing O_2 levels leads to decreasing inhibition (Losi and Frankenberger, unpublished data).
4. Reduction of SeO_4^{2-} occurs at low SeO_4^{2-} levels in rich media, without inhibiting growth (Losi and Frankenberger, unpublished data), which indicates that "detoxification mechanisms" would not be a likely explanation.
5. Another *E. cloacae* strain, HO1, has been shown to reduce chromate as result of respiratory metabolism (Wang et al., 1991).
6. In assimilatory S reduction, bacteria assimilate only that which they need to make proteins (Brock and Madigan, 1991) during growth. In washed resting cell suspensions, no growth (no increase in OD_{600} and biomass/volume) occurred for up to 10 hours (Losi and Frankenberger, unpublished data), while Se was being reduced.

The nature of the reductases, or the precise mechanisms by which they function under microaerophilic conditions, are not known at this time. The observations that SeO_4^{2-} and NO_3^- are reduced concomitantly and that interference by NO_3^- is minimal (Fig. 5) suggest the presence of separate reductases (e.g., SeO_4^{2-} and NO_3^- reductases). Alternatively, evidence that dissimilatory reduction of SeO_4^{2-} occurs under microaerophilic conditions, combined with the constitutive ability of this organism to reduce SeO_4^{2-}, suggests that reductases vary in specificity, and that during periods of vigorous activity multiple substrates may be reduced concurrently by a single enzyme or enzyme system. It is also possible that various other oxidoreductase systems may be functioning, alone or in tandem. However, SeO_4^{2-} reduction by SLD1a-1 under microaerophilic conditions is clearly enzymatic and cannot be explained in terms of assimilatory reduction or "detoxification mechanisms."

Since SLD1a-1 is a facultative anaerobe (which can grow vigorously under aerobic conditions) and reduction of SeO_4^{2-} occurs under microaerophilic conditions as a result of constitutive enzymes, the biochemistry appears to differ from that of SES-3, an organism recently isolated by Oremland et al. (1994) which is an obligate anaerobe and reduces SeO_4^{2-} via induced reductases. As mentioned, the observation that NO_3^- may be necessary for complete reduction of SeO_4^{2-} to Se^0 during anaerobic growth suggests a similarity to *T. selenatis* (Macy, 1994). However, complete reduction to Se^0 is routinely observed by SLD1a-1 without the addition of NO_3^- in washed cell suspensions that are open to the atmosphere. The effect of O_2 on reduction of SeO_3^{2-} by *T. selenatis* has not been reported (as far as we are aware), and in our studies with SLD1a-1, it appears equally plausible that O_2 is needed for reduction of SeO_3^{2-} to occur. It is clear that the precise enzymology and biochemistry involved in these processes remain to be elucidated.

D. Electron Microscopy

The washed cell suspension was viewed after reduction of SeO_4^{2-} to Se^0 using transmission electron microscopy (TEM) in an effort to examine the relationship of the Se precipitate to the cells, and to estimate the size of the precipitate particles. Electron micrographs of the washed cell suspension containing SLD1a-1 and precipitated particles are presented in Figure 6. Selenium was confirmed by electron diffraction spectroscopy (EDS) as the only element heavier than

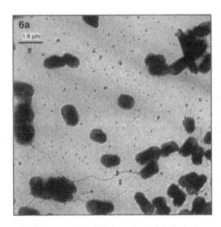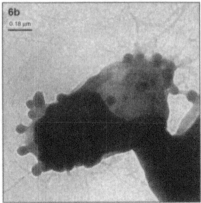

FIGURE 6   Transmission electron micrographs of whole mounts of *E. cloacae* strain SLD1a-1 from a washed cell suspension that was incubated with 633 μM SeO_4^{2-}. (a) Typical field in which precipitated particles are visible outside the cells. The particles were determined through EDS to contain Se as the only element heavier than sodium. EDS analysis of cells revealed no intracellular Se present. At higher magnification (b), particles appear to be associated with the outside of the cell membrane. This evidence suggests that the reaction occurs very near the cell surface, with particles subsequently expelled to the outside of the cell.

sodium (Na) present in the precipitate particles (Fig. 7a). The EDS spectrum shown in Figure 7b is typical of those collected with the beam focused on a cell, indicating that no Se is present.

Particles were less than 0.1 μm in diameter and were observed free in the medium (Fig. 6a) and protruding from the outer surfaces of cells (Fig. 6b). We examined eight grids containing many fields and hundreds of cells, and found no evidence of the precipitate being sequestered intracellularly. EDS analysis probing characteristic cells from about 10 fields yielded spectra identical to that shown in Figure 7b (no Se). This evidence suggests that the precipitate is probably produced near the cell membrane and rapidly expelled. There was no washing procedure after incubation with SeO_4^{2-} to disrupt the cells, and no evidence of cell lysis or fragmentation, which could have led to the release of intracellular precipitate particles. It appears likely that the reaction occurs very near the cell surface, possibly as a result of membrane-associated reductase(s), whereupon the precipitate is rapidly expelled in a manner analogous to the exportation of intracellular arsenic by membrane-associated efflux pumps (Mobley and Rosen, 1982).

This is somewhat in contrast to earlier studies in which other Se-reducing bacteria were determined to contain intracellular Se^0 particles. Selenate reduction during growth of *Desulfovibrio desulfuricans* yielded amorphous (as determined by X-ray diffraction) particles that were deposited inside the cell near the periphery (Tomei et al., 1992). These particles were observed only in cells that had reached the stationary phase. Similarly, *Wolinella succinogenes* growing in medium containing SeO_4^{2-} accumulated intracellular Se-rich granules (Tomei et al., 1995). However, in both these studies free Se^0 granules were also observed in the medium. A facultative anaerobe, *Pseudomonas alcaligenes*, also was observed to accumulate Se intracellularly to concentrations above that in the surrounding environment (Altringer et al., 1989). Unlike the earlier studies, the cells we analyzed were reducing SeO_4^{2-} as a washed cell suspension. Thus no observations could be made regarding effects of SeO_4^{2-} on cell morphology and development; but it is unlikely that cell metabolic processes are affected in a way that would alter the nature of the precipitation reaction. We also surveyed whole mounts of cells prepared while the cells were actively growing, and the trends were identical to those noted in the washed cell preparations. Micrographs of these cells are not shown because of obfuscation by medium components during the drying process.

With regard to analysis of the precipitate particles, EDS does not specifically identify the Se species present. However, the spectrum shown in Figure 7a along with X-ray diffraction data (Fig. 1), and the insolubility and brick-red color of the particles, provide ample evidence that the precipitate is primarily Se^0. The size of the particles is significant because it shows that final filtration must account for particles smaller than 0.1 μm. Interestingly, Barton et al. (1994) studied

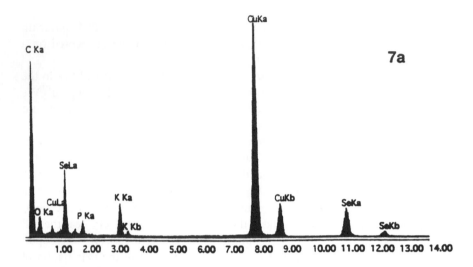

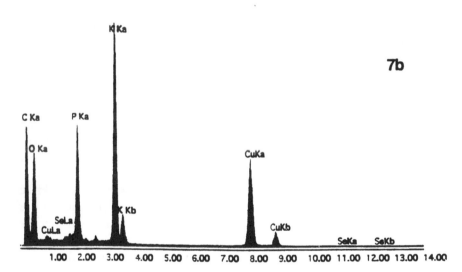

FIGURE 7 EDS spectra of a precipitate particle (a) and a cell (b) from a washed cell suspension of SLD1a-1 incubated with 633 μM SeO_4^{2-}. It is of note that in (a), Se is the only heavy element present (K and P are probably artifacts due to the buffer); and in (b), no Se is present, which was true of virtually all cells analyzed and was consistent with visual observations.

bacterially produced colloidal Se^0 and discovered that the particles retain a slight surface charge, which may be amenable to electrofiltration techniques.

V. USE OF 2,4-DINITROPHENOL TO ASSESS:
$SeO_4^{2-} \rightarrow SeO_3^{2-}$

The bioreduction of SeO_4^{2-} presumably can proceed as follows:

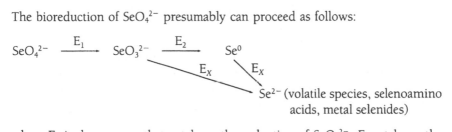

Se^{2-} (volatile species, selenoamino acids, metal selenides)

where E_1 is the enzyme that catalyzes the reduction of SeO_4^{2-}, E_2 catalyzes the reduction of SeO_3^{2-}, and other enzymes (E_x) catalyze reactions yielding assimilated and volatile organic forms. Reduction of SeO_4^{2-} and SeO_3^{2-} is analogous to the first two steps in denitrification, in which nitrate (NO_3^-) is reduced to nitrite (NO_2^-) by NO_3^- reductase, and NO_2^- to N_2O by NO_2^- reductase. By applying techniques used in plant physiology, Abdelmagid and Tabatabai (1987) were able to measure NO_3^- reductase activity in soils by employing the use of 2,4-dinitrophenol (DNP), an uncoupler of oxidative phosphorylation. DNP inhibits NO_2^- reductase but not NO_3^- reductase, thus allowing NO_2^- to accumulate long enough to permit NO_3^- reductase activity to be measured.

Experiments in our laboratory have revealed that this technique is also applicable to SeO_4^{2-} reduction by SLD1a-1 (Losi and Frankenberger, 1997b). When SeO_4^{2-} is added to washed cell suspensions treated with DNP, SeO_4^{2-} reduction proceeds virtually unimpeded, but SeO_3^{2-} reduction is inhibited and SeO_3^{2-} accumulates in solution. In untreated samples, SeO_3^{2-} is more rapidly removed from solution as further reduction occurs with greater efficiency. This provides evidence that two different enzymes are responsible and also gives us a tool for the independent measurement of the activity of the enzyme responsible for the first reduction step. In the following experiments, the optimum concentration of DNP that inhibits SeO_3^{2-} reduction was determined. The influence of various environmental parameters on the reduction of SeO_4^{2-} was then evaluated by measuring the production of SeO_3^{2-}.

This system was used because it provides a means to investigate reduction mechanisms and to rapidly assess the influence of a variety of environmental parameters that are relevant to the potential applicability of this organism for removing Se oxyanions from contaminated water. Within the course of this discussion, references to "SeO_4^{2-} reduction" and "reduction of SeO_4^{2-}" refer only to the transformation: $SeO_4^{2-} \rightarrow SeO_3^{2-}$. Therefore, unless otherwise noted, these expressions are used interchangeably with "SeO_3^{2-} production."

A. Optimal Concentration of DNP

The concentration of DNP that yielded the maximum production of SeO_3^{2-} from SeO_4^{2-} reduction by SLD1a-1 was determined first over a reaction time of 8 hours. The maximum amount of SeO_3^{2-} produced and remaining in solution after 8 hours was observed at a DNP concentration of 1.09 mM (Fig. 8). Although DNP is known to uncouple oxidative phosphorylation, its effect on NO_3^- reduction is not clearly understood. Plant scientists have observed that reduction of both NO_3^- and NO_2^- in plant leaves occurs in light, but only slightly if at all in darkness (Beevers and Hageman, 1969; Canvin and Atkins, 1974). However,

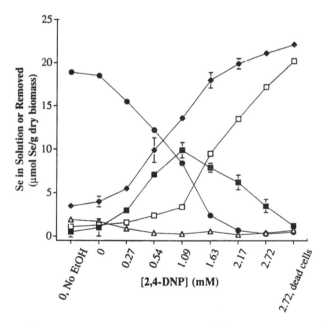

FIGURE 8   Soluble Se species produced or removed following reduction of SeO_4^{2-} by washed cell preparations of *E. cloacae* strain SLD1a-1 in the presence of various concentrations of DNP. Se species represented by solid symbols were measured directly, and those represented by open symbols were calculated. Initial SeO_4^{2-} concentration was 22.4 μmol Se/g dry biomass. Incubation time was 8 hours, and error bars represent standard error of the mean, $n = 3$.

Klepper (1976) reported the accumulation of NO_2^- in DNP-treated leaves in darkness, suggesting that under these conditions, NO_3^- but not NO_2^- reductase(s) was stimulated by DNP. Klepper (1976) postulated that uncoupling of respiration (brought about by DNP) accelerated glycolysis, increasing production of nicotin-amide adenine dinucleotide (NADH), which was used in NO_3^- reduction. Why this sequence stimulated reduction of NO_3^- and not NO_2^- was not discussed.

Abdelmagid and Tabatabai (1987), who observed inhibition of NO_2^- reduc-tase but not NO_3^- reductase in DNP-treated soils, cited Klepper's work in explain-ing that inhibition of NO_2^- reductase occurs because uncoupling of oxidative phosphorylation during NO_3^- reduction blocks production of ATP; thus the energy necessary for NO_2^- reduction is depleted. Our data suggest that analogous to Abdelmagid and Tabatabai's (1987) work, SeO_4^{2-} reduction proceeds relatively unimpeded, while reduction of SeO_3^{2-} is inhibited. Figure 8 shows that at DNP levels lower than 1.09 mM, the uncoupling effect may have been less effective, and thus more energy was available for reduction of SeO_3^{2-}. At DNP concentra-tions greater than 1.09 mM an inhibitory effect was observed such that instead of being reduced, SeO_3^{2-} was not being formed, as suggested by the presence of unreacted SeO_4^{2-}. This effect could possibly be due to the toxicity of DNP.

Another possible explanation for the selective inhibition of NO_2^- reduction by DNP is that the NO_2^- reductase sites are bound by the nitro groups of the DNP. If this occurs, NO_2^- reductase would be inhibited but not NO_3^- reductase, a result consistent with the observations. This supports other data (Section II) suggesting that SeO_3^{2-} reduction in SLD1a-1 may occur via NO_2^- reductase, as is observed in *T. selenatis* (DeMoll-Decker and Macy, 1993). Laboratory studies with *T. selenatis* demonstrated that SeO_4^{2-} and SeO_3^{2-} were reduced concomi-tantly, but mutants lacking NO_2^- reductase activity were unable to reduce NO_2^- and SeO_3^{2-}, which accumulated in the medium. At this time, the exact inhibitory mechanism of DNP is unclear. For subsequent experiments in our study, DNP was added at a level of 1.09 mM, with the goal of measuring the activity of the enzyme responsible for reduction of SeO_4^{2-} to SeO_3^{2-}.

B. Effectiveness of DNP; Time Course of SeO_3^{2-} Production

The time course for the reduction of SeO_4^{2-} to SeO_3^{2-} by SLD1a-1 was measured to permit observation of the reaction rate and selection of a standardized time period over which to assay SeO_3^{2-} production in subsequent experiments, as well as to assess the overall effect of DNP. An increase in the SeO_3^{2-} concentration was observed in live cell suspensions both with and without DNP, peaking at about 100 minutes and dropping thereafter (Fig. 9). Figure 9 reveals the effectiveness of DNP in blocking the reduction of SeO_3^{2-}, the first product in the reduction of SeO_4^{2-}. In treatments receiving DNP, SeO_3^{2-} was produced in greater quantities and persisted in solution longer than treatments not receiving DNP. The reaction

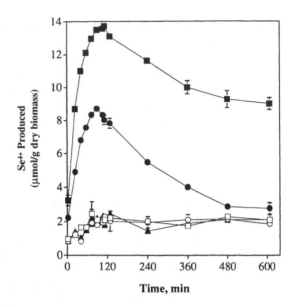

FIGURE 9 Production of SeO_3^{2-} from reduction of SeO_4^{2-} over time by live (solid symbols) and dead (open symbols) washed cell preparations of *E. cloacae* strain SLD1a-1, with (solid) and without (open) 2,4-dinitrophenol. Initial Se level was 127 μM (20.75 μmol Se/g dry biomass) as SeO_4^{2-}. Error bars represent standard error of the mean, $n = 3$.

was nearly linear up to 60 minutes, and thus 30 minutes (approximately the midpoint of this linear range) was selected as the standard time over which environmental parameters would be assessed.

C. Influence of Environmental Parameters

1. pH

The optimum pH for reduction of SeO_4^{2-} to SeO_3^{2-} by SLD1a-1 was observed to be between 6.5 and 7.0, with only a slight drop at 7.5 (Fig. 10a). Lortie et al. (1992) assessed pH effects on aerobic reduction of SeO_4^{2-} and SeO_3^{2-} by a *Pseudomonas stutzeri* isolate and reported a higher pH optimum, from 7 to about 9. Rech and Macy (1992) determined that the pH optimum for SeO_4^{2-} reductase in *T. selenatis* was 6.0. This study and others suggest that there may be significant differences in the pH optima for SeO_4^{2-} reduction by various organisms.

The San Luis Drain water from which the organism was isolated had a pH of 8.0, which is typical of agricultural drainage water in this region. Macy et al. (1993b) reported that the pH values of 17 different drainwater samples from the Westlands Water District (San Joaquin Valley) averaged from 7.4 to 8.1. In their studies, which involved using a bioreactor to remove of SeO_4^{2-} from agricultural

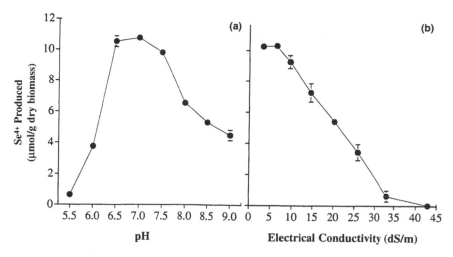

FIGURE 10   Production of SeO_3^{2-} through reduction of SeO_4^{2-} by washed cell suspensions of *E. cloacae* strain SLD1a-1 as a function of increasing pH (a) and salinity (electrical conductivity) (b). Error bars represent standard error of the mean, $n = 3$.

drainwater, the pH of the water was easily lowered before treatment, but an increase was observed during the treatment process. This information may indicate that for waters with a high initial pH, periodic monitoring and adjustment may be necessary to achieve optimal reduction rates.

2.   Electrical Conductivity

Since salinity strongly influences cell metabolism and can vary widely in agricultural drainage water, the effect of electrical conductivity (EC) on SeO_4^{2-} reduction is an important parameter in assessing organisms for potential use in bioremediation. Figure 10b plots production of SeO_3^{2-} as a function of EC. The EC of the San Luis Drain water from which this organism was isolated was 9.3 decisiemens per meter (dS/m). Published EC values may range from 10 to 15 dS/m in agricultural drainage water to around 40 and as high as 200 dS/m in evaporation ponds (Larsen et al., 1989; Chilcott et al., 1990; Thompson-Eagle and Frankenberger, 1990; Euliss et al., 1991). Results depicted in Figure 10b reveal only a 9% loss in SeO_4^{2-} reduction efficiency due to an increase in EC from 6 to 10 dS/m; but the reaction is about 55% less efficient at an EC of 20 dS/m. Thus the organism probably is best suited for treating agricultural drainage water prior to discharge into evaporation ponds.

3.   Alternate Electron Acceptors

The influence of alternate electron acceptors on reduction of SeO_4^{2-} and on growth was variable (Table 4). Respiration has been identified as a mechanism

Table 4 Influence of Various Terminal Electron Acceptors on Growth and Reduction of SeO_3^{2-} to SeO_3^{2-} by *E. cloacae* Strain SLD1a-1[a]

Electron acceptor	Concentration tested (mM)	SeO_3^{2-} yield	Growth yield as % of control[b]
Control	None	100	1.81 (0.02)
NO_3^-	0.13	99.9	1.79 (0.03)
	0.25	94.2	1.84 (0.05)
	0.63	85.5[c]	1.87 (0.09)
	1.27	73.0[c]	1.95 (0.07)[c]
	6.33	60.1[c]	2.03 (0.05)[c]
	12.7	59.0[c]	1.96 (0.05)[c]
NO_2^-	0.13	99.8	1.77 (0.02)
SO_4^{2-}	0.13	97.9	1.79 (0.03)
	1.27	99.7	1.78 (0.02)
	12.7	98.6	1.76 (0.03)
	30.0	86.9[c]	1.74 (0.08)
SO_3^{2-}	0.13	74.2[c]	1.42 (0.05)[c]
AsO_4^{3-}	0.13	95.7	1.74 (0.03)
Fe^{3+}	0.13	95.9	1.78 (0.02)
CrO_4^{2-}	0.13	10.5[c]	0.59 (0.03)[c]

[a]Growth (optical density at a wavelength of 600 nm) was quantified in liquid medium (tryptic soy broth) after 15 hours and reduction in 0.05 M sodium phosphate buffer by washed cell suspensions after 30 minutes. Each treatment contained SeO_4^{2-} at 0.127 mM and the electron acceptor at the indicated concentration. Controls contained only SeO_4^{2-} at 0.13 mM. Treatments were replicated in triplicate.
[b]Standard error of the mean is shown in parentheses; $n = 3$.
[c]Statistically different from the control for each column (Duncan's multiple range test, $P > 0.05$).

by which bacteria reduce Se oxyanions (Oremland et al., 1990, 1994; Rech and Macy, 1992). As noted, work in our laboratory with SLD1a-1 has shown that reduction of SeO_4^{2-} occurs via terminal reductase(s) as a result of respiration (Losi and Frankenberger, 1997a). Alternate electron acceptors may interfere with SeO_4^{2-} reduction and/or diminish reaction efficiency if toxicity occurs, if reductases specific to anions other than SeO_4^{2-} are operating, or, more theoretically, if SeO_4^{2-} reductase or nonspecific reductases are operating and are inhibited by the presence of concentrations of competing anions well in excess of SeO_4^{2-}. The first instance was observed by Oremland et al. (1989), who measured a decrease in reduction of SeO_4^{2-} by sediments and mixed cultures subjected to CrO_4^{2-}, which has been shown to have toxic and mutagenic effects on various bacteria (Lester et al., 1979; Ross et al., 1981; Coleman, 1988). The second was shown to occur in other selenate-respiring organisms that reduce NO_3^- preferentially over SeO_4^{2-} at equimolar concentrations (Steinberg et al., 1992). It was suggested that for these organisms, reduction of SeO_4^{2-} occurs via NO_3^- reductase. Conversely, the organism isolated by Rech and Macy (1992), *T. selenatis*, reduced SeO_4^{2-} at the same rate regardless of whether NO_3^- was present at equimolar concentrations, and this was viewed as evidence for a separate SeO_4^{2-} reductase.

In our study, NO_3^- (at concentrations > 5 times that of SeO_4^{2-}), SO_4^{2-} (at a concentration > 236 times that of SeO_4^{2-}), and SO_3^{2-} and CrO_4^{2-} (at equimolar concentrations) had an inhibitory effect on reduction of SeO_4^{2-} by SLD1a-1 (Table 4). The SO_3^{2-} and CrO_4^{2-} anions also inhibited growth, which suggests toxicity as the cause for decreased reduction efficiency: SO_3^{2-} is a well-known biostatic agent used in food preservation, and toxicity of CrO_4^{2-} to bacteria has been documented (Lester et al., 1979; Ross et al., 1981; Coleman, 1988). However, these properties do not explain the lower reduction efficiency at high NO_3^- and SO_4^{2-} levels. The presence of NO_3^- at ≥ 1.27 mM yielded significantly more growth than the control, and SO_4^{2-} at 30 mM had no significant effect.

The observation that SO_4^{2-} had no effect at substantially higher concentrations (up to 100-fold) suggests that SeO_4^{2-} and SO_4^{2-} have different reductive pathways, which is in agreement with other studies (Oremland et al., 1989). However, at 30 mM SO_4^{2-} (236 times the SeO_4^{2-} level), a 13.1% decrease in SeO_4^{2-} reduction efficiency was measured. Data from a similar study (Lortie et al., 1992) showed a decrease in the SeO_4^{2-} reduction rate by a *Pseudomonas* isolate of 9.5% when the SO_4^{2-} level was 20-fold that of SeO_4^{2-}, but this drop was apparently not significant. The EC of the buffer in our study was about 2.9 dS/m, and that of the buffer + 30 mM SO_4^{2-} (as K_2SO_4) was 8.6 dS/m. Therefore the decrease in SeO_4^{2-} reduction at 30 mM SO_4^{2-} is probably due largely to a salting-out effect, whereby an imbalance in the osmotic pressure of the cell is created, inhibiting the reductases.

It has been demonstrated that SLD1a-1 is able to reduce SeO_4^{2-} and NO_3^- concomitantly, and that NO_3^- reduction proceeds at a faster rate (see Section IV.B) . In addition, although equimolar amounts of NO_3^- did not interfere with SeO_4^{2-} reduction in growth experiments in a rich medium, small but significant inhibition was observed in a washed cell suspension (see Sections III.C and IV.B). In these prior experiments, the effect of NO_3^- was assessed on the removal of Se (added as SeO_4^{2-}) from solution, which actually encompasses both reduction of SeO_4^{2-} and SeO_3^{2-}. Therefore, it was not clear which step was being inhibited. Results presented here (Table 4) show that NO_3^- levels equimolar and 2× equimolar with those of SeO_4^{2-} did not inhibit SeO_3^{2-} production in a washed cell suspension. This suggests that the inhibition noted earlier may occur during the reduction of SeO_3^{2-} to Se^0.

The lack of SeO_4^{2-} reduction interference by NO_3^- at levels equimolar and 2× equimolar with those of SeO_4^{2-} also suggests that enzymes responsible for SeO_4^{2-} reduction may be independent of those that reduce NO_3^-. Nitrate interfered at a much lower concentration than did SO_4^{2-}, but the SeO_4^{2-}/SeO_3^{2-} redox couple has a potential similar to that of NO_3^-/NO_2^- (0.44 and 0.42 V, respectively), while that of SO_4^{2-}/SO_3^{2-} is much lower, -0.52 (Thauer et al., 1977; Doran, 1982). Thus NO_3^- is more easily reduced than SO_4^{2-}.

Interestingly, NO_3^- levels of 1.27 mM and above enhanced growth. We have demonstrated in experiments with SLD1a-1 that during growth, rapid removal of O_2 exceeds its diffusion back into solution, leading to microaerophilic (≤ 0.2 mg/L O_2) conditions (Losi and Frankenberger, 1997a). Since NO_3^- can serve as a terminal electron acceptor in respiration for SLD1a-1, high NO_3^- levels can increase energy and growth compared to cultures with less alternate electron acceptor or none.

With regard to bioremediation applications, typical NO_3^- and SO_4^{2-} levels in agricultural drainage water are about 4.5 and 45 mM, respectively (Chilcott et al., 1990; Macy et al., 1993b). Our results suggest that substantial SeO_4^{2-} reduction occurs at these levels; however, pretreatment lowering NO_3^- levels may be necessary to achieve maximum Se removal efficiency. Alternatively, since SLD1a-1 also reduces NO_3^-, enough carbon can be added to support reduction of both SeO_4^{2-} and NO_3^-.

4. Carbon Sources

Various carbohydrates were tested to determine the response of the reaction to different electron sources and to provide a basis for assessing possible supplements to stimulate the reaction. Monosaccharides containing 5 carbons (arabinose and xylose) and 6 carbons (glucose, mannose, and galactose), disaccharides (lactose and maltose), and a ketose (fructose) promoted reduction of SeO_4^{2-} by SLD1a-1 to varying degrees (Fig. 11a). These results indicate that the more easily an electron donor enters the glycolytic pathway, the greater its propensity to promote SeO_4^{2-} reduction.

Glucose, which enters the pathway directly, promoted the greatest SeO_4^{2-} reduction. Mannose, an epimer of glucose that differs in configuration only at C-2, behaved as glucose did, which suggests that mannose can also directly enter the pathway, or only minor modifications are required to convert mannose to glucose. Galactose is also an epimer of glucose, differing in configuration only at C-4. Galactose promoted less SeO_4^{2-} reduction than glucose or mannose, but whereas glucose is converted to glucose-6-phosphate (the first step in glycolysis) in one step, galactose uses four steps and four enzymes (Stryer, 1988), which may require more time for synthesis. A decrease in SeO_4^{2-} reduction relative to glucose was also observed with fructose as the electron donor. Fructose can be converted to glucose-6-phosphate by the same mechanism as glucose, but the process is much less efficient. Alternatively, it can enter glycolysis through the fructose-1-phosphate pathway, requiring additional enzymes which, for this or-

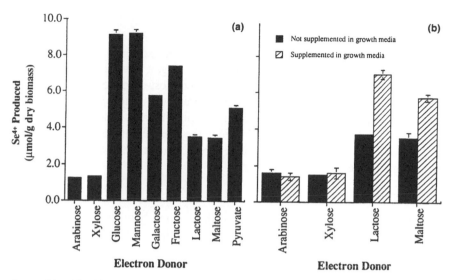

FIGURE 11 (a) Influence of various electron donors and (b) induction of catabolic enzymes on the reduction of SeO_4^{2-} to SeO_3^{2-} by washed cell suspensions of *E. cloacae* strain SLD1a-1. Incubation time was 30 minutes, and error bars represent standard error of the mean, $n = 3$.

ganism, may not be induced or may be constitutive at low levels while growing in tryptic soy broth. Pyruvate, the end product of glycolysis, was about half as efficient as glucose in promoting the reaction. This was expected, since each glucose molecule yields two pyruvate molecules, and pyruvate enters directly into the citric acid cycle. Therefore if these processes were operating, and uptake rates were similar, glucose should yield twofold the activity of pyruvate on a molar basis, which is very close to what was observed.

Arabinose and xylose (pentoses), and lactose and maltose (disaccharides) promoted about 13 and 40%, respectively, of SeO_4^{2-} reduction compared to glucose. In some cases, enzymes necessary for bacteria to utilize arabinose and lactose as an energy source are known to be induced (Stryer, 1988). The experiment in which cells were grown in the presence and absence of a sugar and then tested for SeO_4^{2-}-reducing activity using the corresponding sugar as the electron donor was designed to determine whether SeO_4^{2-} reduction could be enhanced by this process. Figure 11b shows that there was no difference in production of SeO_3^{2-} with arabinose or xylose as the electron donor by cells grown with or without the respective sugar. However, prior exposure of cells to galactose or maltose increased the ability of SLD1a-1 to use these compounds as electron donors in SeO_4^{2-} reduction. Reduction by means of lactose of SeO_4^{2-} by cells that had previously been exposed to lactose was roughly intermediate to that promoted by glucose and galactose, which are the subunits of lactose. In light of this evidence, it can be said that electron donors exert a strong influence on SeO_4^{2-} reduction by this organism. The data suggest that glucose should be provided as the electron donor if possible. Other cost-effective sources that are easily obtained, and contain blends of carbohydrates to promote SeO_4^{2-} reduction, should be screened. In terms of bioreactor applications, the data in Figure 11b suggest that initially growing the organism in the presence of a particular electron source prior to introduction of the influent may promote increased utilization of that source in the reduction of SeO_4^{2-}.

5. Temperature

Temperature is an important parameter with regard to bioremediation. To assess the effect of temperature, removal of Se from solution was measured instead of SeO_3^{2-} production. This is because results of preliminary experiments were anomalous and suggested that high temperatures affected the stability and/or effectiveness of the DNP. Figure 12 depicts the effect of temperature on SeO_4^{2-} removal. The optimum was found to be 30 to 35°C, indicating that SeO_4^{2-} removal will be most efficient in warmer months. In cooler periods, a heating element/thermostat may be necessary to maintain bioreactor temperatures of at least 30°C to optimize Se reduction.

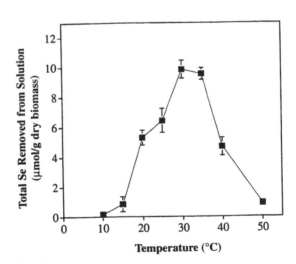

FIGURE 12   Effect of temperature on removal of SeO_4^{2-} from solution by a washed cell suspension of *E. cloacae* strain SLD1a-1 with an incubation time of 4 hours. Initial SeO_4^{2-} content was 22.4 μmol Se/g dry biomass. Error bars represent standard error of the mean, $n = 3$.

6. Oxygen Content

The effect of O_2 concentration on reduction of SeO_4^{2-} by SLD1a-1 was evaluated to define optimal levels for operation of a bioreactor, and to obtain supplementary information regarding the metabolic mechanisms involved. The assessment was carried out by continuously sparging flasks containing the washed cell suspension with gases or gas mixtures of varying O_2 content, including 100% O_2; room air (21% O_2); 75% N_2, 25% room air (5% O_2); 100% N_2 (0% O_2); and no sparging. Our results (Table 5) indicate that reduction rates of SeO_4^{2-} to SeO_3^{2-} and removal of Se from solution by SLD1a-1 are increased as O_2 becomes more limiting. Response of the unsparged treatment, which is known to rapidly consume O_2, generally fell in between those of the 0% and 5% O_2 levels. These results are in agreement with the previously discussed study which found that SLD1a-1 uses SeO_4^{2-} as a terminal electron acceptor (Losi and Frankenberger, 1997a).

Prior experimentation involving reduction of SeO_4^{2-} during growth of mixed pseudomonad cultures has shown that optimal reduction occurs when O_2 is limiting, but not absent (Larsen et al., 1989). However in the study just cited, treatments in which O_2 was present but limiting supported higher cell numbers than treatments in which O_2 was absent, and thus the combination of higher cell numbers and limiting O_2 promoted more efficient SeO_4^{2-} reduction than lower cell numbers and no O_2. In our experiment, the number of cells was constant,

TABLE 5 Production of SeO_3^{2-} and Removal of Se from Solution Due to Reduction of SeO_4^{2-} by Washed and Sparged Cell Suspensions of *E. cloacae* Strain SLD1a-1[a]

	Amount produced or removed (μmol/g dry biomass)[b]		
O_2 content of gas sparged	SeO_3^{2-} produced[c]	Total Se remaining in solution[d]	Total Se removed from solution (%)
100% O_2	1.67 (0.05)A	20.9 (0.25)A	0
21% O_2 (room air)	1.80 (0.16)A	19.6 (0.47)A	6.22
5% O_2 (75% N_2, 25% room air)	4.08 (0.20)B	5.9 (0.42)B	71.8
0% O_2 (100% N_2)	16.6 (1.43)D	1.40 (0.11)C	93.3
No sparging	9.99 (0.12)C	2.61 (0.29)C	87.5

[a]2,4-DNP was added at 1.09 mM in the experiment to measure SeO_3^{2-} production and not added in the experiment to measure Se removal. The initial SeO_4^{2-} concentration was 0.127 mM (20.9 μmol Se/g dry biomass).

[b]Standard error of the mean is shown in parentheses; $n = 3$. Values followed by different letters were statistically different within each column (Duncan's multiple range test, $P > 0.05$).

[c]Analyzed after 30 minutes.

[d]Analyzed after 8 hours.

suggesting that for SLD1a-1, lower O_2 levels promote greater SeO_4^{2-} reduction. These were resting cells, however, and during growth small amounts of O_2 may stimulate the process through enhanced bioactivity and biomass production. In practice, the bioreactor influent will contain O_2. It appears possible that if high enough growth rates can be achieved, O_2 may be consumed at a rate that will lead to efficient reduction of SeO_4^{2-}. To determine whether this is the case will require future bioreactor studies, and it is also possible that pretreatment may be necessary to lower influent O_2 levels.

With regard to practical applications, experiments performed by Larsen et al. (1989) showed that a *Pseudomonas* isolate reduced SeO_4^{2-} more efficiently in solution culture when O_2 was limiting (bubbled with 5% O_2) than when O_2 was plentiful or absent (bubbled with 21% or 0% O_2, respectively), and that cell numbers were greater at increased O_2 content. Greater reduction efficiency at 5% O_2 may be attributable to a point at which a balance is achieved between higher cell numbers and less interference of reductase(s) by O_2 occurring under fully aerobic conditions. This idea merits further investigation and may represent an advantage of working with facultative anaerobes.

Another major potential advantage is that the SeO_4^{2-} reduction efficiency in a bioreactor depends, in part, on the size of the Se-reducing population. Theoretically, high initial cell numbers can be achieved with a facultative organism during an initial aerobic growth phase. This approach would take advantage of the greater energy yield achieved while respiring O_2 versus alternate electron acceptors, as a result of the higher reduction potential of O_2. Additionally, organisms able to reduce SeO_4^{2-} and NO_3^- oxyanions in the presence of even small amounts of O_2 would offer the advantage of diminishing the need for strict anaerobicity in a given treatment system, and enough carbon could be added to support reduction of both oxyanions. A final point with regard to applied aspects: if pure or mixed bacterial cultures are to be used in Se bioremediation systems, facultative anaerobes capable of vigorous aerobic growth have a distinct advantage over strict anaerobes in terms of general ease of handling, processing, and maintenance of the cultures.

VI. CONCLUSIONS

Enterobacter cloacae SLD1a-1 may be useful in a variety of treatment schemes designed to remove Se oxyanions from agricultural drainage water. Potential advantages include the following:

> High cell densities may be achieved rapidly in an aerobic growth phase.
> Reduction occurs in microaerophilic environments.
> No induction or conditioning process is necessary.

SeO_4^{2-}, SeO_3^{2-}, and NO_3^- are reduced concomitantly.
The organism is easy to culture and work with.

Experimentation has demonstrated the ability of SLD1a-1 to reduce SeO_4^{2-} under environmental conditions similar to those commonly observed in San Joaquin Valley drainage water. The development of a biological flow reactor is compatible with the construction of drainage water flow systems through wetland cells. In addition, the organism may be useful as a component in integrated remediation systems—for example, as a pretreatment to be followed by microphytic organisms (algae), which can lower Se concentrations to below 2 μg/L (Teresa Fan, 1996, personal communication) but may have slower removal rates. Other methods that may be used in a treatment train with bioremediation to further decrease the Se concentration include ferric sulfate coagulation, activated alumina, ion exchange with strong-based resins, and cogeneration (Thompson-Eagle and Frankenberger, 1992),

ACKNOWLEDGMENTS

This work was supported by a grant from the University of California Salinity/Drainage Task Force. The authors acknowledge Krassimir Bozhilov, microscopist for the Electron Microscopy Facility at the Institute of Geophysics and Planetary Physics, University of California, Riverside for transmission electron microscopy work, and K. L. Rose, Department of Soil and Environmental Sciences, University of Calif., Riverside, for performing the X-ray diffraction analysis.

REFERENCES

Abdelmagid, H. M., and M. A. Tabatabai. 1987. Nitrate reductase activity in soils. *Soil Biol. Biochem.* 19:421–427.

Altringer, P. B., D. M. Larsen, and K. R. Gardner. 1989. Bench-scale process development of selenium removal from wastewater using facultative bacteria. In J. Salley, R. G. L. McCready, and P. L. Wichlacz (eds.), *Biohydrometallurgy* (*Proceedings of the International Symposium, Jackson Hole, Wyoming, 13–18 August 1989*), pp. 643–657. CANMET, Ottawa, Ont., Canada.

Barton, L. L., H. E. Nuttall, W. C. Lindeman, and R. C. Blake II. 1994. Biocolloid formation: An approach to bioremediation of toxic metal wastes. In D. L. Wise and D. J. Tarantolo (eds.), *Remediation of Hazardous Waste-Contaminated Soils*, pp.481–496. Dekker, New York.

Beevers, L., and R. H. Hageman. 1969. Nitrate reduction in higher plants. *Annu. Rev. Plant Physiol.* 20:495–522.

Brock, T. D., and M. T. Madigan. 1991. *Biology of Microorganisms*. Prentice-Hall, Englewood Cliffs, NJ.

Canvin, D. T., and C. V. Atkins. 1974. Nitrate, nitrite and ammonia assimilation: Effects of light, carbon dioxide and oxygen. *Planta* 116:207–224.

Chilcott, J. E., D. W Westcot, A. L. Toto, and C. A. Enos. 1990. Water quality in evaporation basins used for the disposal of agricultural subsurface drainage water in the San Joaquin Valley, California, 1988 and 1989. California Regional Water Quality Control Board, Central Valley Region, Sacramento.

Coleman, R. N. 1988. Chromium toxicity: Effects on microorganisms with special reference to the soil matrix. *In* J. O. Nriagu and E. Nieboer, (eds.), *Chromium in Natural and Human Environments*, pp. 335–351. Wiley, New York.

DeMoll-Decker, H., and J. M. Macy. 1993. The periplasmic nitrite reductase *of Thauera selenatis* may catalyze the reduction of selenite to elemental selenium. *Arch. Microbiol.* 160:241–247.

Doran, J. W. 1982. Microorganisms and the biological cycling of selenium. *Adv. Microbiol. Ecol.* 6:1–32.

Ehrlich, H. L. 1996. *Geomicrobiology*, 3rd ed. Dekker. New York.

Euliss, N. E. Jr., R. L. Jarvis, and D. S. Gilmer. 1991. Standing crops and ecology of aquatic invertebrates in agricultural drainwater ponds in California. *Wetlands* 11:179–190.

Fujii, R. S., J. Deverel, and D. B. Hatfield. 1988. Distribution of selenium in soils of agricultural fields, western San Joaquin Valley, California. *Soil Sci. Soc. Am. J.* 52:1274–1283.

Klepper, L. A. 1976. Nitrate accumulation in herbicide-treated leaves. *Weed Sci.* 24: 533–535.

Larsen, D. M., Gardiner, K. R., and P. B. Altringer. 1989. Biologically assisted control of selenium in process waste waters. *In* B. J. Scheiner, F. M. Doyle, and S. K. Kawatra (eds.), *Biotechnology in Minerals and Metal Processing*, pp. 177–186. Society of Mining Engineers, Littleton, CO.

Lester, J. N., R. Perry, and A. H. Dadd. 1979. The influence of heavy metals on a mixed bacterial population of sewage origin in the chemostat. *Water Res.* 13:1055–1063.

Lortie, L., W. D. Gould, S. Rajan, R. G. L. McCready, and K.-J. Cheng. 1992. Reduction of selenate and selenite by a *Pseudomonas stutzeri* isolate. *Appl. Environ. Microbiol.* 58:4043–4044.

Losi, M. E., and W. T. Frankenberger, Jr. 1997a. Reduction of selenium by *Enterobacter cloacae* SLD1a-1: Isolation and growth of the bacterium and its expulsion of selenium particles. *Appl. Environ. Microbiol.* 63:3079–3084.

Losi, M. E., and W. T. Frankenberger, Jr. 1997b. Reduction of selenium by *Enterobacter cloacae* strain SLD1a-1: Reduction of selenate to selenite. *Environ. Toxicol. Chem.* 16:1851–1858.

Macy, J. M. 1994. Biochemistry of selenium metabolism by *Thauera selenatis* gen. nov. sp. nov. and use of the organism for bioremediation of selenium oxyanions in San Joaquin Valley drainage water. *In* W. T. Frankenberger, Jr. and S. Benson (eds.), *Selenium in the Environment*, pp. 421–444. Dekker, New York.

Macy, J. M., S. Rech, G. Auling, M. Dorsch, E. Stackebrandt, and L. Sly. 1993a. *Thauera selenatis* gen. nov. sp. nov., a member of the beta sub-class of Proteobacteria with a novel type of respiration. *Int. J. Syst. Bacteriol.* 43:135–142.

Macy, J. M., S. Lawson, and H Demoll-Decker. 1993b. Bioremediation of selenium oxyanions in San Joaquin drainage water using *Thauera selenatis* in a biological reactor system. *Appl. Microbiol. Biotechnol.* 40:588–594.

Maiers, D. T. , P. L. Wichlacz, D. L. Thompson, and D. F. Bruhn. 1988. Selenate reduction by bacteria from a selenium-rich environments. *Appl. Environ. Microbiol.* 54:2591-2593.

Mobley, H. T., and B. P. Rosen. 1982. Energetics of plasmid-determined arsenate resistance in *Escherichia coli*. *Proc. Natl. Acad. Sci. USA* 79:6119–6122.

National Research Council. 1989. *Irrigation-Induced Water Quality Problems: What Can Be Learned from the San Joaquin Valley Experience?* National Academy of Sciences Press, Washington DC.

Ohlendorf, H. M., D. J. Hoffman, M. K. Saiki, and T. W. Aldrich. 1986. Embryonic mortality and abnormalities of aquatic birds: Apparent impact of selenium from irrigation drain water. *Sci. Total Environ.* 52:49–63.

Oremland, R. S., J. T. Hollibaugh, A. S. Maerst, T. S. Presser, L. G. Miller, and C. W. Culbertson. 1989. Selenate reduction to elemental selenium by anaerobic bacteria in sediments and culture: Biogeochemical significance of a novel, sulfate-independent respiration. *Appl. Environ. Microbiol.* 55:2333–2343.

Oremland, R. S., N. A. Steinberg, A. S. Maerst, L. G. Miller, and J. T. Hollibaugh. 1990. Measurement of in situ of selenate removal by dissimilatory bacterial reduction in sediments. *Environ. Sci. Technol.* 24:1157–1164.

Oremland, R. S., J. S. Blum, C. W. Culbertson, P. T. Visscher, L. G. Miller, P. Dowdle, and F. Strohmaier. 1994. Isolation, growth, and metabolism of an obligately anaerobic, selenate-respiring bacterium, strain SES-3. *Appl. Environ. Microbiol.* 8:3011–3019.

Presser, T. S., and H. M. Ohlendorf. 1987. Biogeochemical cycling of selenium in the San Joaquin Valley, California. *Environ. Manage.* 11:805–821.

Presser, T. S., M. A. Sylvester, and W. H. Low. 1994. Bioaccumulation of selenium from natural geologic sources in western states and its potential consequences. *Environ. Manage.* 18:423–436.

Rech, S. A., and J. M. Macy. 1992. The terminal reductases for selenate and nitrate respiration in *Thauera selenatis* are two distinct enzymes. *J. Bacteriol.* 174:7316–7320.

Ross, D. S., R. E. Sjogren, and R. J. Bartlett. 1981. Behavior of chromium in soils: IV. Toxicity to microorganisms. *J. Environ. Qual.* 2:145–148.

State of California. 1987. Regulation of agricultural drainage to the San Joaquin River. Water Quality Control Board Order no. W. Q. 85-1, Sacramento.

Steinberg, N. A., J. S. Blum, L. Hochstein, and R. S. Oremland. 1992. Nitrate is a preferred electron acceptor for growth of fresh water selenate-respiring bacteria. *Appl. Environ. Microbiol.* 58:426–428.

Stryer, L. 1988. *Biochemistry*, 3rd ed., pp. 357–358, 801–803. Freeman. New York.

Sylvester, M. A. 1990. Overview of the salt and agricultural drainage problem in the western San Joaquin Valley, California. U.S. Geological Survey Circular no. 1033c.

Thauer, R. K., K. Jungermann, and K. Decker. 1977. Energy conservation in chemotrophic anaerobic bacteria. *Bacteriol. Rev.* 41:100–180.

Thompson-Eagle, E. T., and W. T. Frankenberger, Jr. 1990. Protein-mediated selenium biomethylation in evaporation pond water. *Environ. Toxicol. Chem.* 6:1453–1462.

Thompson-Eagle, E. T., and W. T. Frankenberger, Jr. 1992. Bioremediation of soils contaminated with selenium. *In* R. Lal and B. A. Stewart (eds)., *Advances in Soil Science*, pp. 261–310. Springer-Verlag, New York.

Tomei, F. A., L. L. Barton, C. L. Lemanski, and T. G. Zocco. 1992. Reduction of selenate and selenite to elemental selenium by *Wolinella succinogenes*. *Can. J. Microbiol.* 38:1328–1333.

Tomei, F. A., L. L. Barton, C. L. Lemanski, T. G. Zocco, N. H. Fink, and L. O. Sillerud. 1995. Transformation of selenate and selenite to elemental selenium by *Desulfovibrio desulfuricans. J. Ind. Microbiol.* 14:329–336.

Wang, P.-C., K. Toda, H. Ohtake, I. Kusaka, and I. Yabe. 1991. Membrane-bound respiratory system of *Enterobacter cloacae* strain HO1 grown anaerobically with chromate. *FEMS Microbiol. Lett.* 78:11–16.

Weres, O., A.-R. Jaouni, and L. Tsao. 1989. The distribution, speciation and geochemical cycling of selenium in a sedimentary environment, Kesterson Reservoir, California, U.S.A. *Appl. Geochem.* 4:543–563.

27

Biochemical Fate of Selenium in Microphytes: Natural Bioremediation by Volatilization and Sedimentation in Aquatic Environments

Teresa W.-M. Fan and Richard M. Higashi
University of California at Davis, Davis, California

I. INTRODUCTION

The unusually narrow margin between nutritional requirement and toxicity, complex biogeochemical cycling, and heterogeneous distribution all contribute to the difficult problem of selenium (Se) contamination. This is well illustrated by the wildlife deformities that occurred in California's Kesterson Reservoir in the early 1980s. The toxic symptoms observed have since been attributed to Se bioaccumulation and biotransformation through the food chain (Skorupa and Ohlendorf, 1991; Maier and Knight, 1994). However, Se biotransformation through the aquatic food web is largely unknown. This ecotoxicological effect is becoming a major concern for agricultural operations in seleniferous soils and for industrial discharges throughout the United States (Skorupa and Ohlendorf, 1991; Fairbrother et al., 1996; Reash et al., 1996). Long-term solutions to the selenium problem need to be explored to make these agricultural and industrial activities sustainable.

Various physical and chemical removal schemes for Se contamination have been tested but were shown to be prohibitively expensive, or ineffective for waterborne Se concentration of 10 μg/L or lower (Mudder, 1997). Some biological remediation schemes based on precipitation and/or volatilization of selenium by soil bacteria and vascular plants have also been proposed (Gerhardt et al., 1991; Thompson-Eagle and Frankenberger, 1991; Terry et al., 1992). These schemes are discussed in other chapters of this book. This chapter describes an approach that utilizes the intrinsic biotransformation activities of aquatic microphytes in Se-laden agricultural evaporation basins for "natural" remediation. Since this process is solar-driven and has been in full-scale operation for over 20 years, the economical advantage is self-evident. We will also illustrate the importance of acquiring a detailed understanding of the biochemical fate of Se to achieve not just Se removal but, more importantly, reduction in ecotoxic risk, referred to by some as "natural attenuation."

II. BACKGROUND AND CONCEPT

Evaporation basins have been employed for the disposal of large volumes of agricultural drainage waters in the San Joaquin Valley of California for more than two decades. An example is a multibasin system that channels drainage waters through a sequence of shallow (e.g., 0.7–1.5 m) basins to optimize water evaporation. Since the drainage water itself can be moderately saline [e.g., 7 parts per thousand (‰)], salinity (primarily Na_2SO_4 and NaCl) in successive basins can reach as high as 300‰, a result of evaporation (see, Fig. 1B). In many cases, contaminant salts including Se oxyanions can be predicted to increase based on evaporite chemistry (Tanji, 1989), as observed in systems such as the Peck Pond basin, located in the Fresno County, California (Fig. 1A).

However, waterborne Se concentrations can exhibit a *decreasing* trend with increasing salinities in a sequential multibasin system. An example is provided by the Tulare Lake Drainage District (TLDD) basin, located in the Tulare County, California; this system has been in continuous operation for over 20 years. The calculated amount of Se dissipated annually is on the order of millions of grams (Fig. 1B). Since this disappearance of Se cannot be readily accounted for by abiotic physical and chemical mechanisms, we are investigating the biochemical processes that may be the main driving force.

An initial survey of the TLDD basins indicates that vascular plants and aquatic macrophytes are absent but the basin waters are rich in aquatic microphytes, including various cyanobacteria, diatoms, and green algae. Except for the most saline basin, these microphytes are abundant in the waters regardless of the season. We have recently reported that a green coccoid microalga (*Chlorella* sp.) isolated from the Pryse Pond system in Tulare County is very active in transforming Se oxyanions including Se volatilization (Fan et al., 1997). The

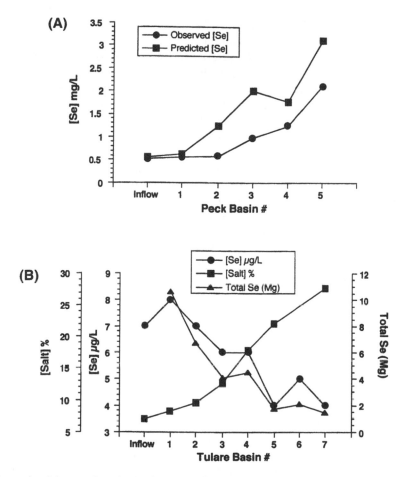

FIGURE I   Selenium fate observed at saline drainage evaporation basins deviates from that predicted based on increasing salinities. The observed concentration [Se] in multibasin agricultural evaporation systems often increases progressively with increasing salinities (i.e., increasing basin no.) as in (A), the Peck Pond system, in California (Tanji, 1989). This behavior is consistent with that predicted from evaporite chemistry as shown (Tanji, 1989). However, an opposite behavior has been observed for the Tulare basins (B), where waterborne [Se] declined with increasing salinities through the sequential basins. (Data from personal communication, D. Davis, 1996.) Also shown is the calculated Se mass megagram (Mg) in the water of each basin at the time of sampling (Data from personal communication, D. Davis, 1996.)

microphytes at the TLDD basins, where Se may be dissipated against the salinity gradient, warrants a similar investigation.

We expect that a detailed characterization of Se biotransformation activities by these microphytes will help elucidate the mechanism(s) underlying the dissipation of Se from the TLDD basins. In addition, by tracing the environmental fate of the biotransformed products, particularly through the food web, it may be possible to develop an in situ bioremediation approach with ecotoxicological compatibility. Since it is unlikely that the natural phenomenon is, by pure chance, optimal in Se removal, we project that the evaporation basin may be readily managed for a more efficient removal of Se or other contaminants, provided the mechanism or mechanisms of removal are understood. Furthermore, these investigations should contribute to a general understanding of the role(s) of aquatic microphytes in the biogeochemical cycling of Se or other elements in surface waters. These organisms have been postulated to be major drivers of Se biogeochemistry (Cutter and Bruland, 1984; Cooke and Bruland, 1987), as they are in the case of sulfur (e.g. Andreae, 1986; Bates et al., 1987). To date, however, there is very little direct evidence available.

III. METHODOLOGIES

A. Microphyte Isolation

Microphytes were isolated from the evaporation basin waters according to the procedure described by Fan et al. (1997). Briefly, a small amount of water was streaked onto a 1% agarose (in f/2 seawater medium, Anderson et al., 1991) plate and incubated at 22°C under a light/dark cycle of 16/8 hours. Distinct and isolated colonies were then inoculated separately into f/2 seawater broth medium and incubated similarly to establish stock cultures. Under microscopic examination, one microphyte from the TLDD basin waters appeared to be a monoculture of a filamentous cyanophyte, while the other was coccoid with two cells frequently joined; the latter assumes the morphology of a *Synechocystis* sp. in the TLDD basins. The species of *Chlorella* that was abundant in the Pryse Pond system was not as prevalent in the TLDD basins. The species identity of two cyanophytes from TLDD is currently under investigation.

B. Microphyte Pigment Composition

Biomass from monocultures, water column particulate matter from basin waters, and basin sediment were harvested by centrifugation, then lyophilized and extracted with methanol in the dark for 1 hour, and centrifuged again. The methanolic extract was then filtered through a 0.22 μm Teflon filter before high performance liquid chromatography (HPLC) or spectrophotometric analysis.

HPLC analysis and the tentative identification of pigments essentially followed the procedure of the Scientific Committee for Oceanic Research (SCOR) Working Group 78 on Measurement of Photosynthetic Pigments in Seawater (see Wright et al., 1991).

C. Measurement of Selenium Volatilization Kinetics

The rate of Se volatilization from growing microphyte cultures was measured as described in detail by Fan et al. (1997). Briefly, 0.8 L of f/2 seawater medium supplemented with a given concentration of Se was inoculated with microphyte stocks and constantly bubbled with 0.22 μm filtered air at 30°C under continuous fluorescent light. The air aeration served to provide CO_2 for microphyte growth and allowed volatile Se compounds (e.g., alkyl selenides) to be purged from the medium, which was subsequently trapped by oxidation to selenate in alkaline peroxide [50 mM NaOH and 30% H_2O_2, 4 : 1 (v/v)] solutions for 18 to 22 hours. Alternatively, volatile Se compounds were trapped in their original forms into a 0.25 in. Teflon tube immersed in liquid nitrogen. Total Se in the trap was then measured by means of a gas chromatograph–electron capture detector (GC-ECD) or by fluorescence, while the chemical form(s) of trapped Se were analyzed by gas chromatography–mass spectrometry (GC-MS) (see below). The rate of Se volatilization was calculated from the total Se and the duration of the trapping.

D. Total Selenium Analysis

Two analytical methods were used to measure total Se in water: the trap and biomass samples. The GC-ECD method is described in details in Fan et al. (1997), while the fluorescence method was modified from the Analytical Methods Committee (1979). Briefly, both methods employed nitric acid or alkaline peroxide microdigestion to convert organic Se into Se oxyanions, which were then reduced to selenite by 6 N HCl at 105°C, followed by derivatization with 4-nitrophenylene-o-diamine or 2,3-diaminonaphthalene to form the corresponding piazselenol derivatives, which were quantified by GC-ECD or fluorescence, respectively. The detection limit for the GC-ECD method was 1 μg/L, while that for the fluorescence method was in the submicrogram-per-liter range, both based on 250 μL of water sample. The typical correlation coefficient obtained for a standard curve of 12 Se standards ranging from 0 to 200 μg/L was better than 0.99. In addition, samples with Se spike were analyzed to determine the bias of the analysis, which was consistently within 10% of predicted values.

E. Analysis of Selenium Biotransformation Products

We have developed the GC-MS and nuclear magnetic resonance (NMR) methodologies to analyze three major classes of Se metabolites: volatile alkyl selenides, selenonium compounds, and selenoamino acids. The detailed procedures for the analysis of these metabolites are described by Fan et al. (1997). Alkyl selenides trapped by liquid nitrogen were analyzed on a capillary DB-1 column in a Varian 3400 gas chromatograph coupled with a Finnegan ITD 806 mass spectrometer. Selenonium compounds and selenoamino acids were extracted from microphyte biomass using 5% perchloric acid (PCA) or 10% trichloroacetic acid (TCA). Selenoamino acids in the extract was then silylated with N-methyl-N-[tert-butyldimethylsilyl]trifluoroacetamide) (MTBSTFA) before GC-MS analysis, or removed of paramagnetic ions by passing through Chelex-100 resin column before NMR analysis. Selenonium compounds in biomass or TCA extracts were inferred from the release of DMSe by 5 M NaOH treatment at room temperature or at 105°C (Fan et al., 1997). Selenonium compounds including dimethylselenonium propionate (DMSeP) and methylselenomethionine (CH$_3$-Se-Met) are considered to be the biological precursors of dimethyl selenide (DMSe), while selenoamino acids such as selenomethionine (Se-Met) are currently thought to be most toxic to wildlife (e.g., Heinz et al., 1996).

IV. RESULTS AND DISCUSSION

A. Characterization of Evaporation Basin Microphytes

One common trait of the three isolated microphytes was capability of growth over a wide range of salinities (e.g., 7–33‰) in culture (Fan et al., 1997). This is presumably because they can adapt to the fluctuating salinities of the basin waters from which they were isolated.

Following the initial morphologic characterization, we used HPLC analysis of pigment composition to help survey the types and amounts of microphytes in basin water communities. Since microphyte classes can be distinguished by their types and/or ratios of pigments, this approach makes it possible to obtain a gross composition of microphytes from a more representative (e.g., a liter) water sample (e.g., Wright et al., 1991) than that practical for microscopic examination (e.g., 10 μL). Examples of pigment analysis are shown in Figure 2 for the isolated Synechocystis and filamentous strains, which revealed, as expected, phycobillin in their pigment extracts, while the Pryse Chlorella sp. did not. This is consistent with the first two being cyanobacteria. Also as expected, the Chlorella pigment profile was similar to the profile for terrestrial vascular plants, which has been observed for many green phytoplankton species (Wright et al., 1991).

An examination of the pigments in the "C" sequential basin systems at TLDD revealed some interesting trends (Fig. 3). In general, the concentration of most pigments declined with increasing salinity, but there were also large differences in relative amounts of pigments. This represents changes in community structure in the water along the salinity gradient. For example, several forms of

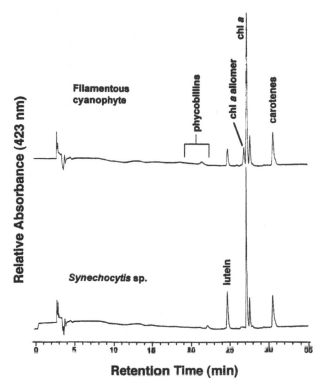

FIGURE 2 HPLC determination of pigment profiles of cyanophytes isolated from TLDD basins (chl a, chlorophyll a). Pigments were analyzed as described in the text. Different classes of microphytes exhibit different pigment profiles, as illustrated for two of the TLDD cyanophytes capable of volatilizing Se.

chlorophyll (Chl) c, which are markers of eukaryotic algae such as diatoms, were the highest in basin C3 (about twice the salinity of seawater); however, another diatom marker, fucoxanthin, steadily declined with increasing salinity. One simple explanation for this pattern is that the diatom species in basin C3 have higher ratios of Chl c to fucoxanthin than those in other basins. Hence, the diatom species composition in basin C3 may differ from that in other basins. Cyanobacterial markers such as phycobillins were present throughout and declined with increasing salinity. Some other pigments such as Chl a are not very useful for compositional analysis because they are common to most microphytes, while the relationships for carotenes are taxonomically complex (Wright et al., 1991; Hirschberg and Chamovitz, 1994).

In addition, the microphyte community in the top sediment layer in relation to that of the water column may be important for understanding the mechanisms of Se removal from water. Figure 3 compares the pigment profile of the least saline water (basin C1) with that of the corresponding top sediment layer, which revealed surprisingly similar pigment signatures. Thus, it appears that diatoms (i.e., as judged by fucoxanthin) and cyanophytes (i.e., based on phycobillins) were present in the top sediment layer and in the water column in similar ratios. In contrast, the saline basin C3 showed significant differences in pigment composition between the water column and top sediment layer (Fig. 3). Here, the diatoms present in the water column were absent from the top sediment layer, while some cyanophytes were still present in both media.

Of course, interpretation of community structure by pigment composition can be complicated by many factors; for example, pigment composition could vary even within a species in response to salinity or other factors. This is known to occur in cyanobacteria (Hirschberg and Chamovitz, 1994). Nevertheless, it is clear from the analyses that the biochemistry of the aquatic and bottom sediment microphyte communities are different in each basin and that these differences can be readily monitored for the purpose of relating to the Se biogeochemistry in the system.

B. Se Volatilization by Isolated Microphytes

Similar to the Pryse Pond *Chlorella* sp., the two cyanophytes isolated from the TLDD basin waters volatilized Se from selenite-supplemented f/2 seawater media, as shown in Figure 4. The rate of volatilization for all three microphytes was dependent on the suspended vegetative growth (optical density at 680 nm), but the time course of volatilization varied among the three species. Volatilization of Se by the *Chlorella* sp. and filamentous cyanophyte tracked the vegetative growth, while that by the *Synechocystis* sp. exhibited a more complex pattern. The peak rate of volatilization for the filamentous cyanophyte was about two fold higher than that for the *Chlorella* sp. at 1 mg/L Se supplement and may be comparable to that for the *Synechocystis* sp., assuming that the volatilization rate is proportional to the Se concentration in the medium.

At a much lower selenite treatment (10 μg/L), the rate of Se volatilization by the filamentous cyanophyte was proportionally lower than that for the 1 mg/ L treatment, as shown in Figure 5. Note that the time course of the suspended vegetative growth (absorbance at 680 nm, A_{680}) was comparable between the two treatments. Also shown in Figure 5 is the decline curve of the Se concentration in the medium, which appears to be in response to the changes in Se volatilization rates. However, volatilization accounted for only a part of the Se depletion from the treatment medium, while Se incorporation into the biomass accounts for an important fraction of the total Se loss (see below). Thus, both processes are candidates for the mechanism of the decrease in waterborne Se concentrations of the TLDD basins (Fig. 1).

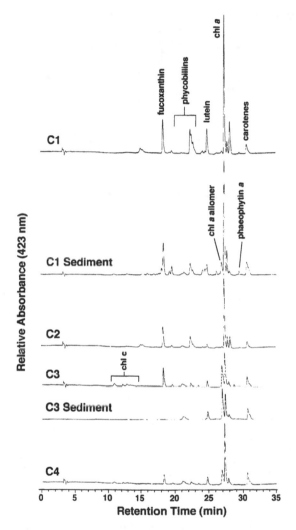

FIGURE 3 HPLC determination of pigment profiles of basin waters and corresponding sediments from TLDD basins. Labels refer to basins of the "C" system of sequential evaporation basins at the facility, with salinity increasing from C1 to C4; basin C4 is terminal and no water is discharged from the system. Pigments were analyzed as described in the text. Water pigment profiles on the ordinate scale are normalized to sample volume, while the sediment profile ordinates are arbitrarily scaled to illustrate pigment composition. The differences in the pigment profiles are indicative of changes in community structure along the salinity gradient of the sequential basins.

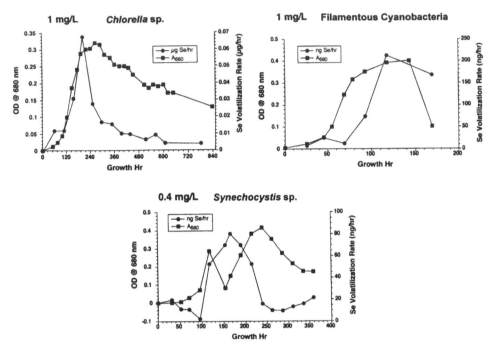

FIGURE 4 Time courses of Se volatilization by three selenite-treated microphyte species isolated from agricultural drainage waters. The *Chlorella* species was isolated from the Pryse Pond system (reprinted with permission from Fan et al., 1997. Copyright 1997, American Chemical Society), while the two cyanophytes (filamentous and *Synechocystis*) were isolated from the TLDD basins. The three microphytes were cultured in f/2 seawater medium supplemented with 1 mg/L (*Chlorella* and filamentous) or 0.4 mg/L (*Synechocystis*) Se (as Na_2SeO_3). Volatile Se was air-purged from the media and trapped in an NaOH/ H_2O_2, solution while the optical density (OD) at 680 nm of the media was monitored for vegetative growth. Total Se in the trap was analyzed by GC-ECD or fluorescence.

The chemical form(s) of the volatilized Se were also investigated by GC-MS analysis of a liquid nitrogen headspace trap, in place of the alkaline peroxide trap. Figure 6 illustrates the total ion chromatogram of volatile compounds released by a *Chlorella* culture grown at 100 mg/L Se (as selenite). Three prominent volatile Se compounds were identified, namely, DMSe, DMDSe (dimethyl diselenide), and DMSeS (dimethylselenenyl sulfide), along with the sulfur analog of DMSe (DMS) (Fan et al., 1997). A similar analysis of a liquid nitrogen trap of the filamentous cyanophyte grown at 1 mg/L Se (as selenite) revealed the presence of DMSe and DMS (data not shown). We currently hypothesize that DMSe and DMS are common—possibly the dominant—forms of selenium and sulfur volatilized by microphytes from the evaporation basins.

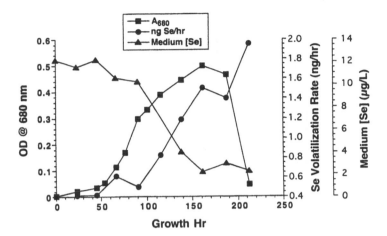

FIGURE 5   Time courses of Se volatilization and depletion from medium by selenite-treated filamentous cyanophyte isolated from the TLDD basins. The filamentous cyanophyte was cultured in f/2 seawater medium supplemented with 10 μg/L Se (as Na₂SeO₃). Volatilized Se and suspended vegetative growth were monitored as in Figure 4. Medium Se concentration was analyzed by the fluorescence method (see Sect. III, Methodologies).

C. Se Fractionation in Isolated Microphytes

To follow the fate of Se in the microphyte biomass, we tested a fractionation scheme on the *Chlorella* biomass, which involved extraction of the biomass with 10% trichloroacetic acid (TCA) to generate the TCA-soluble and residue fractions (Fan et al., 1997). The TCA-soluble fraction generally includes water-soluble, low molecular weight metabolites including small peptides, while the residue fraction comprises of proteins, lipids, and other macromolecules such as cell wall constituents. The total Se distribution in the whole biomass and two TCA fractions is summarized in Figure 7. It is clear that the *Chlorella* biomass contained a significant amount of Se when grown in the 100 mg/L selenite medium. A major fraction (60%) of this Se resided in the TCA residue fraction, while the TCA-soluble fraction accounted for 14 to 16%. It should be noted that Se⁰ is expected to be present in the TCA residue, since a red amorphous material codeposited with the biomass and was not extracted by TCA. Whether the protein fraction in the TCA residue contained a significant amount of Se, and if so, its chemical form(s), is under investigation.

In addition to total Se, we also analyzed for selenonium forms of Se in the biomass and TCA fractions (Fan et al., 1997). This was accomplished by measuring the volatile Se released from the hydroelimination reaction of selenonium compounds with NaOH, which was confirmed as DMSe by headspace GC-MS

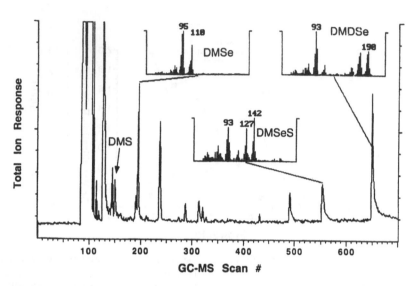

FIGURE 6 GC-MS analysis of liquid N_2 headspace trap of selenite-treated *Chlorella* culture. The liquid N_2 trap was obtained from the culture treated with 100 mg/L Se in the late exponential phase, and the GC-MS analysis was performed as described by Fan et al. (1997). Identification of DMSe, DMDSe, and DMS was based on GC retention times, molecular ions, and mass fragmentation patterns (insets) as compared with authentic standards. The assignment of DMSeS was rationalized from the molecular ion, mass fragmentation pattern, and known GC elution order. (Reprinted with permission from Fan et al., 1997, American Chemical Society.)

analysis (Fig. 8). Also released by this treatment was dimethyl sulfide (DMS), which is diagnostic of the presence of the precursor, dimethylsulfonium propionate (DMSP) (White, 1982). By analogy, the NaOH-induced release of DMSe may be diagnostic for the presence of the selenonium compound DMSeP (Fan et al., 1997). The Se present as selenonium form(s) accounted for all the Se in the *Chlorella* TCA extract (Fig. 7). In fact, the TCA extract contained all the selenonium metabolite(s) in the whole biomass, which would be expected of DMSeP, since it should be very soluble in 10% TCA. We are currently utilizing two-dimensional NMR techniques (e.g., Fan, 1996) to determine the detailed structure of this precursor.

A preliminary analysis of the filamentous cyanophyte biomass indicated that this microphyte also contained a DMSeP-like metabolite (data not shown), while this compound was absent from the leaf tissues of several vascular plants (*Salicornia,* alfalfa, canola, *Atriplex,* purslane, and duckweed) grown with Se supplement, based on an initial survey.

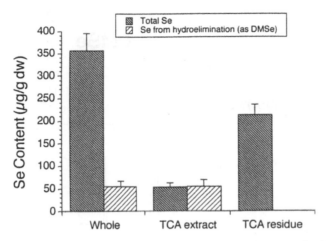

FIGURE 7 Distribution of total Se and Se as DMSe in *Chlorella* biomass and TCA fractions. Two separate *Chlorella* cultures were treated with 100 mg/L Se (as Na_2SeO_3) for 1.5 and 3 months. The algal biomass was collected by centrifugation, lyophilized, and extracted with 10% TCA. Total Se was measured by GC-ECD as the piazselenol derivative of 4-NPD. Selenium as DMSe was calculated from GC-MS headspace analysis of DMSe released from 5 M NaOH/105°C-treated samples. The amount of DMSe, in micrograms per gram, dry weight, was multiplied by 0.73 to convert to Se in the same units. No DMSe was detected in the NaOH-treated TCA residue fraction. (Reprinted with permission from Fan et al., 1997, American Chemical Society.)

D. Biochemical Forms of Selenium in Isolated Microphytes

Since the biological impact and environmental fate of Se is critically dependent on its chemical forms, we want to develop methods for the analysis of some of these forms. Our present focus is on selenonium metabolites (as described above) and selenoamino acids, since the former may be indicative of the Se volatilization potential for a given organism while the latter may have ecotoxic significance to wildlife. Volatilization and ecotoxic effects are not necessarily mutually exclusive, however, because of the possibility that parts of the biochemical pathways are shared for both processes.

To obtain structural information for the selenonium compound from the *Chlorella* TCA extract, we have employed several NMR methods (including ^1H and ^{77}Se NMR) in addition to the headspace GC-MS analysis described above. From two-dimensional ^1H total correlation spectroscopic (TOCSY) analysis, the structure of DMSP was identified directly from the crude *Chlorella* TCA extract (Fan et al., 1997). Without a DMSeP standard, however, it was difficult to confirm the ^1H resonances of DMSeP. When the same extract was analyzed by one-

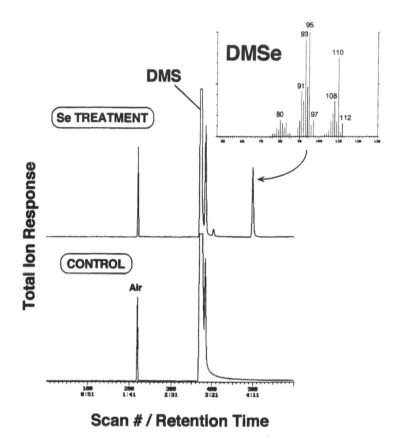

FIGURE 8 GC-MS analysis of headspace alkaline hydroelimination products of *Chlorella* biomass from the 100 mg/L Se and control treatments. Lyophilized *Chlorella* cells from the culture of Figure 6 (treated with 100 mg/L Se) were subjected to alkaline treatment as described in Figure 7, and the headspace content was analyzed by GC-MS as described by Fan et al. (1997). The identity of DMSe and DMS was confirmed by GC retention time, molecular ion, and mass fragmentation pattern (inset), and their content in *Chlorella* biomass was approximately 5 ng DMSe/mg and 185 ng DMS/mg.

FIGURE 9 One-dimensional ^{77}Se NMR analysis of *Chlorella* extract and Se standards. The TCA extract of a *Chlorella* biomass treated with 100 mg/L Se was passed through a Chelex 100 column before analysis on the Bruker AM-400 NMR spectrometer operating at 76.3 MHz using 90° pulse width (6 μs), 20,000 to 125,000 Hz sweep width, and 0.262 second acquisition time. The ^{77}Se NMR spectrum of the *Chlorella* extract is shown along with that of the Se-Met, Se-cystine, and trimethylselenonium ion (TMSe$^+$) standards in D$_2$O. The major resonance at a chemical shift of −845 ppm in the microphyte spectrum did not correspond to that of selenate, selenite, Se-Met, Se-cystine, or TMSe$^+$. However, this chemical shift is characteristic of a selenonium compound (Duddeck, 1994). (Adapted with permission from Fan et al., 1997, American Chemical Society.)

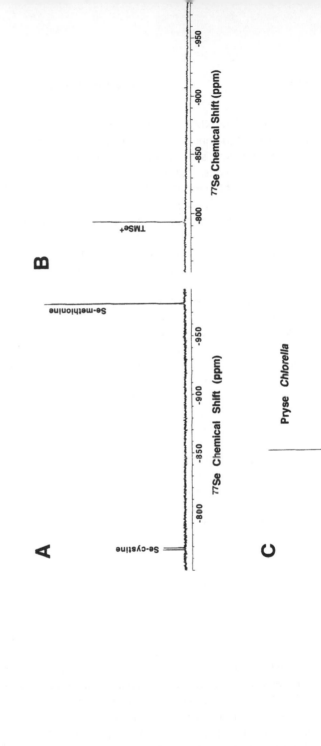

B

TMSe+

⁷⁷Se Chemical Shift (ppm)

-950   -900   -850   -800

A

Se-methionine

Se-cystine

⁷⁷Se Chemical Shift (ppm)

-950   -900   -850   -800

C

Pryse *Chlorella*

dimensional ^{77}Se NMR, a major resonance was observed at -845 ppm, which corresponded to none of the resonances arising from available standards (trimethylselenonium ion, selenocystine, Se-Met) (Fig. 9C). The chemical shift of this resonance was consistent with that of a selenonium compound. We are working with Dr. Dean Martens of the U.S. Department of Agriculture Salinity Laboratory (Riverside, CA), on the synthesis of DMSeP and CH_3-Se-Met, which should facilitate the identification of this Se metabolite from *Chlorella* as well as those from the filamentous cyanophyte.

From the preceding NMR analysis, it is clear that selenoamino acids were not present at any appreciable concentration in the *Chlorella* TCA extract, which was the impetus for employing GC-MS for trace level analysis (Fan et al., 1997). Following silylation of the TCA extract with MTBSTFA, Se-Met in *Chlorella* was analyzed by GC-MS as shown in Figure 10. Under electron ionization mode, no Se-Met peak was observable in either the total or selected ion chromatogram of *Chlorella*. However, using the 40-fold more sensitive isobutane chemical ionization mode, a peak with molecular mass of 427 appearing at the GC retention time of Se-Met was detected, providing evidence for the presence of Se-Met in *Chlorella* tissue at the submicrogram-per-liter level. Using this approach, a number of other amino acids, organic acids, and phosphate metabolites were also simultaneously measured. This procedure greatly simplifies the sample preparation—only one-step extraction is required—while maximizing the number of metabolites analyzed. More importantly, the approach is based on GC–ion trap MS analysis, widely accepted as one of the few reliable methods of structure confirmation and quantification when faced with trace levels of organic compounds. Recently we successfully adapted this method for the simultaneous analysis of selenocysteine, selenocystine, and methylselenocysteine in addition to Se-Met, and are adapting it to the analysis of other forms of organic Se.

V. CONCLUDING REMARKS

As recipients of agricultural drainage waters, it is not surprising that evaporation basins of the San Joaquin Valley will readily support primary productivity. The euryhaline nature of these basins tends toward the predominance of euryhaline microphytes, possibly consisting largely of marine estuarine species. From the initial surveys we conducted, it appears that the microphyte community structure in the Tulare Lake Drainage District basin is strongly regulated by salinity. We found that at least three of the microphytes from evaporation basins actively biotransform Se oxyanions into organic Se metabolites, including the volatile DMSe, DMDSe, and DMSeS. The biochemical characterization supports the hypothesis that microphytes are important drivers of the Se biogeochemical cycling in these saline surface waters. We postulate that microphytes may play an analogous role in the oceans.

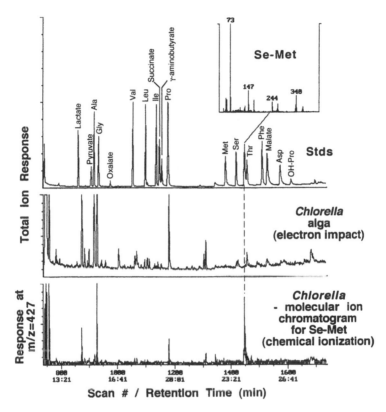

FIGURE 10   GC-MS analysis of silylated *Chlorella* extract and standards. The same micro-phyte biomass as in Figure 9 was extracted with 5% perchloric acid, silylated using MTBSTFA, and analyzed by GC-MS as described by Fan et al. (1997). The silylated derivatives of a mixed standard were also analyzed by GC-MS for comparison (top panel). The identity of Se-Met and other metabolites in the *Chlorella* extract was confirmed by GC retention time (dotted line for Se-Met) and molecular ion detection using chemical ionization (bottom panel). Noted that Se-Met was present only at trace (submicrogram-per-liter) levels in the *Chlorella* extract, and thus its corresponding peak in the electron impact ion chromatogram (middle panel) was not detectable. However, using the 40-fold more sensitive (for Se-Met) chemical ionization method, a peak with the molecular ion of silylated Se-Met (m/z = 427) was evident (bottom panel). (Reprinted with permission from Fan et al., 1997, American Chemical Society.)

Thus, volatilization of Se by microphytes may be an important contributor to the declining, instead of accumulative, trend of waterborne Se concentrations at evaporation basins such as at TLDD. It should be possible to enhance such naturally occurring Se removal process(es) and apply them to other Se-contaminated environments, but only if the mechanism or mechanisms involved are understood in detail. This is because removal of "selenium"—the element—does not equate with remediation. The real goal of Se remediation—that is, to minimize ecotoxic risk to wildlife—is likely to be intertwined with the specific forms (i.e., the biochemical pathways and their branchpoints). If so, true remediation will be difficult to achieve without a mechanistic understanding of the biochemical fate of Se across multiple trophic levels.

ACKNOWLEDGMENT

We thank Dr. Andrew Lane for his NMR expertise and valuable discussion, and Dr. Dean Martens for synthesizing the trimethylselenonium standard. We are grateful to Mr. Douglas Davis and his staff at Tulare Lake Drainage District for the water samples and general assistance in the field. This work was supported in part by the UC Salinity/Drainage program and U.S. EPA (R819658) Center for Ecological Health Research. The authors also acknowledge the MRC Biomedical NMR Centre at Mill Hill, U.K., for providing valuable NMR instrumentation.

REFERENCES

Analytical Methods Committee. 1979. Determination of small amounts of selenium in organic matter. *Analyst* 104:778–787.

Anderson, R. A., D. M. Jacobson, and J. P. Sexton. 1991. *In Catalog of Strains*. Provasoli–Guillard Center for Culture of Marine Phytoplankton, West Boothbay Harbor, ME.

Andreae, M. O. 1986. The ocean as a source of atmospheric sulfur compounds. *In* P. Buat-Ménard (ed.), *The Role of Air–Sea Exchange in Geochemical Cycling*, pp. 331–362. D. Reidel, Dordrecht.

Bates, T. S., R. J. Charlson, and R. H. Gammon. 1987. Evidence for the climate role of marine biogenic sulphur. *Nature* 329:319–321.

Cooke, T. D., and K. W. Bruland. 1987. Aquatic chemistry of selenium: Evidence of biomethylation. *Environ. Sci. Technol.* 21:1214–1219.

Cutter, G. A., and K. W. Bruland. 1984. The marine biogeochemistry of selenium: A re-evaluation. *Limnol. Oceanogr.* 29:1179–1192.

Duddeck, H. 1994. Selenium-77 nuclear magnetic resonance spectroscopy. *Prog. NMR Spectrosc.* 27:1–323.

Fairbrother, A., R. S. Bennett, L. A. Kapustka, E. J. Dorward-King, and W. J. Adams. 1996. Risk from mining activities to birds in southshore wetlands of Great Salt Lake, Utah. *In Abstracts of the 17th Annual Meeting of the Society of Environmental*

Toxicology and Chemistry, 1996. p. 97. Society of Environmental Toxicology and Chemistry, Pensacola, FL.

Fan, T. W.-M. 1996. Metablite profiling by one and two-dimensional NMR analysis of complex mixtures. *Prog. Nuclear Magn. Resonance Spectrosc.* 28:161–219.

Fan, T. W.-M., A. N. Lane, and R. M. Higashi. 1997. Selenium biotransformations by a euryhaline microalga isolated from a saline evaporation pond. *Environ. Sci. Technol.* 31:569–576.

Gerhardt, M. B., F. B. Green, R. D. Newman, T. J. Lundquist, R. B. Tresan, and W. J. Oswald. 1991. Removal of selenium using a novel algal–bacterial process. *Res. J. Water Pollut. Control Fed.* 63:799–805.

Heinz, G. H., D. J. Hoffman, and L. J. LeCaptain. 1996. Toxicity of seleno-L-methionine, seleno-DL-methionine, high selenium wheat, and selenized yeast to mallard ducklings. *Arch. Environ. Contam. Toxicol.* 30:93–99.

Hirschberg, J., and D. Chamovitz. 1994. Carotenoids in cyanobacteria. *In* D. A. Bryant (ed.), *The Molecular Biology of Cyanobacteria*, pp. 559–579. Kluwer Academic Publishers, Dordrecht.

Maier, K. J., and A. W. Knight. 1994. Ecotoxicology of selenium in freshwater systems. *Rev. Environ. Contam. Toxicol.* 134:31.

Mudder, T. I. 1997. Selenium treatment technologies. *In Abstracts of a Workshop on Understanding Selenium in the Aquatic Environment.* Kennecott Utah Copper, Salt Lake City, UT.

Reash, R., T. Lohner, K. Wood, and R. Leveille. 1996. Selenium in fish inhabiting a fly ash receiving stream: Implications for national water quality criteria. *In Abstracts of the 17th Annual Meeting of the Society of Environmental Toxicology and Chemistry*, p. 12. Society of Environmental Toxicology and Chemistry, Pensacola, FL.

Skorupa, J. P., and H. M. Ohlendorf. 1991. Contaminants in drainage water and avian risk thresholds. *In* A. Dinar and D. Zilberman (eds.), *The Economy and Management of Water and Drainage in Agriculture*, p. 345. Kluwer Academic Publishers, Norwell, MA.

Tanji, K. K. 1989. Chemistry of toxic elements (As, B, Mo, Se) accumulating in agricultural evaporation ponds. *In* J. B. Summers (ed.), *Toxic Substances in Agricultural Water Supply and Drainage. An International Environmental Perspective*, pp. 109–121. Abstracts of the Second Pan-American Regional Conference of the International Commission on Irrigation and Drainage, Ottawa, Canada. U.S. Committee on Irrigation and Drainage, Denver.

Terry, N., C. Carlson, T. K. Raab, and A. M. Zayed. 1992. Rates of selenium volatilization among crop species. *J. Environ. Qual.* 21:341–344.

Thompson-Eagle, E. T., and W. T. Frankenberger, Jr. 1991. Selenium biomethylation in an alkaline, saline environment. *Water Res.* 25:231–240.

White, R. H. 1982. Analysis of dimethyl sulfonium compounds in marine algae. *J. Mar. Res.* 40:529–536.

Wright, S. W., S. W. Jeffrey, R. F. C. Mantoura, C. A. Llewellyn, T. Bjørnland, D. Repeta, and N. Welschmeyer. 1991. Improved HPLC method for the analysis of chlorophylls and carotenoids from marine phytoplankton. *Mar. Ecol. Prog. Ser.* 77:183–196.

28

Microbial Deselenification of Agricultural Drainage Water in Flash Evaporation Treatment Systems

WILLIAM T. FRANKENBERGER, JR., and GEORGE P. HANNA, JR.†
University of California at Riverside, Riverside, California

One of the major concerns in the San Joaquin Valley, California, is the hazardous levels of selenium (Se) in evaporation ponds comprised of agricultural drainage water. Seleniferous salts from drainage waters accumulate in the sediments of these evaporation ponds. Drainage water carries primarily selenate (SeO_4^{2-}) and selenite (SeO_3^{2-}) as oxidized species of Se. Their residence time varies in evaporation ponds, but the fate is most likely controlled by various microbial transformations including reduction with subsequent precipitation of elemental selenium (Se^0) in the sediment and methylation/volatilization resulting in a volatile gaseous Se^{2-} species.

The U.S. Department of the Interior, Bureau of Reclamation, is considering microbial volatilization of Se as a method of detoxification for sites such as Kesterson Reservoir, which are contaminated by seleniferous salts (Frankenberger and Karlson, 1988). The alkylselenides of microbial volatilization have high vapor pressures (Karlson et al., 1994) and thus are readily diluted and dispersed into the atmosphere. The transformation of nonvolatile Se species into volatile products is an important link in the geochemical cycle of this element (Atkinson et al., 1990; Rael et al., 1996). The primary volatile Se compound that evolves from soils and water is dimethyl selenide (DMSe) (Barkes and Fleming, 1974; Doran and Alexander, 1977; Karlson and Frankenberger, 1988, 1989; Thompson-Eagle

---

† Deceased.

and Frankenberger, 1990), although other Se compounds such as dimethyl diselenide (DMDSe) and dimethylselenenyl sulfide (CH_3SeSCH_3) also may be produced (Shrift, 1973; Chau et al., 1976; Chasteen, 1993). The toxicity of DMSe is approximately three orders of magnitude less than that of inorganic Se (McConnell and Portman, 1952; Wilber, 1980).

Only recently has a thorough investigation been made to determine factors affecting methyl selenide production from soil (Karlson and Frankenberger, 1989; Frankenberger and Karlson, 1989, 1994, 1995). The primary factors include carbon sources, activators, temperature, moisture, and aeration. Microorganisms active in volatilization of Se include fungi and bacteria (Challenger, 1945; Abu-Eirreish et al., 1968; Barkes and Fleming, 1974; Karlson and Frankenberger, 1988). Both inorganic and organic Se compounds can be methylated/volatilized into gaseous Se. For supplementary information, the reader may consult any of several additional review articles (Doran, 1982; Thompson-Eagle and Frankenberger, 1992; Karlson and Frankenberger, 1993; Frankenberger and Losi, 1995).

The objective of this project was to develop a flash evaporation treatment cell treated with drainage water from an adjacent agricultural field. The drainage water is applied to the treatment cell by means of sprinkler irrigation. The Se-laden drainage water is allowed to fully evaporate as the soil Se concentration increases. Methylating microorganisms were acclimated to the seleniferous environment and maintained active by keeping the soil moist. A readily available carbon source provided methyl donors and an energy source for growth and proliferation of the methylating microorganisms. The Se inventory in the soil as well as the volatilization rates were continuously monitored to determine whether Se was accumulating. This study reveals the feasibility of implementing a treatment bed both to remove Se and to minimize the possibility of Se entering the aquatic food chain.

I. MATERIALS AND METHODS

A. Field Design

1. Adams Avenue Agricultural Drainage Center

A 2.8 ha treatment cell (91.2 m wide $\times$ 304 m long) was installed at the Adams Avenue Agricultural Drainage Research Center near Tranquility, California, on April 1, 1995, as a flash evaporation site to optimize microbial volatilization of Se (Fig. 1). This site was selected in a field with the native soil exposed. Cattle manure obtained from Harris Ranch near Coalinga, California, was applied at a rate of 3.1 Mg/ha. The soil was then tilled to a 6-inch depth to maximize soil porosity. Tillage was necessary to promote aeration within the soil and to support the aerobic methylating–volatilizing organisms. Also, tillage allows DMSe to escape readily into the atmosphere.

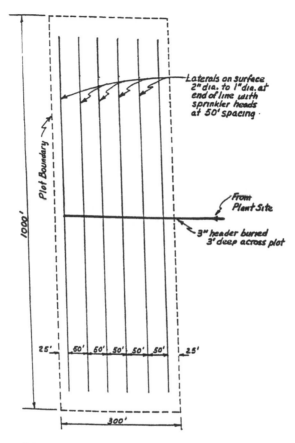

FIGURE I Layout of the sprinkler system on the treatment plot at the Adams Avenue Agricultural Drainage Research Center.

Agricultural drainage feedwater to the site was pumped through a 15 cm pressure main from the Westlands Water District drainage collector no. 136.0 sump located at the intersection of Lincoln Avenue and the San Luis Drain. Feedwater to the Se volatilization test plot was metered through a 3-inch header that discharged into six laterals. The laterals were spaced at 15 m intervals and fitted with sprinkler heads at 15 m spacings. Figure 1 shows the layout of the sprinkler system on the test plot. The sprinkler irrigation system was used to distribute the drainage water containing an average of 530 μg/L Se water to the fallow area to keep the microorganisms in an active state. During the summer months, water was applied daily for short intervals. Despite attempts to keep the soil moist but not saturated; some of the soluble Se fraction did move downward into the soil profile. The soil microbiota in this treatment bed had an extremely high tolerance for saline conditions, and therefore the saline drainage water had little effect in terms of osmotic tension.

2. Peck Ranch

A 4 ha treatment cell, adjacent to an evaporation pond at the Sumner Peck Ranch (Fresno County, CA) was selected as the other site for a flash evaporation cell and was optimized for microbial volatilization of Se. Cattle manure was obtained from the Harris Ranch near Coalinga and applied at a rate of 4.5 Mg/ha. The soil was then tilled to a 15 cm depth to maximize soil porosity. A sprinkler irrigation system was used to distribute tailwater containing approximately 150 μg/ L of Se to the fallow area to keep the microorganisms in an active state. During the summer months, water was applied daily for short intervals.

B. Soil Sampling

At the initiation and at the end of this project, several soil samples were taken from both cells at a depth of 60 cm to determine the concentration gradient of Se within the soil profile. The Se inventory also was determined to assess spatial variability of Se on site. These samples were collected from each cell as 10 composite samples from random sites throughout the fallow flash evaporation treatment cell. Soil samples were taken with a 2.5 cm diameter probe in a five-point pattern. All samples were kept refrigerated at 4°C during storage. Samples were extracted with the sample digestion procedure of EPA method 3050 and analyzed by atomic absorption spectrometry (AAS/hydride generation) (EPA method 270.3) for determination of Se and reported on a dry weight basis.

C. Monitoring Selenium Volatilization in the Field

Figure 2 illustrates the gas sampling system used in the field. Each line consisted of 0.64 cm air tubing inside 1.9 cm schedule 40 PVC pipe to protect the air tubing and electrical service line. Figure 3 shows the details of the gas sampler. An inverted aluminum box (56 cm × 56 cm × 10 cm) with a brass tubing connector at the top center was used as a flux chamber. Additional equipment required for this operation consisted of a vacuum pump (Fisher Scientific, Tustin, CA), 250 mL gas wash bottle (40–60 μm porosity, Fisher Scientific), six-place manifold, 0.6 cm i.d. vinyl tubing, and surgical tubing. Reagents used for the trapping solution were hydrogen peroxide (H_2O_2) (30%) and 0.05 N sodium hydroxide (NaOH). The purpose of the H_2O_2 was to oxidize the volatile Se into the SeO_4^{2-} species.

The flux chamber was placed in the center of each plot to be sampled and pushed into the soil approximately 2.5 cm. The vacuum pump was connected

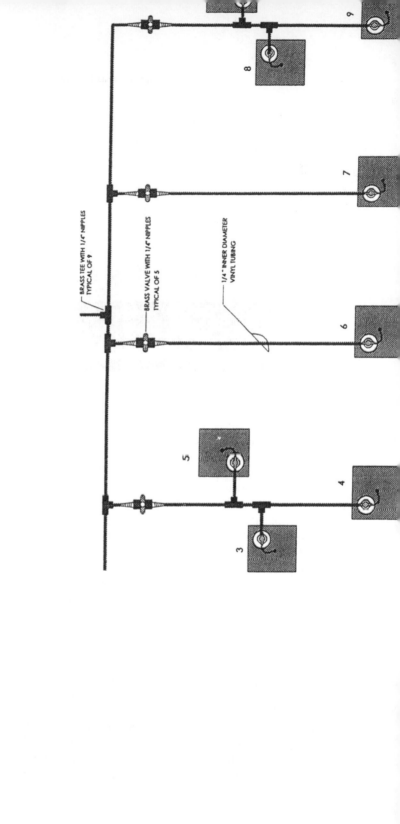

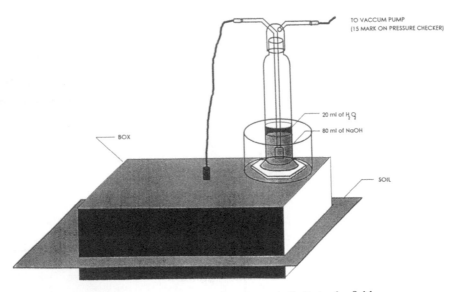

TO VACCUM PUMP
(15 MARK ON PRESSURE CHECKER)

20 ml of H₂O

80 ml of NaOH

BOX

SOIL

FIGURE 3   Inverted box used to trap volatile Se in the field.

to an electrical supply. The manifold was then connected to the pump with vinyl tubing. A trap was inserted between the manifold and pump to prevent the scrubbing solution from being drawn into the pump in case the gas wash bottle was tipped or blown over by the wind. From the manifold, 0 to 5.6 m sections of vinyl tubing were connected to the sampling boxes. The trapping solution consisted of 80 mL 0.05 N NaOH (premeasured) plus 20 mL of H_2O_2 (final concentration, 6%), which was kept cold in an ice chest until used. Because vinyl tubing binds to glass, hindering attachment and removal from the wash bottle, a small piece of surgical tubing was placed on the end of the vinyl tubing. When the trapping solution had been added and all the tubing was connected, the pump was started. At the port of each chamber, the flow rate of the manifold was adjusted with a flow gauge to 2 L/min, and this setting was checked repeatedly. The duration of the measurements was one hour, and the starting and finishing times were recorded. In most cases, measurements were made at midday. At the end of the sampling, the pump was turned off, the wash bottle was opened, and the alkali–peroxide solution containing the oxidized Se was poured directly into a 125 mL polyethylene bottle and placed into an ice chest packed with blue ice. The wash bottles were then rinsed twice with deionized water and filled with 80 mL of 0.05 N NaOH to be used for the next sampling.

The alkali–peroxide sample was boiled to drive off the residual H_2O_2, using a large hot plate with the capacity to hold a dozen 250 mL beakers. Each sample was boiled for 15 minutes, allowed to cool, brought up to volume (100 mL), and stored in 125 mL polyethylene bottles. Analyses for determination of Se were carried out with EPA Method 270.3 (AAS/hydride generation).

D. Quality Control

Quality control was maintained on both soil and Se gas analyses through the use of matrix spikes, matrix spike duplicates, and a National Institute of Standards and Technology standard reference material. Recovery of analytical spikes was monitored at two levels including both "low" (100 μg/L) and "high" level (500 μg/L) spikes. Average percentage recoveries for analytical spikes were well within the $\pm$20% range, and reference sample recoveries averaged 95.8%.

II. RESULTS

A. Adams Avenue Agricultural Drainage Center

The objective of this remedial technology was to test the feasibility of a land treatment bed to deselenify drainage water collected from irrigation agriculture. The chemical composition of the agricultural drainage water to the Adams Avenue Agricultural Drainage Research Center bed is shown in Table 1. The chemical components of the drainage water that are of particular relevance are the sodium content, with an average concentration of 1714 mg/L; chloride, 997 mg/L, sulfate, 3932 mg/L; boron, 12.6 mg/L; total dissolved solids, 8216 mg/L; soluble selenium, 513 μg/L; and total selenium, 534 μg/L.

Table 2 provides a record of the applied water to the flash evaporation treatment system at the Adams Avenue Agricultural Drainage Research Center. The total amount of water delivered was over 75 × 10^4 L. More water was applied in the summer of 1995 than in the fall of 1994 because of warmer temperatures.

The average Se concentration in the land treatment bed at the initiation of this project was 0.48 mg/kg. The flux of gaseous Se was monitored from September 1994 to November 1995 at 10 random subplots. In September 1994, volatilization rates varied from 0.23 to 0.95 μg Se/m^2/h (Table 3). In November, the temperatures began to cool, and volatilization rates were at lower levels, averaging 0.55 in November 1994 and 0.47 μg Se/m^2/h in December 1994. In September–October 1995, the emission flux of gaseous Se increased with increasing temperature, but began to decline in November as the temperatures decreased. The average microbial volatilization rate in September 1995 was 0.77, October 1995, 0.98, and November 1995, 0.54 mg Se/m^2/h.

Table 4 shows the soil selenium inventory for the 10 plots within the test cell. In monitoring the Se content in the soil, it was evident that there was very little change from April 1994 to July 1995. However, after this time, we observed an increase in the Se inventory within the topsoil increasing dramatically in

Chemical components (mg/L)

Sodium	Calcium	Magnesium	Potassium	Chloride	Sulfate	Boron	TSS	VSS	TDS	TOC	NO$_3$-N
1620	540	235	1.9	958	4150	12.0	28	12	7840	13	ND
1500	565	275	1.9	963	4150	11.4	32	26	8260	10	ND
1750	590	245	1.9	1020	4180	11.4	30	6	8180	12	ND
1750	555	258	1.1	983	3916	11.3	8	5	8038	9	ND
1750	565	260	1.9	981	3989	11.3	3	1	8255	11	ND
1750	555	260	1.7	1130	4000	11.6	36	0	8132	14	ND
1750	555	270	1.6	1038	3727	13.8	5	4	8055	9	ND
2000	563	240	1.1	1014	3826	11.5	25	11	8080	10	ND
1290	560	234	6.1	996	3802	12.5	18	17	8120	14	ND
1375	550	243	6.1	1005	3764	12.1	7	2	8090	12	ND
1750	280	124	6.3	967	3639	13.2	28	0	ND	13	64
1750	540	228	7.7	950	3796	13.2	28	0	8354	16	66
2000	530	232	7.4	948	3935	14.0	20	7	8595	18	65
2000	525	258	8.4	1008	4179	17.6	23	1	8810	29	66
1714	534	240	3.9	997	3932	12.6	19.9	6.8	8216	13.6	65.3

pended solids; VSS, volatile suspended solids; TDS, total dissolved solids; TOC, total organic carbon, ND, not determined.

October 1995. The Se content in October 1995 was 3.1-fold greater than that of April 1994 at the onset of this investigation. The Se level in the topsoil increased from 0.48 mg/kg in April 1994 to 1.48 mg/kg in October 1995.

Four profile samples were taken from sites 1.5 m outside the four corners of the test site in April 1994 and again in October 1995 to determine the distribution of Se with soil depth. Spatial variability of Se in each of the soil profiles was evident. Samples were taken from five depths (0–10, 10–20, 20–30, 30–45, and 45–60 cm). At the onset of this investigation, approximately 20.8% of the Se was concentrated in the topsoil (0–10 cm) of four profile samples (Table 5). At the end of this study it was apparent that a considerable amount (31.2%) of the added Se had accumulated in the surface soil (0–10 cm). There was little evidence showing that the soluble Se was leached into the subsurface.

B. Peck Ranch

The Peck Ranch was uniquely different from the Adams Avenue Agricultural Drainage Research Center in that we used tailwater that had considerably less Se, averaging 150 μg/L. The Se content in the Peck soil (2.78 mg/kg) was 5.8-fold higher than the Adams Avenue soil (0.48 mg/kg). Thus, greater volatilization rates would be expected from the Peck matrix. Monitoring of the volatilization rates revealed variability throughout the test plots, ranging from 0.48 to 6.61 μg Se/m²/h during September 1992 (Table 6). As the temperature decreased during the winter months, volatilization rates also decreased, particularly in January 1993. With increasing temperatures in the spring, Se volatilization rates averaged 15.8 μg Se/m²/h for April 1993. With continuous monitoring of the volatilization rates, it was obvious that Se was becoming limited in June 1993 and thereafter. The gaseous emission flux of Se was minimal in the summer of 1993, averaging 0.27 μg Se/m²/h.

The Se content in the Peck soil also varied throughout this experimental project, with the average ranging from 2.15 to 2.84 mg/kg (Table 7). There was no significant change in the Se content in the topsoil over the 1992–93 term of this study, indicating that Se was not accumulating. However, one must note that the Se load was approximately four times less than that of the Adams Avenue Project.

At the beginning of the Peck study, a considerable fraction of the Se was concentrated in the upper topsoil, (0–15 cm) (Table 8). At the end of this study, the Se content had decreased 58.6% in the topsoil after treatment, primarily as a result of dissipation by volatilization. The Se load in the entire profile of the cell decreased by an average of 40% after this study.

III. DISCUSSION

The primary objective of this research was to evaluate microbial volatilization as a remedial technology to remove soluble Se in the drainage water as a gaseous

[Water Application to the Flash Evaporation Treatment System at the Adams Avenue Agricultural Drainage

Flow meter reading (L)		Application time			Water applied (
Start	End	Total delivered (L)	(min/row/day)	Total time (min/day)	
0	18.7×10^3	18.7×10^3	10	60	312
18.7×10^3	33.3×10^3	14.6×10^3	10	60	243
33.3×10^3	48.8×10^4	15.5×10^3	10	60	258
		0 (rain)	0		
		0			
		0 (rain)	0		
		0 (rain)	0		
		0			
49.0×10^3	64.4×10^3	15.4×10^3	10	60	256
54.4×10^3	79.3×10^3	15.0×10^3	10	60	254
79.3×10^3	10.2×10^4	22.9×10^3	15	90	255
		0			
		0			

10.2×10^4	13.2×10^4	0	20	120	248
13.2×10^4	14.8×10^4	29.7×10^3	10	60	273
		16.4×10^3			
14.8×10^4	17.8×10^4	0	20	120	249
17.8×10^4	19.4×10^4	29.9×10^3	10	60	257
19.4×10^4	20.9×10^4	15.4×10^3	10	60	252
20.9×10^4	22.4×10^4	15.1×10^3	20	60	253
22.4×10^4	23.9×10^4	15.1×10^3	10	60	253
23.9×10^4	25.4×10^4	15.2×10^3	10	60	246
		14.8×10^3			
26.7×10^4	31.7×10^4	50.3×10^3	30	180	280
31.7×10^4	36.0×10^4	43.1×10^3	30	180	240
36.0×10^4	40.6×10^4	45.7×10^3	30	180	254
40.6×10^4	44.9×10^4	43.2×10^3	30	180	240
44.9×10^4	49.2×10^4	42.6×10^3	30	180	237
49.2×10^4	55.7×10^4	64.8×10^3	45	270	240
55.7×10^4	62.2×10^4	65.2×10^3	45	270	242
62.2×10^4	70.9×10^4	87.3×10^3	60	360	243
70.9×10^4	77.5×10^4	65.7×10^3	45	270	243

Volatilization Rates at the Adams Avenue Agricultural Drainage Research Center

	Dimethyl selenide evolution (μg Se/m^2/h)					
September 4, 1994	November 26, 1994	December 26, 1994	September 25, 1995	October 16, 1995	Novemb... 199...
0.39	3.75	1.27	1.12	1.37	1.5
0.23	0.27	0.20	0.91	1.22	0.4
0.34	0.56	1.25	1.36	1.91	0.7
0.28	<0.05	0.15	0.84	0.95	0.3
0.41	0.06	<0.05	1.01	0.93	0.2
0.26	0.09	0.09	1.06	1.47	0.3
0.52	0.18	0.31	0.34	0.39	0.3
0.95	<0.05	0.27	0.38	0.67	0.4
0.38	0.60	1.02	0.45	0.55	0.9
0.51	<0.05	0.15	0.26	0.34	0.1
0.43	0.55	0.47	0.77	0.98	0.5

pril	1994				1995				O
	September	October	November	December	June	July	August	September	
.45	0.46	0.63	0.60	0.63	0.58	0.73	0.74	0.82	
.50	0.34	0.52	0.44	0.51	0.50	0.52	0.89	0.92	
.47	0.31	0.44	0.37	0.53	0.34	0.41	0.70	0.75	
.51	0.38	0.61	0.58	0.47	0.48	0.50	0.88	1.12	
.60	0.39	0.43	0.37	0.43	0.32	0.46	0.56	0.87	
.39	0.35	0.50	0.44	0.56	0.32	0.56	0.52	0.79	
.56	0.58	0.74	0.66	0.72	0.61	0.68	0.84	1.04	
.51	0.35	0.49	0.38	0.48	0.41	0.42	0.59	0.89	
.39	0.37	0.51	0.62	0.64	0.42	0.33	0.68	0.81	
.40	0.28	0.45	0.36	0.46	0.35	0.44	0.53	0.69	
.48	0.38	0.53	0.48	0.54	0.43	0.51	0.69	0.87	

Distribution in Soil Profiles at the Beginning and End of This Study at the Adams Avenue Agricultural Drainage

Selenium distribution (mg/kg) in soil profiles

Beginning of project (outside of cell)				End of project (plots in cell)			
NW	NE	SE	SW	1	3	5	6
0.60	0.38	0.71	0.42	0.59	0.85	1.65	1.43
0.52	0.39	0.64	0.31	0.46	0.55	0.51	0.81
0.50	0.41	0.66	0.33	0.41	0.54	0.44	0.50
0.41	0.53	0.70	0.45	0.51	0.41	0.48	0.51
0.41	0.44	0.72	0.56	0.53	0.45	0.73	0.56

Volatilization Rates at the Peck Ranch

	October 4, 1992	October 16, 1992	November 14, 1992	November 27, 1992	December 13, 1992	December 27, 1992	January 30, 1993	April 17, 1993	May 28, 1993	June 26, 1993	July 4, 1993	July 17, 1993	3
	1.91	4.32	8.57	5.93	1.35	0.42	0.51	2.09	0.56	0.23	0.27	0.06	
	1.02	3.67	8.26	23.74	4.08	3.13	<0.05	78.80	10.60	0.12	0.21	<0.05	
	0.35	2.97	5.98	1.74	2.31	1.56	0.20	10.54	3.16	<0.05	0.06	<0.05	
	0.31	3.27	5.73	3.41	3.72	2.78	<0.05	2.38	2.50	0.37	0.13	2.02	
	0.44	2.64	9.72	2.99	1.93	1.49	<0.05	8.69	17.80	<0.05	<0.05	0.17	
	3.78	2.32	35.24	19.15	0.49	0.19	1.65	55.60	19.93	<0.05	<0.05	0.77	
	0.84	1.96	0.62	0.44	0.34	0.16	0.18	<0.05	0.67	0.42	0.22	0.64	
	0.58	2.41	0.62	0.46	0.49	0.18	0.37	<0.05	0.29	0.37	<0.05	1.32	
	0.43	1.42	1.22	2.84	0.42	0.30	<0.05	<0.05	0.09	<0.05	<0.05	1.40	
	0.08	0.82	1.52	2.84	0.56	<0.05	<0.05	<0.05	<0.05	<0.05	0.08	<0.05	
	0.97	2.58	7.65	6.35	1.57	1.02	0.30	15.81	5.56	0.15	0.10	0.64	

Dimethyl selenide evolution (μg Se/m^2/h)

enium (mg/kg) in Soil at the Peck Ranch

	1992					1993			
October	November	December	January	February	March	April	May	June
2.88	2.48	2.69	2.30	2.53	2.88	3.79	2.53	2.68
3.66	2.56	3.64	2.19	2.54	3.63	3.75	2.66	3.02
3.31	2.47	3.86	2.26	2.28	3.02	2.64	1.99	2.66
3.01	2.56	3.63	2.15	2.29	2.29	3.16	2.23	2.56
3.26	3.44	2.65	2.12	2.47	2.68	2.87	1.91	2.17
2.47	2.53	2.13	2.04	2.50	2.25	2.98	2.08	2.13
2.42	2.05	2.03	1.94	2.08	2.05	2.19	1.67	2.32
1.96	2.08	2.29	2.05	1.93	2.29	2.29	2.08	2.66
2.26	2.11	2.56	1.86	2.37	2.36	2.23	1.57	2.57
2.62	2.40	2.36	2.54	2.26	2.56	2.45	1.88	2.77
2.78	2.47	2.78	2.15	2.33	2.60	2.84	2.06	2.55

Distribution in Soil Profiles at the Beginning and End of This Study at the Peck Ranch

| Selenium distribution (mg/kg) in soil profiles | | | | | | | | |
| Beginning of project (plot in cell) | | | | | End of project (plot in cell) | | | |
1	2	3	4	5	1	2	3	4
6.41	6.36	5.51	6.12	7.12	2.32	1.89	2.20	2.33
3.35	4.19	3.89	3.72	2.52	2.12	1.98	2.25	1.99
3.89	3.30	3.14	3.65	2.70	2.39	2.17	2.16	1.84
2.15	2.25	2.26	2.01	2.05	2.99	1.86	2.94	1.89

species. The principle behind this technology was to maintain the indigenous methylating microorganisms in an active state to convert SeO_4^{2-} and SeO_3^{2-} into DMSe. The land treatment beds were supplied with sufficient carbon to provide a methyl donor under moist and aerated conditions. The most readily available Se species for microbial uptake in the drainage water was SeO_4^{2-} with less SeO_3^{2-}. Selenate is soluble in soils at most pH values and weakly adsorbed by soil particles (Arlich and Hossner, 1987). Most SeO_3^{2-} salts are less soluble than the corresponding SeO_4^{2-} salts (Elrashidi et al., 1987). Selenite is rapidly adsorbed at all pH values to soil particles such as clays (Bar-Yosef and Meek, 1987) and in particular, Fe oxyhydroxides (Balistrieri and Chao, 1987). To methylate Se, microorganisms have to convert SeO_4^{2-} into SeO_3^{2-} and subsequently into the volatile Se^{2-} fraction. However, seleno-oxyanions can also be reduced into Se^0, which is not readily available for methylation, probably as a result of its low solubility (Reamer and Zoller, 1980). Reduction into Se^0 can occur under oxidized conditions, influencing the effectiveness of this dissipation technology.

The Se volatilization rates from the Peck soil were higher than those from the Adams Avenue Agriculture Drainage Research Center soil. This result is most likely due to the higher residual Se content in the Peck soil. Although greater Se concentrations within the drainage water were applied at the Adams Avenue site compared to the Peck Ranch, microorganisms in the soil matrix were utilizing the residual Se present in the Peck soil. Studies have shown that Se volatilization is a first-order reaction, highly dependent on Se concentration. As the concentration of Se decreases, the volatilization reaction also tends to decrease. However, the Se volatilization capacity of the soil is highly dependent on the water-soluble Se fraction that directly governs this process (Karlson and Frankenberger, 1988). Zieve and Peterson (1981) were able to correlate a decrease in water-soluble Se of soil as volatilization increased. Calderone et al. (1990) found an interaction ($r = 0.81$) between the decrease in the cumulative soluble Se collected upon leaching in soil columns and an increase in the cumulative volatilization upon amendments with organic materials.

Cattle manure was used as an organic carbon (C) source to stimulate the population of methylating microorganisms within the treatment bed. Previous studies had shown that with the Peck sediments, cattle manure stimulated microbial volatilization of Se (Frankenberger and Karlson, 1995). In a laboratory study, as much as 29% of the total Se was volatilized from the Peck sediment after 273 days of incubation upon treatment of cattle manure compared to 14% without C addition (Karlson and Frankenberger, 1990). The Peck soil is low in organic carbon and nitrogen. The application of a rich organic C source provides carbon and energy for microorganisms to proliferate and remain active. The rate of methylation and subsequent volatilization can be increased up to three-fold upon incorporation of C amendments (Frankenberger and Karlson, 1988, 1989). One must consider that microorganisms are in a continuous state of starvation. When

C is provided, this stimulates microbial growth with the uptake of nutrients and elements in their surrounding environment. With the uptake of Se, specific organisms can methylate Se into a volatile species that dissipates into the atmosphere. Using radioactive ^{75}Se in the laboratory, we have demonstrated that as much as 50% (50 mg/kg) of the Se inventory in spiked seleniferous soils was removed in 4 months upon added C amendments (Karlson and Frankenberger, 1989). In another study, as much as 30% of the Se inventory (30 mg/kg) was volatilized in 37 days upon the addition of pectin to soil (Karlson and Frankenberger, 1988).

The addition of organic carbon provides a methyl donor for the transmethylation reaction. There are many one-carbon-unit degradation reactions in soil which may serve as methyl donors in transmethylation reactions involving metals and metalloids. With the addition of cattle manure, degradation reactions yielding one-carbon units will be promoted. Thus, the addition of supplemental organic carbon can dramatically stimulate microbial volatilization of Se.

The treatment bed was rototilled weekly to promote aeration of the plot. This is necessary to keep microorganisms maintained under aerobic respiration and in active state for the methylation reaction. In earlier studies, we found that fungi comprise a predominant group of methylating organisms in soils and sediments active in volatilization of Se (Karlson and Frankenberger, 1988). Soil fungi, which are heterotrophic and utilize organic materials for carbon and energy, constitute a very important fraction of biomass. The genera isolated in our studies that are prolific producers of DMSe include *Acremonium falciforme, Penicillium citrinum,* and *Ulocladium tuberculatum* (Karlson and Frankenberger, 1988). Since the fungi are strict aerobes, it is important to keep the porosity relatively high to maintain aerobic respiration in the topsoil. Each of the land treatment beds was tilled within the upper few centimeters, since this is the active zone for methylation of Se. The soil should have enough porosity to allow the gas to disperse into the atmosphere. Tillage also prevents a crust from forming on the surface as a result of irrigation.

Moisture affects this biotransformation, since microorganisms active in methylation require water to remain in a viable state. Excess water can promote leaching of the water-soluble Se fraction into the lower profile in which the readily available Se for biomethylation becomes unaccessable. Frankenberger and Karlson (1989) reported that volatilization of Se is much less efficient in soils that become either desiccated or water-saturated. Maximum Se volatilization from seleniferous dewatered sediments occurs at about 50 to 70% of the water-holding capacity (field moist soil) (Frankenberger and Karlson, 1989, 1995), while in a heavy clay soil between 18 and 25% moisture was optimum (Abu-Eirreish et al., 1968). In a loam soil, Zieve and Peterson (1981) found that 28% was optimum for volatilization, while 16 and 40% moisture gave rise to considerable less volatile Se. Fluctuations in the soil water content may stimulate Se volatilization, since sequential drying and rewetting of soil promotes the release of volatile Se (Hamdy and Gissel-Nielsen, 1976). Decomposition of soil organic matter is directly related to repeated drying and rewetting cycles, hence nutrients may become more available for soil microflora and increase their metabolic activity under these conditions (Sørensen, 1974).

Greater emission rates of volatile Se were recorded in the hot summer versus the cooler winter. Selenium volatilization is temperature dependent. The maximum release of DMSe from lake sediments occurred at 23°C (Chau et al., 1976), from California pond water at 35°C (Thompson-Eagle and Frankenberger, 1990), from British loamy soil at 20°C (Zieve and Peterson, 1981), and from a California sandy textured soil at 35°C (Frankenberger and Karlson, 1989). Soil temperatures in the field seasonably varies between 4 and 50°C in the Central Valley of California (Frankenberger and Karlson, 1988). During the winter months, Se emission is relatively low because of less production and low vapor pressure, but it increases during the spring and summer months with warmer temperatures. Karlson et al. (1994) used an isotenoscope method to determine the vapor pressure of DMSe. They found that the vapor pressure of DMSe at 25°C was 32 kPa and enthalpy of vaporization was 31.9 kJ/mol. The Henry's law constant for DMSe was calculated at 143 kPa·kg/mol.

A linear regression analysis was performed on the volatile alkylselenide gas being released from the Peck Ranch upon treatment of cattle manure versus soil temperature (Frankenberger and Karlson, 1995). A significant correlation between soil temperature and volatile Se released ($r = 0.75$, significant at the 95% level) was found. Between 8 and 18°C there was a linear increase in Se volatilization, with temperature and volatilization rates approximately doubled. Also Se produced from the seleniferous soil at the Peck Ranch was related to the diurnal variation of temperature. The diurnal peak of volatile Se emission occurred during midday, an event that correlates with soil temperature (Frankenberger and Karlson, 1988).

The sampling technology used relies on gas exchange with a net release of DMSe from soil pores to the atmosphere. Two important factors affecting volatilization of Se include temperature and wind velocity. The thermal differentiation permits gas exchange of DMSe between the atmosphere and soil air at the immediate soil surface. The vapor pressure of DMSe increases with temperature. Raising the temperature from 10°C to 25°C doubles the vapor pressure of DMSe while raising it from 25°C to 40°C doubles it again (Karlson et al., 1994). At temperatures above the optimum for microbial activity, the increased vapor pressure may more than compensate for reduced microbial production, thus explaining high emissions rates at the extreme summer temperatures. The pressure and suction effects of high wind result in renewal of soil air, particularly with barren soils. The effect of air turbulence on the transfer of vapor in soils suggests that mass air flow should be high. A wind speed of 25 to 50 mph can penetrate several centimeters into soils (Farrell et al., 1966). Even without mass flow, fluctuation in air pressure at the soil surface results in considerable mixing, affecting gas fluxes more markedly than the effects of diffusion.

In summary, this technology may have applications for irrigation drainage water at relatively low levels of Se (< 150 μg/L). This field study has shown that with elevated levels (> 500 μl/L) of seleniferous drainage water, the Se inventory will accumulate over time in the topsoil, even though the Se inventory in the water itself is mostly in the water-soluble fraction. Microorganisms active in reduction of Se not only produce DMSe, but also convert some of the soluble Se into elemental Se as a detoxification process. Elemental Se is known to be expelled from the microbial cells as a process of elimination (see Losi and Frankenberger, Chapter 26, this volume). Once the Se species has been transferred to Se^0, it is very difficult to methylate because of its low solubility. At the Adams Avenue Agricultural Drainage Center test plot, it was also evident that some of the water-soluble Se had leached further down into the soil profile. Obviously when this occurs, Se is not readily available for microbial volatilization. This study shows that under two different sets of field conditions, flash evaporation treatment systems may be applicable to sites receiving relatively low concentrations of Se in the drainage water.

ACKNOWLEDGMENTS

We thank the California State Water Resources Control Board for support of this project.

George P. Hanna, Jr.
1918–1997

George P. Hanna, Jr., 79, Professor Emeritus at California State University, Fresno, died April 5, 1997 during the production stages of this book. Dr. Hanna was the Director of Engineering Research Institute at CSUF and was active in research at Kesterson Reservoir. George was a professor at Ohio State University and the Director of the Water Resources Center from 1959–69. He was Chairman of the Department of Civil Engineering at the University of Nebraska from 1969–71 and then Dean of the College of Engineering and Technology from 1971–79. He then retired and was a Professor in the Department of Civil Engineering from 1979–89. From 1986 to 1997, George was active in the San Joaquin Valley Agricultural Drainage Program as a member of the Technical Advisory Committee and Chair of the Treatment and Disposal Subcommittee. Dr. Hanna served as Past President of the American Academy of Environmental Engineers, Past President of the California Society of Professional Engineers, and is a Fellow of the American Society of Civil Engineers. This chapter is dedicated to George P. Hanna, Jr. who has provided an important contribution to the understanding of selenium geochemistry.

REFERENCES

Abu-Erreish, G. M., E. I. Whitehead, and O. E. Olson. 1968. Evolution of volatile selenium from soils. *Soil Sci.* 106:415–420.

Arlich, J. R., and L. R. Hossner. 1987. Selenate and selenite mobility in overburden by saturated flow. *J. Environ. Qual.* 16:95–98.

Atkinson, R., S. M. Aschmann, D. Hasegawa, E. T. Thompson-Eagle, and W. T. Frankenberger, Jr. 1990. Kinetics of the atmospherically important reactions of dimethyl selenide. *Environ. Sci. Technol.* 24:1326–1332.

Balistrieri, L. S., and T. T. Chao. 1987. Selenium adsorption by goethite. *Soil Sci. Soc. Am. J.* 51:1145–1151.

Barkes, L., and R. W. Fleming. 1974. Production of dimethylselenide gas from inorganic selenium by eleven soil fungi. *Bull. Environ. Contam. Toxicol.* 12:308–311.

Bar-Yosef, B., and D. Meek. 1987. Selenium adsorption by kaolinite and montmorillonite. *Soil Sci.* 144:11–19.

Calderone, S. J., W. T. Frankenberger, Jr., D. R. Parker, and U. Karlson. 1990. Influence of temperature and organic amendments on the mobilization of selenium in sediments. *Soil Biol. Biochem.* 22:615–620.

Challenger, F. 1945. Biological methylation. *Chem. Rev.* 36:315–361.

Chasteen, T. G. 1993. Confusion between dimethyl selenenyl sulfide and dimethyl selenonone released by bacteria. *Appl. Organomet. Chem.* 7:335–342.

Chau, Y. K., P. T. S. Wong, B. A. Silverberg, P. L. Luxon, and G. A. Bengert. 1976. Methylation of selenium in the aquatic environment. *Science* 912:1130–1131.

Doran, J. W. 1982. Microorganisms and the biological cycling of selenium. *Adv. Microb. Ecol.* 6:1–32.

Doran, J. W., and M. Alexander. 1977. Microbial formation of volatile selenium compounds in soil. *Soil Sci. Soc. Am. J.* 41:70–73.

Elrashidi, M. A., D. C. Adriano, S. M. Workman, and W. L. Lindsay. 1987. Chemical equilibria of selenium in soils: A theoretical development. *Soil Sci.* 144:141–152.

Farrell, D. A., E. L. Greacen, and C. G. Curr. 1966. Vapor transfer in soil due to air turbulence. *Soil Sci.* 102:305–313.

Frankenberger, W. T., Jr., and U. Karlson. 1988. Dissipation of soil selenium by microbial volatilization at Kesterson Reservoir. December. Prepared for the U.S. Department of Interior, Bureau of Reclamation, Contract No. 7-FC-20-05240.

Frankenberger, W. T., Jr., and U. Karlson. 1989. Environmental factors affecting microbial production of dimethylselenide in a selenium-contaminated sediment. *Soil Sci. Soc. Am. J.* 53:1435–1442.

Frankenberger, W. T., Jr., and U. Karlson. 1994. Soil management factors affecting volatilization of selenium from dewatered sediments. *Geomicrobiol. J.* 12:265–278.

Frankenberger, W. T., Jr., and U. Karlson. 1995. Volatilization of selenium from a dewatered seleniferous sediment: A field study. *J. Ind. Microbiol.* 14:226–232.

Frankenberger, W. T., Jr., and M. E. Losi. 1995. Applications of bioremediation in the cleanup of heavy metals and metalloids. In H. D. Skipper and R. F. Turco (eds.), *Bioremediation: Science and Application*, pp. 173–210. Soil Science Society of America. Special Publication No. 43. SSSA, Madison, WI.

Hamdy, A. A., and G. Gissel-Nielsen. 1976. Volatilization of selenium from soils. *Z. Pflanzen. Boden.* 6:671–678.

Karlson, U., and W. T. Frankenberger, Jr. 1988. Effects of carbon and trace element addition on alkylselenide production by soil. *Soil Sci. Soc. Am. J.* 52:1640–1644.

Karlson, U., and W. T. Frankenberger, Jr. 1989. Accelerated rates of selenium volatilization from California soils. *Soil Sci. Soc. Am. J.* 53:749–753.

Karlson, U., and W. T. Frankenberger, Jr. 1990. Volatilization of selenium from agricultural evaporation pond sediments. *Sci. Total Environ.* 92:41–54.

Karlson, U., and W. T. Frankenberger, Jr. 1993. Biological alkylation of selenium and tellurium. In H. Sigel and A. Sigel (eds.), *Metal Ions in Biological Systems*, Vol. 29, pp. 185–227. Dekker, New York.

Karlson, U., W. T. Frankenberger, Jr., and W. F. Spencer. 1994. Physicochemical properties of dimethylselenide and dimethyldiselenide. *J. Chem. Eng. Data* 39:608–610.

McConnell, K. P., and O. W. Portman. 1952. Toxicity of dimethylselenide in the rat and mouse. *Proc. Soc. Exp. Biol. Med.* 79:230–231.

Rael, R. M., E. C. Tuazon, and W. T. Frankenberger, Jr. 1996. Gas-phase reactions of dimethyl selenide with ozone and the hydroxyl and nitrate radicals. *Atmos. Environ.* 30:1221–1232.

Reamer, D. C., and W. H. Zoller. 1980. Selenium biomethylation products from soil and sewage sludge. *Science* 208:500–502.

Shrift, A. 1973. Metabolism of selenium by plants and microorganisms. In D. L. Klayman and W. H. H. Gunther (eds.), *Organic Selenium Compounds: Their Chemistry and Biology*, pp. 763–814. Wiley, New York.

Sørensen, J. H. 1974. Rate of decomposition of organic matter in soil as influenced by repeated air drying–rewetting and repeated additions of organic material. *Soil Biol. Biochem.* 6:287–292.

Thompson-Eagle, E. T., and W. T. Frankenberger, Jr. 1990. Volatilization of selenium from agricultural evaporation pond water. *J. Environ. Qual.* 19:125–131.

Thompson-Eagle, E. T., and W. T. Frankenberger, Jr. 1992. Bioremediation of soils contaminated with selenium. *Adv. Soil Sci.* 17:261–310.

Wilber, C. G. 1980. Toxicology of selenium: A review. *Clin. Toxicol.* 17:171–230.

Zieve, R., and P. J. Peterson. 1981. Factors influencing the volatilization of selenium from soil. *Sci. Total Environ.* 19:277–284.

29

Volatile Chemical Species of Selenium

Thomas G. Chasteen
Sam Houston State University, Huntsville, Texas

I. SELENIUM VOLATILITY

Because the major oxyanions of selenium (Se) in the environment, selenate
(SeO_4^{2-}) and selenite (SeO_3^{2-})—and their associated acids in solution—have
negligible vapor pressure at normal biospheric temperatures, the volatile chemical
species of Se are the reduced and methylated forms. Five different volatile forms
of reduced Se have been detected in laboratory experiments, and three of these
have been determined in the environment. They are hydrogen selenide (H_2Se),
methaneselenol (CH_3SeH), dimethyl selenide (CH_3SeCH_3), dimethyl selenenyl
sulfide (CH_3SeSCH_3), and dimethyl diselenide ($CH_3SeSeCH_3$). The relatively high
vapor pressures of these organometalloidal compounds, in theory, make them
significant in the biogeochemical cycling of Se, where they can play a part in the
global processes of Se distribution from sediments to aqueous and vapor phases
and back; however, the rapid oxidation of the first two in this list and the lower
vapor pressure—at normal temperatures—of the last two probably leave dimethyl
selenide as the most significant contributor to environmental Se mobility. The
available physicochemical data for these five compounds are listed in Table 1,
ordered by boiling point. One more compound is included, dimethylselenone,
$CH_3SeO_2CH_3$. With a melting point of 153°C, this compound is not a significant
vapor phase Se-containing compound; however, its historical importance in the
discussion will become clear below. A few other mixed sulfur–selenium alkylated
species have recently been detected in laboratory experiments with Se-rich plants
such as garlic; however, these experiments were carried out at temperatures
substantially above room temperature, and except for high temperature environments like hot springs or ocean vents, the presence of these compounds in the
environment probably is unimportant.

TABLE I   Physicochemical Data for Volatile Selenium Compounds

Chemical species	Boiling point (°C)	Melting point (°C)	Vapor pressure at 25°C (kPa)
H_2Se	−41	−66	—
CH_3SeH	26[a]	—	—
CH_3SeCH_3	58	—	32.03[e]
CH_3SeSCH_3	132[b, c]	—	—
$CH_3SeSeCH_3$	156	60	0.38[e]
$CH_3SeO_2CH_3$	—	153[d]	—

[a]Kaufman (1961).
[b]Chasteen (1993).
[c]Potapov et al. (1992).
[d]Rebane (1974).
[e]Karlson et al. (1994).

This chapter is structured chronologically, and, although there are probably better ways to organize information of this sort, chronological development allows the reader to see why certain volatile Se compounds were determined very early and others not until more recently. To a large degree this two-phase research effort was due to the development of separation science. With early analytical work involving compound derivatization, collection, and identification by melting point, it was only the later development of gas chromatography that allowed for the identification of more organoselenium species. Subsections include the earliest report of each organoselenium compound. That said, the chronological order has been disrupted by a division between Se volatilization by animals (rats), microorganisms (bacteria and fungi), and plants. Finally, the particulate Se content of the atmosphere and gas phase reactions of volatile Se species are discussed.

II. REPORTS OF VOLATILE SELENIUM SPECIES RELEASED BY MICROBES AND RATS

A. Dimethyl Selenide

The scientific reports of volatile organoselenium (and organotellurium) began in the nineteenth century based to a large degree on the smell of laboratories working with Se-amended (i.e., Se-added) bacterial cultures or the breath of animals and humans dosed with metalloidal salts (Challenger, 1945, 1978, and references therein). The postulate that dimethyl selenide (DMSe), was released from microbial environments was confirmed in 1934 in the laboratory of Frederick Challenger at the University of Leeds: fungi grown on bread in the presence of Se salts produced DMSe, which could be detected via gas phase trapping and subsequent derivatization (Challenger and North, 1934). The seminal organometalloidal research in this chemistry group encompassed arsenic, selenium, tellurium, and even antimony (Challenger, 1945, 1951, 1978). Table 2 lists the volatile Se species that have been detected.

After World War II the ensuing research in this field included experiments that determined simply "volatile selenium" produced by beef liver, spleen, whole blood, and plasma extracts amended with selenite or selenate (Rosenfeld and Beath, 1948). Autoclaving destroyed these tissues' abilities to produce the unknown volatile Se compounds, and therefore these authors concluded that the agent of this process was "enzymatic in nature." In later work by McConnell and Portman (1952), DMSe was detected in the breath of rats injected with DMSe or sodium selenate. Exhaled DMSe was trapped in saturated mercury(II) chloride solutions and determined as the mercuric complex or, for smaller amounts, determined by radiolabeled ^{75}Se. Results of this nature were also later replicated with selenate and selenite injections in rats (Hirooka and Galambos, 1966; Hassoun et al., 1995) and by selenocystine added to the drinking water of rats (see Jiang et al., 1983b, and references therein). The methylation of added H_2Se and methaneselenol by S-adenosylmethionine (SAM) to form DMSe was also described by Bremer and Natori (1960). This last report among others helped to support the then relatively new hypothesis that SAM was a strong contender for a biologically active methyl donor to Se in some systems. This hypothesis had been put forth by Cantoni, who (apparently) coined the name for this ATP-activated form of methionine, and others (Cantoni, 1952, Challenger, 1955, 1959).

Ganther's work in the early and mid-1960s strengthened the evidence for the SAM methylation of Se oxyanions (Ganther, 1966). His comparison of the methylating ability of SAM and methionine in in vitro studies showed that SAM was more successful in this regard as measured by the production of DMSe, and furthermore that DMSe production was inhibited by arsenic at extremely low concentrations (10^{-6} to 10^{-8} M As). Work by Hsieh and Ganther (1977) 10 years later extended this research. Biomethylating agents that possibly play a part in this process with Se are methionine (Thompson-Eagle et al., 1989, Thompson-Eagle and Frankenberger, 1991), selenomethionine (Zayed and Terry, 1992), methylcobalamin (Thompson-Eagle et al., 1989, Gadd, 1993), and N^5-methyltetrahydrofolate (Wood, 1975; Doran and Alexander, 1977; Ridley et al., 1977; Voet and Voet, 1995). Transmethylation of Se from methylated metal or metalloids (not inherently in biological enzymes) has also been proposed (see, e.g., Craig, 1986).

Work has also established the production of mixed selenium/sulfur species—which probably play a part in the subsequent Se reduction and methylation—resulting from the interaction of glutathione/glutathione reductase with Se oxyanions: RSSeSR, RSSeSH, and RSSeH (Ganther and Corcoran, 1969; Hsieh and Ganther, 1975); although this type of biologically produced, mixed Se/S

Source	Organism	
Liver extracts	Rat	3
Bacterial headspace	*Citrobacter freundii* KS8	
	Pseudomonas aeruginosa VS7	
	Pseudomonas cepacia KS5	
	Pseudomonas sp. VW1	
Fungal headspace	*Scopulariopsis brevicaulis*	4
Tissue extracts	Mammals	5
Exhalation product	Rat	6–
S-adenosylmethionine/liver microsomes mixture	Rat	9
Liver extracts	Mouse	10
Fungal headspace	*Penicillium* strain	12
Aerated soil headspace	Mixed aerobes	13
Fungal headspace	*Penicillium* strain	14
	Fusarium strain	
	Cephalosporium sp.	
	Scopulariopsis sp.	
Bacterial headspace	*Corynebacterium* sp.	1
Lake water–sediment	Mixed anaerobes	15
Soil, sewage sludge headspace	Mixed microbes	16
Soil	Mixed microbes	18
Atmospheric air	Mixed microbes	19
Surface seawater	Mixed microbes	2
Fungal headspace	*Penicillium* sp.	2
Plant extract	*Allium ampeloprasum* (elephant garlic)	24
Human breath	*Allium* sp. ingestion	2
Human breath	*Homo sapiens*	20

	Source	Organism	
	Soil, sewage sludge headspace	Mixed microbes	16
	Atmospheric air	Mixed microbes	19
	Bacterial headspace	Aeromonas sp. VS6	3,
		Citrobacter freundii KS8	
		Pseudomonas aeruginosa VS7	
		Pseudomonas fluorescens K27	
		Pseudomonas sp. VW1	
	Plant extract	Allium ampeloprasum (elephant garlic)	24
	Human breath	Allium sp. ingestion	25
	Plant extract	Allium ampeloprasum (elephant garlic)	24
	Human breath	Allium sp. ingestion	25
	Lake water–sediment	Mixed anaerobes	15
	Bacterial headspace	Corynebacterium sp.	1
	Soil, sewage sludge headspace	Mixed microbes	16
	Atmospheric air	Mixed microbes	19
	Groundwater	Mixed microbes	29
	Soil	Mixed microbes	18
I_2	Plant extract	Allium ampeloprasum (elephant garlic)	24
	Human breath	Allium sp. ingestion	25
	Plant extract	Allium ampeloprasum (elephant garlic)	24
I_3	Plant extract	Allium ampeloprasum (elephant garlic)	24
	Human breath	Allium sp. ingestion	25
H	Plant extract	Allium ampeloprasum (elephant garlic)	24

species had been described much earlier (Painter, 1941). Finally, Goeger and Ganther (1994) determined in vitro that rat liver microsomes and monoxygenase from pig liver will oxidize DMSe to dimethylselenoxide, and furthermore this work suggested that in vivo Se redox cycling might occur.

Abu-Erreish et al. (1968) determined the evolution of volatile Se from soils collected from seleniferous regions of South Dakota. The trapping of purged volatile Se species in nitric acid and subsequent determination via spectrophotometric analysis showed that soil microbes were responsible for the volatilization. Radiolabeling (^{75}Se) experiments also were carried out. No specific Se species determination was made; however, Se volatilization varied with the amounts of organic matter these researchers added and the water-soluble Se content. Emissions decreased substantially from soils that were autoclaved. The continued, but highly attenuated, evolution of Se from autoclaved soils was attributed to molds that could be observed growing even after efforts at soil sterilization. Others have found the volatilization of Se from soils with mixed bacterial and fungal cultures to be dependent on moisture content and temperature (Zieve and Peterson, 1981).

B. Dimethyl Diselenide

Research in the early 1970s involved DMSe (and dimethyl telluride) production from fungal cultures amended with selenate or selenite (Fleming and Alexander, 1972). The fungus used by these investigators (a strain of *Penicillium*) was isolated from sewage, and organoselenium production was to a degree dependent on pH. DMSe was determined by gas chromatography with flame ionization detection (Evans and Johnson, 1966). Again using this type of detection, Barkes and Fleming (1974) reported DMSe production from fungal cultures amended with selenate and selenite. Eleven different fungi were examined, and although all produced DMSe when amended with selenite, only six fungi responded similarly when exposed to selenate. In Doran and Alexander's work (1977) with cell extracts from a strain of *Corynebacterium*, when either elemental Se, selenocystine, or methaneseleninate was added, DMSe was detected. Dimethyl diselenide (DMDSe) was also detected in some systems examined in this work. This is the first report of microbial DMDSe production extant in the literature, although plant production had been noted 10 years before (Evans et al., 1968; see below). Time course experiments later showed that the production of DMDSe depended on the growth of a *Pseudomonas* sp. when grown with selenocystine (Doran and Alexander, 1977).

C. Hydrogen Selenide

Volatile Se species that had been detected up to this point (only DMSe and DMDSe) were broadened in the 1970s. Diplock et al. (1973) reported the production of a volatile Se species, released from rat liver amended with selenite, that acted like hydrogen selenide (H_2Se). These authors suggested a mechanism for the reduction and methylation of selenite that could, depending on the conditions, yield DMSe, cationic trimethylselenonium, or, under acidic conditions, H_2Se. Using a similar gas trapping method, Doran and Alexander (1977) also reported H_2Se detection in microbial headspace experiments in which a black precipitate formed upon headspace gas trapping with silver nitrate; however, microbial H_2S production from these soil samples would presumably have given the same color precipitate in that system but was not discussed. Byard and others have also determined trimethylselenonium as a detoxification metabolite in rats (Byard, 1969; Palmer et al., 1970) and in lake water samples (Tanzer and Heumann, 1990). Additional work in H. E. Ganther's laboratory suggested that H_2Se might yield methaneselenol (CH_3SeH), DMSe, and trimethylselenonium in the presence of SAM in rat metabolism (Ganther, 1971; Hsieh and Ganther, 1975; Ip et al., 1991); however, actual determination of CH_3SeH was not performed.

The gas chromatographic methods of detecting alkyl selenides before the mid-1970s usually involved flame ionization or electron capture detection (Evans and Johnson, 1966; Vlasáková et al., 1972; Doran and Alexander, 1977); gas chromatography–mass spectrometry (GC-MS) was also becoming available for this purpose (Francis et al., 1974; Reamer, 1978), but this instrumentation would not become widely available until the 1980s. Radiolabeled Se also was used for the detection of alkyl selenides (Hirooka and Galambos, 1966; Diplock et al., 1973). Chau and coworkers described a gas chromatographic–atomic absorption method (GC-AAS) for the determination of DMSe and DMDSe in gas phase samples (Chau et al., 1975). This analytical method was later used to determine DMSe and DMDSe released in laboratory experiments with 12 Canadian lake sediments and waters incubated for a week with one of the following Se-containing nutrients added: sodium selenite, sodium selenate, selenocystine, selenourea, seleno-DL-methionine (Chau et al., 1976). The Se compounds were added individually to the lake water–sediment system to investigate the possibility of methylation (Y. K. Chau, personal communication, 1996). An unknown volatile compound was detected in this work also. This unknown compound also contained Se, since it and the other two compounds detected were produced only with the addition of Se compounds to the cultures and, most importantly, all were detected using selenium's atomic absorption line at 196 nm (Chau et al., 1975).

The first measurement of volatile organoselenium in the atmosphere was performed by Reamer (1978). Large volumes of ambient air were passed through a Teflon tube packed with 0.25 g of a commercial hydrophobic adsorbent, Spherocarb (carbon-based molecular sieve) after particles had been filtered from the gas stream using disk filters (Nuclepore). An average flow rate of 0.3 m^3/h of atmospheric air was used. Although seven sites were sampled (in and around the University of Maryland, College Park, and a remote site in Arizona) only a sampling site "next to an aerobic digestion tank" at the Piscataway, Maryland, sewage treatment plant yielded gas phase samples that tested above the detection limit for DMSe (> 0.02 ng DMSe/m³). The highest value reported from that site was 5.4 ng/m³ DMSe. No other organoselenide was detected.

Reamer's work also included an excellent laboratory simulation of environmental conditions in which microbes might reduce and/or methylate Se (Reamer, 1978; Reamer and Zoller, 1980). In these experiments, sewage sludge or soil was maintained in 500 mL flasks at 21°C, amended with either sodium selenite or elemental selenium (Se⁰) but no added nutrients, and then purged with either high purity nitrogen or air (flow rates 50–100 mL/min). Amendment concentrations ranged from 1 μg added Se compound per gram of sample to 1000 μg/g. Volatile organoselenium products were collected on a room temperature adsorbent (Spherocarb), extracted from that adsorbent using methanol, and then analyzed by coupling a gas chromatograph to a microwave discharge detector (Reamer, 1975, 1978; Reamer et al., 1978; Reamer and Zoller, 1980). This device determined Se emissions at 196 nm. Over a 30-day period, DMSe and DMDSe were detected in the purge gas. In experiments with sewage sludge amended with sodium selenite, the production rate of volatile Se varied with the amount of selenite added. And except for the lowest Se concentration experiments (1 μg/ g amendments), methylation was initially high (day 1 to approximately day 10) and then decreased gradually. The lowest amendment experiments produced relatively constant amounts of volatile Se compounds. The largest amount of added Se (as selenite) that was released as volatile compounds was 7.9% from a sewage sludge augmented with 1000 μg $SeO_3{}^{2-}$/g sample and swept with air for 30 days. Nitrogen-purged versions of this experiment produced less volatile Se, and soil and sewage sludge amended with Se⁰ released the smallest amounts of volatile Se. Reddish deposits of Se⁰ were apparent in all selenite-amended cultures by the end of the 30-day period, and at higher $SeO_3{}^{2-}$ concentrations, Se⁰ deposits were evident within a few days. Soil and sewage samples with no amended Se or samples that had been sterilized by autoclave and then amended with Se produced no detectable amounts of volatile Se compounds. Another chemical species identified by Reamer and Zoller (1980) in this research as dimethylselenone (see dimethyl selenenyl sulfide discussion in Sect. III), was also detected in these experiments as a volatile compound collected by the Spherocarb gas trap.

According to the literature, the second atmospheric (gas phase) determination of organoselenium was by Jiang and coworkers (1982, 1983a). Using GC coupled with graphite furnace atomic absorption detection at 196.1 nm, gas phase samples were collected on glass wool (augmented with 1 mm diameter glass beads) maintained at −130 °C at gas sampling flow rates of approximately 3 L/min. A later report of the same work, interestingly, reported the trapping temperature as −140°C (Jiang et al., 1989). After sampling periods as long as 4 hours, the trapped volatile components were desorbed, chromatographed, and analyzed by GC-AAS. Identification of DMSe and DMDSe was based on retention times of commercial standards. A third Se-containing compound was tentatively identified as dimethylselenone (see preceding discussion of Reamer and Zoller's work and Sect. III). Triplicate gas phase samples were taken at sites in north central Belgium in and around Antwerp and on one North Sea site approximately 100 km to the west. Particulate and aqueous samples were taken simultaneously at these sites. Though samples from many of the sites showed no detectable organoselenium compounds, (approximately < 0.2 ng compound/m^3), samples near a sewage treatment plant, a smelter, a lake at the University of Antwerp, and downwind from a coal-fired power plant showed at least two of the three Se-containing compounds determined. The North Sea sample, taken with wind blowing in from the sea, showed no detectable volatile Se compounds (Jiang et al., 1983a).

One of the only other studies to have detected DMSe in the field was conducted by German researchers (Tanzer and Heumann, 1990, 1992). These workers analyzed for volatile Se in Atlantic Ocean seawaters taken during a voyage from Cape Town, South Africa, to Bremerhaven, Germany, in 1989. Surface water samples (down to 30 cm) analyzed for organoselenium were stored at −10°C and analyzed at the end of the voyage. Volatile Se compounds were subjected to purge-and-trap GC analysis, followed by atomic emission detection (see below). Three samples taken near the equator showed DMSe concentrations of 1 to 5 ng Se/L seawater. The identity of this chemical species was confirmed by comparison to the retention time of a commercial DMSe standard spiked into degassed seawater samples that were then analyzed in exactly the same manner.

An extensive body of work in this field beginning in the mid-1980s came from the University of California at Riverside under the direction of William T. Frankenberger, Jr. This research encompassed analytical methods for organoselenium detection using radiolabeled Se (Karlson and Frankenberger, 1988a), GC-AAS (Thompson-Eagle et al., 1989), and GC with flame ionization detection (FID) or GC-MS (Frankenberger and Karlson, 1989). In addition to establishing the applicability of these methods to organoselenium emissions from soils and sediments, further work involved the investigation of the effects of carbon source, trace elements (Karlson and Frankenberger, 1988b), and environmental conditions (pH, temperature, moisture, etc.) on microbial production of volatile Se compounds (Frankenberger and Karlson, 1989, Karlson and Frankenberger, 1989). In the last two papers, these workers reported that Se-containing soil from California's Kesterson Reservoir (Weres et al., 1989; Presser and Piper, chapter 10, this volume) produced only DMSe (no DMDSe) and illustrated this volatilization process with time course plots at different pH, moisture contents, temperatures, Se sources (e.g., Se-containing amino acids), and carbon sources. For these soil systems, optimum Se volatilization occurred at pH 8.0, 70% water holding capacity (field capacity), and 35°C (the highest temperature studied), with an Se substrate of selenomethionine, a carbon source of glucose, and an additional protein amendment such as casein (Frankenberger and Karlson, 1989). The increased yield from casein amendments may have been due to the additional methionine content of this complex milk protein (Thompson-Eagle and Franken-berger, 1990, 1991). In closed greenhouse-based studies with these sediments over a period of 273 days, two sites' soils that were examined (with 61 and 9 mg/kg Se burden) showed Se volatilization of 44 and 29%, respectively (Karlson and Frankenberger, 1990). Sterilized and nonsterile soils from this region that were inoculated with fungal isolates from the soils and amended with selenate or selenite produced DMSe and small amounts of DMDSe. The identity of these species was established by GC-MS. Again, organoselenium production varied with the added carbon source and Se oxyanion oxidation state; however, the optimum emission levels for a given oxidation state (SeO_4^{2-} or SeO_3^{2-}) depended on whether a C source was added to the systems: in general, the addition of a carbon source increased the level of added Se content that would give an optimum methylated Se yield (Karlson and Frankenberger, 1989).

Though this fungal work was initially promising, subsequent studies showed that mixed cultures of native microbes from this region that were most successful at Se volatilization in amendment experiments were mostly bacteria (Thompson-Eagle and Frankenberger, 1991). These studies cumulatively showed the relative economic viability of using microbial bioremediation on pollution sites of this nature (Thompson-Eagle and Frankenberger, 1992). Recent work in this field has shown in a 22-month-long study that soils augmented with cattle manure and regularly moistened and tilled (dewatered sediments, containing a mean 11.4 mg/kg Se) removed 57.8% of the initial Se during that period (Frankenberger and Karlson, 1995).

It may also be of interest that *Penicillium citrinum* and *Acremonium falciforme*, two of the fungi examined in this work that reduced and methylated Se, have also shown the ability to reduce and methylate tellurite oxyanion (TeO_3^{2-}) and have produced dimethyl telluride (CH_3TeCH_3) and dimethyl ditelluride ($CH_3TeTeCH_3$) in work reported by a different laboratory. The identity of these chemical species in (aerobic) fungal headspace was confirmed by GC-MS and comparison to the retention time of a commercial standard (dimethyl telluride) or, for dimethyl ditelluride, by GC-MS alone (Chasteen, 1990; Chasteen et al., 1990).

D. Methaneselenol

A survey of 17 selenium-resistant microbes collected at Kesterson and Volta Reservoirs (Burton et al., 1987) and at the University Pond, University of Colorado, Boulder yielded the following results (Chasteen, 1990): 14 of the 17 cultures analyzed emitted methylated sulfur species (e.g., methanethiol, dimethyl sulfide, dimethyl disulfide); however, only 6 strains, when amended with selenate, produced volatile Se species including methaneselenol, DMSe, dimethyl selenenyl sulfide (CH$_3$SeSCH$_3$—see discussion below), and DMDSe. The detection of (very small amounts of) methaneselenol (in only four strains) in this work is the only known report on the biological production of this compound. Though Bremer and Natori suggested that H$_2$Se when acted upon by SAM and rat liver microsomes in vitro produced methaneselenol and DMSe, the tentative confirmation of CH$_3$SeH production in that work was based on determination of the melting point of the organoselenium–HgCl$_2$ derivative, and in the case of their mixture of CH$_3$SeCH$_3$ and CH$_3$SeH, "the separation of the mono- and dimethyl compound was not attempted owing to the insolubility and heat lability of the methyl selenol mercury salt" (Bremer and Natori, 1960).

III. CONFUSION BETWEEN DIMETHYLSELENONE AND DIMETHYL SELENENYL SULFIDE

The identity of the volatile compounds of Se that have been detected in either microbial culture or via gas phase sampling of the environment (in modern reports of the last 30 years) have been determined by comparison of chromatographic retention times with those of commercial standards (Chau et al., 1976; Reamer, 1978; Reamer and Zoller, 1980; Jiang et al., 1983a, 1989; Chasteen et al., 1990; Tanzer and Heumann, 1992) or via mass spectra generated by GC-MS (Reamer, 1978; Reamer and Zoller, 1980; Chasteen, 1990; Chasteen et al., 1990). In the cases of DMSe, DMDSe, and methaneselenol, this identification is, on the whole, clear and uncontroversial. However, another microbially produced, volatile selenium-containing compound that has been reported in the literature, dimethylselenone, presents a more interesting situation.

Frederick Challenger's postulate in 1945 that the reduction and methylation steps leading from selenate or selenite to dimethyl selenide pass through a chemical intermediate, dimethylselenone, CH$_3$SeO$_2$CH$_3$ (Challenger, 1945, 1951), has kept researchers on the lookout for this compound in microbial cultures because this chemical species, like DMSe and DMDSe, might be detectable in culture headspace. The other compounds in the proposed pathway are presumably readily ionizable salts (even at neutral pH) and therefore would not have significant vapor pressure at ambient temperatures in aqueous solutions. Karlson et al. (1994) determined the vapor pressures for DMSe and DMDSe at 25°C: 32.03 kPa for DMSe and 0.38 kPa for DMDSe. These workers also determined a relatively high solubility of DMSe in water (2.44 g DMSe/100 mL H$_2$O).

Indeed, a selenium-containing, gas phase component of live microbial cultures that had been amended with Se was detected by Reamer and Zoller (see discussion above) and reported to be dimethylselenone (Reamer and Zoller, 1980). The authors' identification was based ostensibly on comparison of this compound's GC-MS spectrum with an authentic dimethylselenone spectrum from the literature (Rebane, 1974). In this same work, Reamer and Zoller referred to an earlier report (Chau et al., 1976) that had noted an unidentified volatile selenium-containing compound found in the headspace of lake water–sediment systems. At that time Reamer and Zoller (1980) and later Chau and another coworker on the 1976 paper described that unknown as dimethylselenone (Chau and Wong, 1986). Work by Jiang and coresearchers in the 1980s also reported $CH_3SeO_2CH_3$ detection in gas phase samples taken at sites in Belgium (see above). Identification of dimethylselenone was not based on either GC-MS or a commercial reagent but instead on boiling point considerations (F. Adams, personal communication, 1990). This then comprises in toto the evidence for the detection of "volatile" dimethylselenone since 1945 when Frederick Challenger proposed it.

The problem arises with the conclusions drawn from these confirmations as reported, and an account of these mistakes in the literature has been published (Chasteen, 1993). Those arguments will be briefly recounted. Dimethylselenone has been incorrectly identified as a gas phase component in the biological reduction and methylation of Se. The actual compound is dimethyl selenenyl sulfide, CH_3SeSCH_3 (methanesulfenoselenoic acid methyl ester). The compound that Reamer and Zoller (1980) reported to be dimethylselenone eluted in their chromatographic analysis between DMSe and DMDSe (Reamer, 1978). This elution order was also true in Jiang's work (S. Jiang, personal communication, 1990) and in the work of Chau et al. 1976 (Y. K. Chau, personal communication, 1996), though the chromatographic elution position of this analyte was not given in either earlier publication. All these gas chromatographic methods yielded analyte elution orders based on compound boiling point, low boilers eluting first, and so on, and all used the emission or absorption from the Se 196 nm line as a means of detection, Reamer via microwave discharge spectroscopy and Chau et al. and Jiang et al. via atomic absorption. The main problem with this research design is that while dimethylselenone's boiling point in unknown, its melting point is known to be approximately 153°C with decomposition (Rebane, 1974). Since DMSe and DMDSe have boiling points of 58 and 156°C, respectively, it is impossible for dimethylselenone to elute between these two compounds as was reported in these studies (Chau et al., 1976; Reamer, 1978; Reamer and Zoller, 1980; Jiang et al., 1983a, 1989). The formula, melting point, boiling point, and vapor pressure of these relevant compounds (when available) are listed in Table 1. It must be noted that the 196 nm Se emission line on which the discharge and AAS work is based would not preclude the detection of CH_3SeSCH_3, since it too contains Se.

However, as noted above, the evidence for the proported detection of dimethylselenone also includes GC-MS data. Yet here, too, a mistake was made in the earliest work. The coincidental datum that starts this confusion is that, given the naturally occurring Se isotopic distributions, dimethylselenone and dimethyl selenenyl sulfide have the same molecular weight, 142 Da, and therefore, with MS, the same molecular ion. Reamer and Zoller's mass spectrum of the compound that eluted between DMSe and DMDSe in their chromatography was favorably compared in Reamer's dissertation (Reamer, 1978) with Rebane's mass spectrum of synthesized dimethylselenone (Rebane, 1974). Rebane also published the mass spectrum of deuterated dimethylselenone in that same work. Reamer's interpretation aside, significant differences between these spectra are apparent, some of which stem from (Reamer's) clearly incorrect peak (i.e., mass) assignments, but the differences are most apparent in comparing the most prominent peaks in each spectrum beginning with the most intense peak (base peak). Data adapted from Rebane's mass spectrum of authentic dimethylselenone (Rebane, 1974), Reamer's mass spectrum of the compound eluting between DMSe and DMDSe in his experiments (Reamer, 1978), and Chasteen's dimethyl selenenyl sulfide mass spectrum (Chasteen, 1990) are shown in Figure 1 for comparison. Clearly dimethylselenone has a significantly different mass spectrum from the other two (all are electron impact MS). Prominent mass fragment assignments are noted in these spectra. The failure of Reamer's spectrum to show mass peaks below about 76 was probably due to his mass spectrometer's response or the possibility that the low mass region was simply not included in the original figure (D. C. Reamer, personal communication, 1996).

Dimethyl disulfide (DMDS) was another bacterial metabolite that Reamer detected in microbial headspace via GC/MS. Comparison of his mass spectrum of DMDS to that of Chasteen's DMDS GC-MS spectrum taken about 10 years later (T. Chasteen, unpublished data, 1990) also shows a reduced response in the lower mass range of Reamer's DMDS spectrum (Reamer, 1978).

More recent work also noted the detection of an organoselenide that eluted between DMSe and DMDSe in similar GC analysis of bacterial headspace followed by fluorine-induced chemiluminescence detection. These authors identified this selenium-containing compound as dimethyl selenenyl sulfide based on its GC-MS mass spectrum (Fig. 1C) (Chasteen et al., 1990, Chasteen, 1993). Besides being detected in microbial headspace, dimethyl selenenyl sulfide was synthesized in that work by mixing DMDS and DMDSe in the presence of a zinc/HCl mixture that caused the reduction of the sulfur–sulfur and selenium–selenium bonds and formation of the corresponding thiol and selenol. Therefore methanethiol and methaneselenol were produced in this mixture and, when these compounds were examined by GC-MS, their mass spectra were recorded along with those of CH_3SeSCH_3 and some unreacted DMDS and DMDSe. Exchange/reduction between thiol/selenols and disulfides and/or diselenides is well known (Eldjarn and Pihl, 1957; Günther, 1967; McFarlane, 1969; Killa and Rabenstein, 1988; Potapov et al., 1992) and is therefore the source of dimethyl selenenyl sulfide in that reaction mixture. The elution order for these compounds in this capillary GC method was methanethiol, methaneselenol, DMDS, dimethyl selenenyl sulfide,

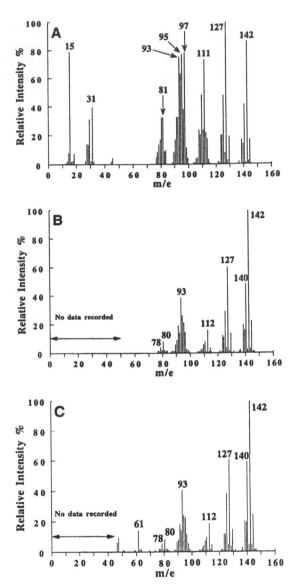

FIGURE I Mass spectra of organoselenium species: relative intensity of mass fragments plotted as a function of mass-to-charge ratio m/e. (A) $CH_3SeO_2CH_3$. (From Rebane, 1974). (B) Peak eluting between DMDS and DMDSe in Reamer's chromatography. (From Reamer, 1978.) (C) CH_3SeSCH_3. (From Chasteen, 1990.)

DMDSe. This elution order has been confirmed by subsequent work with a similar GC stationary phase and atomic emission detection (Cai et al., 1994). The estimation of the boiling point of CH_3SeSCH_3 from strictly GC retention time was 132°C (Chasteen, 1993). The synthesis and subsequent boiling point determination of dimethyl selenenyl sulfide, published elsewhere (Potapov et al., 1992), yielded a boiling point very close to this (boiling point 31°C/11 mmHg ≈ 132°C/760 mmHg).

Further work with the fluorine-induced chemiluminescence detection system has yielded reports of DMSe, DMDSe, and dimethyl selenenyl sulfide released by photosynthetic bacteria (McCarty et al., 1993; Stalder et al., 1995). Time course experiments with a lab-based dynamic headspace sampling system detail the evolution of organoselenium and -sulfur species over a 100-hour experiment (McCarty et al., 1995). Most recently, when the cultures were amended with sodium selenite (1000 μm/L), *Penicillium* sp. maintained aerobically in a 5 L bioreactor at 25°C produced an organoselenium compound thought by the authors to be DMSe. This volatile Se compound was trapped on charcoal, oxidized with HCl/HNO_3 and determined by AAS (Brady et al., 1996).

IV. REPORTS OF VOLATILE SELENIUM RELEASED BY PLANTS

Lewis et al. (1974) reported that DMSe was the only detectable volatile Se species produced by cabbage amended with either potassium selenate or selenite. These results built on Lewis's earlier work with *Astragalus racemosus* in which unidentified volatile Se had been detected (Lewis et al., 1966). In the latter work, determination of DMSe was performed by GC with FID detection using a commercial standard for confirmation. Although the authors looked for DMDSe in their experiments with this plant, none was detected; however DMDS was found. This contrasts with work performed earlier by Evans and her coworkers (1968), who examined extracts of ^{75}Se spiked *Astragalus racemosus* and found DMDSe using GC with FID and/or electron capture detection. Although these authors looked for DMSe in these plant extracts, it was not detected by their methods; however a higher boiling compound containing Se that eluted after DMSe yet before DMDSe was detected but not identified. Examination of the chromatography in that report shows that the unknown compound could have been dimethyl selenenyl sulfide (see above). Evans et al. (1968) determination of DMDSe in gas-purged plant chambers is the first determination of biogenic dimethyl diselenide reported.

A recent extensive survey of common crop species (rice, alfalfa, cabbage, onion, etc.) compared the relative Se volatilization by these plants grown in enclosures and hydroponically amended with selenate (Terry et al., 1992). The results showed that plants that were the best at accumulating Se in leaf and stem tissue also volatilized Se best. This has, in general, been confirmed elsewhere (Duckart et al., 1992). A slightly later report by Zayed and Terry (1992) noted the effects of sulfate on Se accumulation and volatilization by broccoli and found total volatile Se production to be inversely related to hydroponic sulfate concentration, and they later suggested that at least some of the volatilization of Se by plants "may involve microbes, i.e., bacteria" present in the rhizosphere or the plant itself (Zayed and Terry, 1994).

One of the most recent plant-based determinations of volatile Se species involves an odoriferous species: garlic (*Allium sativum*). In crushed homogenates of elephant garlic, Cai et al. (1994) found one of the most complex organoselenium gas phase mixtures yet reported. Their detection system involved capillary GC followed by atomic emission detection (AED). Sampling was performed by an autosampler that purged (with helium) vials containing crushed garlic. To promote the volatilization of the rather low vapor pressure components reported, "homogenized" garlic was shaken in a saturated, aqueous sodium sulfate solution at 90°C for 45 minutes before headspace sampling. The AED produces excited state emissions from microwave plasma excitation of eluting GC analytes. Emission wavelengths from many elements are simultaneously detected by a photodiode array. The identity of the following compounds was determined by GC-MS or by comparison of emission spectra to those of chemical standards. In addition to organosulfur compounds such as allylmethyl sulfide ($CH_3SCHCH{=}CH_2$) and DMDS, these authors reported the determination of DMSe and DMDSe. They also recounted the determination of mixed selenium/sulfur species: dimethyl selenenyl sulfide, CH_3SeSCH_3 (methanesulfenoselenoic acid methyl ester); bis(methylthio)selenide, $CH_3SSeSCH_3$; 2-propenesulfenoselenoic acid methyl ester, $CH_3SeSCHCH{=}CH_2$; 1-propenesulfenoselenoic acid methyl ester, $CH_3SeSCH{=}CHCH_3$; and (allylthio)(methylthio)selenide, $CH_3SSeSCHCH{=}CH_2$. The complex nature of the gas phase mixtures reported here is due partly to the elevated temperature of the sample extract: compounds that elute beyond DMDSe (boiling point 158°C) in this kind of GC have very low vapor pressures at room temperature and, as a result, have a very small presence in the headspaces of room temperature samples. Therefore, the effect of these particular "volatile" Se species on environmental Se cycling is expected to be minor. That said, the headspace of common garlic that had been treated in the same manner but held at only 30°C produced trace amounts of $CH_3SeSCHCH{=}CH_2$ as determined by capillary GC with fluorine-induced chemiluminescence detection (E. Becer and T. G. Chasteen, unpublished results, 1996).

In a subsequent paper using similar analytical methods, Cai and coworkers describe Se-containing amino acids in "ordinary and selenium-enriched garlic, onion, and broccoli" via derivatization (Cai et al., 1995b) and the detection of many volatile Se species in the breath of a human subject fed 3 g of garlic (Cai et al., 1995a). An even more recent report of DMSe determination in human breath has been made by Feldmann et al. (1996). Of six human subjects examined (one woman and five men), all showed DMSe "outgassing" as detected by cryogenic trapping of 20 to 30 L of their breath followed by GC with inductively coupled plasma/mass spectrometric detection. The average DMSe gas phase concentration detected in these exhalations was 0.45 μg/m^{-3}. No other organometal(loid)s (organoarsenic, -tellurium, -mercury, or -diselenide) for which these subjects were screened with this method were detected, and no information about these human subjects' diets was recorded (Feldmann et al., 1996). Though the volatile Se determined here probably came from the garlic itself, reduction and methylation of inorganic salts of Se by humans or enteric microbes is quite probable. Nevertheless, confirmatory data apparently are available only for tellurium salts (Challenger, 1945).

V. PARTICULATE PHASE SELENIUM

Though not strictly within the purview of a chapter on volatile Se species, particulate phase Se is as important a component of the geochemical cycling of Se as it is for sulfur, and, as seen below, vapor phase/particle phase exchange is the fate of most volatile Se. The atmospheric determinations of organoselenium compounds are few and far between, probably because of the relative reactivity of organoselenium compounds in the gas phase (see below). Particulate Se contents have been measured in ice at the South Pole (Zoller et al., 1974), over the ocean in the remote Northern Hemisphere (Duce et al., 1975), and at island sites in the North Pacific (Mosher et al., 1987). A historical record (800 B.C.–A.D. 1965) of Se (and sulfur) depositions has been extracted from cores of remote ice sheets from Greenland and Antarctica (Weiss et al., 1971). Though the chemical species that transported Se to these sites were unknown to these workers, comparison of the determined amounts of particulate Se with well-known Se crustal abundances suggests that volatile Se species did contribute to the Se content of these remote regions. Mosher and coworkers concluded that biological productivity in local oceanic waters influenced aerosol Se (Mosher and Duce, 1981; Mosher et al., 1987). Their calculations showed an annual Se cycling (of all kinds, particulate and volatile) of 13 to 19 × 10^9 g Se with marine flux approximately twice that of Se released from coal combustion (Mosher and Duce, 1987). Earlier particulate Se estimates by Mackenzie et al. (1979) were in the same order of magnitude, and these workers estimated an Se residence time of 9.6 days in the atmosphere, about the same as atmospheric water. An atmospheric Se budget has been constructed for the region 30°N to 90°N, (Ross, 1985).

VI. THE FATE OF VOLATILE SELENIUM SPECIES IN THE ATMOSPHERE

The absence of H_2Se from any atmospheric samples ever analyzed, and the scarcity of reports of it even in reducing, bacterial headspace (Doran and Alexander, 1977) probably can be explained by the ease with which this substance is oxidized (Doran, 1982). Sampling and analytical methods for H_2Se that involve even traces of O_2 or another oxidizing species probably oxidize H_2Se to Se^0 (Diplock et al., 1973). Methaneselenol, in a manner analogous to methanethiol, can be expected to quickly oxidize to DMDSe in the atmosphere (Günther, 1967). DMSe no doubt has a longer atmospheric lifetime, although an atmospheric residence time for this species has not been estimated. The gas phase oxidation products of DMSe are probably less volatile selenoxides, diselenides, or maybe selenones. In 1990 Atkinson et al. reported that DMSe did not undergo gas phase photolysis but was approximately an order of magnitude more reactive than dimethyl sulfide (DMS) toward hydroxyl (OH) and nitrate (NO_3) radicals. DMSe also reacted with ozone (O_3), whereas DMS probably does not (Atkinson et al., 1990). More recent investigation of the O_3 gas phase oxidation of DMSe suggests that dimethyl selenoxide was produced at approximately 90% yield at room temperature and about 1 atm pressure (Rael et al., 1996). The oxidation of DMSe by OH and NO_3 radicals was decidedly more complex because of the breakage of the Se—C bond (and subsequent formation of formaldehyde). The ultimate result of this oxidation, however, probably was the formation of "salt-type compounds" of very low volatility, which would likely be incorporated into aerosols in the atmosphere. An example of one of the suggested salts was $[(CH_3)_2SeOH]^+ NO_3^-$. Dimethyl selenoxide was tentatively identified by infrared absorption bands as an intermediate and/or product in this process, which continued via reaction of methaneseleninic acid $[CH_3Se(O)OH]$ with nitrate radicals present in that experimental system (Rael et al., 1996). As a comparison, in another oxidative study of DMSe with MnO_2, the major product was also dimethyl selenoxide, confirmed by mass spectrometry (Wang and Burau, 1995).

VII. CONCLUSIONS

Volatile organoselenium species are created by microorganisms, plants, animals, and probably humans and released into the environment. DMSe, DMDSe, dimethyl selenenyl sulfide, methaneselenol, and probably H_2Se have all been detected as produced by biological sources. Much less volatile –SSeS– and –SSe– allyl-type organoselenium compounds also have been detected in plants, but their impact on geochemical cycling is probably insignificant. Biogenic emissions of volatile Se rival those from anthropogenic sources. Release of volatile Se by human activity is probably mainly in the form of selenium oxides and small-diameter, Se-containing particulates that quickly fall out of the gas phase. The stable products of organoselenium reactions in the atmosphere are low or non-volatile oxides or diselenides that are scavenged by aerosols and returned to the biosphere via wet and dry gas phase deposition. The estimated residence time of Se in the atmosphere is on the same time scale as water and suggests that particulate scrubbing would be an efficient means of removal.

REFERENCES

Abu-Erreish, G. M., E. I. Whitehead, and O. E. Olson. 1968. Evolution of volatile selenium from soils. *Soil Sci.* 106:415–420.

Atkinson, R., S. M. Aschmann, D. Hasegawa, E. T. Thompson-Eagle, and W. T. Frankenberger, Jr. 1990. Kinetics of the atmospherically important reactions of dimethyl selenide. *Environ. Sci. Technol.* 24:1326–1332.

Barkes, L., and R. W. Fleming. 1974. Production of dimethylselenide gas from inorganic selenium by eleven soil fungi. *Bull. Environ. Contam. Toxicol.* 12:308–311.

Brady, J. M., J. M. Tobin, and G. M. Gadd. 1996. Volatilization of selenite in aqueous medium by a *Penicillium* species. *Mycol. Res.* 100:955–961.

Bremer, J., and Y. Natori. 1960. Behavior of some selenium compounds in transmethylation. *Biochim. Biophys. Acta* 44:367–370.

Burton, G.A., T. H. Giddings, P. De Brine, and R. Fall. 1987. High incidence of selenite-resistant bacteria from a site polluted with selenium. *Appl. Environ. Microbiol.* 53:185–188.

Byard, J. L. 1969. Trimethyl selenide a urinary metabolite of selenite. *Arch. Biochem. Biophys.* 130:556–560.

Cai, X.-J., P. C. Uden, E. Block, X. Zhang, B. D. Quimby, and J. J. Sullivan. 1994. *Allium* chemistry: Identification of natural abundance organoselenium volatiles from garlic, elephant garlic, onion, and Chinese chive using headspace gas chromatography with atomic emission detection. *J. Agric. Food Chem.* 42:2081–2084.

Cai, X.-J., E. Block, P. C. Uden, B. D. Quimby, and J. J. Sullivan. 1995a. *Allium* chemistry: Identification of natural abundance organoselenium compounds in human breath after ingestion of garlic using gas chromatography with atomic emission detection. *J. Agric. Food Chem.* 43:1751–1753.

Cai, X.-J., E. Block, P. C. Uden, X. Zhang, B. D. Quimby, and J. J. Sullivan. 1995b. *Allium* chemistry: Identification of selenoamino acids in ordinary and selenium-enriched garlic, onion, and broccoli using gas chromatography with atomic emission detection. *J. Agric. Food Chem.* 43:1754–1757.

Cantoni, G. L. 1952. The nature of the active methyl donor formed enzymatically from L-methionine and adenosine triphosphate. *J. Am. Chem. Soc.* 74:2942–2943.

Challenger, F. 1945. Biological methylation. *Chem. Rev.* 36:315–361.

Challenger, F. 1951. Biological methylation. *Adv. Enzymol.* 12:429–491.

Challenger, F. 1955. Biological methylation. *Q. Rev. Chem. Soc.* 9:255–286.

Challenger, F. 1959. Biological methylation with particular reference to compounds of sulfur. In *Aspects of the Organic Chemistry of Sulfur*, pp. 162–206. Buttersworths Scientific Publications, London.

Challenger, F. 1978. Biosynthesis of organometallic and organometalloidal compounds, In F. E. Brinckman and J. M. Bellama (eds.), *Organometals and Organometalloids— Occurrence and Fate in the Environment*, pp. 1–22. ACS Symposium Series No. 82, American Chemical Society, Washington, DC.

Challenger, F., and H. E. North. 1934. The production of organo-metalloidal compounds by microorganisms, II. Dimethyl selenide. *J. Chem. Soc.* 68–71.

Chasteen, T. G. 1990. Ph.D. thesis. University of Colorado, Boulder.

Chasteen, T. G. 1993. Confusion between dimethyl selenenyl sulfide and dimethyl selenone released by bacteria. *Appl. Organomet. Chem.* 7:335–342.

Chasteen, T. G., G. M. Silver, J. W. Birks, and R. Fall. 1990. Fluorine-induced chemiluminescence detection of biologically methylated tellurium, selenium, and sulfur compounds. *Chromatographia* 30:181–185.

Chau, Y. K., and P. T. S. Wong. 1986. Organic group VI elements in the environment. In P. J. Craig (ed.), *Organometallic Compounds in the Environment: Principles and Reactions*, pp. 254–278. Wiley-Interscience, New York.

Chau, Y. K., P. T. S. Wong, and P. D. Goulden. 1975. Gas chromatography–atomic absorption method for the determination of dimethyl selenide and dimethyl diselenide. *Anal. Chem.* 47:2279–2281.

Chau, Y. K. , P. T. S. Wong, B. A Silverberg, P. L. Luxon, and G. A. Bengert. 1976. Methylation of selenium in the aquatic environment. *Science.* 192:1130–1131.

Cooke, T. D., and K. W. Bruland. 1987. Aquatic chemistry of selenium: Evidence of biomethylation. *Environ. Sci. Technol.* 21:1214–1219.

Craig, P. J. 1986. *Organometallic Compounds in the Environment: Principles and Reactions.* Wiley-Interscience, New York.

Diplock, A. T., C. P. J. Caygill, E. H. Jeffery, and C. Thomas. 1973. The nature of the acid-volatile selenium in the liver of the male rat. *Biochem. J.* 134:283–293.

Doran, J. W. 1982. Microorganisms and the biological cycling of selenium. In K. C. Marshall (ed.), *Advances in Microbial Ecology*, Vol 6. Plenum Press, New York.

Doran, J. W., and M. Alexander. 1976. Microbial formation of volatile selenium compounds in soil. *Soil Sci. Soc. Am. J.* 40:687–690.

Doran, J. W., and M. Alexander. 1977. Microbial transformation of selenium. *Appl. Environ. Microbiol.* 33:31–37.

Duce, R. A., G. L. Hoffman, and W. H. Zoller. 1975. Atmospheric trace metals at remote Northern and Southern Hemisphere sites: Pollution or natural? *Science* 187:59–61.

Duckart, E. C., L. J. Waldron, and H. E. Donner. 1992. Selenium uptake and volatilization from plants growing in soil. *Soil Sci.* 53:94–99.

Eldjarn, L., and A. Pihl. 1957. The equilibrium constants and oxidation–reduction potentials of some thiol–disulfide systems. *J. Am. Chem. Soc.* 79:4589–4593.

Evans, C. S., and C. M. Johnson. 1966. The separation of some alkyl–selenium compounds by gas chromatography. *J. Chromatogr.* 21:202–206.

Evans, C. S., C. J. Asher, and C. M. Johnson. 1968. Isolation of dimethyl diselenide and other volatile selenium compounds from *Astragalus racemosus* (Pursh.). *Aust. J. Biol. Sci.* 21:13–20.

Feldmann, J., T. Riechmann, and A. V. Hirner. 1996. Determination of organometallics in intra-oral air by LT-GC/ICP-MS. *Fresenius J. Anal. Chem.* 354:620–623.

Fleming, R. W., and M. Alexander. 1972. Dimethylselenide and dimethyltelluride formation by a strain of *Penicillium*. *Appl. Microbiol.* 24:424–429.

Francis, A. J., J. M. Duxbury, and M. Alexander. 1974. Evolution of dimethylselenide from soils. *Appl. Microbiol.* 28:248–250.

Frankenberger, W. T., Jr., and U. Karlson 1989. Environmental factors affecting microbial production of dimethylselenide in a selenium-contaminated sediment. *Soil Sci. Soc. Am. J.* 53:1435–1442.

Frankenberger, W. T., Jr., and U. Karlson. 1995. Volatilization of selenium from a dewatered seleniferous sediment: A field study. *J. Indust. Micriobiol.* 14:226–232.

Gadd, G. M. 1993. Microbial formation and transformation of organometallic and organometalloid compounds. *FEMS Microbiol. Rev.* 11:297–316.

Ganther, H. E. 1966. Enzymic synthesis of dimethyl selenide from sodium selenite in mouse liver extracts. *Biochemistry* 5:1089–1098.

Ganther, H. E. 1971. Reduction of the selenotrisulfide derivative of glutathione to a persulfide analog by glutathione reductase. *Biochemistry* 10:4089–4098.

Ganther, H. E., and C. Corcoran. 1969. Selenotrisulfides. II. Crosslinking of reduced pancreatic ribonuclease with selenium. *Biochemistry* 8:2557–2563.

Goeger, D. E., and H. E. Ganther. 1994. Oxidation of dimethylselenide to dimethylselenoxide by microsomes from rat liver and lung and by flavin-containing monooxygenase from pig liver. *Arch. Biochem. Biophys.* 310:448–451.

Günther, W. H. H. 1967. Methods in selenium chemistry. II. The reduction of diselenides with dithiothreitol. *J. Org. Chem.* 32:3931–3933.

Hassoun, B. S., I. S. Palmer, and C. Dwivedi. 1995. Selenium detoxification by methylation. *Res. Commun. Mol. Pathol. Pharmacol.* 90:133–142.

Hirooka, T., and J. T. Galambos. 1966. Selenium: I. Respiratory excretions. *Biochim. Biophys. Acta* 130:313–320.

Hsieh, H. S., and H. E. Ganther. 1975. Acid-volatile selenium formation catalyzed by glutathione reductase. *Biochemistry* 14:1632–1636.

Hsieh, H. S., and H. E. Ganther. 1977. Biosynthesis of dimethyl selenide from sodium selenide in rat liver and kidney cell-free systems. *Biochim Biophys. Acta* 497:205–217.

Ip, C., C. Hayes, R. M. Budnick, and H. E. Ganther. 1991. Chemical form of selenium, critical metabolites, and cancer prevention. *Cancer Res.* 51:595–600.

Jiang, S., W. De Jonghe, and F. Adams. 1982. Determination of alkylselenide compounds in air by gas chromatography–atomic absorption spectrometry. *Anal. Chim. Acta.* 136:183–190.

Jiang, S., H. Robberecht, and F. Adams. 1983a. Identification and determination of alkylselenide compounds in environmental air. *Atmos. Environ.* 17:111–114.

Jiang, S., H. Robberecht, and D. Vanden Berghe. 1983b. Elimination of selenium compounds by mice through formation of different volatile selenides. *Experientia* 39:293–294.

Jiang, S., H. Robberecht, and F. Adams. 1989. Studies of the naturally occurring biomethylation of selenium and the determination of the products. *Appl. Organomet. Chem.* 3:99–104.

Karlson, U., and W. T. Frankenberger, Jr. 1988a. Determination of gaseous selenium-75 evolved from soil. *Soil Sci. Soc. Am. J.* 52:678–681.

Karlson, U., and W. T. Frankenberger, Jr. 1988b. Effects of carbon and trace element addition on alkylselenide production by soil. *Soil Sci. Soc. Am. J.* 52:1640–1644.

Karlson, U., and W. T. Frankenberger, Jr. 1989. Accelerated rates of selenium volatilization from California soils. *Soil Sci. Soc. Am. J.* 53:749–753.

Karlson, U., and W. T. Frankenberger, Jr. 1990. Volatilization of selenium from agricultural evaporation pond sediments. *Sci. Total Environ.* 92:41–54.

Karlson, U., W. T. Frankenberger, Jr., and W. F. Spencer. 1994. Physicochemical properties of dimethyl selenide and dimethyl diselenide. *J. Chem. Eng. Data* 39:608–610.

Kaufman, H. C. 1961. *Handbook of Organometallic Compounds*. Van Nostrand, New York.

Killa, H. M. A., and D. L. Rabenstein. 1988. Determination of selenols, diselenides, and selenenyl sulfides by reverse-phase liquid chromatography with electrochemical detection. *Anal. Chem.* 60:2283–2287.

Lewis, B. G., C. M. Johnson, and C. C. Delwiche. 1966. Release of volatile selenium compounds by plants. Collection procedures and preliminary observations. *J. Agric. Food Chem.* 14:638–640.

Lewis, B. G., C. M. Johnson, and T. C. Broyer. 1974. Volatile selenium in higher plants: The production of dimethyl selenide in cabbage leaves by enzymatic cleavage of Se-methyl selenomethionine selenonium salt. *Plant Soil* 40:107–118.

Mackenzie, F. T., R. J. Lantzy, and V. Paterson. 1979. Global trace metal cycles and prediction. *Math. Geol.* 11:99–142.

McCarty, S. L., T. G. Chasteen, M. Marshall, R. Fall, and R. Bachofen. 1993. Phototrophic bacteria produce volatile, methylated sulfur and selenium compounds. *FEMS Lett.* 112:93–98.

McCarty, S. L., T. G. Chasteen, V. Stalder, and R. Bachofen. 1995. Bacterial bioremediation of selenium oxyanions using a dynamic flow bioreactor and headspace analysis. In R. E. Hinchee, J. L. Means, and D. R. Burris (eds.), *Bioremediation of Inorganics*, pp. 95–102. Battelle Press, Columbus, OH.

McConnell, K. P., and O. W. Portman. 1952. Excretion of dimethyl selenide by the rat. *J. Biol. Chem.* 195:277–282.

McFarlane, W. 1969. Magnetic double resonance study of selenium-77 spin coupling in diselenides. *J. Chem. Soc.* (A). 670–672.

Mosher, B., and R. Duce. 1981. Vapor phase selenium in the atmosphere. *SEAREX News.* 4:9–10.

Mosher, B. W., and R. A. Duce. 1987. A global atmospheric selenium budget. *J. Geophys. Res.* 92:13289–13298.

Mosher, B. W., R. A. Duce, J. M. Prospero, and D. L. Savoie. 1987. Atmospheric selenium: Geographical distribution and ocean to atmosphere flux in the Pacific. *J. Geophys. Res.* 92:13277–13287.

Painter, E. P. 1941. The chemistry and toxicity of selenium compounds with special reference to the selenium problem. *Chem. Rev.* 28:179–213.

Palmer, I. S., R. P. Gunsalus, A. W. Haverson, and O. E. Olson. 1970. Trimethylselenonium ion as a general excretory product from selenium metabolism in the rat. *Biochim. Biophys Acta* 208:260–266.

Potapov, V. A., S. V. Amosova, P. A. Petrov, L. S. Romanenko, and V. V. Keiko. 1992. Exchange reactions of dialkyl dichalcogens. *Sulfur Lett.* 15:121–126.

Rael, R. M., E. C. Tuazon, and W. T. Frankenberger, Jr. 1996. Gas-phase reactions of dimethyl selenide with ozone and the hydroxyl and nitrate radicals. *Atmos. Environ.* 30:1221–1232.

Reamer, D. C. 1975. M.S. thesis. University of Maryland, College Park.

Reamer, D. C. 1978. Ph.D. thesis. University of Maryland, College Park.

Reamer, D. C., and W. H. Zoller. 1980. Selenium biomethylation products from soil and sewage sludge. *Science* 208:500–502.

Reamer, D. C., W. H. Zoller, and T. C. O'Haver. 1978. Gas chromatography–microwave plasma detector for the determination of tetraalkyllead species in the atmosphere. *Anal. Chem.* 50:1449–1453.

Rebane, E. 1974. Mass spectrometric studies on organic selenium–oxygen compounds. *Chem. Scripta* 5:65–73.

Ridley, W. P., L. J. Dizikes, and J. M. Wood. 1977. Biomethylation of toxic elements in the environment. *Science* 197:329–332.

Rosenfeld, I., and O. A. Beath. 1948. Metabolism of sodium selenate and selenite by the tissues. *J. Biol. Chem.* 172:333–341.

Ross, H. B. 1985. An atmospheric selenium budget for the region 30°N to 90°N. *Tellus* 37B:78–90.

Stalder, V., N. Bernard, K. W. Hanselmann, R. Bachofen, and T. G. Chasteen. 1995. A method of repeated sampling of static headspace above anaerobic bacterial cultures with fluorine-induced chemiluminescence detection. *Anal. Chim. Acta* 303:91–97.

Tanzer, D., and K. G. Heumann. 1990. GC determination of dimethyl selenide and trimethyl selenonium ions in aquatic systems using element specific detection. *Atmos. Environ.* 24A:3099–3102.

Tanzer, D., and K. G. Heumann. 1992. Gas chromatographic trace-level determination of volatile organic sulfides and selenides and of methyl iodide in Atlantic surface waters. *Int. J. Anal. Chem.* 48:17–31.

Terry, N., C. Carlson, T. K. Raab, and A. M. Zayed. 1992. Rates of selenium volatilization among crop species. *J. Environ. Qual.* 21:341–344.

Thompson-Eagle, E. T., and W. T. Frankenberger, Jr. 1990. Volatilization of selenium from agricultural evaporation pond water. *J. Environ. Qual.* 19:125–131.

Thompson-Eagle, E. T., and W. T. Frankenberger, Jr. 1991. Selenium biomethylation in an alkaline saline environment. *Water Res.* 25:231–240.

Thompson-Eagle, E T., and W. T. Frankenberger, Jr. 1992. Bioremediation of soils contaminated with selenium. In R. Lal and B. A. Stewart (eds.), *Advances in Soil Science*, Vol. 8, *Soil Restoration*, pp. 261–310. Springer-Verlag, New York.

Thompson-Eagle, E. T., W. T. Frankenberger, Jr., and U. Karlson. 1989. Volatilization of selenium by *Alternaria alternata*. *Appl. Environ. Microbiol.* 55:1406–1413.

Vlasáková, V., J. Benes, and J. Parizek. 1972. Application of gas chromatography for the analysis of trace amounts of volatile ^{75}Se metabolites in expired air. *Radiochem. Radioanal. Lett.* 10:251–258.

Voet, D., and J. G. Voet. 1995. *Biochemistry*, pp. 761–764. Wiley, New York.

Wang, B., and R. G. Burau. 1995. Oxidation of dimethylselenide by MnO_2: Oxidation product and factors affecting oxidation rate. *Environ. Sci. Technol.* 29:1504–1510.

Weiss, H. V., M. Koide, and E. D. Goldberg. 1971. Selenium and sulfur in a Greenland ice sheet: Relation to fossil fuel combustion. *Science* 172:261–263.

Weres, O., A.-R. Jaouni, and L. Tsao. 1989. The distribution, speciation and geochemical cycling of selenium in a sedimentary environment, Kesterson Reservoir, California, U.S.A. *Appl. Geochem.* 4:543–563.

Wood, J. M. 1975. Biological cycles for elements in the environment. *Naturwissenchaften* 62:357–364.

Zayed, A. M., and N. Terry. 1992. Selenium volatilization in broccoli as influenced by sulfate supply. *J. Plant Physiol.* 140:646–652.

Zayed, A. M., and N. Terry. 1994. Selenium volatilization in roots and shoots: Effects of shoot removal and sulfate level. *J. Plant Physiol.* 143:8–14.

Zieve, R., and P J. Peterson. 1981. Factors influencing the volatilization of selenium from soil. *Sci. Total Environ.* 19:277–283.

Zoller, W. H., E. S. Gladney, and R. A. Duce. 1974. Atmospheric concentrations and sources of trace metals at the South Pole. *Science* 183:198–200.

30

Particulate Selenium in the Atmosphere

THOMAS A. CAHILL and ROBERT A. ELDRED
University of California at Davis, Davis, California

I. INTRODUCTION

Many physical and biological processes, both natural and anthropogenic, result in the evolution of selenium compounds into the atmosphere. Mosher and Duce (1987) estimated that one-half to two-thirds of the worldwide selenium emissions come from natural sources. Approximately 90% of all natural emissions are biogenic, with the marine biosphere accounting for about 70%, and the continental biosphere about 20%. The biogenic emissions are mainly of dimethyl selenide gas. Nonbiogenic natural sources include volcanoes (8%), sea salt (2%), and crustal weathering (< 1%). Two-thirds of the anthropogenic selenium emissions come from combustion: 50% from coal, 9% from oil combustion, and 10% from other combustion. Metals production accounts for most of the rest: 20% from copper, 4% from zinc and lead, and 4% from selenium. Other manufacturing accounts for the remaining 4%, primarily from glass and ceramic manufacturing.

Selenium is predominantly emitted as a gas, although a small fraction may be emitted directly as primary particles. The gas may be removed by dry deposition onto surfaces or by wet deposition after being incorporated into water droplets. The gas not deposited is converted into secondary particles. For both anthropogenic and biogenic emissions, the conversion is rapid enough so that most of the gas is converted. In coal-fired power plants, and possibly also in oil-fired power plants, the selenium vapor condenses to particles very rapidly (Andren et al., 1975; Ondov et al., 1989). The biogenic dimethyl selenide is converted to particles rapidly in the presence of ozone, OH, and NO_3 (Rael et al., 1996).

Primary and secondary particles may be removed by dry and wet deposition. Particles larger than about 2 μm are effectively removed by gravitational settling. Particles smaller than about 0.05 μm diffuse easily to surfaces or other particles; thus, those that are not deposited tend to form larger particles. The result is that the particles tend to be in the "accumulation mode," centered around 0.3 μm in diameter. Accumulation mode particles have a much longer lifetime in the atmosphere than do gases and can remain in the atmosphere for a week or more. During this time they may be transported thousands of kilometers. Accumulation mode particles are important for visibility, health, and climate, as they scatter light effectively, penetrate deeply into the lung, catalyze rainfall, and modify climate. For simplicity in sampling, most measurements include all particles smaller than 2.5 μm in aerodynamic diameter, referred to as $PM_{2.5}$.

There is a significant difference in the elevation at which anthropogenic and biogenic particles are produced. Most anthropogenic selenium emissions occur at high temperatures, with the result that the particles are lofted far enough above the surface to minimize dry deposition and maximize transport. Since biogenic gas is emitted at ambient temperature, the gas and converted particles will remain near the surface, increasing dry deposition. Thus, while the biogenic selenium gas emission is large, the contribution of particles to the atmosphere may be small.

Major efforts have been made during the past two decades to measure the concentration and composition of $PM_{2.5}$ particles, but little effort has been directed specifically to selenium compounds. The most important reason for the lack of information is the low concentration of selenium in the atmosphere, with annual average concentrations ranging from less than 0.1 ng/m^3 at some remote sites in northwestern United States to 1 ng/m^3 in the Appalachian Mountains. At these low concentrations, selenium has little potential to affect health or visibility. However, selenium is a key tracer for sulfate aerosols; selenium and sulfur tend to be emitted by common sources, since they are chemical analogs in the periodic table. Sulfur is the largest or second largest component of the $PM_{2.5}$ mass budget and of the light extinction budget at remote sites in the United States (Malm et al., 1994). At remote sites in the Appalachian and Northeast regions, sulfate accounts for 50 to 60% of the fine mass budget and 60 to 70% of the reconstructed light extinction budget. The remainder of this chapter focuses on the relationship of selenium and sulfur.

II. THE IMPROVE NETWORK

The largest source of data on atmospheric selenium and sulfur at remote sites is the Interagency Monitoring of Protected Visual Environments (IMPROVE) network (Eldred, 1997). This cooperative program involves the National Park Service, the

Forest Service, the Fish and Wildlife Service, the Bureau of Land Management, and the Environmental Protection Agency. IMPROVE was a response to the Clean Air Act Amendments of 1977, which required protection of visibility at national parks, monuments, and wilderness areas, and mandated measurements to find the sources of haze (Malm et al., 1994). Since monitoring began in 1988, the particulate portion of the network has been operated by the Air Quality Group, Crocker Nuclear Laboratory, at the University of California, Davis. The network includes sites operated for other agencies, such as the Tahoe Regional Planning Agency and the Northeast States for Air Use Management, using the IMPROVE protocols. The 76 remote aerosol sampling sites of the IMPROVE network discussed in this chapter are shown in Figure 1. Most of the sites are in or near EPA class I visibility areas. Two 24-hour samples are collected each week. Validated samples were collected for 95% of the possible periods from 1992 to 1996.

The IMPROVE particulate sampler was designed for the IMPROVE network by Crocker Nuclear Laboratory (Eldred et al., 1990). The sampler consists of three independent PM$_{2.5}$ modules with Teflon, nylon, and quartz filters, and one PM$_{10}$ module with a Teflon filter. A denuder in the nylon module removes nitric acid vapors. Each PM$_{2.5}$ module has a cyclone with an aerodynamic diameter cut point of 2.5 mm for a flow rate of 22.8 L/min. The flow rate is measured by two independent gauges before and after each sample is collected. The precision of the volume of air is estimated to be 3%. All filter handling is performed at

FIGURE 1   Remote aerosol sampling sites in the IMPROVE network included in the discussion. The sites Hopi Point and Indian Gardens are both in Grand Canyon National Park; Hopi Point is on the rim, and Indian Gardens is 1000 m below.

Davis with the filters shipped to and from the sites in sealed cassettes and insulated boxes.

The Teflon filters are analyzed for mass, absorption, elemental hydrogen, and all elements between sodium to lead that are present in concentrations above the analytical minimum detectable limit (MDL). Sulfur is among the elements measured by particle-induced X-ray emission (PIXE). Since summer 1992, selenium and other trace elements have been measured by X-ray fluorescence (XRF). The nylon filters are analyzed by ion chromatography for nitrate and sulfate. The prefired quartz filters are analyzed for elemental and organic carbon by the thermal/optical reflectance combustion method. The PM_{10} Teflon filter is analyzed for mass. A small fraction of the PM_{10} Teflon filters are analyzed by PIXE and XRF. In addition, SO_2 is measured at some of the sites by means of a carbonate-impregnated filter inserted after the PM_{10} filter. This filter is analyzed by ion chromatography.

The PIXE and XRF analytical systems are calibrated by means of a series of elemental standards; validation consists of reanalyzing a series of filters from the preceding analytical session. The normal replicate precision for elements present in high concentrations by either system is 4%. The uncertainty for each measurement is calculated as the quadratic sum of the 3% uncertainty from collection, the 4% uncertainty from analytical calibration, and the statistical uncertainty associated with the number of X-rays observed. For sulfur, the calculated uncertainty is very close to 5% for all samples, with the largest uncertainty in summer 1993 being 8%. For selenium concentrations above 0.5 ng/m^3, the calculated uncertainty is approximately 7%, while for concentrations below this, the uncertainty is approximately 0.04 ng/m^3. The validity of the uncertainty estimates has been verified by the reanalysis samples and collocated samples.

Approximately 15% of the samples collected in the IMPROVE network had no statistically significant selenium peak in the X-ray spectrum. The selenium MDL, which is calculated from the background counts under a typical peak at the location of the selenium X-ray line, corresponds to a 95% probability that the peak will be identified. The typical selenium MDL for the IMPROVE network is 0.03 ng/m^3. In calculating mean concentrations, whenever the peak was not identified, half of the MDL is used as the ambient concentration. At several sites in the Pacific Northwest and Alaska, less than half the concentrations were above the MDL. In these cases, the relative uncertainty in the mean is large. Nevertheless, the means are retained because they provide valuable qualitative information that the selenium concentrations are extremely low.

The data validation procedures include routine intercomparisons between the various filters and analytical methods. Sulfur collected on the Teflon filter and measured by PIXE is generally within 5% of the sulfate collected on the nylon filter and analyzed by ion chromatography. The selenium measured by

XRF is compared to that measured by PIXE on the same filter. Good agreement is obtained for samples with enough selenium to be above the PIXE MDL.

III. NATURAL CONTRIBUTIONS TO SELENIUM IN THE IMPROVE NETWORK

Although marine sources contribute half or more of the atmospheric selenium, the marine contribution to particulate selenium from sea salt and biological activity appears to be significant at only four sites in the IMPROVE network: Point Reyes National Seashore, Redwood National Park, Haleakala National Park, and Virgin Islands National Park. Even at these sites, the largest selenium concentrations occur on samples with low Na. If we eliminate samples with Na less than 1 $\mu g/m^3$, there is a moderate correlation between Se and Na ($r^2 = 0.5$). The slope and the ratio of means (Se/Na) are both about 200 mg/kg. This is 10^4 times the ratio of dissolved minerals in sea-water (Kennish, 1994), substantiating the argument that most of the selenium comes from the marine biosphere rather than from dissolved minerals in the water (Mosher and Duce, 1983; Ellis et al., 1993). Over an annual cycle, approximately half the fine particle selenium at these four sites is of marine origin. A similar comparison of sodium and sulfur shows that some of the sulfur is also of marine origin. The observed S/Na ratio is only twice the ratio for dissolved minerals. This indicates that while the marine biosphere contributes almost all the marine selenium, it contributes very little marine sulfur. In addition, a single PM_{10} measurement at Point Reyes in summer 1993 indicates that the selenium is much coarser than at continental sites but has about the same $PM_{2.5} / PM_{10}$ ratio as sodium (Eldred et al. 1997).

The large particle size of marine selenium was also observed by Mosher and Duce (1989). Since most marine selenium is biogenic, this suggests that the biogenic gas or secondary particles must become attached to larger particles of sea salt. Because of the large size of marine particles, the marine influence drops off rapidly, since these particles will settle rapidly compared to fine particles. Moderate marine influences were observed at Pinnacles National Monument, 60 km from the ocean. From the measured Na and an Se/Na ratio of 200 mg/kg, it may be estimated that marine selenium accounts for 10% of the total annual selenium. Because of the higher anthropogenic sulfur and selenium concentrations at the four IMPROVE sites mentioned above, the marine contribution to the annual selenium mean is estimated to be less than 10%. Based on Na concentrations, the contribution of marine selenium to the annual budget at all other sites would be negligible.

Natural continental selenium sources include the continental biosphere, volcanoes, and soil. Particulate Se concentrations from all natural sources must be less than about 0.05 ng/m^3, the annual average at Crater Lake National Park.

This is small compared to the 0.5 to 1.5 ng/m^3 average concentration throughout the eastern United States. The smallest of these sources, soil, can be estimated from the measured concentrations of the soil-derived elements. Selenium is present in soil with a concentration of approximately 0.05 mg/kg for upper continental crust (Taylor and McLennan, 1985). Assuming this composition for the fine aerosol, and calculating soil as the sum of soil-derived elements with their oxides, then for 25 μg/m^3, the highest soil concentration observed between 1992 and 1996, the corresponding selenium concentration would be 0.001 ng/m^3, well below the MDL of 0.03 ng/m^3. In addition, soil-derived selenium would be coarser than selenium from high temperature combustion. From a small number of PM$_{10}$ samples from the network that have been analyzed for elemental composition, the PM$_{2.5}$ / PM$_{10}$ ratio for selenium (0.88) is equal to that for sulfur (0.88), but much higher than that for the soil elements (0.16).

Comparison of the sulfur/selenium ratios for Sequoia and Yosemite National Parks in California provides a test for soil-derived selenium. Both sites are east of the San Joaquin Valley, but Sequoia should experience a much greater impact from the Kesterson area, which has extremely high selenium content. However, the S/Se ratios for Sequoia and Yosemite are equal, indicating no major impact of selenium from the Kesterson area at Sequoia. Some selenium could come from biogenic activity in the lakes and ponds of the basin, but this source would also have a very small S/Se ratio (Fan et al., 1997). The conclusion is that soil is not a significant source of particulate selenium, except possibly at locations with highly enriched soil.

The marine data indicate that the marine biosphere emits far more selenium than sulfur, relative to dissolved minerals. If this is also the case for the continental biosphere, then the S/Se ratio for biogenic selenium should be much less than for anthropogenic selenium. Thus, in regions of low sulfur, the S/Se should drop considerably if there were significant biogenic selenium. This is not observed; sites with low sulfur tend to have correspondingly low selenium. Most sites with low sulfur have relatively high S/Se ratios.

IV.   MEAN SELENIUM AND SULFUR AT IMPROVE SITES

This section analyzes the relationship between selenium and sulfur by examining the mean concentrations at each IMPROVE site for all summer and winter seasons between summer 1992 and summer 1996. The seasonal mean concentrations are appropriate for determining regional patterns, since variations over a shorter time scale are averaged out. The major source of uncertainty in the mean is the variation in ambient concentrations rather than analytical uncertainty, except when the selenium means are below about 0.10 ng/m^3. Figure 2 shows maps of mean concentrations for selenium for both summer and winter. The concentra-

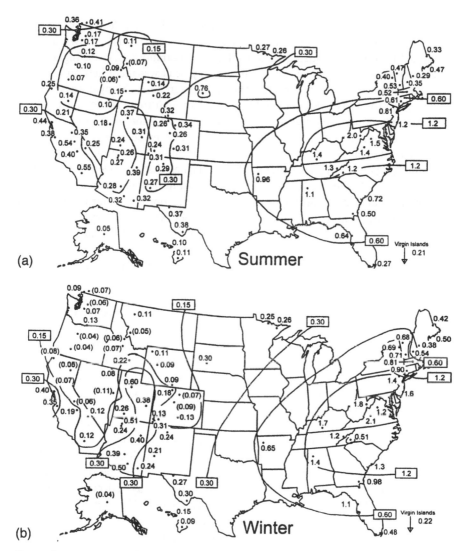

FIGURE 2 Maps of mean selenium, in nanograms per cubic meter, for all data collected at IMPROVE sites between June 1992 and August 1996: (a) summer (June–August) and (b) winter (December–February). The means in parentheses indicate that selenium was identified on less than half the samples. The concentrations are standardized to sea level.

tions on the maps in this chapter are standardized to sea level, to allow for comparison of sites. The contours are based solely on the data shown and are intended to aid in visualizing regions. The concentrations may be higher near selenium sources. Whenever the selenium X-ray peak was identified on less than half the samples at a site, the mean is shown in parentheses. Many of these means are close to the typical MDL of 0.03 ng/m^3. Although these means have a high relative uncertainty, they give a valuable indication that the selenium concentrations are extremely low. Several observations can be made from these maps.

The first observation is that the mean selenium concentrations generally exhibit well-defined spatial patterns. This indicates that the selenium observed at these generally remote sites primarily comes from regional sources. There are a few cases of individual sites not following the normal pattern. The major anomaly during summer is Badlands, South Dakota, which has very high concentrations. Most anomalies occur during winter, when lower boundary layers can inhibit long-range transport. Hopi Point on the Grand Canyon rim and Indian Gardens, 1000 m below, have similar standardized concentrations in summer but differ by a factor of 2 in winter. The most likely explanation is that during winter selenium from a large coal-fired power plant 130 km up the river descends into the canyon but does not always rise to the rim. Shining Rock Wilderness, North Carolina, is near Great Smoky Mountains National Park, but is on the east side of the Appalachian crest. During summer there is good transportation from sites to the west to both Great Smoky Mountains and Shining Rock, so that the concentrations are nearly equal. During winter, the selenium is transported to the mountains as well as across them, so that the mean selenium concentration at Shining Rock is less than half that at Great Smoky Mountains. Two sites show enhanced winter concentrations probably associated with sources in the same air basin. Lone Peak Wilderness, Utah, which has much higher concentrations in winter than at nearby sites, lies east of the industrialized Salt Lake–Provo region. The site in Jefferson National Forest in Virginia has higher winter concentrations than nearby sites. In this case, the increase may reflect contributions from coal-fired power plants in the James River basin. The conclusion is that generally selenium follows regional patterns, but during winter local sources may play a key role because transportation is often inhibited.

The second observation is that the mean concentrations are often less than 0.1 ng/m^3 at sites throughout the Pacific Northwest, the northern Rocky Mountains, Denali, and Mauna Loa. These averages are similar to the concentrations of 0.04 to 0.06 ng/m^3 observed at remote sites at Barrow, Alaska; Northwest Territories, Canada; Bolivia; and American Samoa (Mosher and Duce, 1983). Thus, the concentrations at several regions in the United States are near the world baseline.

The third observation is that the mean selenium concentrations in the East are generally much larger than in the West. The median East site has a concentration four times that at the median West site in summer and nine times in winter. The northern Minnesota sites are more like western sites than eastern sites. The lowest concentrations in the East are in the Northeast and southern Florida and the largest in the Appalachian mountains. The concentrations in the Appalachian mountains are generally 1 to 2 ng/m^3.

Figure 3 shows the mean sulfur concentrations for the same sites and seasons. The spatial patterns for sulfur and selenium are similar. For both elements, the average eastern concentrations are much larger than those in the West. In the East, the concentrations are highest in the Appalachian region. In the West, the concentrations are lowest in the portions of the Northwest away from Puget Sound and Portland and highest in the southern Southwest.

The ratio of the S and Se means at a given site reflects two factors: the emissions ratio of total sulfur (gaseous and particulate) to Se and the amount of transformation of sulfur from SO$_2$ (gas) to SO$_4$ (particle). For a given emissions ratio, a higher S/Se ratio at the site indicates more transformation. Figure 4 shows the S/Se ratio for summer and for winter. The ratios are not shown for sites at which selenium was identified on less than half the samples. Both maps show relatively uniform ratios throughout the network. In summer, the median ratios are slightly lower in the West than in the East (1800 vs. 2300). In winter, the median ratios are slightly higher in the West than in the East (1300 vs. 1000). At the eastern sites, summer ratios are more than twice the winter ratios. The highest S/Se ratios are at the sites in the Pacific Northwest, suggesting that the sulfur sources in this region are low in selenium. If the winter S/Se ratio of 3000 obtained at Columbia River Gorge is assumed to be appropriate at the sites with low sulfur, then a typical winter mean sulfur concentration in the northwest region of 70 ng/m^3 would yield a mean selenium concentration of 0.02 ng/m^3, which is below the typical MDL of 0.03 ng/m^3. The low summer ratio at Badlands suggests that there is either a source with high selenium or a source that is so near that not much sulfur transformation is allowed. In New England, the summer ratios are all large and fairly uniform, which is consistent with sources that have uniform emissions ratios and are far enough away to permit most of the SO$_2$ to be converted. Low winter ratios of 500 at Dolly Sods, West Virginia, Jefferson, Virginia, Indian Gardens, Arizona, and Pinnacles, California, would be consistent with relatively little conversion of SO$_2$.

The concentration of selenium at the sampling site depends on two factors: the emission rate of the various sources and the transport from the sources to the site. The transport factor depends on wind trajectories, deposition during transport, dispersion of the air mass, and the height of the boundary layer. The concentration of sulfur depends on these two factors plus a third, the amount of conversion of SO$_2$ (gas) to SO$_4$ particles. This conversion would depend on the transport time, the temperature, and relative humidity, as well as increased deposition while in the gas phase. The relative importance of these processes can

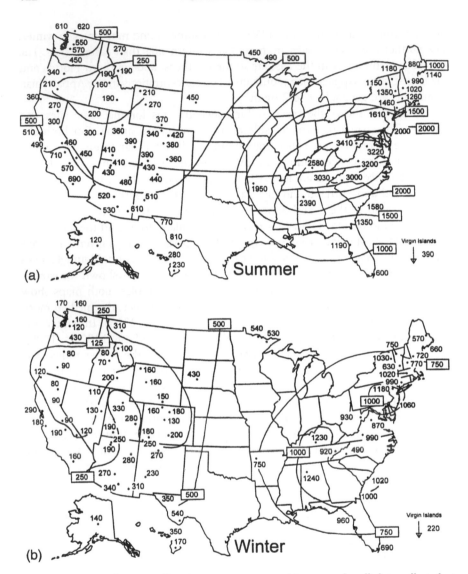

FIGURE 3   Maps of mean sulfur, in nanograms per cubic meter, for all data collected at
IMPROVE sites between June 1992 and August 1996: (a) summer (June–August) and
(b) winter (December–February). The concentrations are standardized to sea level.

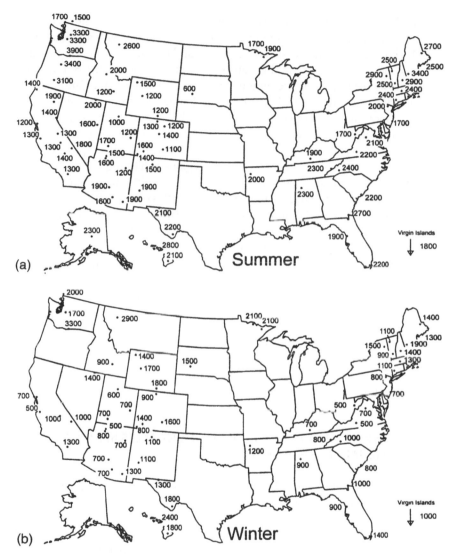

FIGURE 4 Maps of the ratio of mean sulfur to mean selenium for (a) summer and (b) winter. The ratios are not shown for sites at which selenium was identified on less than half the samples.

be seen by comparing the summer and winter means. Figure 5a shows that the summer/winter ratios of selenium means are close to unity, with a median ratio of 0.9. This indicates that the average emission and transport factors are the same summer and winter. There are three areas with ratios higher than 1.8, indicating more emission/transport of selenium in summer: Rocky Mountain National Park, in Colorado; Badlands, South Dakota; Mount Rainier, Washington; and Sequoia and San Gorgonio, California. Sites in the Northeast and Southeast have ratios around 0.6, suggesting lower emission/transport in summer. Figure 5b shows that the summer/winter ratio for sulfur is greater than 1 at all except three sites, with a median ratio of 1.9. There are three regions exhibiting a ratio of means consistently higher than 3: Appalachia (3.5 ± 0.5); California, excluding Point Reyes (3.6 ± 1.3); and Puget Sound (4.9 ± 1.2). If we compare the two maps, of Figure 5, we find that the sulfur ratio is around two to three times the selenium ratio at most sites in the East, at many sites in the Southwest, and at Mount Rainier. At most of the sites in the network, the sulfur concentrations are higher in summer than in winter. If we assume that sulfur and selenium come from common sources, and that the emission ratios of SO_2/Se do not change from summer to winter, then the difference in summer/winter ratios for sulfur and selenium at these sites reflects an increase in the conversion of SO_2 to SO_4 during the summer compared to the winter. This analysis suggests that conversion plays the key role for higher summer sulfur in much of eastern United States, in Arizona, and at Pinnacles, California, while emission transport plays the main role at San Gorgonio and Sequoia, California, and at three sites near the eastern edge of the Rocky Mountains. Both mechanisms are involved at Mount Rainier.

V. SELENIUM AND SULFUR ON INDIVIDUAL SAMPLES IN THE IMPROVE NETWORK

The relationships for individual samples will differ from those for seasonal mean samples if there are variations in the S/Se ratio from day to day that are smoothed out. Figure 6 shows the correlation coefficients (r^2) at each site for the regression of selenium and sulfur, based on samples in which selenium was identified. A high correlation for a given site over a season requires (1) that the sources have relatively uniform SO_2/Se ratios and (2) that the fraction of SO_2 converted to sulfate during transport be constant. The correlation is much better at eastern sites than at western sites. The highest coefficients are found at all the sites in the Northeast, with r^2 of 0.8 to 0.95. A high correlation is also found at the northernmost site in the Appalachian region, at Shenandoah, and at Brigantine, on the New Jersey coast. The coefficients rarely exceed 0.5 at sites in the West. The high correlation coefficients at all the sites in the Northeast and the similar S/Se ratios of means (Figure 4) suggest that the sulfate throughout the Northeast

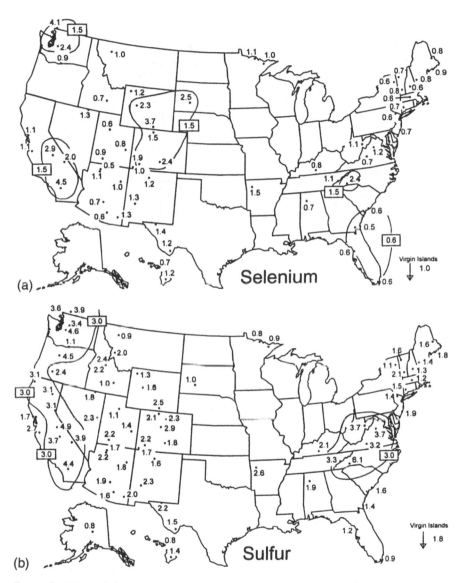

FIGURE 5   Maps of the summer/winter ratios for (a) selenium and (b) sulfur. The ratios are not shown for sites at which selenium was identified on less than half the samples.

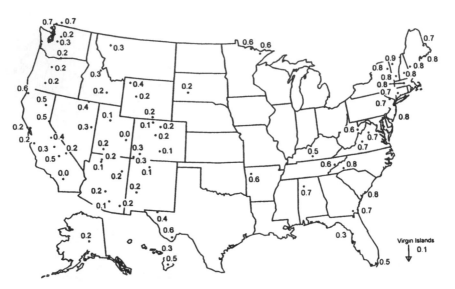

FIGURE 6 Map of correlation coefficients (r^2) for selenium versus sulfur for all summer (June–August) data collected between June 1992 and August 1996. The coefficients are based only on cases in which selenium was identified. The coefficients are not shown for sites at which selenium was identified on less than half the samples.

comes from a single set of sources. To achieve a uniform conversion at all sites, the sources probably would have to be far enough away to permit most of the SO_2 to convert. This is consistent with the hypothesis that the primary sulfate sources for the Northeast are coal-fired power plants in the Ohio Valley.

Figure 7 compares the sulfur and selenium concentrations for all eastern and most western remote sites in summer 1993. (Badlands and the western marine sites are excluded.) This again shows that the correlation in the East is much higher than in the West ($r^2 = 0.76$ vs. $r^2 = 0.39$). In the East, most of the samples have Se/S ratios between 1300 and 3000, with 5 cases out of 565 having ratios around 700. There are no samples with high sulfur or high selenium concentrations that do not also have high concentrations of the other element.

In the West, the situation is somewhat different. There are several samples with high S/Se ratios, indicating that some of the sulfur has low or no selenium. There is also a large group of samples with low ratios, indicating less particulate sulfur relative to selenium. Note that the low correlation is true not only for the West as a whole, but also at each western site. Any explanation must account for the fact that while the individual S/Se ratios are more variable at western sites, the seasonal average gives a ratio that does not vary significantly from site to site. One hypothesis proposes a more variable SO_2 to sulfate conversion rate

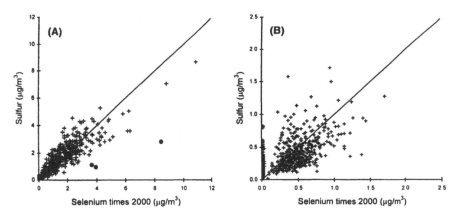

FIGURE 7 Comparison of selenium and sulfur for summer 1993 (June–August). (A) for 514 samples at the sites in the East; (B) for 732 samples at the sites in the West, excluding data from Badlands and the marine sites. Samples with selenium concentrations near zero are cases in which selenium was below the MDL. In the West, 32% of the samples, primarily in the Northwest, were below the MDL. The diagonal lines represent a S/Se ratio of 2000. The correlation coefficient (r^2) is 0.76 in the East and 0.39 in the West. If the three circled samples from Dolly Sods are deleted, the r^2 increases from 0.75 to 0.80.

in the West. If the conversion rate were the sole factor, selenium would correlate with total sulfur [particulate sulfur + (0.5 × SO$_2$)] better than with sulfate alone. However, at the sites where SO$_2$ was also measured, there was no improvement in correlation. The most likely explanation is that there are fewer but more diverse sulfur sources in the West. With fewer sources, the impact of a single source for a given 24-hour period will be more significant. The S/Se ratio at the site would depend on the emission ratios and on the distance from the source. A very high S/Se ratio in the western regression plot suggests a selenium-poor source. A very low S/Se ratio suggests either a selenium-rich source or a source with a normal emission factor combined with low conversion during transport. The net result is to give more variation from sample to sample, but a uniform S/Se ratio when averaged over a season.

VI. SELENIUM AND SULFUR AT GREAT SMOKY MOUNTAINS NATIONAL PARK

Figure 8 shows 12-hour concentrations of selenium and sulfur at Great Smoky Mountains National Park during the summer of 1995. In this region most sulfur and selenium is derived from coal combustion. There are numerous coal-fired

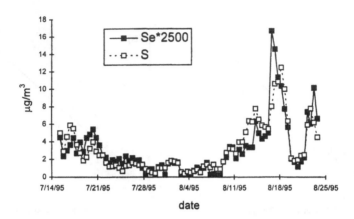

FIGURE 8 Time plot of selenium and sulfur concentrations at Great Smoky Mountains National Park between July 14 to August 25, 1995. The ratio of mean S to mean Se is 2500.

power plants directly upwind, several within 100 miles of the site. Overall, the selenium and sulfur concentrations correlate reasonably well ($r^2 = 0.74$). During the sulfate episode from August 16–19, the sulfur is slightly larger than 2500 times selenium, probably indicating greater conversion of sulfur from gas to particle. The episode from days 228 to 231 is even more interesting. During the night of August 16–17, a plume with elevated selenium, SO_2, and organics, reached the site and remained because of stagnant meteorological conditions. The selenium decreased over a period of 72 hours. The peak in the particulate sulfur concentrations was about 12 hours after the peak in the selenium concentrations, as the stagnant SO_2 was transformed into particulate sulfur. Both elements then decreased until the air mass was removed. This example illustrates why sulfur and selenium do not necessarily correlate even when they probably originate from a common source. Note that selenium provided information on the sulfur sources that could not be obtained by sulfur alone.

VII. SELENIUM IN THE SIERRA NEVADA

The roles of selenium in the Sierra Nevada and in the Appalachians are quite different. The California Central Valley has industrial sources of sulfur and selenium, as well as regions with soil of very high selenium content. However, coal-fired power plants are totally absent in California. Extensive studies indicate that particles can be efficiently transported from the Central Valley into the Sierra Nevada terrestrial and aquatic ecosystems. These studies were included in the Sierra Nevada Ecosystem Project (SNEP), a large congressionally mandated study of all aspects of the Sierra Nevada ecosystem. Table 1 gives the mean values of fine particulate matter and its major and minor constituents including selenium, in the Cascades and Sierra Nevada from five IMPROVE sites. There is a north-to-south increase in most parameters, especially for the anthropogenically dominated nitrate and sulfate aerosols and several trace elements including selenium. This suggests that these trace elements also have strong anthropogenic sources. The concentrations of sulfur and selenium at Sequoia in the summer are the highest among the western IMPROVE sites. The concentrations of sulfur, nitrate, and trace elements in the southern Cascades (Crater Lake and Lassen Volcanic) are close to global background levels, and similar to those at our sites at the global baseline Mauna Loa Observatory, Hawaii, and Denali National Park, Alaska.

Figure 9 shows that that there is a moderate correlation ($r^2 = 0.57$) between selenium and sulfur at Sequoia National Park. (Figure 6 shows that the other Sierra Nevada sites have a similar correlation.) The plot on the left compares the

TABLE I   Average $PM_{2.5}$ Concentrations of Mass (All Particles), Major Constituents, and Selected Trace Elements for Five Remote Sites in the Sierra Nevadas and Southern Cascades for Samples Collected March 1993 to February 1994

	Sites (arranged north → south)				
	Sequoia	Yosemite	Bliss	Lassen Volcanic	Crater Lake
Mass (μg/m³)	9.58	4.38	3.33	3.07	2.71
Major Constituents (μg/m³)					
Organic[a]	3.16	1.64	1.32	1.26	0.94
Sulfate[b]	2.06	1.03	0.73	0.61	0.54
Nitrate[c]	2.08	0.44	0.31	0.26	0.19
Soil[d]	0.99	0.55	0.47	0.49	0.45
Trace elements (ng/m³)[e]					
Selenium	0.33	0.16	0.09	0.08	0.05
Zinc	3.52	1.45	1.09	1.13	4.92
Bromine	2.71	1.44	1.25	0.93	0.73
Lead	1.17	0.65	0.54	0.45	0.51
Copper	1.79	0.48	0.32	0.34	0.40
Nickel	0.13	0.06	0.05	0.05	0.12

[a] Total organic material, including carbon, hydrogen, oxygen, and nitrogen: estimated to be 1.4 times the total organic carbon measured on the quartz filter.
[b] Sulfate is 4.125 times the sulfur measured on the Teflon filter and represents ammonium sulfate.
[c] Nitrate is 1.29 times the nitrate ion measured on the nylon filter following a denuder to remove nitric acid vapor and represents the particulate ammonium nitrate.
[d] Sum of soil-related elements measured on the Teflon filter, plus their normol oxides.
[e] Measured on the Teflon filter.

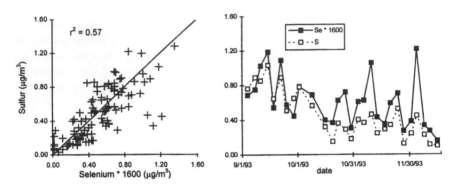

FIGURE 9   Comparison of selenium and sulfur concentrations at Sequoia National Park. *Left:* comparison of all samples collected between March 1993 and February 1994 (correlation coefficient r^2-0.57); *right:* time plots for a limited period, September 1 through December 15, 1993

concentrations for an entire year, while the right-hand plot shows concentration versus time for the part of 1993 that includes the samples with the largest differences between sulfur and selenium. The elevated selenium concentrations could indicate either a selenium-rich source or lower conversion from SO_2. A possible selenium-rich source would be the Central Valley soil, which has elevated selenium. However, these samples did not have elevated soil elements. The largest differences occurred between October 20 and December 4, 1993. During this time, selenium and sulfur still correlate, but with a S/Se ratio of around 800, rather than the normal 1600. The most probable cause is a shorter transport time.

VIII.   CONCLUSIONS

The mean fine particle selenium concentrations at remote sites throughout the United States range from approximately 0.05 ng/m^3 at several regions in the West during winter to 1 ng/m^3 in the Appalachian Mountains. Selenium from the marine biosphere is significant at sites with major marine influences, but because of the large particle size of marine selenium, it settles rapidly, so the contribution at most continental sites is negligible. The continental biosphere does not appear to contribute a significant fraction of the particulate selenium at any site. Sites with low sulfur have correspondingly low selenium. Nearly all the particulate selenium appears to be associated with sulfur. This is especially evident in the East, where the correlation between the two elements for individual samples is very high ($r^2 = 0.8$) and the ratio of sulfur and selenium mean concentrations is relatively uniform. In the West, the correlation for individual sample is lower

($r^2 = 0.4$), but the ratio of means is similar. This suggests that even in the West there are no major selenium sources without sulfur.

The mean concentrations exhibit reasonable geographic patterns, suggesting that the selenium observed at remote sites is from regional rather than local sources, especially in summer. The IMPROVE network has a few cases of local sources that have a greater influence in winter than in summer because of inhibited transport.

The measurement of both sulfur and selenium provides valuable information on sulfur behavior. Examination of the summer/winter ratios of selenium and sulfur indicates that the summer sulfate concentrations at all sites in the East and many in the West are higher than those in winter because of greater transformation of SO_2 to SO_4 rather than because of better transport. Examination of 12-hour samples at Great Smoky Mountains shows that the maximum concentrations of particulate sulfur may occur several hours after a plume of SO_2, SO_4, and Se has reached a site, provided the air mass remains in the vicinity, as SO_2 is converted to particulate sulfur.

REFERENCES

Andren, A. W., D. H. Klein, and Y. Talmi. 1975. Selenium in coal-fired steam plant emissions. *Environ. Sci. Technol.* 9: 856–858.

Eldred, R. A. 1997. Comparison of selenium and sulfur at remote sites throughout the United States. *J. Air Waste Manage. Assoc.* 47:204–211.

Eldred, R. A., T. A. Cahill, L. K. Wilkinson, P. J. Feeney, J. C. Chow, and W. C. Malm. 1990. Measurement of fine particles and their chemical components in the IMPROVE/NPS networks. In C. V. Mathai (ed.), *Visibility and Fine Particles,* pp. 187–196. Air & Waste Management Association, Pittsburgh.

Eldred, R. A., T. A. Cahill, and R. G. Flocchini. 1997. Composition of $PM_{2.5}$ and PM_{10} aerosols in the IMPROVE network. *J. Air Waste Manage. Assoc.* 47:194–203.

Ellis, W. G., R. Arimoto, D. L. Savoie, J. T. Merrill, R. A. Duce, and J. M. Prospero. 1993. Aerosol selenium at Bermuda and Barbados. *J. Geophys. Res.* 98:12,673–12,685.

Fan, T. W.-M., A. N. Lane, and R. M. Higashi. 1997. Selenium biotransformations by a euryhaline microalga isolated from a saline evaporative pond. *Environ. Sci. Technol.* 31:569–576.

Kennish, M. J. 1994. *Practical Handbook of Marine Science.* CRC Press, Boca Raton, FL.

Malm, W. C., J. F. Sisler, D. Huffman, R. A. Eldred, and T. A. Cahill. 1994. Spatial and seasonal trends in particle concentration and optical extinction in the United States. *J. Geophys. Res.* 99:1347–1370.

Mosher, B. W., and R. A. Duce. 1983. Vapor phase and particulate selenium in the marine atmosphere. *J. Geophys. Res.* 88:6761–6768.

Mosher, B. W., and R. A. Duce. 1987. A global atmospheric selenium budget. *J. Geophys. Res.* 92:13,289–13,298.

Mosher, B. W., and R. A. Duce. 1989. The atmosphere. In M. Ihnat (ed.), *Occurrence and Distribution of Selenium,* pp. 295–325. CRC Press, Boca Raton, FL.

Ondov, J. M., C. E. Choquette, W. H. Zoller, G. E. Gordon, A. H. Biermann, and R. E. Heft. 1989. Atmospheric behavior of trace elements on particles emitted from a coal-fired power plant. *Atmos. Environ.* 23:2193–2204.

Rael, R. M., E. C. Tuazon, and W. T. Frankenberger, Jr. 1996. Gas-phase reactions of dimethyl selenide with ozone and the hydroxyl and nitrate radicals. *Atmos. Environ.* 30:1221–1232.

Taylor, S. R., and S. M. McLennan. 1985. *The Continental Crust: Its Composition and Evolution.* Oxford University Press, Oxford.

31

Phytoremediation of Selenium

NORMAN TERRY and ADEL ZAYED

University of California at Berkeley, Berkeley, California

I. INTRODUCTION

Selenium (Se) pollution of soil and water is a major environmental problem in many areas of the world. High concentrations of Se occur naturally in some soils, especially those derived from Cretaceous shale parent materials. In well-aerated alkaline soils, oxidized forms of Se are easily mobilized into solution. With the irrigation of crops, Se concentrations in the agricultural drainage waters may reach hazardous levels. Other anthropogenic activities also lead to Se pollution of water. Selenium contamination of coastal waters is a major problem in areas close to oil refineries, which release substantial amounts of Se in their wastewater discharge. Electric utilities' aqueous discharges, which result from the storage of coal and other by-products (e.g., coal pile runoff, coal pile seepage, coal ash landfill discharges) present other significant sources of environmental Se pollution. Unfortunately, the few available technologies for wastewater treatment (e.g., chemical, microbiological, or electrochemical treatments) are not cost-effective in many cases and produce a large amount of unwanted by-product in the form of hazardous waste that must be shipped to a toxic landfill.

An inexpensive, environmentally friendly alternative to cleaning up toxic trace elements in soils is phytoremediation, an emerging technology that is receiving increasing recognition (Ernst, 1988; McGrath et al., 1993; Wentzel et al., 1993; Baker et al., 1994; Kumar et al., 1995; Salt et al., 1995). Plants, especially Se accumulators, are very effective in removing Se from soil and water. Once in the plant, the Se is transported to stems and leaves, where it may be harvested and removed from the site. Another approach is plant volatilization of Se: in this case, Se absorbed by the plant may be metabolized (mostly in the roots) and released to the atmosphere in relatively nontoxic volatile forms (e.g., dimethyl selenide). In this process, volatilization by plants may be assisted by the presence of microbes, particularly bacteria, either in the rhizosphere or inside the root.

In this chapter we review the research efforts that have been made to establish phytoremediation as a new technology for the cleanup of Se pollution in soil and water. We include research dealing with Se removal from soils by plant uptake and accumulation, Se removal by plant volatilization, Se removal in the rhizosphere, Se removal by wetlands, and advances in the genetic engineering of Se-volatilizing plants.

II. PHYTOREMEDIATION BY PLANT UPTAKE AND ACCUMULATION OF SELENIUM

A. Availability of Selenium for Plant Uptake

Selenium is a group VI metalloid that is found directly below sulfur (S) in the periodic table of the elements. Like S, Se can exist in a variety of oxidation states: selenide (Se^{2-}), elemental or "colloidal" Se (Se^0), selenite (Se^{4+}), selenate (Se^{6+}), and several organic and volatile Se compounds (Rosenfeld and Beath, 1964). Selenium removal from polluted soils and waters by uptake and accumulation in plant tissues is mostly governed by the chemical species of the element (Mikkelsen et al., 1989; Blaylock and James, 1994). While selenate and selenite, the oxidized forms of Se, are highly available to plants because of their high solubility, selenide and elemental Se (reduced forms of Se) are insoluble and therefore generally unavailable to plants. Oxidized forms are typically found in well-aerated alkaline soils, while reduced forms are found in waterlogged acid soils (Allaway, 1971). Most research regarding the availability of different forms of Se for plant uptake and accumulation has focused on the two oxidized forms: selenate and selenite. Generally, Se concentrations in plants grown in selenate-rich soils are an order of magnitude higher than plants grown in selenite-rich soils (Bañuelos and Meek, 1990).

B. Uptake and Metabolism of Selenium by Plants

Since Se and S share very similar chemical properties, Se compounds are absorbed and metabolized by mechanisms similar to those for their S analogs. This results in competitive interactions between the two elements at both the uptake and assimilation levels. Selenium is taken up by plants as selenate, selenite, and organic selenium. The uptake of selenate and organic Se is driven metabolically, whereas the uptake of selenite may have a passive component (Ulrich and Shrift, 1968; Abrams et al., 1990). Furthermore, organic forms of Se may be more readily available to plant uptake than inorganic forms (Table 1). Once inside the plant, Se is almost certainly metabolized by the enzymes of the S assimilation pathway (Brown and Shrift, 1982; Terry and Zayed, 1994).

The chemical species of Se also influences the translocation and distribution of Se among different plant parts. Unpublished data from our laboratory show that Se accumulation in shoots and roots of broccoli was affected substantially by the form of Se supplied; when plants were supplied with 20 μM Se as selenate, selenite, or Se-methionine, Se accumulation in leaves was best when the element came from selenate, followed by Se-methionine and then by selenite, while in roots, it was best when supplied as Se-methionine, followed in order by selenite and selenate (Fig. 1). Thus, more Se accumulates in aboveground tissues when the available form of Se is selenate versus selenite or Se-methionine.

Plant species are not equally effective in removing Se by uptake and accumulation in their tissues. Initial research by Hurd-Karrer (1935) demonstrated that plant species with a high S requirement (such as members of the Brassicaceae family) tend to absorb more Se than other plant species. This conclusion has been confirmed repeatedly by other researchers (Bañuelos and Schrale, 1989; Bañuelos and Meek, 1990; Terry et al., 1992). For instance, recent research by Bañuelos and his colleagues showed that *Brassica juncea*, an S-loving plant, has a better capacity to remove Se from soil, probably because of its inability to discriminate between absorbing selenate (SeO_4^{2-}) and sulfate (SO_4^{2-}) (Bañuelos and Schrale, 1989; Bañuelos and Meek, 1990). In our laboratory we compared 14 crop species for their ability to absorb Se from a nutrient solution containing 20 μM Se and found that rice, broccoli, and cabbage accumulated Se to a much higher degree than did all the other 11 plant species (Table 2) (Terry et al., 1992).

Generally, plants may be divided into three groups, depending on their ability to accumulate Se in their tissues (Rosenfeld and Beath, 1964; Brown and Shrift, 1982). Plant species that preferentially grow on soils high in Se and accumulate Se in the order of several thousand milligrams per kilogram dry

TABLE I Bioavailability of Different Chemical Species of Se for Plant Uptake

Se species	Amount of Se added (μg)	Relative recovery (%)	
		A. bisulcatus	Wheat grass
Selenite	7000	6	25
Se-methionine	4200	23	93
Se-cysteine	4200	29	27

Source: Adapted from Williams and Mayland (1992).

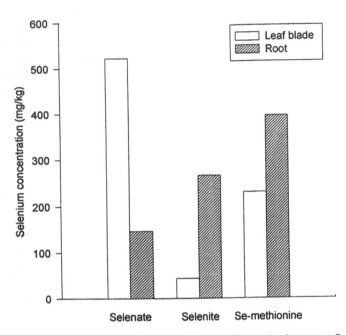

FIGURE 1   Selenium concentration in leaves and roots of broccoli plants as influenced by the form of Se supplied at 20 μM in a half-strength Hoagland's solution.

weight, are referred to as group I or primary Se accumulators (or indicators). Group I plants include several species of the legume genus *Astragalus* and some *Neptunia* species. Group II plant species are not confined to seleniferous soils but may accumulate up to 1000 mg of Se per kilogram dry weight. Termed secondary Se accumulators (or Se absorbers), these plants include some *Aster*, *Atriplex*, *Castilleja*, and *Grindelia* species. Group III plant species do not accumulate Se beyond 50 mg/kg dry weight under field conditions. These species, which include many grains and grasses, are known as nonaccumulators of Se.

C.   Phytoremediation Potential of Different Plant Species

For a plant species to be effective in phytoremediation, it must be capable of achieving a significant reduction in Se concentration in soil or water over a reasonably short period of time. Ideally, one would choose a plant species that exhibits a high rate of shoot growth combined with high Se concentrations in shoot tissues. Parker and Page (1994) proposed that vegetation management of the Se contamination problem would be feasible if the selected plant species could attain a minimum tissue Se concentration of 100 mg/kg and produce a dry matter yield of 4 to 16 t/ha. They based their assumption on an estimated soil Se burden of 15 kg/ha.

TABLE 2 Selenium Concentrations in Plant Tissues for 14 Crop Plants Grown in Quarter-Strength Hoagland's Solution with Se supplied at 20 μM as Sodium Selenate

Plant species	Selenium concentration (mg/kg dry weight)		
	Leaf	Stem	Root
Alfalfa	248	55	59
Barley	393	107	136
Bean	103	91	124
Broccoli	435	254	493
Cabbage	203	198	470
Carrot	128	57	104
Cotton	144	68	141
Cucumber	181	76	81
Eggplant	184	118	104
Lettuce	77	44	71
Maize	93	55	61
Sugar beet	28	23	137
Rice	502	155	114
Tomato	236	86	129

Source: Adapted from Terry et al. (1992).

Utilizing existing information about Se-accumulating plants, researchers evaluated the ability of primary (group I), secondary (group II) and non-Se-accumulating plants (group III) to absorb Se under a variety of conditions (Rosenfeld and Beath, 1964; Wu et al., 1988; Mayland et al., 1989; Bañuelos and Meek, 1990; Parker et al., 1991; Bell et al., 1992; Retana et al., 1993). Certain species of *Astragalus* (e.g., *A. bisulcatus*) accumulated the most Se; unfortunately, *Astragalus* spp. grow slowly and have undefined growth requirements, which limits their use for effective phytoremediation. Moreover, *Astragalus* spp. have not been genetically improved, and their general traits (vigor, resistance to disease and pests, responsiveness to fertilization, etc.) are neither superior nor well characterized.

Indian mustard (*Brassica juncea*) appears to be a plant species that possesses all the characteristics needed to be an excellent phytoremediator of Se (Bañuelos and Schrale, 1989; Bañuelos and Meek, 1990). This group II plant species has a typical plant Se content of 350 mg/kg (Bañuelos and Schrale, 1989). Bañuelos and his colleagues estimated that at a plant density of 247,000 Indian mustard plants per hectare, and with a dry weight per plant averaging 8 g, 800 kg of hay would be produced at each harvest. Thus, 690 g of Se could be removed per hectare per harvest. With five harvests, a total of 1134 g of Se would be removed in the hay if the rate of Se uptake observed in the greenhouse were sustained in the field. In follow-up multiyear studies, Bañuelos et al. (1995) used *Brassica juncea* and some other plant species to determine their effectiveness in lowering soil Se concentrations under field conditions. The results showed that after three years *Brassica juncea* had lowered the total soil Se inventory between 0 and 75 cm by almost 50%. Other plant species also significantly reduced total soil Se content but to a lesser extent (Table 3). These investigators also demonstrated that in each year of the three years of the field study, *Brassica juncea* accumulated substantially higher concentrations of Se in its shoot and root tissues than the other plant species. Thus, *Brassica juncea* seems to be an appropriate plant species for use in the vegetation management of Se.

D. Salinity

The removal of Se from soil by plants may be considerably impacted by the magnitude and type of salinity. The impact of sulfate salinity on Se uptake by plants exceeds that of chloride salinity because of the competitive interaction between Se and S. It has been established for a number of plant species that saline levels of sulfate inhibit plant uptake of selenate drastically (Mikkelsen et al., 1989; Shennan et al., 1990; Wu and Huang, 1991b; Zayed and Terry, 1992). Fortunately, not all plant species are affected to the same extent by sulfate salinity. Selenium accumulator species may exhibit preferential uptake of Se relative to S. The Se accumulator *Astragalus bisulcatus* accumulates very high levels of Se even when growing in gypsiferous soils high in soluble sulfate (Rosenfeld and Beath, 1964) or at high external sulfate levels provided hydroponically (Bell et al., 1992). Furthermore, much of the S in field soils may not be present as soluble sulfate. Thus, the inhibition of selenate uptake by S is often less pronounced in field soils than in water culture studies where both selenate and sulfate are added as soluble salts.

TABLE 3 Removal of Native Soil Selenium by Some Plant Species After Three Croppings Under Field Conditions

Plant species	Soil Se (mg/kg)		Soil Se removed (%)
	Preplant	Postharvest	
Control	0.99	0.82	17
Brassica juncea	1.12	0.60	46
Tall fescue	1.25	0.87	30
Birdsfoot trefoil	0.98	0.56	43
Kenaf	0.96	0.66	31

Source: Adapted from Bañuelos et al. (1995).

Plant species that have been identified as fast-growing Se accumulators should also be tested for their salt tolerance, since high Se levels in soil are often associated with high salinity. The effect of saline concentrations on Se uptake by plants has been investigated only in recent years. Some salt/Se research has been conducted with different plant species of *Festuca* (Wu et al., 1988), *Atriplex* (Watson et al., 1993), *Astragalus* (Bell et al., 1992), and *Brassica* (Bañuelos et al., 1993, 1997). This research showed that Indian mustard, canola, tall fescue, and in particular the Se accumulators *Astragalus bisulcatus* and *A. racemosus,* are good candidates for Se removal from saline soils (Parker et al., 1991; Wu and Huang, 1991a; Bañuelos et al., 1993, 1997). Generally, there is a decrease in shoot accumulation of Se with increasing salt levels. In our research, we compared the feasibility of using four different plant species including canola, tall fescue, kenaf, and birdsfoot trefoil to remove soil Se under increasing salt regimes. We found that canola was the best of the four in removing Se from selenate-contaminated soils under saline conditions (Fig. 2). Canola accumulated three times more Se than kenaf, and birdsfoot trefoil and at least 10 times as much Se as tall fescue.

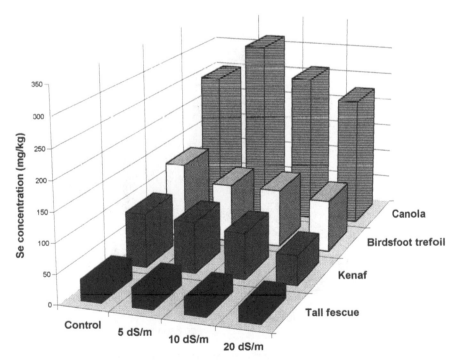

FIGURE 2 Selenium concentration in aboveground tissues of four plant species grown in Se-enriched soil as affected by the level of chloride salinity in the soil.

Canola also tolerated high levels of salinity (up to 20 dS/m), although there was some growth reduction (Fig. 3).

E. Disposal of Selenium-Laden Plant Material

The disposal of Se-laden plant material is an important concern in any phytoremediation scheme based on the uptake and accumulation of Se in plant tissue. Such plant tissue must be harvested, removed from the site, and disposed in a safe manner. There are a number of options for the safe disposal of plant materials rich in Se. Because Se is an essential trace element for adequate nutrition and health in mammals, one disposal option is to use selenized plant materials as a forage blend in Se-deficient regions to improve the Se status of animals. Another option is to add Se as a source of organic Se fertilizer to soils supporting forage crops. Bañuelos et al. (1991) demonstrated that the addition of Se-enriched plant tissues to soil caused increased accumulation of Se in alfalfa. Alternatively, if concentrations of other toxic trace elements in plant tissues (cadmium, chromium,

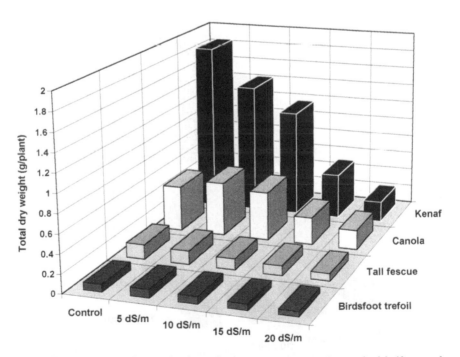

FIGURE 3 Biomass production by four plant species grown in Se-enriched half-strength Hoagland's solution as affected by the level of chloride salinity in the solution. (Adapted from Bañuelos et al., 1997.)

arsenic, mercury) exceed the safe limits for animal consumption, the Se-laden plant material may serve instead as fuel for the generation of electricity.

III. PHYTOREMEDIATION BY PHYTOVOLATILIZATION

The ability of plants, or plant–microbe associations, to take up Se and metabolize it to volatile forms (e.g., dimethyl selenide) may be referred to as "phytovolatiliza-tion." The idea that biological volatilization of Se could be used in the cleanup of Se from Se-contaminated soils was first proposed by Frankenberger and Karlson (1988), who suggested the use of soil microbes, especially soil fungi, for this purpose. Later, it was demonstrated that plants too possess the capacity to volatil-ize Se (Duckart et al., 1992; Terry et al., 1992; Biggar and Jayaweera, 1993). Plant (and microbial) volatilization of Se is an especially attractive method for the bioremediation of Se pollution because it removes Se completely from the local ecosystem, thereby minimizing its entry into the food chain (Terry et al., 1992; Terry and Zayed, 1994).

Several studies have shown that the addition of plants to soil increases the rate of Se volatilization above that for soil alone. Zieve and Peterson (1984) observed that planting barley improved volatilization of ^{75}Se from soil. Biggar and Jayaweera (1993) found that soil plus barley volatilized 19.6 times faster than soil alone. Duckart et al. (1992) obtained relative volatilization rates as high as 225% (expressed as a percentage of that of fallow soil) for *Astragalus bisulcatus*. They also calculated the time required for an amount of Se equal to half the selenate added to the soil to be removed by both plant uptake and volatilization. This time was 461 days for the soil alone, and ranged from 5.9 to 12 days when different plants were added to the soil (Duckart et al., 1992).

The first indication that volatile Se can be released from plant tissues was reported by Beath et al. (1935), who noticed that volatile Se compounds were released from *Astragalus bisulcatus* plant materials during storage. Later, Lewis et al. (1966) demonstrated that volatile Se compounds are released from intact higher plants when grown in Se-rich media. They showed that the Se volatilization process is not restricted to Se-accumulators but also occurs in nonaccumulator plant species. However, the chemical form of the volatile Se released by nonaccu-mulators was found to be different from that produced by Se accumulator plants. Evans et al. (1968) found that Se accumulators mainly volatilize dimethyl diselen-ide (DMDSe), while Lewis (1971) showed that Se nonaccumulators typically release dimethyl selenide (DMSe) and do not produce DMDSe.

A. Selenium Volatilization by Different Plant Species

Selenium volatilization is a widespread phenomenon among different plant spe-cies. Terry and coworkers compared volatilization rates for 18 plant species grown

under standardized environmental conditions (Terry et al., 1992). Their data
show that plant species from the family Brassicaceae and from rice were superior
volatilizers of Se. These species (viz., rice, broccoli, cabbage, cauliflower, Indian
mustard, Chinese mustard) volatilized Se at a rate exceeding 1500 μg of Se per
kilogram of dry weight per day when supplied with 20 μM Se as sodium selenate
(Table 4). Sugar beet, bean, lettuce, and onion exhibited very low rates of Se
volatilization (< 250 μg/kg/day). Other plant species tested (carrot, barley, alfalfa,
tomato, cucumber, cotton, eggplant, maize) showed intermediate rates of 280 to
750 μg/kg/day (Table 4). The rate of Se volatilization by different plant species
was found to be highly correlated with Se concentrations in plant tissues. However,
no correlation was obvious between salt tolerance of various plants and their
ability to take up and volatilize Se (Terry et al., 1992).

Duckart et al. (1992) measured volatilization rates for five plant species
(tomato, alfalfa, broccoli, tall fescue, and *Astragalus bisulcatus*) grown in a greenhouse, in pots filled with Panoche fine sandy loam soil containing 16.5 μg/kg
soil extractable Se and amended with 500 μg/kg Se as sodium selenate. Their
data show large differences among the plant species tested. *Astragalus bisulcatus*

TABLE 4   Rates of Selenium Volatilization by 18 Crop Plants Grown in
Quarter-Strength Hoagland's Solution with Se Supplied at 20 μM as
Sodium Selenate

Plant species	Rate of selenium volatilization (μg/kg/day)
Alfalfa	280
Barley	489
Bean	217
Broccoli	2393
Cauliflower	2286
Cabbage	2309
Carrot	548
Chinese mustard	2015
Cotton	478
Cucumber	752
Eggplant	462
Indian mustard	2429
Lettuce	179
Maize	420
Onion	229
Sugar beet	240
Rice	1495
Tomato	742

Source: Adapted from Terry et al. (1992).

and broccoli posted the highest rates of Se volatilization, both on a leaf area and soil dry weight basis, followed by tomato, tall fescue, and alfalfa, respectively. These results are in agreement with those obtained by Terry et al. (1992) for plants cultured hydroponically, indicating the importance of the solution culture experiments in identifying crops that appear to be superior volatilizers of Se.

B. Volatilization by Roots and Shoots

For some time it was believed that plant foliage was the main site for plant volatilization of Se. This assumption came from data of Lewis et al. (1966), who concluded that Se is mostly volatilized from shoots on the basis of their finding that considerably less Se was volatilized by alfalfa plants after the tops were removed. Zayed and Terry (1994) carefully measured the rate of volatilization of roots separately from shoots and found that for six different plant species, roots volatilize most of the Se despite having a much smaller mass than shoots. For instance, their results with broccoli plants indicate that on a dry weight basis, the roots volatilized Se 13 to 26 times faster than shoots (Table 5), even though Se concentration in shoots was two- to five-fold higher than in roots. Furthermore, they also showed that shoot removal increased Se volatilization by the remaining root up to five-fold in the following 24 hours. Selenium volatilization by the detopped root continued to increase progressively for 72 hours after shoot removal, attaining rates that were up to 30 times higher than the rate of Se volatilization by the intact root (Table 6). It is not known why rates of Se volatilization are enhanced after shoot removal. The increased availability of free amino acids in plant tissues may contribute to such improved volatilization; recent results indicate that most free amino acids in plant tissues increased by 13.2 to 942% upon detopping (Zayed and Terry, unpublished). Thus, volatilization may be enhanced because of an increased concentration of selenoamino

TABLE 5 Rates of Se Volatilization by Roots and Shoots of Broccoli Plants Supplied with 20μM Se in Half-Strength Hoagland's Solution as Affected by Shoot Removal and Sulfate Supply

Sulfate level (mM)	Rate of Se volatilization (mg/kg/day)		
	Shoot	Root	Detopped root
0.00	0.13	3.24	36.3
0.25	0.32	7.71	17.5
1.00	0.12	1.78	5.2
10.00	0.02	0.20	1.3

Source: Adapted from Zayed and Terry (1994).

TABLE 6   Changes in Se Volatilization Rate with Time After Shoot Removal for Broccoli Plants Grown in Half-Strength Hoagland's Solution with 20μM Se

Days after shoot removal	Rate of Se volatilization (mg/kg/day)	Percent of intact root
0	1.35	100
1	5.35	399
2	19.10	1415
3	39.55	2930
7	2.69	199
14	0.83	61

Source: Adapted from Zayed and Terry (1994).

acids, such as selenomethionine, which has been shown to be very rapidly volatilized. An alternative possibility is that the enhanced volatilization following detopping is due to increased microbial activity (see below, Sect. IV).

C. Effect of Chemical Form of Selenium Supplied

When Se is supplied as selenate, it competes with sulfate with respect to plant uptake and metabolism. Increased levels of sulfate in the soil will diminish uptake, metabolism, and volatilization of selenate. This is illustrated by the research of Zayed and Terry (1994), who showed that increasing the concentration of sulfate in the culture solution led to a progressive decrease in the Se volatilization rate by broccoli plants. With an increase in the sulfate concentration from 0.25 mM to 10 mM, the rate of Se volatilization decreased from 97 to 14 μg of Se per square meter of leaf area per day. The decrease in Se volatilization rate was correlated with a decrease in the ratio of Se to S in plant tissues (Table 7). These results suggest that Se analogs of S compounds were outcompeted by the S compounds for the active sites of enzymes responsible for the uptake and conversion of inorganic Se to volatile forms. The competitive effect of S on Se volatilization may be less important with Se accumulators, since these plants may exhibit preferential uptake of Se relative to S (Bell et al., 1992).

Several lines of research suggest that plant volatilization proceeds more rapidly if Se is supplied in more reduced or organic forms such as selenite or selenomethionine. Early research by Lewis et al. (1974) showed that cabbage leaves from plants supplied with selenite released 10 to 16 times more volatile Se than those taken from plants supplied with selenate. Similarly, roots of selenite-grown plants released 11 times more volatile Se during oven-drying than did roots of selenate-grown plants (Asher et al., 1967). In our research, we examined the influence of the chemical form of Se in the root medium on the rate of Se volatilization by plants. We supplied broccoli plants with 20 μM Se as selenate,

TABLE 7 Rate of Se Volatilization and Ratio of Se to S in Plant Tissues of Broccoli Plants Grown in Half-Strength Hoagland's Solution at Five Levels of Sulfate and Supplied with Se at 20 μM as Sodium Selenate

Sulfate level (mM)	Rate of Se volatilization (μg/m²/day)	Ratio of Se to S (dry weight basis)	
		Leaf	Root
0.25	96.7	0.101	0.155
0.50	61.6	0.063	0.100
1.00	52.2	0.035	0.064
5.00	16.0	0.009	0.012
10.00	13.8	0.005	0.006

Source: Adapted from Zayed and Terry (1994).

selenite, or Se-methionine in a solution culture and measured the rate of Se volatilization. Interestingly, we found that when plants were supplied with Se-methionine, the roots volatilized Se 13 times faster and the shoots volatilized Se 3 times faster than the roots and shoots of plants supplied with selenate (Table 8). When plants were supplied with selenite, Se was volatilized 4 times faster in roots and 60% faster in shoots compared to the selenate-grown plants. These findings have been extended to three other plant species, namely, Indian mustard, sugar beet, and rice. It should be noted that the same trend is observed with microbes in that the following substrates are more effective in terms of stimulating volatilization: selenomethionine $> SeO_3^{2-} > SeO_4^{2-}$ (Azaizeh et al., 1997).

D. Proposed Mechanism of Selenium Volatilization in Plants

The biomethylation of Se by plants involves several metabolic processes, a detailed description of which is given elsewhere (Terry and Zayed, 1994). Briefly, once

TABLE 8 Rates of Selenium Volatilization of Broccoli Roots and Shoots as Influenced by the Form of Se Supplied at 20 μM in Half-Strength Hoagland's Solution

Selenium form supplied	Rate of Selenium volatilization (mg/kg/day)		
	Shoot	Root	Detopped root
Selenate	0.068	1.124	2.118
Selenite	0.110	5.278	9.876
Se-methionine	0.226	14.827	22.118

Source: Adapted from Zayed and Terry (1994).

selenate has entered the plant, it is activated by the enzyme ATP sulfurylase (for which sulfate is the normal substrate) to form adenosine 5'-phosphoselenate (APSe) (Burnell, 1981). APSe apparently undergoes reduction to selenite, though no conclusive data exist on this step. Selenite is then reduced to selenide, followed by the incorporation of selenide into Se-cysteine (Ng and Anderson, 1978, 1979); selenocysteine is converted to Se-methionine via Se-cystathionine and Se-homocysteine (Dawson and Anderson, 1988). Se-methionine is the most likely precursor, eventually leading to the formation of volatile Se, dimethyl selenide (DMSe) (Fig. 4). This is the form of volatile Se produced by the majority of plant species and microorganisms.

E. Toxicity and Fate of Volatile Selenium in the Atmosphere

Some concerns have been expressed that biomethylation of Se may lead to the production of toxic forms of gaseous Se, or that Se may be redeposited in other areas and become toxic there. Dimethyl selenide is reported to be 500 to 700 times less toxic than selenate and selenite (LD_{50} of DMSe is 1600–2200 mg Se/

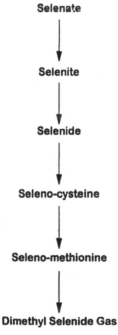

FIGURE 4 Suggested major steps in the metabolic pathway for the biomethylation of inorganic Se into volatile forms of Se. (Adapted from Zayed and Terry, 1992.)

kg rat) (Wilber, 1980; Ganther et al., 1966; McConnell and Portman, 1952). Using a volatilization rate of 250 $\mu g/m^2/h$, the highest 24-hour average exposure of Se under stagnant conditions was computed to be 837 ng/m^3 (Frankenberger and Karlson, 1988). An acceptable intake level documented by the U.S. Environmental Protection Agency (EPA) in guidance for Superfund sites is considered to be 3500 ng/m^3, which is substantially higher than that calculated for Se bioremediation.

With regard to the fate of the gaseous Se in the atmosphere, recent studies on the tropospheric transformation of DMSe on the west side of San Joaquin Valley of California indicate that during the short lifetime of gaseous DMSe (9.6 days), the gas will be dispersed and diluted by air currents directly away from the contaminated areas, with deposition possibly occurring in the Se-deficient areas (Atkinson et al., 1990). At the highest annual deposition flux (4.5 g/h Se) mixed in the upper 10 cm of soil, the soil Se content would be increased by about 0.005 mg/kg (Frankenberger and Karlson, 1988).

IV. THE ROLE OF RHIZOSPHERE MICROBES IN SELENIUM VOLATILIZATION

One explanation for the increase in Se volatilization by roots after the removal of shoots is that once the shoots are gone, reduced carbon compounds (including free amino acids) leak into the rhizosphere, thereby accelerating the production of volatile Se by rhizosphere microorganisms. This hypothesis triggered us to study the role of rhizosphere microbes in the process of Se volatilization by plants. Zayed and Terry (1994) showed that Se volatilization by detopped broccoli roots in nutrient solution was substantially inhibited when bacterial (but not fungal) antibiotics were added. The volatilization rate of the nutrient solution alone (detopped roots removed) was only about 10% of the rate when plants were included. Since the contribution of the microbes in the nutrient solution to volatilization represents only a small fraction of the total Se volatilized, the inhibition of volatilization of detopped roots by the antibiotics was most likely due to the action of antibiotics on bacteria present in the rhizosphere, or in the root itself.

What is the role of rhizosphere or other root bacteria in phytovolatilization? We have obtained some evidence to suggest that bacteria may be involved in the reduction and metabolism of inorganic Se (selenate or selenite) to organic forms of Se (e.g., selenoamino acids), which can be volatilized by plants. Nonaxenically grown broccoli plants can volatilize Se when supplied with Se as selenate; volatilization is faster if Se is supplied as selenite, and even faster if supplied with Se-methionine (Terry and Zayed, 1994). However, axenically grown plants volatilized very little Se from selenate or selenite, although they could volatilize Se at near-normal rates from selenomethionine. This pair of findings suggests that microbes may be needed to convert selenate and selenite to selenoamino acids such as selenomethionine. Similar results have been obtained with wetland plants. Bulrush plants were grown on selenate, selenite, and Se-methionine in hydroponic solutions with and without the bacterial antibiotic ampicillin (de Souza and Terry, unpublished). Ampicillin strongly inhibited volatile Se production from selenate and selenite but not from Se-methionine.

A recent Se volatilization study performed in vitro on samples taken from a constructed wetland contaminated with selenite showed that microbial cultures prepared from rhizosphere soils had higher rates of Se volatilization than cultures prepared from bulk soil (Azaizeh et al., 1997). Most of the Se volatilized by rhizosphere microbes was due to bacteria, rather than fungi, in the sediments. Selenium volatilization by these bacterial cultures was greatly enhanced by aeration and the addition of an energy source, which, in situ, is probably derived from wetland plants.

V. CONSTRUCTED WETLANDS

Constructed wetlands offer tremendous promise for the cleanup of agricultural and industrial wastewaters contaminated with Se and many other toxic trace elements. Wetlands exhibit some of the highest rates of biomass production per land area of any known ecosystem ($\sim$2500 g/m^2/yr: Dennison and Berry, 1993). When wastewater is fed into a constructed wetland, many organic and inorganic contaminants are filtered out so that water released from the wetland, after perhaps a 7- to 10-day period, is substantially cleaner than before. The capacity of many wetlands to reduce the levels of biochemical oxygen demand (BOD), nutrients, and metal concentrations and other pollutants in water passing through them has been well documented (Kadlec and Kadlec, 1979; Nixon and Lee, 1986). All over the world, constructed wetlands are being utilized increasingly to treat wastewater discharges of many different types (Suzuki et al., 1989; Bastian and Hammer, 1993). Wetlands consume less energy, are more reliable, require less operation and maintenance, and cost less than more conventional chemical and mechanical water treatment systems. Constructed wetlands are especially efficient in removing low concentrations of contaminants present in large volumes of wastewater.

In constructed wetlands, Se is reduced to insoluble forms, which are deposited in the sediments, by accumulation into plant tissues and by volatilization to the atmosphere (i.e., through biological volatilization of plants, plant–microbe associations, and microbes alone). Cooke and Bruland (1987) estimated that as much as 30% of Se introduced into the ponds of Kesterson Reservoir in the San Joaquin Valley of California was lost by volatilization into the atmosphere through biomethylation and that the process varied seasonally. Velinsky and Cutter (1991) calculated Se volatilization rates of a salt marsh from measurements of total Se and Se/S ratios, and from published rates of S emission. They found that the calculated rate of Se loss through volatilization was of the same order of magnitude as the measured loss of total Se from the salt marsh. Allen (1991) tested the ability of a constructed wetland to remove Se from water spiked with selenite. She showed that the rate of Se removal by wetland plants exceeded 90% of the total Se introduced and credited biological volatilization as being an important component of this loss.

The efficacy of constructed wetlands in the cleanup of Se-polluted industrial wastewaters is exemplified by the Chevron Water Enhancement Wetland at Richmond, California. This 36-hectare constructed wetland was found to remove over 70% of the Se in the 10 million liters of oil refinery wastewater supplied to the wetland each day (Duda, 1992). In a recent research project at the Chevron site, we found that over a period of 4 months (June 23 to October 13, 1995), the wetland removed Se at an average rate of 156 g/day, which is equivalent to 89.4% of the rate of Se entering the wetland (Hansen et al., unpublished). Most of this removal occurred in the first pass (about 12 ha in area). Preliminary measurements of Se volatilization over a small area of the wetland indicated rates from 10 to 330 μg Se/m^2/day. If, for the sake of argument, one were to assume an average rate of volatilization over the entire wetland of 100 μg Se/m^2/day, Se volatilization would account for about 23.4% of the Se removed by the wetland. Before we can provide an accurate estimate of the extent of volatilization as a pathway for Se removal, however, more research is needed. It is clear that the wetland ecosystem acts as a very efficient filter to remove Se, with a significant, but as yet unknown, portion being removed by biological volatilization.

To maximize the efficiency of wetlands in Se removal, we must learn more about which plant species are most efficient in removing this trace element. Some plants have natural attributes that enable them to absorb and hyperaccumulate toxic elements in their tissues. Such plants can tolerate having several percent of their dry weight of metals and other ions. Unfortunately, little information is available on the uptake and accumulation of Se by aquatic plants. In one study, salt grass accumulated 429 mg/kg Se when supplied with 1 mg/L Se in a solution culture (Wu and Huang, 1991a). In another study, Allen (1991) grew seven wetland plant species (arrowhead, canary grass, cattail, duckweed, elodea, pond-weed, and coontail) in an experimental wetland with an Se concentration of 10 μg/L in the water. She found that duckweed had the highest concentration among the wetland plants, with a maximum of 115 mg of Se per dry weight kilogram. In our research, we found that duckweed bioconcentrated Se from 200- to 400-fold the initial supply concentrations, depending on the supply level (Zayed and Terry, unpublished). Duckweed, therefore, shows promise for the removal of Se from Se-laden wastewater.

VI. GENETIC ENGINEERING OF PLANTS FOR SELENIUM PHYTOREMEDIATION

For Se phytoremediation, the ideal plant species should be able to accumulate and volatilize large amounts of Se, grow rapidly and produce a large biomass, tolerate salinity and other toxic conditions, and provide a safe source of forage for Se-deficient livestock. Indian mustard (*Brassica juncea*), an excellent volatilizer, is such a plant. Indian mustard is also easy to transform. We are focusing on Indian mustard as our target phytoremediating plant species for genetic improvement. Using modern biotechnological approaches, we are genetically engineering plants to accumulate and volatilize Se at enhanced rates. Our goal is to overexpress rate-limiting enzymes involved in Se volatilization. In the Se volatilization pathway proposed by Zayed and Terry (1994), selenate is metabolized via the sulfate assimilation pathway; inorganic forms of Se are then reduced and incorporated into the amino acids Se-cysteine and Se-methionine, and finally methylated to the volatile compound dimethyl selenide. To date, we have succeeded in obtaining transgenic plants that overexpress five different enzymes associated with the Se volatilization pathway. These are ATP-sulfurylase, glutathione reductase, cysteine synthetase, cystathionine-β-lyase, and S-adenosylmethionine-synthetase. Glutathione reductase and cysteine synthetase transgenic lines have been shown to have substantially enhanced levels of activity of their respective enzymes and are being tested for increased volatilization.

Selenium accumulation and volatilization may also be rate-limited by the uptake of selenate into plant roots (Terry et al., 1992). The protein mediating the uptake of selenate is the sulfate transporter sulfate permease. The gene for a high affinity sulfate transporter has been cloned from the legume *Stylosanthes hamata* and was shown to be induced by S starvation (Smith et al., 1995). When the amount of sulfate transporter protein in the root membrane is increased through S starvation, subsequent addition of nutrient sulfate results in a substantial increase in the accumulation of sulfate in the plant. Thus, the sulfate transporter is rate-limiting for sulfate uptake and accumulation and may well rate-limit selenate uptake as well. Unpublished data from this laboratory support this hypothesis: when plants were preconditioned on low sulfate to increase the amount of sulfate transporter protein, and subsequently transferred back to a normal growth medium spiked with selenate, they accumulated and volatilized more Se than plants that were not preconditioned (Table 9). Thus, we are genetically engineering plants to have increased amounts of sulfate transporter (sulfate permease) to test the hypothesis that volatilization is rate-limited by this protein. So far we have developed eight transgenic lines that overexpress this gene and we are testing them for Se accumulation and volatilization.

TABLE 9 Rates of Selenium Volatilization (mg/kg/day) of Broccoli Plants Grown on Half-Strength Hoagland's Solution Containing 20 μM of Selenate and Different Levels of Sulfate

| | Rate of selenium volatilization (mg/kg/day) | |
Sulfate supply (mM)	Preconditioned[a]	Control[b]
0.00	0.90	0.38
0.25	1.99	1.26
1.00	1.58	0.41
5.00	1.23	0.09

[a] Preconditioned on 0.25 mM sulfate for one week.
[b] Grown on 1 mM sulfate.

VII. FUTURE OUTLOOK

Within the next decade, the phytoremediation of Se-contaminated soils and waters must surely become a standard technology. Plant species such as Indian mustard show tremendous promise for the removal of Se from upland soils. Genetic improvement of Indian mustard (and other plant species) using molecular biology approaches should pave the way for the rapid development of superior transgenic lines of phytoremediating plants. The technology of using constructed wetlands to clean up Se from agricultural and industrial wastewaters is already available and will be applied increasingly in the next few years. Our laboratory, in cooperation with the University of California Salinity/Drainage Task Force and the Tulare Lake Drainage District, has constructed 10 quarter-acre experimental wetland cells at Corcoran, California, to develop the use of constructed wetlands for the removal of Se from agricultural irrigation drainage waters. If this experiment is successful, it will revolutionize the treatment of Se-contaminated drainage water and increase the acreage of wetlands throughout the western United States dramatically.

REFERENCES

Abrams, M. M., C. Shennan, J. Zasoski, and R. G. Burau. 1990. Selenomethionine uptake by wheat seedlings. *Agron. J.* 82:1127–1130.

Allaway, W. H. 1971. Distribution of selenium. *In* Selenium in nutrition. Report of the Subcommittee on selenium, Committee on Animal Nutrition, Agricultural Board, National Research Council. National Academy of Sciences, Washington, DC.

Allen, K. N. 1991. Seasonal variation of selenium in outdoor experimental stream–wetland systems. *J. Environ. Qual.* 20:865–868.

Asher, C. J., C. S. Evans, and C. M. Johnson. 1967. Collection and partial characterization of volatile selenium compounds from *Medicago sativa* L. *Aust. J. Biol. Sci.* 20:737–748.

Atkinson, R., S. M. Aschmann, D. Hasegawa, E. T. Thompson-Eagle, and W. T. Frankenberger, Jr. 1990. Kinetics of the atmospherically important reactions of dimethyl selenide. *Environ. Sci. Technol.* 24:1326–1332.

Azaizeh, H. A., S. Gowthaman, and N. Terry. 1997. Microbial selenium volatilization in rhizosphere and bulk soils from a constructed wetland. *J. Environ. Qual.* 26:666–672.

Baker, A. J. M., S. P. McGrath, C. M. D. Sidoli, and R. D. Reeves. 1994. The possibility of in situ heavy metal decontamination of polluted soils using crops of metal-accumulation crops. *Resource Conserv. Recycl.* 11:41–49.

Bañuelos, G. S., and D. W. Meek. 1990. Accumulation of selenium in plants grown on selenium-treated soil. *J. Environ. Qual.* 19:772–777.

Bañuelos, G. S., and G. Schrale. 1989. Plants that remove selenium from soils. *California Agric.* May/June:19–20.

Bañuelos, G. S., R. Mead, and S. Akohoue. 1991. Adding selenium-enriched plant tissue to soil causes the accumulation of selenium in alfalfa. *J. Plant Nutr.* 14:701–713.

Bañuelos, G .S., G. E. Cardon, C. J. Phene, L. Wu, S. Akohoue, and S. Zambrzuski. 1993. Soil boron and selenium removal by three plant species. *Plant Soil.* 148:253–263.

Bañuelos, G. S., N. Terry, A. Zayed, and L. Wu. 1995. Managing high soil selenium with phytoremediation. In G. E. Schuman and G. F. Vance (eds.), *Selenium: Mining, Reclamation, and Environmental Impact*, pp. 394–405. *Proceedings of the 12th Annual National Meeting of the American Society of Surface Mining and Reclamation.* June 5–8, Gillette, WY.

Bañuelos, G. S., A. Zayed, N. Terry, B. Mackey, L. Wu, S. Akohoue, and S. Zambrzuski. 1996. Accumulation of selenium by different plant species grown under increasing salt-regimes. *Plant Soil.* 183:49–59.

Bastian, R. K., and D. A. Hammer. 1993. The use of constructed wetlands for wastewater treatment and recycling. In G. Moshiri (ed.), *Constructed Wetlands for Water Quality Improvement*, pp. 59–68. Lewis Publishers, London.

Beath, O. A., H. F. Eppson, and C. S. Gilbert. 1935. Selenium and other toxic minerals in soils and vegetation. *Wyoming Agric. Sta. Bull.* 206:1–9.

Bell, P. F., D. R. Parker, and A. L. Page. 1992. Contrasting selenate–sulfate interactions in selenium-accumulating and nonaccumulating plant species. *Soil Sci. Soc. Am. J.* 56:1818–1824.

Biggar, J. W., and G. R. Jayaweera. 1993. Measurement of selenium volatilization in the field. *Soil Sci.* 155:31–35.

Blaylock, M. J., and B. R. James. 1994. Redox transformation and plant uptake of Se resulting from root–soil interactions. *Plant Soil* 158:1–17.

Brown, T. A., and A. Shrift. 1982. Selenium: Toxicity and tolerance in higher plants. *Biol. Rev.* 57:59–84.

Burnell, J. N. 1981. Selenium metabolism in *Neptunia amplexicaulis*. *Plant Physiol.* 67:316–324.

Cooke, T. C., and K. W. Bruland. 1987. Aquatic chemistry of selenium: Evidence of biomethylation. *Environ. Sci. Technol.* 21:1214–1219.

Dawson, J. C., and J. W. Anderson. 1988. Incorporation of cysteine and selenocysteine into cystathionine and selenocystathionine by crude extracts of spinach. *Phytochemistry* 27:3453–3460.

Dennison, M. S., and J. F. Berry. 1993. *Wetlands: Guide to Science, Law, and Technology.* Noyes Publications. Park Ridge, NJ.

Duckart, E. C., L. J. Waldron, and H. E. Doner. 1992. Selenium uptake and volatilization from plants growing in soil. *Soil Sci.* 153:94–99.

Duda, P. J. 1992. Chevron's Richmond Refinery Water Enhancement Wetland. Technical report submitted to the Regional Water Quality Control Board, Oakland, CA, December, 1992.

Ernst, W. H. U. 1988. Decontamination of mine sites by plants: An analysis of the efficiency, In *Proceedings of the International Conference on Environmental Contamination, Venice,* pp. 305–310. CEP Consultants, Ltd., Edinburgh.

Evans, C. S., C. J. Asher, and C. M. Johnson. 1968. Isolation of dimethyl diselenide and other volatile selenium compounds from *Astragalus racemosus* (Pursh.). *Aust. J. Biol. Sci.* 21:13–20.

Frankenberger, W.T., Jr., and U. Karlson. 1988. Dissipation of soil selenium by microbial volatilization at Kesterson Reservoir. Final report submitted to U.S. Department of the Interior, Project No. 7-F.C-20-05240, November. U.S. Bureau of Reclamation, Sacramento, CA.

Ganther, H. E., O. A. Levander, and C. A. Saumann. 1966. Dietary control of selenium volatilization in the rat. *J. Nutr.* 88:55–60.

Hurd-Karrer, A. M. 1935. Factors affecting the absorption of selenium from soils by plants. *J. Agric. Res.* 54:601–608.

Kadlec, R. H., and J. A. Kadlec. 1979. Welands and water quality. In P. E. Greeson and J. R. Clark (eds.), *Wetland Functions and Values: The State of Our Understanding,* pp. 436–456. American Water Resources Association, Bethesda, MD.

Kumar, N., V. Dushenkov, H. Motto, and I. Raskin. 1995. Phytoextraction: The use of plants to remove heavy metals from soils. *Environ. Sci. Technol.* 29:1232–1238.

Lewis, B. G. 1971. Ph.D. thesis. University of California, Berkeley.

Lewis, B. G., C. M. Johnson, and C. C. Delwiche. 1966. Release of volatile selenium compounds by plants: Collection procedures and preliminary observations. *J. Agric. Food Chem.* 14:638–640.

Lewis, B. G., C. M. Johnson, and T. C. Broyer. 1974. Volatile selenium in higher plants. The production of dimethyl selenide in cabbage leaves by enzymatic cleavage of Se-methyl selenomethionine selenonium salt. *Plant Soil.* 40:107–118.

Mayland, H. F., L. F. James, K. E. Panter, and J. L. Sonderegger. 1989. Selenium in seleniferous environments. In L. W. Jacobs (ed.), *Selenium in Agriculture and the Environment,* pp. 15–50. American Society Agronomy, Madison, WI.

McConnell, K. P., and O. W. Portman. 1952. Toxicity of dimethyl selenide in the rat and mouse. *Proc. Soc. Exp. Biol. Med.* 79:230–231.

McGrath, S. P., C. M. D. Sidoli, A. J. M. Baker, and R. D. Reeves. 1993. The potential for the use of metal-accumulation plants for the in situ decontamination of metal-polluted soils. In J.P. Eysakers and T. Hamers (eds.), *Integrated Soil and Sediment Research: A Basis for Proper Protection,* pp. 673–676. Kluwer Academic Publishers, Dordrecht.

Mikkelsen, R. L., A. L. Page, and F. T. Bingham. 1989. Factors affecting selenium accumulation by agricultural crops. In L.W. Jacobs (ed.), *Selenium in Agriculture and the Environment.* pp. 65–94. Soil Science Society of America Special Publication No. 23. SSSA, Madison, WI.

Ng, B. H., and J. W. Anderson. 1978. Synthesis of selenocysteine by cysteine synthases from selenium accumulator and non-accumulator plants. *Phytochemistry* 17:2074.

Ng, B. H., and J. W. Anderson. 1979. Light-dependent incorporation of selenite and sulphite into selenocysteine and cysteine by isolated pea chloroplasts. *Phytochemistry.* 18:573–580.

Nixon, S. W., and V. Lee. 1986. Wetland and water quality: A regional view of recent research in the United States on the role of freshwater and saltwater wetlands as sources, sinks, and transformers of nitrogen, phosphorus, and various heavy metals. Technical report No. Y-86-2. U.S. Army Corps of Engineers, Vicksburg, MI.

Parker, D. R., and A. L. Page. 1994. Vegetation management strategies for remediation of selenium-contaminated soils. *In* W.T. Frankenberger, Jr., and S. Benson (eds.), *Selenium in the Environment,* pp. 327–347. Dekker, New York.

Parker, D. R., A. L. Page, and D. N. Thomas. 1991. Salinity and boron tolerances of candidate plants for the removal of selenium from soils. *J. Environ. Qual.* 20:157–164.

Retana, J., D. R. Parker, C. Amerhein, and A. L. Page. 1993. Growth and trace element concentrations of five plant species grown in highly saline soil. *J. Environ. Qual.* 22:805–811.

Rosenfeld, I., and O. A. Beath. 1964. *Selenium, Geobotany, Biochemistry, Toxicity, and Nutrition.* Academic Press, New York.

Salt, D .E., M. J. Blaylock, D. B. A Kumar, V. Dushenkov, B. D. Ensley, I. Chet, and I. Raskin. 1995. Phytoremediation: A novel strategy for the removal of toxic metals from the environment using plants. *Environ. Sci. Technol.* 13:468–474.

Shennan, C., D. Schachtman, and G. R. Cramer. 1990. Variation in [^{75}Se]selenate uptake and partitioning among tomato cultivars and wild species. *New Phytol.* 115:523–530.

Smith F. W., P. M. Ealing, M. J. Hawkesford, and D. T. Clarkson. 1995. Plant members of a family of sulfate transporters reveal functional subtypes. *Proc. Natl. Acad. Sci. USA.* 92:9373–9377.

Suzuki, T., W. G. A. Nissanka, and Y. Kurihara. 1989. Constructed wetlands for wastewater treatment: municipal, industrial and agricultural. In D. A. Hammer (ed.), *Proceedings from the First International Conference on Constructed Wetlands for Wastewater Treatment,* pp. 530–535. Chattanooga, TN, June 13–17, 1989. Lewis Publishers, Chelsea, MI.

Terry, N. and A. M. Zayed. 1994. Selenium volatilization by plants. In W.T. Frankenberger, Jr., and S. Benson (eds.), *Selenium in the Environment,* pp. 343–367. Dekker, New York.

Terry, N., C. Carlson, T. K. Raab, and A. M. Zayed. 1992. Rates of selenium volatilization among crop species. *J. Environ. Qual.* 21:341–344.

Ulrich, J. M., and A. Shrift. 1968. Selenium absorption by excised *Astragalus* roots. *Plant Physiol.* 43:14–19.

Velinsky, D., and G. A. Cutter. 1991. Geochemistry of selenium in a coastal salt march. *Geochim. Cosmochim. Acta* 55:179–191.

Watson, F., G. Bañuelos, and J. O'Leary. 1993. Trace element composition of *Atriplex* species. *J. Environ. Qual.* 48(2):157–162.

Wentzel, W. W., H. Stattler, and F. Jockwer. 1993. Metal hyperaccumulator plants: A survey on species to be potentially used for soil remediation. *Agron. Abstr.* 1993:52.

Wilber, C. G. 1980. Toxicology of selenium: A review. *Clin. Toxicol.* 17:171–230.

Williams, M. C,. and H. F. Mayland. 1992. Selenium absorption by twogrooved milkvetch and western wheatgrass from selenomethionine, selenocystine, and selenite. *J. Range Manage.* 45:374–378.

Wu, L. and Z. H. Huang. 1991a. Selenium tolerance, salt tolerance, and selenium accumulation in tall fescue lines. *Ecotoxicol. Environ. Saf.* 21:47–56.

Wu, L., and Z. H. Huang. 1991b. Chloride and sulfate salinity effects on selenium accumulation by tall fescue. *Crop Sci.* 31:114–118.

Wu, L., Z. H. Huang, and R. G. Burau. 1988. Selenium accumulation and selenium–salt cotolerance in five grass species. *Crop Sci.* 28:517–522.

Zayed, A., and N. Terry, 1992. Selenium volatilization in broccoli as influenced by sulfate supply. *J. Plant Physiol.* 140:646–652.

Zayed, A., and N. Terry, 1994. Selenium volatilization in roots and shoots: Effects of shoot removal and sulfate level. *J. Plant Physiol.* 143:8–14.

Zieve, R., and P. J. Peterson. 1984. The accumulation and assimilation of dimethyl selenide by four plant species. *Planta* 160:180–184.

32

Selenium Accumulation and Uptake by Crop and Grassland Plant Species

Lin L. Wu
University of California at Davis, Davis, California

I. INTRODUCTION

The uptake of selenium (Se) by plants has received considerable attention since the 1930s, when toxic effects where reported in animals that had consumed forage containing selenium in amounts greater than 1 mg Se/kg dry matter (Robinson,1933, Beath et al.,1934). Selenium once was considered a highly toxic carcinogenic element and was used primarily for industrial purposes (Nelson et al.,1943; Volgary and Tscherkes,1967). This perception changed in 1957, when Se was found to prevent dietary hepatic necrosis in rats (Schwarz and Foltz,1957). Research on Se uptake by plants has emphasized plant species considered as Se accumulators (Brown and Shrift,1982). Selenium deficiency and the importance of Se in nutritional and metabolic functions in livestock (Paterson et al., 1957; Schwarz and Foltz, 1957; Drake et al.,1960; Sharman, 1960; Tagwerker, 1960) became the focus of widespread interest. In the 1980s, the discovery of high concentrations of Se in agricultural drainwater (Presser and Barnes, 1984) and the resultant embryonic mortality of waterfowl at Kesterson Reservoir, Merced County, California (Ohlendorf et al., 1986a, 1986b), renewed interest in the environmental toxicity of Se and in food chain contamination. Recent research on plant uptake of Se has placed greater emphasis on Se nonaccumulators, including pasture plants, vegetable and cereal crops, and naturally established field plants. It is not the purpose of this chapter to review extensively the literature on uptake of Se by plants, but rather to focus on the recent research on Se uptake and accumulation by crop and grassland plant species.

II. SELENIUM ACCUMULATION

Among different plant species, the ability to accumulate Se differs strikingly. Generally, plants are classified into three groups on the basis of their ability to accumulate Se when grown on seleniferous soils (Rosenfeld and Beath,1964). Primary indicators, or accumulators, can accumulate several thousand milligrams of se per kilogram dry weight. (Unless otherwise stated, all plant species and soil weights are given as dry weight.) A number of different species in the genera *Astragalus, Stanleya, Haplopappus,* and *Machaeranthera* belong to this category. Secondary accumulators may accumulate from several hundred up to 1000 mg Se/kg (Barneby, 1964). Some plant species in the genera of *Astragalus, Aster, Atriplex, Castilleja, Comandra, Grayia, Grindelia, Gutierrezia, Machaeranthera,* and *Mentzelia* are considered as secondary accumulators. The third group of plants, the Se nonaccumulators, comprise forage and crop plants, grasses, and most open field plant species. Nonaccumulator plant species generally contain less than 25 mg Se/kg. Between 1930 and 1960, Se accumulation by the primary indicator species was the subject of numerous studies. The principal concern at that time was livestock poisoning due to the toxic effects of Se (Gissel-Nielsen et al., 1984). In the last 10 years research interest has become focused on food crops, forage plants, and natural grassland plant species. Selenium accumulation by these plants has been found to be profoundly influenced by the plant species, soil Se concentration, soil salinity, and field management strategies.

A. Selenium Concentrations in Crops from California Farmland

The recognition that elevated Se concentrations in agricultural drainage water caused toxic effects on wildlife in California's San Joaquin Valley prompted the desire to determine the degree of transfer of Se from the soil in this region into animals and humans. A wide range of food crop samples have been analyzed for tissue Se concentrations. Burau et al. (1988) collected 381 edible tissue samples of 17 different crops from farms on the west side of the San Joaquin Valley and analyzed them for tissue Se concentrations. A wide range of Se concentrations were detected among the plant samples within crop plant species: in alfalfa leaf, the Se concentrations ranged from nondetectable to 1460 μg Se/kg; in bell pepper, 89 to 1640 μg/kg; in broccoli, 60 to 1530 μg/kg; in cottonseed, 98 to 3380 μg/ kg; and in tomato, 15 to 770 μg/kg. The data indicated that the crop samples from higher elevations on alluvial fans had lower Se concentrations, which was not unexpected, since the soluble soil Se had been well leached. On the other hand, plants from the lower elevations tended to have higher tissue Se concentrations.

Figure 1 shows the ranked geometric mean values of Se concentrations in edible tissues of 16 food crop species. Among the different crop species, broccoli, cotton, and cauliflower had the highest tissue Se concentrations. Broccoli and cauliflower are in the genus *Brassica* and are high sulfur uptake species. Plants

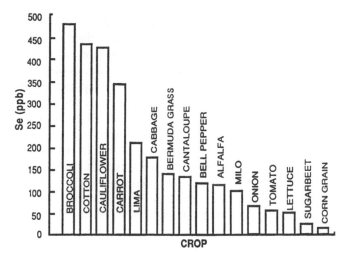

FIGURE I Ranked geometric mean values of Se concentrations (dry weight) in edible tissues of crops sampled from the west side of the San Joaquin Valley, California. (From Burau et al. 1988.)

that contain high concentrations of sulfur accumulate more Se than those that contain low concentrations of sulfur (Stadtman, 1974). Cotton is grown in the valley at lower elevations, where soil salinity is relatively high and higher soil Se concentrations are expected. Overall, the crop tissue Se concentration data fall in a nonseleniferous category, except the highest Se concentration of cotton that was at a threshold of potential toxicity to animals.

The potential toxic effects and the degree of transfer of Se from the soil to the crop plants should not be overlooked because even among the Se nonaccumulator plant species, Se accumulation and Se tolerance have been found to be quite different. Wheat (*Triticum vulgare* L.), for example, can accumulate moderate amounts of Se, up to 39 μg/g, without growth retardation. Selenium-sensitive pasture crops, such as white clover (*Trifolium repens* L.), red clover (*Trifolium pretense* L.), and perennial ryegrass (*Lolium perenne* L.), may show growth inhibition in the presence of Se accumulations exceeding 5 μg/g (Peterson and Butler, 1966; Smith and Watkinson, 1984). In vegetables, the ability to accumulate Se was found to be positively correlated with the sulfur content of the species (Bañuelos and Meek, 1988).

B. Selenium Accumulation by Naturally Established Grassland Plants in Seleniferous Soils Contaminated by Agricultural Drainwater Sediment

The California State Water Resources Control Board initiated a Kesterson Cleanup Action plan in 1988 to remediate the Se-laden soil. The Se-contaminated evaporation ponds were transformed into an upland grassland by filling with a layer of uncontaminated soil. To examine the potential food chain contamination and the environmental impact of the reconstructed grassland, naturally established grassland plant species have been monitored for Se accumulation (Huang and Wu, 1991; Wu et al.,1995) since the closure of the evaporation ponds.

The soil Se-concentrations monitored at four field sites over six collections from May 1989 to May 1992 are presented in Table 1. The total soil Se concentrations for the fresh soil fill sites in Ponds 2 and 6 were higher for the subsurface soil (30–45 cm deep) than for the top 15 cm of soil. In the Pond 2 fresh soil fill site, an average of 5372 μg Se/kg dry soil was found in the subsurface soil at depths of 30 to 45 cm. This was over seven times higher than the total soil Se concentration detected for the top 15 cm of the soil horizon (an average of 732 mg Se/kg). The average water-extractable soil Se was 30 μg/kg for the top 15 cm soil and 241 μg/kg for the subsurface soil; it was about 5% of the total soil Se concentration for each horizon.

The soil Se concentrations were found to vary greatly among soil samples within a site. For the native soil sites (Ponds 6 and 7), greater soil Se concentrations were found in the top 15 cm soil than in the subsurface soil. The top 15 cm soil of the Pond 6 native soil site had Se concentrations ranging from approximately 6400 to 10,000 μg/kg. This was about two to six times the soil Se concentration detected in the subsurface soil. For the Pond 7 site, the total soil Se and water-extractable Se concentrations of the top 15 cm soil detected over 3 years were about 50% of the soil Se concentrations found in the Pond 6 native soil site. The spatial distribution of soil Se concentrations within sites fluctuated widely and could vary by as much as a factor of 100 between samples.

Earlier studies indicated that only a small fraction of total Se in soils is in the soluble form. Adriano (1986) reported that water-soluble Se varied between 0.3 and 7% of the total Se in soil. Byers et al. (1933), who measured water-soluble Se in more than 100 soil samples, found water-soluble Se present at 0.1 mg/kg for a majority of the samples. Workman and Soltanpour (1980) stated that many cultivated soils contain 0.05 mg/kg. Cary and Allaway (1969) indicated that when elemental Se is added to soils, a portion of it rapidly oxidizes to selenite. After the rapid initial oxidation, the remaining Se is quite inert, and its oxidation then proceeds very slowly. Therefore, the Se in soils of the existing grassland at Kesterson may represent a long-term release of Se into the environment. A conceivable adverse environmental impact of elevated soil Se is bioconcentration to toxic levels, as occurred in the evaporation ponds (Ohlendorf et al., 1986b).

Table 2 shows that the Pond 2 fresh soil fill site had a considerable reduction of the plant species richness for the May 1990 survey (taken right after filling the ponds with fresh soil) compared to the May 1989 survey. The species that did not exist in the May 1990 survey are winter annuals such as *Avena, Bromus,*

Soil depth (cm)	Se concentration (mg/kg) at six sampling times					
	May 1989	May 1990	September 1990	May 1991	September 1991	May 1
ll site						
0–15	720 ± 630[a]	648 ± 648	763 ± 469	740 ± 470	799 ± 659	721 ±
30–45	5001 ± 1550	4174 ± 4115	6174 ± 3116	5672 ± 4178	5772 ± 2636	5441 ±
e 1–15	29 ± 20	32 ± 16	20 ± 33	20 ± 4	29 ± 27	17 ±
30–45	303 ± 300	205 ± 325	215 ± 312	128 ± 195	42 ± 53	552 ±
ll site						
0–15	209 ± 95	158 ± 100	140 ± 84	177 ± 134	253 ± 242	171 ±
30–45	3320 ± 1200	3290 ± 1180	3142 ± 1100	2287 ± 1320	2375 ± 1431	3190 ±
e 0–5	53 ± 50	51 ± 54	10 ± 11	45 ± 105	21 ± 36	7 ±
30–45	220 ± 180	138 ± 200	136 ± 173	195 ± 265	174 ± 222	718 ±
site						
0–15	6634 ± 4000	6843 ± 3800	6724 ± 2283	9919 ± 4980	9855 ± 4462	6428 ±
30–45	2219 ± 1200	1091 ± 930	1951 ± 901	1505 ± 560	2417 ± 980	1660 ±
e 0–15	290 ± 90	274 ± 101	224 ± 180	529 ± 299	150 ± 42	163 ±
30–45	95 ± 65	86 ± 53	66 ± 33	226 ± 133	107 ± 51	35 ±
site						
0–15	3010 ± 3200	3518 ± 2565	3966 ± 1037	4726 ± 3442	3040 ± 2334	4830 ±
30–45	1120 ± 1200	1051 ± 1003	1445 ± 760	926 ± 434	1672 ± 764	1446 ±

12.33 ± 1.25	—	17.71 ± 2.0	—	13.02 ± 9.75	—	—
12.86 ± 3.90	1.46 ± 0.95[a]	—	24.52 ± 1.25	1.75 ± 0.14	—	—
3.36 ± 3.05	—	—	—	—	—	—
4.18 ± 3.52	—	—	—	—	—	—
9.31 ± 0.77	2.23 ± 0.83	14.86 ± 8.45	10.92 ± 4.50	3.04 ± 0.99	22.98 ± 11.9	10.53 ±
3.31 ± 3.21	—	—	—	0.87 ± 0.38	—	—
—	—	—	—	1.56 ± 0.48	—	1.35 ±
1.57 ± 0.42	—	3.28 ± 2.01	—	0.96 ± 0.54	—	1.49 ±
16.15 ± 1.22	5.84 ± 4.30	—	—	—	—	—
—	—	—	—	3.01 ± 1.95	—	—
18.47 ± 3.05	—	19.34 ± 3.56	3.48 ± 0.91	—	—	1.30 ±
2.03 ± 0.06	—	—	—	0.97 ± 0.50	—	6.75 ±
—	—	—	—	—	—	—
12.02 ± 2.54	24.74 ± 12.25	—	—	—	—	6.51 ±
2.78 ± 1.85	—	—	—	2.66 ± 1.80	—	3.67 ±
—	—	—	—	3.53 ± 2.26	—	—
3.36 ± 1.26	—	—	—	—	—	—
1.06 ± 1.00	—	4.60 ± 2.10	1.47 ± 1.14	1.69 ± 1.06	—	5.78 ±
19.01 ± 8.51	—	12.01 ± 5.34	4.03 ± 2.03	11.71 ± 6.96	—	—
—	—	—	—	—	14.92 ± 4.50	—
8.12 ± 6.49	11.73 ± 10.30	12.81 ± 7.33	7.31 ± 9.67	5.39 ± 6.91	18.95 ± 19.1	4.41 ±

eviation of tissue Se concentration.

5).

Festuca, and *Melilotus.* This reduction in species richness may be attributable both to the increase of Se and salinity and to summer drought in the fresh soil. The plant species richness increased again for the May 1991 survey, possibly as a result of the increased rainfall during the winter months. Species richness was quite variable between years. Generally, fewer plant species were found in the fall (an average of about four species) compared to the spring (an average of nine species). Plant samples collected in the fall generally had higher tissue Se concentrations than the species collected in the spring. For example, *Atriplex canescens* and *A. patula* had average shoot tissue Se concentrations ranging from 1.5 mg/kg (May 1990) to 17.7 and 24.5 mg/kg (September 1990 and December 1990). *Bassia hyssopifolia* had average tissue Se concentrations ranging from 2.2 mg/kg (May 1990) to 22.9 mg/kg (September 1991). The winter weed *Bromus rubens* had average shoot tissue Se concentrations ranging from 0.96 (May 1991) to 1.49 mg Se/kg (May 1992). The average tissue Se concentrations found in *M. indica* were from 1.06 to 5.78 mg/kg. The overall mean plant tissue Se concentrations were also much higher for the September survey (ranging from 12.81 to 18.95 mg/kg) than for May survey (ranging from 4.41 to 11.73 mg/kg) (Table 2).

Table 3 presents the distribution of plant tissue Se concentrations from the Pond 6 native soil site over a period from May 1989 to May 1992. The predominant plant species was salt grass (*Distichlis spicata*). *Atriplex patula* and *Frankenia grandifolia* contributed about 20% of the vegetation coverage. The average plant tissue Se concentrations in salt grass ranged from 1.07 μg/g (September 1991) to 4.6 mg/kg (May 1990). Selenium concentrations in *F. grandifolia* ranged from 2.3 (May 1992) to 10.5 mg Se/kg (May 1990); and in *Atriplex* from 2.1 (September 1991) to 12.57 mg/kg (May 1989). Plant species richness was considerably increased in the May 1992 survey. This increase was attributable to increased precipitation in the winter of 1991 and to the increase in colonization of certain species, such as *Melilotus indica* and *Cressa truxilensis.* No consistent reduction or increase of tissue Se concentration was found over time except that the average shoot tissue Se concentration of salt grass was reduced from an average of 4.3 mg/kg in the May 1989 and 1990 surveys to 1.2 mg/kg in May 1991 and 1992.

Factors that could affect Se accumulation by the grassland plants at Kesterson Reservoir are plant species, season, filling with fresh topsoil, soil moisture, soil salinity, and soil sulfate and Se concentrations. The vegetation in the Kesterson grassland is represented by annual herbaceous plant species or by perennial halophyte salt grass. Filling with fresh topsoil reduced Se accumulation by the shallow-rooted plants. Among the annual species, the shallow-rooted winter and spring grass species had relatively low tissue Se concentrations at the fresh soil fill sites, and these plants may not contribute significant amounts of Se in bioaccumulation. However, most of the summer glycophyte weeds, such as *Bassia hyssopifolia, Lactuca serriola,* and *Salsola kali,* were brought into the pond soil with the fresh fill soil. Because these species are deep rooted, elevated concentrations of

	May 1989	May 1990	September 1990	December 1990	May 1991	September 1991	May
	12.57 ± 6.51[a]	4.33 ± 3.01	6.32 ± 3.50	2.39 ± 0.53	5.63 ± 1.22	2.10 ± 1.12	2.43 ±
	—	—	—	—	7.43 ± 4.61	—	1.81 ±
	—	—	—	—	—	—	1.75 ±
	—	—	—	—	—	—	3.27 ±
	—	—	—	—	1.68 ± 0.33	8.88 ± 9.53	2.36 ±
	4.06 ± 0.86	4.61 ± 0.86	2.80 ± 0.30	1.14 ± 0.69	1.16 ± 0.41	1.07 ± 0.50	1.18 ±
a	3.05 ± 2.21	10.50 ± 4.01	4.32 ± 2.50	2.77 ± 1.27	5.09 ± 2.05	6.79 ± 7.20	2.30 ±
	—	—	—	—	—	—	1.22 ±
	—	—	—	—	7.82 ± 3.61	—	2.13 ±
	—	—	—	—	—	—	2.70 ±
	9.39 ± 0.27	—	—	—	4.26 ± 1.21	—	2.50 ±
	—	—	—	—	—	—	
	7.38 ± 4.36	6.48 ± 3.48	4.48 ± 1.77	2.10 ± 0.58	4.71 ± 2.57	4.71 ± 3.73	2.22 ±

eviation.
)5).

Se (>30 mg/kg) were accumulated even in the fresh soil fill sites, and these plants may contribute significant amounts of Se to the food chain. The tissue Se concentrations of the halophilous *Atriplex* sp. found on the Kesterson Reservoir fresh soil fill sites had tissue Se concentrations greater than 13 mg/kg. Salt grass formed a large percentage of the vegetation coverage in the grassland. It contained relatively low tissue Se concentrations even when it grew at the native soil sites (Huang and Wu, 1991). It appears to be a plant species that could reduce the risk of bioconcentration of Se.

III. UPTAKE AND TRANSLOCATION OF SELENATE AND SELENITE IN PLANTS

Of the different forms of Se present in soils, selenate and selenite are the major forms that may be toxic to plants. Both anions are readily absorbed by the plant and metabolized to organic Se compounds that act as analogs of essential sulfur compounds. Studies of differences in selenate and selenite uptake using excised roots (Shrift and Ulrich, 1976; Ulrich and Shrift, 1986) indicate that uptake of selenate requires energy, whereas uptake of selenite seems to be energy-independent. When roots were provided with selenite, the concentration in the roots never exceeded the external concentration. However, selenate was able to accumulate in the roots. This accumulation was suggested to be evidence of active transport across the cell wall and the cell membranes. Differences of movement of selenate and selenite from root to shoot were examined by measuring Se in the xylem exudate of detopped plants (Asher et al.,1977). Since selenate was unchanged during absorption and translocation, the movement of this anion can be said to resemble translocation of sulfate (Tolbert and Wiebe,1955). When selenite was supplied to the roots, however, very little could be detected in the exudate, suggesting that selenite is readily metabolized. Whether this mechanism bears any relationship to that responsible for sulfite assimilation has not yet determined.

Wu et al. (1988) tested the effects of selenate and selenite on Se uptake and plant growth using whole-plant and long-term growth conditions. Tall fescue plants were grown in nutrient solution culture supplemented with 2 mg Se/L, either as sodium selenate or as sodium selenite, over 5 weeks. Table 4 shows that by the end of 3 weeks' growth, selenate had caused greater growth inhibition than selenite, and this difference increased up to 5 weeks of growth. During the fourth and fifth weeks of exposure, the tolerance ratio was reduced from 74 to 48% by selenite to a reduction from 86 to 25% by selenate. After 5 weeks of growth, the roots and shoots were harvested and the tissue Se concentrations were measured. The root tissue treated with selenate had a slightly lower Se concentration than the roots treated with selenite (384 and 433 mg/kg, respec-

TABLE 4 Selenium Tolerance Ratio and Selenium Accumulation in Tall Fescue Grown in Culture Solution Supplemented with 4 mg Se/L as Either Selenate or Selenite

	Selenium tolerance ratio (%)		Selenium accumulation (μg/g)	
Treatment	3 weeks	5 weeks	Roots	Shoots
Control	100 ± 0.0[a]	100 ± 0	0.6 ± 0.1	0.2 ± 0.0
Selenate	86 ± 2	25 ± 1	384 ± 14	883 ± 35
Selenite	74 ± 2	48 ± 3	433 ± 14	142 ± 13

[a]Mean ± standard deviation.
Source: Wu et al. (1988).

tively). However, the shoot tissue treated with selenate had an Se concentration more than four times greater than that of the shoot tissue treated with selenite. The greater growth inhibition caused by selenate as compared to selenite apparently was attributable to the translocation of greater amounts of Se to the shoot tissue of the plants (883 and 142 mg/kg, respectively). Selenium has an adverse effect on the production of porphobilinogen synthetase, an enzyme required for chlorophyll biosynthesis (Panmaja et al., 1989), and is damaging to plant growth. The relationship between tissue Se concentration and root zone selenate and selenite concentration was investigated recently for field-grown plants. Mantgem et al. (1996) found that plant tissue Se concentration was least significantly correlated with soil selenite concentration; greater significant correlations were found for soil selenate, soil organic Se, and total water-extractable Se (Fig. 2).

IV. ASSIMILATION AND TOLERANCE OF SELENIUM

Selenium is not known to be a required nutrient for plant growth, but its concentration in forage and hay crops is important to animal health. Plants may contain Se in amounts below detectable limits up to several thousand milligrams per kilograms. In the last two decades, the mechanism of Se tolerance in Se accumulator plant species has been studied extensively. In contrast to nonaccumulating species, an increasingly clearer picture of the biochemical events that underlie both selenium toxicity and selenium tolerance has emerged. The exclusion of Se from protein is believed to be the mechanism of Se tolerance in Se accumulator plants (Brown and Shrift, 1981). There are several possible ways in which exclusion of Se can occur:

1. In accumulators, Se metabolism may not result in synthesis of the selenoamino acids that would be incorporated into proteins. This

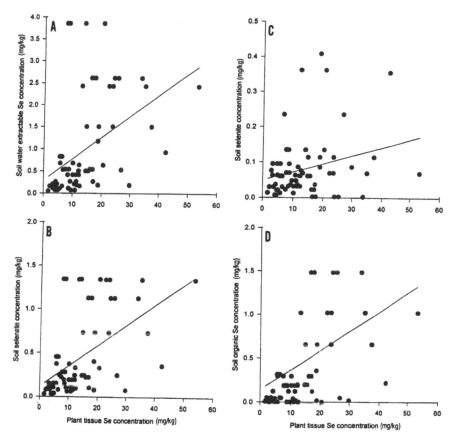

FIGURE 2 Joint distribution and correlations between plant tissue Se concentrations and (A) water-extractable, (B) selenate, (C) selenite, or (D) organic selenium concentrations found in soil samples at the immediate root zone of the plant samples collected from the forage management field plots. (From Mantgem et al. 1996.)

possibility is supported by the identification of nonprotein seleno-amino acids, such as Se-methylselenocysteine, selenocystathionine, Semethylselenomethionine, γ-glutamyl-Se-methylselenocysteine, γ-glutamylselenocystathionine, selenopeptide, and selenohomocysteine (Brown and Shrift,1982). Synthesis of these compounds perhaps would divert Se from the formation of selenocysteine and selenomethionine.

2. There may be a mechanism that discriminates against Se compounds during protein synthesis. The cysteinyl–tRNA synthetase of the accumulator *A. bisculatus* was unable to use the Se analog as a substrate (Burnell and Shrift 1979).

3. The Se content in proteins of accumulator plants may be reduced by enzymatic removal. Such a reaction would be comparable to posttranslational deamination of glutamine and asparagine residues to form glutamatic acid and aspartic acid. Therefore, the replacement of cysteine by selenocysteine in proteins is considered the major cause of the Se toxicity in nonaccumulator plants (Wold,1977).

Among the nonaccumulator plant species, such as crop plants and natural grassland field plants, considerable differences in Se tolerance exist. Wu et al. (1988) tested five forage and turf grass species in solution culture supplemented with Se or salt concentrations as sodium selenate and sodium chloride. It was found that the root and shoot growth of Bermuda grass (*Cynodon dactylon* L.) was severely inhibited by 1 mg/L Se, and the growth of buffalo grass [*Buchloe dactyloides* (Nutt.)], creeping bent grass (*Agrostis stolonifera* L.), and crested wheat grass (*Agropyron desertorum* Fich) was inhibited by 2 mg/L Se. Tall fescue (*Festuca arundinacea* Schred.) displayed the greatest Se tolerance and did not reduce its growth in 2 mg Se/L treatment. Selenium uptake by the five grass species grown in three concentrations of Se (selenate) showed that Se accumulation in both root and shoot tissues corresponded positively with Se concentrations in the culture solution (Fig. 3). Shoot tissue accumulated greater amounts of Se than did root tissue. The amounts of Se accumulated by the plants appear to be inversely related to the Se tolerance of the plant species (Fig. 3).

Wu and Huang (1992) further studied Se assimilation and tolerance in the crop plants, tall fescue and white clover (*Trifolium. repens* L.). These plants, which display substantial differences in their Se tolerance, were grown in sand culture and irrigated with 2 mg Se/L as sodium selenate in nutrient solution containing either 0, 1, or 5 mM Na_2SO_4. Selenium tolerance was measured based on the percentage of shoot dry weight produced under Se treatment to the control treatment. In the presence of Na_2SO_4, tall fescue produced an average of 91% dry weight of its control treatment, but the white clover only had 53% growth. Increased SO_4^{2-} concentrations increased Se tolerance and reduced Se accumulation for both species. When the tissue protein and Se concentrations were measured, the white clover had significantly higher protein Se concentrations than tall fescue (Table 5). plant tissue protein concentrations were not affected by Se and sulfate treatments. Selenium concentrations in nonprotein fractions (converted from the percentage of protein Se in total tissue Se) were much higher in white clover than in tall fescue and higher in the shoot tissue without the sulfate treatment than in the tissue with sulfate treatment. The percentage of total protein Se in total tissue Se in tall fescue was much higher, because less total tissue Se was accumulated in tall fescue than in white clover (Table 5). The correlation analysis between tissue Se concentration and protein Se concentration indicates that both tall fescue ($r = 0.94$, $P < 0.01$) and white clover ($r = 0.98$, $P <$

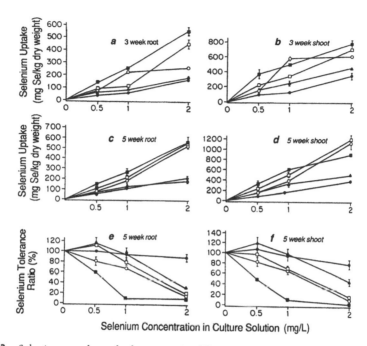

Selenium Concentration in Culture Solution (mg/L)

FIGURE 3   Selenium uptake and tolerance ratio of five grass species after 3 weeks of growth in culture solution supplemented with selenium as sodium selenate: ■, bermuda grass; ▲, buffalo grass; O, crested wheat grass; □, seaside bent grass; ●, tall fescue. Each point represents the mean of three replicates. (From Wu et al. (1988.)

0.01) were significantly and positively correlated. No indication of Se exclusion at the protein level, such as is found in Se accumulators, was detected in these two species. However, the results do suggest an Se exclusion mechanism that restricts Se uptake by plants with greater Se tolerances.

The inverse relationship between Se tolerance and Se uptake has also been found in green algae (Foe and Knight, 1985). This study indicates a mechanism that restricts Se uptake by the plant and, consequently, reduces Se incorporation into its protein, rather than a major mechanism excluding Se from incorporation into proteins as in Se accumulator plants (Brown and Shrift, 1981). Both short-term kinetic uptake studies (Vange et al., 1974, Asher et al., 1977) and long-term Se accumulation experiments (Mikkelsen et al., 1988; Wu and Huang, 1991) indicated that an antagonistic effect exists between Se and sulfate. The present study further confirms that this antagonistic effect acts at the protein assimilation level.

...dium Selenate and Varied Concentrations of Na_2SO_4

nts	Total tissue Se concentration (mg/kg)	Protein Se concentration (mg/g)	Protein Se in total tissue Se concentration (%)	Se tolera...
	455.3 ± 74.8[a]	243.2 ± 23.3	61.4 ± 4	91
	88.1 ± 12.4	27.6 ± 1.5	55.0 ± 17	102
	43.0 ± 4.6	20.1 ± 2.8	32.0 ± 6	112
	0.9 ± 0.2	0.2 ± 0.1	12.0 ± 2	
	789.1 ± 102.1	388.7 ± 18.3	38.3 ± 11.8	53
	242.6 ± 40.2	94.2 ± 5.7	27.6 ± 15.8	81
	53.7 ± 1.1	14.3 ± 3.2	17.0 ± 6.0	95
	1.0 ± 0.3	1.3 ± 1.0	6.0 ± 1.0	10

ercentages represent means ± one standard deviation.
g (1992).

V. ACCUMULATION OF SELENOAMINO ACIDS

Accumulation of the selenium analogs of methionine, sulfur-methylcysteine, γ-glutamyl Se-methylselenocysteine, and cystathionine appears to be common in Se accumulator plants such as *Astragalus* (Stadtman, 1974). The major nonprotein selenoamino acid, Se-methylselenocysteine, first isolated from *Astragalus bisulcatus,* accounted for 80% of the plant tissue Se. This selenoamino acid either was not found in Se nonaccumulator plant species (Nigam et al., 1969; Virupaksha et al., 1966) or was detected only in very small concentrations (Martin et al., 1971; Nigam and McConnell, 1973).

  Melilotus indica, an annual legume species, expanded its range of colonization since being brought into the Kesterson Reservoir with the uncontaminated fill soils by the Kesterson Cleanup Action in the fall of 1988 (Huang and Wu, 1991). This species can accumulate much higher tissue Se concentrations (>200 mg/kg) than the other field plant species growing in the same seleniferous soils (Wu et al., 1993). A study was conducted to determine whether this plant has any characteristics that resemble those of Se accumulator plants at the selenoamino acid metabolism level. The plant materials of *M. indica* were collected from three sites at Kesterson where the average total soil Se concentrations were 0.4, 8, and 15 mg/kg. The average plant tissue Se concentrations were 5.07, 22.02, and 177.35 mg/kg, respectively. The tissue-free selenoamino acids and their corresponding sulfuramino acids were analyzed by Guo and Wu (1997).

  Table 6 presents the distribution of selenoamino acid concentrations and selenoamino acid Se concentrations in the plant tissue. For the five selenoamino acids, selenocysteine (SeCys), selenomethyl-Se-cysteine (SeMSeCys), selenomethionine (SeMet), selenocystine (SeCys$_2$), and γ-glutamyl-Se-methylselenocysteine (γ-GSeMSeCys), there was a 3- (SeMet) to 12- (SeCys$_2$) fold increase of the tissue selenoamino acid concentration when the tissue total Se concentration increased from 5.07 mg/kg to 22.02 mg/kg. When the tissue total Se concentration further increased from 22.02 mg/kg to 117.35 mg/kg, the SeMet concentration increased from 317.1 μmol/kg to 536.8 μmol/kg and SeMSeCys concentration increased from 53.6 μmol/kg to 109.8 μmol/kg. The other three selenoamino acids did not respond significantly to the increase in the total tissue Se concentration. For the tissue concentrations of the five corresponding sulfuramino acids, methionine (Met) and methylcysteine (MCys) were 3779.1 and 3491.1 μmol/kg, respectively, and the other three ranged from 213.6 for γ-glutamylmethylcysteine (γ-GMCys) to 610.2 μmol/kg for cysteine (Cys). However, these sulfur-amino acids did not exhibit any consistent pattern of increase or decrease of their concentration with the increase of total tissue Se concentration.

  Among the 14 nonsulfur amino acids (Table 6), over the three different total tissue Se concentrations, alanine (Ala), tyrosine (Tyr), and lysine (Lys) had concentrations ranging from 2958.7 to over 6495.6 μmol/kg. Glycine (Gly), ion of Selenoamino Acid Concentrations in Plant Tissues of *Melilotus indica* Grown in Selenium-Laden Soils

Total tissue Se, dry weight

5.07 mg/kg		22.02 mg/kg		117.35 mg/kg	
AA concentration (μmol/kg)	AA-Se in plant tissue (μg/kg)	AA concentration (μmol/kg)	AA-Se in plant tissue (μg/kg)	AA concentration (μmol/kg)	AA-Se tissue
20.5 ± 2.3[a]	9.6 ± 1.1	94.0 ± 5.9	44.1 ± 2.8	97.2 ± 2.8	45.6
15.3 ± 2.1	6.6 ± 0.9	53.6 ± 1.6	23.2 ± 0.7	109.8 ± 4.0	47.2
115.6 ± 5.9	46.5 ± 2.4	317.1 ± 16.2	127.7 ± 6.5	536.8 ± 6.6	216.1
12.4 ± 1.1	2.9 ± 0.3	148.9 ± 7.0	35.2 ± 1.7	154.6 ± 4.5	36.6
2.1 ± 0.8	0.2 ± 0.1	3.2 ± 1.4	0.8 ± 0.1	4.1 ± 1.1	1.2
610.2 ± 18.0		229.9 ± 16.7		571.3 ± 18.9	
3491.1 ± 55.7		2429.0 ± 96.3		3479.7 ± 29.3	
3779.1 ± 93.0		1115.0 ± 26.7		2917.0 ± 32.5	
223.9 ± 6.0		1507.9 ± 49.6		696.6 ± 56.4	
213.6 ± 5.7		350.9 ± 6.9		486.5 ± 12.2	
3151.0 ± 131.7		3918.8 ± 143.1		2958.7 ± 79.1	
1716.1 ± 27.4		1679.6 ± 38.5		1282.7 ± 18.3	
290.2 ± 6.9		348.5 ± 8.1		354.7 ± 12.1	

%Se-AA in total AA	%AA-Se in total tissue selenium	%Se-AA in total AA	%AA-Se in total tissue selenium	%Se-AA in total AA	%AA total selenium
336.2 ± 11.9		407.7 ± 3.6		485.5 ± 4.4	
309.5 ± 8.3		455.0 ± 5.3		614.1 ± 12.7	
1056.1 ± 51.8		1107.8 ± 13.6		1445.9 ± 12.8	
948.2 ± 36.2		607.1 ± 62.8		739.4 ± 160.0	
110.0 ± 20.0		110.0 ± 10.0		110.0 ± 10.5	
371.4 ± 15.4		217.4 ± 17.1		112.4 ± 8.3	
1110.1 ± 136.1		877.9 ± 101.7		1029.2 ± 134.3	
4733.5 ± 43.7		6495.8 ± 83.0		5713.2 ± 64.8	
3130.4 ± 43.2		4219.3 ± 64.3		3845.2 ± 51.7	
1418.3 ± 138.7		1607.1 ± 121.2		1000.7 ± 113.5	
853.1 ± 35.5		1692.1 ± 17.5		1140.17 ± 15.7	
24,916.2 A[b]	65.7 C	26,487.5 A	230.2 B	25,654.0 A	345.6
0.66 B[b]	1.30 A	2.33 A	1.05 B	3.38 A	0.29

Se, amino acid selenium concentration; SE-AA, selenoamino acid.

, means followed by the same letter are not significantly different at $P = 0.01$, according to Duncan's new multiple range test.

(1997).

leucine (Leu), glutamine (Glu), histidine (His), and tryptophan (Typ) had concentrations ranged from 877.9 to 1716.1 μmol/kg, and the remaining six nonsulfur amino acids had concentrations ranging from 110 to about 600 μmol/kg. Glycine showed a slight decrease and Leu showed a slight increase with the increase of tissue Se concentrations, but the rest of the nonsulfur amino acids had no consistent response of tissue concentration to the increase of plant total tissue Se concentration.

The mean values of the total amino acid concentrations, including selenoamino acids, were 24,916.2, 26,487.5, and 25,654.0 μmol/kg for the plants having total tissue Se concentrations of 5.07, 22.02, and 117.35 mg/kg, respectively. The total tissue amino acid concentrations of the plants grown in different soil Se concentrations were not significantly different ($P > 0.05$), even though their total tissue Se concentrations were significantly different ($P < 0.01$). Selenoamino acid concentrations increased from 0.66% for the plants having total tissue Se concentrations of 5.07 mg/kg to 2.33% for the plants having total tissue Se of 22.02 mg/kg, and to 3.38% for the plants having a total tissue Se concentration of 117.35 mg/kg. Among the five selenoamino acids, selenomethionine had the highest concentration. It constituted 70, 51, and 59% of the total selenoamino acid content measured for the plants having total tissue selenium concentrations of 5.07, 22.02, and 117.35 mg/kg, respectively.

The total selenium concentration in the selenoamino acids increased by 3.41 times (from 6.57 μg/kg to 230.2 μg/kg) when there was a 4.3-fold increase of the plant tissue total Se concentration (from 5.07 μg/kg to 22.02 mg/kg dry weight), but it only had a 1.5-fold increase (from 230.2 μg/kg to 345.6 μg/kg) when the tissue total Se concentration increased over 5-fold (from 22.07 mg/kg to 117.35 mg/kg). The percentage of selenoamino acid Se decreased with the increase of total tissue Se concentration; values were 1.30, 1.05, and 0.29% for plants having tissue Se concentrations of 5.07, 22.02, and 117.35 mg/kg, respectively.

Despite the similarities between the chemical properties of Se and sulfur, the substitution selenocysteine for cysteine may have significant effects on the properties and functions of proteins. In particular, Se and sulfur differ in two important respects that could affect protein function:

1. Selenium atoms are larger than sulfur atoms, and the diselenide bond is approximately one-seventh longer and one-fifth weaker than the disulfide bond.
2. The ionization properties of the seleno (—SeH) and sulfhydryl (—SH) radicals are different over the pH range 7.0 to 7.5 (Huber and Criddle, 1967a).

In M. indica, there was a five-fold increase of selenocysteine when the total tissue Se increased from 5.07 mg/kg to 22.02 mg/kg, but there was no increase in tissue selenocysteine concentration when the total tissue Se concentration increased from 22.02 mg/kg to 117.35 mg/kg. This phenomenon indicates that there is a control mechanism at the level of selenocysteine metabolism, keeping the tissue selenocysteine concentration below the level that becomes toxic to the plant.

Methionine seems play a less specific role in protein structure than cysteine. Therefore, introduction of selenomethionine into a protein might not have a serious effect on conformation, despite the greater size of the Se atom. However, selenomethionine is less soluble in water than methionine, and a substitution may affect normal enzyme function by its hydrophobic interactions (Shepherd and Huber, 1969). Despite this difference in hydrophobic character, replacement of methionine by selenomethionine in more than 50% of methionine residues in bacteria β-galactosidase by selenomethionine had no negative effect (Huber and Criddle, 1967b). The present study showed that in M. indica, selenomethinine comprised more than 50% of the total selenoamino acid measured. This may be an overestimate, because not all the selenoamino acids in the plant tissue were identified and measured, but it is reasonable to suggest that the accumulation of selenomethionine may serve as a buffer against excessive amounts of Se in the plant.

More research is needed to reveal the role of selenomethinone in Se accumulation by plants. The Se in the selenoamino acids accounted for 1.3% or less of the total tissue Se concentration. The protein Se concentration of the plant was not measured in the present study. It has been suggested that in Se nonaccumulator plants, most of the tissue Se is incorporated in the proteins (Brown and Shrift, 1982). Wu and Huang (1992) reported that in tall fescue (F. arundinacea Schreb.) and white clover (T. repens L.), the percentage of protein Se in total tissue Se concentrations ranged from 32 to 61% for tall fescue and 17 to 38% for white clover. Thus, for the M. indica selenoamino acid analysis, approximately half the plant tissue Se is not accounted for. Horn and Jones (1941) suggested that about 80% of Se in plants of Astragalus pectinatus is present in organic compounds with an aromatic character; but so far, Se-containing aromatic compounds have not been isolated from plant materials. Further research on the distribution of Se in different plant organic compounds should be an important area in the study of physiology and chemistry of Se accumulation by plants.

Se-methylselenocysteine was found in the plant tissue of M. indica. The accumulation of this nonprotein selenoamino acid in this species resembles that found in Se accumulator plants. A compound known to be toxic to animals, γ-glutamyl-Se-methylselenocysteine (γ-GSeMSeCys), was detected in this species. However, the detected concentration was very low, (<5 μmol/kg), and it might not be a toxic constituent in the food chain at this concentration.

Selenoamino acid accumulation was further examined by Wu et al. (1997) for two legume and two grass species that differ in their Se tolerance. Naturally established plants of yellow sour clover (M. indica) and rabbitfoot grass (Polypogon *monspeliensis*) were collected from the Se-laden soil site in Pond 2 at Kesterson Reservoir and from an uncontaminated field at Davis, California. Alfalfa (*Medicago sativa*) and tall fescue (*F. arundinacea*) were either grown in the Se-laden soil that came from Kesterson Pond 2 (a 2-acre field) or an uncontaminated field in Davis, California. The distribution of selenoamino acids and the total tissue Se concentrations of the four plant species were analyzed. The results (Table 7) show that tissue selenoamino acid concentrations were significantly and positively correlated with the increase of total tissue Se concentrations.

Sulfur amino acids mostly were negatively correlated with the increase of total tissue Se concentration. Highly significant negative correlations between tissue Se concentration and tissue cysteine (Cys) concentration were found in all four plant species. However, the selenocysteine concentration was positively correlated with the increase of the tissue Se concentration. Except in the sour clover, the tissue methionine (Met) concentrations of the plant species studied were significantly and negatively correlated with the total tissue Se concentrations. Generally, the tissue Se amino acid concentrations were higher than their correspondings amino acid concentrations. Increase of total plant tissue Se concentration clearly elevated selenoamino acid concentrations but showed only slightly reduction of their corresponding amino acid concentrations. In sour clover, tissue selenomethionine (SeMet) concentrations increased from an average of 1.658 nmol in the Davis field plant sample to 6.165 nmol in the Kesterson field plant sample. The greenhouse-grown Kesterson soil plants had the highest methionine (Met) concentration (0.767 nmol) and the Kesterson field-collected plants had the lowest methionine (0.198 nmol) concentration. The increase of tissue methionine concentration was less affected by the increase of total tissue amino acid concentration compared to the increase of tissue selenomethionine concentration. Cysteine (Cys) decreased from an average of 0.784 nmol in the Davis field plant samples to 0.048 nmol in the Kesterson field plant samples, while Se-cysteine (SeCys) increased from an average of 1.147 to 2.556 nmol respectively. The increase of Se-methylselenocysteine was also accompanied by the decrease of methylcysteine (MCys) in the plant tissue.

The study shows that there is an antagonistic relationship between tissue selenoamino acid levels and their corresponding sulfur amino acid concentrations. This is a demonstration of competitive interaction between Se and sulfur at the amino acid synthesis level. Bañuelos and Meek (1988) and Bañuelos et al. (1990) found a strong negative correlation between tissue Se concentration and sulfate concentration in sulfur-accumulating plants (Brassicaceae), and they suggest that the plants of this family are able to accumulate high concentrations of Se because they do not discriminate between selenate and sulfate. Among the four plant species tested, sour clover was higher in Se tolerance and Se accumulation than the other three species. The tissue methionine concentration of sour clover was less affected by an increased tissue selenomethionine concentration than in alfalfa, and in alfalfa this relationship was significantly negatively correlated (Fig. 4). This discrepancy suggests that a less antagonistic effect under the increase of selenoamino acid analogs in plant tissue might be able to minimize Se toxicity to the plant. However, more research is needed to elucidate this presumption.

In tall fescue, an Se avoidance mechanism was found that prevents Se from being taken up by the plant (Wu and Huang, 1992). Therefore, the plant can tolerate relatively high Se concentrations in the soil. Different mechanisms of Se tolerance may exist among the Se nonaccumulator plant species. Both Se-methylselenocysteine (a nonprotein amino acid analog) and selenomethinone (a protein amino acid analog) were accumulated in the plant tissue when the plants were grown in Se-laden soils. Concentrations of both these selenoamino acids were greater than those of their corresponding sulfur amino acid analogs. This phenomenon suggests that these selenoamino acid analogs are preferentially accumulated when Se is present. The non-sulfur-containing amino acids were not substantially affected by an increase in tissue Se concentration.

VI. SELENIUM ACCUMULATION AND PHYTOREMEDIATION

A. Selenium Accumulation and Soil Selenium Removal

The advantage of using crop and/or grassland plant species rather than Se accumulator plant species for remediation of Se-laden soils is that the Se-enriched plant materials may be harvested and used as an Se supplement for Se-deficient forage or used as a soil amendment for Se-deficient rangeland in the western United States. Bañuelos et al. (1996) examined Se uptake and soil Se removal by canola (*Brassica napus*), kenaf (*Hibiscus cannabinus* L.), and tall fescue (*F. arundinacea*). The plants were grown in a greenhouse in seleniferous Turlock soil. The soil had a total Se concentration of 40 mg/kg, with a water-extractable Se concentration of 10 mg/kg. The plants were harvested and replanted three times, and the plant tissue Se and soil Se concentrations were measured. For the first harvest, the plant tissue Se concentrations were 470, 45, and 50 mg/kg for canola, kenaf, and tall fescue, respectively. By the third harvest, the plant tissue Se concentrations were 288, 36, and 17 mg/kg, respectively. The cultivation of the three crop plants led to a significant reduction of total soil Se between preplant and final harvest: 47% for canola, 23% for kenaf, and 21% for tall fescue.

Selenium removal by the crop plants also was studied by Bañuelos et al. (1992) for Se-laden soil enriched by Se-rich plant tissues. This study was conducted under field conditions. Selenium-enriched (250 mg Se/kg) wild brown mustard (*B. juncea*) plant tissue was incorporated into the field soil at a rate of 1500 kg/ha. Alfalfa (*Medicago sativa*), tall fescue (*F. arundinacea*), birdsfoot trifoil (*Lotus corniculatus*), canola (*B. napus*), and wild brown mustard (*B. juncea*) were

ino Acid Concentrations and Corresponding Sulfur Amino Acid Concentrations Deteced from Plant Tissue of Four P...
oils with Different Selenium Concentrations and Correlations Tested Between Seleno Amino Acid Concentration and
...ncentration

| | Mean tissue amino acid concentration (nmol/g dry weight)[b] | | | | | | | Total ti... |
	Se MSeCys	Met	Se Met	Cys	SeCys	Cys2	SeCys2	Se(mg...
Mcys								N...
−0.575	0.957	−0.356	0.983	−0.650	0.908	−0.982	0.925	
$P < 0.01$	$P < 0.01$	$P > 0.10$	$P < 0.01$	$P < 0.05$	$P < 0.01$	$P < 0.01$	$P < 0.01$	
1.421	2.003	0.261	1.658	0.784	1.147	7.564	4.538	0.94
0.420	2.329	0.458	2.475	0.245	1.743	6.498	10.650	6.00
0.382	2.195	0.767	3.242	0.202	1.825	6.569	16.550	22.02
2)								117.35 +
0.409	3.373	0.198	6.165	0.048	2.556	3.146	15.999	N...
−0.960	0.948	−0.588	0.980	−0.987	0.978	0.956	0.973	
$P < 0.01$	$P < 0.01$	$P < 0.05$	$P < 0.01$	$P < 0.01$	$P < 0.01$	$P < 0.01$	$P < 0.01$	
1.200	2.008	0.840	2.062	0.824	0.134	3.090	3.412	0.52

0.967	4.016	0.692	6.709	0.198	0.970	3.604	7.194	6.43
0.242	0.946	−0.747	0.929	−0.149	0.906	0.912	0.798	NV
$P > 0.01$	$P < 0.01$	$P < 0.02$	$P < 0.01$	$P > 0.10$	$P < 0.01$	$P < 0.01$	$P < 0.01$	
0.035	0.809	0.278	0.428	0.128	0.284	1.314	2.179	0.24 =
0.035	2.107	0.117	2.214	0.150	0.380	3.332	13.247	5.84 =
0.045	2.732	0.134	2.845	0.115	0.639	3.875	12.419	15.53 ±
−0.148	0.737	0.734	0.578	−0.104	0.890	0.768	0.841	NV
$P > 0.10$	$P < 0.05$	$P < 0.05$	$P < 0.05$	$P > 0.10$	$P < 0.01$	$P < 0.05$	$P < 0.01$	
0.010	0.884	0.112	0.382	0.147	0.010	0.488	0.503	0.23 ±
0.375	1.743	0.123	3.272	0.035	0.370	0.704	4.260	0.26 ±
0.059	1.982	0.146	2.791	0.085	1.072	2.256	7.223	10.68 ±

parentheses.
...t and significance level are given for each amino acid and plant species on the appropriate line
...leviation; NV, not valid.
...97).

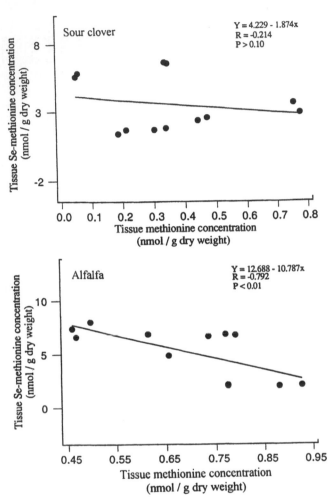

FIGURE 4 Joint distribution and linear regression between tissue selenomethinonine concentration and methionine concentration detected in leaf tissue of sour clover (*Melilotus indica*) and alfalfa (*Medicago sativa*). Correlation coefficient and significance level are indicated on the right corner of the graphs. (From Wu et al., 1996.)

grown in the Se-enriched soil. The soil Se concentration at the time of planting was 0.7 mg Se/kg.

The plants were clipped at 60 and 145 days after planting, and plant tissue dry weight, tissue Se concentration, and tissue crude protein were measured. Tissue Se concentrations ranged from 1.3 mg/kg in the tall fescue to 50 mg Se/

kg in the mustard. The relative Se concentration detected in the plant tissue among the plant species was wild mustard = canola > alfalfa > birdsfoot trefoil > tall fescue. Crude protein ranged from 16 to 27% for all species. Digestible dry matter of the forage materials was found at least 90%.

This study demonstrated that soil Se enriched by organic matter can be effectively taken up by plants. Substantial differences in the rate of Se accumulation were detected among the plant species even though they were grown under identical soil conditions. Based on the suggested safe Se concentration for forage use (0.05–2 mg Se/kg dry matter), tall fescue grown at the soil Se level enriched by the Se-laden plant materials could be fed directly to the animals (Bañuelos et al., 1992). Alfalfa and birdfoot trefoil grown under the same soil Se conditions appear to be useful as Se supplement forage for areas with a Se deficiency. Wild mustard and canola accumulated much higher tissue Se concentrations (50 and 34 mg/kg, respectively). It may be risky to use these plants for forage purposes, but they could serve as a soil amendment for Se-deficient rangeland. For the application and disposal of Se-rich plant materials, long-term studies are necessary for the development of agromanagement strategies, including frequency of harvesting, monitoring Se content in biomass, forage quality, yield, palatability for animals, residual soil Se, and irrigation strategies.

B. Effects of Plants on Redistribution, Leaching, and Bioextraction of Soil Selenium

The effects of forage plants on soil Se redistribution, leaching, and bioextraction were studied by Wu et al. (1996), using soil columns under simulated field soil conditions. The bottom 65 cm of the column was filled with uncontaminated field soil. On top of the uncontaminated soil, was placed 20 cm of the Kesterson Se-laden soil. Tall fescue (F. arundinacea, a forage plant) and yellow sour clover (Melilotus indica, a naturally occurring legume species in Se-contaminated soils) were used for the study. Important questions directly addressed by this study were:

1. Does plant Se accumulation change over repeated harvests?
2. Do the volume and Se concentration of leachate change over the course of plant growth and harvests?
3. Do the selected plant species increase the rate of Se dissipation compared to bare soil?
4. Are there changes in soil Se remobilization over depth depending on crop planting strategy and age of plant growth?

The results of this study indicate that both leachate volume and Se concentration in the leachate were greatly influenced by the presence of vegetation. The volume of leachate was considerably lower for soil having either tall fescue or yellow sour clover plantings. Tall fescue had a higher water use rate and greater rooting density than did yellow sour clover, indicating that the tall fescue will be more practical for bioremediation of Se-laden soil than yellow sour clover.

Soil Se distribution analyses yielded three trends for this study:

1. Transport of Se down to the lower soil profile of the uncontaminated soil occurred at the first harvest, but the Se concentrations at these depths were not at high levels.
2. Except for selenite, all forms of water-extractable soil Se concentrations showed a clear reduction over time upon vegetation harvest.
3. The reduction of total soil Se was significant, but the difference between the forage plantings and the control treatment was relatively small. This is because a relatively small fraction (<10%) of the total Se inventory was water soluble and available to plants at any one time. A large reduction may become apparent after a longer period of forage plant management.

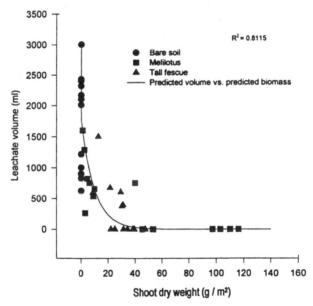

Figure 5 Predicted relationship between leachate volume (liters) and combined vegetation biomass (grams dry weight per square meter) of three planting treatments (tall fescue, yellow sour clover, bare soil) of the greenhouse soil column experiment using the equation $L = A \exp(-k\,W)$, where A (mL) decreases as shoot dry weight W (g) increases in a nonlinear relationship. The parameter k (1 /g) affects the rate at which predicted leachate would reach zero as soon as dry weight increased up to about 40 g/m^2.
Source: Wu, et al., 1996.

4. The volume of leachate collected from the column, under an irrigation rate of 8 cm of water per week, is strongly dependant on the presence of biomass. A nonlinear regression (Fig. 5) shows that leachate at zero biomass was estimated at a mean of 1770 mL, and the minimum biomass required to arrive at a total absence of leachate was 22 g/m^2 for tall fescue and 47 g/m^2 for yellow sour clover.

The rate and efficiency of soil Se removal by plants may vary greatly depending on the source of soil Se, the plant species, the physical and chemical soil conditions, and the culture management strategies. More research is needed for the successful application of Se nonaccumulator plant species in the phytoremediation of Se-contaminated soils.

REFERENCES

Adriano, D. C. 1986. *Trace Elements in the Terrestrial Environment*, pp. 51–76. Springer-Verlag, New York.

Asher, C. J., G. W. Butler and P. J. Peterson. 1977. Selenium transport in root systems of tomato. *J. Exp. Bot.* 28:279–291.

Bañuelos, G. S., and D. W. Meek. 1988. Crop selection for removing selenium from soil. *Calif. Agric.* 43:19–20.

Bañuelos, G. S., D. W. Meek, and G. J. Hoffman. 1990. Influence of selenium and salinity, and boron on selenium accumulation in wild mustard. *Plant Soil* 127:201–206.

Bañuelos, G. S., R. Meed, L. Wu, P. Beuselinck, and S. Akohoue. 1992. Differential accumulation among forage plant species grown in soils amended with selenium-enriched plant tissue. *J. Soil Water Conserv.* 47:338–342.

Bañuelos, G. S., H. A. Ajwa, B. Mackey, L. Wu, C. Cook, S. Akohoue, and S. Zambrzuski. 1996. Evaluation of different plant species used for phytoremedation of high soil selenium. *J. Environ. Qual.* 26:639–646.

Barneby, R. C. 1964. Atlas of North American *Astragalus*. *Mem. New York Bot. Garden* 13:1–188.

Beath, O. A., J. H. Draize, H. F. Eppson, C. S. Gilbert, and O. C. McCreary. 1934. Certain poisonous plants of Wyoming activated by selenium and their association with respect to soil types. *J. Am. Pharm. Assoc.* 23:94–97.

Brown, T. A., and A. Shrift. 1981. Exclusion of selenium from proteins of selenium-tolerant *Astragalus* species. *Plant Physiol.* 67:1051–1053.

Brown, T. A., and A. Shrift. 1982. Selenium: Toxicity and tolerance in higher plants. *Biol. Rev.* 57:59–84.

Burau R. G., A. McDonald, A. Jacobson, D. May, S. Grattan, C. Shennan, B. Swanton, D. Scherer, M. Abrams, E. Epstein, and V. Rendig. 1988. Selenium in tissues of crops sampled from the West Side of the San Joaquin, California. In *Selenium Contents in Animal and Human Food Crops Grown in California*, pp. 61–66. ANR Special Publication no. 3330.

Burnell, J. N., and A. Shrift. 1979. Cysteinyl–tRNA synthetase from *Astragalus* species. *Plant Physiol.* 63:1095–1097.

Byers, H. G., J. T. Miller, K. T. Williams, and H. W. Lakin. 1933. Selenium occurrence in certain soils in the United States with a discussion of related topics. U.S. Department of Agriculture Tech. Bull. no. 601, pp. 1–74. Government Printing Office, Washington, DC.

Cary, E. E., and W. H. Allaway. 1969. The stability of different forms of selenium applied to low-selenium soils. *Soil Sci. Soc. Am. Proc.* 33:571–574.

Drake, C., A. B. Grant, and W. J. Hartley. 1960. Selenium and animal health. Parts I and II. *N. Z. Vet. J.* 8:4–10.

Foe, C., and A. W. Knight. 1985. Selenium bioaccumulation and toxicity in the green alga *Selenatrum capricornutum*, and dietary toxicity of the contaminated alga to *Daphnia magna*. In *Proceedings of the First Annual Environmental Symposium on Selenium in the Environment*, Fresno, CA, pp. 76–88. June 10–12, 1985.

Gissel-Nielsen, G., U. C. Gupta, M. Lamand, and T. Westermarck. 1984 Selenium in soils and plants and its importance in livestock and human nutrition. *Adv. Agron.* 37:397–460.

Guo, X., and L. Wu. 1997. Distribution of seleno-amino acids in plant tissue of *Melilotus indica* L. grown in selenium laden soils. *Environ. Exp. Bot.*

Horn, M. J., and D. B. Jones. 1941. Isolation from *Astragalus pectinatus* of a crystalline amino acid complex containing selenium and sulfur. *J. Biol. Chem.* 139:649–660.

Huang, Z. Z., and L. Wu. 1991. Species richness and selenium accumulation of plants in soils with elevated concentration of selenium and salinity. *Ecotoxicol. Environ. Saf.* 22:251–266.

Huber, R. E., and R. S. Criddle. 1967a. Comparison of the chemical properties of selenocysteine and selenocystine with their sulfur analogs. *Arch. Biochem. Biophys.* 122:164–173.

Huber, R. E., and R. S. Criddle. 1967b. The isolation and properties of β-galactosidase from *Escherichia coli* grown on sodium selenite. *Biochim. Biophys. Acta* 141:587–599.

Mantgem, P. J., L. Wu, and G. S. Bañuelos. 1996. Bioextraction of selenium by forage and selected field legume species in selenium laden soils under minimal field management conditions. *Ecotoxicol. Environ. Saf.* 34:228–238.

Martin, J. L., A. Shrift, and M. L. Gerlach. 1971. Use of ^{75}Se-selenite for the study of selenium metabolism in *Astragalus*. *Phytochemistry* 10:945–952.

Mikkelsen, R. L., G. H. Haghnia, A. L. Page, and F. T. Bingham. 1988. The influence of selenium, salinity, and boron on alfalfa tissue composition and yield. *J. Environ. Qual.* 17:85–88.

Nelson, A. A., O. G. Fitzhugh, and O. H. Calvery. 1943. Liver tumor following cirrhosis caused by selenium in rats. *Cancer Res.* 3, 230–236.

Nigam, S. N., and W. B. McConnell. 1973. Biosynthesis of Se-methylselenocysteine in lima beans. *Phytochemistry* 12:359–362.

Nigam, S. N., J. I. Tu, and W. B. McConnell. 1969. Distribution of selenomethylselenocysteine and some other amono acids in species *Astragalus*, with special reference to their distribution during the growth of *A. bisulcatus*. *Phytochemistry* 8, 1161–1165.

Ohlendorf, H. M., D. J. Hoffman, D. J. Saiki, and T. W. Aldrich. 1986a. Embryonic mortality and abnormalities of aquatic birds. *Sci. Total Environ.* 52:49–63.

Ohlendorf, H. M., D. J. Hoffman, C. M. Bunck, T. W. Aldrich, and J. F. Moore 1986b. Relationships between selenium concentrations and avian reproduction. *Trans. North Am. Wild. Nat. Resour. Conf.* 51:330.

Panmaja, K., D. D. K. Prasadand, and A. R. K. Prasad. 1989. Effects of selenium on chlorophyll biosynthesis in munseedlings. *Phytochemistry* 28: 3321–3324.

Paterson, E. L., R. Milstery, and E. L. R. Stokstad. 1957. Effects of selenium in preventing exudative diathesis in chicks. *Proc. Soc. of Exp. Biol. Med.* 95:617–620.

Peterson, P. J., and G. W. Butler. 1966. Colloidal selenium availability to three pasture species in pot culture. *Nature* 212:961–962.

Presser, T. S., and I. Barnes. 1984. Selenium concentration in water tributary and in vicinity of the Kesterson National Wildlife Refuge, Fresno and Merced Counties, California. U.S. Geological Survey Water Resources Invest. Rep. No. 84–4112. USGC, Sacramento.

Robinson, W. O. 1933. Determination of selenium in wheat and soils. *J. Assoc. Off. Agric. Chem.* 16:423–424.

Rosenfeld, I., and O. A. Beath. 1964. *Selenium, Geobotony, Biochemistry, Toxicity and Nutrition.* Academic Press, New York.

Schwarz, K., and C. Foltz. 1957. Selenium as an integral part of factor 3 against dietary necrotic liver degeneration. *J. Am. Chem. Soc.* 79:3292–3293.

Sharman, G. A. M. 1960. Selenium in animal health. *Proc. Nutr. Soc.* 19:169–172.

Shepherd, L., and R. E. Huber. 1969. Some chemical and biochemical properties of selenomethionine. *Can. J. Biochem.* 47:877–881.

Shrift A. and J. M. Ulrich. 1976. Transport of selenate and selenite into *Astragalus* roots. *Plant Physiol.* 44:893–896.

Smith, G. S., and J. H. Watkinson. 1984. Selenium toxicity in perennial ryegrass and white clover. *New Phytol.* 97:557–564.

Stadtman T. C. 1974. Selenium biochemistry. *Science* 183:915–922.

Tagwerker, F. J. 1960. Vitamin E and selenium in animal and poultry nutrition. Parts 1 and 11. *Agric. Vet. Chem.* 1:23-(6):78–84.

Tolbert, N. E., and H. M. Wiebe. 1955. Phosphorus and sulfur compounds in plant xylem sap. *Plant Physiol.* 30:499–594.

Ulrich, J. M., and A. Shrift. 1986. Selenium absorption by excised *Astragalus* roots. *Plant Physiol.* 43:14–20.

Vange, M. S., K. Holmern, and P. Nissen. 1974. Multiphasic uptake of sulphate by barley roots. I. Effects of analogues, phosphate and pH. *Physiol. Plant,* 31:292–301.

Virupaksha, T. K., A. Shrift and H. Tarver. 1966. Metabolism of selenomethionine in selenium accumulator and non-accumulator *Astragalus* species. *Biochim. Biophys. Acta* 130:45–55.

Volgary, M. N., and L. A. Tscherkes. 1967. Further studies in tissue changes associated with sodium selenate. In O. H. Nuth (ed.) *Selenium in Biomedicine,* pp. 179–184. AVI Publishers, Westport, CT.

Wold, U. R. F. 1977. Posttranslational covalent modofication of proteins. *Science* 198:890–896.

Workman, S. W., and P. N. Soltanpour. 1980. Importance of prereducing selenium(VI) to selenium(IV) and decomposing organic matter in oil extracts prior to determination of selenium using hydride generation. *Soil Sci. Soc. Am. J.* 44:1331–1333.

Wu, L., and Z. Z. Huang. 1991. Selenium tolerance, salt tolerance, and selenium accumulation in tall fescue lines. *Ecotoxicol. Environ. Saf.* 22:232–247.

Wu, L., and Z. Z. Huang. 1992. Selenium assimilation and nutrient element uptake in white clover and tall fescue under the influence of sulphate concentration and selenium tolerance of the plants. *J. Exp. Bot.* 43:549–555.

Wu, L., Z. Z. Huang, and G. R. Burau. 1988. Selenium accumulation and selenium–salt cotolerance in five grass species. *Crop Sci.* 28:517–522.

Wu, L., A. Enberg, and K. K. Tanji. 1993. Natural establishment and selenium accumulation of herbacious plant species in soils with elevated concentrations of selenium and salinity under irrigation and tillage practices. *Ecotoxicol. Environ. Saf.* 25:127–140.

Wu, L., J. Chen, K. K. Tanji, and G. S. Bañuelos. 1995. Distribution and biomagnification of selenium in a restored upland grassland contaminated by selenium from agricultural drain water. *Environ. Toxicol. Chem.* 14:733–742.

Wu, L., P. J. Mantgem, and X. Guo. 1996. Effects of forage plant and field legume species on soil selenium redistribution, leaching, and bioextration in soils contaminated by agricultural drain water sediment. *Arch. Environ. Contam. Toxicol.* 31:329–338.

Wu, L., X. Guo, and G. S. Bañuelos. 1997. Accumulation of seleno-amino acids in legume and grass plant species in selenium soils. *Environ. Toxicol. Chem.* 16:491–497.

Index